RNA

A LABORATORY MANUAL

ALSO FROM COLD SPRING HARBOR LABORATORY PRESS

RELATED TITLES

Basic Methods in Microscopy: Protocols and Concepts from Cells: A Laboratory Manual

The Condensed Protocols from Molecular Cloning: A Laboratory Manual

Emerging Model Organisms: A Laboratory Manual, Volume 1 and Volume 2

Gene Transfer: Delivery and Expression of DNA and RNA, A Laboratory Manual

Imaging in Neuroscience and Development: A Laboratory Manual

Live Cell Imaging: A Laboratory Manual, Second Edition

Manipulating the Mouse Embryo: A Laboratory Manual, Third Edition

Molecular Cloning: A Laboratory Manual, Third Edition

RNAi: A Guide to Gene Silencing

RNA Worlds: From Life's Origins to Diversity in Gene Regulation

HANDBOOKS

At the Bench: A Laboratory Navigator, Updated Edition

At the Helm: Leading Your Laboratory, Second Edition

Binding and Kinetics for Molecular Biologists

Experimental Design for Biologists

Lab Math: A Handbook of Measurements, Calculations, and Other Quantitative Skills for Use at the Bench

Lab Ref: A Handbook of Recipes, Reagents, and Other Reference Tools for Use at the Bench, Volume 1 and Volume 2

Statistics at the Bench: A Step-by-Step Handbook for Biologists

WEBSITE

Cold Spring Harbor Protocols
http://www.cshprotocols.org

RNA

A LABORATORY MANUAL

Donald C. Rio
University of California, Berkeley

Gregory J. Hannon
Cold Spring Harbor Laboratory

Manuel Ares, Jr.
University of California, Santa Cruz

Timothy W. Nilsen
Case Western Reserve University School of Medicine

COLD SPRING HARBOR LABORATORY PRESS
Cold Spring Harbor, New York • www.cshlpress.com

RNA: A Laboratory Manual

Printed in the United States of America

Publisher	John Inglis
Acquisition Editor	Alexander Gann
Managing Editor	Judy Cuddihy
Developmental Editors	Judy Cuddihy and Maria Smit
Project Manager	Inez Sialiano
Permissions Coordinator	Carol Brown
Production Editor	Rena Steuer
Desktop Editor	Susan Schaefer
Production Manager	Denise Weiss
Book Marketing Manager	Ingrid Bernirschke
Sales Account Managers	Jane Carter and Elizabeth Powers
Cover Designer	Mike Albano

Library of Congress Cataloging-in-Publication Data

RNA : a laboratory manual / Donald C. Rio ... [et al.].
p. ; cm.
Includes bibliographical references and index.
ISBN 978-0-87969-890-4 (hardcover : alk. paper) -- ISBN 978-0-87969-891-1 (pbk. : alk. paper)
1. RNA--Analysis--Laboratory manuals. I. Rio, Donald Charles.
[DNLM: 1. RNA--Laboratory Manuals. 2. Genetic Techniques--Laboratory Manuals. QU 25]
QP623.R5692 2010
572.8'8--dc22

2010030576

10 9 8 7 6 5 4 3 2 1

All Cold Spring Harbor Laboratory Press publications may be ordered directly from Cold Spring Harbor Laboratory Press, 500 Sunnyside Blvd., Woodbury, New York 11797-2924. Phone: 1-800-843-4388 in the Continental U.S. and Canada. All other locations: (516) 422-4100. FAX: (516) 422-4097. E-mail: cshpress@cshl.edu. For a complete catalog of all Cold Spring Harbor Laboratory Press publications, visit our website at http://www.cshlpress.com/.

Contents

Preface, xi
Acknowledgments, xiii

CHAPTER 1 ■ The Fundamentals, 1

Introduction, 1
Common Sense in Dealing with RNA, 1
Purification Methods and When to Use Them, 5
Recovering Purified RNA: Guidelines for Precipitation, 6
Resuspending Pelleted RNA, 8
Assessing the Quantity and Quality of Isolated RNA after Purification, 8
Removing Ribosomal RNAs, 9
Enrichment of mRNA Using Oligo(dT) Cellulose, 10
Enriching for Species of RNA Other Than mRNAs, 10
Miscellaneous (but Important) General Items, 11
Summary, 14

CHAPTER 2 ■ Purification of RNA from Natural Sources, 17

Overview, 17
Protocols
1 SDS Solubilization and Phenol Extraction, 19
2 Purification of RNA Using TRIzol (TRI Reagent), 23
3 Ethanol Precipitation of RNA and the Use of Carriers, 27
4 Guidelines for the Use of RNA Purification Kits, 30
5 Preparation of Cytoplasmic and Nuclear RNA from Tissue Culture Cells, 33
6 Removal of Ribosomal Subunits (and rRNA) from Cytoplasmic Extracts before Solubilization with SDS and Deproteinization, 36
7 Isolation of Total RNA from Yeast Cell Cultures, 38
Method 1: Vegetative Cells, 38
Method 2: Meiotic Cells, 40
8 Bacterial RNA Isolation (*E. coli*), 44

9 Removing rRNA from Deproteinized, Phenol-extracted Total RNA, 48
Method 1: Enzymatic Digestion of rRNA, 48
Method 2: Selective Precipitation of Large RNAs with LiCl or PEG/NaCl, 49
Method 3: Removal of rRNA by Hybrid Selection, 51
10 Enrichment of Poly(A)$^+$ mRNA Using Immobilized Oligo(dT), 53
11 Removal of DNA from RNA, 56
12 Gel Electrophoresis, 58
Method 1: Polyacrylamide Gel Electrophoresis, 58
Method 2: Agarose Gel Electrophoresis (Nondenaturing), 63
13 Determining the Yield and Quality of Purified RNA, 67

CHAPTER 3 ■ Detection and Characterization of Specific RNAs, 75

Overview, 75
Probe Preparation, 75
Northern Blotting, 77
Nuclease Protection, 78
Primer Extension, 78
Other Methods, 80
PCR-based Approaches, 80
Specialized Methods for Detecting Small RNAs, 86
Protocols
1 Labeling of Oligonucleotide Probes (DNA, LNA, RNA) by Polynucleotide Kinase and [γ-^{32}P]ATP, 89
2 DNA Oligonucleotide Radiolabeling by Terminal Deoxynucleotidyl Transferase (TdT), 92
3 Asymmetric Polymerase Chain Reaction to Generate Single-stranded Probes, 95
4 Random Hexamer ^{32}P Radiolabeling of DNA Fragments as Hybridization Probes, 97
5 Nick Translation of Double-stranded DNA for the Preparation of ^{32}P-Labeled Hybridization Probes, 100
6 Enzymatic Dephosphorylation of RNA Using Calf Intestine Alkaline Phosphatase or Shrimp Alkaline Phosphatase, 103
7 Northern Blots: Denaturation and Electrophoresis of RNA with Formaldehyde, 106
8 Northern Blots: Denaturation and Electrophoresis of RNA with Glyoxal, 110
9 Northern Blots: Capillary Transfer of RNA from Agarose Gels and Filter Hybridization Using Standard Stringency Hybridization Conditions, 114
10 Northern Blots: Alternative Method for Processing Northern Blots after Capillary Transfer, 121
11 Northern Blots for Small RNAs and MicroRNAs, 126
12 RNase Protection Assay, 131
13 Nuclease S1 Protection Mapping, 136
14 Primer Extension Analysis of RNA, 141

15 Reverse Primer Extension, 145
16 Poisoned Primer Extension, 149
17 Detection of RNAs by 3′-end Labeling and RNase H Digestion, 152
18 Reverse Transcription-Polymerase Chain Reaction, 155
Method 1: Reverse Transcription (RT): cDNA Priming and Synthesis, 158
Method 2: Polymerase Chain Reaction and Detection of Products, 160
19 Basic Quantitative PCR Using Real-time Fluorescence Measurements, 165
20 5′ Rapid Amplification of cDNA Ends, 171
21 3′ Rapid Amplification of cDNA Ends, 180
22 RACE on Uncapped or Nonpolyadenylated RNAs, 186
23 RNA Sequencing by Primer Extension, 189
24 Direct Chemical Sequencing of End-labeled RNA, 192
25 Splinted Ligation Method to Detect Small RNAs, 200

CHAPTER 4 ▪ Synthesis, Purification Labeling, and Substitution of Transcripts Synthesized In Vitro, 205

Overview, 205
In Vitro Transcription: The Basics, 205
Yield of In Vitro–Transcribed RNAs, 205
Labeling of In Vitro–Synthesized RNAs, 207
Protocols
1 In Vitro Transcription of RNA: Synthesis, Labeling, and Substitution, 210
2 High-yield Synthesis of RNA Using T7 RNA Polymerase and Plasmid DNA or Oligonucleotide Templates, 216
3 Determining the Yield of RNA Synthesized In Vitro, 220
4 Gel Purification of RNA, 222
5 3′-End Labeling of RNA with [^{32}P]pCp and T4 RNA Ligase 1, 225
6 3′-End Labeling of RNA with Yeast Poly(A) Polymerase and 3′-Deoxyadenosine 5′-[α-^{32}P]Triphosphate (Cordycepin 5′-[α-^{32}P]Triphosphate, 228
7 5′-End Labeling of RNA with [γ-^{32}P]ATP and T4 Polynucleotide Kinase, 231
8 Site-specific Labeling and Substitution of RNA, 234

CHAPTER 5 ▪ Detecting and Defining RNA–Protein and RNA–RNA Interactions, 241

Introduction, 241
General Methods for Detecting RNA–Protein Interactions, 241
RNA Structure Probing, 243
Mapping Sites of RNA–Protein Interactions, 244
Summary, 252

Protocols

1 Filter-binding Assay for Analysis of RNA–Protein Interactions, 254
2 Native Polyacrylamide-gel-electrophoresis-binding Assay for RNA–Protein Complexes: Gel Mobility Shift Assay or Electrophoretic Mobility Shift Assay, 257
3 SELEX to Identify Protein-binding Sites on RNA, 262
4 RNA Structure Determination Using Chemical and Nuclease Digestion Methods, 269
Method 1: RNA Structure Determination Using Chemical Methods, 269
Method 2: RNA Structure Determination Using Nuclease Digestion Methods, 273
5 Mapping Sites of RNA–Protein Interactions Using Chemical Methods, 276
6 Mapping Sites of RNA–Protein Interactions Using Nuclease-based Methods: RNase Footprinting and Reverse Footprinting, 284
Method 1: RNase Footprinting, 284
Method 2: Reverse Footprinting, 288
7 Modification Interference and Nucleotide Analog Interference Mapping, 291
Method 1: Chemical Modification Interference, 291
Method 2: Nucleotide Analog Interference Mapping, 294
8 Boundary Analysis, 298
9 Ultraviolet Photochemical Cross-linking to Detect RNA-binding Proteins, 301
10 Photocross-linking Using Uridine Analogs, 304
11 Detecting RNA–RNA Interactions Using Psoralen Derivatives, 307
12 Cross-linking with Formaldehyde, 311
13 Basic Affinity Selection Methods, 314
Method 1: Basic Selection with Beads, 314
Method 2: Selection with a Biotinylated Oligonucleotide, 316
Method 3: Selection of a Ribonucleoprotein Using a Complementary Biotinylated Oligonucleotide, 317

CHAPTER 6 ■ Analysis of RNA-processing Reactions Using Cell-free Systems, 321

Introduction, 321
Guidelines for the Preparation of Active Cell-free Systems, 321
Guidelines for Testing and Optimizing Extracts, 322
Developing Cell-free Systems from Poorly Studied Organisms, 323
Protocols

1 Preparation of Nuclear Extracts from HeLa Cells, 325
2 Analysis of Pre-mRNA Splicing Using HeLa Cell Nuclear Extracts, 330
3 Preparation of *Drosophila* Kc Cell Nuclear Extracts for In Vitro Splicing, 334
4 In Vitro Splicing Reactions in *Drosophila* Kc Nuclear Extracts, 339
5 Preparation and Analysis of Cell-free Splicing Extracts from *Saccharomyces cerevisiae*, 343
6 Analysis of Splicing In Vitro Using Extracts of *Saccharomyces cerevisiae*, 348
7 Analysis of Splicing Complexes on Native Gels, 352

CHAPTER 7 ■ RNA Interference, 355

Introduction, 355
RNAi: The Basics, 356
The Design of RNAi Experiments: General Considerations and Strategies, 357
RNAi with Long dsRNA, 359
Protocols
1 dsRNA-induced RNAi in *Drosophila* Cells by Soaking, 363
2 dsRNA-induced RNAi in *Drosophila* Cells by Transfection, 366
3 dsRNA-induced RNAi in *C. elegans* by Feeding with dsRNA-expressing *E. coli*, 370
4 RNAi in *C. elegans* by Injection of dsRNA, 374
5 RNAi in Cultured Mammalian Cells Using Synthetic siRNAs, 380
6 Construction of Transgenic *Drosophila* Expressing shRNAs in the miR-1 Backbone, 387
7 Creating an miR30-based shRNA Vector, 395
8 Packaging shRNA Retroviruses and Host Cell Infection, 399
9 Infection of Mammalian Cells with Retroviral shRNAs, 403
10 Creating Transgenic shRNA Mice by Recombinase-mediated Cassette Exchange, 407

CHAPTER 8 ■ Genomic Approaches, 419

Overview, 419
Microarrays, 420
Sequencing of RNAs by High-throughout Methods, 424
Identifying RNAs Associated with Specific Proteins, 428
Analysis of Genomic Data, 431
Protocols
1 Preparation of Fluorescent-dye-labeled cDNA from RNA for Microarray Hybridization, 435
2 Microarray Slide Hybridization, 440
3 Scanning Microarray Slides, 446
4 Tips on Hybridizing, Washing, and Scanning Affymetrix Microarrays, 451
5 Fragmentation of Whole-transcriptome RNA Using *E. coli* RNase III, 457
6 Preparation of Small RNA Libraries for High-throughput Sequencing, 460
7 Large-scale Immunopurification of RNP Complexes from *Drosophila* Nucleoplasmic Extracts for Tiling Microarrays (RNA Immunoprecpitation Microarray; RIP-Chip), 471
8 CLIP (Cross-linking and Immunoprecipitation) Identification of RNAs Bound by a Specific Protein, 476
9 Preparation of mRNA-Seq Libraries from Poly(A)$^+$ mRNA for Illumina Transcriptome High-throughput Sequencing, 491
10 Methods for Processing Microarray Data, 496
11 Methods for Processing High-throughput Sequencing Data, 503

APPENDICES

Common Recipes, 513

Quick Reference for Enzymes Commonly Used in RNA Research, 523

Protocols, 525

Protocols

1 RNA-friendly Plasmid Preparation, 526

2 Preparing Cellular DNA from Nuclei or Whole Cells, 532

3 Expression and Purification of Active Recombinant T7 RNA Polymerase from *E. coli*, 535

4 Preparing Size Markers for Gel Electrophoresis, 541

5 Toeprinting, 544

6 Cell Fractionation, 547

7 hnRNP-Enriched Nucleoplasmic Extracts from *Drosophila* or Mammalial Tissue Culture Cells, 548

8 Measuring the Length of Poly(A) Tails, 554

Method 1: Low-resolution Measurement of Poly(A) Tails Using Northern Blotting, 555

Method 2: High-resolution Measurement of Poly(A) Tails Using Northern Blotting, 555

Method 3: PCR-based and Ligation-mediated Approaches for Measuring Poly(A) Tail Length, 557

Cautions, 559

Index, 567

Preface

RNA MOLECULES PARTICIPATE IN AND REGULATE a vast array of cellular processes, and the scientific community is now well into a new era in which some aspect of RNA biology—as a tool, a therapeutic, a diagnostic, or a central player in a fundamental biological process—is becoming increasingly important. Nevertheless, beginning RNA research can be daunting, and without a thorough understanding of the challenges and complexities inherent in handling this fragile nucleic acid, forays into the RNA world can be quite frustrating. For example, the introduction of many kits designed to perform everything from purification to high-throughput sequencing has been both a boon and a bane to RNA science. Although these kits have made RNA analysis accessible to nearly every laboratory, they give many researchers a false sense of security—they believe that whatever comes out of the final tube is exactly what they expect. Unfortunately, unbeknownst to the investigator, this is often not the case.

We have written this manual with all RNA scientists—novices and more advanced researchers — in mind. Chapters 1 through 3 are specifically targeted to those new to the area or those needing some "refreshing" of fundamental principles and techniques. We have emphasized "common sense" approaches and highlight the fact that it is essential to monitor the yield and quality of any RNA preparation after each step. Anyone contemplating RNA research should be comfortable with the material in these chapters before moving on to the more complex procedures in the chapters that follow.

Chapters 4 through 6 are written for those who wish to investigate RNA-related topics in more depth. Here, we describe a variety of techniques, some quite complex, designed primarily to explore specific RNA–protein interactions and elucidate the mechanisms of RNA-processing reactions. Chapter 7 describes current methodologies for using RNA interference as a tool for determining gene function, and Chapter 8 presents methods for performing increasingly popular genome-wide investigations. Finally, the Appendix includes a few methods that we have found to be useful but did not fit into any particular area, as well as recipes for commonly used reagents and some general reference information.

Almost all of the methods described in this manual are routinely used in one or more of our laboratories. In the few cases where we do not have direct experience, we have enlisted the expertise of trusted colleagues. Although we have attempted to be comprehensive, we realize that this is an impossible task. New approaches, especially in the area of genome-wide analyses, appear with high frequency. Moreover, no one method is ideally suited to any particular question. We fully realize that there are alternative approaches available for nearly every method that we describe. In addition, we acknowledge that many specialized techniques, such as circular permutation and direct approaches to RNA analysis (e.g., fingerprinting and thin-layer chromatography), have not been described at all because we believe that these methods would be of limited appeal to most RNA researchers.

It is often difficult to identify with any precision the source of particular methods because as they are adopted by different laboratories, they "evolve." Accordingly, we have not attempted to provide extensive referencing for the protocols; only a few representative publications are cited. We apologize in advance to all of those authors who have published methods or improvements thereon who have not been cited in this manual.

Finally, we note that many "standard" molecular biology techniques, such as protein gels, western blotting, cloning, etc., are not covered in this manual. Such material is covered in great detail in the latest edition of *Molecular Cloning: A Laboratory Manual*, and readers requiring this information are strongly encouraged to consult that manual in conjunction with this one.

In *RNA: A Laboratory Manual*, we have attempted to describe a broad scope of current techniques used in RNA research. We hope that this manual will enable any researcher to approach a wide variety of problems from the most fundamental to the highly sophisticated with confidence and high expectation of success.

DONALD C. RIO
MANUEL ARES, JR.
GREGORY J. HANNON
TIMOTHY W. NILSEN

Acknowledgments

ALL OF THE AUTHORS THANK Alex Gann, Cold Spring Harbor Laboratory Press, for suggesting and initiating this project and for his patience through many delays. We also thank John Inglis, Publisher, for his advice and support. Inez Sialiano, Project Manager, and Rena Steuer, Production Editor, provided guidance through the production process. We also are especially grateful to Judy Cuddihy for her exceptional efforts in pulling the book together from start to finish. Without her hard work and extraordinary patience, this manual would not exist.

Don Rio would like to thank the current and previous graduate students and postdocs who have contributed to developing and/or perfecting the protocols that his lab contributed to this manual. These individuals are Melissa Adams, Asoka Amarasinghe, Julie Aspden, Eileen Beall, Marco Blanchette, Lucille Fresco, Amanda Hannaford, Roland Kanaar, Paul Kaufman, Emmanuel Labourier, Charles Lee, Enbo Ma, Bosun Min, Sima Misra, Siobhan Roche, David Rudner, Kathleen Rusilas, Alex Sabogal, Christian Siebel, Mei Tang, and Brian Weinert. Don Rio would also like to thank the following investigators who sent protocols from their labs: David Bartel, Douglas Black, Marco Blanchette, Kathy Collins, Bob Darnell, Xiang-Dong Fu, Brenton Graveley, Adrian Krainer, Robin Reed, Jeremy Sanford, and Shalini Sharma.

Manny Ares thanks Jeremy Sanford, Brent Graveley, Lily Shiue, Melissa Cline, John Paul Donohue, and Lisa Munding for their help.

Greg Hannon thanks Chris Hammel (CSHL) and Rene Ketting (Hubrecht Laboratories) for help in preparation of protocols for *C. elegans* RNAi. Rui Zhou, Jian Quan Ni, Norbert Perrimon, the *Drosophila* RNAi Screening Center (DRSC), and the Transgenic RNAi Project (TRiP, Harvard) provided many protocols for RNAi in *Drosophila*. Prem Premsrirut, Young Park, and Scott Lowe (CSHL) developed and provided the RMCE protocol for shRNA mice. Amy Valentine, Ken Chang, and Krista Marran provided protocols for RNAi in mammalian cells. Jim Duffy (CSHL) drew the figures for Chapter 7.

Tim Nilsen is extremely grateful to Pat Maroney who wrote (and perfected) all of the methods from the Nilsen lab. In addition, she read and in some cases revised methods from other laboratories. Her efforts on this book cannot be overstated. He would also like to thank Ann Marie Micenmacher for her patience and exceptional talents in preparing the manuscript. In addition, he thanks John Denker and Jesse Fisher for their input and the late Hugh Robertson who introduced Tim to RNA research 30 years ago.

All of the authors are grateful to the following investigators who supplied protocols that appear in the book: Richard Davis (University of Colorado Health Sciences), RACE protocols; Jim Manley (Columbia), SELEX; Bob Darnell (The Rockefeller University), CLIP; Mark Caprara (Case Western Reserve University), structure determination; and Brent Graveley (University of Connecticut), high-throughput sequencing.

1 The Fundamentals

Introduction, 1
Common Sense in Dealing with RNA, 1
Purification Methods and When to Use Them, 5
Recovering Purified RNA: Guidelines for Precipitation, 6
Resuspending Pelleted RNA, 8
Assessing the Quantity and Quality of Isolated RNA after Purification, 8
Removing Ribosomal RNAs, 9
Enrichment of mRNA Using Oligo(dT) Cellulose, 10
Enriching for Species of RNA Other Than mRNAs, 10
Miscellaneous (but Important) General Items, 11
Summary, 14
Reading List, 14

WHY A BOOK DEVOTED TO techniques used in RNA research now? There has been an explosion of interest in RNA due to multiple discoveries, especially during the last 10 years, that have revealed the wide-ranging roles that RNA has in gene expression and regulation. It is now clear that much of the genome is transcribed into RNA, even though only a small fraction of these transcripts code for proteins. And it has become apparent that a majority of genetic diseases are caused by mutations that cause mistakes in splicing, alternative splicing, or other RNA-processing reactions. Moreover, microRNAs (miRNAs) are now known to regulate more than two-thirds of all genes, and these tiny RNAs have been shown to function as both oncogenes and tumor suppressors. As a result, it is difficult to imagine a biological question that does not involve some aspect of RNA metabolism. The ability to purify, analyze, and manipulate RNA is now essential for any laboratory working in the life sciences; however, the skills and practices required to work with RNA are not present in every laboratory, and initiating RNA research can be intimidating. Our intent in this manual is to provide enough information so that any competent researcher can confidently handle RNA and perform meaningful experiments from the most basic to the highly sophisticated.

Before describing the nuts and bolts of laboratory RNA research, we remind the reader of a few basic facts. In any cell or tissue, the vast majority (~95%) of the mass of RNA is composed of ribosomal RNAs (rRNAs). Messenger RNAs (mRNAs) account for ~3% of total RNA, and the remaining 1%–2% of total RNA contains a remarkably complex array of noncoding RNAs, including transfer RNAs (tRNAs), small nuclear RNAs (snRNAs), large noncoding RNAs, and tiny RNAs such as miRNAs, just to name a few. Several fundamental considerations should be kept in mind in all RNA work, and these issues are discussed in this chapter.

COMMON SENSE IN DEALING WITH RNA

The major challenge in any aspect of RNA research is preparing intact RNA molecules and keeping them intact through any subsequent manipulations. Because of the 2′-hydroxyl group attached to the pentose ring, the RNA backbone is intrinsically more sensitive to breakage than DNA. This is because the 2′-hydroxyl group adjacent to phosphodiester linkages can act as an intramolecular nucleophile attacking the adjacent phosphodiester bond, creating a 2′, 3′ cyclic phosphate and cleaving the RNA backbone (Fig. 1-1). There are three major causes of RNA degradation—alkaline pH, divalent metal ions, and ribonucleases—that are all easily preventable.

pH

The higher the pH of a solution, the more readily the hydrogen of the 2′-hydroxyl group will be abstracted by equilibrium deprotonation, causing the 2′ oxygen to become more electronegative. The deprotonated 2′ oxygen is well positioned for nucleophilic attack on the phosphate, thus breaking the RNA strand, a reaction termed alkaline hydrolysis (Fig. 1-2). Subsequent attack by a hydroxide on the 2′, 3′ cyclic phosphate results in a mixture of 2′ and 3′ nucleoside phosphates.

Just a few minutes of RNA in pH 9 or more can cause significant alkaline hydrolysis; thus, it is essential to keep RNA in solutions that are buffered to near neutrality, ideally pH 6–7. Higher pHs may be required for some applications, but it is never advisable to keep RNA in solutions above pH 8 for long periods of time. Obviously, the longer the RNA molecule, the more phosphates it has and the more chance that alkaline hydrolysis will occur. Thus, hydrolysis is more readily observed on large RNAs than on smaller RNAs.

Metal Ions

A second common cause of RNA degradation comes from exposure to divalent metal ions, such as Mg^{++}, Mn^{++}, Fe^{++}, Zn^{++}, and Pb^{++}. These ions are positively charged and so can interact well with the negatively charged phosphodiester backbone. At neutral pH, divalent metal ions can form inner-sphere interactions with phosphate oxygens and position a metal hydoxide that abstracts the proton more efficiently from the 2′-hydroxyl group, causing strand scission via hydrolysis, as described above for high-pH solutions. Binding of divalent metals is often essential for proper RNA folding and function, and not all metal-binding events trigger hydrolysis. The relationship between RNA and metals is complex; therefore, the RNA experimenter must manage metals carefully.

To prevent general metal damage to RNA, ensure that solutions of RNA are made in deionized or distilled water and contain a metal chelator such as EDTA. The main source of divalent metals will be contaminants of other buffer components such as sodium acetate. Normally, 0.1–1 mM EDTA is sufficient to chelate contaminating divalent metal ions. For applications in which metals are required, it is essential to ensure that the metal concentrations used are in appropriate excess of any EDTA pres-

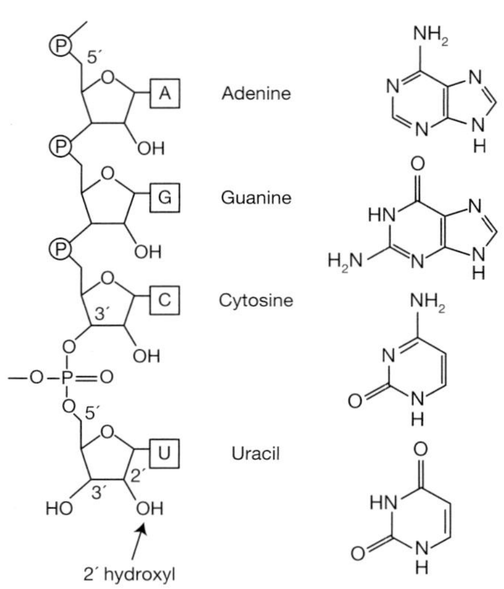

FIGURE 1-1. Chemical structure of RNA. RNA is similar to DNA in that it contains four bases linked by 3′ to 5′ phosphodiester bonds. However, the fourth base in RNA is uracil (instead of thymidine, as in DNA) and the sugar phosphate of its backbone is ribose (instead of deoxyribose, as in DNA).

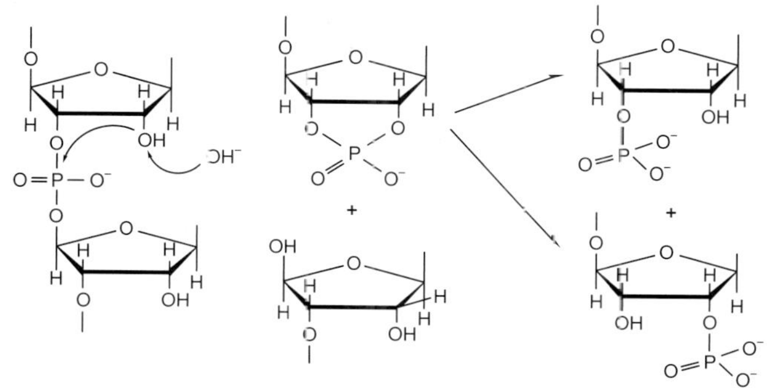

FIGURE 1-2. Alkaline hydrolysis of RNA. (Redrawn from http://www.biochem.arizona.edu/classes/bioc462/462a/NOTES/Nucleic_Acids/RNA_Hydrolysis.gif>.)

ent so that the required effective concentration of the metal is available for the reaction. Various chelators with different affinities for different divalent ions are available. EDTA is widely used because of its affinity for the most common contaminant Mg^{++} and its good general affinity for other metals. EGTA is a good chelator of Ca^{2+} but not Mg^{++}, and trinitrilcacetate (TNA) is particularly good for chelating Zn^{++}.

Ribonucleases

By far, the most vexing and pernicious cause of RNA degradation is the presence of ribonuclease contamination. Unfortunately, RNases are everywhere and can be present in dust, improperly treated glassware, contaminated pipettes, our own skin, bacteria, molds, or anything that has been in contact with a contaminated surface. It is worth noting that many RNases, e.g., RNase A, are resistant to metal-chelating agents and can survive prolonged autoclaving or boiling. In addition, many bacterial RNases are resistant to autoclaving. It is of utmost importance to recognize potential sources of RNase and pay strict attention to detail in avoiding them. The use of RNase inhibitors, such as RNasin (Promega) and many other commercial preparations, may give a false sense of security. Not only are these inhibitors expensive, but they also only inactivate certain RNases and only under reducing (>5 mM dithiothreitol [DTT]) conditions. The bottom line is that RNase inhibitors have several useful applications, but it is best to avoid RNase contamination in the first place.

All solutions used in RNA work must be RNase-free. This begins with RNase-free water. Many commercial suppliers sell RNase-free water, and for laboratories without a dedicated still, this is probably the best way to begin. If highly deionized water is used in place of distilled water, the deionizing systems must be well maintained. Filter units can become contaminated with molds and bacteria that secrete RNases into the deionized water, and autoclaving will not inactive them. Of course, the water is only as clean as the container in which it is kept. It is essential to bake all glassware for 4 h at 200°C before use. Several sources mention using diethyl pyrocarbonate (DEPC) for treatment of glassware and solutions. We do not recommend this because DEPC is not completely decomposed by autoclaving, and residual DEPC will react with adenosine residues in the RNA. Do not bake plastic caps; they will melt. Caps can be separately wrapped in aluminum foil and autoclaved. A more expensive alternative is to store RNase-free water in sterile plasticware. Obviously, exposure to air (and dust) should be kept to a minimum.

Once RNase-free water has been prepared, the next step is to prepare RNase-free solutions such as buffer and salt stock solutions. All chemicals to be used for RNA work should be dedicated for that purpose. For example, Tris, NaCl, $MgCl_2$, etc. should be purchased solely for RNA-related solutions, and, if available, they should be certified to be RNase-free. When possible, prepare concentrated stocks (e.g., nucleoside triphosphates [NTPs] and enzymes) directly in the bottle in which they arrive by adding RNase-free water to achieve the desired stock concentration. NTP stocks prepared in water from powder need to be neutralized (pH ~7–8) with NaOH for long-term storage at –80°C. Working stock aliquots can be kept at –20°C.

When it is not possible to prepare the stock solutions directly, it is important to practice common sense when weighing and dissolving chemicals. For example, never remove a dry chemical using a spatula of any kind—even a clean metal spatula will promote the contamination of the chemical stock with dust and metals. Always pour the chemical into a clean weight boat, determine the weight, transfer it to a baked beaker, and then add RNase-free water in the appropriate amount to achieve the desired final concentration.

Finally, a common source of RNase contamination in the laboratory is dirty stir bars. Stir bars can accumulate tiny iron and steel filings that clump together with nonmagnetic dust and detritus. Always use stir bars that have been extensively washed in sodium dodecyl sulfate (SDS), rinsed extensively with RNase-free water, and baked for at least 4 h at not more than 200°C. Even when all of these steps are followed carefully, some RNase contamination is still possible. If this is the case, perform a final filtration step with a 0.2-µm Nalgene filter.

Once RNase-free solutions have been made, it is critical not to introduce RNase at subsequent steps. Any laboratory planning to do RNA work should have pipetting aids (e.g., Pipetman) dedicated to this work. Many a failed experiment can be traced to the use of a contaminated Pipetman. Imagine spending the morning doing DNA minipreps that involve the pipetting of high concentrations of pancreatic ribonucleases. If the same Pipetman is used for RNA work in the afternoon, it is possible that those carefully prepared solutions will become contaminated with pancreatic RNase and the RNA will be degraded and useless. Even though no obvious contamination may be apparent, aerosols created during pipetting may coat the end and the dead-space surfaces of the pipettor, only to be released upon subsequent pipetting into the dead space and onto the sample in the tip. Restricting a special set of pipettors for RNA work is a simple and effective method of preventing such cross-contamination (see also Proper Use of Micropipettors, below).

All RNA work should be done with gloves to avoid contamination with "fingerases." Common sense is important here as well. Gloves are worthless if they have been used to open doors, talk on the telephone, or handle other items already exposed to RNases.

The foregoing precautions may seem at first to be extreme, but they become easy with practice and attention to detail. In our own labs, we routinely have beginning students pass the "RNase test." We ask them to make their own solutions, prepare and purify a labeled RNA of several hundred nucleotides, and then incubate the labeled RNA with each of their solutions at their working concentrations for at least 1 h at 37°C. If the RNA remains intact under these conditions, the students pass the test and gain confidence that they can work with RNA. We recommend that any laboratory beginning RNA research of any kind perform a similar test before doing actual experiments. Even experienced labs can benefit from these simple RNase assays. They can be quite useful in pinpointing the source of RNase if contamination is suspected. Of course, this test is not useful for solutions such as Tris base or 1 M $MgSO_4$ because the pH and metal content of these solutions degrade RNA in the absence of enzymatic nucleases.

Read the Protocol!

Another "good habit" that should be acquired for all experiments with RNA (or anything else for that matter) is to review each step in the protocol thoroughly before beginning. Many RNA protocols are exceptionally time-sensitive and require immediate transitions from step to step. Thus, it is imperative that all solutions be prepared beforehand, water baths or incubators be at the right temperature, and centrifuges and rotors be precooled (or warmed) as needed. It is also advisable to go through a "dry run" with readily available material to identify any potential glitches before risking valuable (and expensive) samples.

PURIFICATION METHODS AND WHEN TO USE THEM

The introduction above described in a very general way the "good habits" that must be adopted before beginning RNA work. Here, several basic techniques that are used in all RNA experiments are described in detail, starting with methods for RNA purification. Adopting and understanding these basics will allow even the most inexperienced researcher to perform a wide spectrum of experimental approaches from the very simple to the most sophisticated. There are several different approaches to purifying RNA and they all work well when used appropriately. Each technique will be described in this section, as will the advantages and disadvantages of each. Specific protocols for many of these basic techniques can be found in Chapter 2.

Solubilizing and Deproteinizing

The first step in any purification is solubilization of the material from which the RNA is to be extracted. The basic goal is to minimize nuclease activity while optimizing recovery. With some exceptions, it is desirable to inactivate nucleases as quickly as possible, and this is accomplished by solubilizing biological material in a chaotropic salt such as guanidine isothiocyanate or a denaturing detergent such as SDS. The following techniques can be used to prepare total cellular RNA.

TRIzol

TRIzol reagent (Invitrogen) is a monophasic solution of phenol and guanidine isothiocyanate that disrupts cells and dissolves cellular components. It is used to quickly dissolve cells or tissues for preparation of RNA of all sizes. For all applications involving cell pellets or tissues, TRIzol solution is used at 10:1 volume/volume of material to be extracted. Following solubilization, add 0.2 mL of chloroform per milliliter of TRIzol solution, followed by vigorous mixing (shaking or vortexing) for 15 sec. The chloroform causes the solution to separate into two phases, and after centrifugation, the organic (red) phase is at the bottom of the tube and the aqueous (clear) phase, which contains the extracted RNA, is at the top. In some cases, there will be a visible interface that contains DNA and some denatured protein. When the aqueous phase is transferred to a fresh tube, it is important to avoid disturbing this interface or pipetting any of this material. To be absolutely certain of avoiding contamination of the aqueous phase, leave a small amount of aqueous phase behind when pipetting. The minimum extraction volume should be 200 µL to minimize loss of the aqueous phase.

The TRIzol approach can also be used to extract RNA from liquid samples (e.g., cellular extracts or enzyme reactions). For these purposes, a different formulation of TRIzol (TRIzol LS) designed for liquid samples is used (see Chapter 2, Protocol 2).

The advantages of using TRIzol are that the reagents are premade and the procedure is quick and straightforward. Disadvantages include possible contamination with DNA and the exceptionally high salt concentrations in the aqueous phase (see Recovering Purified RNA: Guidelines for Precipitation, below).

Guanidine Cesium Chloride

Another common and simple way to purify RNA involves solubilizing cells and tissues with guanidine isothiocyanate with no phenol, followed by centrifugation through a cushion of cesium chloride (CsCl). This procedure relies on the differences in buoyant density among protein, RNA, and DNA. Because of the 2′-hydroxyl group, RNA is significantly more dense than DNA, and both RNA and DNA are significantly more dense than proteins. The density of the CsCl cushion is adjusted so that the DNA does not enter the cushion, whereas the RNA traverses the cushion and pellets at the bottom of the tube. Although straightforward and an excellent way to produce rigorously DNA-free RNA, we do not recommend this procedure for general use, mainly because the RNA pellets can be very difficult to resuspend and small RNAs (e.g., miRNAs) are poorly recovered.

Column Chromatography

A third method for purifying RNA is hydrophobic interaction chromatography. This technique provides the basis for RNA purification kits (see Chapter 2, Protocol 4). Here, cells or tissues are dissolved in guanidine isothiocyanate and the extract is passed over a silica gel column in a nonpolar (organic solvent-containing) buffer. RNA adheres to the silica resin while DNA and protein are removed by washes with increasingly polar buffers. After washing, the RNA is eluted with H_2O or low-salt buffers.

SDS/Proteinase K/Phenol:Chloroform

For "routine" deproteinizations (e.g., extracting RNA from processing reactions) and removing enzymes, traditional SDS solubilization, followed by proteinase K digestion and by phenol:chloroform extraction, should be used. In brief, solutions are made with 0.25% SDS (which denatures protein and separates it from RNA). EDTA is added to ~1 mM in excess of Mg^{++} in the solution, and the sample is then digested with 100 µg/mL proteinase K for 15 min at 42°C. The proteinase K step can be skipped if the total protein concentration is low, i.e., when removing enzymes from labeling reactions. Digesting proteins greatly reduces interfaces and makes extractions cleaner. Following proteinase K digestion, the solution is made 0.3 M in sodium acetate (NaOAc), two volumes of phenol SEVAG (SEVAG is a solution of 24 parts chloroform and 1 part isoamyl alcohol; see Chapter 2, Protocol 1) are added, and the solution is mixed vigorously by vortexing. Following centrifugation, the organic phase is at the bottom and the aqueous phase containing the extracted RNA is at the top. The aqueous phase is removed (do not disturb the interface) and reextracted with chloroform isoamyl alcohol. The reextracted phase is centrifuged again and the aqueous (top) phase is removed for precipitation (see also Recovering Purified RNA: Guidelines for Precipitation, below). The advantage of this technique is that phenol extraction is much more cost effective than TRIzol or kits and it avoids the high-salt problem.

RNA Extraction from Tissues

All of the methods described above are appropriate for extraction from tissue culture cells. The following special precautions must be taken when RNA extraction is to be performed from tissues. Because tissues are often enriched for nucleases, it is essential that they be diced and snap frozen in liquid nitrogen or treated with a stabilizer such as the commercial reagent RNA*later* (Ambion).

Snap freezing followed by powdering on dry ice is an effective method for extracting RNA from tissue samples. In brief, the tissue is quickly diced or minced (with a scalpel) to fragments ≤1 mm^3 and snap frozen using liquid nitrogen. The frozen material is then transferred to a precooled mortar and pestle. Note that it is important to prevent any thawing of the tissue; thus, the mortar and pestle should be precooled in an ice bucket with dry ice and then flooded with liquid N_2 just before use. Using the cold pestle, the tissue is ground into a fine powder and liquid N_2 is added as needed. After powdering, TRIzol is added in sufficient amounts to solubilize the material. The TRIzol will freeze on contact with the frozen mortar, but upon thawing at room temperature, the tissue will be completely dissolved and ready for extraction.

RECOVERING PURIFIED RNA: GUIDELINES FOR PRECIPITATION

After any purification procedure, the RNA will be present in a CLEAR aqueous solution. If there is any turbidity (other than precipitated SDS, which should redissolve upon warming), the RNA is contaminated and should be extracted again with phenol. The standard way to recover purified RNA is to use ethanol precipitation (Chapter 2, Protocol 3). Although this procedure is deceptively simple, care must be taken to ensure optimal recovery. For example, the salt concentration should be increased to 0.3 M, usually in NaOAc at pH 5.2, before adding ethanol. It is also important that a minimal concentration of RNA be present; RNA solutions below 10 μg/mL will not precipitate efficiently. If the RNA concentration is below 10 μg/mL, add carrier RNA or another polymer, depending on the nature of the subsequent manipulations. For many applications, tRNA is suitable as a carrier; for others (e.g., when the recovered RNA is to be end-labeled or used for cloning), glycogen should be used as a carrier. Carriers serve multiple purposes: They promote efficient precipitation, prevent small amounts of RNA from sticking to the tube (an often overlooked problem), and aid in detecting pellets.

Once the RNA solution is of the appropriate concentration, it (and the carrier, if needed) can be precipitated by the addition of 2 volumes of ethanol, frozen on dry ice for ≥1.5 h or overnight at −20°C. Another simple means to chill samples for ethanol precipitation is to place a metal rack in a styrofoam container or ice bucket with dry ice and 95% ethanol; this is referred to as a dry ice–ethanol bath. If the RNA is sufficiently concentrated (~100 μg/mL), freezing may not be necessary. If a precipitate is not observed, the solution should be frozen. Following freezing, thaw the solution at room temperature and recover the precipitated RNA by centrifugation. Precipitation is not readily reversible, so it is not necessary to keep the solution cold, nor is it necessary to centrifuge in the cold. Many laboratories have a misconception about using cold centrifuges and keep microfuges in a cold room for RNA precipitation. This not only is unnecessary, but can be counterproductive as well because cold rooms can cultivate mold, a rich source of RNase.

Following centrifugation, the RNA pellets must be identified and resuspended. It can be difficult to see the pellet if you are working with small amounts of material. To make this easier, a colored carrier, e.g., GlycoBlue (Ambion), can be used for precipitations in which the anticipated yield is at or below 10 μg of RNA. The blue pellet is easily seen and readily identified. When pure RNA is precipitated without carrier, it should appear as a bluish translucent pellet, which may disappear completely when dry (Fig. 1-3). Once RNA, or RNA plus carrier, is pelleted, the ethanol must be removed. This can be accomplished by several means including aspiration with a baked Pasteur pipette or simply inverting the tube and pouring off the ethanol. Inevitably, however, some ethanol will remain in the tube and must be removed before proceeding. Ethanol per se is not the problem; it is the salt that is dissolved in it. When ethanol evaporates, the salt remains behind and this concentrated salt can make resuspension of RNA pellets quite difficult, not to mention contributing to solution composition in uncontrolled ways in later procedures. If the pelleted RNA is to be run on a gel (see below), the presence of salt can lead to distorted bands or hyperfocusing artifacts, in which the bands appear to be of normal width at the top but tiny at the bottom of the lane; in other words, an ugly gel (see RNA Analysis: Never Run an Ugly Gel, below).

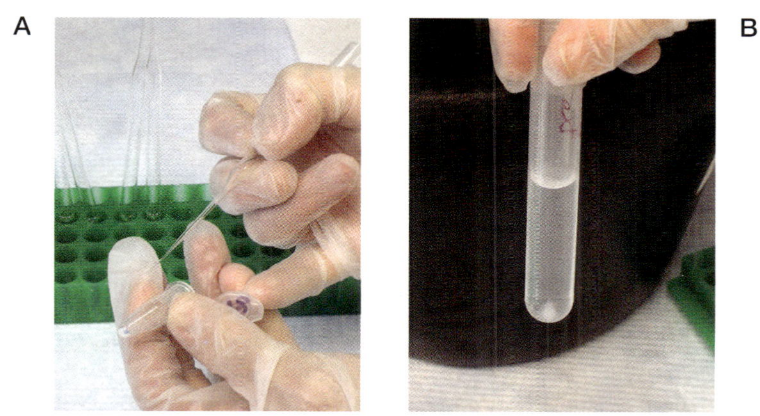

FIGURE 1-3. RNA pellets after ethanol precipitation and centrifugation. (*A*) Pellet obtained from 25 μg of total RNA with 20 μg of glycoblue added as carrier. The pulled Pasteur pipette is used to remove any excess ethanol. (*B*) Pellet obtained from 500 μg of total RNA precipitated without carrier.

The first method for removing excess ethanol is to rinse the pellet with 70% ethanol (lacking salt). The idea here is that the "fresh" ethanol will dilute the residual salt and remove most of it. A good technique is to remove the original supernatant, add cold (–20°C) 70% ethanol, chill again, centrifuge again, and remove the supernatant.

Another method is to remove the excess ethanol after the 70% wash with a drawn pasteur pipette and either briefly air dry or use a SpeedVac Concentrator for 1–2 min. However, DO NOT SpeedVac to dryness because excessively dried nucleic acids can be extremely difficult to resuspend. It is important to note that when using a SpeedVac, it should be dedicated to "drying" nucleic acids. If the apparatus is used for other purposes (e.g., for evaporating alkaline solutions), RNA degradation may be encountered.

An alternative and very reliable method for removing excess ethanol is to remove it manually using "ethanol grabbers," finely pulled Pasteur pipettes that suck up ethanol droplets by capillary action. This method takes only a few seconds to remove excess ethanol and its accompanying salt completely.

RESUSPENDING PELLETED RNA

If the procedures described above have been followed, the RNA will be pure and intact and it can be resuspended in RNase-free water. If everything is clean, RNA in solution is quite stable at room temperature and can remain on the benchtop. Nevertheless, most researchers prefer to store valuable samples in a –80°C freezer. For long-term storage (indefinite), RNA can be resuspended in 0.5% SDS at a concentration of ≥100 µg/mL RNA or RNA plus carrier. RNA stored in this way must be reprecipitated before use. For immediate use, we recommend that RNA (or RNA plus carrier) be resuspended in RNase-free water or in Tris-EDTA (TE) buffer to a final concentration of ≥100 µg/mL. RNA can then be aliquoted to appropriate concentrations and stored frozen for further analysis.

ASSESSING THE QUANTITY AND QUALITY OF ISOLATED RNA AFTER PURIFICATION

It is obviously a good idea to have an approximate estimate of the amount of RNA expected given the amount of starting material. For example, a "typical" eukaryotic cell contains ~15–30 pg of RNA, so that an RNA preparation from 10^7 cells should be about 150–300 µg; if you consider RNA that is not rRNA, then the expected amount is ~7.5–15 µg. A reasonable estimate of nuclear RNA is 2% of total RNA, so a nuclear RNA preparation from 10^7 cells would be expected to have 3–6 µg of RNA. Similar calculations can be made for the expected amount of total RNA from tissues, bacteria, or yeast.

To determine the amount of RNA prepared, take a small aliquot of the resuspended RNA and determine the absorbance at 260 nm using a spectrophotometer (see Chapter 2, Protocol 13). If the absorbance is unexpectedly high, the sample may be contaminated with phenol or DNA, both of which absorb UV light at this wavelength. Read the absorbance at 260 nm, 270 nm, and 280 nm. If the optical density (OD) at 270 nm is greater than that at 260, this indicates phenol contamination of the RNA. If the ratio of OD at 260–280 nm is less than 1.9, this may indicate protein, which absorbs light at 280 nm due to tryptophan residues. In the former case, reprecipitate with ethanol and the absorbance should come down because the phenol should be removed. In the case of DNA contamination, the absorbance at 260 nm will remain unchanged after reprecipitation, and the sample may need to be treated with DNase before proceeding (see Chapter 2, Protocol 11). If at the end of these procedures the RNA sample is far below expected yield (≤30%) or significantly more than expected (≥ threefold), something is wrong with either the calculations or the preparation. In the latter case, we recommend starting over.

Although the foregoing procedure tells the amount of RNA, it does not tell anything about the quality of the preparation; i.e., is the RNA intact or degraded? A quick and valuable assay for RNA integrity is to subject a small aliquot (~1 μg) of RNA (total or cytoplasmic) to electrophoresis on a nondenaturing agarose gel along with a known intact RNA sample and/or RNA molecular-weight markers. After ethidium bromide or SYBR staining, two bright and distinct bands corresponding to 28S and 18S rRNAs should be seen. A general "fuzziness" or smearing down from the major rRNA bands indicates that substantial degradation has occurred. In addition, a molar ratio of large-to-small subunit RNAs of less than 1 may indicate preferential degradation of the large subunit RNA, suggesting that the sample has been compromised. If this is the case, it may be advisable to repurify RNA from the beginning.

Such gels are also useful for "normalizing" different RNA preparations; e.g., samples prepared from cells exposed to different treatments. For example, to determine if the level of a specific mRNA changes in abundance in different cell types or tissues, it is always advisable to load the equivalent amount of total RNA; normalizing to rRNA content is a valuable first step.

A second, but more cumbersome, way to assess RNA integrity (and to normalize RNA content) is to perform northern blotting by probing for an abundant "housekeeping" mRNA such as actin or tubulin. This procedure should be used either when rRNA has been removed (see below) or when rRNA is not abundantly present (e.g., nuclear RNA).

More recently, Agilent has introduced the Bioanalyzer, a capillary electrophoresis "chip" that measures RNA amounts and integrity (see Chapter 2, Protocol 13). The traces produced by this machine are convenient to analyze for relative yield and quality of large and small rRNAs and RNA amount relative to a standard. These data are archivable and can be analyzed in a variety of ways that allow many RNA samples to be compared under uniform standards.

REMOVING RIBOSOMAL RNAS

Although rRNAs can be useful for estimating the quality (and quantity) of total RNA purifications, they can also be quite problematic. It is important to remember that rRNAs (28S, 18S, 5.8S, and 5S in most higher eukaryotes) comprise >95% of total cellular RNA and thus are abundant components of the preparation that can interfere with subsequent analysis. The large rRNAs can prevent hybridization to mRNAs on a northern blot when the mRNA is of similar size, "trap" and otherwise disturb the migration of other RNA species during electrophoresis, and "soak up" (compete for) label during 5′- or 3′-end-labeling reactions.

It is often desirable to remove rRNAs either before (preferable) or after RNA purification. The fact that rRNAs are present in massive (>30S) particles makes it possible to remove them before denaturation using sedimentation. As detailed in Chapter 2, Protocol 6, centrifugation of cytoplasmic extracts can effectively remove the bulk of ribosomal subunits and therefore rRNA. The advantage of this approach is that it is quick and easy. Its disadvantages are that it can only be used for

RNA ANALYSIS: NEVER RUN AN UGLY GEL

By far, the most common and useful way of fractionating and analyzing RNA is by gel electrophoresis. There is nothing more satisfying than looking at a pretty gel where all of the bands are straight and crisp, the lanes are uniform (not fat at the top and narrow at the bottom), the gel does not "smile," and there are no streaks or distortions. Pretty gels should be the norm, but the literature is heavily contaminated with "ugly" gels. The most common causes of poor-quality gels are overloading, excess salt in the RNA to be analyzed, badly prepared gels, gels running too quickly, and inadequate sample preparation. It is imperative to avoid shortcuts in both the preparation and running of the gel. It is also important to choose the appropriate gel matrix (agarose or acrylamide), as well as the appropriate percentage and the appropriate amount of cross-linker. Also note that gel boxes, combs, etc., must be ones dedicated to RNA work; never assess the integrity of RNA using gel boxes that have been exposed to nuclease. Chapter 2, Protocol 12 describes how to make gels for RNA work and their most appropriate uses.

RNA preparations from cells lysed and subjected to subcellular fractionation procedures that may compromise RNA integrity. In addition, some large nonribosomal ribonucleoproteins (RNPs) (e.g., mRNPs) of mass >30S may be removed along with the ribosomal subunits.

There are two methods for removing rRNAs after purification. The first is enzymatic degradation using a processive exonuclease that cleaves 5′→3′ and requires a 5′ monophosphate terminus. Because all mRNAs and many small RNAs contain a GpppG 5′ cap, they are resistant to 5′ exonuclease digestion. This technique is very appealing in principle. However, the exonuclease is extremely expensive, and large amounts of it are necessary to convert all rRNA sequences into mononucleotides. Furthermore, because miRNAs are uncapped, they will also be degraded by the nuclease. Thus, this procedure should be used sparingly or when the situation warrants (see Chapter 2, Protocol 9).

The second method for removing rRNA after purification is to use hybrid selection. In this procedure, biotinylated DNA oligonucleotides complementary to rRNA are mixed with the sample, and the oligonucleotides hybridize to the rRNAs in solution. The hybrids are removed by avidin affinity chromatography via biotinylated DNA. This procedure can be effective in removing bulk rRNA (see Chapter 2, Protocol 9). Commercial kits are available for yeast, mouse, and human rRNAs, but they are expensive.

ENRICHMENT OF MRNA USING IMMOBILIZED OLIGO(DT) CELLULOSE

Although removal of rRNA is often desirable, it is also possible to enrich for a specific class of RNA while leaving rRNA behind. For mRNAs, affinity purification via oligo(dT) is the best and easiest way. Almost all eukaryotic mRNAs possess a poly(A) tail (rRNAs and other RNAs do not). Therefore, it is possible to take advantage of the ability of these poly(A) stretches to hybridize to strings of oligo(dT). Fortunately, these hybrids are only stable in high salt (0.5 M), and they dissociate at room temperature when salt is absent. Passing total RNA over immobilized oligo(dT) (either on magnetic beads or on cellulose) in high salt retains the mRNA while allowing other RNAs to remain in solution and flow through the column. Following removal of the other RNAs, mRNAs can be recovered by elution with water (see Chapter 2, Protocol 10). This technique is valuable for enriching for mRNAs, but it does have limitations; for example, mRNAs that are not naturally polyadenylated (e.g., histone) and mRNAs that are deadenylated will not be retained. The latter situation may be especially problematic because of the recent demonstration that miRNAs may promote deadenylation, and thus it is possible that oligo(dT) selection may potentially result in the loss of miRNA-targeted mRNAs.

ENRICHING FOR SPECIES OF RNA OTHER THAN MRNAS

It is possible to enrich for certain other types of RNAs as well. For example, snRNAs associated with Sm proteins invariably contain a trimethylated guanosine cap structure, and there is a commercially available monoclonal antibody that specifically recognizes this cap. Therefore, immunoprecipitation with this antiserum can effectively enrich for Sm snRNAs as well as certain small nucleolar RNAs (snoRNAs).

The anti-2,2,7-trimethylguanosine (anti-TMG) antibody works well with deproteinized RNAs, but immunoprecipitation can also be used to effectively recover classes of RNA that are bound to common proteins. For example, snoRNAs of the Box C/D type are associated with a common core of protein constituents, as are snoRNAs of the Box H/ACA type with a different set of proteins. Similarly, miRNA-protein complexes contain a member of the Argonaute protein family. By using appropriate antisera (before exposure to any denaturant), it is possible to highly enrich for the desired RNAs. Such enrichment can be very useful for many applications, including deep sequencing.

MISCELLANEOUS (BUT IMPORTANT) GENERAL ITEMS

The Importance of Master Mixes

Most protocols in this manual require assembling multiple components in small-volume reactions. We have observed many beginning researchers attempt to add each ingredient separately to a tube. This is problematic for at least two reasons. First, for example, if a specific reaction calls for 3 mM $MgCl_2$ in total volume of 10 µL, one would have to add 0.03 µL of 1 M $MgCl_2$ or 0.3 µL of a 100 mM stock. Some pipetting aids can, in theory, pipette such small amounts, but it is highly unlikely if not impossible for an average person with an average Pipetman to achieve such accuracy, because the variance in the sample obtained is near the total volume attempted. In fact, a major source of error is inaccurate pipetting. Viscous solutions such as glycerol are especially problematic. Thus, the absolute minimum volume to pipette should be 1 µL. (For pipetting guidelines, see Proper Use of Micropipettors, below.) A second major problem with pipetting small amounts is successful mixing of the assembled ingredients. It is extremely difficult to successfully mix small volumes of concentrated solutions, especially if the solutions have vastly different densities and viscosities.

To avoid both of these problems, master mixes should be used whenever possible. The concept of master mixes is straightforward. All of the "inactive" ingredients (salt, buffer, glycerol, nucleotides, etc.) are mixed in a separate tube to 10x or 5x of their final concentrations. It is then possible to add a reasonable amount of master mix to the actual tube to be used; i.e., 1 µL of 10x mix or 2 µL of 5x mix for a 10-µL reaction.

Additionally, reactions should be assembled in a specific order: (1) water, (2) master mix, (3) RNA, (4) and extract or enzyme. Following these simple steps will serve to increase reproducibility and thus reduce experimental error.

Proper Use of Micropipettors

The use of micropipettors in molecular biology research is now entrenched to the point where instruments such as Gilson Pipetman, Eppendorf Reference, Oxford Benchmate, or Rainin Classic pipettes are used in a wide variety of critical applications. Despite the high impact on research quality that the use of these instruments has, it is also common to observe them being used improperly, even by veteran researchers or, more entertainingly, by actors portraying forensic scientists on TV. Training in proper use of micropipettors is almost certain to improve the quality of research in any lab.

Micropipettors share a common construction, consisting of a plunger that is moved by the user's thumb or index finger a specific distance that can be adjusted by the user. The distance that the plunger moves (the "throw") and the cross-sectional area of the plunger represent the volume that the investigator intends to pipette. This volume is calibrated so that the adjustment dial shows the volume, and adjusting for the volume actually adjusts the throw of the plunger. A short throw will provide a small volume, and a longer throw will provide a larger volume. Thus, the mechanical adjustment that controls the throw must be calibrated properly for the pipettor to be accurate. This leads to the first and second rules. The first rule is to use a pipettor that has been properly calibrated by an expert. The second is to *never* set the pipettor to below zero or above the maximum capacity of the pipettor. Doing so ensures that the calibration will be compromised, due to the forces placed on the mechanical stops that maintain the proper throw of the mechanism.

The range of volumes that can be accurately and safely pipetted is smaller than the range of adjustment of the instrument. Here is where the investigator must apply judgment. Although a 1000-µL pipettor can be adjusted to deliver 10 µL, it will not do so accurately. A 20-µL pipettor should be used. Most 20-µL pipettors can be used down to 2 µL with confidence; a 200-µL pipettor will be fine for 20 µL but not less, and a 1000-µL pipettor is not reliable below 100 µL. At the high end, the problem to be avoided is contamination of the pipettor, because the surface of the sample comes quite close to the barrel, affording an opportunity for splashing and bubbling to wash material from one sample to the other in undesirable ways. Use the correct-size pipettor for the volume needed.

Let's say that the pipettor is calibrated and set properly, and the disposable tip is placed snuggly on the end of the barrel of the pipettor. To obtain the sample, the plunger is depressed to the first of two "stops," or positions in the throw that can be felt by the user. The tip is inserted vertically 2–3 mm (not more) into the solution, and the plunger is slowly released. What just happened? As the plunger was released, a vacuum was created in the chamber between the end of the plunger and the surface of the liquid sample at the opening of the tip (the "dead space"). Under this reduced pressure, the liquid sample moved into the tip to equalize the pressure inside the chamber with the atmospheric pressure, resulting in a volume of liquid equal to the volume described by the movement of the plunger. If done correctly, this will be true; however, there are several ways to go wrong.

Inserting the tip >2–3 mm into the solution will create additional pressure against the dead space, resulting in more than the desired amount of liquid. Holding the pipettor off from vertical will also result in more sample, because the tip will support sample on the sides in excess of sample that is being supported by the pressure equilibration. Snapping the plunger up instead of allowing it to return slowly is incorrect for several reasons. First, snapping creates unnecessary forces on the calibration stops, and continued snapping will rapidly cause the instrument to be incorrectly calibrated. Second, a rapid snap creates a more extreme transient pressure gradient in the dead space that will strain the critical pressure seals between the disposable tip and the end of the barrel, as well as in the O-ring at the end of the plunger. When these seals are strained by an extreme pressure gradient, they leak air into the dead space, and that air, rather than the desired volume of liquid, will accomplish the pressure equalization. In addition, snapping can generate aerosols that will cross-contaminate reagents and reactions (see below).

Dispense the sample by inserting the tip 2–3 mm into the diluent liquid, or by holding the tip against the side surface of the container near the bottom, and slowly depress the plunger to the first stop. Wait a second and then depress the plunger to the second stop to force out any residual liquid adhering to the inside of the tip. Do not try to dispense the volume through midair, because the surface tension of the liquid will likely prevent clean separation of the liquid from the tip. Squirting is not appropriate. Holding the tip in the diluent or against the side of the surface prevents droplets from being flung and lost.

Problematic Solutions: Viscosity, Volatility, and Wetting

Many solutions are made with glycerol, which can make accurate pipetting a challenge. The viscosity of glycerol, especially cold glycerol, creates resistance to flow and increases transient extreme pressure gradients in the dead space. This means that slower plunger movement and increased patience for equilibration may be needed. It is wise to avoid pipetting small volumes of high-percentage glycerol solutions if accuracy is important. For best accuracy, use the master mix concept (see above) and make the critical measurements using larger volumes of lower-percentage glycerol. Glycerol also clings to the outside of the tip. Obviously, the volume inside the tip is the desired amount and yet several microliters of a viscous solution might cling to the outside. It may require extra attention to determine how deep the tip is inserted when pipetting such solutions; alternatively, the outside of the tip may be carefully contacted with the edge of a clean Kimwipe to reduce the amount of carryover.

Sometimes, organic chemicals and acids need to be pipetted. Volatile materials of this type require extra attention both for safety and to protect the instrument. The consequence of snapping the plunger while pipetting a volatile acid such as hydrochloric acid can be severe, because acid aerosol may permeate the instrument and corrode the metal pieces. We do not recommend using micropipettors for concentrated acid solutions. Organics such as phenol and chloroform have very different surface tension properties from aqueous solutions and also form aerosols readily. Although organics do not generally harm the metal pieces of the pipettor, the integrity of the plastic and rubber fittings inside may be compromised by organics. To reduce the chance of contaminating the barrel, do not approach the maximum volume of the pipettor with organics. For example, pipetting 200 μL of an organic should be done using a 1000-μL pipettor rather than a 200-μL pipettor. Again, do not snap the plunger or the organics will leap into the barrel and make the experiments go poorly.

Some solutions can successfully "wet" a surface and some cannot. A solution with SDS will generally be a good wetting agent, because the detergent disrupts the forces that create surface tension in the liquid. Reduced surface tension can also occur in protein solutions, because proteins can also be good surfactants. Bubbles can easily form in such solutions, and bubbles are unwanted. First, bubbles in a solution volume close to the maximum volume of the pipettor can easily contact the barrel, even if the plunger is moved slowly. Thus, consider using the larger of two equally qualified pipettors to maintain sample distance. In addition, bubbles pop, and when they do, tiny droplets of sample are flung into and throughout the dead space, contaminating the barrel, plunger, and O-rings. Snapping the plunger or otherwise creating an extreme pressure gradient in the dead space can pop many bubbles at once due to the pressure change, leading to massive contamination. Extra caution and care are required for solutions that may bubble.

Checking Pipettor Accuracy

If attention is paid to the above rules and common sense is applied, using a micropipettor should produce accurate and reproducible results. To check a micropipettor, it is simple to use a balance and pure water. Water weighs 1 gm/mL, so 1 μL of water should weigh 1 mg. The accuracy (mean difference between the expected and the observed) and precision (standard deviation from the mean) of pipettors can be tested without spending a lot of time. Be aware that weighing 1 μL of water can be tricky due to the rate of evaporation.

The Importance of K_m: Remembering Enzyme Kinetics

In any enzymatic reaction, the rate (and extent) of product formation is determined by the concentration of substrate. Each enzyme has a characteristic affinity for its substrate, and for many enzymes required for procedures described in this book, the K_m (concentration of substrates required for half-maximal enzyme velocity) can be quite high. Most enzymes are sold in "units of activity." It is important to note that units are normally determined under optimum reaction conditions where substrates are at saturating concentrations.

Many investigators make the mistake of thinking that if they add enough units of enzyme to their reactions, the reaction will go to completion. This is often not the case because the substrate concentration may be a fraction of the K_m of the enzyme. For example, the K_m of T7 RNA polymerase for NTPs is ~160 μM. If synthesis of in vitro transcripts is attempted using low (≤10 μM) concentrations of NTPs (e.g., to increase specific activity; see below), the rate of incorporation will be very low, and vanishingly small amounts of RNA will be made regardless of the amount of units of the enzyme. It is extremely important (especially for radioactive applications) to make certain that substrate concentrations are sufficient to ensure that the enzymes used will be active for the application. It is often desirable (and necessary) to mix radioactively labeled substrate with nonradioactive substrate to reach the appropriate concentrations (see Chapter 4, Protocol 1).

Specific Activity: Important Considerations

Specific activity is defined as the amount of radioactivity per mass unit of material and is usually expressed as Ci/mole or Ci/mmole. A Ci (Curie) is defined as 2.22×10^{12} radioactive disintegrations per minute. As presented in detail in several chapters, many RNA analyses rely on the use of ^{32}P. We trust that radiation safety officers have provided appropriate training in the proper handling and disposal of radioactive materials. Our purpose here is to provide guidance for the practical use of ^{32}P in RNA experiments. In our experience, specific activity is one of the most misunderstood concepts, and improper use of radiolabeling is commonplace.

Many researchers appear to believe "the hotter the better." This is often not the case. Consider radiolabeled nucleotides. Most α-labeled nucleotides are delivered at 400–800 Ci/mmole. Therefore, 800 μCi is equivalent to 800 nmole and 80 μCi has a mass equivalent of 0.1 nmole. If 0.1 nmole is suspended in 50 μL (a typical in vitro transcription reaction size), the concentration will be 2 μM, far below the K_m of the enzyme. Therefore, if RNA synthesis is attempted with only radiolabeled material, investigators will be highly disappointed. Most transcription products will be less than full length (abortive transcripts), and the full-length transcripts (if any) will be subject to radiolytic decay, which can cause strand scission. Similar considerations apply to the synthesis of RNA and DNA probes.

Intuitively, many investigators believe that the "hotter" the probe, the more sensitivity. This is not always correct because hybridization is a bimolecular reaction, and the mass of the probe can drive the reaction forward. In addition, if the RNA is a substrate for an enzyme with a K_m for RNA that is higher than the very low RNA concentration achieved when high-specific-activity RNA is used, the reaction may not proceed. We discuss appropriate specific activities for each application as they come up in specific protocols, but, in general, whenever a probe or transcript is made by polymerization, it is advisable to include unlabeled nucleotides along with the labeled material. An exception to this is 3′- or 5′-end labeling with T4 RNA ligase or kinase. These are not polymerization reactions, and in most cases, it is possible to achieve the appropriate concentrations using only labeled nucleotide.

SUMMARY

This chapter covers many basic aspects of working with RNA. We cannot stress enough that a fundamentally sound technique is essential for successful execution of the protocols detailed in this manual. We believe that anyone can perform high-quality experiments using a combination of good habits, common sense, and attention to detail.

In succeeding chapters, we provide step-by-step methods for a large variety of RNA-related experimental approaches. We begin in Chapter 2 by describing basic methods for purifying and fractionating RNA, followed by methods in Chapter 3 for detecting and quantitating individual RNAs in a purified preparation. We detail several approaches to detect both large and small RNAs, and highlight the advantages and disadvantages of each.

We then move to more mechanistic types of analyses that require the use not of natural transcripts but of those that are synthesized in vitro. In Chapter 4, we first detail how to synthesize such transcripts and then describe a variety of specialized methods to label and modify synthetic RNAs for subsequent functional studies. Chapter 5 describes the many uses of synthetic transcripts, including the analysis of RNA–protein interactions, how RNAs can be used as affinity reagents, and approaches that use cross-linking reagents. Chapter 6 discusses and presents methods for using synthetic RNAs to monitor processing reactions such as splicing. Also included are methods to prepare active subcellular extracts.

In Chapter 7, we turn to some basic methods used in RNA interference, realizing that such methods can (and do) occupy entire books. We then conclude the main text with a suite of methodologies in Chapter 8 used for genome-wide studies of RNA expression, including arrays and deep sequencing and how to analyze the voluminous data that are generated by these approaches.

Finally, a variety of methods that we have found to be useful in our laboratories but do not mesh well with the major topics above can be found in the Appendix. The following books and articles listed as references will provide the interested reader with an introduction to the complex world of RNA.

READING LIST

Moore MJ. 2005. From birth to death: The complex lives of eukaryotic mRNAs. *Science* **309:** 1514–1518.

Nilsen TW. 2003. The spliceosome: The most complex macromolecular machine in the cell? *Bioessays* **25:** 1147–1149.

Nilsen TW. 2007. RNA 1997–2007: A remarkable decade of discovery. *Mol Cell* **14:** 715–720.

Regulatory RNAs. 2006. Cold Spring Harbor Symposia on Quantitative Biology LXXI (ed. B Stillman and D Stewart), p. 574. Cold Spring Harbor Laboratory Press, Cold Spring Harbor, New York.

The Ribosome. 2001. Cold Spring Harbor Symposia on Quantitative Biology LXVI. Cold Spring Harbor Laboratory Press, Cold Spring Harbor, New York.

RNA Structure and Function. 1998. Cold Spring Harbor Monograph Series 35 (ed. RW Simons and M Grunberg-Manago). Cold Spring Harbor Laboratory Press, Cold Spring Harbor, New York.

The RNA World. 2006. Third Edition. Cold Spring Harbor Monograph Series 43 (ed. RF Gesteland et al.). Cold Spring Harbor Laboratory Press, Cold Spring Harbor, New York.

Translational Control in Biology and Medicine. 2007. Cold Spring Harbor Monograph Series 48 (ed. MB Mathews et al.), p. 934. Cold Spring Harbor Laboratory Press, Cold Spring Harbor, New York.

2 Purification of RNA from Natural Sources

PROTOCOLS

1 SDS Solubilization and Phenol Extraction, 19
2 Purification of RNA Using TRIzol (TRI Reagent), 23
3 Ethanol Precipitation of RNA and the Use of Carriers, 27
4 Guidelines for the Use of RNA Purification Kits, 30
5 Preparation of Cytoplasmic and Nuclear RNA from Tissue Culture Cells, 33
6 Removal of Ribosomal Subunits (and rRNA) from Cytoplasmic Extracts before Solubilization with SDS and Deproteinization, 36
7 Isolation of Total RNA from Yeast Cell Cultures, 38
8 Bacterial RNA Isolation (*E. coli*), 44
9 Removing rRNA from Deproteinized, Phenol-extracted Total RNA, 48
10 Enrichment of Poly(A)$^+$ mRNA Using Immobilized Oligo(dT), 53
11 Removal of DNA from RNA, 56
12 Gel Electrophoresis, 58
13 Determining the Yield and Quality of Purified RNA, 67

CHAPTER 1 OUTLINED A VARIETY of fundamental and practical considerations applicable to all experiments dealing with RNA. Chapter 2 presents detailed protocols for several basic techniques. It is impossible to envision every experimental scenario, but the procedures described in this chapter should have general utility. It is important to remember to adjust the scale of each approach to the scale of your particular applications.

In general, all RNA purification approaches involve three steps: solubilization, deproteinization (i.e., removing protein and sometimes DNA from RNA, also called extraction), and recovery of RNA free from other contaminants. Some protocols are designed to remove DNA during extraction, whereas in others contaminating DNA is removed after recovery of protein-free nucleic acids. All purification methods are designed to minimize loss of RNA due to endogenous (or exogenous) nucleases and to maximize recovery. In many cases, it is advantageous to enrich for specific classes of RNAs such as messenger RNAs (mRNAs) or small RNAs either before or after purification. Solubilization is achieved by dissolving biological material in either ionic detergents (e.g., sodium dodecyl sulfate [SDS]) or chaotropic salts (e.g., guanidinium isothiocyanate). For soft tissues (e.g., liver, blood) and tissue culture cells, solubilization is accomplished simply by adding a solution of detergent or chaotropic salt. In other cases, e.g., bone, kidney, most plant cells, and other cells with rigid cell walls, it is necessary to disrupt the material enzymatically or mechanically before solubilization.

In this chapter, we first discuss widely applicable general methods for solubilization and extraction of RNA; we then discuss general methods for recovering extracted RNA. We show how these methods can be used to purify RNA from tissue culture cells, yeast, and bacteria, and subsequently outline a variety of methods designed to either enrich or deplete specific types of RNA once total RNA is purified. The strategies for RNA purification are outlined in Figure 2-1.

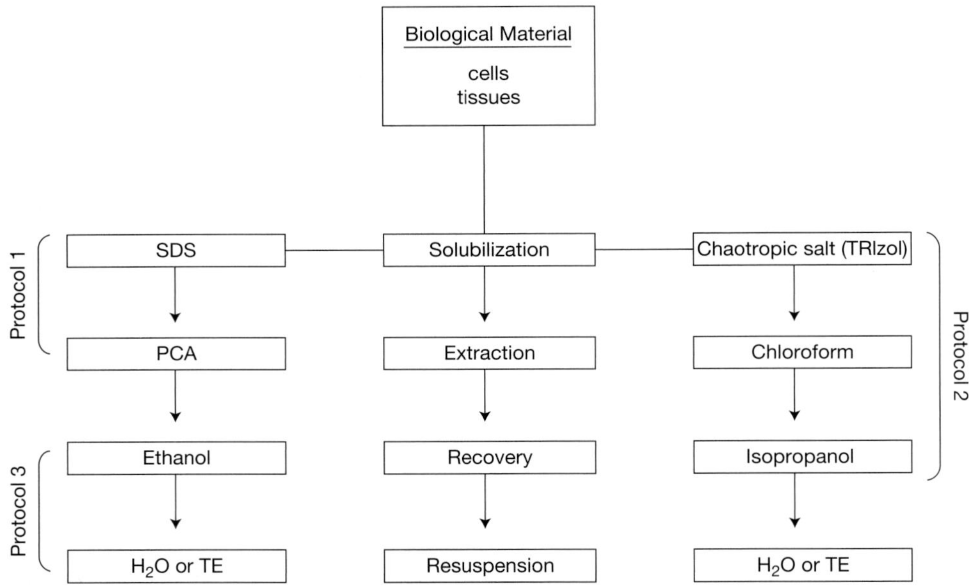

FIGURE 2-1. Flow chart outlining purification strategies.

WWW RESOURCES

http://www.ambion.com/techlib/append/rna_yields.html RNA yields from tissues and cells, technical resource appendix, Applied Biosystems.

http://www1.qiagen.com/HB/RNeasyMini/ 2006. *RNeasy Mini Handbook*, 4th ed. QIAGEN.

Protocol 1

SDS Solubilization and Phenol Extraction

This method is of wide utility and is routinely used to deproteinize RNAs in biological material that has been solubilized in SDS. SDS (at least 0.25%) is an ionic detergent that dissolves membranes, disrupts protein–nucleic acid interactions, and inactivates ribonucleases. Once solubilized, addition of phenol (or phenol:chloroform:isoamyl alcohol [PCA] 25:24:1) completely denatures the protein, and it becomes insoluble in aqueous solution. Throughout the rest of the book, this method is referred to as PCA extraction. PCA extraction is the method of choice for preparing cytoplasmic RNA from tissue culture cells or in any other situation (e.g., enzyme reactions) where solubilization in SDS is easily achievable.

Note that SEVAG is 24:1 chloroform:isoamyl alcohol, which is often mixed 1:1 with TE-saturated neutralized phenol to give PCA (Sevag et al. 1938).

MATERIALS

The recipe for the item marked with <R> is at the end of the protocol or in the Common Recipes Appendix.

CAUTION: See the Cautions Appendix for appropriate handling of materials marked with <!>.

Reagents

Biological material or sample containing RNA and protein (e.g., analytical reactions)
Chloroform <!>
Isoamyl alcohol <!>
Phenol (purchased), saturated and equilibrated <!>
Proteinase K (Roche) (10 mg/mL in 20 mM Tris, pH 7.5) <!>
SDS extraction buffer (10x) <R>
Sodium acetate (NaOAc) (3 M, pH 5.2, RNase-free)

Equipment

Centrifuge (large scale, Beckman J6B or equivalent) or microcentrifuge (small scale)
Freezer (–80°C) (optional; see Step 2)
Incubator or water bath (42°C) (optional; see Step 2)
Pipettes according to scale, e.g., Pipetman tips or baked 5-mL and 10-mL glass pipettes
Tubes (1.5-mL microcentrifuge or 12-mL polypropylene [Sarstedt] with lids)
Vortex mixer

OVERVIEW

Before beginning this protocol, prepare the PCA 25:24:1 solution, 10x SDS extraction buffer, 3 M NaOAc, and proteinase K. In addition, it is necessary to have the RNA-containing sample prepared and ready for solubilization and extraction. Such samples can be cytoplasmic extracts from tissue culture cells or "soft" tissues (see Protocol 5), analytical RNA-processing reactions, or enzyme reactions. It is never advisable to use SDS solubilization and phenol extraction to prepare RNA from material that contains large amounts of DNA (e.g., whole mammalian cells or mammalian cell nuclei). Such

FIGURE 2-2. Phases formed after centrifugation in phenol extraction procedure.

samples become highly viscous when exposed to SDS. In these situations, we recommend using TRIzol (see Protocol 2).

In general, the aqueous RNA- and protein-containing solution will be brought up to 0.3 M NaOH, 0.5% SDS, 1 mM EDTA, and 20 mM Tris-HCl, pH 7.5, and emulsified with the organic mixture. The SDS disrupts protein–nucleic acid interactions and, together with the chloroform and phenol, denatures the proteins so that they become insoluble in the aqueous solution. The charged nucleic acids will remain in the salty aqueous phase, whereas the proteins, with their aromatic and aliphatic side chains, will partition to the organic phase. Because phenol:chloroform is immiscible with water, centrifugation results in the formation of two phases: a lower organic phase that contains denatured proteins and an upper aqueous phase that contains nucleic acid (Fig. 2-2). The isoamyl alcohol facilitates the separation of the two phases. Removing the aqueous phase recovers the RNA from the protein, and a second organic extraction with chloroform removes the considerable amount (~6% by volume) of phenol that dissolves in the aqueous phase with the RNA and that can contaminate the RNA. When protein concentration is high (i.e., when cytoplasmic preparations are extracted), some denatured protein can form an insoluble gel at the organic–aqueous interface. Because this gel can trap some nucleic acid, it is recommended that samples containing high concentrations of proteins be predigested with proteinase K (this is one of the few enzymes that is active in SDS). Treatment with proteinase K greatly reduces or eliminates the interface. When appropriately performed, PCA extraction results in a quantitative yield of RNA from SDS-solubilized material. The procedure is outlined in the flow chart in Figure 2-3.

METHOD

For specific methods for processing of biological materials (e.g., cells and tissues) see Protocols 5, 7, and 8.

1. Starting with a solution of the RNA and protein to be processed, estimate the volume to be extracted (i.e., the total volume of the aqueous RNA solution should be ≥200 µL; if not, increase to 200 µL with RNase-free water). Note this starting volume and whether the solution has a large amount of protein (as in a cell extract) or a small amount (≤100 µg/mL, as in an enzyme reaction or for RNA that has been partially purified already).
2. Add a 1/9 volume 10x SDS buffer to the solution of RNA and protein, e.g., cytoplasmic extract (see Protocol 5), to be extracted. (We recommend 200 µL as the minimum volume to be extracted.) Vortex and keep at room temperature for immediate extraction; otherwise, freeze at –80°C until extraction. If frozen, thaw quickly and proceed to next steps.
3. In all cases, except those with minimal protein concentration (e.g., enzyme reactions such as removing DNase or ligase), add proteinase K to a concentration of 100 µg/mL and incubate the

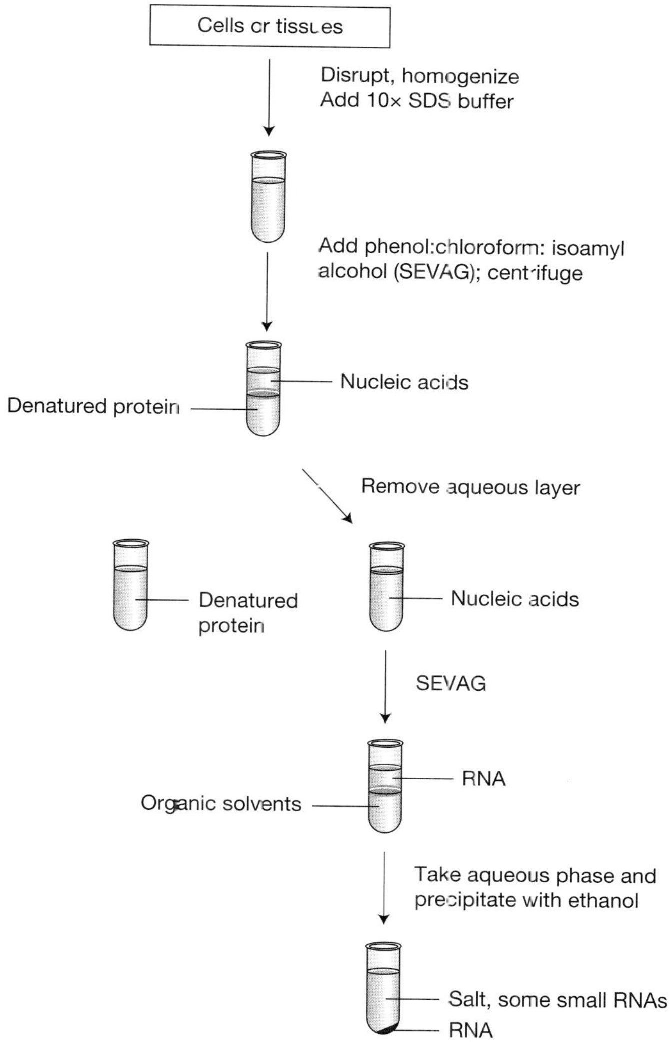

FIGURE 2-3. Flow chart for phenol extraction procedure.

mixture for 20 min at 42°C. If the protein concentration is negligible, skip this step and move to Step 4.

When in doubt regarding protein concentration, use proteinase K.

4. Make the solution 0.3 M NaOAc, pH 5.2. For every volume of solution to be extracted, add 2 volumes of PCA and vortex vigorously for 1 min. Centrifuge (maximum speed in a microcentrifuge for 2 min or at 2000 rpm in a J6B or clinical centrifuge for 5 min at room temperature) to separate the phases. The organic phase will be at the bottom (see Fig. 2-2). If protein is present, it will appear as a white precipitate at the interface between the phases.
5. Remove the aqueous (top) phase, which contains the RNA, to a fresh tube. Be careful not to disturb the interface and if necessary, leave a drop of the aqueous phase behind to avoid contamination of the RNA sample with phenol.
6. Add 2 volumes of chloroform:isoamyl alcohol without phenol to the aqueous phase. Vortex vigorously to remove residual phenol from the aqueous phase.
7. Centrifuge as previously in Step 4. Remove the aqueous phase for subsequent ethanol precipitation (see Protocol 3). We recommend that samples be ethanol precipitated immediately.

8. Resuspend RNA in H_2O, quantitate, and check quality on an agarose gel (see Protocol 13).

TIPS AND TROUBLESHOOTING

1. SDS and potassium salts (e.g., KCl or KOAc) are not compatible; this combination forms an insoluble potassium dodecyl sulfate precipitate. Accordingly, never use potassium salts ≥20 mM for phenol/SDS extraction.
2. SDS and chloroform can form a white emulsion that prevents phase separation. This occurs when the salt concentration is too low; ensure that the salt concentration in the aqueous solution to be extracted is at least 0.3 M.
3. After PCA extraction, the aqueous phase should be clear. If not, dilute it with an equal volume of 1x SDS buffer and repeat the extractions.

SOLUTION AND REAGENT RECIPE

CAUTION: See the Cautions Appendix for appropriate handling of materials marked with <!>.

SDS Buffer (10x)

Reagent	Quantity (for 500 mL)	Final concentration (10x)
SDS (10%) <!>	250 mL	5%
EDTA (500 mM, pH 8.0) <!>	10 mL	10 mM
Tris-HCl (1 M, pH 7.5) <!>	100 mL	200 mM
H_2O	140 mL	

Store indefinitely at room temperature.

READING LIST

Sevag MG, Lackman DB, Smolens J. 1938. The isolation of the components of streptococcal nucleoproteins in serologically active form. *J Biol Chem* **124:** 425–436.

Protocol 2

Purification of RNA Using TRIzol (TRI Reagent)

TRIzol solubilization and extraction is a relatively recently developed second general method for deproteinizing RNA. This method is particularly advantageous in situations where cells or tissues are enriched for endogenous RNases or when separation of cytoplasmic RNA from nuclear RNA is impractical. TRIzol (or TRI Reagent) is a monophasic solution of phenol and guanidinium isothiocyanate that simultaneously solubilizes biological material and denatures protein. After solubilization, the addition of chloroform causes phase separation (organic and aqueous), much like PCA extraction, where protein is extracted to the organic phase, DNA goes to the interface, and RNA remains in the aqueous phase. Therefore, RNA, DNA, and protein can be purified from a single sample (hence the name TRIzol). Nevertheless, TRIzol is expensive and RNA pellets can be difficult to resuspend. We do not recommend the use of TRIzol instead of regular phenol extraction when the latter technique is practical. In addition, TRIzol extraction is an effective method for isolating small RNAs, such as microRNAs (miRNAs), piwi-associated RNAs (piRNAs), or endogeneous, small interfering RNAs (endo-siRNAs).

MATERIALS

The recipe for the item marked with <R> is at the end of the protocol or in the Common Recipes Appendix.

CAUTION: See the Cautions Appendix for appropriate handling of materials marked with <!>.

Reagents

Cells or tissues for extraction
Chloroform <!>
ddH_2O
Ethanol (100%) <!>
Glycogen or GlycoBlue (Ambion, Applied Biosystems)
Isopropanol (isopropyl alcohol) <!>
NaOAc
Phenol <!>
SDS buffer (1×) <R>
TRIzol (Invitrogen) or TRI Reagent (Molecular Research Center, Inc.) <!>

Equipment

Centrifuge or microcentrifuge
Dry ice <!>
Freezer (–20°C and –80°C)
Homogenizer
Hood
Liquid nitrogen <!>
Microcentrifuge tubes (1.5-mL or 12-mL polypropylene tubes [Sarstedt] with lids)
Mortar and pestle, precooled
Pasteur pipettes (glass, 5 mL or 10 mL)
Tube rocker
Vortex mixer

OVERVIEW

TRIzol extraction is a straightforward way to prepare RNA from cells or tissues. TRIzol LS can also be used to extract RNA from liquid samples or from yeast and bacteria. When used appropriately, the RNA yield is quantitative.

METHOD

Preparation of Samples

1. For tissues:
   i. Snap-freeze the tissue to be extracted and powder on dry ice. Alternatively, using a pre-cooled mortar and pestle, grind the frozen tissue under liquid nitrogen. A less desirable alternative is to use a mechanical homogenizer.
   ii. Add 1 mL of TRIzol per 50–100 mg of tissue. For larger amounts, scale up accordingly.
   Never let the sample volume exceed 10% of the TRIzol volume. It is essential to dissolve the tissues as quickly as possible in TRIzol.
   iii. Once the sample has been completely solubilized, it can be kept for a few hours at room temperature or frozen for later use at –80°C. The RNA is stable indefinitely at –80°C.
2. For cells:
   i. Add 1 mL of TRIzol per 10^6 tissue culture cells. For larger amounts, scale up accordingly.
   Never let the sample volume exceed 10% of the TRIzol volume. It is essential to dissolve the cells as quickly as possible in TRIzol.
   ii. Collect the cells by centrifugation and resuspend the cell pellets (for cell preparation, see Protocol 5) in TRIzol by pipetting with a pasteur pipette, inverting the tubes, or vortexing.
   iii. Once the sample has been completely solubilized, it can be kept for a few hours at room temperature or frozen for later use at –80°C.
3. For liquid samples:
   i. Add 3 volumes of TRIzol LS to 1 volume of liquid sample.

For All Samples after Solubilization

4. Following solubilization for 5 min at room temperature, add 0.2 mL of chloroform per milliliter of TRIzol.
   It is important that the chloroform does not contain any additives such as isoamyl alcohol used in Protocol 1. The addition of chloroform promotes phase separation.
5. Mix vigorously by shaking or vortex for 15 sec. Allow the mixture to stand for 2–3 min at room temperature.
6. Centrifuge at maximum speed in a microcentrifuge or 10,000*g* in a centrifuge for 10 min at room temperature.
   At this point, there should be two phases: an upper clear one that contains RNA and a lower red one that contains protein. DNA is at the interface.
7. Remove the upper clear phase to a fresh tube taking care not to disturb the interface and add 0.5 mL of isopropanol per each milliliter of the clear phase.

The isopropanol will precipitate the RNA. If the expected RNA concentration is ≤10 µg/mL, it is necessary to add a carrier (i.e., glycogen or GlycoBlue) (see Protocol 3) before adding isopropanol.

8. Mix vigorously by rapid shaking or vortexing and then let stand for 10 min.
9. Collect the precipitated RNA by centrifugation at maximum speed in a microcentrifuge or 10,000*g* in a centrifuge for 10 min at 4°C.

   The use of GlycoBlue as a carrier will greatly aid in visualizing the pellet.
10. Carefully pour off the supernatant; the RNA pellet (white, or blue if GlycoBlue has been used) will appear as a triangle extending from the bottom of the tube slightly upward. Remove any remaining liquid with a pulled pasteur pipette.
11. Because of the extremely high salt concentration present in TRIzol, the pellet may be difficult to resuspend. We recommend immediately resuspending the pellet (without drying) in 1x SDS extraction buffer. Resuspension may take several minutes, and if GlycoBlue has been used, it is easy to visualize this process.
12. After resuspension, add NaOAc to 3 M. Then reextract the solution with PCA and precipitate with ethanol (see Protocols 1 and 3). To assess the quantity and quality of the RNA, see Protocol 13.
13. The purified RNA can be stored in ethanol at –20°C or recovered by centrifugation and resuspended in H_2O. The resuspended RNA should be stored at –20°C or at –80°C for long-term storage (i.e., indefinitely).

TIPS AND TROUBLESHOOTING

1. The suppliers of TRI Reagent and TRIzol recommend washing the isopropanol pellet with 75% ethanol and vortexing. We advise against this step; we have found that immediate resuspension followed by phenol extraction gives much cleaner RNA.
2. Although TRIzol removes large DNA molecules efficiently, it does not remove plasmid DNA or DNA fragments efficiently. These can be problematic for subsequent PCR applications.
3. Unexpectedly low RNA yield can occur when the initial solubilization step is not effective; i.e., it is too short or not enough TRIzol has been used. Make sure that the solution is clear and of uniform viscosity.

SOLUTION AND REAGENT RECIPE

CAUTION: See the Cautions Appendix for appropriate handling of materials marked with <!>.

SDS Extraction Buffer

Reagent	Quantity (for 500 mL)	Final concentration
SDS (10%) <!>	25 mL	0.5%
EDTA (500 mM, pH 8.0) <!>	1 mL	1 mM
Tris-HCl (1 M, pH 7.5) <!>	10 mL	20 mM
H_2O	464 mL	

Store for up to 1 yr at room temperature.

READING LIST

Chirgwin JJ, Przbyla AE, MacDonald RJ, Rutter WJ. 1979. Isolation of biologically active ribonucleic acid from sources enriched in ribonuclease. *Biochemistry* **18:** 5294.

Chomczynski P, Sacchi N. 1987. Single-step method of RNA isolation by acid guanidinium thiocyanate-phenol-chloroform extraction. *Anal Biochem* **162:** 156–159.

Chomczynski P, Sacchi N. 2006. Single-step method of RNA isolation by acid guanidinium thiocyanate-phenol-chloroform extract: Twenty-something years on. *Nature Prot* **1:** 581–585.

Protocol 3

Ethanol Precipitation of RNA and the Use of Carriers

Recovery of deproteinized RNA is performed almost always via precipitation with ethanol (but see Protocol 2 for precipitation from TRIzol extractions). If the RNA is free of contaminants (i.e., it is well extracted) (see Protocol 13), it will resuspend easily and be ready for any subsequent manipulation. Ethanol precipitation is quite useful because it provides quantitative recovery of any sized RNA, from several kilobases to ~20 nucleotides. Major considerations in this protocol are RNA concentration, pellet identification, and resuspension.

MATERIALS

The recipe for the item marked with <R> is at the end of the protocol or in the Common Recipes Appendix.

CAUTION: See the Cautions Appendix for appropriate handling of materials marked with <!>.

Reagents

Carriers
  GlycoBlue (10 mg/mL) or Glycogen, RNase-free (20 mg/mL)
  Escherichia coli tRNA (10 mg/mL)
Ethanol (95% or 100%) <!>
H_2O, RNase-free
RNA (from Protocol 1 or 2)
Sodium acetate (NaOAc) (3 M, pH 5.2), RNase-free
TE buffer (optional; see Step 9) <R> <!>

Equipment

Centrifuge (fixed-angle rotor) or microcentrifuge
Freezer (–20°C)
Pasteur pipettes, baked
Tubes (12 mL [Sarstedt] or 1.5 mL [microcentrifuge])
Vortex mixer

OVERVIEW

The procedures described here are important for recovering RNA after deproteinization. RNA is inherently hydrophilic; i.e., it dissolves readily in water. However, the hydrophilicity of RNA is reduced by the presence of salt (0.3 M) at acidic pH and further reduced (to the point of precipitation) by addition of ethanol to 66%. Thus, ethanol precipitation is the preferred method for recovering RNA from aqueous solutions. It is important to note that efficient precipitation only occurs when a minimum (~10 µg/mL) concentration of RNA is present. If solutions are less concentrated than 10 µg/mL, it is necessary to add a "carrier": either additional RNA or a polysaccharide with limited solubility in ethanol, such as glycogen. Ethanol precipitation is not size dependent until chain lengths become very short; hence, any RNA of 20 nucleotides or more is precipitated efficiently.

METHOD

1. Because ethanol precipitation is concentration dependent, it is necessary to know the approximate concentration of the RNA sample. This can be estimated from the amount of starting material (see Table 2-1). When in doubt, add a carrier, but ensure that the carrier will be compatible with subsequent manipulations to be performed on the RNA (see Step 10).

   If the RNA concentration is ≥10 µL/mL, it can be precipitated directly by ethanol. If it is ≤10 µL/mL, see Step 10 before proceeding.

2. Make sure that the solution (i.e., the aqueous phase containing RNA) is at least 0.3 M NaOAc. If not, make the solution 0.3 M NaOAc by adding concentrated (3 M) NaOAc as appropriate. Add 2.5 volumes of 95% or 100% ethanol and mix vigorously by vortexing.
3. Freeze-thaw the ethanol solution on dry ice or leave in a freezer overnight at –20°C.

   For long-term storage, keep the ethanol solution at –80°C.

4. Thaw the frozen solution and pellet the precipitated RNA by centrifugation for 10 min at maximum speed in a microcentrifuge or 20 min at 10,000*g* in a fixed-angle rotor for larger volumes. Make sure that tubes are oriented for subsequent pellet identification.

   On the top of each tube, mark the outside (relative to the center of the rotor). Pellets will be at the bottom of the tube on the same side as the mark.

5. Remove the tubes from the centrifuge immediately after the centrifuge stops so that the pellets do not dislodge. Clean RNA will appear as a translucent bluish triangle at the bottom of the tube.
6. Carefully pour off the ethanol into a fresh tube, taking care not to disturb the pellet.
7. Remove residual ethanol/salt droplets from the RNA pellet with a pulled pasteur pipette.

   Traces of salt left behind will make the pellet difficult to resuspend, and traces of SDS will inhibit subsequent enzyme reactions.

8. Resuspend the pellet in RNase-free H_2O so that the concentration of RNA or RNA plus carrier will be ≥100 µg/mL.

TABLE 2-1. Expected Amount of Total RNA and Specific RNA from Various Tissues and Cells

Tissues/cells	Total RNA	Specific RNA
Tissue/cells amounts		
5 mg tissue or 2.5 x 10^6 cells	10 µg	0.3 µg mRNA, 8 µg rRNA
1 g tissue or 5 x 10^8 cells	2 mg	60 µg mRNA, 1.6 mg rRNA
Cell cultures (1 x 10^6 cells)		
NIH/3T3	10 µg	
HeLa	15 µg	
COS-7	35 µg	
Mouse/rat tissues (10 mg)		
13-d embryo	25 µg	
Kidney	23–30 µg	
Liver	40–60 µg	
Spleen	30–40 µg	
Thymus	40–50 µg	
Lung	10–20 µg	
Yeast (1 x 10^7 cells)		
Saccharomyces cerevisiae	25 µg	
Plants (100 mg leaves)		
Arabidopsis	35 µg	
Maize	25 µg	

Data modified from www.ambion.com/techlib/append/rna_yields.html; www1.qiagen.com/HB/RNeasyMin.

9. If precipitating RNA from SDS-containing solutions, the RNA should be diluted with TE buffer and reprecipitated (i.e., follow Steps 1–7 again).
10. If the starting concentration of RNA is ≤10 µg/mL, it is necessary to add a carrier (RNase-free glycogen, GlycoBlue, or transfer RNA [tRNA]) before precipitation. Add tRNA, glycogen, or GlycoBlue to at least 20 µg/mL. Add 1.5 µL of 15 mg/mL GlycoBlue to small-scale precipitations to facilitate pellet visualization.

    tRNA is advantageous as a carrier when the goal is simply RNA recovery. It should never be used when the recovered RNA is going to be manipulated (e.g., 5′- or 3′-end labeled) subsequently or quantitated by A_{260} spectrophotometry. Glycogen is recommended for large-scale precipitations. GlycoBlue is recommended for small-scale precipitations where its blue color greatly aids visualization of small pellets.
11. Following the addition of appropriate carriers, precipitation is exactly as described in Steps 1–7 above.
12. The concentration of resuspended RNA can be determined by spectrophotometry or dye-based methods (see Protocol 13).

TIPS AND TROUBLESHOOTING

1. When possible, add the carrier before deproteinization and extraction for samples ≤10 µg/mL. When working with minute amounts of RNA, this prevents adherence of the RNA to the walls of tubes.
2. Many protocols advocate rinsing pellets with 70% ethanol to remove salt. We do not recommend this procedure for small amounts of RNA because it can disturb the pellets. If excess salt is a concern, we recommend reprecipitation and removing drops of liquid from the tube with a pulled pasteur pipette. However, if you want to wash RNA pellets with 70% ethanol, use cold (–20°C) ethanol and then chill and centrifuge the tubes for 5 min. Carefully remove the supernatant after the centrifugation.

Protocol 4

Guidelines for the Use of RNA Purification Kits

With the proliferation of researchers investigating various aspects of RNA biology has come a corresponding proliferation of commercially available kits for RNA purification. Some of these kits have specialized uses and applications, and it is not always possible to simply mix and match the different protocols. Although the instructions are often written with a minimum of specifics concerning the chemical basis for the separation characteristics that the kits use, the competent researcher will want to understand these important features. Unfortunately, commercial sources work hard to hide the specifics of their methods. Fortunately, patent documents are publicly available, and the government mandates material safety data disclosures of the composition of at least those components that represent a threat to environmental health and safety. Thus, it is possible to infer many aspects of such "proprietary" materials, which in most cases are only slightly and unimportantly modified from published techniques. We do not describe in detail the specific steps involved in the use of particular kits and their components (e.g., proprietary buffers, etc.); they are detailed in the literature that comes with each kit. Instead, we detail some advantages and disadvantages of using RNA purification kits.

HOW RNA ISOLATION KITS WORK

The first reason to purchase RNA isolation kits is the convenience of not having to prepare the buffers and columns already provided in the kit. The second reason is claimed consistency in performance of the kit components due to automation and control in manufacturing, as well as quality testing of the materials, although in practice, companies make no guarantee that their kits work as claimed. A contributing factor to continuing kit use is that, owing to the hidden elements of the protocols, it is difficult for a researcher who has used one kit to change manufacturers without adding unknown variables. Thus, kits engender an unscientific atmosphere in two ways: (1) Knowable aspects of the experimental procedure are kept secret in violation of good scientific practice and (2) principal investigators are unable to explain to their students why such unknowns should be tolerated scientifically in an experiment, leading students to accept a lower standard in their science. Nonetheless, RNA isolation kits are based on chemical principles that are known in general, if not in detail, and they tend to work more or less routinely.

There are several kinds of RNA isolation kits on the market. Some are designed to purify total RNA from cell lysates, and others are designed to select for or enrich subclasses of RNA. Some contain materials that are compatible with extraction buffers, and for others RNA must be purified at least partially and dissolved in aqueous buffers. Table 2-2 shows a comparison of several RNA isolation kits available from some common sources. Many companies (e.g., Invitrogen, QIAGEN, Ambion/Applied Biosystems, Exiqon, Zymo Research, Molecular Devices, Agilent/Stratagene, GE Healthcare/Amersham, Roche Applied Biosciences, Sigma-Aldrich, New England BioLabs, Macherey-Nagel, and others) produce fundamentally identical products, and we do not recommend the use of any one of these over another. Rather, we encourage the researcher to read carefully whatever literature comes with the kit and, if insufficient information is given, to check material safety data sheets (MSDSs) to determine whether hazardous materials are used in the kit. If more detail is needed, U.S. Patent Applications and Patent documents are searchable using Google (http://www.google.com/patents?hl =en).

Table 2-2 indicates the different applications and approaches used in RNA isolation kits. Many of those designed to isolate total RNA rely on the fact that nucleic acids (both RNA and DNA) bind tightly to silica-based materials (e.g., membranes or beads composed of silica glass) in the presence

TABLE 2-2. RNA Isolation Kits

Kit name	Lysis buffer	Selectivity	Binding chemistry
RiboPure (Ambion)	TRIzol	Total RNA (not selective)	Silica gel/glass fiber in high salt and ethanol
TRIzol-Plus (Invitrogen)	TRIzol	Total RNA (not selective)	Silica gel/glass fiber in high salt and ethanol
ChargeSwitch (Invitrogen)	Urea-LiCl	Total RNA (not selective)	Organic pH-dependent anion exchange matrix coated on paramagnetic beads
RNaqueous (Ambion)	GITC	Total RNA (not selective)	Silica gel/glass fiber in high salt and ethanol
RNeasy (QIAGEN)	GITC	Large RNAs (>200 nt)	Silica gel/glass fiber in high salt and ethanol
*mir*Vana (Ambion)	GITC	Small RNAs	Silica gel/glass fiber in high salt and ethanol
miRNeasy (QIAGEN)	GITC	Inclusive of small RNAs, not selective (total RNA)	Silica gel/glass fiber in high salt and ethanol
Purelink miRNA (Invitrogen)	GITC	Small RNAs	Silica gel/glass fiber in high salt and ethanol
Oligotex poly(A)$^+$ (QIAGEN)	GITC, urea-SDS, or purified total RNA	Poly(A)$^+$ RNAs	dC10dT30 covalently linked to polystyrene-latex beads
DynaBeads mRNA DIRECT (Invitrogen)	0.5 M LiCl, lithium DS	Poly(A)$^+$ RNAs	dT25 attached to paramagnetic beads
RiboMinus (Invitrogen)	Purified total RNA in solution	Non-rRNAs	5′-Biotinylated LNA oligonucleotides bound to streptavidin-linked paramagnetic beads

(GITC) Guanidinium isothiocyanate, (lithium DS) lithium dodecyl sulfate, (LNA) locked nucleic acid, (nt) nucleotide(s), (TRIzol) a monophasic mixture of phenol and guanidinium isothiocyanate.

of chaotropic salts, such as guanidinium isothiocyanate, whereas other less-charged cellular components do not. A cell or tissue lysate is passed through a silica-based material and washed; the RNA and DNA are bound to the silica while protein and other cellular constituents pass through. After washing, the DNA is eluted selectively, although not completely removed, with low salt in the presence of 35% ethanol. Subsequently, RNA is eluted with water. Alternatively, DNA is removed by on-column DNase I treatment under high salt conditions that preserve binding, and after washing away the DNase and salt, RNA is eluted. There is a dependence on ethanol concentration in the wash buffers for retaining small RNAs (e.g., miRNAs, <200 nucleotides) such that at 50% but not 35% ethanol, small RNAs are retained on the silica gel/glass fiber matrix. Thus, attention must be paid if it is important to have small RNAs remain in the sample. Some kits (e.g., RNeasy, Table 2-2) are designed so that small RNAs do not bind to the column and flow through instead. These kits are obviously not useful for recovering miRNAs. However, miRNAs can be recovered via the use of a second column where conditions are optimized for binding and subsequent elution of small RNAs. Such two-column kits are commercially available (Table 2-2).

At least one kit (ChargeSwitch, Invitrogen) does not use a silica-based matrix to bind RNA but instead uses magnetic beads coated with an organic matrix with protonatable groups that have a pK_a in the 6.8 range (e.g., piperazine-*N,N′*-bis[2-ethanesulfonic acid] [PIPES]). At low pH (6.0), the RNA binds because of the positively charged, protonated matrix, and, upon washing with more neutral pH buffers, contaminants are removed. Finally, RNA is eluted with Tris-HCl, pH 9.0, by deprotonating and neutralizing the matrix. This chemistry is not compatible with phenol or guanidinium thiocyanate, and thus cells must be lysed in a different buffer. As described elsewhere in this book, the type of lysis buffer and the conditions of lysis must be matched carefully to the kind of biological material being extracted, regardless of the downstream features of the kit (or other procedure steps).

Several kits are designed to bypass the isolation of total RNA to capture poly(A)$^+$ RNA directly from cell lysates. These kits rely on the ability of the poly(A) tail on the RNA to anneal with an oligo(dT) tract immobilized on a solid matrix. After annealing in high salt and washing, low salt at room temperature melts the unstable rA-dT tracts and releases the mRNA. Some matrix materials are reasonably resistant to the agents used to lyse cells, but others require gentler cell lysis agents (thus increasing the opportunity for RNase attack), and still others can only be performed using purified aqueous solutions of RNA. Isolation of poly(A)$^+$ RNA often leaves residual ribosomal (r)RNA. Another method for removing rRNA from total RNA uses a similar hybridization–elution approach but captures the rRNA rather than the mRNA. Custom oligonucleotides complementary to the rRNAs are bound to beads, and the rRNA is annealed and binds while the non-rRNA flows through. Note that the RNA populations selected in these two methods are not strictly complementary. Poly(A)$^+$ selection enriches essentially only mRNAs, the primary class of polyadenylated RNAs in the cell, whereas removing rRNA enriches for tRNAs, small nuclear RNAs (snRNAs), small nucleolar RNAs (snoRNAs), miRNAs, and siRNAs, as well as mRNAs.

In general, RNA purification kits perform as described by their manufacturers (if the directions are followed carefully), and they can be quite useful for RNA isolation from small numbers of cells or amounts of tissues. Nevertheless, it is important to select the right kit for the application and to be aware of potential problems.

SOLVING SOME POTENTIAL PROBLEMS WITH KITS

Inadequate Solubilization of Starting Material

If the starting material (i.e., cells or tissues) is not well solubilized, the yield will be low. To avoid this common problem, monitor solubilization by microscopy, if possible, and be sure not to exceed the amount of starting material recommended by the manufacturer. In some cases, treatment with digestive enzymes before solubilization is recommended. A drawback to this approach is that RNAs can be degraded during enzyme treatment.

Low or No Yield

A common and frequently overlooked problem is the fact that extremely small amounts of starting material are often processed using RNA purification kits; thus, the yield of RNA under perfect conditions will be very low and the concentration after elution negligible. In such situations, the RNA that is present can adhere to the walls of the tube. To prevent this, perform elution with a carrier in the collection tube (e.g., GlycoBlue or glycogen; see Protocol 3). This not only will prevent loss of RNA to the tube walls but also permit concentration of the RNA by ethanol precipitation (see Protocol 3).

Contamination with DNA or Ethanol

The presence of DNA in the eluant can cause problems, especially for reverse transcriptase–polymerase chain reaction (RT-PCR) applications. The presence of residual ethanol in the eluant can inhibit subsequent analyses (e.g., reverse transcriptase, ligase, or kinase). These problems can be easily overcome by treating the RNA with DNase for DNA (see Protocol 11) or ethanol precipitation and resuspension (see Protocol 3). Indeed, we recommend ethanol precipitation (with appropriate carrier; see Protocol 3) and resuspension after recovering RNA, using any kit, if such problems arise.

In summary, RNA purification kits can be valuable tools. In principle, they are easy to use. Nevertheless, they are expensive and, if used improperly, they can yield disappointing (or no) results. It is important that good RNA handling techniques are used when using kits. Even kits are not immune to RNase contamination! Moreover, a sound understanding of the fundamentals of dealing with RNA (see Chapter 1) will suggest logical ways to troubleshoot any problems that occur.

Protocol 5

Preparation of Cytoplasmic and Nuclear RNA from Tissue Culture Cells

As discussed in Chapter 1, it is often desirable to "prefractionate" RNA before subsequent analysis. Ordinarily, this can be done only with tissue culture cells, although it is possible to isolate nuclei and cytoplasm from certain "soft" tissues such as liver and white blood cells. Here, we describe a method for separating nuclei from the cytoplasm that can be used for many tissue culture types. This procedure is useful for cells grown in suspension or for adherent cells. It relies on swelling in hypotonic buffer, subsequent gentle homogenization, and centrifugation. This method is not appropriate for material (e.g., bacteria, yeast, etc.) that has high intrinsic RNase activity, or tissues that are difficult to solubilize, such as muscle, or plant material.

MATERIALS

The recipe for the item marked with <R> is at the end of the protocol or in the Common Recipes Appendix.

CAUTION: See the Cautions Appendix for appropriate handling of materials marked with <!>.

Reagents

Buffer A <R>
Cells of interest in medium
Dry ice <!>
Emetine (5 mg/mL) (optional) <!>
Phosphate-buffered saline (PBS) (1×) (4°C)
RNasin (optional)
Triton X-100 (10%) <!>
TRIzol (optional; see Step 9) <!>

Equipment

Cell culture dishes (10 cm)
Cell scraper
Centrifuge tubes (15-mL conical or 50 mL)
Dounce homogenizer (Kontes) with type B pestle, baked
Ice
Microscope slides
Phase-contrast microscope
Pipettes
Refrigerated high-speed centrifuge (e.g., Sorvall SS34 or equivalent)
Refrigerated low-speed centrifuge (e.g., Beckman JS 4.2 or equivalent)
Vortex mixer

OVERVIEW

A key factor in isolating intact RNA from tissue culture cells or in preparing extracts is speed in avoiding degradation of RNA by endogenous RNases. Accordingly, it is essential that all solutions are premade and chilled to 4°C, and that centrifuges and rotors are ready and precooled. It is also essential that all tubes, glassware, pipettes, etc., are RNase-free (see Chapter 1). Use a separate Dounce homogenizer for each sample. However, if many samples are to be processed, rinse the homogenizers thoroughly with ddH_2O and aspirate any remaining drops of water. Do not add detergent to the Dounce homogenizer. The method described has many applications that are specified in the notes at the end of the protocol.

METHOD

Harvesting Cells

1. For suspension or nonadherent cells:
   i. Centrifuge the cells plus medium at 1000*g* for 5 min.
   ii. Remove all of the medium, being careful not to disturb the pellet.
   iii. Wash the pellet three times with ice-cold PBS to remove any serum that contains RNases. For the final wash, use 10 mL of PBS and transfer the pellet to a 15-mL conical tube.
   iv. After the final wash, aspirate the PBS and immediately add at least 3 volumes (relative to the cell pellet) of cold Buffer A.

      If the cell pellet is ≤0.3 mL, add Buffer A to 1 mL. Never try to process less than 1 mL.

   v. Mix by pipetting and vortexing briefly to resuspend the cells. Let it sit on ice for 10 min.
2. For adherent cells:
   i. Wash the cells on the plate three times (10 mL per 10-cm dish) with cold PBS. A few rocks back and forth should suffice.
   ii. Add 10 mL of PBS and then, using a cell scraper, scrape the cells from the plate. Transfer the cells to a centrifuge tube (15-mL or 50-mL conical as appropriate). If more than one plate is used, pool cells in one 50-mL conical tube.
   iii. Centrifuge the cells at 1000*g* for 5 min.
   iv. Aspirate the PBS and then add at least 3 volumes (relative to the cell pellet) of cold Buffer A. Let sit on ice for 10 min.

      If the cell pellet is ≤0.3 mL, add Buffer A to 1 mL. Never try to process less than 1 mL.

Breaking the Cells

3. Immediately after adding Buffer A, transfer a small amount of the suspension to a microscope slide using a Pipetman and check the cells under a phase-contrast microscope.

   The cells should be well suspended (an occasional sheet of cells will appear if working with adherent cells). This examination will reveal the preswollen shape of the cells.

4. After 9 min in Buffer A, examine another aliquot of cells under the microscope.

   At this point, the cells should be completely swollen with well-defined nuclei and expanded (~3x) in size relative to unswollen cells, i.e., the cells examined immediately after adding Buffer A.

5. At 10 min, transfer the cells to a baked Dounce homogenizer and apply 15 strokes with a baked, type B pestle.

   The strokes should be firm but not so vigorous as to cause bubbles.

6. Check the cells again under the microscope. There may be a few intact cells, but look for ≥90% breakage (i.e., 9:1 free nuclei to intact cells).

   The field should appear as free nuclei with many small particulates. The particulates are fragments of membranes, mitochondria, etc. There should be very few broken nuclei. If there are many fragmented nuclei, the homogenizing was too vigorous. If cells are inadequately broken, try five more strokes in the Dounce homogenizer and examine again.

7. Transfer the homogenate to a fresh tube. Add Triton X-100 from a 10% stock solution to make the homogenate 0.1% Triton X-100. Mix by inversion (four or five times).

   The addition of this nonionic detergent strips residual cytoplasmic material from the nuclei.

8. Immediately centrifuge the homogenate at 1500*g* for 5 min to pellet the cell nuclei.

   The pellet should be white and compact.

9. Transfer the supernatant, which contains the cytoplasm, to a fresh tube without disturbing the nuclear pellet. If desired, immediately add an appropriate amount of TRIzol to the tube with the nuclear pellet for preparation of nuclear RNA (see Protocol 2).

10. Isolate cytoplasmic RNA by phenol:chloroform extraction (see Protocol 1), remove ribosomal subunits (see Protocol 9), or freeze it on dry ice for further analysis and store at –80°C.

    The cytoplasmic extract is stable indefinitely at –80°C, but some nuclease activity can occur upon thawing.

TIPS AND TROUBLESHOOTING

1. If you wish to isolate mRNAs associated with ribosomes or perform polysome analysis, treat the cells with 100 µg/mL of emetine (from 5 mg/mL stock) for 10 min before harvesting. This will prevent ribosome runoff during extract preparation.
2. If the cells of choice have significant RNase, it may be advisable to add RNasin to the Buffer A during swelling and homogenization.
3. A low yield of cytoplasmic RNA can be a result of inadequate cell breakage; i.e., the RNA remains trapped in the intact cells. Be sure to monitor breakage under a microscope (see Step 6).
4. Degraded RNA can result if the steps are not performed in a timely fashion, something was allowed to warm during the procedure, or buffers or water have become contaminated with ribonuclease. Make sure everything is cold and the steps are performed as described. Make new solutions if necessary.
5. If the nuclear RNA is heavily contaminated with cytoplasmic RNA or vice versa, it is likely that the cells were overhomogenized or underhomogenized. Monitor cell (and nuclear) damage carefully under the microscope.

SOLUTION AND REAGENT RECIPE

CAUTION: See the Cautions Appendix for appropriate handling of materials marked with <!>.

Buffer A

Reagent	Quantity (for 500 mL)	Final concentration
KCl (2 M) <!>	2.5 mL	10 mM
$MgCl_2$ (1 M) <!>	0.75 mL	1.5 mM
Tris-HCl (1 M, pH 7.5) <!>	10 mL	20 mM
H_2O	486.75 mL	

Before using this buffer, it is necessary to add fresh DTT <!>. Prepare a working stock by adding DTT from a 1 M stock to make 10 mL of 1 mM. Discard leftovers.

Protocol 6

Removal of Ribosomal Subunits (and rRNA) from Cytoplasmic Extracts before Solubilization with SDS and Deproteinization

As discussed in Chapter 1, more than 95% of total RNA is composed of ribosomal RNAs (rRNAs) (28S, 18S, 5.8S, and 5S). Here, we present a method that is effective in removing rRNA before extraction and purification of total RNA. Subsequently (Protocol 9), we present methods that remove rRNA after RNA purification. If you choose to prepare cytoplasmic RNA (see Protocol 5) and wish to analyze any RNA other than rRNA, it is desirable to eliminate the rRNAs by taking advantage of the fact that the ribosomal subunits are very large (40S and 60S). Few, if any, other cellular RNAs are present in such large macromolecular complexes. The vast majority of rRNAs can be removed by sedimentation. Of course, steps must be taken to avoid cosedimentation of desired RNAs. Cosedimentation can be greatly reduced by first dissociating ribosomes into their respective subunits by EDTA treatment. The subunits are then "cleaned" by treatment with high salt and nonionic detergent. Ribosomal subunits remain intact under these conditions and can be sedimented free of other RNAs. Subsequently, the remaining RNAs (mRNAs and all other RNAs) can be purified and analyzed by a variety of methods.

MATERIALS

The recipe for the item marked with <R> is at the end of the protocol or in the Common Recipes Appendix.

CAUTION: See the Cautions Appendix for appropriate handling of material marked with <!>.

Reagents

Buffer A <R>
Cytoplasmic extract (see Protocol 5)
EDTA (0.5 M, pH 8.0) stock solution
NaCl (5 M)
Triton X-100 (10%) <!>

Equipment

Ice
Pasteur pipette
Polycarbonate tubes with caps (10 mL)
Ti50 rotor, 70.1 Ti rotor (Beckman), or equivalent
Ultracentrifuge

OVERVIEW

In this method, cytoplasmic extract from tissue culture cells (Protocol 5) is depleted of ribosomal rRNAs via sedimentation of intact ribosomal subunits. Intact ribosomes are first separated into subunits by treatment with EDTA and loosely bound (nonribosomal) proteins are dissociated by treatment with high salt. Subunits are then pelleted by ultracentrifugation, leaving all (or most) other RNAs in the supernatant.

METHOD

1. Dilute the cytoplasmic extract to 1 mL with Buffer A (if necessary) and adjust to 0.1% Triton X-100 or NP-40.
2. Add a 1/9 volume of 125 mM EDTA and incubate the mixture on ice for 15 min.
3. Add a 1/9 volume of 5 M NaCl and incubate the mixture on ice for 10 min.
4. Transfer the mixture to a 10-mL polycarbonate centrifuge tube.

   It is not necessary to fill the polycarbonate tubes. Although we recommend a volume of at least 1 mL, volumes of as little as 0.5 mL can be centrifuged or diluted before centrifugation.
5. Centrifuge at 49,000 rpm (~200,000*g*) for 2 h at 4°C in a 70.1 Ti rotor.

   For smaller or more dilute volumes, a tabletop ultracentrifuge can be used with 1-mL open-cap polycarbonate tubes in a TLA 120 rotor adjusted to the same run conditions. This rotor uses a tabletop Beckman ultracentrifuge

   It is important not to overcentrifuge. This will cause the large ribonucleoproteins (RNPs) to pellet.
6. Remove the supernatant, which contains all non-rRNAs, with a pasteur pipette to a fresh tube without disturbing the pellet.

   The pellet, which contains 5S, 5.8S, 18S, and 28S rRNAs in the form of intact subunits, will appear as a glassy tight circle at the bottom of the tube (toward the top of the rotor angle).
7. The supernatant is now suitable for further analysis (e.g., deproteinization by phenol:chloroform extraction [Protocol 1] or immunoprecipitation). Unless rRNA is to be analyzed, the pellet can be discarded.

TIPS AND TROUBLESHOOTING

1. This protocol is not suitable for deproteinized RNAs.
2. It is best to use enough material (at least 0.2 mL of extract) so that the resulting pellets can be seen.
3. It is a good idea to monitor the supernatant and pellet by northern blotting or semiquantitative PCR (Chapter 5, Protocols 7–10 and 18) to assess the partitioning of RNAs of interest, i.e., to make sure that they have not pelleted.
4. For subsequent immunoprecipitation analysis, it may be necessary to dilute or dialyze the supernatant to reduce salt concentration.

SOLUTION AND REAGENT RECIPE

CAUTION: See the Cautions Appendix for appropriate handling of materials marked with <!>.

Buffer A

Reagent	Quantity (for 500 mL)	Final concentration
KCl (2 M) <!>	2.5 mL	10 mM
$MgCl_2$ (1 M) <!>	0.75 mL	1.5 mM
Tris-HCl (1 M, pH 7.5) <!>	10 mL	20 mM
H_2O	486.75 mL	

Before using this buffer, it is necessary to add fresh DTT <!>. Prepare a working stock by adding DTT from a 1 M stock to make 10 mL of 1 mM. Discard leftovers.

Protocol 7

Isolation of Total RNA from Yeast Cell Cultures

Method 1: Vegetative Cells

This procedure allows the convenient isolation of total RNA from early log-phase cultures of yeast cells. RNA isolated in this way is intact and sufficiently pure for use in microarray experiments, primer extension, RNase protection mapping, etc. With additional treatment to remove contaminating genomic DNA, the preparation here is suitable for RT-PCR, quantitative PCR (qPCR), cDNA library construction, high-throughput sequencing of RNA, or other manipulations. Extraction of RNA from cells late in meiosis (asci and ascospores) requires additional effort to achieve cell disruption and efficient extraction (see Method 2 on p. 39). Method 2 may be preferable for cells from stationary cultures or from yeast strains and other fungal species isolated from the environment.

MATERIALS

The recipe for the item marked with <R> is at the end of the protocol or in the Common Recipes Appendix.

CAUTION: See the Cautions Appendix for appropriate handling of materials marked with <!>.

Reagents

AE buffer <R>
Chloroform <!>
Ethanol (100% and 70%) <!>
H_2O
PCA 25:24:1 <!>
SDS (10%, w/v) <!>
Sodium acetate (NaOAc) (3 M, pH 5.2)
Yeast culture (an aliquot of cells no larger than 10 mL at $A_{600} = 0.5$)

Equipment

Ice bucket (with ice)
Microcentrifuge
Microcentrifuge tubes (2 mL)
Phase Lock Gel tubes (2 mL, heavy [Eppendorf, Catalog no. 955154045])
Spectrophotometer or nanospectrophotometer
Vortex mixer
Water bath or heat block (65°C)

OVERVIEW

This is a hot PCA extraction with SDS followed by separation of the organic phase using Phase Lock Gel, an inert material with a density intermediate between the organic and aqueous samples. The sample is split into three phases: from bottom to top, these are phenol and chloroform, the inert gel, and the aqueous phase with the RNA. The gel acts as a physical barrier between the sample and the organic phase. Following organic extraction, the RNA is concentrated by ethanol precipitation.

METHOD

1. Read the A_{600} of the yeast culture using a spectrophotometer. Use an aliquot of cells no larger than 10 mL at A_{600} = 0.5 (meaning 1 mL at A_{600} = 5, or 50 mL at A_{600} = 0.1, etc.); otherwise, the Phase Lock Gel will be overloaded.
2. Centrifuge the yeast culture to pellet the cells and then remove the culture fluid by decanting it carefully.
3. Resuspend the pelleted cells in 400 µL of AE buffer and transfer them to a 2-mL microcentrifuge tube.
4. Add 40 µL of 10% SDS and 400 µL of PCA. Vortex the mixture quickly and then place the tube into a water bath or heat block at 65°C.
5. Incubate the cells for 10 min at 65°C. Vortex each tube about once a minute during this 10-min period, for 5–10 sec each time.
6. Incubate the cells for 5 min on ice. While the cells are incubating, centrifuge the gel to the bottom of the appropriate number of Phase Lock Gel tubes with a 10-sec spin.

   During shipping and storage, the gel can settle anywhere in the tube; this step returns the gel to the bottom of the tube.
7. Pour the entire chilled mixture of PCA and aqueous sample into a Phase Lock Gel tube. Once the sample is in the gel tube, DO NOT VORTEX because the gel will become fragmented.
8. Centrifuge the sample with the gel at 14,000 rpm in a microcentrifuge for 5 min.

   After centrifugation, the gel will float on the organic phase, separating it from the aqueous phase. The RNA will be in the upper aqueous phase.
9. Leave everything in the tube and add 400 µL of chloroform to the top (aqueous) phase. Mix by gently shaking, but do not vortex.

   This step removes water-dissolved organic solvents from the aqueous phase.
10. Centrifuge again at 14,000 rpm for 5 min.
11. Add another 400 µL of chloroform to the tube and gently shake again. Centrifuge one last time at 14,000 rpm for 5 min.
12. Pour the top aqueous phase into a new 2-mL microcentrifuge tube. The Phase Lock Gel should hold the organic phase in the tube while the aqueous phase is poured off (if it has not been fragmented).
13. Add 50 µL of 3 M sodium acetate, pH 5.2, to the aqueous phase and add 100% ethanol up to 2 mL. Do not overfill the tube. Make sure that the cap fits and then mix well by inverting several times.

    The RNA should precipitate readily at this concentration so it is not necessary to chill the ethanol.
14. Centrifuge at 14,000 rpm for 5–10 min.
15. Pour off the ethanol and rinse the pellet with 70% ethanol. Centrifuge at 14,000 rpm for 5 min.
16. Pour off the ethanol and allow the pellet to air-dry or dry BRIEFLY in an RNA-dedicated Speed Vac.

    Do not desiccate the RNA because this will make resolubilization troublesome. A small amount of ethanol contamination is likely to cause much less trouble.

17. Resuspend the pellet in 100 μL of RNase-free H_2O by vortexing. Expect at least 10 μg of total RNA per milliliter of cells at A_{600} = 0.5, giving a yield of ~100 μg of total RNA for this protocol. It may take some time for the RNA to come back into solution, so be patient and keep the RNA on ice. Carefully inspect the sample visually to make sure that there are no clear chunks of RNA precipitate still floating and that the side of the tube to which the RNA was pelleted is clean. Do not proceed until you are reasonably convinced that you have a solution, and not a suspension, of RNA.
18. Once the pellet is completely in solution, read the A_{260}, A_{270}, A_{280}, and A_{320} in a spectrophotometer or take a spectrum in a nanospectrophotometer (see Protocol 13).

    If the RNA is pure, the A_{260}:A_{280} ratio should be 1.8–2.2, the A_{320} (particulates) should be very low, and the A_{270} should never be higher than the A_{260} value; otherwise, there is phenol contamination.
19. Store the RNA in H_2O at –80°C. It should be stable for several years.

Method 2: Meiotic Cells

During meiosis and ascospore development, a tough cell wall composed of heavily cross-linked polysaccharides and proteins is built around the four spores. One consequence is that isolation of RNA from late meiotic stages of yeast becomes less reliable when only Method 1 is used. Method 2 presents an alternative protocol for meiotic cells that leads to uniform extraction of RNA. The last part of this method is identical to Method 1.

MATERIALS

The recipes for the items marked with <R> are at the end of the protocol or in the Common Recipes Appendix.

CAUTION: See the Cautions Appendix for appropriate handling of material marked with <!>.

All of the materials from Method 1 plus:

Reagents

AE buffer (2X) <R>
Cells of interest
Lysis buffer <R>
PCA 25:24:1 <!>
Water-saturated phenol (neutral pH) (purchased)

Equipment

Centrifuge tubes (15 mL, plastic [phenol resistant, e.g., Nunc EZ top polypropylene tubes])
Glass beads (acid-washed, 0.45–0.5-mm diameter)
Ice water slurry

OVERVIEW

To achieve uniform RNA extraction from yeast and other cell-wall-producing microorganisms, it is vital to ensure uniform and complete breakage of the cells. A common method is vortexing with glass beads. SDS and phenol without chloroform is a more potent protein denaturant, and the presence of the glass beads during vortexing helps this process. After removal of the glass beads and reextraction of the sample with PCA, a crude ethanol precipitation step concentrates the RNA. Finally, the sample is reextracted using Method 1. This longer, more elaborate process is required to release all of the RNA from meiotic cells.

METHOD

1. Pellet the cells (no more than 10 mL of cells at $A_{600} = 0.5$) in a 15-mL phenol-resistant disposable centrifuge tube.
2. Resuspend the pellet in 0.5 mL of lysis buffer and add 1 mL of water-saturated phenol. Add one full 1.5-mL microcentrifuge tube of acid-washed glass beads. Vortex on a benchtop vortex machine for 7 min in 20-sec bursts followed by at least 20 sec in an ice water slurry after each burst.

   Ice alone provides insufficient contact to keep the sample cold.
3. Add another 0.5 mL of phenol, 0.5 mL of lysis buffer, and 200 µL of 10% SDS. Vortex for 1 min more (20 sec vortex, then 20 sec on ice; repeat three times).
4. Centrifuge at 4000*g* for 10 min at 4°C to separate the phases. Remove the aqueous phase containing the RNA to a new 15-mL tube by pipetting and add 1 mL of PCA 25:24:1. Vortex for 1 min and then centrifuge at 4000*g* for 10 min at 4°C.
5. Remove the aqueous phase to two 2-mL microcentrifuge tubes. Estimate the volume recovered and add 2.5 volumes of 100% ethanol. Mix well and place at –20°C for ≥90 min.
6. Centrifuge the ethanol precipitates at 14,000 rpm for 10 min in a microcentrifuge. Rinse the pellets with 70% ethanol, centrifuge for 1 min, and discard the ethanol. Do not dry the pellets.
7. Completely resuspend each pellet in 100 µL of RNase-free H_2O. Combine the two pellets for each of the samples from Step 6 into one of the 2-mL microcentrifuge tubes. Add 200 µL of 2x AE buffer.
8. Perform Method 1, beginning with Step 3.

DISCUSSION

For the first level of quality control, it is important to have spectrophotometric readings on dilutions of the sample that are between an A_{260} of 0.1 and 0.7. Readings at A_{270} that are greater than those at A_{260} indicate phenol contamination. Readings at A_{280} that are too high relative to A_{260} give low ratios (optimal A_{260}:A_{280} ratios are 1.8–2.2 for RNA) and indicate residual protein contamination. Readings at A_{320} provide an estimate of the light scattering from particulate debris in the sample and should be close to zero (see Protocol 13 for more details).

For a second level of quality control, the intactness of the RNA sample should be evaluated. Conventionally, this is done by measuring the amount of large rRNA (23S–28S) relative to small rRNA (16S–18S) in the sample, assuming equimolar ratios of the two RNAs in the cell. (Note: This may not be true in yeast mutants unable to process one or the other rRNA.) Because the larger RNA will show slight degradation of the sample more readily than its smaller partner, the molar ratio of the two RNAs makes a good measure. Molar ratios of 28S–18S rRNA lower than 1 may indicate degradation. This ratio can be measured absolutely on a gel using dyes that bind rRNA proportionately to its length (e.g., toluidine blue, not ethidium bromide); however, relative changes in the ratio from sample to sample can be determined even with ethidium bromide. The Agilent Bioanalyzer is a capillary electrophoresis device that has software allowing easy determination of these numbers, together with profiles that can be analyzed for degradation of samples (see Protocol 13).

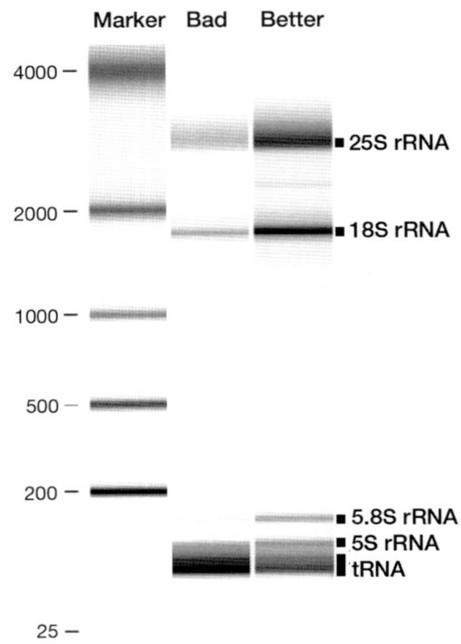

FIGURE 2-4. Bioanylzer trace of RNA purified from yeast meiotic cells. (Bad) Poor extraction, (Better) good extraction results.

Finally, extraction of perfectly intact RNA may occur but with variable relative efficiencies due to differences in the "extractability" of the RNA. In yeast, the obvious measure to notice is variation from preparation to preparation in the amount of rRNA relative to tRNA (Figure 2-4). In samples where extraction is poor because of incomplete cell breakage or other problems, tRNA will represent a much larger proportion of the extracted RNA than with samples in which there is complete extraction, which releases the large and small ribosomal subunits (compare lane "Bad" to lane "Better" in Figure 2-4). Although there may be biological reasons for variation in the rRNA-to-tRNA content in cells grown under different physiological regimes, comparison of extractions of the same cells under the same physiological state should provide constant ratios if extraction is uniform.

SOLUTION AND REAGENT RECIPES

CAUTION: See the Cautions Appendix for appropriate handling of material marked with <!>.

AE Buffer

Reagent	Quantity (for 10 mL)	Final concentration
NaOAc (3 M, pH 5.2)	0.167 mL	50 mM
EDTA (0.5 M, pH 8.0) <!>	0.2 mL	10 mM
ddH_2O	to 10 mL	

Prepare fresh; store for no longer than 1 wk at room temperature.

Lysis Buffer

Reagent	Quantity (for 100 mL)	Final concentration
Tris-HCl (1 M, pH 6.8) <!>	5 mL	50 mM
NaCl (1 M)	10 mL	100 mM
EDTA (0.5 M, pH 8.0) <!>	2 mL	10 mM
ddH$_2$O	to 100 mL	

Buffer can be prepared well in advance; store indefinitely at room temperature.

Protocol 8

Bacterial RNA Isolation (*E. coli*)

Bacteria are an extremely diverse group of organisms, whose apparent morphological simplicity belies an incredible elaboration of biochemical capabilities and compositions. Although RNA contained inside bacteria is comparatively uniform across the kingdom, bacterial cell walls and capsules, which contain the RNA, are decidedly not so. In addition, microbial biofilms can be as complex as eukaryotic tissues with respect to the physical properties of individual cell types and their impact on the extractability of intact RNA that accurately reflects the intracellular pools.

There are numerous ways to break open bacteria: freeze–thaw cycles, sonication, milling with 0.1-mm or 0.2-mm glass beads, enzymatic digestion of the cell wall, and treatment with detergents. Methods such as chilling, centrifugation, and cell-wall digestion may lead to physiological changes that interfere with gene expression regulation events of interest. Rifampicin or other transcriptional inhibitors can be added to block new gene expression, but these have their own effects. Some steps may unleash RNA degradation that reflects the procedure rather than the biology. Because bacterial RNAs tend to have short half-lives and low steady-state levels of any one mRNA in the first place, any loss caused by poor extraction can easily render them undetectable.

A recent practice involves adding "RNA-protective" solutions of quaternary amine-based cationic detergents (Macfarlane and Dahle 1993) such as RNAProtect from QIAGEN (similar to CTAB [hexadecyltrimethylammonium bromide]) or Ambion's RNALater (essentially, 70% ammonium sulfate; e.g., see Lader 2001) directly to the culture. These formulations permeate cells and precipitate the RNA in a form sequestered from nucleases until extraction can be performed.

MATERIALS

CAUTION: See the Cautions Appendix for appropriate handling of materials marked with <!>.

Reagents

Chloroform <!>
E. coli culture (5 mL, 2×10^9 cells, $OD_{600} = 0.2$)
Ethanol (100%) <!>
Lysozyme dissolved in 10 mM Tris-HCl (pH 8.0), 10 mM EDTA <!>
NaOAc (3 M, pH 5.2)
PCA 25:24:1 <!>
RNAProtect Bacteria Reagent (10% tetradecyltrimethylammonium oxalate <!>) (QIAGEN)
SDS (10%, w/v) <!>

Equipment

Ice bucket (with ice)
Incubator
Liquid nitrogen or dry ice/ethanol <!>
Microcentrifuge
Microcentrifuge tubes (2 mL)
Phase Lock Gel tubes (2 mL, heavy [Eppendorf, Catalog no. 955154045])
Pipettor

Polypropylene conical tubes (15 mL)
Pulled capillary
Spectrophotometer
Vortex mixer
Water bath or heat block (65°C)

OVERVIEW

This protocol uses an "RNA-protective" treatment followed by lysozyme digestion of the peptidoglycan component of the cell wall. Cells begin to lyse during the digestion in the hypotonic lysozyme buffer and lysis is completed by SDS and the hot PCA extraction. The separation of the organic phase from the aqueous phase is achieved using Phase Lock Gel, an inert material with a density intermediate between the organic and aqueous samples. The sample is split into three phases: from bottom to top, these are phenol and chloroform (organic phase), the inert gel with the interface material, and the aqueous phase with the RNA. The gel acts as a physical barrier between the sample and the organic phase plus interface. Following organic extraction, the RNA is concentrated by ethanol precipitation. EDTA promotes the loss of the outer membrane of Gram-negative bacteria and allows the lysozyme better access to the peptidoglycan. SDS and hot phenol disrupt membranes, denature protein (including RNase), and strip proteins from RNA.

METHOD

1. Determine the optical density (OD) of the *E. coli* culture by spectrophotometry. The starting material should be 5 mL of *E. coli* culture containing 2.5 x 10^9 cells (OD_{600} = 0.2). If the spectrophotometer reading indicates a higher concentration, dilute the cells with medium or sterile water or scale up the reagents.
2. Pipette 10 mL of RNAProtect Bacteria into a 15-mL polypropylene conical tube and 5 mL of the *E. coli* culture into the same tube. Mix immediately by inverting and vortexing for 5 sec. Incubate for 5 min at room temperature.
3. Centrifuge the cells at 8000 rpm for 2 min at 4°C. Completely pour off the excess supernatant. The pelleted cells can be snap-frozen in liquid nitrogen or on dry ice/ethanol and stored for 1 month at –80°C. Alternatively, RNA extraction can proceed immediately.
4. Digest the bacterial cell walls by resuspending the pellet in 400 µL of freshly prepared 0.5 mg/mL lysozyme, 10 mM Tris-HCl (pH 8.0), and 10 mM EDTA. Incubate for 2 min at room temperature.
5. Lyse the cells by adding 15 µL of 3 M NaOAc (pH 5.2) and 45 µL of 10% SDS. Mix by vortexing for 10 sec, after which the suspension should clear. Add 400 µL of PCA 25:24:1 and quickly vortex to emulsify. Then place the tube in an incubator at 65°C.
6. Incubate the mixture for 10 min at 65°C, vortexing each tube about once a minute for 5–10 sec, returning the tube to 65°C each time. Incubate the tubes for 5 min on ice.
7. During this time, centrifuge an appropriate number of Phase Lock Gel tubes to reorient the gel to the tube bottom with a brief spin.

   During shipping and storage, the gel can settle anywhere in the tube; this step returns the gel to the bottom of the tube.
8. Pour the entire chilled mixture of PCA 25:24:1 and the aqueous sample from Step 6 into a Phase Lock Gel tube. Once the sample is in the gel tube, DO NOT VORTEX because the gel will become fragmented.

9. Centrifuge the sample with the gel at 14,000 rpm in a microcentrifuge for 5 min at room temperature.

   After centrifugation, the gel will float on the organic phase, trapping the interface, which contains denatured proteins. This separates them from the aqueous phase containing the RNA.

10. Leave everything in the tube and add 400 μL of chloroform to the top (aqueous) phase. Mix by gently shaking, but do not vortex.

    This step removes water-dissolved organic solvents from the aqueous phase.

11. Centrifuge again at 14,000 rpm for 5 min at room temperature.
12. Add another 400 μL of chloroform and gently shake again. Centrifuge one last time at room temperature.
13. Pour the top phase into a new 2-mL microcentrifuge tube. The Phase Lock Gel should hold the organic phase in the "old" tube (if the gel has not been fragmented).
14. Add 50 μL of 3 M sodium acetate (pH 5.2) to the aqueous phase and add 100% ethanol up to 2 mL. Do not overfill the tube, make sure that the cap fits, and then mix well by inverting several times.

    The RNA should precipitate readily at this concentration so it is not necessary to chill the ethanol.

15. Centrifuge the tube at 14,000 rpm for 5–10 min at room temperature.
16. Pour off the ethanol and remove any residual ethanol with pulled capillaries or add 1 mL of 70% ethanol.
17. Rechill and recentrifuge.
18. Carefully remove the ethanol and allow the pellet to air-dry or dry BRIEFLY in a RNA-dedicated Speed Vac.

    Do not desiccate the RNA because this will make resolubilization troublesome; a small amount of ethanol contamination is likely to cause much less trouble.

19. Resuspend the pellet in 100 μL of H_2O by vortexing. Store at –80°C. Analyze the quality of RNA by spectrophotometry and gel electrophoresis or with the Bioanalyzer (see Protocols 12 and 13).

TIPS AND TROUBLESHOOTING

1. For bacteria other than *E. coli*, care must be taken to ensure that the cells are lysed. This may require modification of the protocol to include alternative methods that replace lysozyme treatment with bead milling, freeze–thaw cycles, etc.
2. This procedure has been designed for laboratory strains of *E. coli*, a well-studied Gram-negative bacterium that is quite unlike many other bacteria in numerous respects. The effectiveness of this protocol for other bacteria with different cell walls or capsules should not be assumed.
3. DNA contamination can occur as a result of prolonged or vigorous cell lysis. It can be removed by subsequent DNase treatment (see Protocol 11).

DISCUSSION

As with any extraction and purification protocol, monitoring the quality of the product is key. Intactness of the RNA can be evaluated by the proportion of larger to smaller rRNAs. If 23S rRNA is present in reduced molar ratios to 16S or 5S rRNA, degradation may be suspected. If tRNAs are highly represented relative to rRNAs, or their representation fluctuates relative to rRNA, steps to increase the effort or alter the method of cell disruption should be taken. With respect to mRNA recovery, use of protective solutions and rapid processing of the samples to denaturing conditions could improve recovery of labile mRNAs.

REFERENCES

Lader E. 2001. Methods and reagents for preserving RNA in cell and tissue samples, U.S. Patent 6204375.

Macfarlane D, Dahle C. 1993. Isolating RNA from whole blood—The dawn of RNA-based diagnosis? *Nature* **326:** 186–188.

Protocol 9

Removing rRNA from Deproteinized, Phenol-extracted Total RNA

As discussed previously in Chapter 1, rRNAs (28S, 18S, 5.8S, and 5S) comprise the vast majority of total cellular RNAs (>95%). When other RNAs are to be analyzed, it is very advantageous to remove the rRNA species. For example, if cellular RNA is 3′-end labeled (see Chapter 4, Protocol 5), the only visible bands in the small range of <200 nucleotides will be tRNA, 5S, and 5.8S rRNAs. More-over, the large rRNAs inevitably "electrophoretically trap" and obscure detection by northern blotting of mRNAs in the 28S and 18S regions. As discussed in Protocol 6, the best way to eliminate rRNA "contamination" is to pellet subunits before deproteinization. When this is not possible, certain RNAs (e.g., mRNAs, 2,2,7-trimethylguanosine (TMG)-capped RNAs; see Chapter 4, Protocol 1) can be separated from bulk rRNA after extraction. Nevertheless, many RNAs cannot be enriched by these methods. In these cases, it is still desirable to remove rRNAs. We are aware of three methods that can be used to accomplish this: enzymatic digestion, selective precipitation, and hybrid selection. Each method, as well as its respective advantages and disadvantages, is discussed in this protocol.

Method 1: Enzymatic Digestion of rRNA

This procedure relies on the properties of certain exonucleases that digest RNA in a 5′→3′ direction and require a 5′ phosphate for activity. Because rRNAs, with the exception of 5S rRNA, contain a 5′-phosphate terminus, they are substrates for these nucleases. It is important to remember that other RNAs, such as miRNAs and intronic snoRNAs, also contain these termini and they are substrates as well. Thus, this procedure should never be used when non-rRNAs that contain 5′-phosphate ends are to be analyzed. Because the 5′ ends of mRNAs and many snRNAs are protected by caps, they are resistant to degradation by these exonucleases. Therefore, this procedure will enrich for capped RNAs and RNAs with triphosphorylated 5′ ends. A 5′-phosphate-dependent exonuclease (Terminator) is commercially available from Epicentre, although it is quite expensive ($179 for 20 units). One unit can digest 0.2–10 µg of 5′-phosphate-terminated RNA, so this procedure becomes prohibitively expensive if one is working with milligram quantities of total RNA.

MATERIALS

Reagents

ddH_2O, RNase-free
EDTA (100 mM) <!>
Exonuclease reaction buffer (Epicentre) (10x)
Terminator exonuclease (1 unit/µL)
Total RNA, resuspended (200 ng–10 µg)

Equipment

Incubator
Microcentrifuge
Microcentrifuge tubes (1.5 mL)

OVERVIEW

For this method, it is necessary to have purified total RNA free of SDS. This method relies on the ability of a 5′→3′-exonuclease to digest RNAs with 5′-phosphate termini such as rRNAs, while leaving RNAs with 5′-terminal caps or triphosphates intact. The method is not appropriate when analyzing non-rRNAs, such as miRNAs or many snoRNAs, that also contain 5′-phosphate termini.

METHOD

1. Add the following ingredients in this order to a 1.5-mL tube:

Resuspended total RNA	200 ng–10 µg
ddH_2O	to 17 µL
Exonuclease reaction buffer (10×)	2 µL
Terminator exonuclease	1 µL

2. Mix by pipeting up and down and incubate for 30 min at 30°C.
3. Stop the reaction by adding 1 µL of 100 mM EDTA.
4. Perform phenol:chloroform extraction and ethanol precipitation (see Protocols 1 and 3).

TIPS AND TROUBLESHOOTING

1. If there is enough RNA, it is a good idea to check the precipitated RNA before and after digestion on an agarose gel (see Protocol 12) to determine whether the rRNA has disappeared.
2. The RNA concentration must be at least 12 ng/µL.
3. Longer incubation times may be required to achieve complete digestion.
4. Do not use this method if any 5′-phosphate-containing RNAs are to be analyzed; they will be digested by this procedure.

Method 2: Selective Precipitation of Large RNAs with LiCl or PEG/NaCl

For reasons that are somewhat mysterious, "large" RNAs ≥100 nucleotides long can be selectively separated from small RNAs <100 nucleotides long by precipitation with either LiCl or PEG/NaCl. Both techniques are very straightforward and have the major advantage that nothing gets lost (compared to the nuclease digestion method). These methods are very valuable for enriching for very small RNAs (i.e., miRNAs) in unprecipitated material while retaining all other RNAs in the precipitate. Accordingly, it is possible to isolate and concentrate small RNAs from large amounts of total RNA for further analysis.

MATERIALS

CAUTION: See the Cautions Appendix for appropriate handling of material marked with <!>.

Reagents

LiCl stock (5 M) <!>
NaCl stock (5 M)

PEG 8000 stock (50%)
Total RNA at 1 µg/µL in RNase-free ddH_2O or TE buffer

Equipment

Ice
Microcentrifuge
Microcentrifuge tubes (1.5 mL)

OVERVIEW

This method relies on the ability of high concentrations of LiCl or PEG/NaCl to precipitate large RNAs (≥100 nucleotides) while leaving small RNAs in the supernatant, which is useful when enrichment of small RNAs is desired. Both LiCl and PEG/NaCl work equally well, although we prefer to use LiCl because it is easier to resuspend LiCl pellets. This is advantageous for analyzing both small and large RNAs from the same preparation of total RNA.

METHOD

1. (a) Mix RNA (at least 1 µg/µL) with 5 M LiCl to a final concentration of 2.5 M.

   OR

   (b) Mix RNA (at least 1 µg/µL) with 50% PEG 8000 and 5 M NaCl to final concentrations of 5% PEG and 0.5 M NaCl.
2. Incubate the mixture for 30 min on ice.
3. Centrifuge the mixture at maximum speed in a microcentrifuge for 30 min or at 10,000 rpm in a centrifuge for 30 min. The resulting pellet contains the large RNAs (≥100 nucleotides long).
4. Remove the supernatant, which contains the small RNAs, to a fresh tube and dilute the supernatant to 0.5 M LiCl with TE buffer (a) or 0.1 M NaCl (b).
5. Ethanol precipitate the diluted supernatant using GlycoBlue carrier following the method used in Protocol 3 to obtain small RNAs.
6. The pellet from Step 3, which contains RNAs ≥100 nucleotides long, should be resuspended in H_2O and precipitated with ethanol. Store for days at –20°C temporarily or indefinitely at –80°C.

TIPS AND TROUBLESHOOTING

1. We have not observed any noticeable differences between performances of the LiCl and PEG/NaCl methods.
2. It is a good idea to check the yield and integrity of small RNAs by running an aliquot of the RNA on an acrylamide gel and staining with ethidium bromide or SYBR dye. The tRNAs at ~75 nucleotides should be very visible.
3. This protocol is not useful for separating rRNA from mRNA because all mRNAs will precipitate along with the rRNA.
4. The only potential problem is that the large RNAs fail to precipitate. In this case, the starting RNA concentration was most likely too low. If this occurs, reprecipitate with ethanol (see Protocol 3) and repeat the procedure. Resuspend in a smaller volume and using an agarose gel, check that the large RNAs are intact in the starting material.

Method 3: Removal of rRNA by Hybrid Selection

This protocol is essentially the inverse of oligo(dT) (see Protocol 10) to enrich for mRNAs. The basic idea is to remove rRNA by hybridizing to complementary oligonucleotides that can be used to remove the rRNA selectively, leaving all other RNAs behind. This is accomplished by using biotinylated oligodeoxynucleotides that can then be affinity purified using streptavidin beads. Although commercially available kits can be used to perform this procedure, they are expensive. We recommend that investigators order their own species-specific oligodeoxynucleotides for their own applications. There are well-established secondary structure predictions for all rRNA. Oligonucleotides should be designed to be complementary to regions that are predicted to be single stranded. Several commercially available kits with custom-designed oligodeoxynucleotides are available from Invitrogen.

MATERIALS

Reagents

Biotinylated oligonucleotides
ddH$_2$O, RNase-free
NP40 (10%)
SSC (1x and 10x)
Streptividin beads, agarose, 50-μL washed slurry 1:1, or magnetic bead solution such as 50 μL Dynol
Total RNA (100 μg)

Equipment

Centrifuge
Disposable spin columns
Freezer
Heat blocks (95°C, 65°C, and 30°C) or thermal cycler
Microcentrifuge tubes (1.5 mL)
Pipettor
Rotator

OVERVIEW

This method efficiently removes rRNA from total RNA preparations by hybrid selection and is used when enrichment of the RNA(s) of interest cannot be obtained by other procedures. The starting RNA must be purified and free of protein.

METHOD

1. Anneal total RNA to the biotinylated oligonucleotides complementary to 18S, 28S, 5S, and 5.8S. Use 1 μg of each oligonucleotide per 10 μg of total RNA. For example, to remove rRNA from 100 μg of total RNA, add 50 μL of RNA (2 μg/μL) to a 1.5-mL microcentrifuge tube. Then add 10.25 μL of each complementary oligonucleotide (2 μg/μL), 10 μL of 10x SSC, and H$_2$O to 100 μL.
2. Heat the mixture in a heat block for 2 min at 95°C, then for 10 min at 65°C , and finally for 10 min at 30°C.

3. Dilute the mixture to 500 μL with 1x SSC and add 50 μL of streptavidin beads (washed at least three times with 0.5 mL of 1x SSC). Add 2.5 μL of 10% NP40 to make the solution 0.05% NP40.
4. Rotate the mixture for 1 h and then centrifuge out the beads using a low-speed setting (1000*g*) for 30 seconds at room temperature (see Tips and Troubleshooting).
5. Carefully pipette off the supernatant. Recover ribosomal-depleted RNA by ethanol precipitation as in Protocol 3, using GlycoBlue as a carrier. Alternatively, the bead slurry in Step 4 can be transferred to an empty spin column. Set in an empty collection tube and centrifuge at 1000*g* for 30 seconds to recover the liquid and minimize any contamination with the beads.

TIPS AND TROUBLESHOOTING

1. If magnetic streptavidin beads are used, use a magnet to recover them instead of centrifugation.
2. It is a good idea to monitor removal of rRNA by comparing samples before and after depletion by agarose gel electrophoresis (see Protocol 12).

Protocol 10

Enrichment of Poly(A)$^+$ mRNA Using Immobilized Oligo(dT)

This protocol takes advantage of the fact that the vast majority of eukaryotic mRNAs contain a 3′ tail of polyadenylic acid added posttranscriptionally. It relies on the ability of these tails to form stable hybrids with oligo(dT) under high-salt conditions. The hybrids are destabilized when salt is absent, allowing mRNAs to be recovered. This protocol is effective in removing the bulk of non-mRNA RNAs. A drawback of this technique is that it does not enrich for mRNAs that lack poly(A) tails (e.g., normal histone mRNAs or deadenylated mRNAs whose poly[A] tails have been removed during mRNA turnover). Several commercially available kits can be used for selection of poly(A)$^+$ RNAs. However, they are expensive and do not perform any better than "homemade" reagents using (dT) cellulose. One obvious advantage of this method is that mRNAs are enriched approximately 50-fold. Accordingly, subsequent detection by northern blotting or RT-PCR is considerably more sensitive.

MATERIALS

The recipes for the items marked with <R> are at the end of the protocol or in the Common Recipes Appendix.

CAUTION: See the Cautions Appendix for appropriate handling of materials marked with <!>.

Reagents

Binding buffer <R>
Elution buffer <R>
Glycogen or GlycoBlue (20 mg/mL)
LiCl (5 M) <!>
RNA sample dissolved in RNase-free H_2O, or in 1× elution buffer (pH 7.5)
SDS (10%) <!>

Equipment

Oligo(dT), cellulose or alternatives
Pasteur pipette
Poly-Prep columns (0.8 × 4 cm) or Econo Columns for larger preparations
Tubes (12 mL [Sarstedt])

OVERVIEW

Before beginning this procedure, prepare total RNA from tissue or tissue culture cells as described in Protocol 5. When this RNA is passed over immobilized oligo(dT) in high salt, RNAs containing poly(A) stretches (mRNAs) will anneal to the oligo(dT), whereas other RNAs will not hybridize. Because the dT:A hybrids are stable only in high salt, washing with solutions that do not contain salt destabilizes the hybrids, and poly(A)$^+$ RNA is recovered.

METHOD

1. Dilute purified RNA (if necessary) to obtain samples in at least 2 mL of 1x binding buffer by adding water and concentrated stock solution (10x elution buffer and 5 M LiCl).
2. Prepare oligo(dT) cellulose by suspending dry powder in binding buffer without LiCl (elution buffer). Use 100 µL of hydrated cellulose for up to 100 µg of total RNA; use 1 mL of cellulose for ≥1 mg of total RNA. Using a pasteur pipette, pipette the suspended cellulose into the Poly-Prep column and let it drip dry. Perform all steps at room temperature.
3. Equilibrate the column with binding buffer. Use 2 x 10 mL for a small-scale preparation and 2 x 15 mL for a large-scale preparation. Allow all of the buffer to run through completely before adding the second aliquot. The column is now ready to run.
4. Place the RNA sample in 2 mL (small scale) or ≥10 mL (large scale) of binding buffer and pass the sample through the column. Collect the flowthrough in a clean tube. Repeat this procedure two more times so that each sample has passed through the column three times. Save the flowthrough (this contains poly[A]$^-$ RNA) for subsequent analysis if desired.
5. Wash the column with 3 x 5 mL (small scale) or 4 x 10 mL (large scale) binding buffer and let the column drip dry between washes.
6. Add the elution buffer (3 x 0.5 mL for small scale or 3 x 1.0 mL for large scale) into the column and collect the flowthrough in a fresh tube. The eluted material will contain poly(A)$^+$ RNA. Because poly(A)$^+$ RNA represents only about 2%–3% of the total RNA, the expected yield will be 20–30 µg/mg of the starting RNA.
7. Add glycogen or GlycoBlue and then ethanol precipitate as described in Protocol 3.

   Remember that polyadenylated RNA is ~2% of the total RNA, so a carrier is almost always necessary.

TIPS AND TROUBLESHOOTING

1. The procedure can be done in batch (i.e., not in a column); however, we do not recommend this because cellulose pellets are easily disturbed.
2. Alternatives include oligo(dT) affixed to magnetic beads. These beads usually have a reduced capacity relative to cellulose, so make sure to use enough beads to capture all of the polyadenylated RNA. Some companies recommend these for isolation of mRNA for RNA sequencing libraries.
3. Although this procedure greatly enriches for mRNA, there is inevitably some rRNA that remains. If removing this rRNA is critical, the previous procedure can be repeated as many times as necessary to deplete the rRNA.
4. If the yield is far below that expected, as assessed by OD or northern blotting, it is likely that the starting material was degraded, and it is necessary to start again from the beginning. It is unlikely that the poly(A)$^+$ RNA will be contaminated with DNA, but if this is the case (as assessed by RT-PCR), DNase treatment (Protocol 11) is recommended.
5. Oligo(dT) is expensive. It is important to note that it can be reused multiple times. For storage, keep the columns in elution buffer (stoppered at both ends). Before reuse, wash again with elution buffer then reequilibrate with binding buffer as described previously. Manufacturers provide instructions for reusing oligo(dT) cellulose.

SOLUTION AND REAGENT RECIPES

CAUTION: See the Cautions Appendix for appropriate handling of materials marked with <!>.

Binding Buffer (1x)

Reagent	Quantity (for 500 mL)	Final concentration
LiCl (5 M) <!>	50 mL	0.5 M
EDTA (0.5 M) <!>	10 mL	10 mM
SDS (10%) <!>	250 mL	5%
Tris-HCl (1 M, pH 7.5) <!>	100 mL	200 mM
Elution buffer (10x)	50 mL	1x
H_2O	90 mL	

Store for up to 1 yr at room temperature.

Elution Buffer (10x)

Reagent	Quantity (for 500 mL)	Final concentration
EDTA (0.5 M) <!>	10 mL	10 mM
SDS (10%) <!>	250 mL	5%
Tris-HCl (1 M, pH 7.5) <!>	100 mL	200 mM
H_2O	140 mL	

Store for up to 1 yr at room temperature. Dilute to 1:10 as needed to make 1x.

READING LIST

Aviv H, Leder P. 1972. Purification of biologically active globin messenger RNA by chromatography on oligothymidylic acid-cellulose. *Proc Natl Acad Sci* **69:** 1408–1412.

Chirgwin JM, Przybyla AE, MacDonald RJ. 1979. Isolation of biologically active ribonucleic acid from sources enriched in ribonuclease. *Biochemistry* **18:** 5294–5299.

Protocol 11

Removal of DNA from RNA

DNA contamination of RNA preparations can present significant problems, especially for PCR-based applications. No RNA extraction procedure excludes DNA entirely. Cytoplasmic RNA (see Protocol 5) can be contaminated with DNA because of nuclear breakage during preparation. Moreover, TRIzol preparations do not exclude plasmids or other small DNA fragments. We have found that eliminating DNA via DNase digestion is the most reliable and effective method.

MATERIALS

The recipe for the item marked with <R> is at the end of the protocol or in the Common Recipes Appendix.

Reagents

DNase buffer (10×) (see Step 1)
DNase I, RNase-free
H_2O, RNase-free
NaOAc (3 M)
RNA (total) preparation
RNasin RNase inhibitor
SDS extraction buffer (2×) <R>

Equipment

Incubator
Microcentrifuge (for phenol extraction, Step 5)
Microcentrifuge tubes (1.5 mL)
Pipettor

OVERVIEW

This protocol efficiently removes contaminating DNA from RNA preparations. DNase I specifically digests DNA into small oligonucleotides, leaving RNA intact. After DNase treatment, it is necessary to remove the enzyme by phenol extraction (see Protocol 1).

METHOD

1. Suspend the RNA sample in RNase-free H_2O. To 25 µg of RNA resuspended in RNase-free H_2O (50 µL), add the following:

DNase buffer (10×)	10 µL
RNasin	40 units/µL, 2 µL
DNase I	10 units/µL, 2 µL
H_2O	36 µL

2. Mix by pipetting up and down.
3. Incubate the mixture for 30 min at 37°C.
4. Bring up to 200 μL with an equal volume (100 μL) of 2x SDS buffer and add NaOAc to 0.3 M by adding 20 μL of 3 M NaOAc.
5. The RNA sample is now ready for phenol extraction (Protocol 1), ethanol precipitation (Protocol 3), and resuspension.

TIPS AND TROUBLESHOOTING

1. DNase denatures very easily, so do not vortex. Mix by pipetting up and down.
2. Any residual SDS in the RNA sample will inhibit DNase activity. If SDS is a problem, another round of ethanol precipitation (Protocol 3) should remove it.

SOLUTION AND REAGENT RECIPE

CAUTION: See the Cautions Appendix for appropriate handling of materials marked with <!>.

SDS Extraction Buffer (2x)

Reagent	Quantity (for 10 mL)	Final concentration
SDS (10%) <!>	1 mL	1%
EDTA (0.5 M) <!>	20 μL	1 mM
Tris-HCl (1 M, pH 7.5) <!>	200 μL	20 mM
H_2O	8.78 mL	

Store indefinitely at room temperature, but remake if cloudy or there is visible dust.

Protocol 12

Gel Electrophoresis

Perhaps the most important and certainly the most often used technique in RNA analysis is gel electrophoresis. This technique (or, more appropriately, techniques) is generally applicable for RNA detection, quantification, purification by size, quality assessment, etc. Because RNAs are negatively charged, they migrate toward the anode in the presence of electric current. The gel acts as a sieve to selectively impede the migration of the RNA in proportion to its mass, given that its mass is generally proportional to its charge. Because mass is approximately related to chain length, the length of an RNA is more generally determined by its migration. In addition, topology (i.e., circularity) can affect migration, making RNAs appear longer on the gel than they actually are.

Gels are used in almost all protocols described in this book including northern blotting, primer extension, footprinting, analyzing processing reactions, etc. They are invaluable as preparative and fractionating tools. It is imperative that good practices are used to avoid misshapen, streaky, or otherwise ugly gels. This protocol describes how to prepare various types of gels, load them, and run them. In addition, guidelines are given for the appropriate use of each gel type. There are two common types of gel: polyacrylamide and agarose. We describe polyacrylamide gels first. For many applications, denaturing acrylamide gels are most appropriate. These gels are extremely versatile and can resolve RNAs from ~600 to ≤20 nucleotides. In certain circumstances, e.g., resolving different conformers of RNAs or RNA–protein complexes, native acrylamide gels are appropriate. Agarose gels are normally used for assessing the quality of RNA preparations and for detecting high-molecular-weight RNAs by northern blotting.

Method 1: Polyacrylamide Gel Electrophoresis

Acrylamide gels are often the gel of choice because of their superior resolving power (in many cases, to single-nucleotide resolution) and their versatility. The only disadvantage to acrylamide gels is that they are not suitable for analyzing large RNAs (≥600 nucleotides).

MATERIALS

The recipes for the items marked with <R> are at the end of the protocol or in the Common Recipes Appendix.

CAUTION: See the Cautions Appendix for appropriate handling of materials marked with <!>.

Reagents

Acrylamide stock solution (40%, 19:1 acrylamide:bisacrylamide) <R> <!>
Ammonium persulfate (10%, w/v, freshly prepared) <!>
Distilled water
Dye (ethidium bromide OTC SYBR-type dye) <!>
Ethanol <!>
N,N,N′,N′-Tetramethylethylenediamine (TEMED), electrophoresis grade <!>
RNA gel-loading buffer <R>
RNA sample
Siliconizing agent (e.g., Rainex)
TBE electrophoresis buffer <R>
Urea, ultrapure <!>

Equipment

Beaker (200 mL), baked
Binder clips
Combs and spacers (0.4 mm)
Gel drier or vacuum drier
Gel-loading tips
Gel plates (20 x 40 cm)
Heat block
High-voltage power supplies
Ice
Phosphorimaging or autoradiography equipment
Pipetman
Plastic wrap
Spatula
Squirt bottle
Stir bar, baked
Syringe (60 mL)
Thermometer
Tray
Vacuum system for degassing (optional) (see notes)
Vertical gel box
Water bath (60°C)
Whatman 3MM paper
Windex
Yellow tape (Scotch 3M electrical, 1.5 in)

OVERVIEW

In this method, we describe one specific example of a denaturing polyacrylamide gel. However, this example can be easily extrapolated to any size gel. In this regard, we prefer to run the same size (large) gels for all applications, whereas other investigators prefer to run smaller gels. Because good habits are essential for obtaining good results, we recommend that the beginning investigator practice running gels before investing valuable samples.

METHOD

1. Before beginning, assemble all equipment and solutions.
2. Prepare the gel plates; gel plates must be cleaned before each use.
   i. For first-time use, wash with soap and water, then dry. Mark the outside of the plate and then clean the inside surface with Windex or ethanol. If there is any particulate material or dried acrylamide evident, scrape clean with a new single-edge razor blade, but do not scratch the glass plates.
   ii. Treat the inside of the shorter plate with a siliconizing agent or preferably a commercial substitute such as Rainex. Apply Rainex to the plate with a paper towel or large Kimwipe. Cover evenly but do not allow Rainex to get on the edges or the other side of the plate. Air-dry and wipe clean with Kimwipes. Rinse with water and dry for use.

      The water should bead up, indicating that the surface has been properly treated. This treatment should last for many gel runs. Re-treat when the water fails to bead on the surface of the plate.

3. Prepare the gel solutions. The following protocol is for a 6% gel. This is enough gel solution to make a 20-cm × 40-cm × 0.4-mm-thick gel with enough left to use in the event that some solution gets lost or is required to be used as refill after any leakage. Adjust as necessary for other percentages.

   i. For 100 mL, weigh out 48 g of urea (electrophoresis grade) in a 200-mL baked beaker; add 15 mL of 40% 19:1 acrylamide:bis-acrylamide, 10 mL of 10× TBE, and water to 100 mL. Note that the final concentration of urea is 8 M. Stir to dissolve with a baked stirbar.

      We recommend using premade solutions of acrylamide:bis-acrylamide to avoid exposure to dry chemicals. If dry acrylamide is to be weighed out, the balance should be moved to the hood and leveled, and the investigator should wear a mask to avoid inhaling particulate acrylamide.

      A little heating in a 60°C water bath helps the process, but do not heat for long because the solution should be at room temperature for the polymerization step.

      If you prepare your own acrylamide solutions, filter the gel mix through either a 0.45-µm bottle-top filter or three layers of Whatman No. 1 paper using a Buchner funnel and a 250-mL side-arm Erlenmeyer flask. Remove the funnel, place a rubber stopper over the top of the flask, and de-gas briefly under vacuum.

4. Assemble the plates (Fig. 2-5).

   i. Place spacers on the edges of the inside surface of the long plate. Cover with a second plate, with the treated side down; be careful not to move the spacers. Hold in place with a binder clip on one side (Fig. 2-5A).

   ii. Tape the unclipped side and part of the bottom with yellow tape; use the nonsharp side of a single-edge razor to smooth the tape. (Razor blades are the most useful for cutting tape.)

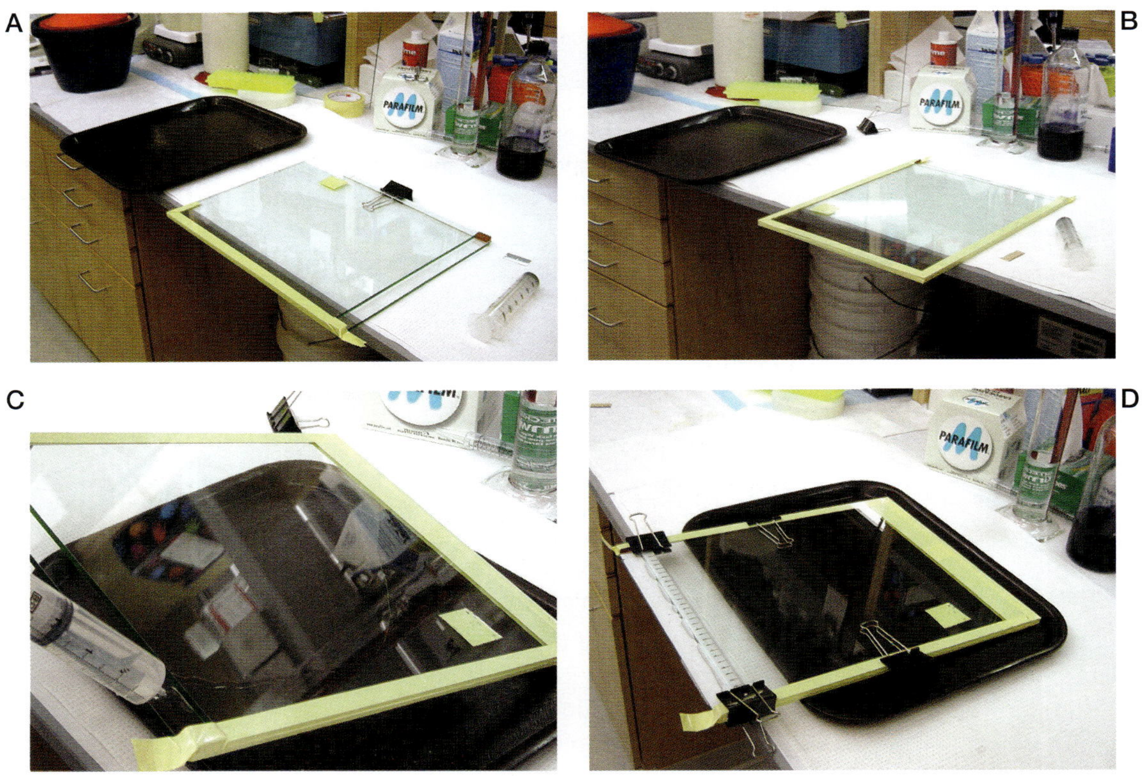

FIGURE 2-5. Preparing the gel plates and pouring the gel. (*A*) Partially taped plates. (*B*) Fully taped plates. (*C*) Pouring the gel. (*D*) Clamped comb in place.

If the small plate does not have "ears," extend the tape over gel spacer tabs. Clip the taped side one-third of the way up from the bottom. Make sure that the end of the clip is positioned over the tape and spacer.

iii. Unclip the other side and tape the plate as described above. Tape the bottom of the plates overlapping a few inches up the sides. Check to ensure that the comb fits securely (Fig. 2-5B).

5. You are now ready to pour the gel (see Fig. 2-5C). Have everything ready for pouring because the solution will polymerize quickly; the higher percentage the gel, the more quickly it will polymerize. Use a cafeteria tray or its equivalent to catch any spilled or overflow acrylamide solution.

   i. Add 900 µL of 10% ammonium persulfate and 80 µL of TEMED and mix briefly with a stirbar. Fill a 60-mL syringe with the gel solution.

   ii. Tip the assembled plates up and balance on one bottom corner. Using the syringe, slowly and steadily run the gel solution down the side on which the plates are balanced. As the solution reaches the bottom corner, rotate the bottom plates so that the solution spreads across the bottom to the other corner. As you continue to supply solution, slowly move the plates to a more level (horizontal) position so that the gel fills the plates from bottom to top.

   iii. When the solution reaches the top of the short plate (with no bubbles), lay the gel down in the tray with the top resting on the edge. Insert the comb and clamp both top corners so that the comb is held tightly between the two plates (Fig. 2-5D).

   If bubbles form around the wells, remove the comb and try again to ensure that the acrylamide solution covers the wells.

   If the fit is not tight, a thin film of polymerized acrylamide will form between the wells and the plates and will interfere with sample loading in Step 7.

   iv. Leave the remainder of the gel solution in the beaker to check for polymerization. Once the solution has completely polymerized (~15–20 min), leave the gel for another 5–10 min. Make sure that the partitions between the teeth of the comb are completely formed.

   As polymerization proceeds, the refractive index of the gel changes and this can be visualized as a "Schlieren line" near the edge of the gel plate and comb. This effect is a result of a change in refractive index between the polymerized and unpolymerized acrylamide.

6. Set up the gel. After the gel is polymerized,

   i. Remove all clips and the tape from the bottom edge (or slice with a razor blade) and rinse the top of the gel around the comb using a squirt bottle filled with distilled water.

   ii. Remove the comb and rinse the wells with water (using a squirt bottle, syringe, or pasteur pipette).

   iii. Attach the gel plates to the gel apparatus, with the short side in. Clamp in place with equal pressure on both sides.

   iv. Add TBE to the top and bottom chambers. Use a pasteur pipette or syringe to rinse the wells again, this time with TBE; make sure that there are no air bubbles in the wells and that no air bubbles are trapped under the bottom of the gel.

   v. Close the lids, attach the electrodes (red on the bottom, black on the top), and prerun the gel at 45 mA for 20–45 min.

   Prerunning removes excess persulfate and eliminates hyperfocusing. Longer preruns (45 min) are recommended for analyzing short RNAs of ≤50 nucleotides.

7. Load the gel. After the prerun is completed,

   i. Turn the power supply off.

   ii. Heat the RNA samples at 90°C–95°C and then chill on ice.

iii. Because urea will diffuse out of the gel and remain in the bottom of the wells, rinse the wells again to disperse the urea. This will allow the sample to settle with as little upward spread as possible.

iv. Using a Pipetman, load the RNA samples quickly so that the wells remain free of urea. Avoid creating air bubbles, which will disturb the samples.

Thin gel-loading flat Pipetman tips that fit between the 0.4-mm spacers and reach the bottom of the well are available. In general, it is advisable to load minimal volumes of ≤10 µL per 0.8-cm-wide well with a 0.4-mm-thick gel. Adjust the loading volume according to thickness and width of the well. For preparative applications, it is advisable to run thicker gels. For information on loading buffers and dyes, see individual protocols regarding sample preparation and appropriate loading dyes.

8. Run the gel.

i. For a 20-cm-wide x 40-cm-long x 0.4-mm-thick gel, run the gel either at constant current (45 mA) or constant voltage (1600 V).

It is important not to run the gels too fast. Haste makes waste! A thermometer to monitor heating can be attached to the front plate with silicone grease. The temperature must remain under ~58°C. Increasing the current or the voltage runs the risk of overheating the gel or even cracking the gel plates. Overheating or running too fast can greatly distort the appearance of bands and accentuates any irregularities in the gel matrix.

ii. Run the appropriate gel distance to optimize the resolution of the bands of interest.

The tracking dyes in the loading buffers (xylene cyanol and bromophenol blue) ALWAYS run to the same position, but depending on the percentage gel, they comigrate with different-sized nucleic acids. Gels are highly reproducible when prepared and run under the same conditions; therefore, it is advisable to become accustomed to the behavior of the specific gel you use.

9. Disassemble the gel.

i. Turn the power supply off and drain the buffer. If the RNA sample run was radioactive, the bottom buffer in particular may be radioactive and should be handled accordingly.

ii. Detach the plates and lay them down on bench paper, with the large plate down. Remove the yellow tape, either with a razor blade or by pulling it off if possible. Slide the spacers out of the sides.

iii. Using a thin spatula (dedicated to this purpose only), pry the top plate up, starting from a corner. Do this slowly until it is clear that the gel remains on the bottom plate. Wear safety glasses in the event that the corner of the glass chips.

iv. Trim a piece of Whatman 3MM paper to the size of the gel and carefully lay it over the gel. Smooth the paper with a gloved hand to ensure uniform contact with the gel. Pick up the paper (and gel) and lay them gel side up on clean bench paper.

Gels can be covered by plastic wrap and exposed to X-ray film at –80°C without drying. However, to prevent damage to phosphor screens, drying is recommended.

v. Cover the gel with plastic wrap and dry at 80°C using a vacuum-driven gel dryer.

Make sure that the gel is completely dry before opening the drier. If the drier is opened too soon, the gel may crack or shrivel.

10. The gel is now ready for autoradiography or phosphorimaging. Expose the gel to the phosphorimager.

Gels dried without fixing and washing in methanol–acetic acid remain tacky after drying because of the urea. Leaving the plastic wrap on them will help to protect the phosphor screen or prevent the gel from sticking to the film.

TIPS AND TROUBLESHOOTING

1. If running preparative gels, use thicker spacers and combs, e.g., ≥0.8 mm.
2. For large-scale purification (e.g., from in vitro transcription reactions), pellets contain high concentrations of nucleotides and salt. Do not overload the gel with these samples; it is advisable to run such samples in multiple lanes.
3. High-percentage gels must run at constant current to avoid overheating. The milliamps are determined by the size and percentage of the gel.
4. We run 20-cm x 40-cm gels for almost every acrylamide application because these large gels are very versatile and afford maximal resolution for RNAs of any size ≤600 nucleotides.
5. Loading dyes for denaturing and native gels are commercially available and we recommend using them.
6. "Smiling" of marker dyes is caused by uneven heat distribution when using a gel apparatus without a metal backing plate. Other causes are improper taping or clamping of plates, a bowed plate, a bad spacer (unmatched), and running the gel too fast. To avoid the latter problem, run the gel more slowly or use a gel apparatus with a metal diffuser plate.
7. Distortion of lanes is caused by inappropriate pulling of combs or acrylamide in wells, salt in samples, overloading the gel, or bubbles in the gel. Make sure that the tops of the wells are wet when pulling the comb and rinse sample wells well. For the problem of salt in the sample, reprecipitate with ethanol. If overloading is the culprit, reduce the amount of sample loaded.
8. A hyperfocused tracking dye indicates that the gel was not prerun for enough time. Prerun gels adequately, making sure that running conditions are the same as those of the prerun.
9. If the sample is trapped in the wells, this may be the result of inadequate resuspension of the sample, aggregates in the sample, or inadequate denaturation. Make sure that the sample is well dissolved and not overconcentrated. Denature properly.
10. Streaking of radioactivity down the lanes can be the result of any of the above situations.
11. The gel sticking to both plates can be caused by inadequate "waterproofing" of the short plate or overheating during the run. Re-treat the top plate with Rainex and make sure that run conditions are correct.
12. If your gel cracks after drying, the gel was not completely dry. Dry gels for a longer amount of time and make sure that the gel drier vacuum is sealed.
13. Acrylamide gels can be strained after running with ethidium SYBR dyes.

READING LIST

Maxam AM, Gilbert W. 1980. Sequencing end-labeled DNA with base-specific chemical cleavages. *Methods Enzymol* **65:** 499–560.

Sambrook J, Russell DW. 2001. *Molecular cloning: A laboratory manual*, 3rd ed. Cold Spring Harbor Laboratory Press, Cold Spring Harbor, New York.

Method 2: Agarose Gel Electrophoresis (Nondenaturing)

Agarose gels are valuable for fractionation of large RNAs. They are generally used to analyze RNAs of ≥600 nucleotides, such as mRNAs (e.g., by northern blotting). RNA analysis on agarose gels is essentially identical to DNA analysis (except that the gel boxes used must be dedicated to RNA work or to other ribonuclease-free work). Here we describe straightforward Tris borate, EDTA (TBE) gels for routine analysis. For particular applications such as northern blotting, refer to specific Protocols

7–10 in Chapter 5. These gels are appropriate for determining the integrity of RNA before using it for other applications (Protocol 13).

MATERIALS

The recipe for the item marked with <R> is at the end of the protocol or in the Common Recipes Appendix.

CAUTION: See the Cautions Appendix for appropriate handling of materials marked with <!>.

Reagents

Agarose
Bromophenol blue <!>
Ethidium bromide or SYBR Gold (Invitrogen) <!>
RNA sample (at least 200 ng)
TBE electrophoresis buffer (10x) <R>

Equipment

Gel combs (purchased with the gel system)
Gel tray
Horizontal gel box (dedicated to RNA work)
Microwave oven
Power supply
UV transilluminator
Water bath (55°C)

OVERVIEW

Analysis of RNA by agarose gel electrophoresis is the same as analysis of DNA. Here we describe the use of these gels to evaluate the quantity and quality of RNA preparations.

METHOD

1. Before beginning, assemble the necessary equipment and solutions.
2. Prepare the gel matrix.
   i. Suspend preweighed agarose in 1x TBE to the desired final percentage; e.g., for a 1% gel, suspend 1 g of agarose in 100 mL of 1x TBE.
   ii. Boil the suspension to dissolve the agarose, then cool for 5 min in a 55°C water bath.
   iii. If the gel is to be prestained, add 0.5 µg/mL of ethidium bromide to the gel solution before pouring the gel.
   iv. Pour the suspension into the gel tray to the desired thickness. Insert the comb and let the gel solidify.
3. Place the gel tray into the gel box and add TBE, ensuring that the gel is submerged. Carefully remove the comb.
4. Load the gel by carefully pipetting the RNA sample into the bottom of the well.

   The buffer level must be high enough to prevent surface tension from disturbing the sample. RNA samples should be at a concentration of 0.2–1 µg/µL in 10 µL of TE buffer. Add nondenaturing loading dye.

5. Place the electrodes on the gel box and turn on the power supply. Most gels can be run at 100–150 V. Run until the bromophenol blue is two-thirds of the way to the bottom.

   See the Tips and Troubleshooting section below for visualizing RNA.

6. For poststaining, which is usually necessary to visualize small RNAs, soak the gel for 10 min in 0.5 μg/mL ethidium bromide. Rinse or destain with water for 15 min and visualize on a UV transilluminator. If there is too much background ethidium bromide, destain for a longer amount of time.

   For greater sensitivity, stain with SYBR Gold according to the suppliers' instructions.

TIPS AND TROUBLESHOOTING

1. This procedure should not be used to determine size with accuracy, because the migration of the RNA will be influenced by secondary structure.
2. The RNA sample must be clean, disaggregated, and denatured at the start or it will not run properly.
3. Because these gels are nondenaturing, the bands may be broader and aggregates may be present. In addition, differently folded RNAs of identical chain length may migrate in different positions even though they are the same molecule.

SOLUTION AND REAGENT RECIPES

CAUTION: See the Cautions Appendix for appropriate handling of materials marked with <!>.

Acrylamide Stock (40%; 19:1 acrylamide:bis-acrylamide)

Note: We recommend purchasing premade solutions (e.g., BioRad or Amresco).

Reagent	Quantity (for 500 mL)	Final concentration
Acrylamide (electrophoresis grade) <!>	190 g	38% (w/v)
Bisacrylamide (electrophoresis grade) <!>	10 g	2% (w/v)
H_2O	to 500 mL	

Filter with a 0.45-μm filter and store at 4°C in a dark bottle.

RNA Gel-loading Buffer (1.5x)

Formamide + dyes

Reagent	Quantity (for 10 mL of 1.5x)	Final concentration
Ultrapure formamide <!>	9.5 mL	95%
Bromophenol blue (2.5%, w/v) <!>	100 μL	0.025%
Xylene cyanol FF (2.5%, w/v) <!>	100 μL	0.025%
EDTA (0.25 M, pH 8.0) <!>	200 μL	5 mM EDTA

Use 5 μL for a 2.5-μL sample. Purchase a distilled, deionized preparation of formamide and the above loading dyes. Store in small aliquots at –20°C. This solution is available commercially (Ambion) and is recommended over homemade. Make 1-mL aliquots. Store for up to 1 yr at –20°C.

TBE Electrophoresis Buffer (10x)

Reagent	Quantity (for 1 L)	Final concentration
Tris base	121.1 g	1 M
Boric acid <!>	61.8 g	1 M
EDTA (disodium salt) <!>	7.4 g	0.02 M

Dilute 50 mL to 1 L to make gel running buffer. Store for up to 6 mo at room temperature.

Protocol 13

Determining the Yield and Quality of Purified RNA

After any RNA purification by any method, it is necessary to ascertain both the amount (yield) and the integrity (quality) of the RNA obtained. This is essential for any subsequent analysis and critical for any comparative analyses.

METHODS FOR DETERMINING YIELD

Before attempting to use any of the following procedures, it is important to have a "ballpark" idea of what you expect the yield to be (see Table 2-1).

Determining Yield by Spectrophotometry or Fluorometry

The easiest way to determine the quantity of RNA in a sample is to measure the absorbance at 260 nm (A_{260}) using a spectrophotometer. Because the bases in RNA absorb UV light in the 250–265-nm range, one can use this property to quantitatively measure the concentration of an RNA solution, using an average absorbance for the four nucleotide bases. A solution of RNA at 40 µg/mL will have an absorbance of ~1. Accordingly, if 50 µL of the same solution is diluted in 1 mL of H_2O and read in a 1-mL cuvette, the absorbance will be 0.05; this is the absolute minimum that we recommend for accurate OD readings. The advent of nanospectrophotometers, such as the Thermo-Fisher NanoDrop and GE Healthcare NanoVue, has greatly increased the ease, sensitivity, and accuracy of determining the concentration of low-volume (microliters) samples by UV absorbance (see below). With this equipment, one can measure the A_{260} without dilution and with minimal waste of the sample. If you do not have access to these spectrophotometers, we recommend that OD be used as a measure of concentration, only when you have a significant amount of RNA present in a concentrated solution, e.g., at least 1.0 µg/µL. We have often observed students interpret A_{260} readings of 0.003 or less, but measurements as low as this are essentially meaningless.

Although measuring A_{260} is generally a reliable way to quantitate RNA concentrations, this method can be confounded if the RNA is contaminated with DNA, protein, or phenol, all of which absorb some UV light at 260 nm. A diagnostic for protein contamination is absorbance at 280 nm. Phenol and TRIzol Reagent both absorb at 270 nm and 230 nm. The chaotropic agents guanidine-HCl and guanidinium isothiocyanate, commonly used for RNA purification, absorb at ~230 nm and ~260 nm, respectively. For purified DNA, the A_{260}:A_{280} ratio should be ~1.8, and for purified RNA, this ratio should be ~2.0, because the unpaired bases in RNA absorb more UV light than the base-paired bases in duplex DNA. However, these ratios are "rules of thumb" and assume an average base composition of the RNA sample, because different bases each have different A_{260}:A_{280} ratios. If the A_{260}:A_{280} ratio is <2.0, protein contamination is probable and reextraction with phenol is recommended. If the A_{260} is high, phenol contamination is probable and another round of ethanol precipitation and resuspension is recommended. Contamination with DNA is harder to detect because RNA and DNA essentially have identical absorbance spectra. Therefore, if the A_{260} of the sample is higher than expected, and protein and phenol contamination have been ruled out, the sample is likely contaminated with DNA. In this case, DNase treatment followed by phenol extraction and ethanol precipitation is recommended (see Protocols 1 and 3). In addition, it is possible to use the A_{260}:A_{230} ratio as an indicator of nucleic acid purity, with ratios commonly in the range of 2.0–2.2. Note that absorbance is pH dependent, so for accurate readings, keep the pH constant and near 7.5.

UV Absorbance Determination of RNA Concentrations Using a Nanospectrophotometer

The development of small-volume (0.5–2 µL) UV-visible spectrophotometers can now be used to sensitively measure RNA or DNA concentrations and fluorochrome (e.g., Cy3/Cy5) dye coupling to allylamine-modified cDNA, or for protein concentration determination. These instruments (e.g., the Thermo-Fisher NanoDrop or GE Healthcare NanoVue nanospectrophotometers) use fiber optic technology and surface tension to hold a 1-µL sample in place; thus, they eliminate the need for traditional sample holding by cuvettes. The dynamic range of these instruments is high, ranging from 2 ng/µL to 3700 ng/µL. The following discussion provides practical information about the use of this equipment and interpretation of results. Step-by-step instructions for operation are supplied in the manufacturers' instrument manuals.

Procedure

This procedure is simple. Place 1.0–1.5 µL of the RNA sample onto the sample pedestal. The UV absorbance of the sample is then read either at a fixed wavelength or in a UV-visible scan. Typically, pure RNA (or DNA) samples are read at A_{260} and A_{280}. It is possible also to use these instruments to scan a full UV and visible wavelength absorbance spectrum, from 220 nm to 750 nm. The UV-visible-wavelength scanning procedure is more useful in microarray studies when one wants to quantitate fluorochrome dye coupling to cDNA (see Chapter 8, Protocol 1). Figure 2-6 shows some results obtained using a nanospectrophotometer and how these results are interpreted.

TIPS AND TROUBLESHOOTING

1. As with the UV absorbance methods described previously, any compounds (e.g., free nucleotides, phenol, or other organic compounds such as guanidinium isothiocyanate) that absorb UV light near 260 nm will interfere. Often, a simple ethanol precipitation can be used to remove the contaminating substance.
2. When adding a sample to the pedestal, ensure that the sample is placed over the "eye" (the eye is the little metal circle with the tiny hole in the NanoDrop).
3. Bubbles are incompatible with accurate readings. Although the instructions may claim that only 1 µL is needed, in practice a larger volume (1.3–1.5 µL) will produce more reliable readings because the droplet will have better optical characteristics. Small-volume droplets can give incomplete coverage across the pedestal.
4. It is possible to saturate the spectrophotometer with high concentration solutions of RNA or DNA. This will cause an underestimate of the true concentration. Try to obtain readings using solutions at ≤2.5 µg/µL. For samples that give higher readings (2500 ng/µL), it is a good idea to dilute 1:10 and read the dilution to get the most accurate reading.

Fluorescent Dye Binding for RNA and DNA Quantitation

As an alternative to spectrophotometry, RNA can be quantitated using fluorescent dye binding. This is a sensitive assay for detecting and determining the quantity of RNA (and contaminating DNA) present in a purified RNA sample or in crude extracts or chromatographic fractions. It is ~1000 times more sensitive than using UV absorbance and can detect RNA at 1 ng/mL. In addition, this method can be used for quantitation of in vitro–transcribed RNA samples or for determining RNA concentrations before northern blotting, RNase protection, RT-PCR, or cDNA library preparation. It is similar in concept to fluorescent dye binding to DNA (e.g., SYBR Green) used in quantitative PCR (qPCR) (see Chapter 4, Protocol 19). The RiboGreen RNA reagent, a proprietary fluorescent

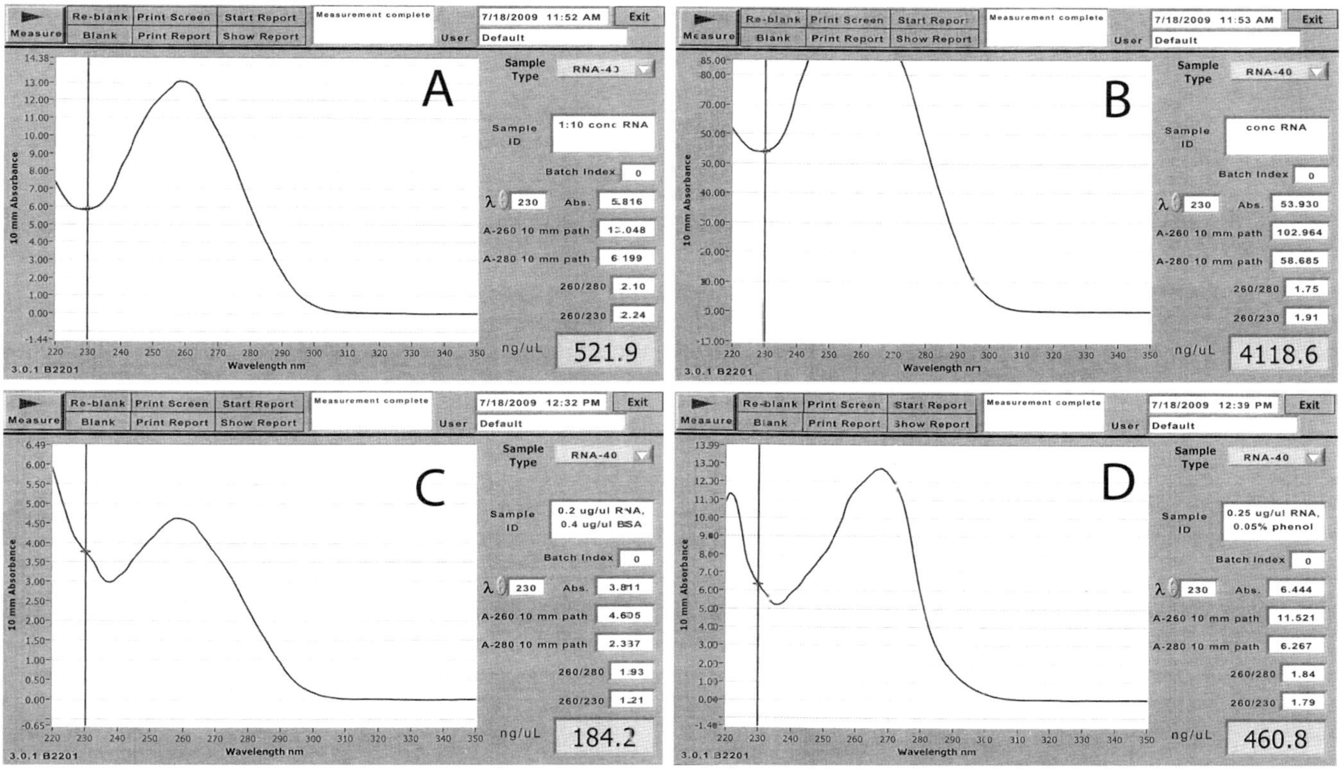

FIGURE 2-6. Using the microspectrophotometer (e.g., Nanodrop), according to the manufacturer's instructions, purified RNA was evaluated. (*A*) Purified yeast RNA was read at a concentration estimated at 0.5 µg/µL. (*B*) Purified yeast RNA was read at a concentration to be 5 µg/mL. Note that the detector has been saturated at this concentration and is underestimating the RNA content by 20%. (*C*) Mixture of yeast RNA and bovine serum albumin (BSA). Note the increased absorption at 220 nm due to the presence of peptide bonds; however, the increase in 280 is less noticeable, owing to the fact that BSA has relatively few tryptophans (3/607) or tyrosines (21/607). In this case, the peptide bond absorption at 220 is a better indicator (BSA has 606 peptide bonds/mole), as shown by the 260:230 ratio (230 is used because RNA has a local minimum at 230, whereas the peptide bond contributes at this wavelength; see *A*). The 260:230 ratio in the BSA-contaminated sample is 1.21, whereas it is 2.24 for the purified RNA. The 260:280 ratio has dropped from 2.1 for pure RNA to 1.9 in this sample. Thus, although this RNA passes the >1.8 test for 260:280, it should be rejected or reextracted on the basis of its 260:230 ratio. (*D*) Mixture of 250 ng/µL purified yeast RNA and 0.05% v/v of water-saturated phenol in water (monophasic). Note that absorption is at 270>260 and that the RNA amount is overestimated. Phenol has a second peak near 224 that is obvious in this spectrum, and the RNA minimum at 230 has shifted up toward 240, which is a phenol minimum. Recognizing the presence of phenol in the spectrum by noting these features is more valuable than simply evaluating the 260:280 and 260:230 ratios, but it can be seen that these have dropped in comparison to the purified sample.

dye, preferentially binds to RNA, but it can also detect DNA (see the following Tips and Troubleshooting section). This method is useful when making RNA from nuclear fractions that might be contaminated with DNA and for assaying very small quantities of RNA prepared from limited quantities of starting material. Step-by-step instructions for using this reagent are supplied in the manufacturer's user manual.

TIPS AND TROUBLESHOOTING

1. If you suspect that the RNA sample of interest is contaminated with DNA (perhaps cellular genomic DNA in a total RNA preparation), you can DNase-I treat the sample (Protocol 11) or use PicoGreen, which detects double-stranded DNA only.

2. The assay remains linear in the presence of several compounds that commonly contaminate nucleic acid preparations, although the signal intensity may be affected. Thus, to serve as an effective control, treat the RNA solution used to prepare the standard curve the same way as the experimental samples; it should contain similar levels of such compounds.
3. There can be some interference with the fluorescence assay by salts, organic solvents, detergents, proteins, or other compounds. If these are present in the sample, control for this by placing them in the standard curve RNA dilutions.

Determining Yield by Gel Electrophoresis

For samples of total or cytoplasmic RNA (with rRNA), a simple and straightforward way to determine yield is to separate a small aliquot of the RNA on an agarose gel and stain with ethidium bromide or SYBR Gold (see Protocol 12). Bands of rRNA (28S and 18S) are visualized, and their intensity is compared to that of a preparation of known quantity. In our laboratories, we keep a "stock" of high-quality cytoplasmic RNA prepared from tissue culture cells as a reference. This stock can be diluted as appropriate to obtain the equivalent amount of RNA that you expect from the sample. Because SYBR Gold is 10-fold more sensitive than ethidium bromide, it should be used for small amounts of RNA.

An advantage of this technique is that it measures both the quantity and the quality of RNA; i.e., sharp and distinct rRNA bands without a pronounced haze below them is a good sign that the preparation is not significantly degraded. The only disadvantage of this approach is sensitivity; at least 500 ng of RNA must be available to sacrifice for this assessment. To assess quality, sufficient quantity must be loaded to detect degraded material.

Use of Rapid Capillary Electrophoresis on an Agilent Bioanalyzer for RNA Sample Quality Control

A bioanalyzer provides a convenient and more sensitive way to assess RNA quality. The Agilent 2100 Bioanalyzer basically performs small-volume electrophoresis, similar to capillary DNA sequencing but in a "chip" format that allows analysis of 12 samples at once. The bioanalyzer can be used for analysis of the quality of total RNA preparations by visualizing the rRNA bands and intact mRNA, and to detect and quantify RT-PCR products (using DNA chips; see Klinck et al. 2008; Venables et al. 2008, 2009 in the Reading List at the end of this protocol). Agilent RNA kits, which are designed for use with the Agilent 2100 Bioanalyzer only, contain chips and reagents designed for analysis of RNAs. Each RNA chip contains an interconnected set of microchannels (capillaries) that is used for separation of nucleic acid fragments based on their size as they are driven through the microchannels electrophoretically. The following discussion provides practical information about the use of this equipment and interpretation of results obtained running RNA samples on a bioanalyzer. Step-by-step instructions for operation of this equipment are supplied in the manufacturer's user manuals.

A computer runs the machine, analyzes the traces, and presents the data. Of interest is the trace that estimates the amount of each rRNA that is present and their ratios. Because the larger rRNA is more susceptible to degradation because of its greater length, the ratio of 28S to 18S rRNA is a convenient measure of the integrity of the sample. This information is displayed by the software. The amount of RNA in the sample is determined by comparison to a fixed known amount of the RNA ladder RNAs. Thus, it is possible to obtain both concentration and integrity of the sample. Because the data are simple to archive and compare, many laboratories that handle and compare many samples believe that this approach is superior to gel analysis, primarily because of its data archiving and comparison ability.

Expected Results of Bioanalyzer Analysis of RNA Quality

The quality of purified total RNA can be analyzed using a bioanalyzer, such as the Agilent 2100. In Figure 2-7, this bioanalyzer was used to show the presence of 18S and 28S rRNA (mammalian), 18S

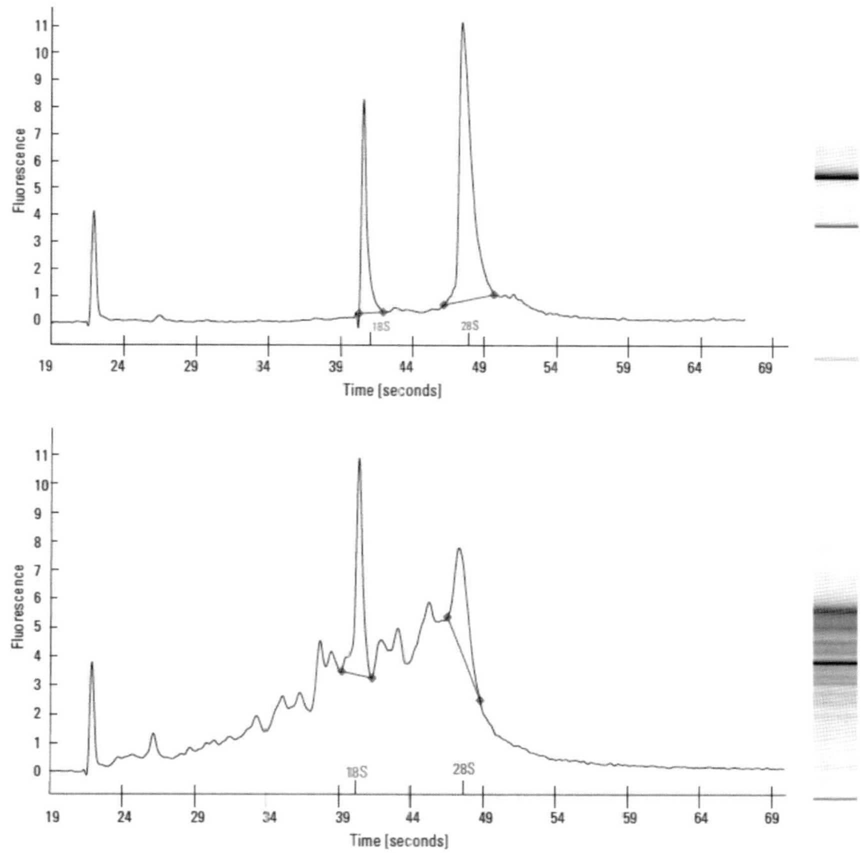

FIGURE 2-7. Analysis of intact and degraded total RNA. Electropherograms and gel-like images of intact total RNA (*top*) and partially degraded total RNA (*bottom*). Partially degraded RNA shows a decrease in the 18S/28S ribosomal band ratio and an increase in fragmentation products. The RNA ladder markers give discrete peaks following electrophoresis and the instrument software can generate a "virtual gel" pattern (*right*). (Reprinted, with permission, from http://www.chem.agilent.com [publication 5988-7650EN. ©Agilent Technologies 2002].)

and 25S/26S rRNA (yeast), 16S and 23S rRNA (bacteria), and small RNA species in total RNA. Typically, for total RNA samples, two major rRNA peaks are observed (because these account for >90% of the total RNA in cells: for eukaryotic total RNA, the two rRNA peaks correspond to the 18S and 28S rRNAs, for yeast total RNA the two large rRNA peaks correspond to the 18S and 25S/26S rRNAs, for bacterial total RNA, the two large rRNA peaks correspond to the 16S and 23S rRNAs) (see Fig. 2-7).

TIPS AND TROUBLESHOOTING

1. Sample preparation for the RNA Nano Chip:
   i. Ideally, sample concentrations should be 100–200 ng/μL; RNA concentrations as low as 50 ng/μL can be used.
   ii. Each sample well must contain a total of 6 μL (1 μL for the RNA Pico Chip).
   iii. The nano marker must be placed in every sample well and the ladder well.
   iv. Add water or nano marker to unused wells to bring the volume up to 6 μL.

   v. Use the chip within 5 min of preparation to prevent evaporation. Cover the chip with plastic wrap or parafilm if it will be left standing for any length of time.
   vi. RNA samples may be denatured to remove secondary structure. Denature for 2 min at 70°C before placing the samples in the wells of the chip.
   vii. Genomic DNA can produce stray bands or clog the capillaries in the chip. To check for genomic DNA contamination, treat the samples with DNase I. Run a DNase-I–treated sample next to an untreated sample.

2. It is important not to leave the used chip in the bioanalyzer after the run is complete because this will dry out the electrodes and make them difficult to clean.
3. A critical part of the assay is preparing the chip. This involves loading the capillaries with the "gel" matrix material and the dyes that will stain the RNA so that the detector in the machine can measure it. The chip is loaded with a syringe system that is fairly easy to use. Fill each well (make sure that there are no bubbles!), add the sample, vortex, and load into the machine according to the instructions.
4. Because this protocol uses small-volume electrophoresis, the samples must be in a low-ionic-strength solution (preferably in RNase-free water or 10 mM TE buffer).
5. Accurate pipetting is very important for reproducible results. Make sure to use properly calibrated pipettes and to place the tip into the center and bottom of each well in the chip when dispensing. To avoid bubbles, do not push past the first resistance point on the pipette. You may pipette up and down gently to mix samples in the wells of the chip.
6. Protect the gel-dye mix from light by covering the tube with foil. Return the reagents to the cold room when you are finished.
7. When using the priming station, press down slowly and steadily on the plunger when priming. After releasing, the plunger should come up to at least 0.7 mL in 1–2 sec. If this does not occur, check that the gasket is clean and retry. If it still does not prime well, change the gasket (see Agilent 2100 instrument manual).

Determining Yield by Quantitative or Semiquantitative PCR

When it is not possible to quantitate the amount of RNA in a preparation by any of the methods previously described (e.g., when the amount of expected RNA is very small), reverse transcription followed by qPCR or semi-qPCR is recommended (see Chapter 3, Protocols 18 and 19). These techniques are extremely sensitive and provide information with regard to both the quantity and quality of RNA. In addition, these approaches allow a determination of whether the sample is significantly contaminated with DNA. We recommend assaying a housekeeping mRNA (e.g., actin, glyceraldehyde 3-phosphate dehydrogenase [GAPDH], tubulin, etc.). Again, as with gel electrophoresis, we recommend using a "stock" RNA preparation (diluted as appropriate) as a reference. Parallel assessment with and without reverse transcription will show any DNA contamination (i.e., amplification without reverse transcription indicates the presence of DNA, unless the PCR primer pairs are in different exons separated by a large intron; in this case, the genomic DNA contamination product will be too large to show up). For RNA preparations in which large RNAs have been removed, it is necessary to assay a small RNA (e.g., a constitutively expressed miRNA) by RT-PCR.

All of the techniques described above are useful methods for assessing the quantity of RNA that has been prepared; gel electrophoresis and qPCR or semi-qPCR also provide information regarding quality. Common sense indicates that if the yield is far less (≥30%) than expected, something has gone wrong with the preparation and the RNA is suspect. Rather than proceeding with a suspect preparation, start over from the beginning.

NORTHERN BLOTTING TO ASSESS THE QUALITY OF RNA PREPARATIONS

As discussed previously, gel electrophoresis, visualization of RNA bands, and qPCR or semi-qPCR (in all cases compared to a reference) are valuable methods for assessing quantity as well as quality of RNA. Nevertheless, each of these methods can be potentially misleading. PCR will amplify fragments of RNA, and rRNA visualization is a somewhat crude method for assessing integrity of an RNA preparation. Moreover, spectrophotometry tells nothing about the integrity of the RNA preparation. By far, the best method for assessing the quality of an RNA preparation is northern blotting (see Chapter 3, Protocols 7–10). Because this technique visualizes the entire RNA, it is diagnostic for any degradation. Therefore, if you are working with specific RNAs, we recommend assaying each preparation by northern blotting. Again, this is best done by comparing the quality (i.e., sharp band) and quantity of the signal obtained to those obtained from a reference northern blot performed on RNA of known quantity and quality.

READING LIST

Nanospectrophotometry

Gallagher SR, Desjardins PR. 2008. Quantitation of DNA and RNA with absorption and fluorescence spectroscopy. *Curr Protoc Mol Biol* **52:** A.4K.1–A.4K.21.

Fluorescent Dye Binding

Gallagher SR, Desjardins PR. 2008. Quantitation of DNA and RNA with absorption and fluorescence spectroscopy. *Curr Protoc Mol Biol* **52:** A.4K.1–A.4K.21.

Jones LJ, Yue ST, Cheung C-Y, Singer VL. 1998. RNA quantitation by fluorescence-based solution assay: RiboGreen reagent characterization. *Anal Biochem* **265:** 368–374.

Bioanalyzer

Klinck R, Bramard A, Inke L, Dufresne-Martin G, Gervais-Bird J, Madden R, Paquet ER, Koh C, Venables JP, Prinos P, et al. 2008. Multiple alternative splicing markers for ovarian cancer. *Cancer Res* **68:** 657–663.

Venables JP, Koh CS, Froehlich U, Lapointe E, Couture S, Inkel L, Bramard A, Paquet ER, Watier V, Durand M, et al. 2008. Multiple and specific mRNA processing targets for the major human hnRNP proteins. *Mol Cell Biol* **28:** 6033–6043.

Venables JP, Klinck R, Koh C, Gervais-Bird J, Bramard A, Inkel L, Durand M, Couture S, Froehlich U, Lapointe E, et al. 2009. Cancer-associated regulation of alternative splicing. *Nat Struct Mol Biol* **16:** 717–724.

WWW RESOURCES

Nanospectrophotometry

http://www1.gelifesciences.com/ GE Healthcare NanoVue User Manual 28944299AA.pdf.

http://www.nanodrop.com/Support.aspx?Type=User%20Guides&Cat=NanoDrop%201000 NanoDrop User Bulletin T042, Nucleic Acid Purity Ratios.

http://www.nanodrop.com/ Thermo Fisher Scientific Inc. NanoDrop User Manual (nd-1000-v3.7 User's Manual).

Fluorescent Dye Binding

http://www.invitrogen.com/site/us/en/home/References/Molecular-Probes-The-Handbook/Nucleic-Acid-Detection-and-Genomics-Technology/Nucleic-Acid-Detection-and-Quantitation-in-Solution.html/ Quant-iT™ RiboGreen® RNA Reagent and Kit: mp11490.pdf.

Bioanalyzer

http://www.chem.agilent.com/ Agilent 2100 Bioanalyzer User Guide.pdf; Agilent RNA 6000 Nano Kit Guide G2938-90034_KitRNA6000Nano_ebook.pdf; Quantitation comparison of total RNA using the Agilent 2100 bioanalyzer, RiboGreen analysis and UV spectrometry.pdf; Agilent 2100 Bioanalyzer User Guide.pdf.

3 Detection and Characterization of Specific RNAs

IN ANY GIVEN CELL, THE NUMBER OF distinct RNA species is remarkably complex. Several thousand pre-mRNAs and their corresponding mRNAs are expressed, and myriad noncoding mRNA-like transcripts, small nucleolar RNAs (snoRNAs), small nuclear RNAs (snRNAs), small cytoplasmic RNAs (scRNAs), and small noncoding RNAs such as micro RNAs (miRNAs) are present. The sizes of RNAs can vary from 20 nucleotides to several kilobases, and abundances can vary from a few molecules per cell (e.g., in the case of rare mRNAs) to hundreds of thousands of molecules per cell (e.g., in the case of certain snRNAs). It is of obvious interest to determine the abundance and integrity of specific RNA species within these complex mixtures. In this chapter, we discuss commonly (and less commonly) used methods for detecting, characterizing, and quantifying RNAs both known and unknown. We also present methods for determining the sequence of RNAs by direct and indirect approaches. Because many of the approaches require the use of radiolabeled probes, we first outline general considerations for probe synthesis. We then describe the methods in general terms before presenting detailed protocols.

PROTOCOLS

1 Labeling of Oligonucleotide Probes (DNA, LNA, RNA) by Polynucleotide Kinase and [γ-^{32}P]ATP, 89
2 DNA Oligonucleotide Radiolabeling by Terminal Deoxynucleotidyl Transferase (TdT), 92
3 Asymmetric Polymerase Chain Reaction to Generate Single-stranded Probes, 95
4 Random Hexamer ^{32}P Radiolabeling of DNA Fragments as Hybridization Probes, 97
5 Nick Translation of Double-stranded DNA for the Preparation of ^{32}P-labeled Hybridization Probes, 100
6 Enzymatic Dephosphorylation of RNA Using Calf Intestine Alkaline Phosphatase or Shrimp Alkaline Phosphatase, 103
7 Northern Blots: Denaturation and Electrophoresis of RNA with Formaldehyde, 106
8 Northern Blots: Denaturation and Electrophoresis of RNA with Glyoxal, 110
9 Northern Blots: Capillary Transfer of RNA from Agarose Gels and Filter Hybridization Using Standard Stringency Hybridization Conditions, 114
10 Northern Blots: Alternative Method for Processing Northern Blots after Capillary Transfer, 121
11 Northern Blots for Small RNAs and MicroRNAs, 126
12 RNase Protection Assay, 131
13 Nuclease S1 Protection Mapping, 136
14 Primer Extension Analysis of RNA, 141
15 Reverse Primer Extension, 145
16 Poisoned Primer Extension, 149
17 Detection of RNAs by 3′-end Labeling and RNase H Digestion, 152
18 Reverse Transcription–Polymerase Chain Reaction, 155
19 Basic Quantitative PCR Using Real-time Fluorescence Measurements, 165
20 5′ Rapid Amplification of cDNA Ends, 171

(continued)

PROBE PREPARATION

Radiolabeled DNA probes are convenient for detecting RNA molecules in hybridization experiments such as S1 nuclease protection mapping and RNA blot filter hybridization. There are several ways to radiolabel DNA probes (Table 3-1). The standard methods include kinasing the 5′-hydroxyl end with [γ-^{32}P]ATP and T4 polynucleotide kinase (Protocol 1), tailing the 3′-hydroxyl end with terminal deoxynucleotidyl transferase (TdT) (Protocol 2), and uniformly labeling one strand of the DNA by incorporation of [α-^{32}P]dCTP using asymmetric polymerase chain reaction (PCR) (Protocol 3). Chemically synthesized DNA oligonucleotides do not have a 5′-phosphate group and so can be subjected to the T4-kinase-labeling reaction without phosphatase treatment. It is also possible to uniformly radiolabel long DNA molecules (e.g., restriction enzyme fragments or PCR products) for use in RNA blot filter hybridizations (i.e., northern blots) using either random hexamer priming in the presence of [α-^{32}P]dCTP (Protocol 4) or nick translation (Protocol 5). Finally, for detecting RNAs by hybridization in solution, by hybridization on membranes, or by in situ hybridization, chemically synthesized oligonucleotides called locked nucleic acids (LNAs) (Protocol 1) provide a more sensitive (up to 10x)—although also more expensive—method to detect both long and short RNA molecules via improved hybridization characteristics.

PROTOCOLS (*continued*)

21 3′ Rapid Amplification of cDNA Ends, 180

22 RACE on Uncapped or Nonpolyadenylated RNAs, 186

23 RNA Sequencing by Primer Extension, 189

24 Direct Chemical Sequencing of End-labeled RNA, 192

25 Splinted Ligation Method to Detect Small RNAs, 200

Labeled RNAs can also be used as probes and are superior for some applications (see Chapter 4, Protocol 1). RNA-based probes are generated by in vitro transcription, typically using recombinant bacteriophage (T7, SP6, or T3) RNA polymerases. These uniformly or "body"-labeled RNAs are generated by adding [α-^{32}P]UTP or [α-^{32}P]GTP to the nucleotide mix during the in vitro transcription reaction. These are high-specific-activity (~5–6 x 10^8 cpm/μg) radioactive RNAs that are suitable for use as hybridization probes for filter hybridizations (northern blots) or for probes in solution (RNase protection assays). They can also be used for in vitro RNA-processing reactions, such as for pre-mRNA splicing or 3′-end processing or polyadenylation assays (see Chapter 6). Often, RNA molecules, either chemically or enzymatically synthesized, can be labeled at their 5′ ends with T4 polynucleotide kinase and [γ-^{32}P]ATP (Protocol 1). In this case, as with synthetic DNA, RNA is chemically synthesized without a 5′ phosphate and so can be treated with T4 polynucleotide kinase and [γ-^{32}P]ATP. In contrast, RNAs prepared by in vitro transcription contain a 5′-triphosphate terminus that must be removed by treatment with calf intestine or shrimp alkaline phosphatase before being subjected to the T4-kinase-labeling reaction (Protocol 6). These 5′-end-labeled RNAs are useful as probes

TABLE 3-1. Techniques for Generating Probes for RNA Detection

	Application	
	Northern blotting	Nuclease protection
Probe type	*Kinased oligo LNA	5′-labeled DNA
	*Kinased oligo DNA	*5′-labeled RNA
	Kinased oligo RNA	Asymmetric PCR primed with kinased DNA oligonucleotide
	5′-labeled RNA	*Body-labeled RNA
	5′-labeled DNA	
	3′-labeled RNA (ligase)	
	3′-labeled RNA (cordycepin)	
	3′-labeled oligo DNA (TdT)	
	*Asymmetric PCR (DNA)	
	Nick translated (DNA)	
	Random primed (DNA)	
	Body labeled (RNA)	

As described in the text, there are many ways to generate labeled probes for RNA detection. Each is listed above under the technique where each is appropriate. Techniques marked with an asterisk are preferred.

for RNA-binding assays (such as gel-shift or nitrocellulose filter binding) or for use in "footprinting" assays using either chemical or enzymatic probes that identify where a protein interacts with the RNA (see Chapter 5). RNA molecules, either natural or synthetic, can also be labeled at their 3′ ends either with [5′-^{32}P]pCp and RNA ligase or with poly(A) polymerase and [α-^{32}P]ddATP (cordycepin). 3′-End labeling is also useful for generating RNA probes for footprinting or chemical modification–interference experiments (see Chapter 5).

NORTHERN BLOTTING

Northern blotting is perhaps the most popular of all techniques for detecting specific RNAs. In this technique, an RNA sample is denatured in the presence of formaldehyde or glyoxal and fractionated by gel electrophoresis (Protocols 7–10); it is then transferred and fixed to a membrane. Specific RNAs are detected by hybridizing radioactive probes to the fixed RNA and performing autoradiography or phosphorimaging. Every laboratory uses slightly different procedures for prehybridization, hybridization, and washing. Therefore, we present two different procedures for these steps. Investigators are encouraged to follow whichever procedure they are more comfortable with, but it is important to follow either procedure exactly; "mixing and matching" could lead to poor results.

In general, northern blotting has several advantages over other techniques. RNAs of any size can be detected (including small RNAs; see Protocol 11), so the technique is "universally" applicable. In addition, northern blotting allows for an estimate of the size of the detected RNA, and this can be important when measuring levels of intact mRNAs while avoiding measurement of their degradation products. The accuracy of the size estimate varies with the length of the detected RNA: The smaller the RNA, the more precise the estimate.

Disadvantages of northern blotting include the fact that it is relatively insensitive; rare transcripts are difficult to detect with this technique. It can also be difficult to prepare probes of high enough specific activity, and because the RNA target is immobilized on the membrane and not free in solution, both the kinetics and the extent of hybridization are reduced. Only a small fraction of the probe will hybridize to the target, leaving the unhybridized material to contribute to background. Increasing the background radioactivity on the filter reduces the dynamic range and sensitivity of the method. Moreover, very large RNAs and very small RNAs can be problematic in that large RNAs may not transfer efficiently and very small RNAs can be lost during transfer. Finally, the northern blotting procedure is inherently labor-intensive, quite slow, and somewhat cumbersome. Nevertheless, it remains the "gold standard" for direct analysis of the expression of moderately abundant and abundant mRNAs as well as numerous other RNAs (Fig. 3-1).

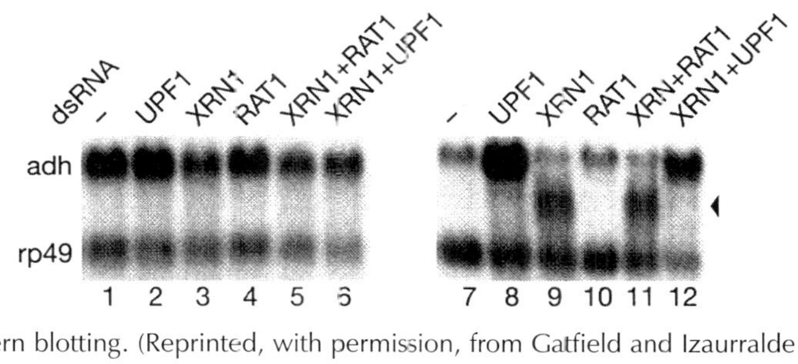

FIGURE 3-1. Northern blotting. (Reprinted, with permission, from Gatfield and Izaurralde 2004 [©Macmillan].)

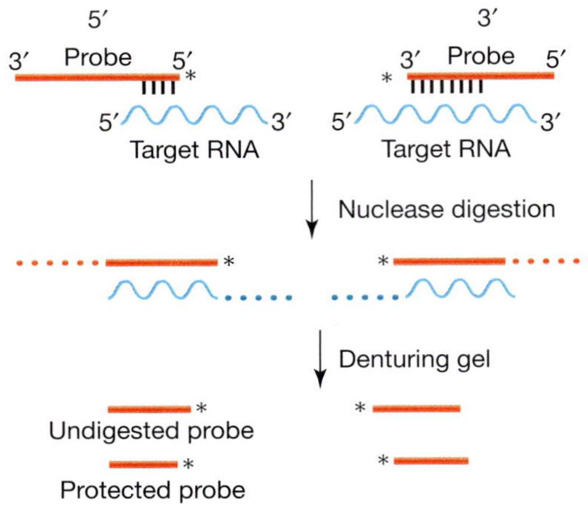

FIGURE 3-2. Nuclease protection.

NUCLEASE PROTECTION

Nuclease protection is another common approach for detecting and quantifying specific species of RNA (Protocols 12 and 13). This technique relies on the properties of certain nucleases that digest single-stranded nucleic acids but do not digest double-stranded structures. The approach is conceptually straightforward. A radiolabeled probe that partially overlaps the RNA of interest is designed (see Fig. 3-2), the probe is then hybridized to the target RNA, and the hybrids are digested with single-strand-specific nucleases. During nuclease digestion, the single-stranded regions are degraded and the probe is shortened (the unhybridized probe is digested completely). After digestion, the nuclease is removed by phenol:chloroform:isoamyl alcohol (PCA) extraction (Chapter 2, Protocol 1), and the remaining nucleic acids are fractionated on denaturing acrylamide gels followed by phosphorimaging or autoradiography.

In principle, nuclease protection has several advantages over northern blotting. First, hybridization takes place in solution, allowing saturation of the target with the probe and thus more sensitive detection. Second, rRNAs are not a factor in nuclease protection. (Note though that they can present a problem during northern blotting if total RNA is used because rRNAs are present at high concentrations and compete for filter binding with similarly sized mRNAs.) Despite these advantages, we have found that nuclease protection can be a touchy technique. It often takes a considerable amount of nuclease to achieve complete digestion of the unprotected probe. Furthermore, any duplex structure—including the target-probe hybrid—is subject to thermal "breathing" or end fraying, which allows the nuclease to nibble at the ends of the probe, generating a heterogeneous set of bands that may complicate interpretation. Although fraying and breathing can be minimized by raising the monovalent cation (Na^+) concentration and lowering the digestion temperature, these conditions reduce nuclease activity, which may lead to incomplete digestion. It may be necessary to titrate the nuclease(s) to find the balance at which nuclease activity is adequate to digest the probe and thermal breathing and fraying are reduced. Moreover, because the technique relies on probe shortening, any spurious nuclease nicks might appear to be products. Another limitation of nuclease protection compared to northern blotting is that only a portion of the target RNA is assayed; therefore, degradation products of the target such as 5′-end-containing pieces may be detected.

PRIMER EXTENSION

Primer extension is a very powerful method that in principle can be used to detect any RNA. Like northern blotting and nuclease protection, it is conceptually straightforward. In a standard primer

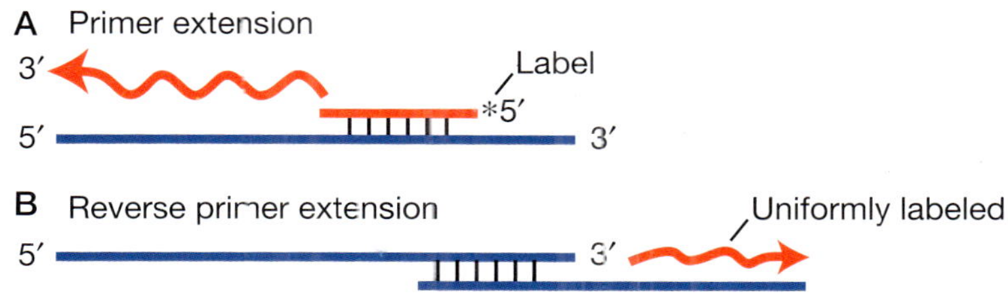

FIGURE 3-3. Standard primer extension and reverse primer extension. (*A*) Standard primer extension design wherein a 5′-end-labeled oligodeoxynucleotide is annealed to the RNA of interest. Extension by reverse transcriptase proceeds to the 5′ end of the target molecule. (*B*) Reverse primer extension. In this case, an oligodeoxynucleotide is designed that overlaps the 3′ end of the RNA of interest. The free 3′ hydroxyl of the RNA is extended to the 5′ end of the oligodeoxynucleotide.

extension assay (Protocol 14; Fig. 3-3A), a radiolabeled probe (almost always a 5′-end-labeled DNA oligonucleotide) is annealed to the target RNA of interest. After hybridization, the free 3′-hydroxyl group of the oligonucleotide primes cDNA synthesis by reverse transcriptase. Extension by reverse transcriptase to the 5′ end of the target RNA produces a product of a diagnostic size that is detected after fractionation on denaturing polyacrylamide gels and subsequent phosphorimaging or autoradiography.

In reverse primer extension (Protocol 15; Fig. 3-3B), the target RNA is lengthened, not the probe. An unlabeled oligonucleotide is annealed to the 3′ end of the target RNA, and the free 3′-hydroxyl group of the target RNA is extended with labeled nucleotides. The extended RNA is then detected after gel electrophoresis and phosphorimaging or autoradiography.

Both of these methods are extremely sensitive, highly specific, and versatile. Regular primer extension can be used for RNA sequencing (see Protocol 23), modification studies (see Chapter 4), and footprinting analyses (see Chapter 4). Reverse primer extension can be used for detection of specific RNAs or affinity tagging of specific RNAs or ribonucleoproteins (RNPs) (see Protocol 15).

Poisoned Primer Extension

Poisoned primer extension (Protocol 16; Fig. 3-4) is a variation of regular primer extension that is primarily used to distinguish between RNAs that are nearly identical in sequence (e.g., edited vs. nonedited transcripts). In such situations, the RNAs are of the same size and cannot be distinguished by regular primer extension. This method requires the design of primers whose 3′ end is positioned so that the base that is detected is the first of that type encountered as the primer is extended (Fig. 3-4).

FIGURE 3-4. Poisoned primer extension used to distinguish between edited and nonedited RNAs. In this example of A-to-I editing (conversion of adenosine residues to inosine residues by adenosine deaminase), regular primer extension would yield identical results. However, when a chain-terminating nucleotide (the "poison") is included (in this case, ddTTP), two distinguishable products are produced. In both cases, extension is stopped at the first A by incorporation of ddT. With the upper sequence, the primer is extended by two nucleotides, and with the lower sequence, the primer is extended by seven nucleotides. The extent of editing can be directly measured by the ratio of the +2 and +7 extension products.

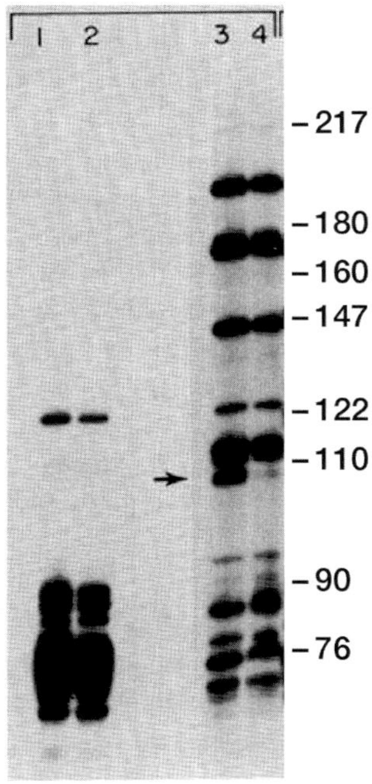

FIGURE 3-5. An example of targeted RNase H digestion to identify an RNA. (Lanes *1* and *2*) Total RNA labeled with [^{32}P]pCp either untreated or treated with an oligodeoxynucleotide complementary to the RNA of interest and RNase H. (Lanes *3* and *4*) Same samples after prior enrichment for RNAs containing a trimethylguanosine cap structure. The absence of the band (indicated by an arrow) in the treated sample indicates that the RNA of interest was present. For details of the experiment, see Nilsen et al. (1989). (Reprinted, with permission, from Nilsen et al. 1989 [©American Society for Microbiology].)

OTHER METHODS

The methods described above are "indirect" in that they rely on using hybridization to labeled probes or incorporation of label via polymerization to identify an RNA species. It is sometimes advantageous to directly label the RNA of interest to high specific activity by either kinasing the 5′ end or ligating labeled pCp to the 3′ end. After labeling, specific RNAs can be identified by their susceptibility to digestion in the presence of RNase H and sequence-specific oligodeoxynucleotides (Protocol 17; Fig. 3-5). This is an extremely sensitive procedure that is only useful and practical when the RNAs of interest have been highly enriched via biochemical purification *before* deproteinization. For example, if total RNA is labeled with pCp, the only RNAs that will be visualized by autoradiography or phosphorimaging will be tRNA and 5.8S and 5S rRNAs because of their high abundance. Therefore, this technique is valuable when ribonucleoprotein complexes such as spliceosomes, snoRNPs, and miRNPs have been purified first and then deproteinized. Because of its sensitivity, the direct labeling approach is invaluable for detecting RNAs that may be present in a biochemical by purified (e.g., immunoprecipitated) samples; the sequence of such RNAs can be determined by direct chemical sequencing.

PCR-BASED APPROACHES

Because of their exquisite sensitivity, PCR-based approaches have become exceptionally popular methods for detecting and quantifying RNA—from mRNA to miRNA. In general, there are three types of PCR assays (each with many technical variations): end point (Protocol 18), semiquantitative, and quantitative (Protocol 19). Although these methods can be extremely powerful, they are often misused and the results they produce are frequently overinterpreted. *Before using any PCR-based method, it is critical to keep in mind that the quality of the data obtained is only as good as the quality of the RNA sample analyzed.*

End-point PCR

End-point PCR is used to determine whether a specific RNA is present in a sample; it cannot be used to quantitate the abundance of an RNA. Detection is achieved by amplifying the cDNA that has been copied from an RNA template by reverse transcriptase (RT). Reverse transcription is primed with an RNA-specific primer, oligo(dT), or random hexamers (Protocol 18); we have found that a mixture of oligo(dT) and random hexamers is the preferred method because the resulting cDNA can be amplified with any set of primers. It is always necessary to run a parallel RT-PCR without reverse transcriptase to detect genomic DNA contamination of the RNA sample. Any amplified products observed in the absence of the enzyme arise from DNA contamination, and it is necessary to remove the DNA from the RNA sample by DNase I treatment (see Chapter 2, Protocol 11) before proceeding further.

An obvious but amazingly often overlooked problem is inappropriate or nonspecific amplification. We have observed students and postdocs who see sharp bands after a PCR and assume that they are the correct RT-PCR products, ignoring the fact that the bands are not the expected size. Therefore, first make sure that the PCR product is of the anticipated size. It is also helpful to perform restriction enzyme digestions or sequencing on the PCR products to confirm that the correct fragment has been amplified.

Semiquantitative PCR

Once the RT-PCR conditions for an RNA of interest have been established, it is often desirable to determine the relative abundance of that RNA in one or more samples (e.g., in samples prepared from different tissues, cell lines, or cell fractions). A convenient method for accomplishing this is called semiquantitative PCR, which does not require any specialized equipment other than a thermal cycler. In this approach, it is necessary to determine the number of amplification cycles that are needed both to visualize a product (which is almost always radioactive) and to determine a point (cycle number) where the amount of product is proportional to the amount of input cDNA—the so-called "linear" portion of the product accumulation curve (Fig. 3-6A). Any number of cycles under the linear range will yield no observable product, whereas excessive amplification produces skewed results where the amount of product is no longer proportional to the input (Fig. 3-6B).

When comparing RNA samples, it is always necessary to normalize them based on the level of an RNA that is known to be constant (e.g., an abundant housekeeping mRNA such as glyceraldehyde

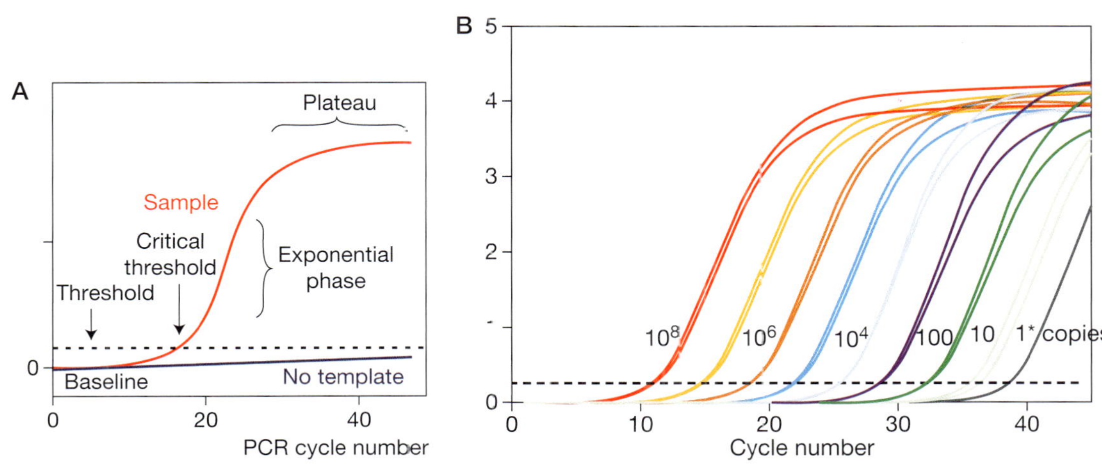

FIGURE 3-6. Product accumulation curves in PCRs. (*A*) Typical PCR accumulation curve illustrating the phases of product accumulation. (*B*) Illustration of accumulation curves (and cycle numbers) when different amounts of the same template are being ampified. Comparison of such curves using different samples provides a measurement of the relative amounts of template (i.e., specific mRNAs in a given sample).

phosphate dehydrogenase [GAPDH], rp49, actin, or tubulin). Once the linear range for this control transcript is established in each sample, the most concentrated sample should be diluted to match the concentration of the least concentrated. It is often desirable to "prenormalize" the RNA concentrations of the samples using spectrophotometry, if possible, so that each sample gives the same signal at the same cycle number. Once these conditions have been met, one can proceed to quantitate the transcript of interest, keeping in mind that the number of cycles required to reach the linear range for a less abundant transcript is going to be significantly greater (e.g., by 5–15 cycles) (see Fig. 3-6B) than the number needed for the abundant housekeeping gene. In sum, semiquantitative PCR is an accepted and powerful method for quantifying the relative abundance of specific RNAs in different preparations. It is important to realize and remember, however, that this approach is not useful for absolute quantitation.

Quantitative PCR

A third PCR-based approach is quantitative PCR (qPCR) or real-time PCR (which is not abbreviated RT-PCR) (Protocol 19). This method is exceptionally sensitive; in theory, it can detect a single RNA molecule. qPCR is advantageous when exact quantitation is necessary. Multiplexing (measuring several RNAs simultaneously) is also possible with this technique.

qPCR requires specialized (and often quite expensive) equipment. There are two basic methods by which amplification is measured: TaqMan and SYBR Green. In TaqMan measurements, a quenched fluorescent oligonucleotide probe is included in the PCR (in addition to the forward and reverse amplification primers). The probe contains a fluorescent dye at one end and a fluorescent quencher at the other end. When the probe is intact, the fluor is quenched. During amplification, the probe anneals to the middle of one of the denatured strands of DNA. When *Taq* polymerase extends the amplicon during the synthesis phase of a PCR cycle, its 5′ to 3′ exonuclease digests the TaqMan probe, releasing the fluorescent dye from the quencher. The amount of fluor detected is proportional to the number of DNA molecules synthesized (see Fig. 3-7). The TaqMan probes are often expensive because fluorochromes must be coupled to custom-designed probes; therefore, this method is most useful for analytical methods when the same RT-PCR product is routinely analyzed.

SYBR Green detection also monitors an increase in fluorescence during synthesis, except that the dye only fluoresces when it is bound to double-stranded DNA (dsDNA). As dsDNA accumulates in the reaction tube during amplification, the fluorescence signal increases in proportion to the amount of DNA that is synthesized (Fig. 3-8). Both methods are reliable, but TaqMan detection

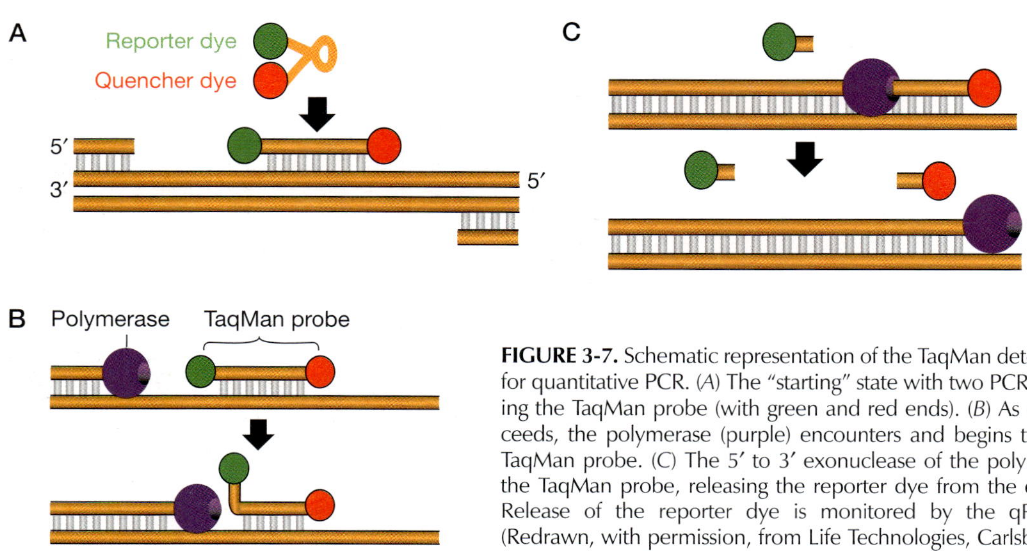

FIGURE 3-7. Schematic representation of the TaqMan detection method for quantitative PCR. *(A)* The "starting" state with two PCR primers flanking the TaqMan probe (with green and red ends). *(B)* As extension proceeds, the polymerase (purple) encounters and begins to displace the TaqMan probe. *(C)* The 5′ to 3′ exonuclease of the polymerase digests the TaqMan probe, releasing the reporter dye from the quencher dye. Release of the reporter dye is monitored by the qPCR machine. (Redrawn, with permission, from Life Technologies, Carlsbad, CA.)

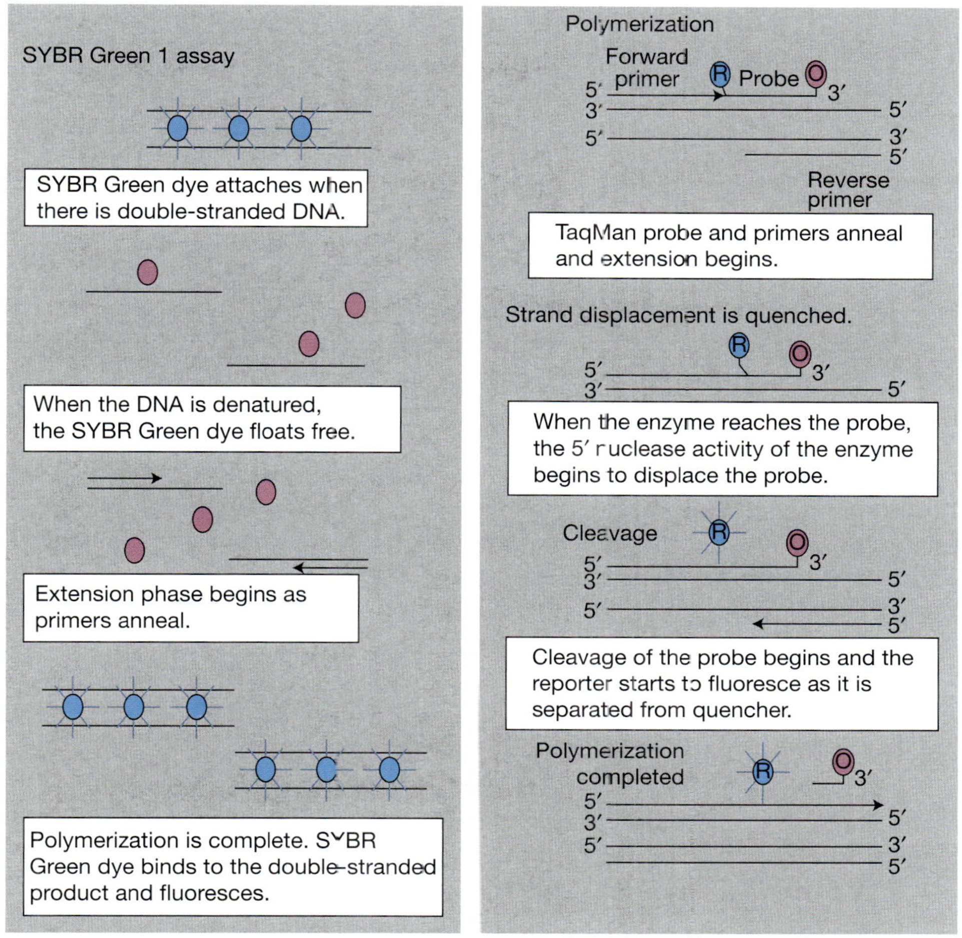

FIGURE 3-8. Comparison of SYBR and TaqMan detection of amplification for qPCR analysis. For discussion, see the text. (Redrawn, with permission, from Life Technologies, Carlsbad, CA.)

offers an additional layer of specificity because a signal is observed only when the probe can bind to the desired amplification product via hybridization.

Although qPCR is an extremely powerful and valuable technique, it can be misused. It is a "blind" technique, unless the investigator takes time to analyze the product by gel electrophoresis or by constructing a melting curve. Thus, it is always advisable to use validated primer sets (i.e., those that do not form primer dimers and do not amplify sequences other than the target of choice). The efficacy of the primer pairs should first be tested using standard RT-PCR where the amplification products are detected by agarose or acrylamide gel electrophoresis.

Finally, before leaving PCR-based methods, we note that there are several specialized techniques that rely on PCR but whose primary purpose is not detection. This includes 5′ and 3′ rapid amplification of cDNA ends (RACEs) (Protocols 20, 21, and 22; Figs. 3-9 and 3-10).

5′ and 3′ RACE

RACEs are specialized PCR-based techniques that are used to determine the 5′ or 3′ ends of mRNAs. These approaches are particularly useful if an investigator wishes to assemble a complete cDNA clone for a mRNA of interest or when only partial cDNA sequences are available. These approaches can have important applications, e.g., in characterizing unknown RNAs or for characterizing transcripts in organisms for which abundant sequence information is absent. The basic steps for 5′ and 3′ RACE are described below.

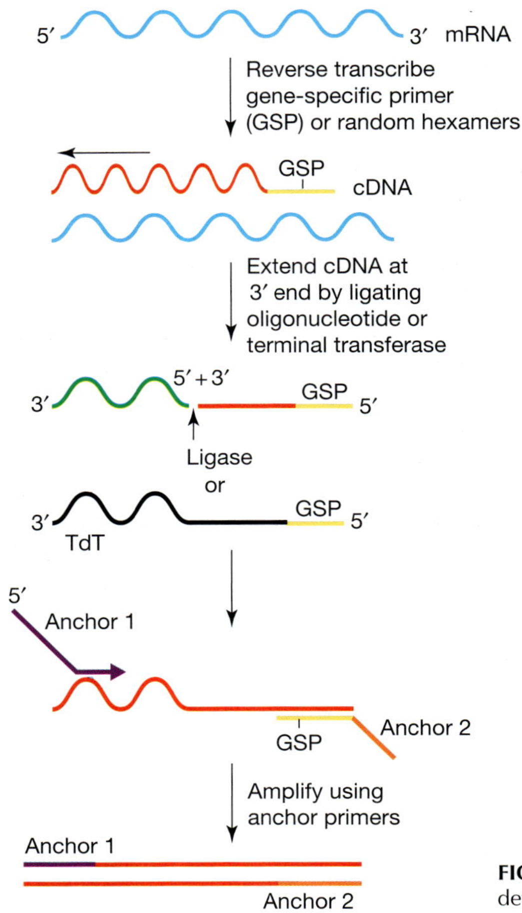

FIGURE 3-9. Outline of 5′ RACE procedure. See the text for detailed description.

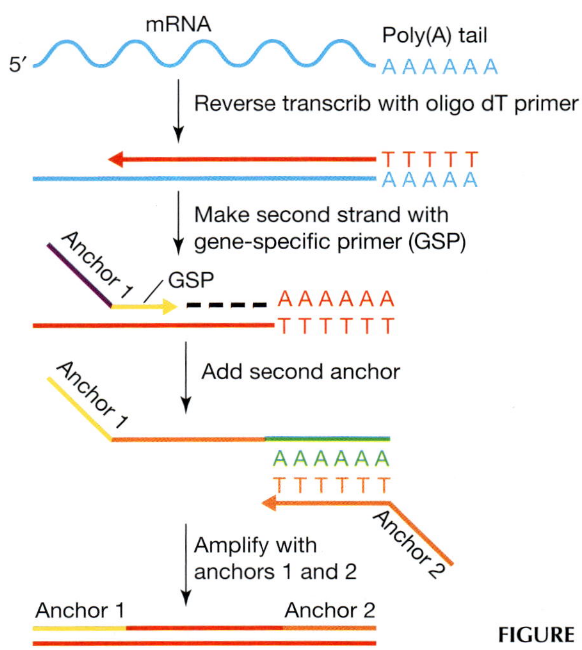

FIGURE 3-10. Outline of 3′ RACE procedure.

5′ RACE

In 5′ RACE (see Fig. 3-9), first-strand cDNA is made using either random hexamer primers or a gene-specific primer. RNA is removed by the RNase H activity of reverse transcriptase and the cDNA is "tailed" by TdT to create a binding site and "anchor" sequence for PCR primers. Newer methods of 5′ RACE (such as the protocol described in this chapter) include ligation of an "anchor" oligonucleotide to decapped mRNAs. The ligated anchor sequence is copied to become the 3′ end of the cDNA.

In both cases, the 3′-tail extension of the cDNA serves as a primer-binding site for second-strand synthesis. The primer has a 5′ extension for later rounds of amplification (anchor). Finally, the dsDNA is amplified by PCR using a forward primer complementary to the anchor and a gene-specific reverse primer. The resultant PCR product can be cloned and sequenced or sequenced directly by cycle sequencing.

In theory, this approach will yield the complete 5′ end of the RNA of interest. Nevertheless, some problems can be encountered, including incomplete reverse transcription in the first step that may result from stable secondary structure(s) in the 5′ region of the RNA of interest. This potential problem can usually be avoided if reverse transcription is performed at a high temperature using thermostable reverse transcriptase. A second potential problem is mispriming by either the anchor primer or the gene-specific primer. If mispriming is observed with the gene-specific primer (i.e., a lot of background), it is advisable to follow the first PCR with a second in which another gene-specific reverse primer inside the first is used. This is called nested priming.

3′ RACE

3′ RACE (see Fig. 3-10) is conceptually very similar to 5′ RACE. Here, the poly(A) tail of the mRNA serves as a binding site for a dT anchor primer used for first-strand cDNA synthesis. After removal of the RNA, a gene-specific primer is used as the forward primer and the anchor is the reverse primer. As with 5′ RACE, any mispriming by the gene-specific primer can be "cured" by using nested primers. Again, the resulting product can be cloned or sequenced.

Sequencing

With abundant genome and expressed sequence tag (EST) sequences available, it is relatively uncommon that direct RNA sequencing techniques are needed. However, in the fortuitous event that an unknown RNA is discovered (e.g., by pCp labeling of a biochemically purified sample as described above), it is of obvious interest to obtain the sequence of that RNA. It is often easiest and most convenient to use cloning methods to obtain the sequence of such RNA molecules (see Protocols 23 and 24).

In that regard, if the RNA of interest is polyadenylated, it is straightforward to prime first-strand cDNA synthesis with oligo(dT). The resulting cDNA can then be extended with TdT. (To determine whether the RNA is polyadenylated, the 3′-end-labeled RNA can be passed over an oligo[dT] column [Chapter 2, Protocol 10]. If the RNA is not polyadenylated, it can be extended at the 3′ end with yeast poly[A] polymerase.) After amplification (as in 3′ RACE), the double-stranded cDNA can be cloned and sequenced or the PCR product can be sequenced directly.

An alternative to cloning strategies is direct RNA sequencing based on base-specific chemical modifications. This is a particularly valuable method when abundant sequence information (e.g., EST or genomic sequence data) is not available. The approach is also useful if the RNA of interest has been discovered by direct labeling with [^{32}P]pCp. Although it is desirable to obtain as much sequence information as possible, it is usually enough to obtain 20–30 bases of sequence, which enables the design of cDNA oligonucleotides for RACE (see Protocols 20–22) followed by cloning and DNA sequencing or primer extension sequencing (see Protocol 23).

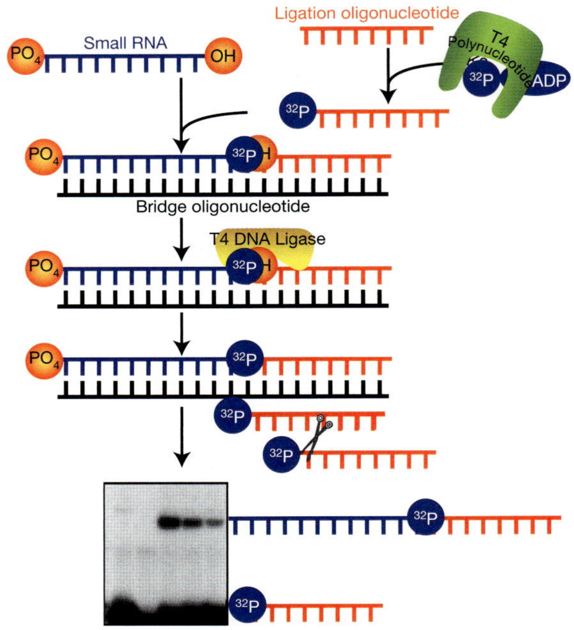

FIGURE 3-11. Schematic depiction of the splinted ligation method for detecting small RNAs The gel shows detection of a specific miRNA expressed in HeLa cells using three different amounts of input RNA. For detailed discussion, see the text and Maroney et al. (2008). (Adapted, with permission, from Maroney et al. 2008 [©Macmillan].)

SPECIALIZED METHODS FOR DETECTING SMALL RNAs

The discovery and subsequent explosion of interest in small noncoding RNAs such as miRNAs, Piwi-interacting RNAs (piRNAs), and endogenous small interfering RNAs (siRNAs) have prompted the development of a variety of sensitive methods designed to detect and quantify the expression of these RNAs. At least 20 distinct methodologies of varying complexities and utilities have been described and each has its own advantages and disadvantages. Somewhat surprisingly, northern blotting (Protocols 7–9) remains the most reliable and popular method for detecting small RNAs, despite its relatively poor sensitivity. The introduction of LNA probes (Protocol 1) has greatly enhanced the sensitivity and specificity of this approach. Here, we highlight three alternatives to northern blotting that we have found to work well in our hands and therefore can be recommended: (1) splinted ligation (Protocol 25), (2) "gel shift of labeled small RNAs," and (3) stem-loop PCR. This is not to say that other methodologies are inferior; the reader should consult the relevant literature for other protocols that might be suitable for their particular application.

Splinted Ligations

Splinted ligation (Protocol 25) relies on the ability of T4 DNA ligase to covalently join the 3′-hydroxyl end of an RNA to the 5′-phosphorylated end of a ^{32}P-radiolabeled DNA oligonucleotide. A splint (or bridge) oligodeoxynucleotide is designed to hybridize to the miRNA of interest (Fig. 3-11). It contains a 5′ overhang that is complementary to a probe (or detection) oligonucleotide. Following hybridization, the miRNA and labeled detection oligonucleotide become juxtaposed on the splint and can be covalently joined by T4 DNA ligase to form a standard 5′ to 3′ phosphodiester linkage. After ligation, the labeled miRNA, lengthened by the covalent addition of the ^{32}P-labeled oligonucleotide probe, is visualized by denaturing gel electrophoresis and phosphorimaging. We have found that this method is significantly more sensitive than northern blotting and is much faster. We recommend this approach for the routine detection and quantification of specific miRNAs.

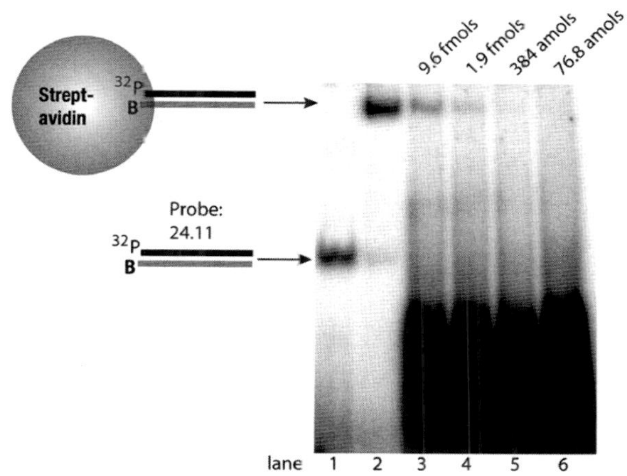

FIGURE 3-12. Illustration of the gel-shift method for detecting small RNAs. The dark "smear" at the bottom of the gel is total kinased small RNAs. The band at the top of the gel (streptavidin plus arrow) indicates the presence of a specific small RNA. For details, see the text and Ebhardt and Unrau (2009). (Reprinted, with permission, from Ebhardt and Unrau 2009.)

Native Gel Electrophoresis

The streptavidin "gel-shift assay" of labeled small RNAs (Fig. 3-12) is a new and highly sensitive method for detecting small RNAs and for discovering novel small RNAs. Size-selected RNAs (18–30 nucleotides) are treated with phosphatase and labeled at their 5′ ends to high specific activity with T4 polynucleotide kinase and [γ-^{32}P]ATP (see Protocol 1). This generates a pool of labeled small RNAs. Biotinylated oligodeoxynucleotide probes are then designed; these are complementary to the small RNA of interest. After hybridization, streptavidin is added. The extremely tight binding between avidin and biotin (10^{-15} M) increases the molecular weight of the hybridized complex, and any labeled RNA that hybridizes to the probe can be visualized on a native polyacrylamide gel.

In addition, a specific advantage of this method is that it can be used to identify small RNAs. For example, if it is suspected that a specific region of the genome is transcribed to yield a small RNA, one can design a biotinylated probe spanning the region and proceed as described. A positive result (i.e., a shifted radiolabeled RNA species) indicates that a small RNA in this region is present.

Stem-Loop PCR

Stem-loop PCR is an extremely sensitive method for detecting and quantitating the expression of small RNAs. It is a variation of real-time or quantitative PCR (qPCR) designed especially for detecting and quantitating the expression of small RNAs (e.g., miRNAs). This method was developed by scientists at Applied Biosystems. Its advantages include the fact that pre-miRNA and genomic DNA do not interfere with the assay, and it has a very broad dynamic range (seven orders of magnitude). A schematic diagram of the assay is shown in Figure 3-13 (Chen et al. 2005).

In this technique, a "probe" is designed that has partial complementarity to the 3′ end of the small RNA of interest and, after annealing, is extended by reverse transcriptase (Fig. 3-13). Following removal of the RNA, a second small RNA-specific primer is added to make dsDNA. This DNA is then amplified using qPCR in the presence of a TaqMan probe, as described in the Quantitative PCR section, above. As with all qPCR applications, it is essential that the probes be validated. We recommend that primers be purchased from commercial suppliers; appropriate primers for a large number of miRNAs for a wide variety of species are available, as are kits for multiplexed analysis of many (hundreds) of miRNAs simultaneously. MiRBase, the microRNA database (www.mirbase.org), is a searchable database of published miRNA sequences and annotation that can also be consulted.

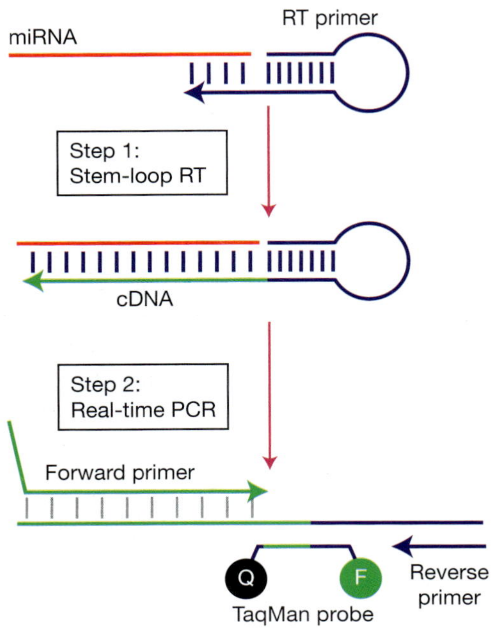

FIGURE 3-13. Schematic diagram of stem-loop PCR. In the first step, an miRNA-specific primer is used to create a cDNA copy, and in the second step, the miRNA is amplified by real-time PCR in the presence of a TaqMan probe (see Fig. 3-7). (Redrawn, with permission, from Chen et al. 2005 [Oxford University Press].)

REFERENCES

Chen C, Ridzon DA, Broomer AJ, Zhou Z, Lee DH, Nguyen JT, Barbisin M, Xu NL, Mahuvakar VR, Andersen MR, et al. 2005. Real-time quantification of microRNAs by stem–loop RT-PCR. *Nucleic Acids Res* **33:** e179.

Ebhardt HA, Unrau PJ. 2009. Characterizing multiple exogenous and endogenous small RNA populations in parallel with subfemtomolar sensitivity using a streptavidin gel-shift assay. *RNA* **15:** 724–731.

Gatfield D, Izaurralde E. 2004. Nonsense-mediated messenger RNA decay is initiated by endonucleolytic cleavage in *Drosophila*. *Nature* **429:** 575–578.

Maroney PA, Chamnongpol S, Souret F, Nilsen TW. 2008. Direct detection of small RNAs using splinted ligation. *Nat Protoc* **3:** 279–287.

Nilsen T, Shambaugh J, Denker J, Chubb G, Faser C, Putnam L, Bennett K. 1989. Characterization and expression of a spliced leader RNA in the parasitic nematode *Ascaris lumbricoides* var. suum. *Mol Cell Biol* **9:** 3543–3547.

READING LIST

Hausner TP, Giglio LM, Weiner AM. 1990. Evidence for base-pairing between mammalian U2 and U6 small nuclear ribonucleoprotein particles. *Genes Dev* **4:** 2146–2156.

Hayes ML, Hanson MR. 2007. Assay of editing of exogenous RNAs in chloroplast extracts of *Arabidopsis*, maize, pea, and tobacco. *Methods Enzymol* **424:** 459–482.

Keegan LP, Rosenthal JJ, Roberson LM, O'Connell MA. 2007. Purification and assay of ADAR activity. *Methods Enzymol* **424:** 301–317.

Maroney PA, Chamnongpol S, Souret F, Nilsen TW. 2007. A rapid, quantitative assay for direct detection of microRNAs and other small RNAs using splinted ligation. *RNA* **13:** 930–936.

WWW RESOURCE

Applied Biosystems tutorial http://www.appliedbiosystems.com/support/tutorials/pdf/rtpcr_vs_tradpcr.pdf
MiRBase, the microRNA database (www.mirbase.org).

Protocol 1

Labeling of Oligonucleotide Probes (DNA, LNA, RNA) by Polynucleotide Kinase and [γ-^{32}P]ATP

5′-labeled oligonucleotides are exceptionally valuable for a number of applications: They can be used as probes for northern blots (Protocols 7–9), primers for primer extension (Protocols 14–16), and primers for radioactive PCR. Oligonucleotides are advantageous over many other probes (see Table 3-1) because they can be labeled to very high specific activity and are long enough to form stable hybrids. LNAs are nucleic acid mimics in which the 2′ oxygen is linked to the 4′ carbon by a methylene group. This chemistry "locks" the nucleoside into the C3′-*endo* conformation, which is more favorable for duplex stability and hybridization. DNA, RNA, and LNA oligonucleotides are synthesized with 5′-hydroxyl groups (Fig. 3-14) that are recognized by polynucleotide kinase and can therefore accept ^{32}P-phosphate from ATP. For most applications, labeled DNA oligonucleotides are preferred because of their stability and cost. However, when extra sensitivity is required, LNA probes are preferred. LNAs provide superior hybridization performance compared to DNA oligonucleotides and are particularly useful for detecting small RNAs such as miRNAs. Depending on the sequence composition of the probe, thermal stability can be raised by as much as 7°C for LNA/RNA duplexes relative to DNA/RNA duplexes. Because LNAs have a free 5′ OH, they can be labeled using kinase exactly as described for DNA oligonucleotides. LNAs are commercially available (e.g., Exiqon) and their only disadvantage is their high cost relative to DNA oligonucleotides.

MATERIALS

The recipes for the items marked with <R> are at the end of the protocol or in the Common Recipes Appendix.

CAUTION: See the Cautions Appendix for appropriate handling of materials marked with <!>.

Reagents

[γ-^{32}P]ATP (7000 Ci/mmol; 167 μCi/μL) <!>
OptiKinase (20 units/μL)
OptiKinase buffer (10x) <R>
SDS (10%) <!>
Synthetic DNA, LNA, or RNA oligonucleotides to be labeled (10 pmol/μL, desalted)
TE buffer <R>

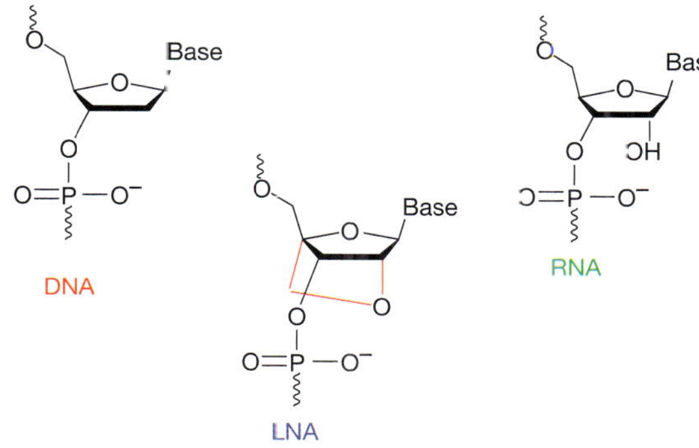

FIGURE 3-14. Depiction of DNA, LNA, and RNA.

Equipment

Geiger counter

Use a handheld counter capable of detecting γ-^{32}P isotopes.

Micro Bio-Spin 6 Column (Bio-Rad)

Water baths (37°C and 65°C)

OVERVIEW

This method uses [γ-^{32}P]ATP and polynucleotide kinase to label oligonucleotides to high specific activity. After the kinase reaction, labeled oligonucleotides are separated from unincorporated [γ-^{32}P]ATP using a spin column.

METHOD

Initial Reaction

1. Set up the reaction with the following reagents in the order listed:

H_2O	4 µL
OptiKinase buffer (10x)	1 µL
Synthetic DNA, LNA, or RNA oligonucleotides (10 pmol/µL, desalted)	1 µL
[γ-^{32}P]ATP (7000 Ci/mmol; 167 µCi/µL)	3 µL
OptiKinase (20 units/µL)	1 µL
Total	10 µL

At least 10 pmol of ATP must be included to kinase 10 pmol of DNA, LNA, or RNA.

2. Incubate for 60 min at 37°C.
3. Add 65 µL of TE buffer and 1.5 µL of 10% SDS.
4. Heat-inactivate the kinase for 5 min at 65°C.
5. Separate the radiolabeled nucleic acids from the unincorporated [γ-^{32}P]ATP using a Micro Bio-Spin 6 Column.

Estimation of Incorporation

6. Compare the radioactivity of the flowthrough to the radioactivity retained on the spin column using a handheld Geiger counter. Discard the column in the radioactive waste!

   The flowthrough should incorporate between 50% and 75% of the ^{32}P from the [γ-^{32}P]ATP. The incorporation can also be determined more accurately by using the TCA or DEAE filter paper methods.

TIPS AND TROUBLESHOOTING

1. The only potential problems with this technique are poor quantitation of the oligonucleotide or "dirty" preparations of oligonucleotides.
2. For poor quantitation of the oligonucleotides, make sure to obtain an accurate measure of amount.
3. For "dirty" preparations of oligonucleotides, we recommend PCA extraction, ethanol precipitation, and resuspension.

SOLUTION AND REAGENT RECIPE

CAUTION: See the Cautions Appendix for appropriate handling of materials marked with <!>.

OptiKinase Buffer (10x)

Reagent	Quantity (for 1 mL)	Final concentration (10x)
Tris-HCl (1 M, pH 7.5)	500 μL	0.5 M
$MgCl_2$ (1 M) <!>	100 μL	100 mM
DTT (1 M) <!>	50 μL	50 mM
H_2O	to 1 mL	

Store for up to 3 mo at –20°C.

Protocol 2

DNA Oligonucleotide Radiolabeling by Terminal Deoxynucleotidyl Transferase (TdT)

This is an alternative procedure for rapidly radiolabeling DNA oligonucleotides for use as probes for northern blots, small RNA blots, or other hybridization detection methods. It uses the template-independent polymerase activity of TdT to add radiolabeled nucleotides to the free 3′-hydroxyl end of a DNA oligonucleotide of a defined sequence (Deng and Wu 1983; Eschenfeldt et al. 1987). It is ideal for labeling synthetic DNA oligonucleotides. The resulting homopolymeric ^{32}P-labeled tails do not interfere with subsequent hybridization steps.

MATERIALS

The recipes for the items marked with <R> are at the end of the protocol or in the Common Recipes Appendix.

CAUTION: See the Cautions Appendix for appropriate handling of materials marked with <!>.

Reagents

[α-^{32}P]dATP (10 mCi/mL; 800–3000 Ci/mmol) <!>
$CoCl_2$ (12.5 mM) <!>
DNA oligonucleotide (1 pmol/µL)
EDTA <!>/glycogen mixture <R>
Ethanol (75% and 100%), prechilled to –20°C <!>
Hybridization buffer appropriate for protocol of interest (see Step 8)
LiCl (4 M) <!>
Tailing buffer (5x) <R>
TdT (10 units/µL)

Equipment

Dry ice <!>/ethanol <!> bath
Ice
Microcentrifuge (4°C)
Scintillation counter
Water bath or temperature block (37°C)

OVERVIEW

A specific DNA oligonucleotide containing a 3′-hydroxyl group is treated with TdT and the 3′ end is extended with [α-^{32}P]dATP. This results in a highly radioactive probe.

METHOD

1. Set up the TdT labeling reaction with the following reagents in the order listed:

H_2O	2.2 μL
Tailing buffer (5x)	3.0 μL
$CoCl_2$ (12.5 mM)	2.0 μL
DNA oligonucleotide (1 pmol/μL)	1.8 μL
[α-^{32}P]dATP (10 mCi/mL; 800–3000 Ci/mmol)	5.0 μL
TdT (10 units/μL)	1.0 μL
Total	15 μL

2. Incubate the reaction for 30 min at 37°C and then place it on ice.
3. Add 6 μL of the EDTA/glycogen mixture.
4. Add 1.9 μL of 4 M LiCl.
5. Add 65 μL of cold 100% ethanol and place it in a dry ice/ethanol bath for 30 min.
6. Centrifuge at maximum speed in a microcentrifuge for 10 min at 4°C. Remove the radioactive ethanol and dispose of it in an appropriate waste container.
7. Wash the pellet twice with 200 μL of cold 75% ethanol. During each wash, centrifuge and discard the radioactive ethanol as described in Step 6.
8. Dissolve the pellet in 200 μL of hybridization buffer.
9. Measure the total radioactivity by counting 0.2 μL using Cerenkov scintillation counting.

   The incorporation of isotope should be between 40% and 60%. See Tips and Troubleshooting.
10. If necessary, store the probe at −20°C.

TIPS AND TROUBLESHOOTING

If no labeled DNA is obtained in Step 9, it may help to gel-purify the DNA oligonucleotide before labeling.

SOLUTION AND REAGENT RECIPES

CAUTION: See the Cautions Appendix for appropriate handling of materials marked with <!>.

Tailing Buffer

Reagent	Quantity (for 1 mL)	Final concentration
Potassium cacodylate (0.5 M, pH 7.2) <!>	200 μL	100 mM
$CoCl_2$ (0.2 M) <!>	10 μL	2 mM
H_2O	790 μL	

Store for up to 2 yr at –20°C.

EDTA/Glycogen Mixture

Reagent	Quantity (for 12 μL)	Final concentration
Glycogen (10 mg/mL)	2.0 μL	2 mg/mL
EDTA (250 mM, pH 8.0) <!>	10 μL	210 mM

Use 6 μL of the mixture for each reaction. Prepare fresh before use; do not store.

REFERENCES

Deng GR, Wu R. 1983. Terminal transferase: Use in the tailing of DNA and for in vitro mutagenesis. *Methods Enzymol* **100:** 96–116.

Eschenfeldt WH, Puskas RS, Berger SL. 1987. Homopolymeric tailing. *Methods Enzymol* **152:** 337–342.

Protocol 3

Asymmetric Polymerase Chain Reaction to Generate Single-stranded Probes

Asymmetric PCR is a convenient method to produce high-specific-activity strand-specific probes for hybridization applications such as northern blotting. The principle of asymmetric PCR, or linear amplification, is straightforward: During each cycle, one primer (the only primer) anneals to its complementary strand and is extended. Because a second primer is not present, there is not an exponential increase in product as observed with standard PCR. Thus, 40 cycles of asymmetric PCR will result in no more than a 40-fold amplification of one strand. This procedure has advantages over nick translation and random priming in that the probes are derived from only one strand and therefore cannot self-hybridize. With nick translation and random priming, half of the labeled nucleic acids in the probe solution cannot hybridize to the target but can potentially hybridize to one another because they are derived from both strands. In addition, in contrast to single-stranded RNA probes, the probes produced by asymmetric PCR are made of DNA and are therefore resistant to ribonuclease activity.

MATERIALS

The recipe for the item marked with <R> is at the end of the protocol or in the Common Recipes Appendix.

CAUTION: See the Cautions Appendix for appropriate handling of material marked with <!>.

Reagents

[α-^{32}P]dCTP (3000 Ci/mmol; 10 µCi/µL) <!>
dATP (5 mM)
dCTP (100 µM)
dGTP (5 mM)
dTTP (5 mM)
PCR buffer (10x) <R>
Primer (25 µM)
Taq polymerase (5 units/µL)
Template DNA (purified PCR product or linearized plasmid)

Equipment

Micro Bio-Spin Columns with Bio-Gel P-6 in Tris buffer (Bio-Rad)
PCR tube
Pipette
Thermal cycler

OVERVIEW

This protocol is for a 50-µL asymmetric PCR. During each cycle, the primer anneals to its complementary strand and is extended, and in each round, a labeled single-stranded probe is made. If more than one blot is to be probed, perform multiple reactions.

METHOD

1. Assemble the following reagents in a PCR tube:

Template DNA (purified PCR product or linearized plasmid)	~5 ng of PCR product or 30–50 ng of linearized plasmid
PCR buffer (10x)	5 µL
dATP (5 mM)	2 µL
dGTP (5 mM)	2 µL
dTTP (5 mM)	2 µL
dCTP (100 µM)	1.5 µL
Primer (25 µM)	2 µL
[α-^{32}P]dCTP (3000 Ci/mmol; 10 µCi/µL)	5 µL (50 µCi)
Taq polymerase (5 units/µL)	1 µL
H_2O	to 50 µL

See Tips and Troubleshooting.

2. Mix the reagents by pipetting up and down.
3. Run 40 cycles of "PCR" at the appropriate hybridization temperature for the primer (see Protocol 18).
4. After the PCR is complete, pass it through a Micro Bio-Spin Column to remove unincorporated label.

   The eluant from the column is ready to use as a probe. DO NOT heat denature. It can be stored for short periods of time (i.e., days, not weeks) at –20°C.

TIPS AND TROUBLESHOOTING

1. A possible alternative to using labeled dCTP in the PCR in Step 1 is to use kinased primer (e.g., from Protocol 1) instead of unlabeled primer and 2 µL of 5 mM dCTP instead of 1.5 µL of 100 µM dCTP.
2. Although this method produces a probe with somewhat lower specific activity, it increases the yield and produces more full-length products. This can be advantageous for some hybridization applications (e.g., nuclease protection analysis; Protocol 13).

SOLUTION AND REAGENT RECIPE

PCR Buffer (10x)

Reagent	Quantity (for 1 mL)	Final concentration (10x)
Tris-HCl (1 M, pH 8.3)	100 µL	100 mM
KCl (2 M) <!>	250 µL	500 mM
$MgCl_2$ (1 M) <!>	15 µL	15 mM
H_2O	635 µL	

Store for up to 1 yr at –20°C.

Protocol 4

Random Hexamer ^{32}P Radiolabeling of DNA Fragments as Hybridization Probes

This method (Feinberg and Vogelstein 1983; Feinberg and Vogelstein 1984) is used to prepare radiolabeled DNA probes for northern (or Southern) blot hybridizations and serves as an alternative to nick translation (Protocol 5). It uses 20–60 ng of purified DNA fragments. These fragments are denatured and annealed to dN_6 (random hexamer) primers, which are then extended with Klenow DNA polymerase and deoxynucleoside triphosphates (dNTPs), including [α-^{32}P]dCTP. The reaction is performed at pH 6.8 to reduce the 3′ to 5′ exonuclease activity of Klenow; alternative variants lacking this exonuclease activity (NE Biolabs or Stratagene) can be used. The procedure routinely generates probes of ~1×10^9 to 4×10^9 cpm/µg of DNA.

MATERIALS

The recipes for the items marked with <R> are at the end of the protocol or in the Common Recipes Appendix.

CAUTION: See the Cautions Appendix for appropriate handling of material marked with <!>.

Reagents

[α-^{32}P]dCTP (3000 Ci/mmol, 10 mCi/mL) <!>
Bovine serum albumin (BSA) (20 µg/mL)
DNA diluted in H_2O (see Step 1)

When considering the quantity of DNA, anything greater than ~5 ng works; theoretically, 20–60 ng produces the hottest probe under the conditions described here. Using more than 100 ng will reduce the final specific activity.

Hexamer mix <R>
Klenow fragment of *Escherichia coli* DNA polymerase I (3–4 units/µL)
Stop mix <R>

Equipment

Gel filtration spin column (e.g., Bio-Gel P-10 [Bio-Rad] or Sephadex G-25)
Ice
Incubator (37°C)
Microcentrifuge tubes (1.5 mL)
Scintillation counter

OVERVIEW

The DNA template molecules are denatured by boiling and annealed to an excess of dN_6 (random hexamer) oligonucleotides, which serve as primers for extension by Klenow fragment DNA polymerase in the presence of dNTPs and [α-^{32}P]dCTP. To label a DNA fragment in low-melting-point agarose, begin the procedure at Step 2; otherwise, perform Step 1 and then proceed to Step 3.

METHOD

DNA Preparation

1. Take 20–6 ng of DNA dissolved in 7 µL of H_2O. Boil for 3–5 min and then place the DNA on ice immediately. Proceed to Step 3.
2. For DNA in low-melting-point agarose, cut the band from a low-melting-point agarose gel and add 1 mL of H_2O for every gram of agarose gel. Boil the gel slice for 10–15 min and then place for 10 min at 37°C.

   Use 2 µL of this mix in the reaction (Step 3). The remaining DNA/gel mix can be frozen at −20°C and reused many times. It must be boiled each time before use.

Labeling Reaction

3. Set up the reaction as follows:

DNA from Step 1 or 2	2–7 µL
Hexamer mix	2.5 µL
BSA (20 µg/mL)	1 µL
[α-^{32}P]dCTP (3000 Ci/mmol, 10 mCi/mL)	3–5 µL
Klenow fragment of *E. coli* DNA polymerase I (3–4 units/µL)	1 µL
dNTP mix	2.5 µL
H_2O	to 25 µL

4. Incubate the reaction for 2–8 h at room temperature.
5. Stop the reaction with 100 µL of stop mix.

Probe Purification

6. Use a small gel-filtration spin column to remove unincorporated nucleotides.
7. Count 0.1–1.0 µL in a scintillation counter.

   There will be incorporation of between 50% and 80%, unless <20–30 ng of DNA was used. Specific activity can reach ~3×10^9 dpm/µg. See Tips and Troubleshooting.
8. Store the probe for up to 7–10 d at –20°C.

TIPS AND TROUBLESHOOTING

1. In general, supercoiled plasmid DNA does not label well. It may help to boil the DNA longer (e.g., 5–10 min in Step 1) or simply to linearize the plasmid with a restriction endonuclease before labeling.
2. Using DNA fragments instead of a plasmid yields probes that are more specific. The label will not be "wasted" synthesizing plasmid sequence.
3. Klenow enzyme should be properly stored to make sure it retains its activity.

SOLUTION AND REAGENT RECIPES

The recipes for the items marked with <R> are at the end of the protocol or in the Common Recipes Appendix. CAUTION: See the Cautions Appendix for appropriate handling of materials marked with <!>.

Hexamer Mix (10x)

Reagent	Quantity (for 100 μL)	Final concentration
HEPES (1 M, pH 7.0)	20 μL	200 mM
Sodium chloride (1 M)	50 μL	500 mM
$MgCl_2$ (1 M) <!>	2 μL	20 mM
Hexamers (10 μg/mL)	20 μL	2 μg/μL
H_2O	to 100 μL	

Store for up to 1 yr at –20°C.

Stop Mix

Reagent	Quantity (for 1 mL)	Final concentration
NaCl (5 M)	4 μL	20 mM
Tris-HCl (1 M, pH 7.5)	20 μL	20 mM
EDTA (100 mM, pH 8.0) <!>	20 μL	2 mM
SDS (10%, w/v) <!>	25 μL	0.25%, w/v
H_2O	931 μL	

Store indefinitely at room temperature.

dNTP Mix (10x)

2 mM each dGTP, dATP, and dTTP.

REFERENCES

Feinberg AP, Vogelstein B. 1983. A technique for radiolabeling DNA restriction endonuclease fragments to high specific activity. *Anal Biochem* **132:** 6–13.

Feinberg AP, Vogelstein B. 1984. A technique for radiolabeling DNA restriction endonuclease fragments to high specific activity. Addendum. *Anal Biochem* **137:** 266–267.

Protocol 5

Nick Translation of Double-stranded DNA for the Preparation of ^{32}P-labeled Hybridization Probes

This procedure uses the combined action of DNase I (which nicks duplex DNA) and the *E. coli* DNA polymerase I 5′ to 3′ exonuclease and polymerase activities to incorporate [α-^{32}P]dNTPs into duplex DNA, making high-specific-activity DNA probes for hybridization assays (Maniatis et al. 1975; Rigby et al. 1977). It uses all four ^{32}P-labeled nucleotides and will label DNA uniformly to ~2 x 10^8 cpm/μg. This specific activity is sufficient to detect rare transcripts on northern blots (or single-copy sequences in genomic Southern blots). The procedure can also be used to generate less radioactive probes using three unlabeled dNTPs.

MATERIALS

The recipes for the items marked with <R> are at the end of the protocol or in the Common Recipes Appendix.

CAUTION: See the Cautions Appendix for appropriate handling of materials marked with <!>.

Reagents

[α-^{32}P]dNTPs (dATP, dCTP, dGTP, and dTTP; each at 800 Ci/mmol; 10 mCi/mL) <!>

Alternatively, use only one [α-^{32}P]dNTP and obtain 20-mM solutions of each of the other three unlabeled dNTPs for use in Step 1.

BSA (1 mg/mL, enzyme grade)

DNA (plasmid or linear fragment)

DNase I (Worthington, LS006330, DPFF grade, >2000 units/mg)

Gently dissolve and then serially dilute this stock in DNase I storage buffer <R> to 100 ng/μL and store at –20°C.

DNase storage buffer <R>

E. coli DNA polymerase I (5 units/μL)

Nick translation buffer (10X) <R>

Trichloroacetic acid (TCA) <!>/sodium pyrophosphate ($NaPP_i$) <!> solution <!> (for Step 5)

Equipment

Sephadex G-50 spin columns (e.g., GE Healthcare ProbeQuant G-50)

Water bath or incubator (15°C)

OVERVIEW

In this protocol, DNase I is used to introduce nicks which then serve as primers for *E. coli* DNA polymerase I. Because the polymerase enzyme has 5′ to 3′ exonuclease activity, it removes nucleotides from the 5′ side of the nick and replaces the preexisting nucleotides with radioactive ones.

METHOD

1. Set up the following reaction on ice:

DNA (plasmid or linear fragment)	1 µg
[α-^{32}P]dNTPs (each at 800 Ci/mmol, 10 mCi/mL)	4 µL each
BSA (1 mg/mL)	2 µL
Nick translation buffer (10x)	4 µL
H_2O	to 40 µL

   For less radioactive probes, use only one [α-^{32}P]dNTP (e.g., [α-^{32}P]dCTP) and add 1 µL of a 20-mM solution of each of the other three unlabeled dNTPs (e.g., if [α-^{32}P]dCTP is used, add unlabeled dATP, dGTP, and dTTP to the reaction instead of the corresponding [α-^{32}P]dNTPs).

2. Add 1–1.5 µL of DNase I (100 ng/µL).
3. Add 0.5 µL of 5 units/µL *E. coli* DNA polymerase I.
4. Incubate for 30 min to 1 h at 15°C.
5. Check incorporation by TCA precipitation.

   Incorporation should be between 30% and 50%. See Tips and Troubleshooting.

6. Remove the unincorporated free nucleotides using a Sephadex G-50 spin column.

   Alternatively, precipitate the probe by adding 3 volumes of ethanol and 0.2 volumes of 10 M ammonium acetate (to a final concentration of 2 M ammonium acetate).

7. Store the probe at −20°C.

TIPS AND TROUBLESHOOTING

It is imperative to keep the reaction at 15°C to avoid strand displacement, which will yield snapback DNA and impede subsequent hybridization of the radiolabeled probe to target sequences.

SOLUTION AND REAGENT RECIPES

CAUTION: See the Cautions Appendix for appropriate handling of materials marked with <!>.

DNase I Storage Buffer (1x)

Reagent	Quantity (for 1 mL)	Final concentration (1x)
Tris-HCl (1 M, pH 7.6)	20 µL	20 mM
NaCl (5 M)	10 µL	50 mM
DTT (0.5 M) <!>	2 µL	1 mM
EDTA (1 mM, pH 8.0) <!>	50 µL	0.05 mM
BSA (20 mg/mL)	5 µL	100 µg/mL
Glycerol	500 µL	50%
H_2O	413 µL	

Store for up to 1 yr at –20°C.

Nick Translation Buffer (10x)

Reagent	Quantity (for 1 mL)	Final concentration (10x)
Tris-HCl (1 M, pH 7.5)	500 µL	500 mM
$MgSO_4$ (1 M) <!>	100 µL	100 mM
DTT (0.5 M) <!>	20 µL	10 mM
H_2O	380 µL	

Store for up to 1 yr at –20°C.

REFERENCES

Maniatis T, Jeffrey A, Kleid DG. 1975. Nucleotide sequence of the rightward operator of phage lambda. *Proc Natl Acad Sci* **72:** 1184–1188.

Rigby PW, Dieckmann M, Rhodes C, Berg P. 1977. Labeling deoxyribonucleic acid to high specific activity in vitro by nick translation with DNA polymerase I. *J Mol Biol* **113:** 237–251.

Protocol 6

Enzymatic Dephosphorylation of RNA Using Calf Intestine Alkaline Phosphatase or Shrimp Alkaline Phosphatase

This procedure is commonly used to dephosphorylate the 5′ ends of RNA or DNA for subsequent radiolabeling with T4 polynucleotide kinase. Two enzymes—calf intestine alkaline phosphatase (CIAP or CIP) and shrimp alkaline phosphatase (SAP)—are commonly used. These enzymes catalyze the hydrolysis of many phosphomonoesters, including those at the 5′ ends of RNA or DNA. In general, phosphatasing is required before most 5′-end-labeling reactions (see protocol) or site-specific labeling (see protocol).

MATERIALS

The recipes for the items marked with <R> are at the end of the protocol or in the Common Recipes Appendix.

CAUTION: See the Cautions Appendix for appropriate handling of materials marked with <!>.

Reagents

Calf intestine alkaline phosphatase (CIP; 20 units/μL) (Roche)
  Alternatively, use SAP (1 unit/μL) (US Biochemicals) (see Step 2).
Ethanol (100%) <!>
Glycogen (20 mg/mL)
$MgCl_2$ (1 M) <!>
Phenol <!>
Reaction buffer (10×) <R>
RNA or DNA sample of interest
SEVAG (chloroform:isoamyl alcohol 24:1) <!>
Sodium acetate (2 M, pH 5.2)
TE buffer <R>

Equipment

Dry ice <!>/ethanol <!> bath
Microcentrifuge capable of cooling to 4°C
Microcentrifuge tubes (1.5 mL)
SpeedVac (*optional*; see Step 6.v)

OVERVIEW

In this procedure, RNA fragments are incubated with CIP or SAP and then recovered by phenol:chloroform extraction and ethanol precipitation.

METHOD

Phosphatase-treat the RNA

1. Calculate the concentration of 5′ ends that need to be dephosphorylated in the RNA or DNA sample. For single-stranded RNA,

   $$5' \text{ ends (pmol)} = [X/Y \times (330 \text{ g/mol})/\text{nucleotide}] \times 10^{12} \text{ pmol/mol},$$

   where X is the mass of the RNA fragment in grams and Y is the length of the RNA fragment in nucleotides. For example, 1 μg of a 100-nucleotide RNA contains 330 pmol of 5′ ends.

2. Set up a reaction to phosphatase-treat the RNA or DNA in a 1.5-mL microcentrifuge tube:

Reaction buffer (10×)	2 μL
RNA sample of interest	~100 pmol of 5′ ends
CIP (20 units/μL)	1 μL
H_2O	to 20 μL

   As a general rule of thumb, use 0.1–1 unit of CIP per pmol of 5′ ends. If using SAP (1 unit/μL), add 2–3 μL to the reaction.

3. Incubate the reaction for 1 h at 37°C if using CIP or 30°C if using SAP. See Tips and Troubleshooting.

4. Add the following:

Sodium acetate (2 M, pH 5.2)	30 μL
Glycogen (20 mg/mL)	0.5 μL
TE buffer	148 μL

5. Phenol:chloroform isoamyl alcohol (PCA)-extract the RNA as described in Chapter 2, Protocol 1.

6. Precipitate the RNA as follows:

   i. Add 500 μL of cold 100% ethanol.

   ii. Chill for at least 10 min in a dry ice/ethanol bath.

   iii. Centrifuge at maximum speed in a microcentrifuge for at least 10 min at room temperature.

   iv. Discard the supernatant. Add 500 μL of cold 100% ethanol to the pellet and repeat Steps 6.ii–6.iii.

   v. Discard the supernatant. Air-dry the RNA pellet or briefly dry the pellet in a SpeedVac.

7. Resuspend the RNA in 5 μL of H_2O.

   The RNA is now ready for radiolabeling with T4 polynucleotide kinase (see Protocol 1). Store the RNA pellet at –80°C.

TIPS AND TROUBLESHOOTING

1. Do not overtreat the RNA with phosphatase in Step 3. In some cases, trace amounts of ribonucleases are present in the reaction; these can degrade RNA while it is being treated with phosphatase.

2. Some manufactuers recommend $ZnCl_2$ in phosphatase reaction buffers. These buffers are not recommended for use with RNA.

3. Heating at 65°C to inactivate the enzyme is not advisable.

SOLUTION AND REAGENT RECIPES

CAUTION: See the Cautions Appendix for appropriate handling of materials marked with <!>.

TE Buffer (10x)

Reagent	Quantity (for 1 mL)	Final concentration (10x)
Tris-HCl (1 M, pH 8.5)	500 µL	500 mM
EDTA (0.25 M) <!>	4 µL	1 mM
H_2O	496 µL	

Store indefinitely at –20°C.

Reaction Buffer (10x)

Reagent	Quantity (for 1 mL)	Final concentration (10x)
Tris-HCl (1 M, pH 8.0)	200 µL	0.2 M
$MgCl_2$ (1 M) <!>	10 µL	0.01 M
H_2O	788 µL	

Store indefinitely.

READING LIST

Sambrook J, Russell DW. 2001. *Molecular cloning: A laboratory manual*, 3rd ed., p. 9.65. Cold Spring Harbor Laboratory Press, Cold Spring Harbor, NY.

Sambrook J, Fritsch EF, Maniatis T. 1989. *Molecular cloning: A laboratory manual*, 2nd ed. Cold Spring Harbor Laboratory Press, Cold Spring Harbor, NY.

Protocol 7

Northern Blots: Denaturation and Electrophoresis of RNA with Formaldehyde

This protocol is used to denature and separate large mRNA molecules (0.5–10 kb) on formaldehyde-containing agarose gels by electrophoretic size fractionation. Formaldehyde contains a carbonyl group that reacts to form Schiff's bases with the imino or amino groups of guanine, adenine, and cytidine. These covalent adducts prevent normal base pairing and maintain the RNA in a denatured state. Because these adducts are unstable, it is necessary to have formaldehyde present in the gel to maintain the RNA in the denatured state. Following electrophoresis, the gel is stained to visualize RNA markers or rRNA using one of several different types of stains. After capillary or vacuum transfer of the separated RNAs from the agarose gel to a nylon or nitrocellulose membrane, the RNA of interest can be detected using the northern hybridization procedure.

MATERIALS

The recipes for the items marked with <R> are at the end of the protocol or in the Common Recipes Appendix.

CAUTION: See the Cautions Appendix for appropriate handling of materials marked with <!>.

Reagents

Agarose (electrophoresis grade)
Formaldehyde (37%; ACS grade) <!>
Formamide (ultrapure, redistilled; e.g., Invitrogen) <!>

If the color of the formamide is yellowish or there is a smell of formic acid in it, do not use it—buy it fresh.

MOPS (morpholinopropanesulfonic acid) buffer (10x) <!> <R>
RNA sample of interest, diluted in H_2O to the appropriate concentration (see Step 4)
Size marker

It is generally best to use a 0.5–10-kb RNA marker ladder (e.g., Ambion 7151, New England Biolabs N0362S, or Invitrogen 15623–200), but a DNA marker can also be used. If using a DNA marker, boil the marker so it is denatured before running on the gel in Step 9.

Stain to detect RNA markers (*optional*; see Step 10)

Ethidium bromide <!>, toluidine blue (0.05%, prepared in H_2O), or a fluorescent dye such as SYBR Green II RNA Gel Stain (Invitrogen S-7564) <!> may be used. Keep in mind that ethidium bromide is less sensitive for single-stranded nucleic acids.

Tracking dye <R>

Equipment

Chemical fume hood
Electrophoresis apparatus for running 15-cm^2 agarose gels
Gel-documentation system (*optional*; see Step 10)
Ice
Microcentrifuge tubes (1.5 mL)
Microwave
Ruler (*optional*; see Step 10)
Water bath (65°C for Step 1.ii) and water bath or temperature block (60°C for Step 7)

OVERVIEW

An agarose gel with formaldehyde is prepared under a chemical fume hood and set up in a horizontal electrophoresis apparatus. RNA samples are prepared and denatured in a solution of formamide and formaldehyde and then, along with 0.5–10-kb size markers, subjected to electrophoresis on the gel.

METHOD

Preparation of a Formaldehyde-agarose Gel

1. Prepare a 1.5%, 15-cm x 15-cm gel (150-mL volume) as follows:
   i. Mix 2.25 g of agarose and 109.5 mL of H_2O.

      A 1.5% agarose gel can separate 0.5–2-kb mRNAs. For larger RNAs, use 0.7% or 1% agarose gels.

   ii. Melt the agarose in a microwave and then cool it to 65°C.

   iii. In a chemical fume hood, add 15 mL of 10x MOPS buffer and 25.5 mL of 37% formaldehyde; mix well.

      MOPS is sensitive to high temperatures and formaldehyde is hazardous, so do not add these to the gel mix until it is slightly cooled.

   iv. In a chemical fume hood, pour the agarose mixture into a 15-cm x 15-cm gel tray and insert the comb. Allow the gel to solidify.

      It is best to prepare the RNA samples (Steps 4–8) while the gel is solidifying in the chemical fume hood.

      Agarose gel boxes come in various sizes. Depending on the surface area of the gel to be poured, prepare enough agarose to make the gel 0.5 cm thick.

2. Prepare sufficient 1x MOPS buffer for the gel tank reservoirs by diluting the 10x stock with H_2O.
3. Put the gel into the tank and add the 1x MOPS buffer. Do not let the gel dry out in the hood.

   Load and run the gel as described in Step 9 as soon as possible so that the formaldehyde does not diffuse out of the gel.

Preparation of RNA Samples for Electrophoresis

4. Add 4.7 µL of the RNA sample to a 1.5-mL microcentrifuge tube.

   Each tube should typically contain 0.5–2 µg of poly(A)$^+$ mRNA or 10–30 µg of total cellular RNA.

5. Prepare a fresh stock of sample buffer containing formamide, 10x MOPS, and formaldehyde as follows:

Ultrapure formamide	660 µL
MOPS buffer (10x)	200 µL
Formaldehyde (37%)	270 µL
Total	1.13 mL

6. Add 11.3 µL of sample buffer to the reaction tube containing 4.7 µL of RNA.
7. Heat the tube for 5 min at 60°C and then cool on ice.
8. Add 4 µL of tracking dye. The total volume is 20 µL.

Electrophoresis and Visualization of RNA

9. Load the gel (from Step 3) with the RNA samples (from Step 8) and size markers and run at low voltage (40–50 V) at room temperature until the bromophenol blue dye is about two-thirds down the gel (usually ~8–9 hours).

   It is important to run size markers on the gel, which will allow the molecular weight of the target transcript to be determined. Do not run at voltages greater than 50 V or band distortion will result. After the RNA has been separated, stain the gel as described in Step 10. If no staining is needed, proceed to Protocol 9 or 10 to detect the transcript by northern hybridization.

10. If desired, stain the RNA in the gel using ethidium bromide, toluidine blue, or another fluorescent dye by following standard procedures. Destain in H_2O for 30 min to 1 h.

    You can use a computer-controlled CCD camera system with a UV transilluminator to photograph the gel; place it next to a ruler to calibrate the relative migration of the marker bands.

    If using poly(A)$^+$-selected mRNA, a smear and faint rRNA bands should be seen. If using total cellular RNA, then two prominent rRNA bands should be seen. The 18S and 28S rRNA species should be visible at ~1800 and 2800 nucleotides, respectively. See Tips and Troubleshooting.

TIPS AND TROUBLESHOOTING

1. Another option for staining in Step 10 is to add 1 µg/µL of ethidium bromide directly to each sample and marker lane before electrophoresis. Run the gel and then photograph it before transfer.
2. A useful alternative to staining the whole gel as described in Step 10 is to run a size marker in an outside lane. This lane can be stained by adding ethidium bromide directly to the marker lane only while loading the lane or by cutting off the marker lane (before transfer) and staining with ethidium bromide. In both cases, photograph only the marker lanes next to a ruler to calibrate relative migration. If staining the marker lane only, stain for 15 min at room temperature with ethidium bromide and then destain in H_2O at least 2 h or overnight. This will eliminate the fluorescence from the formaldehyde and ethidium bromide and yield a nicer photograph.

SOLUTION AND REAGENT RECIPES

CAUTION: See the Cautions Appendix for appropriate handling of materials marked with <!>.

MOPS Buffer (10x)

Reagent	Quantity (for 1 L)	Final concentration (10x)
MOPS free acid, ultrapure (e.g., Calbiochem ULTROL) <!>	41.86 g (or 46.26 g of the sodium salt)	0.2 M
Sodium acetate, anhydrous	4.1 g (or 6.8 g of acetate•3H_2O)	0.05 M
Na_2EDTA (0.5 M, pH 7.5) <!>	20 mL	0.01 M
H_2O	to 1 L	

Bring the pH to 7.0 with NaOH <!> (or if the MOPS sodium salt is used, then use glacial acetic acid <!> to adjust the pH). Do not autoclave. Store in the dark, covered with foil, at room temperature. The solution will turn yellow with time, but this is normal.

Tracking Dye

Reagent	Quantity (for 1 mL)	Final concentration
MOPS buffer (10x) <!>	100 µL	1x
Ficoll 400	150 mg	15%
Bromophenol blue (1%) <!>	200 µL	0.2%
H_2O	to 1 mL	

Store for up to 1 yr at –20°C.

READING LIST

Fritsch EF, Sambrook J, Maniatis T. 1982 *Molecular cloning: A laboratory manual.* Cold Spring Harbor Laboratory, Cold Spring Harbor, NY.

Herrin DL, Schmidt GW. 1988. Rapid, reversible staining of northern blots prior to hybridization. *BioTechniques* **6:** 199–200.

Kroczek RA, Siebert E. 1990. Optimization of northern analysis by vacuum-blotting, RNA-transfer visualization, and ultraviolet fixation. *Anal Biochem* **184:** 90–95.

Sambrook J, Fritsch EF, Maniatis T. 1989. *Molecular cloning: A laboratory manual*, 2nd ed. Cold Spring Harbor Laboratory Press, Cold Spring Harbor, NY.

Thomas PS. 1980. Hybridization of denatured RNA and small DNA fragments transferred to nitrocellulose. *Proc Natl Acad Sci* **77:** 5201–5205.

Protocol 8

Northern Blots: Denaturation and Electrophoresis of RNA with Glyoxal

This protocol is used to denature and separate large mRNA molecules (0.5–10 kb) on agarose gels by electrophoretic size fractionation. It is based on methods developed by Burnett (1997), Carmichael and McMaster (1980), McMaster and Carmichael (1977), and Thomas (1980, 1983). Glyoxal (also called diformyl or ethanedial), the agent responsible for maintaining denaturation, contains two carbonyl groups that react to form a cyclic ring structure with the imino and amino groups of guanine. It can also react with the amino groups of adenine and cytidine. This covalent adduct prevents normal base pairing and maintains the RNA in a denatured state in agarose gels. Once formed, these adducts are stable at room temperature at pH <7.0; thus, no glyoxal needs to be added to the gel to maintain the RNA in the denatured state. Because the RNA is fully denatured and as a result migrates through agarose gels according to its molecular weight, this method can be used to accurately size mRNA molecules. Following electrophoresis and reversal of glyoxalation, the RNA can be detected using a northern hybridization procedure (see Protocols 9 and 10).

MATERIALS

The recipes for the items marked with <R> are at the end of the protocol or in the Common Recipes Appendix.

CAUTION: See the Cautions Appendix for appropriate handling of materials marked with <!>.

Reagents

Agarose, electrophoresis grade

Gel-running buffer (1x PB containing 0.1% SDS) <!> <R>

Alternatively, use 1x BPTE buffer (diluted from the 10x stock <R>) (see Steps 2 and 8).

Glyoxal, deionized <!> <R>

Glyoxal denaturation buffer <!> <R>

This can be prepared without dimethylsulfoxide (DMSO) <!>; the DMSO must then be added separately (see Step 6).

PB buffer (20x) <R>

Alternatively, use 10x BPTE buffer <R> (see Step 1.i).

RNA sample of interest (see Step 4)

SDS (20%, electrophoresis grade) <!>

Size marker (see Step 8)

It is generally best to use a 0.5–10-kb RNA marker ladder (e.g., Ambion 7151, New England Biolabs N0362S, or Invitrogen 15623–200), but a DNA marker can also be used. If using a DNA marker, boil the marker so that it is denatured before running on the gel in Step 8.

Stain (*optional*; see Step 9)

Ethidium bromide <!> or a fluorescent dye such as SYBR Green II RNA Gel Stain (Invitrogen S-7564) <!> may be used.

Equipment

Electrophoresis apparatus for running 15-cm x 15-cm agarose gels

Gel-documentation system (*optional*; see Step 9)

Microcentrifuge tubes (1.5 mL)
Microwave
Peristaltic pump (e.g., Cole Parmer 7543–20 with 7021-24 pump head)
SpeedVac or lyophilizer to dry RNA samples (see Step 5)
Vortex mixer
Water bath (65°C for Step 1.ii) and water bath or temperature block (50°C for Step 7)

OVERVIEW

An agarose gel is prepared and set up in a horizontal electrophoresis apparatus. Unlike formaldehyde gels, glyoxal is not included in the gel or gel buffers. RNA samples are prepared and denatured in glyoxal and are then along with the 0.5–10 kb size markers, subjected to electrophoresis on the gel.

METHOD

Preparation of an Agarose Gel

1. Prepare a 1.5%, 15-cm x 15-cm gel (150-mL volume) as follows:
   i. Mix 2.25 g of agarose, 7.5 mL of 20x PB buffer, and 142.5 mL of H_2O.
      Alternatively, use 15 mL of 10x BPTE buffer in place of the 20x PB buffer and adjust the volume of H_2O to 135 mL. If BPTE is used in the agarose gel, it must also be used in the gel-running buffer (see Step 2). A 1.5% agarose gel can separate 0.5–8 kb mRNAs; for larger RNAs, use 0.7% or 1.0% agarose gels.
      Agarose gel boxes come in various sizes. Depending on the surface area of the gel to be poured, prepare enough agarose to make the gel 0.5 cm thick.
   ii. Melt the agarose in a microwave and then cool it to 65°C.
   iii. Add 0.75 mL of 20% SDS and mix well.
   iv. Pour the agarose mixture into a 15-cm x 15-cm gel. Allow the gel to solidify.
      It is best to prepare the RNA samples (Steps 4–7) while the gel is solidifying. The RNA samples should be ready to run when the gel is placed in the tank (in Step 3).
      Agarose gel boxes come in various sizes. Depending on the surface area of the gel to be poured, prepare enough agarose to make the gel 0.5 cm thick.
2. Prepare sufficient gel-running buffer (1x PB containing 0.1% SDS) for the gel tank reservoirs.
   Alternatively, 1x BPTE may be used as the gel-running buffer (see Step 1.i).
3. Put the gel in the tank and add the gel-running buffer. Do not let the gel dry out.

Preparation of RNA Samples for Electrophoresis

It is important to note that glyoxal reacts with primary amines, so the sample RNA should not be in Tris buffer. This procedure dries the RNA pellet, and in so doing, removes any Tris from the RNA preparation.

4. Dilute the RNA samples to the appropriate concentration in H_2O and add the appropriate volume of RNA sample to a 1.5-mL microcentrifuge tube.
   Each tube should typically contain 0.5–2 µg of poly(A)$^+$ mRNA or 10–30 µg of total cell RNA.
5. Evaporate the H_2O from the RNA sample using a SpeedVac or lyophilizer.
6. Add 20 µL of glyoxal denaturation buffer and vortex to resuspend the RNA.
   Alternatively, dissolve the RNA in 10 µL of the glyoxal denaturation buffer made up without DMSO; when the RNA is dissolved, add 10 µL of DMSO. The benefit in doing it this way is that the RNA will dissolve more readily in H_2O than in organic solvent.
7. Heat the tube for 1 h at 50°C and then cool to room temperature.

Electrophoresis and Visualization of RNA

8. Load the gel (from Step 3) with the RNA samples (from Step 7) and size markers and run at low voltage (40–50 V) at room temperature until the bromophenol blue dye is about two-thirds down the gel (usually ~3–4 h). During electrophoresis, recirculate the buffer between the anode and cathode chambers of the horizontal gel tank using a peristaltic pump.

   It is important to run size markers on the gel; these will allow the molecular weight of the target transcript to be determined. Do not run at voltages greater than 50 V or band distortion will result. If using BPTE buffer, recirculation is not necessary. After the RNA has been separated, stain the gel as described in Step 9. If no staining is needed, proceed to Protocol 9 or 10 to detect the transcript by northern hybridization.

9. If desired, stain the RNA in the gel using ethidium bromide or another fluorescent dye by following standard procedures. Photograph the gel.

   If poly(A)$^+$-selected mRNA is being used, a smear and faint rRNA bands should be seen. If total cell RNA is being used, then two prominent rRNA bands should be seen.

SOLUTION AND REAGENT RECIPES

CAUTION: See the Cautions Appendix for appropriate handling of materials marked with <!>.

PB Buffer (20x)

Reagent	Quantity (for 1800 mL)	Quantity (for 3580 mL)
$NaH_2PO_4 \cdot H_2O$	27.6 g	55.2 g
$Na_2HPO_4 \cdot 7H_2O$	22.45 g	44.9 g

Prepare in H_2O. Autoclave. Store for up to 1 yr at room temperature. See Fritsch et al. (1982), Appendix B, p. 21 regarding the pHs of PO_4 buffers.

BPTE Buffer (10x)

Reagent	Quantity (for 1 L)	Final concentration (10x)
PIPES (free acid)	30 g	100 mM
Bis-Tris (free base) <!>	60 g	300 mM
EDTA (0.5 M, pH 8.0) <!>	2 mL	10 mM
H_2O	to 1 L	

Autoclave. The final pH of the buffer is ~6.5. Store for up to 6 mo at room temperature.

Gel-running Buffer

Reagent	Quantity (for 1 L)	Final concentration
PB buffer (20x)	50 mL	1x
SDS (20%, electrophoresis grade) <!>	5 mL	0.1%
H_2O	to 1 L	

Store for up to 1 yr at room temperature.

Deionized Glyoxal

Commercial stock solutions of 40% glyoxal <!> contain various hydrated forms of glyoxal, as well as oxidation products such as glyocolic acid, formic acid, and other compounds that can degrade RNA and must be removed. Deionize the 40% glyoxal by stirring on a magnetic stirrer with AG 501-X8 mixed bed ion-exchange resin (Bio-Rad 142-6425) <!> for 4 h. Then filter it through Whatman no. 1 paper to separate the resin from the liquid. Store deionized glyoxal in aliquots, under nitrogen <!>, for up to 1–2 yr at –70°C.

Glyoxal Denaturation Buffer

Reagent	Quantity (for 100 µL)	Final concentration
PB buffer (20x)	5 µL	1x
Deionized glyoxal (40%) <!>	14.5 µL	5.8%
DMSO, HPLC grade or higher <!>	50 µL	50%
Bromophenol blue (0.15%, prepared in 1x PB buffer) <!>	2.5 µL	0.00375%
H_2O	28 µL	

It is imperative to deionize the 40% glyoxal stock before preparing this buffer (see Deionized Glyoxal recipe above). Always prepare the sample buffer fresh before use.

REFERENCES

Burnett WV. 1997. Northern blotting of RNA denatured in glyoxal without buffer recirculation. *BioTechniques* **22:** 668–671.

Carmichael GG, McMaster GK. 1980. The analysis of nucleic acids in gels using glyoxal and acridine orange. *Methods Enzymol* **65:** 380–391.

Fritsch EF, Sambrook J, Maniatis T. 1982. *Molecular cloning: A laboratory manual.* Cold Spring Harbor Laboratory, Cold Spring Harbor, New York.

McMaster G, Carmichael GG. 1977. Analysis of single- and double-stranded nucleic acids on polyacrylamide and agarose gels by using glyoxal and acridine orange. *Proc Natl Acad Sci* **74:** 4835–4838.

Thomas PS. 1980. Hybridization of denatured RNA and small DNA fragments transferred to nitrocellulose. *Proc Natl Acad Sci* **77:** 5201–5205.

Thomas PS. 1983. Hybridization of denatured RNA transferred or dotted nitrocellulose paper. *Methods Enzymol* **100:** 255–266.

Protocol 9

Northern Blots: Capillary Transfer of RNA from Agarose Gels and Filter Hybridization Using Standard Stringency Hybridization Conditions

This protocol is used to detect and determine the size of large mRNA molecules (0.5–10 kb) following electrophoretic size fractionation using capillary transfer and northern hybridization (Alwine et al. 1977; Thomas 1980; Reed and Mann 1985). It is suitable for detecting poly(A)$^+$-selected mRNA or mRNA in total cellular RNA if the target transcript is relatively abundant. Using DNA or RNA probes labeled to 1×10^8 to 10×10^8 cpm/µg, it should be possible to detect ~5 pg of a specific RNA.

MATERIALS

The recipes for the items marked with <R> are at the end of the protocol or in the Common Recipes Appendix.

CAUTION: See the Cautions Appendix for appropriate handling of materials marked with <!>.

Reagents

Agarose gel through which RNA samples have been separated (from Protocol 7 or 8)
Formamide (50%)/NaH_2PO_4 (10 mM, pH 6.8) (for stripping the filter only; see Steps 16–18) <!>
HCl (1 M) (for preparing a DNA probe only; see Step 11) <!>
Methylene blue (0.04%), prepared in 0.3 M sodium acetate (pH 5.2) (*optional*; see Step 8) <!>
NaOH (1 M) (for preparing a dsDNA probe only; see Step 11) <!>
Prehybridization/hybridization mix <R>
Radiolabeled dsDNA or RNA probe (100–200 ng, labeled to 1×10^8 to 10×10^8 cpm/µg) <!>

In addition, plan to reprobe the filter using a positive control probe to ensure that the procedure is working properly (see Steps 16–18).

SSC buffer (2x)/SDS (0.1%) (for stripping the filter only; see Steps 16–18) <!>
SSC buffer (20x) <R>

Check the recommended transfer buffer for the transfer membrane and if something other than 20x SSC is recommended, use it instead.

Sodium acetate (0.3 M, pH 5.2) (*optional*; see Step 8)
Tris-HCl (1 M, pH 8.0) (for preparing a DNA probe only; see Step 11)
Wash buffers A, B, C, and D <R>

Equipment

Autoradiography or phosphorimaging apparatus
Gel-documentation system with yellow filter (*optional*; see Step 8)
Glass pipette (5 mL) (*optional*; see Step 4.iv)
Glass plate
Glass trays (*optional*; see Steps 8 and 17)
Hybridization bottle (cylindrical, glass; see Step 9)
Ice (for preparing a DNA probe only; see Step 11)
Kimwipes
Microwave (*optional*; see Step 17)
Nylon transfer membrane (uncharged; e.g., GE Healthcare, Hybond-N RPN303N)

It is also possible to use a positively charged nylon membrane and alkaline transfer conditions if the RNA is large (Reed and Mann 1985). However, alkaline transfer is very tricky because of the sensitivity of RNA to hydrolysis. If a different membrane is chosen, use the transfer buffer recommended by the supplier instead of 20x SSC.

Oven with shaker or hybridization oven with a rotating rack for glass hybridization bottles (42°C)
Paper towels
Plastic container for washes
Plastic tray for capillary gel-transfer system
Polyethylene heat-seal bags or cylindrical glass hybridization bottles

A heat sealer (e.g., National Instruments, Model 310) will be required if using heat-seal bags.

Scalpel
Scissors
Shaking water bath (65°C)
Sponge, clean (2 inches high, and the same length and width as the membrane)
Stratalinker (Stratagene) with 254-nm bulbs

Alternatively, use a 254-nm UV transilluminator <!> (see Step 7).

Water bath or temperature block (90°C, for preparing an RNA probe only; see Step 11)
Weight for capillary gel-transfer system (e.g., a book or 500-mL bottle filled with H_2O)
Whatman 3MM chromatography paper

OVERVIEW

Capillary transfer of RNAs (Reed and Mann 1985) that have been separated through an agarose gel (Protocol 7 or 8) to a nylon membrane is performed in the same manner as DNA is transferred from agarose gels for Southern blots. The membrane is then treated with 254-nm UV light to fix the transferred RNA to the membrane. The transfer membrane is pretreated to block nonspecific probe-binding sites on the membrane, and hybridization of the immobilized RNA to a ^{32}P-labeled DNA or RNA probe specific for the mRNA of interest is performed. The membrane is then washed and subjected to autoradiography or phosphorimaging. Additionally, because the exposure to UV cross-links the RNA to the membrane, it can be stripped and reprobed with other probes.

METHOD

Capillary Transfer

1. Trim a few millimeters of agarose off the edges of the gel from either Protocol 7 or 8 with a scalpel to make it flat.
2. Cut a piece of nylon transfer membrane so that it is ~2 mm larger than the gel on all four sides.
3. Cut 10 pieces of Whatman 3MM paper and a 2-inch stack of paper towels so that they are the same size as the nylon transfer membrane.
4. Set up a capillary gel-transfer system (Fig. 3-15) as follows:
   i. Soak a sponge with 20x SSC buffer and place it into a plastic tray. Add 20x SSC buffer to the bottom of the tray so that the buffer is ~1 cm deep.

      Use the transfer buffer that is recommended by the supplier of the membrane (this may be something other than 20x SSC).

   ii. Prewet three pieces of Whatman 3MM paper from Step 3 with 20x SSC and layer them on top of the sponge.
   iii. Place the gel (upside down relative to how it was in the gel tank) on top of the Whatman 3MM paper. Blot off excess liquid with Kimwipes.

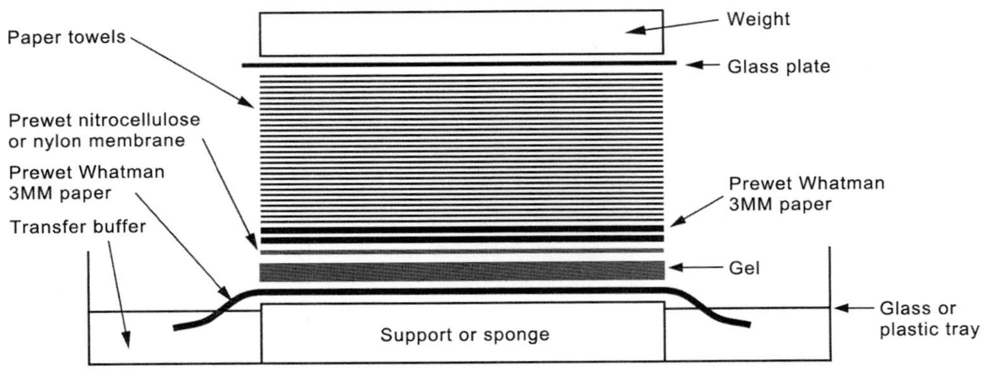

FIGURE 3-15. Capillary gel-transfer system setup. Assembly for the capillary transfer of RNA from agarose gels to nitrocellulose or nylon membranes. Transfer buffer is placed in a glass or plastic tray and a solid support or a sponge is placed in the tray. A prewet piece of Whatman 3MM paper is placed into the tray over the sponge/support to act as a wick to draw liquid from the tray up into the gel. The gel is then placed upside down on the wick and a prewet nitrocellulose membrane is carefully placed on top. Any bubbles are removed and then three sheets of prewet Whatman 3MM paper are placed on top of the membrane. Dry paper towels are cut to size and placed on top of the 3MM paper. Finally, a glass plate and a weight are placed on top of the paper towels. (Adapted, with permission, from Sambrook and Russell 2001.)

iv. Prewet the nylon membrane (from Step 2) with H_2O and then with 20x SSC. Carefully place it on top of the gel. Do not move the membrane once it touches the gel. Use a clean 5-mL glass pipette or a gloved finger to smooth out any bubbles between the gel and the nylon membrane.

v. Prewet another piece of Whatman 3MM paper from Step 3 with 20x SSC and place it on top of the membrane. Place the last six pieces of dry Whatman 3MM paper on top of that.

vi. Place the 2-inch stack of cut paper towels (from Step 3) on top of the paper.

vii. Finally, place a glass plate and a weight on top of the paper towels.

5. Let the sandwich blot overnight.

   At this point, choose between two alternative methods for processing the membrane. Follow Steps 6–15 below for one method or Protocol 10 for the other.

Cross-linking and Hybridization

6. The next day, take the transfer membrane off the sandwich and remove residual agarose by rinsing briefly in 2x SSC. Air-dry the transfer membrane on a fresh paper towel but do not allow it to dry completely.

   Do not allow the membrane to dry out at any stage during the procedure.

7. Place the filter on a sheet of Whatman 3MM paper with the RNA surface facing up. Expose the membrane to short-wave UV light (254 nm; the 1200-mJ auto-cross-link setting in a Stratalinker).

   Alternatively, use a 254-nm UV transilluminator and expose the transfer membrane with the RNA surface facing down against the glass surface for 2 min. This step cross-links the RNA to the nylon transfer membrane.

8. If desired, stain the membrane with 0.04% methylene blue (prepared in 0.3 M sodium acetate [pH 5.2]) in a glass tray for 2–3 min at room temperature. Photograph the membrane with a yellow filter under visible light. Then, destain the transfer membrane in 0.3 M sodium acetate (pH 5.2) for 15 min at room temperature.

9. Place the transfer membrane into a heat-seal bag or cylindrical glass hybridization bottle and add the prehybridization/hybridization mix (use 40–50 mL for a 12-cm^2 filter).
10. Prehybridize with shaking or rotation for 2–4 h at 42°C to block the nonspecific probe-binding sites on the membrane.
11. Prepare the probe as follows:
    - For dsDNA probes, base-denature 200 µL (100–200 ng, or 1×10^7 to 2×10^7 cpm) of the probe by adding 20 µL of 1 M NaOH. Incubate the mixture for 5 min at room temperature and then for 5 min on ice. Then, add 20 µL of 1 M HCl and 20 µL of 1 M Tris-HCl (pH 8.0).
    - For RNA probes, heat-denature 100–200 ng (1×10^7 to 2×10^7 cpm) of the probe for 2 min at 90°C.
12. Mix the probe with ~15 mL of fresh prehybridization/hybridization mix and hybridize for 40–48 h under the same conditions (at 42°C, with shaking or rotation).

    The 42°C hybridization temperature, in the presence of the 50% formamide and 4x SET in the prehybridization/hybridization mix, provides specificity to the probe hybridization without hydrolyzing the RNA immobilized on the nylon membrane (Alwine et al. 1977; Thomas 1980).

Washing

13. Wash the transfer membrane, using agitation, with the following series of increasingly stringent (i.e., lower salt) wash buffers. Washes are performed in a heat-sealed bag or plastic container in a shaking water bath or in a cylindrical glass bottle in a hybridization oven.
    1. Wash once with 50 mL of buffer A for 45 min at 42°C.
    2. Place the container into a shaking water bath or a cylindrical glass bottle in a hybridization oven and wash once with 500 mL of buffer B for 45 min at 65°C.

       Up to four transfer membranes can be washed at once in the plastic container.
    3. Wash twice with 500 mL of buffer C for 15 min at 65°C.
    4. Wash twice with 500 mL of buffer D for 15 min at 65°C.
14. Air-dry the transfer membrane on Whatman 3MM paper.

    When reprobing (Steps 16–18), do not air-dry the membrane completely.
15. Subject the filter to autoradiography or phosphorimaging.

    See Tips and Troubleshooting.

Stripping and Reprobing

Because the RNA is cross-linked to the membrane, it can be stripped and reprobed. Commonly, it is reprobed for a more abundant control transcript such as actin, tubulin, glyceraldehyde-3-phosphate dehydrogenase (GAPDH), or a ribosomal protein. These probe signals can be used for quantitation.

16. Shake the transfer membrane slowly for 1 h at 65°C in 50 mL of 50% formamide/10 mM NaH_2PO_4 (pH 6.8) in a plastic container.
17. Shake the transfer membrane vigorously for 15 min at room temperature in 50 mL of 2x SSC/0.1% SDS in a plastic container.

    Alternatively, boil some H_2O in a glass tray in a microwave and place the membrane in the boiling water for 10 min.
18. Prehybridize and then hybridize with a new radioactive probe, starting with Step 9.

TIPS AND TROUBLESHOOTING

1. If smearing or streaking of the bands is observed in Step 15, the RNA may have been degraded. A control probe may reveal this problem. If the RNA is degraded, obtain a freshly prepared sample and repeat the procedure.
2. If bands are faint or nonexistent in Step 15, consider the following:
   - This northern hybridization procedure can detect ~5 pg of a specific RNA. The mRNA fraction in a typical mammalian cell comprises ~2% of the total RNA. An mRNA is considered abundant if it constitutes >1% of the mRNA fraction, so 5 pg should be present in 100 ng of total RNA. If 10 μg of total RNA is transferred, abundant mRNAs (and less abundant transcripts that constitute greater than 0.01% of the mRNA) should be detectable. For even less abundant mRNAs, a poly$(A)^+$ fraction should be prepared and 2–3 μg of transferred poly$(A)^+$ mRNA should be sufficient to detect an mRNA that constitutes 0.0002% of the poly$(A)^+$ fraction.
   - We have found that staining the gel with either ethidium bromide or SYBR Green II after electrophoresis can reduce the sensitivity of the hybridization and increase the background. However, other investigators have found that if the concentration of ethidium bromide is kept low (<10 mg/mL), it does not interfere with hybridization. In fact, one procedure uses ethidium bromide in the denaturation solution to allow visualization of the RNA in the gel following electrophoresis. If the amount of RNA loaded on the gel is low, however, we do not recommend staining the gel after electrophoresis. As an alternative to staining the gel, stain the membrane with methylene blue after transfer but before prehybridization, as described in Step 8.

SOLUTION AND REAGENT RECIPES

CAUTION: See the Cautions Appendix for appropriate handling of materials marked with <!>.

SSC (20x)

Reagent	Quantity (for 1 L)	Quantity (for 2 L)	Final concentration (20x)
NaCl	175.3 g	350.6 g	3 M
Trisodium citrate <!>	88.3 g	176.5 g	0.3 M
H_2O	to 1 L	to 2 L	

Adjust the pH to 7.0 with NaOH <!> or HCl <!>. Autoclave. Store for more than 1 yr at room temperature.

PB Buffer (20x)

Reagent	Quantity (for 1800 mL)	Quantity (for 3580 mL)
$NaH_2PO_4 \cdot H_2O$	27.6 g	55.2 g
$Na_2HPO_4 \cdot 7H_2O$	22.45 g	44.9 g

Prepare in H_2O. Autoclave. Store for up to 1 yr at room temperature. See Fritsch et al. (1982), Appendix B, p. 21 regarding the pHs of PO_4 buffers.

SET (20x)

Reagent	Quantity (for 1 L)	Quantity (for 2 L)	Final concentration (20x)
NaCl	175.3 g	350.6 g	3 M
EDTA (no sodium) <!>	5.8 g	11.7 g	20 mM
Tris base <!>	24.2 g	48.4 g	200 mM
H_2O	to 1 L	to 2 L	

Adjust the pH to 7.8 with HCl <!> and autoclave. Store for up to 1 yr at room temperature.

Herring Testis Carrier DNA (5 mg/mL)

Add 10 g of herring testis carrier DNA (Type XIV) to sufficient H_2O to make a final volume of 2 L. Stir until all DNA is no longer opaque. This could take overnight. Autoclave. Aliquot in 50-mL tubes and store for more than 1 yr at −20°C.

Denhardt's Solution (100x)

Reagent	Quantity (for 1 L)	Final concentration (100x)
Polyvinylpyrrolidone	20 g	2%
Bovine serum albumin (Type V)	20 g	2%
Ficoll 400	20 g	2%
EDTA (250 mM, pH 8.0) <!>	10 mL	2.5 mM
H_2O	to 1 L	

Store a working stock at 4°C. For long-term storage (more than 1 yr), store at −20°C.

Prehybridization/Hybridization Mix

Reagent	Quantity (for 50 mL)	Final concentration
Formamide <!>	25 mL	50%
NaH_2PO_4 (1 M, pH 7.0)	2.5 mL	50 mM
SET (20x)	10 mL	4x
Denhardt's solution (100x)	0.5 mL	1x
Herring testis carrier DNA (5 mg/mL)	1.5 mL	
H_2O	10.5 mL	

Prepare fresh before use.

Wash Buffer A

Reagent	Quantity (for 25 mL)	Final concentration
Formamide <!>	12.5 mL	50%
SET (20x)	6.25 mL	5x
SDS (10%) <!>	1.25 mL	0.5%
H_2O	5.0 mL	

Prepare fresh before use.

Wash Buffer B

Reagent	Quantity (for 500 mL)	Final concentration
SET (20x)	50 mL	2x
SDS (10%) <!>	25 mL	0.5%
H_2O	425 mL	

Prepare fresh before use.

Wash Buffer C

Reagent	Quantity (for 1 L)	Final concentration
SET (20x)	25 mL	0.5x
SDS (10%) <!>	50 mL	0.5%
H_2O	925 mL	

Prepare fresh before use.

Wash Buffer D

Reagent	Quantity (for 1 L)	Final concentration
SET (20x)	5 mL	0.1x
SDS (10%) <!>	50 mL	0.5%
H_2O	945 mL	

Prepare fresh before use.

REFERENCES

Alwine JC, Kemp DJ, Stark GR. 1977. Method for detection of specific RNAs in agarose gels by transfer to diazobenzyloxymethyl-paper and hybridization with DNA probes. *Proc Natl Acad Sci* **74:** 5350–5354.

Reed KC, Mann DA 1985. Rapid transfer of DNA from agarose gels to nylon membranes. *Nucleic Acids Res* **13:** 7207–7221.

Sambrook J, Russell DW. 2001. *Molecular cloning: A laboratory manual*, 3rd ed. Cold Spring Harbor Laboratory Press, Cold Spring Harbor, NY.

Thomas PS. 1980. Hybridization of denatured RNA and small DNA fragments transferred to nitrocellulose. *Proc Natl Acad Sci* **77:** 5201–5205.

Protocol 10

Northern Blots: Alternative Method for Processing Northern Blots after Capillary Transfer

Different laboratories use different methods for the prehybridization, hybridization, and washing steps of the northern blotting procedure. The methods described in Protocol 9 use one set of solutions. For those who prefer to use fewer ingredients in their solutions, we present here an alternative set of solutions and conditions. Both methods work well; it is simply a matter of "taste."

MATERIALS

The recipes for the items marked with <R> are at the end of the protocol or in the Common Recipes Appendix.

CAUTION: See the Cautions Appendix for appropriate handling of materials marked with <!>.

Reagents

Bovine serum albumin (BSA)
Church and Gilbert hybridization buffer (Step 7; Church and Gilbert 1984) <R>
EDTA (0.5 M, pH 8.0) <!>
Filter from capillary transfer procedure (Step 5 of Protocol 9)
Formamide (for high and moderate stringency only; see Step 7) <!>
Probe prepared by asymmetric PCR (as in Protocol 3)
SDS (20%, w/v) <!>
Sodium phosphate buffer (1 M, pH 7.2) <R>
SSC buffer (20x) <R> (as in Protocol 9)
The 20x stock must be diluted twice for Step 4.
Wash buffer (see Step 22)

Equipment

Aluminum foil
Geiger counter (*optional*; see Step 31)
Incubator with rotating or rocking platform (42°C)
Kimwipes (*optional*; see Step 29)
Oven (80°C)
Paper towels
Pencil
Plastic container with lid to hold and mix 500 mL of liquid
Plastic wrap
Platform shaker at room temperature
Rinse container with lid
Scalpel or razor blade
Scissors
Styrofoam and lead pigs (*optional*; see Step 27)
UV cross-linker with 254-nm bulbs (e.g., Stratalinker, Stratagene)
Vacuum food sealer and bags
Water bath (shaking) at temperature appropriate for desired stringency (see Step 22)
X-ray film and developer (or phosphorimaging equipment)

OVERVIEW

This procedure is performed after capillary transfer (Steps 1–5 of Protocol 9) and is designed to minimize background (nonspecific adherence of probe to membrane and nonspecific hybridization) and maximize specific hybridization to RNAs immobilized on a membrane. The procedure should be followed exactly. It is an alternative to the method described in Steps 6–15 of Protocol 9; the steps here are designed to achieve the same goals.

METHOD

Cross-linking

Perform Steps 1–5 quickly and carefully to prevent the agarose from drying and sticking to the membrane, causing unnecessary background.

1. On the day after performing the capillary transfer procedure (Steps 1–5 of Protocol 9), remove the paper towels and the blot papers from the transfer setup.
2. Mark the locations of the wells with a pencil. Orient the gel and the membrane by slicing off or notching a corner.
3. Peel the gel off of the membrane and discard.
4. Rinse the membrane in 2x SSC to remove any agarose that may be stuck to it. Gently rub the membrane with a gloved hand to be sure that the agarose is removed.
5. Place the membrane on a foil sheet (with the RNA side up).
6. Cross-link at 254 nm (the auto-cross-link setting in a Stratalinker) or for 2 min facing down on a 254-nm light box. Then, place the membrane in an oven and bake for 30 min at 80°C.

   If the RNA was initially separated through gels containing formaldehyde, this will reverse the formaldehyde cross-linking and improve hybridization to the probe.

Prehybridization

7. Prepare ~100 mL of the appropriate Church and Gilbert hybridization buffer as outlined in the table below. After combining the ingredients, heat to 42°C to dissolve and keep the SDS in solution.

	Stringency			Final
	high	moderate	low	concentration
BSA (add the BSA to H_2O first and then add remaining ingredients)	1 g	1 g	1 g	10 mg/mL
H_2O	30 mL	–	15 mL	–
Sodium phosphate buffer (1 M, pH 7.2)	20 mL	50 mL	50 mL	0.2 M (high stringency) or 0.5 M (moderate and low stringency)
Formamide	15 mL	15 mL	–	15% (v/v) (high and moderate stringency)
EDTA (0.5 M, pH 8.0)	0.2 mL	0.2 mL	0.2 mL	1 mM
SDS (20%, w/v)	35 mL	35 mL	35 mL	7% (w/v)

Use high-stringency conditions for long probes with high melting temperatures (T_m). Use moderate-stringency conditions for probes that are ~100 nucleotides long and made with random primers or asymmetric PCR. Use low-stringency conditions for oligonucleotide probes with low T_m.

8. If pieces of the membrane are to be probed separately, cut the membrane into strips. Make orientation marks (such as diagnostic notching of corners) to distinguish the different pieces. If strips from two different gels are to be probed together, place them back to back so that the RNA side is facing out.
9. Cut vacuum seal bags large enough to leave one-half to three-quarters of an inch on either side of the membrane.
10. Seal one side of the bag.
11. Place the membrane (or membrane strips) in the bag. Position the membrane toward the bottom of the bag so that it will be easier to slide the membrane around once it is sealed in the bag.
12. Seal the second side of the bag.
13. Cut a three-quarter-inch slit in the top of the bag.
14. Add warm Church and Gilbert hybridization buffer to the bag, creating as few bubbles as possible. (For a 15 x 15-cm membrane, use 20 mL of buffer.)
15. Remove as many bubbles as possible and seal the top of the bag.
16. Place the bag on a rocking platform in a 42°C incubator. Rock for 3 h to prehybridize. Keep the hybridization buffer in the same incubator to keep the SDS in solution.

Hybridization

17. Drain the buffer from the bag.
18. Add fresh Church and Gilbert hybridization buffer (half as much as was used for prehybridization) and probe to the bag gently and smoothly to avoid bubbles and spilling.
19. Seal the bag. Quarantine as many bubbles as possible by pushing the bubbles to one side of the bag and sealing them off.
20. Place the bag into a larger secondary bag and seal. This helps to contain the radioactivity.

    If there is more than one blot, they can all be placed into the same secondary bag.
21. Place the secondary bag on a rotating platform in a 42°C incubator and incubate overnight or up to 48 h.

Washing Membrane and Visualizing Bands

22. Prepare 2 L of the wash buffer corresponding to the desired hybridization stringency as outlined in the table below. Keep 0.6 L aside for room-temperature washes and prewarm 1.4 L to the desired hybridization temperature.

	SSC (20x)	SDS (20%, w/v)	H_2O	Temperature
High stringency (0.2x SSC, 0.1% SDS)	20 mL	10 mL	1970 mL	65°C
Moderate stringency (0.2x SSC, 0.1% SDS)	20 mL	10 mL	1970 mL	50°C
Low stringency (2x SSC, 1% SDS)	200 mL	100 mL	1700 mL	50°C

23. Pour ~300 mL of wash buffer at room temperature into a wash container (with lid).
24. Remove the primary bag from the secondary bag and place it on a paper towel.

    The paper towel helps to prevent contamination of the bench paper with any radioactivity that may be on the outside of the bag.
25. Remove the blot from the primary bag and place it in the rinse container containing 300 mL of wash buffer. (Place the bag with radioactive hybridization buffer in a secondary bag and seal the

secondary bag. Dispose of this in the radioactive waste.) Rinse and remove excess hybridization buffer from the membrane in the wash buffer with a gloved hand.

The hybridization buffers are thick because of the high concentration of SDS and the residual hybridization buffer must be removed to prevent high background.

26. Pour off the buffer and immediately add another 300 mL of room-temperature wash buffer. Wash the blot on a platform shaker for 10 min at room temperature.

    Never let the membrane dry out. Dispose of *all* wash buffer in the *radioactive sink*.

27. Discard the room-temperature wash buffer and add 500 mL of fresh warm wash buffer. Place the lid on the wash container and place the container in a shaking water bath set at the appropriate temperature for the desired stringency. If necessary, weigh down the container with a few lead pigs on top of a piece of Styrofoam so that it sits on and moves with the shaker's platform. Wash for 15 min.
28. Repeat Step 27 two more times (for a total of three times).
29. Remove excess liquid from the blot by placing it between two paper towels or Kimwipes.
30. Lay the blot out in the original orientation and wrap in plastic wrap. (If the membrane was cut into several pieces for probing, put it back together in the original uncut orientation and make sure the marks denoting the locations of the wells are properly lined up.)
31. Expose the blot to film at –80°C (or to a phosphorimager screen). Develop when the desired exposure has been achieved.

    The typical exposure time is 1–2 h for abundant messages and 16–48 h for nonabundant messages. For abundant messages, monitoring the edges of the blot with a Geiger counter should give little or no signal. The area at which bands are expected should be easily detectable.

SOLUTION AND REAGENT RECIPES

CAUTION: See the Cautions Appendix for appropriate handling of material marked with <!>.

Sodium Phosphate Buffer (1 M, pH 7.2)

Reagent	Quantity (for 1 L)	Final concentration (10X)
$Na_2HPO_4 \cdot 7H_2O$	134 g	500 mM
H_3PO_4 (85%) <!>	4 mL	0.34%
H_2O	to 1 L	

The pH should be at 7.2. Store indefinitely at room temperature.

Church and Gilbert Hybridization Buffer (for 100 mL)

	High	Moderate	Low
BSA (10 mg/mL final)	1 g	1 g	1 g
H_2O	30 mL	—	15 mL
NaP (1 M) <!>	20 mL	50 mL	50 mL
Formamide <!>	15 mL (15% final)	15 mL (15% final)	—
0.5 M EDTA (1 mM) <!>	0.2 mL	0.2 mL	0.2 mL
20% SDS (7% final) <!>	35 mL	35 mL	35 mL

The preparation of this buffer for Church and Gilbert hybridization requires that dibasic sodium phosphate be pHed with phosphoric acid.

REFERENCE

Church GM, Gilbert W. 1984. Genomic sequencing. *Proc Natl Acad Sci* **81:** 1991–1995.

Protocol 11

Northern Blots for Small RNAs and MicroRNAs

This procedure is used for the detection of small RNAs (~10–200 nucleotides) by blot hybridization following electrophoretic separation on polyacrylamide–urea denaturing gels. Because high-percentage polyacrylamide gels are required to separate RNAs in this size range, it is necessary to perform electrophoretic transfer to positively charged nylon membranes. Once transferred, the immobilized RNA can be detected by hybridization using ^{32}P-radiolabeled DNA oligonucleotide probes or antisense single-stranded RNA probes (Pall and Hamilton 2008). This procedure is commonly used to detect small U-rich spliceosomal snRNAs and miRNAs. It should also be possible to detect most miRNAs using high-percentage (e.g., 15%) urea–polyacrylamide gel electrophoresis. However, for greater detection sensitivity, the splinted RNA ligation method with T4 DNA ligase (Protocol 25) is preferred.

MATERIALS

The recipes for the items marked with <R> are at the end of the protocol or in the Common Recipes Appendix.

CAUTION: See the Cautions Appendix for appropriate handling of materials marked with <!>.

Reagents

Bromophenol blue dye <!>

Electrophoretic transfer buffer (20x, pH 8.3) <R>

Dilute the 20x stock to 1x in H_2O and prechill to 4°C before use. The 1x transfer buffer contains 6 mM trisodium citrate <!> and 8 mM sodium phosphate dibasic. Alternatively, use 0.5x TBE electrophoresis buffer.

Formamide/dye loading buffer <R> <!>

Prehybridization/hybridization mix <R>

Radiolabeled DNA or RNA probe, labeled to 2×10^8 to 5×10^8 cpm/μg <!>

The probe is typically a kinased DNA oligonucleotide or body-labeled antisense in vitro RNA transcript.

RNA marker ladder (10–150 nucleotides) (e.g., US Biochemicals Low Molecular Weight Marker or Ambion Decade Marker System)

The marker can be radiolabeled by kinasing (Protocol 1) and run in parallel with the unlabeled RNA samples in Step 2. It also serves as a control for electrophoretic transfer from the gel.

RNA samples (each ~10–15 μg of total RNA)

SSC (20x) <R>

Use this 20x SSC stock and a solution of 10% SDS <!> to prepare three wash solutions just before use (see Step 10): 2x SSC/1% SDS, 1x SSC/1% SDS, and 0.1x SSC.

Equipment

Autoradiography or phosphorimaging equipment

Ballpoint ink pen

Electroblot transfer tank and power supply (e.g., Bio-Rad Trans-Blot cell with wire electrodes and PowerPac HC Power Supply)

Electrophoresis apparatus and a high-percentage (10%–15%) denaturing polyacrylamide <!>–urea <!> gel (19:1 acrylamide:bis-acrylamide)

Typically, this would be an ~18 x 18-cm gel, 1.5 mm thick with a 20-well comb. For details, see the denaturing polyacrylamide–urea gel protocol (Chapter 2, Protocol 12).

Nylon transfer membrane

The membrane should be positively charged (e.g., GE Healthcare, Hybond-N+) and cut the same size as the gel.

Oven with shaker or hybridization oven with a rotating rack for glass hybridization bottles set at 42°C

Plastic wrap

Polyethylene heat-seal bags or cylindrical glass hybridization bottles

A heat sealer (e.g., National Instruments, Model 310) will be required if using heat-seal bags.

Stratalinker (Stratagene) with 254-nm bulbs

Alternatively, use a 254-nm UV transilluminator <!> or a vacuum oven set at 80°C (see Step 7).

Water bath or temperature block (90°C)

Whatman 3MM chromatography paper (four sheets, cut the same size as the gel; see Step 2)

OVERVIEW

The RNA samples are first denatured in formamide and then separated by denaturing polyacrylamide–urea gel electrophoresis. The RNAs are then transferred to a positively charged nylon membrane and fixed by UV cross-linking. The filter is then prehybridized, hybridized to a ^{32}P-radiolabeled DNA or RNA probe, washed, and subjected to phosphorimaging or autoradiography.

METHOD

Denaturing Polyacrylamide–Urea Gel and Electrophoretic Transfer

1. Resuspend each RNA sample (~10–15 µg of total RNA) in 8–12 µL of formamide/dye loading buffer. Heat for 2 min at 90°C.
2. Load the samples and the RNA marker ladder onto a high-percentage (10%–15%) denaturing polyacrylamide–urea gel. Run the gel between 15 W and 20 W, on constant power, until the bromophenol blue is near the bottom of the gel (usually 3–4 h).
3. Remove the gel from the tank and separate the plates. Wet two sheets of Whatman 3MM paper with 20x SSC and place them on top of the gel.
4. Transfer the gel with the paper to a sponge (prewet in transfer buffer) from the electroblot apparatus, so that the paper is on top of the sponge. Next, prewet the nylon transfer membrane in H_2O and then in 20x SSC and place the positively charged nylon membrane on top of the gel and remove any bubbles. Wet two more sheets of Whatman 3MM paper with 20x SSC and place them on top of the membrane. Add the second sponge. Clamp the cassette closed.

   The layers of the sandwich are as follows, starting with the negative (−) pole: sponge, two sheets of Whatman 3MM paper, gel, nylon membrane, two sheets of Whatman 3MM paper, sponge. Remember that the RNA will migrate to the positive (+) pole.

5. Place the assembled gel cassette into the electroblot chamber and fill the chamber with 1x electrophoretic transfer buffer that has been precooled to 4°C. Transfer at 250 mA for 2 h at 4°C and then at 350 mA for another 2 h at 4°C.

Cross-linking, Prehybridization, and Hybridization

6. Disassemble the apparatus and save the nylon membrane. Carefully remove any gel fragments from the membrane. Mark the orientation of the membrane with a ballpoint ink pen. Wrap the membrane in plastic wrap and place it on Whatman 3MM paper.
7. Expose the membrane, with the RNA surface facing up, to short-wave UV light (254 nm; the 1200 mJ auto-cross-linking setting in a Stratalinker) for 1 min.

   Alternatively, use a 254-nm UV transilluminator and expose the membrane with the RNA surface facing down against the glass surface, or bake the membrane in a vacuum oven for 1 h at 80°C. This step cross-links the RNA to the nylon transfer membrane. See Tips and Troubleshooting.
8. Place the membrane in a heat-seal bag or cylindrical glass hybridization bottle and cover with prehybridization/hybridization mix. Use ~3–5 mL for a small hybridization tube or 10 mL for a heat-seal bag. Prehybridize with shaking or rotation for at least 2 h at 42°C.
9. Add the radiolabeled probe to 10^6 cpm/mL in fresh prehybridization/hybridization mix and hybridize overnight at 42°C. See Tips and Troubleshooting.

Washing and Detection

10. Safely dispose of the radioactive liquid. Perform the following 15-min washes on the membrane, all at 42°C with agitation:
    i. 2x SSC/0.1% SDS
    ii. 1x SSC/0.1% SDS
    iii. 0.1x SSC
11. Air-dry the membrane.
12. Cover the membrane with plastic wrap and expose it to a phosphor screen for phosphorimaging or to X-ray film for autoradiography. Exposure for several hours to overnight may be necessary.
13. Image on a phosphorimager or develop the X-ray film. See Tips and Troubleshooting.

TIPS AND TROUBLESHOOTING

1. For miRNAs (RNAs <50 nucleotides), covalent chemical cross-linking (Pall and Hamilton 2008) may be used to fix the small RNAs to the nylon membrane in Step 7.
2. For the detection of miRNAs and other 20–30-nucleotide RNAs, it may be useful to try the low-stringency hybridization conditions in Protocol 10 (and in this case, also omit the high-stringency wash in Step 10.iii).
3. If a high background is observed on the membrane in Step 13, either prehybridize for a longer period of time or add tRNA and/or poly(A) to the prehybridization/hybridization mix at a concentration of 200 µg/mL. Polyacrylamide sticking to the membrane can cause high background by retaining the excess probe or unincorporated label. After transfer, while the membrane is still wet, "wash" the side of the membrane in contact with the gel in 1x transfer buffer using gloved fingers.

SOLUTION AND REAGENT RECIPES

CAUTION: See the Cautions Appendix for appropriate handling of materials marked with <!>.

Formamide/Dye Loading Buffer

Reagent	Quantity (for ~10 mL)	Final concentration
Deionized formamide <!>	9.5 mL	95%
Bromophenol blue <!>	2.5 mg	0.025% (w/v)
Xylene cyanol FF <!>	2.5 mg	0.025% (w/v)
EDTA (0.5 M, pH 8.0) <!>	100 µL	5 mM
H_2O	400 µL	

Store in aliquots at –20°C.

SSC (20x)

Reagent	Quantity (for 1 L)	Quantity (for 2 L)	Final concentration (20x)
NaCl	175.3 g	350.6 g	3 M
Trisodium citrate <!>	88.3 g	176.5 g	0.3 M
H_2O	to 1 L	to 2 L	

Adjust the pH to 7.0 with NaOH <!> or HCl <!>. Autoclave. Store at room temperature for more than 1 yr.

Electrophoretic Transfer Buffer (20x)

Reagent	Quantity (for 1 L)	Final concentration (20x)
Trisodium citrate <!>	35.29 g	120 mM
Sodium phosphate, dibasic (Na_2HPO_4)	22.71 g	160 mM
H_2O	to 1 L	

Autoclave. Store for up to 1–2 yr at room temperature.

Herring Testis Carrier DNA (5 mg/mL)

Add 10 g of herring testis carrier DNA (Type XIV) to sufficient H_2O to make a final volume of 2 L. Stir until all DNA is no longer opaque. This could take overnight. Autoclave. Aliquot in 50-mL tubes and store for >1 yr at –20°C. Alternatively, the carrier DNA can be sonicated and denatured instead of autoclaved.

Denhardt's Solution (100x)

Reagent	Quantity (for 1 L)	Final concentration (100x)
Polyvinylpyrrolidone	20 g	2%
Bovine serum albumin (Type V)	20 g	2%
Ficoll 400	20 g	2%
EDTA (250 mM, pH 8.0) <!>	10 mL	2.5 mM
H_2O	to 1 L	

Store a working stock at 4°C. For long-term storage (>1 yr), store at –20°C.

Prehybridization/Hybridization Mix

Reagent	Quantity (for 50 mL)	Final concentration
Formamide <!>	25 mL	50%
SSC buffer (20x)	12.5 mL	5x
Denhardt's solution (100x)	1.5 mL	3x
Herring testis carrier DNA (5 mg/mL)	2 mL	200 µg/mL
SDS (10%) <!>	0.5 mL	0.1%
H_2O	8.5 mL	

Prepare fresh before use.

REFERENCE

Pall GS, Hamilton AJ. 2008. Improved northern blot method for enhanced detection of small RNA. *Nat Protoc* **3:** 1077–1084.

Protocol 12

RNase Protection Assay

The RNase protection assay can be used to detect RNAs in complex mixtures, to map the positions of introns, and to identify and map the locations of the 5′ and 3′ ends of mRNAs and intron/exon junctions on cloned DNA templates. The assay can be made quantitative by incorporating internal control probes in the same hybridization reaction. Because high-specific-activity probes can be easily generated by in vitro transcription and because RNA:RNA hybrids are much more stable than DNA:RNA hybrids, RNase protection is usually more sensitive than nuclease S1 protection using DNA probes (Casey and Davidson 1977). RNase protection assays are more sensitive than traditional northern blotting and can detect as little as 5 fg of target RNA or 4,000–50,000 copies/sample. RNase protection assays are more tolerant of partially degraded RNA than northern blots; if samples are even slightly degraded, the quality of data from a northern blot is severely compromised. RNase protection assays are able to distinguish among transcripts from multigene families that may comigrate on northern blots. Finally, it is possible to use multiple probes in a single RNase protection assay. Multiprobe RNase protection assays are easy to perform: Up to 10 probes plus one or two internal control probes can routinely be analyzed from a single RNA sample (http://www.ambion.com/techlib/basics/npa/index.html).

MATERIALS

The recipes for the items marked with <R> are at the end of the protocol or in the Common Recipes Appendix.

CAUTION: See the Cautions Appendix for appropriate handling of materials marked with <!>.

Reagents

Carrier RNA (10 mg/mL total yeast RNA or tRNA) or glycogen (20 mg/mL) (see Step 9.ii)

Control probe (tubulin or actin) to serve as internal standard

DNA marker, radiolabeled and denatured

A marker can be prepared from pBR322 DNA by digesting with MspI, filling in with [α-^{32}P]dCTP <!> and Klenow fragment DNA polymerase (see Appendix 3, Protocol 4).

Ethanol (70% and 100%, both prechilled to –20°C) <!>

Formamide hybridization buffer <R> <!>

Hybridization buffer (5x) <R>

Methanol (10%) <!>/acetic acid (10%) <!>

Phenol:chloroform:isoamyl alcohol (50:48:2, v/v/v) <!>

Proteinase K (10 mg/mL) <!>

[^{32}P]UTP-labeled antisense RNA probe, gel-purified (40,000 cpm per sample; see Step 1) <!>

For an in vitro transcription procedure, see Chapter 4, Protocol 1. The probe should usually be 300–900 nucleotides (not too short) and have no long A + T-rich stretches.

RNA sample of interest

Use 0.5 μg of poly(A)$^+$ RNA or 5–30 μg of total cellular RNA.

RNase A/T1 solution <R> <!>

SDS (10%) <!>

Sodium acetate (3 M, pH 5.2)

Urea/dye loading buffer <R> <!>

Equipment

Autoradiography or phosphorimaging equipment
Dry ice <!> or dry ice/ethanol <!> bath (see Steps 2.ii and 9.ii)
Gel dryer
Gel electrophoresis equipment (see Step 12)

A 6%–10% denaturing urea <!>–polyacrylamide <!> gel (20 cm wide x 40 cm high x 0.4 mm thick; 19:1 acrylamide:bis-acrylamide) in 1x TBE is required. See Chapter 12, Protocol 12.

Ice
Incubators or temperature blocks (45°C and 85°C)

A small heated air incubator (e.g., Boekel Scientific CCC Mini Incubator 260700) may be useful (see Step 6).

Microcentrifuge
Microcentrifuge tubes (0.5 or 1.5 mL)
Pipette
Plastic tray
Plastic wrap
SpeedVac
Tube shaker (e.g., Eppendorf Mixer 5432)
Vortex mixer
Water bath (boiling)
Whatman 3MM paper

OVERVIEW

Preparations of total cellular RNA or poly(A)$^+$ RNA containing a target mRNA of interest are hybridized to a single-stranded ^{32}P-labeled RNA probe. Following hybridization, a mixture of RNase A and RNase T1 is used to degrade unhybridized single-strand regions of the probe. The duplex RNase-resistant probe molecules are then separated by denaturing gel electrophoresis and visualized using autoradiography or phosphorimaging.

It is often useful to have a control probe such as tubulin (Hedley and Maniatis 1991) or actin to serve as an internal standard. A control probe serves as a signal with which the experimental probe can be compared; therefore, the two probes (experimental and control) must be different sizes so that they can be distinguished on the denaturing gel.

METHOD

1. Combine the RNA (0.5 µg of poly[A]$^+$ RNA or 5–30 µg of total cellular RNA) and 40,000 cpm of gel-purified [^{32}P]UTP antisense RNA probe in a microcentrifuge tube in 200 µL of H_2O.

   This probe is made to high specific activity (~5 x 10^8 to 6 x 10^8 cpm/µg).

2. Precipitate the RNA as follows:
   i. Add 0.1 volume of 3 M sodium acetate (pH 5.2) and 2.5 volumes of cold (–20°C) 100% ethanol.
   ii. Chill the sample on dry ice or in a dry ice/ethanol bath for 15 min.
   iii. Recover the precipitate by centrifuging at maximum speed in a microcentrifuge for 15 min at room temperature.
   iv. Discard the supernatant. Wash the pellet by adding 1 mL of cold (–20°C) 70% ethanol. Then, chill briefly on dry ice or in the dry ice/ethanol bath and centrifuge again for 2–3 min at 4°C.
   v. Discard the supernatant. Dry the pellet in a SpeedVac briefly (3–5 min) without heat.

3. Resuspend the pellet by adding 1 µL of H_2O by pipetting up and down.
4. Add 10 µL of formamide hybridization buffer to the tube. Vortex thoroughly to mix and centrifuge for 10 sec at room temperature to bring all the liquid to the bottom of the tube.
5. Heat the tube for 10 min at 85°C to denature the RNA.
6. Incubate the tube for 3–4 h at 45°C to hybridize the probe to the target RNA.

   It is useful to use a 45°C incubator rather than a temperature block or water bath so that the sample does not condense on the top of the tube. The sample can be incubated for as long as overnight.
7. Add 300 µL of RNase A/T1 solution to the tube and incubate for 1 h at room temperature.

   This step digests the unprotected RNA probe. Room temperature is used (instead of 37°C) to prevent end-fraying (see chapter Overview) of the RNA:RNA hybrids.
8. Add 20 µL of 10% SDS and 10 µL of 10 mg/mL proteinase K and incubate for 30 min at room temperature.
9. Extract the RNA as follows:
   i. Add 350 µL of phenol:chloroform:isoamyl alcohol (50:48:2, v/v/v) and vortex. Separate the phases by centrifuging at maximum speed in a microcentrifuge for 3–5 min at room temperature.
   ii. Transfer the aqueous phase to a new tube. Precipitate the RNA by adding 1 mL of cold (–20°C) 100% ethanol and 0.5 µL of 10 mg/mL carrier RNA or 20 mg/mL glycogen. Chill the tube on dry ice or in a dry ice/ethanol bath for 15 min.
   iii. Recover the precipitate by centrifuging at maximum speed in a microcentrifuge for 15 min at room temperature.
   iv. Discard the supernatant. Wash the pellet by adding 1 mL of cold (–20°C) 70% ethanol and centrifuging again for 5 min at room temperature.
   v. Discard the supernatant. Dry the pellet briefly (3–5 min) without heat.
10. Resuspend the pellet by adding 8 µL of urea/dye loading buffer and mixing on a shaker for ~5 min at room temperature.
11. Boil the sample for ~1 min in a boiling water bath and then immediately place it on ice.
12. Load half of the reaction (4 µL) on a 6%–10% denaturing urea–polyacrylamide gel in 1x TBE buffer that has been prerun for ~30 min at 20 W. In addition, load 600–2000 cpm of a denatured, radiolabeled DNA marker in a side lane of the gel. Run the gel at 33 W until the bromophenol blue dye is near the bottom of the gel.
13. Separate the gel plates and, while leaving the gel on one plate, fix the gel by soaking it for 10–15 min in a plastic tray containing 10% methanol/10% acetic acid at room temperature.
14. Remove the gel from the fixative and place one sheet of Whatman 3MM paper on the gel. Pick up the paper (and gel) and place paper side down before covering the top of the gel with plastic wrap.
15. Place the gel on a gel dryer and dry with heat for ~1 h.
16. Expose the gel to X-ray film for autoradiography or to a phosphor screen for phosphorimaging. See Tips and Troubleshooting.

TIPS AND TROUBLESHOOTING

1. It is often necessary to titrate the amounts of RNases A and T1. If too little RNase is used, multiple, incompletely cleaved partial digestion products may be observed in Step 16. If too much RNase is used, very little protected fragment is observed because the RNA:RNA hybrids "breathe."

2. Take extra precautions with the RNase A/T1 mix. Do not allow it to contaminate pipettes, racks, or centrifuges.

SOLUTION AND REAGENT RECIPES

CAUTION: See the Cautions Appendix for appropriate handling of materials marked with <!>.

Hybridization Buffer (5x)

Reagent	Quantity (for 1 mL)	Final concentration
NaCl (5 M)	400 µL	2 M
PIPES-NaOH (1 M, pH 6.4)	200 µL	200 mM
EDTA (0.25 M, pH 8.0) <!>	20 µL	5 mM
H_2O	380 µL	

Prepare fresh before use.

Formamide Hybridization Buffer

Reagent	Quantity (for 100 µL)	Final concentration
Deionized formamide <!>	80 µL	80%
Hybridization buffer (5x)	20 µL	1x

Prepare fresh before use.

RNase A/T1 Solution

Reagent	Quantity (for 1 mL)	Final concentration
NaCl (5 M)	60 µL	0.3 M
Tris-HCl (1 M, pH 7.5)	10 µL	10 mM
EDTA (0.25 M, pH 8.0) <!>	20 µL	5 mM
RNase T1 (500 units/µL; Boehringer) <!>	0.1 µL	50 units/mL
RNase A (1 mg/mL, 50 units/mL; Boehringer) <!>	1 µL	0.05 unit/mL
H_2O	to 1 mL	

The amount of both RNases may have to be titrated. We have gone as high as 8x higher for both. Prepare fresh before use.

Urea/Dye Loading Buffer

Reagent	Quantity (for 10 mL)	Final concentration
Urea <!>	4.8 g	8 M
Tris-HCl (1 M, pH 8.0)	200 µL	20 mM
EDTA (0.25 M, pH 8.0) <!>	40 µL	1 mM
Bromophenol blue (0.5%) <!>	1 mL	0.05%
Xylene cyanol (0.5%) <!>	1 mL	0.05%
H_2O	to 10 mL	

Store for 2–4 wk at –20°C.

REFERENCES

Casey J, Davidson N. 1977. Rates of formation and thermal stabilities of RNA:DNA and DNA:DNA duplexes at high concentrations of formamide. *Nucleic Acids Res* **4:** 1539–1552.

Hedley ML, Maniatis T. 1991. Sex-specific binding and polyadenylation of dsx pre-mRNA requires a sequence that binds specifically to tra-2 protein in vitro. *Cell* **65:** 579.

Sambrook J, Fritsch EF, Maniatis T. 1989. *Molecular cloning: A laboratory manual*, 2nd ed. Cold Spring Harbor Laboratory Press, Cold Spring Harbor, NY.

WWW RESOURCE

http://www.ambion.com/techlib/basics/npa/index.html The Basics: What is a nuclease protection assay? Applied Biosystems.

Protocol 13

Nuclease S1 Protection Mapping

This typically involves using end-labeled DNA probes that are hybridized to an RNA sample, and any single-stranded regions, which might correspond to introns, or extensions beyond the 5′ and 3′ ends of the transcript are digested with the single-strand specific nuclease, S1. It is commonly used to find the positions of introns and the 5′ and 3′ ends of transcripts (Calzone et al 1987). This method is quite sensitive and can also be used to detect specific transcription initiation events in reconstituted transcription systems (Fig. 3-16).

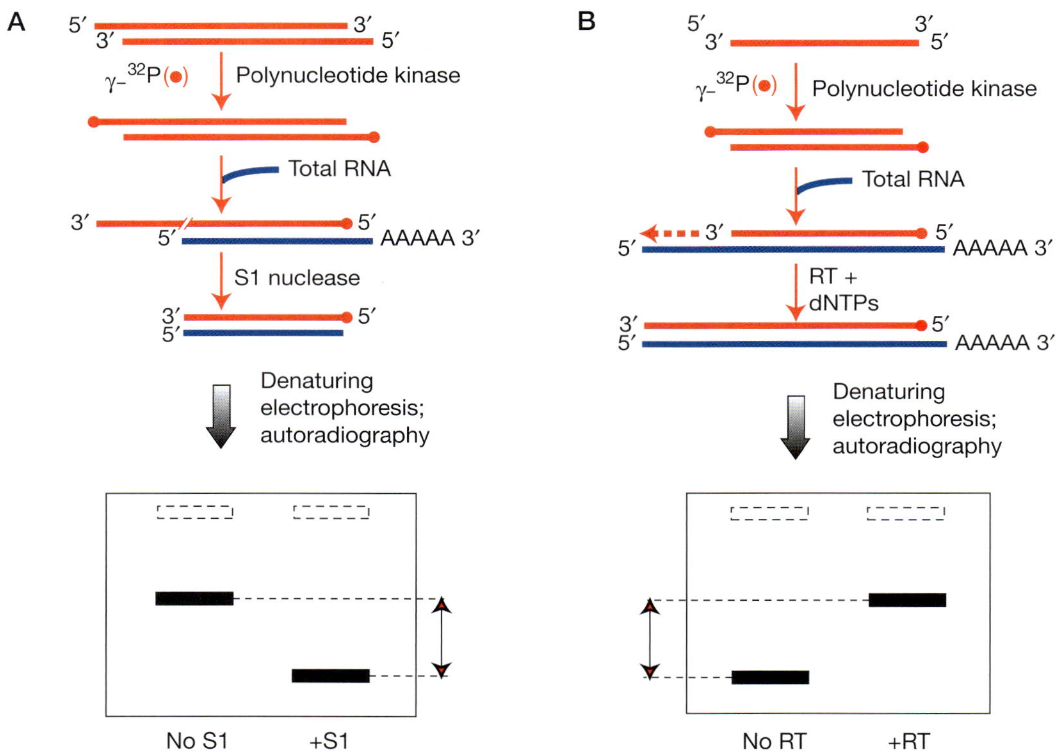

FIGURE 3-16. S1 nuclease protection assay (*A*) compared with primer extension assay (*B*). (Redrawn, with permission, from Strachan and Read 1999 [©Garland Science].)

MATERIALS

The recipes for the items marked with <R> are at the end of the protocol or in the Common Recipes Appendix.

CAUTION: See the Cautions Appendix for appropriate handling of materials marked with <!>.

Reagents

Carrier RNA (10 mg/mL total yeast RNA or tRNA) or glycogen (20 mg/mL)
Deioinized formamide <!> (purchased is preferred)
Denaturing urea <!>–polyacrylamide <!> gel (20 cm wide x 40 cm high x 0.4 mm thick)
 Prepare a polyacrylamide (19:1 acrylamide:bis-acrylamide) gel in 1x TBE buffer and 8.3 M urea. Use an 8%–10% gel for protected fragments 50–150 nucleotides and use a 5%–6% gel for protected fragments >250 nucleotides.
Ethanol (70%, prechilled to –20°C) <!>
Formamide <!>/dyes/NaOH <!>/EDTA <!> <R>
Isopropanol, prechilled to –20°C <!>
Nuclease S1 <!>
Radiolabeled DNA probe (~10^6–10^7 cpm/µg; use 1–20 ng or ~3 x 10^4 to 4 x 10^4 cpm per assay) <!>
 The probe is typically a kinased synthetic oligonucleotide or a kinased duplex DNA generated by PCR or restriction endonuclease cleavage where a single strand is purified on a native acrylamide <!> gel after denaturing a duplex DNA fragment, a so-called strand separation gel (see Sambrook and Russell 2001). The probe can be 50–400 nucleotides.
RNA sample of interest (0.5 µg of poly[A]$^+$ RNA or 5–30 µg of total cellular RNA)
S1 digestion buffer <R>
S1 hybridization buffer (5x) <R>
S1 nuclease enzyme <!>
S1 termination mix <R>
Size markers (see Tips and Troubleshooting, Step 3)
Sodium acetate (0.3 M, pH 5.2)

Equipment

Autoradiography or phosphorimaging equipment
Dry ice <!>/ethanol <!> bath
Gel electrophoresis equipment for denaturing polyacrylamide gel (tank, plates, and spacers)
Ice
Microcentrifuge
Microcentrifuge tubes (1.5 mL)
Micropipette
Small heated air incubator (30°C)
Temperature blocks (12°C–30°C [see Step 10] and 70°C)
Tube shaker (e.g., Eppendorf Mixer 5432) (*optional*; see Step 6)
Water bath (boiling)

OVERVIEW

In this procedure, a radiolabeled single-stranded DNA probe is hybridized to target RNA molecules. These RNA:DNA hybrids are then treated with single-strand-specific nuclease S1 to digest all of the unannealed single-stranded probe and RNA. The remaining duplex S1-resistant hybrids are then denatured and subjected to separation by denaturing polyacrylamide–urea gel electrophoresis.

It is often useful to include an internal control probe. This can be a probe to an endogenous transcript or it can be a spiked-in control in vitro transcript, whose protected RNA:DNA hybrid is different in size from that of the experimental RNA:DNA hybrid signal. It is also useful to perform control samples (1) without nuclease S1 and without target RNA but with probe and (2) with nuclease S1 and probe but without target RNA. These reactions control for the intactness of the starting probe, the activity of the nuclease S1 in the digestion reaction, and any secondary structure in the probe that might give rise to an nuclease S1-resistant band in the absence of target RNA.

METHOD

1. Dilute the RNA (0.5 μg of poly[A]$^+$ RNA or 5–30 μg of total cellular RNA) to 100 μL in 0.3 M sodium acetate (pH 5.2) in an 1.5-mL microcentrifuge tube. Then, add the radioactive probe (generally ~30,000–40,000 cpm). Add 250 μL of 100% ethanol.
2. Chill for 10–15 min in a dry ice/ethanol bath.
3. Centrifuge at maximum speed in a microcentrifuge for 10 min at 4°C.
4. Discard the supernatant. Wash the pellet by adding 0.5 mL of cold (–20°C) 70% ethanol and then repeating Steps 2 and 3.
5. Discard the supernatant. Air-dry the probe–RNA pellet.
6. Add 4 μL of 5x S1 hybridization buffer and 16 μL of deionized formamide to the probe–RNA pellet. Resuspend with manual pipetting or in a tube shaker (10–15 min).
7. Denature the probe–RNA mixture for 5 min at 70°C.
8. Transfer the tube to an incubator and anneal the reaction overnight at 30°C.

   A 30°C heated air incubator will prevent the samples from evaporating, as would occur in a temperature block.
9. Place the tube on ice. Add 300 μL of S1 digestion buffer and 100–1000 units/mL of S1 nuclease enzyme.

   The quantity of S1 nuclease enzyme must be determined empirically. Often, RNA:DNA hybrids undergo thermal "breathing," and if there is a large excess of S1, the protected hybrids may be lost. Once the optimal quantity of S1 enzyme is known, the assay is reproducible as long as the same nuclease S1 (with the same specific activity) is used.
10. Incubate for 1–2 h at 12°C–30°C.

    The temperature must be titrated; generally, shorter or more AT-rich hybrids will require a lower temperature.
11. Stop the nuclease digestion with 75 μL of S1 termination mix and 1 μL of 10 mg/mL carrier RNA or 1 μL of 20 mg/mL glycogen.
12. Add 400 μL of cold (–20°C) isopropanol. Chill in the dry ice/ethanol bath for 10–15 min and then repeat Steps 2 and 3.
13. Discard the supernatant. Wash the pellet by adding 0.5 mL of cold (–20°C) 70% ethanol and then repeat Steps 2 and 3.
14. Carefully discard the supernatant and air-dry the sample.
15. Resuspend the pellet in 3 μL of formamide/dyes/NaOH/EDTA. Place in a boiling water bath for 3 min and then place on ice.
16. Run the sample on a denaturing polyacrylamide gel. Dry the gel and perform autoradiography or phosphorimaging. See Tips and Troubleshooting.

TIPS AND TROUBLESHOOTING

1. If a discrete band is not obtained, this may indicate that the hybrids are being cleaved by nuclease S1. Titrate the nuclease S1 using the least possible amount (often leaving some of the probe intact) and reduce the digestion temperature from room temperature to 15°C. However, lowering the temperature is going to reduce the activity of the enzyme, so this approach may require more nuclease S1.
2. It is important to perform the two controls mentioned above: (1) a reaction of probe alone with no S1 treatment, so the intactness and any potential secondary structures in the probe (that are resistant to 8.3 M urea) can be detected without nuclease cleavage and (2) treating the probe with nuclease S1 in the absence of target RNA so that any S1-resistant secondary structures can be observed.
3. It is imperative to use markers to size the protected probe fragments. DNA markers can be used, e.g., kinased or Klenow-filled pBR322 plasmid DNA cleaved with MspI (see Appendix 3, Protocol 4).

SOLUTION AND REAGENT RECIPES

CAUTION: See the Cautions Appendix for appropriate handling of materials marked with <!>.

S1 Hybridization Buffer (5x)

Reagent	Quantity (for 100 mL)	Final concentration (5x)
PIPES, free acid	6.04 g	0.2 M
NaCl (5 M)	40 mL	2 M
EDTA (250 mM, pH 8.0) <!>	2 mL	5 mM
H_2O	to 100 mL	

Before bringing the final volume to 100 mL with H_2O, adjust the pH to 6.4 with NaOH <!>. Filter with a 0.2-µm filter unit. Store for up to 1–2 yr at room temperature.

S1 Digestion Buffer

Reagent	Quantity (for 100 mL)	Final concentration
NaCl (5 M)	5.6 mL	0.28 M
Sodium acetate (0.5 M, pH 4.6)	10 mL	0.05 M
Zinc acetate (450 mM)	1 mL	4.5 mM

Filter with a 0.2-µm filter unit. Store for up to 1 yr at 4°C.

S1 Termination Mix

Reagent	Quantity (for 100 mL)	Final concentration
Ammonium acetate	30.83 g	4 M
Na_2EDTA <!>	3.72 g	0.1 M

Filter with a 0.2-µm filter unit. Store for up to 1–2 yr at room temperature.

NaOH (0.1 M)/EDTA (1 mM)

Reagent	Quantity (for 100 mL)	Final concentration
NaOH <!>	0.4 g	0.1 M
EDTA (0.25 M, pH 8.0) <!>	0.4 mL	1 mM
H_2O	to 100 mL	

Store in a plastic polypropylene bottle for up to 3 mo at room temperature.

READING LIST

Berk AJ, Sharp PA. 1977. Sizing and mapping of early adenovirus mRNAs by gel electrophoresis of S1 endonuclease-digested hybrids. *Cell* **12:** 721–732.

Calzone FJ, Britten RJ, Davidson EH. 1987. Mapping of gene transcripts by nuclease protection assays and cDNA primer extension. *Meth Enzymol* **157:** 611–632.

Sambrook J, Russell DW. 2001. *Molecular cloning: A laboratory manual*, 3rd ed. Cold Spring Harbor Laboratory Press, Cold Spring Harbor, NY.

Strachan T, Read AP. 1999. *Human molecular genetics*, 2nd ed. Garland, Oxford.

Protocol 14

Primer Extension Analysis of RNA

Primer extension provides a very sensitive method to detect and quantitate specific RNAs. It is particularly useful for the analysis of small RNAs such as snRNAs. In addition, primer extension is often used to detect modified bases in chemical probing analyses. This assay (Calzone et al. 1987; Boorstein and Craig 1989) can also be used to measure the distance from the primer to the 5′ end or a branch point of an RNA molecule. The ideal distance from the 3′ end of the oligonucleotide to the end to be mapped is 50–200 nucleotides, but extensions of several hundred nucleotides are possible depending on the amount of RNA in the sample. The assay can be used with RNA that has been purified from an in vivo source or synthesized in vitro (e.g., from in vitro splicing or in vitro transcription reactions).

MATERIALS

The recipes for the items marked with <R> are at the end of the protocol or in the Common Recipes Appendix.

CAUTION: See the Cautions Appendix for appropriate handling of materials marked with <!>.

Reagents

Ammonium acetate (4 M)
Annealing buffer (5x) <R>
DNA marker, radiolabeled and denatured (*optional*; see Tips and Troubleshooting)

> A marker can be prepared from pBR322 DNA by digesting with MspI, filling in with [α-^{32}P]dCTP <!> and Klenow fragment DNA polymerase, denaturing with formamide dyes <R> <!>, and boiling for 1–2 min (see Appendix 3).

Ethanol (70% and 100%, both chilled to –20°C) <!>
Formamide/dye loading buffer <R> <!>
[γ-^{32}P]ATP (7000 Ci/mmol; 167 mCi/mL) <!>
Glycogen (20 mg/mL)
Kinase buffer (10x) <R>
NaOH (0.1 M)/EDTA (1 mM) <!>
Oligonucleotide primer (~22 mer)
Primer extension (PE) mix (5x) <R>
Reverse transcriptase (e.g., Moloney murine leukemia virus), cloned (15 units/mL; Invitrogen)
RNA, prepared in vivo or in vitro

> Depending on the experiment, the sample can be 20–30 mg of total cytoplasmic or nuclear RNA, 1–2 µg of poly (A)$^+$-selected RNA, or an eluted gel band plus a glycogen carrier.

Sodium acetate (0.3 M, pH 5.2)
T4 polynucleotide kinase (e.g., OptiKinase [20 units/µL] from US Biochemicals)
TE buffer <R>

Equipment

Autoradiography or phosphorimaging equipment
Dry ice <!>/ethanol <!> bath, dry ice <!>, or –80°C freezer (see Steps 4, 10, and 16)
Gel dryer
Ice

Incubators or temperature blocks (37°C, 42°C, 48°C, 65°C, and 80°C)
Microcentrifuge
Microcentrifuge tubes (0.5 or 1.5 mL)
Scintillation counter for Cerenkov counting
Sephadex G-10 or other gel filtration spin column with a molecular weight cutoff (MWCO) of 1000 (*optional*; see Step 3)
Sequencing gel, plates (20 cm wide x 40 cm long x 0.4 mm thick), comb, and gel tank (for Step 19; see Chapter 2, Protocol 12 for details)
SpeedVac
Water bath (boiling)

OVERVIEW

The procedure consists of three steps: (1) radiolabeling the oligonucleotide primer, (2) annealing the primer to the target RNA, and (3) adding dNTPs and extending the primer with M-MuLV reverse transcriptase. All solutions, tubes, and pipette tips must be RNase-free.

METHOD

Radiolabel the Primer Oligonucleotide with T4 Polynucleotide Kinase

It is important to label the olignucleotide to the highest specific activity possible. This is most simply done by labeling the smallest amount possible.

1. Perform the radiolabeling using Protocol 1.

Primer–RNA Annealing and Primer Extension

2. Mix the RNA in 8 µL of H_2O and add 2 µL of 5x annealing buffer.

   Double these volumes (i.e., use 16 µL of H_2O and 4 µL of 5x buffer) if more than 20 µg of RNA is used.

3. Incubate for 5 min at 80°C, 30 min at 65°C, and 30 min at 48°C. Centrifuge for 5 sec after each 30-min incubation to prevent drying the RNA.

   The annealing temperature depends on the length and G + C content of the primer. The conditions specified here (i.e., with an annealing temperature of 48°C) are satisfactory for 20–30-nucleotide primers with 50%–70% G + C content.

4. Add 23 µL of primer extension mix and 0.67 µL of 15 units/µL cloned reverse transcriptase. Mix carefully, centrifuge for 15 sec, and mix again.

   Double these quantities if the annealing was done in 20 µL. The primer extension mix and reverse transcriptase can be first combined as a master mix and then added to the reaction tube.

5. Incubate for 1 h at 42°C.
6. Add 0.3 mL of cold 100% ethanol. Chill the samples on dry ice, in a dry ice/ethanol bath, or for 15 min in a –80°C freezer.
7. Centrifuge in a microcentrifuge for 10 min at 4°C. Wash the pellet with cold 70% ethanol and air-dry (or dry briefly for 2–3 min in the SpeedVac).
8. Dissolve in 2 µL of 0.1 M NaOH/1 mM EDTA and then add 4 µL of formamide/dye loading buffer. Heat for 2 min at 95°C and then chill on ice.
9. Load on a sequencing gel. Perform electrophoresis at 25 W for 1–2 h. See Chapter 2, Protocol 12 for details.

10. Dry the gel and expose to X-ray film or a phosphorimaging screen overnight. Develop or scan. See Tips and Troubleshooting.

TIPS AND TROUBLESHOOTING

1. When working with small amounts of RNA (<10 µg) that are to be precipitated with ethanol, the addition of glycogen (20 µg) as a carrier will facilitate recovery of the RNA.
2. The sensitivity of this assay will depend on the quantity of the target RNA in the sample. If a low but detectable signal is obtained, try adding more input RNA in the primer hybridization step. Remember that a mass excess of primer over target RNA is being used and so what will determine the signal strength is how much target RNA is in the hybridization reaction.
3. A large percentage of oligonucleotides simply do not work well in this assay. This could be due to RNA secondary structure or other reasons. When the assay is not working well with an unproven oligonucleotide, it is always worth trying a different primer, even if it is only shifted by 10 nucleotides or so.
4. To precisely map stops (i.e., to map a branch point), run a dideoxy sequencing ladder alongside the RNA samples. This uses the same primer and 0.5 µg of denatured plasmid DNA containing the target sequence. The annealing and extension reactions are the same for the RNA samples except that the primer extension mix recipe must be adjusted so that the dNTPs have a final concentration of 50 µM, and 20 µM of one ddNTP is also included. After the first 30 min at 42°C in Step 15, bring the dNTP concentration to 500 µM and incubate for another 30 min at 42°C. For mapping branch points, it is also important to run a control reaction with a full-length transcript to determine the strong reverse transcriptase stops present without incubation in splicing extract.

SOLUTION AND REAGENT RECIPES

CAUTION: See the Cautions Appendix for appropriate handling of materials marked with <!>.

Kinase Buffer (10x)

Reagent	Quantity (for 100 µL)	Final concentration
Tris-HCl (2 M, pH 7.6)	33 µL	0.66 M
$MgCl_2$ (1 M) <!>	10 µL	0.1 M
Spermidine (100 mM) <!>	10 µL	10 mM
DTT (0.5 M) <!>	30 µL	150 mM
H_2O	17 µL	

Store at –20°C and make fresh at least once a month (DTT goes bad). It is a good idea to make this buffer fresh each time.

Annealing Buffer (5x)

Reagent	Quantity (for 1 mL)	Final concentration
KCl (3 M) <!>	375 µL	1.25 M
Tris-HCl (1 M, pH 7.9)	10 µL	10 mM
EDTA (0.25 M, pH 8.0) <!>	4 µL	1 mM
H_2O	611 µL	

Store for up to 1–2 yr at –20°C.

Primer Extension (PE) Mix (5x)

Reagent	Quantity (for 500 µL)	Final concentration
Tris-HCl (1 M, pH 8.7 at room temperature or pH 8.3 at 37°C)	10 µL	20 mM
$MgCl_2$ (1 M) <!>	5 µL	10 mM
dNTPs (10 mM each dA,dC,dG, dT)	16.7 µL	0.33 mM
Actinomycin D <!> (1 mg/mL, prepared in 100% DMSO <!>) (*optional*)*	50 µL	100 µg/mL
Dithiothreitol (DTT) (0.5 M) <!>	5 µL	5 mM
H_2O	413.3 µL	

Store for up to 1 yr at –20°C in the dark (aluminum foil–covered tubes).

*Actinomycin D is only recommended for samples containing significant quantities of DNA in the sample (e.g., from in vitro transcription reactions). If the RNA sample does not contain DNA, omit the actinomycin D from the primer extension mix.

Formamide/Dye Loading Buffer

Reagent	Quantity (for ~10 mL)	Final concentration
Deionized formamide <!>	9.5 mL	95%
Bromophenol blue <!>	2.5 mg	0.025% (w/v)
Xylene cyanol FF <!>	2.5 mg	0.025% (w/v)
EDTA (0.5 M, pH 8.0) <!>	100 µL	5 mM
H_2O	400 µL	

Store in aliquots for up to 1 yr at –20°C.

REFERENCES

Boorstein WR, Craig EA. 1989. Primer extension analysis of RNA. *Methods Enzymol* **180:** 347–369.

Calzone FJ, Britten RJ, Davidson EH. 1987. Mapping of gene transcripts by nuclease protection assays and cDNA primer extension. *Methods Enzymol* **152:** 611–632.

Protocol 15

Reverse Primer Extension

Reverse primer extension (i.e., using the 3′ end of an RNA molecule of interest as the primer) is a valuable and versatile method that can be used for RNA detection or affinity selection. The method is straightforward, but it will only work if the 3′ end of the RNA is precisely known. An oligodeoxynucleotide that contains a region of complementarity to the 3′ end of the RNA of interest and an overhanging 5′ region of arbitrary sequence (Fig. 3-17) is annealed to the 3′ end of the RNA of interest. DNA polymerase or reverse transcriptase recognizes the partially duplexed structure as a substrate and "fills in" the single-strand region of the oligonucleotide, using the 3′-hydroxyl group of the RNA to initiate polymerization. Extension in the opposite direction is prohibited if the 3′ end of the oligonucleotide is blocked by a nonextendable end (e.g., a dideoxynucleotide). This protocol incorporates [α-^{32}P]dCTP during extension. It is as sensitive as primer extension (Protocol 14) and can be used if the 5′ end of the RNA of interest contains a significant amount of stable secondary structure. Unlike primer extension, however, the RNA of interest becomes labeled and the RNA is not hydrolyzed after the procedure. Although this method was originally designed for detecting small RNAs (Hausner et al. 1990), it is also quite useful for introducing biotin groups onto the 3′ end of mRNA for subsequent affinity selection. For this application, a "clean" 3′ end must be introduced into the mRNA by site-specific RNase H digestion (Protocol 17). Cleavage by RNase H removes the poly(A) tail and some of the 3′-untranslated region (UTR), resulting in a free 3′ hydroxyl that can then be extended.

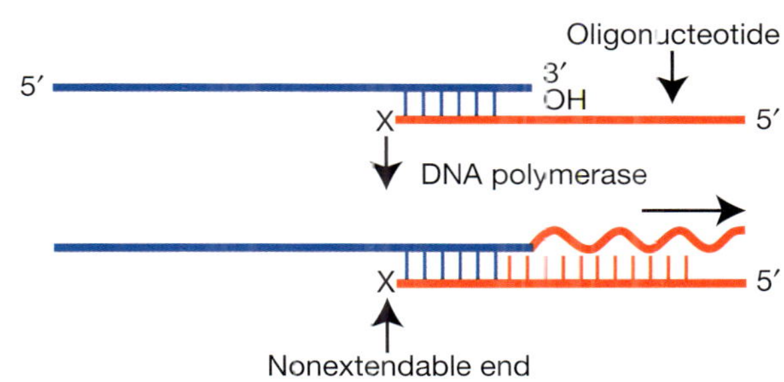

FIGURE 3-17. Reverse primer extension.

MATERIALS

The recipes for the items marked with <R> are at the end of the protocol or in the Common Recipes Appendix.

CAUTION: See the Cautions Appendix for appropriate handling of materials marked with <!>.

Reagents

[α-^{32}P]dCTP (3000 Ci/mmol; 10 µCi/µL) <!>

For biotinylation, use 10 mM biotinylated dUTP instead of [α-^{32}P]dCTP (see Step 3).

Annealing buffer (2x) <R>

dCTP (1 mM)

For biotinylation, use a 10 mM stock of dCTP (see Step 3).

DNA polymerase buffer (5x) <R> or reverse transcriptase buffer (5x) <R> (see Step 3) dTTP (10 mM)

For biotinylation, use a 1 mM stock of dTTP (see Step 3).

Oligodeoxynucleotide

Before beginning this procedure, design an oligodeoxynucleotide with ~20 nucleotides of complementarity to the 3′ end of the RNA of interest, as well as an overhang of 10–15 nucleotides of arbitrary sequence. We have found that an alternating sequence of A and G (e.g., AGAGAGAGAG) is best because it templates dU and dC, thus allowing simultaneous incorporation of [α-^{32}P]dCTP and biotinylated dUTP.

Reagents for ethanol precipitation <!> and denaturing gel electrophoresis (Chapter 2, Protocols 3 and 12, respectively; see Step 6)

RNA of interest (see Step 1)

The RNA sample must be deproteinized. Samples of mRNA can be used, but the mRNA molecules must first be completely deadenylated (see Appendix 3, Protocol 8) or cleaved with site-specific RNase H (see Chapter 4, Protocol 8).

SDS buffer <R> <!>

T7 DNA polymerase (5 units/μL; US Biochemicals) or RNase H^{-} reverse transcriptase (5 units/μL; US Biochemicals)

T7 DNA polymerase is the preferred extension enzyme; if reverse transcriptase is used, it must be RNase H minus.

Equipment

Equipment for ethanol precipitation and denaturing gel electrophoresis (Chapter 3, Protocols 3 and 12, respectively; see Step 6)

Ice

Micro Bio-Spin Columns with Bio-Gel P-6 in Tris buffer (Bio-Rad)

Microcentrifuge

Microcentrifuge tubes (0.5 or 1.5 mL)

Temperature block or water baths (95°C and 65°C; see Step 2, and either 37°C or 42°C; see Step 4)

OVERVIEW

In this protocol, RNA is extended with [α-^{32}P]dCTP, separated from unincorporated nucleotides using a spin column, and analyzed by denaturing gel electrophoresis. It can also be used to introduce biotinylated dUTP for subsequent affinity selection (see Chapter 5, Protocol 13).

METHOD

1. Assemble the following reagents in a microcentrifuge tube:

RNA of interest	up to 4 μL
Oligodeoxynucleotide	100 ng in 1 μL
H_2O	to 5 μL
Annealing buffer (2×)	5 μL
Total	to 10 μL

The appropriate quantity of RNA depends on the sample. It can range from nanogram quantities (e.g., samples enriched for small RNAs) to 1–20 μg (e.g., for total RNA).

2. Heat for 1 min at 95°C and 10 min at 65°C and then quench on ice.

3. Add the following to the reaction:

DNA polymerase buffer (5x) or reverse transcriptase buffer (5x)	5 μL
dTTP (10 mM)	1 μL
dCTP (1 mM)	1.25 μL
[α-^{32}P]dCTP (3000 Ci/mmol; 10 μCi/μL) <!>	5 μL
H_2O	1.75 μL
T7 DNA polymerase (5 units/μL) or RNase H$^-$ reverse transcriptase (5 units/μL)	1 μL

If biotinylating and not radiolabeling, use 2.5 μL of 10 mM dCTP instead of 1.25 μL of 1 mM dCTP, 1 μL of 10 mM biotinylated dUTP instead of the [α-^{32}P]dCTP, and 1 μL of 1 mM dTTP instead of 1 μL of 10 mM dTTP and adjust the volume of H_2O accordingly.

4. Incubate for 30 min at 37°C for DNA polymerase or at 42°C for reverse transcriptase.
5. Dilute to 75 μL with SDS buffer and run through a Micro Bio-Spin Column.
6. Ethanol-precipitate and analyze one-third of the sample by denaturing gel electrophoresis (see Chapter 2, Protocols 3 and 12, respectively).

   It is important not to PCA-extract heavily biotinylated RNAs; they will go to the interface.

7. Store the remainder of the sample at –20°C.

   Biotinylated samples can be stored indefinitely.

TIPS AND TROUBLESHOOTING

The protocol is designed to minimize nonspecific background. Nevertheless, some background from "random priming" on the oligodeoxynucleotide or other RNA molecules may occur. If this is a problem, it may be necessary to gel-purify the RNA of interest.

SOLUTION AND REAGENT RECIPES

CAUTION: See the Cautions Appendix for appropriate handling of materials marked with <!>.

Annealing Buffer (2x)

Reagent	Quantity (for 1 mL)	Final concentration (2x)
KCl (2 M) <!>	150 μL	300 mM
Tris-HCl (1 M, pH 8.0)	200 μL	200 mM
H_2O	650 μL	

Store at −20°C.

DNA Polymerase Buffer (5x)

Reagent	Quantity (for 1 mL)	Final concentration (5x)
Tris-HCl (1 M, pH 7.5)	200 μL	200 mM
$MgCl_2$ (1 M) <!>	100 μL	100 mM
H_2O	700 μL	

Store at −20°C.

Reverse Transcriptase Buffer (5x)

Reagent	Quantity (for 1 mL)	Final concentration (5x)
$MgCl_2$ (1 M) <!>	15 µL	15 mM
Tris-HCl (1 M, pH 8.0)	100 µL	100 mM
Dithiothreitol (DTT) (1 M) <!>	50 µL	50 mM
H_2O	835 µL	

Store for up to 1 wk at −20°C.

SDS Buffer

Reagent	Quantity (for 500 mL)	Final concentration
Tris-HCl (1 M, pH 7.5)	10 mL	20 mM
EDTA (0.5 M, pH 8.0) <!>	1 mL	1 mM
SDS (10%, w/v) <!>	25 mL	0.5% (w/v)
H_2O	464 mL	

Store indefinitely at room temperature.

REFERENCE

Hausner P, Giglio LM, Weiner AM. 1990. Evidence for base-pairing between mammalian U2 and U6 small nuclear ribonucleoprotein particles. *Genes Dev* **4:** 2146–2156.

Protocol 16

Poisoned Primer Extension

This protocol is used to distinguish RNAs that are extremely close but not identical in sequence. It is conceptually identical to primer extension (see Protocol 14), but it involves the use of a nonextendable nucleotide (the poison) in the presence of the other three nucleotides. The protocol is valuable in determining the extent of RNA editing or the accuracy and extent of processing reactions such as splicing (e.g., Hayes and Hanson 2007; Keegan et al. 2007).

MATERIALS

The recipes for the items marked with <R> are at the end of the protocol or in the Common Recipes Appendix.

CAUTION: See the Cautions Appendix for appropriate handling of materials marked with <!>.

Reagents

Annealing buffer (5x) <R>

Denaturing acrylamide <!> gel (7%) (for Step 6; see Chapter 2, Protocol 12)

Dideoxynucleotide triphosphates (ddNTPs) and deoxynucleotide triphosphates (dNTPs)

> A solution of one ddNTP at 0.1 mM is required (e.g., 0.1 mM ddCTP). The specific ddNTP used is determined by the sequence to be analyzed. A second mix of the other three dNTPs at 1 mM each is also required (e.g., 1 mM each of dATP, dGTP, and dTTP).

EDTA (0.5 M, pH 8.0) <!>

Formamide loading dye <!>

Primer, kinased with [γ-^{32}P]ATP (2 pmol/µL) <!>

> To prepare ^{32}P-kinased primer, see Protocol 1. At least 4 pmol (~25 ng) of primer is needed for each reaction.

Reverse transcriptase (e.g., Moloney murine leukemia virus) (200 units/µL)

Reverse transcription mix (10x) <R>

Total cellular RNA (1–10 µg) or a purified in vitro transcript in the presence of a glycogen carrier

Equipment

Gel dryer

Gel electrophoresis apparatus and power supply

Ice

Microcentrifuge tubes (0.5 or 1.5 mL)

Phosphorimager or autoradiography equipment

Water baths or thermal cycler for incubations (42°C, 55°C, and 95°C; see Steps 2, 4, and 6)

OVERVIEW

This method uses a radioactively labeled primer that hybridizes a short distance downstream from the "changed" region of interest. Reverse transcription is used to extend the primer into this region. The reactions contain three of four substrates for extension (e.g., dATP, dGTP, and dTTP) and a chain-terminating substrate (e.g., ddCTP). The extension reaction will stop once it reaches a G in the template; therefore, RNAs that differ in sequence at that position will yield different-sized extension products that can be resolved on a denaturing gel. Note that the nucleotide to be analyzed must be the first nucleotide of its type located after the primer sequence (in the region to be extended); otherwise, primer extension will terminate before the base of interest.

METHOD

1. For a 10-μL final reaction volume, assemble the following reagents in a microcentrifuge tube:

Annealing buffer (5x)	1 μL
Total cellular RNA (1–10 μg)	2 μL
Primer, kinased with [γ-^{32}P]ATP (2 pmol/μL)	2 μL
Total	5 μL

   See Tips and Troubleshooting.

2. Denature for 30 sec at 95°C, anneal for 10 min at 55°C, and then place on ice.
3. Add the following for detecting changes in G (or substitute appropriate mixes for detecting change in the other nucleotides):

Mix of three dNTPs (e.g., 1 mM each of dATP, dGTP, and dTTP)	1 μL
One ddNTP (e.g., 0.1 mM ddCTP)	1 μL
Reverse transcription mix (10x)	1 μL
Reverse transcriptase (200 units/μL) <!>	1 μL
H_2O	1 μL

4. Incubate for 20 min at 42°C.
5. Stop the reaction with 1 μL of 0.5 M EDTA (pH 8.0).
6. Take 2.5 μL of the reaction and add 5 μL of formamide loading dye. Heat for 2 min at 95°C and then fractionate on a 7% denaturing acrylamide gel. Dry the gel and visualize with a phosphorimager or by autoradiography. See Tips and Troubleshooting.

TIPS AND TROUBLESHOOTING

1. It is important to have an excess of primer in Step 1, but too much primer can obscure the signal from very short extensions. Accordingly, it is advisable to titrate the primer to the minimum excess.
2. If the gel is streaky or distorted in Step 6, phenol:chloroform-extract and ethanol-precipitate the remainder of the reaction before electrophoresis.
3. If the bands near the locations of dyes are irregular, run the samples without dye and run the dye in outside lanes for tracking.

SOLUTION AND REAGENT RECIPES

CAUTION: See the Cautions Appendix for appropriate handling of the material marked with <!>.

Annealing Buffer (5x)

Reagent	Quantity (for 1 mL)	Final concentration (5x)
Tris-HCl (1 M, pH 8.3)	80 µL	80 mM
KCl (2 M) <!>	100 µL	200 mM
H_2O	820 µL	

Store indefinitely at room temperature.

Reverse Transcription Mix (10x)

Reagent	Quantity (for 1 mL)	Final concentration (10x)
$MgCl_2$ (1 M) <!>	25 µL	25 mM
DTT (1 M) <!>	200 µL	200 mM
H_2O	775 µL	

Store for up to 1 yr at –20°C.

REFERENCES

Hayes ML, Hanson MR. 2007. Assay of editing of exogenous RNAs in chloroplast extracts of *Arabidopsis*, maize, pea, and tobacco. *Methods Enzymol* **424:** 459–482.

Keegan LP, Rosenthal JJ, Roberson LM, O'Connell MA. 2007. Purification and assay of ADAR activity. *Methods Enzymol* **424:** 301–317.

Protocol 17

Detection of RNAs by 3′-end Labeling and RNase H Digestion

This protocol provides an extremely sensitive procedure for detecting the presence of RNAs (known or unknown). It is useful only when the RNAs of interest have been purified away from bulk cellular RNA. Such enrichment can be achieved by immunoprecipitation when the RNA is present with specific proteins (e.g., snoRNAs can be isolated by virtue of their association with snoRNP-specific proteins, or small RNAs such as miRNAs and piRNAs that are bound by Argonaute proteins). Enrichment can also take advantage of specific properties of the RNA (e.g., a trimethyl-guanosine cap). The method relies on the ability of T4 RNA ligase to attach high-specific-activity [^{32}P]pCp to the terminal 3′-hydroxyl group of an RNA of interest. RNA molecules labeled in this fashion can be identified because they are cleaved by RNase H in the presence of a complementary oligodeoxynucleotide. If the RNA of interest is present, then, after digestion with RNase H, it will exhibit a shift in mobility or will be degraded, and no band will be detected. If the RNA of interest does not cleave with any known complementary oligodeoxynucleotides, then the band will be unaltered; this band could represent a new RNA and may be identified using a variety of cloning techniques or chemical methods (see Protocols 18 and 24).

MATERIALS

The recipes for the items marked with <R> are at the end of the protocol or in the Common Recipes Appendix.

CAUTION: See the Cautions Appendix for appropriate handling of materials marked with <!>.

Reagents

NaOAc

Oligodeoxynucleotide complementary to RNA of interest (50 ng/μL)

Design oligodeoxynucleotides to regions that are predicted to be single stranded; structured regions in RNA are not good substrates for RNase H digestion. The oligodeoxynucleotide should be 15–25 nucleotides long; a length of 20 nucleotides is optimal.

[^{32}P]pCp (6000 Ci/mmol; 150 μCi/μL) <!>

Reagents for phenol:chloroform <!> extraction and ethanol <!> precipitation (for Steps 4 and 9; see Chapter 2, Protocols 1 and 3) <!>

RNA of interest, deproteinized

For this method to be successful, abundant cellular RNAs such as rRNAs and tRNAs must be eliminated from the sample. Enrichment for the RNA of interest can be by size, immunoprecipitation, affinity selection (e.g., splicing complexes), or any other fractionation method. Before using this protocol, the RNA of interest must be completely deproteinized and purified. The concentration of the RNA in the sample (see Step 1) will be dependent on how the RNA is isolated or enriched.

RNase H (2 units/μL)

RNase H buffer (5x) <R>

T4 RNA ligase (1 unit/μL)

T4 RNA ligase buffer (2x) <R>

Tris-HCl (10 mM, pH 7.5)

Equipment

Electrophoresis apparatus for denaturing polyacrylamide (sequencing) gels (see Chapter 2, Protocol 12)
Ice
Micro Bio-Spin Columns with Bio-Gel P-6 in Tris buffer (Bio-Rad)
Microcentrifuge
Microcentrifuge tubes (0.5 to 1.5 mL)
Water baths (37°C and 95°C)

OVERVIEW

Enriched RNAs are labeled with [^{32}P]pCp, separated from unincorporated label, and fractioned on gels before and after digestion with RNase H in the presence and absence of specific oligodeoxynucleotides.

METHOD

1. Combine the following in a microcentrifuge tube:

*RNA of interest, deproteinized	
[^{32}P]pCp (6000 Ci/mmol; 150 µCi/µL)	2 µL
T4 RNA ligase buffer (2x)	5 µL
T4 RNA ligase (1 unit/µL)	1 µL
Total	10 µL

*The quantity of the RNA of interest should be in the nanogram to picogram range (and may be less).

2. Incubate overnight at 4°C.
3. Dilute to 50 µL with 10 mM Tris-HCl (pH 7.5) and pass through a Micro Bio-Spin Column.
4. Collect the flowthrough, dilute to 200 µL with RNase-free H_2O, add NaOAc to 0.3 M, and ethanol-precipitate (as in Chapter 2, Protocol 3).
5. Resuspend the pellet in 10 µL of H_2O. Set aside 1 µL (untreated control) for gel analysis in Step 10.
6. In one microcentrifuge tube, combine 2 µL of labeled RNA, 1 µL of 50 ng/µL complementary oligodeoxynucleotide, and 16 µL of H_2O for a total volume of 19 µL. In another tube, dilute 2 µL of labeled RNA with 17 µL of H_2O for a total volume of 19 µL.
7. Heat both samples for 1 min at 95°C and then quench on ice.
8. Add 5 µL of 5x RNase H buffer and 1 µL of 2 units/µL RNase H to each tube. Incubate for 1 h at 37°C.
9. Perform phenol:chloroform extraction and ethanol precipitation with carrier (see Chapter 2, Protocols 1 and 3). Resuspend each pellet in 2–5 µL of H_2O for gel analysis.
10. Perform gel electrophoresis with a 6%–8% denaturing polyacrylamide gel in TBE as described in Chapter 2, Protocol 12. Compare the migration of the RNA of interest before and after digestion (with and without the oligodeoxynucleotide).

    The disappearance or mobility shift of a labeled band after RNase H digestion with an oligodeoxynucleotide indicates that the RNA is present. RNAs that are resistant to RNase H after exposure to a panel of oligodeoxynucleotides may be novel and their sequence can be determined by direct methods. See Tips and Troubleshooting.

TIPS AND TROUBLESHOOTING

1. If there is no digestion, either there was no RNA complementary to the oligonucleotide or the RNA that was there had traces of inhibitory contaminants. If this occurs, reprecipitate the sample and repeat the procedure with a control sample known to have the RNA of interest.
2. Some RNAs (e.g., U6 snRNA) do not label well with [^{32}P]pCp, but other RNAs (e.g., 5S rRNA) label exceptionally well (it depends on the 3′ end of the RNA). Therefore, this method is not reliable for measuring relative amounts of different RNAs in a sample.

SOLUTION AND REAGENT RECIPES

CAUTION: See the Cautions Appendix for appropriate handling of materials marked with <!>.

T4 RNA Ligase Buffer (2x)

Reagent	Quantity (for 1 mL)	Final concentration (2x)
ATP (100 mM)	10 µL	1 mM
Tris-HCl (1 M, pH 8.0)	200 µL	200 mM
$MgCl_2$ (1 M) <!>	20 µL	20 mM
DTT (1 M) <!>	6.6 µL	6.6 mM
DMSO <!>	200 µL	20% (v/v)
Glycerol (ultrapure)	300 µL	30% (v/v)
H_2O	to 1 mL	

Store for up to 3 mo at –20°C.

RNase H Buffer (5x)

Reagent	Quantity (for 1 mL)	Final concentration (5x)
Tris-HCl (1 M, pH 7.5)	100 µL	100 mM
KCl (2 M) <!>	250 µL	0.5 M
$MgCl_2$ (1 M) <!>	50 µL	50 mM
DTT (1 M) <!>	5 µL	5 mM
Sucrose (ultrapure)	0.25 g	25% (w/v)
H_2O	to 1 mL	

Store for up to 1 mo at –20°C.

READING LIST

Nilsen T, Shambaugh J, Denker J, Chubb G, Faser C, Putnam L, Bennett K. 1989. Characterization and expression of a spliced leader RNA in the parasitic nematode *Ascaris lumbricoides* var. suum. *Mol Cell Biol* **9:** 3543–3547.

Protocol 18

Reverse Transcription–Polymerase Chain Reaction

Reverse transcription coupled to the polymerase chain reaction (RT-PCR) is commonly used to detect and quantitate levels of mRNAs, pre-mRNAs, or other types of RNA such as noncoding RNAs. The method involves using a primer annealed to the RNA of interest. For mRNA, the primer is usually a synthetic oligo(dT)$_{15-18}$, a random hexamer mixture (dN)$_6$, or a synthetic DNA oligonucleotide that is complementary to a specific transcript (a gene-specific primer). This DNA:RNA hybrid serves as a template during reverse transcription, which generates the first strand—a single-stranded cDNA copy of a portion of the target RNA molecule. Using random hexamer priming, it is possible to obtain representative cDNA copies of sequences from the entire length of the mRNAs and pre-mRNAs in a population. This cDNA can then be used as a template for PCR. Because the complexity of cellular cDNA is high and can lead to spurious priming of DNA before or during the denaturation step in the PCR procedure, it is often useful either to do a "hot-start" PCR by adding the *Taq* polymerase after the sample has been heated to >80°C or to use special hot-start thermostable polymerases, which are inactive until an extended (2–15 min) incubation at 95°C. This prevents priming of poorly matched primers (i.e., mispriming) and the formation of primer dimers during the initial phase of the PCR when the temperature ramps to 95°C. Upon addition of gene-specific primers, a specific DNA fragment corresponding to a portion of the RNA of interest is generated. The method can also be used to detect the ratio of different spliced or unspliced mRNA isoforms. This is important when, for instance, one is evaluating or validating microarray expression or spliced isoform patterns or confirming results obtained by RNA-seq experiments.

MATERIALS

The recipes for the items marked with <R> are at the end of the protocol or in the Common Recipes Appendix.

CAUTION: See the Cautions Appendix for appropriate handling of materials marked with <!>.

Reagents

AmpliTaq DNA polymerase (5 units/µL) and 10x buffer (Applied Biosystems)
Dithiothreitol (DTT) (0.1 M) <!>
dNTPs (each 10 mM)
First-strand buffer (5x) <R>
H_2O (RNase-free)
Methanol (10%) <!>/acetic acid (10%) <!> (*optional*; see Step 20)
Mineral oil (for gene-specific priming method only; see Steps 12 and 23)
[^{32}P]dCTP (3000 mCi/mmol, 20 µCi/mL) <!> (for PCR using random hexamer-primed cDNA or oligo[dT]-primed cDNA only; see Step 17)
Primers

For reverse transcription, use random hexamers ([dN]$_6$) (5 mg/mL), oligo(dT)$_{15}$ or oligo(dT)$_{18}$ (100 pmol/µL) or one gene-specific primer (the 3′ primer is used). For PCR, use two gene-specific primers (one 5′ and one 3′). For notes about primer design, see discussion.

RNasin (40 units/µL) (Promega)
SuperScript II reverse transcriptase (20 units/µL) (Invitrogen)

SuperScript II is an engineered version of a Moloney murine leukemia virus (M-MuLV) reverse transcriptase. The recombinant enzyme is purified to near homogeneity from *E. coli.* It has reduced RNase H activity and increased thermal stability. The enzyme can be used to synthesize first-strand cDNA at higher temperatures than conventional M-MuLV reverse transcriptase, providing increased specificity, higher yields of cDNA, and more full-length cDNA product. It can generate cDNAs up to 12.3 kb in length.

Total RNA from sample of interest

The RNA must be treated with DNase to remove contaminating genomic DNA in total RNA samples. Genomic DNA can interfere with RT-PCR assays by producing PCR products in the absence of reverse transcriptase. DNase treatment should be performed and then the RNA is purified and quantitated, depending on the method.

Equipment

Bucket with ice

Gel electrophoresis apparatus, vacuum gel drier, and detection equipment as required (see Steps 19, 20, and 25)

PCR tubes (200 μL)

Thermal cycler

OVERVIEW

RT-PCR proceeds in two stages. First, oligo$(dT)_{15}$, $(dN)_6$, or a gene-specific primer is annealed to the target mRNA population, and reverse transcriptase and dNTPs are added to allow first-strand cDNA synthesis (reverse transcription, or RT). Second, new primers, fresh dNTPs, and *Taq* polymerase are added to generate a PCR to amplify the cDNA of interest. These products are then run on either an agarose gel or a native polyacrylamide gel and analyzed.

Section I (Steps 1–5) uses oligo$(dT)_{15}$ priming, Section II (Steps 6–10) uses random hexamer priming $(dN)_6$, and Section III (Steps 11–15) uses gene-specific priming for the RT reaction for first-strand cDNA synthesis. In the subsequent PCR steps (Sections IV and V), [α-^{32}P]dCTP is added to radioactively label the cDNA products, and the products are then analyzed by native acrylamide gel electrophoresis and quantitative phosphorimaging.

The decision of whether to use oligo$(dT)_{15}$ or $(dN)_6$ priming will depend on the goal. If validating or detecting spliced mRNA isoforms and if the intron(s) of interest is not located near the 3′ end of the transcript, then random-primed cDNA synthesis will be essential. For certain applications, it may be preferable to perform RT using a mixture of oligo(dT) and random hexamer primers in the same reaction tube. For instance, it may be useful when performing RT-PCR to detect alternate splicing patterns for a gene with introns near the 3′ end (close to the oligo[dT] primer on the poly[A] tail) and near the 5′ end (close to the random hexamer primers) (Figs. 3-18 and 3-19). In the protocol by Castle et al. (2003), random hexamer priming was optimized to equally represent all portions of an mRNA as cDNA, whereas oligo(dT) priming obviously biases the cDNA products toward the 3′ end of the gene. That said, a novel mutant and more processive reverse transcriptase called ArrayScript (Ambion), an engineered version of M-MuLV that improves the yield of full-length cDNA using oligo(dT) mRNA priming of reverse transcription, is often used for cDNA synthesis for microarray analysis.

Often, the use of gene-specific primers produces much less background and reduces nonspecific PCR amplification products. For first-strand cDNA synthesis using a gene-specific primer, follow Section III (Steps 11–15) and then proceed to Section V (Steps 21–25). This procedure can be also be used with either oligo$(dT)_{15}$ or $(dN)_6$, but make sure to add 50 pmol of both the upstream and the downstream primers.

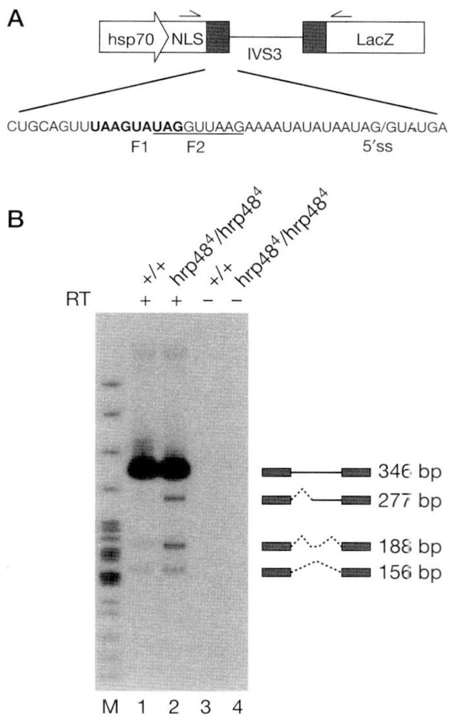

FIGURE 3-18. RT-PCR assay to detect alternative splicing isoforms. (*A*) Transgenic *Drosophila* strains were used to test the effects of mutations in the splicing factor *hrp48* on P-transposable-element third-intron splicing using an hsp70-P-element splicing reporter (Hammond et al. 1997). (*B*) Wild-type and mutant (lanes *1* and *2*) samples were subjected to RT with gene-specific primers and [^{32}P]dCTP in the PCR step and analyzed by native polyacrylamide gel electrophoresis. Control samples prepared without reverse transcriptase (–RT) were run in parallel (lanes *3* and *4*). This control indicates whether there is genomic DNA contamination. (Reprinted, with permission, from Hammond et al. 1997 [American Society for Microbiology].)

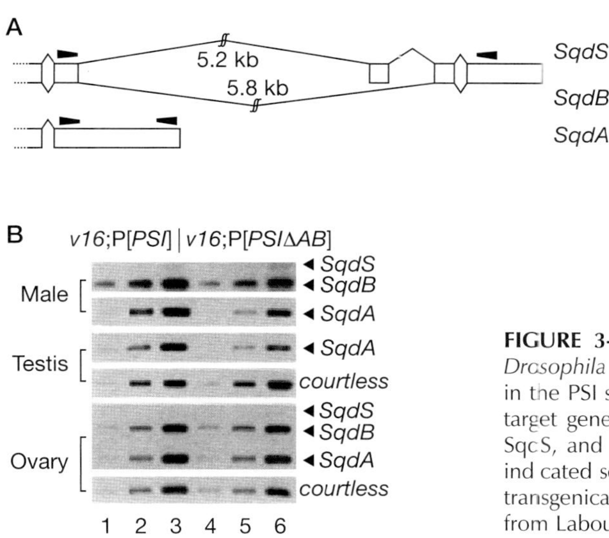

FIGURE 3-19. RT-PCR assay to detect alternative splicing in *Drosophila* PSI mutant flies. Detection of the effects of mutations in the PSI splicing factor gene on the splicing pattern of the PSI target gene squid (*sqd*). (*A*) Primer pairs specific for the SqdB, SqdS, and SqdA isoforms are shown at top. (*B*) RNA from the indicated sex or tissue was analyzed from mutant (lanes *1–3*) or transgenically rescued (lanes *4–6*). (Reprinted, with permission, from Labourier et al. 2002.)

METHOD

Method 1: Reverse Transcription (RT): cDNA Priming and Synthesis

I. Priming with Oligo(dT)$_{15}$

1. Program a thermal cycler with the following program:

Step 1	10 min	70°C
Step 2	at least 2 min	4°C
Step 3	60 min	42°C
Step 4	hold	4°C

2. Mix the following in a 200-µL PCR tube:

Total RNA	5 µg
Oligo(dT$_{15}$) or oligo(dT$_{18}$) (100 pmol/µL)	1 µL
dNTPs (each 10 mM)	1 µL
H_2O (RNase-free)	to 12 µL

3. Run the first two steps of the thermal cycler program from Step 1. Let the reaction cool for at least 2 min at 4°C and halt the program. Move the tube to an ice bucket while adding the reverse transcriptase reagents in Step 2.

4. Add the following components to the PCR tube:

First-strand buffer (5x)	4 µL
DTT (0.1 M)	2 µL
RNasin (40 units/µL)	1 µL
SuperScript II reverse transcriptase (20 units/µL)	1 µL

For multiple reactions, these components can be prepared as a master mix and then 8 µL of the master mix can be added to each tube.

5. Return the tube to the thermal cycler and allow the thermal cycler program to proceed through completion.

Proceed immediately to Section IV (Step 16). Otherwise, store the cDNA indefinitely at –20°C.

II. Priming with Random Hexamers

6. Program a thermal cycler with the following programs.

i. The heat/snap-cool protocol is as follows:

Step 1	5 min	65°C
Step 2	hold	0°C

ii. The RT program is as follows:

Step 1	10 min	4°C
Step 2	10 min	15°C
Step 3	20 min	42°C
Step 4	15 min	70°C
Step 5	hold	4°C

This program maximizes cDNA synthesis by allowing a slow increase in the temperature to 42°C, which allows the random hexamers to be efficiently annealed and extended.

7. Mix the following in a 200-μL PCR tube:

Total RNA	5 μg
$(dN)_6$ (5 mg/mL)	1 μL
dNTPs (each 10 mM)	1 μL
H_2O (RNase-free)	to 12 μL

8. Run the heat/snap-cool protocol on the thermal cycler. Remove the tube and place in an ice bucket.

9. Add the following components to the tube on ice:

First-strand buffer (5x)	4 μL
DTT (0.1 M)	2 μL
RNasin (40 units/μL)	1 μL
SuperScript II reverse transcriptase (20 units/μL)	1 μL

For multiple reactions, these components can be prepared as a master mix and then 8 μL of the master mix can be added to the tube.

10. Return the tubes to the thermal cycler and run the reverse transcription program on the thermal cycler.

Proceed immediately to Section IV (Step 16). Otherwise, store the cDNA indefinitely at –20°C.

III. Priming with Gene-specific Primers

11. Program a thermal cycler with the following programs.

i. The following program will anneal the primer to the template DNA:

Step 1	30 min	80°C to 22°C
Step 2	10 min	22°C
Step 3	hold	4°C

ii. The following program will extend the primers:

Step 1	10 min	4°C
Step 2	5 min	4°C to 15°C
Step 3	10 min	15°C
Step 4	5 min	15°C to 22°C
Step 5	10 min	22°C
Step 6	5 min	22°C to 37°C
Step 7	10 min	37°C
Step 8	5 min	37°C to 42°C
Step 9	60 min	42°C
Step 10	5 min	95°C
Step 11	hold	4°C

12. Mix the following in a 200-μL PCR tube:

Total RNA	1–2 μg
First-strand buffer (5x)	2 μL
Gene-specific primer (3′ only)	25 pmol
RNasin (40 units/μL)	0.3 μL
H_2O (RNase-free)	to 10 μL

Adjust the amount of total RNA according to the relative abundance of the transcript. Overlay the reaction with 30 μL of mineral oil if using a thermal cycler without a heated lid.

13. Place the reaction in the thermal cycler and run the annealing program from Step 11.i. After the program is complete, remove the reaction from the thermal cycler and immediately place it on ice.

14. Add the following components to the tube on ice:

First-strand buffer (5X)	3 µL
RNasin (40 units/µL)	0.3 µL
dNTPs (each 10 mM)	1.5 µL
SuperScript II reverse transciptase (20 units/µL)	1.5 µL
H_2O (RNase-free)	to 15 µL

15. Place the reaction in the thermal cycler and run the cDNA synthesis program from Step 11.ii. After the program is complete, remove the reaction from the PCR machine.

    Mineral oil can be removed by extracting with chloroform using standard procedures. To use the cDNA for PCR, proceed immediately to Section V (Step 21). Otherwise, store the cDNA at –20°C.

Method 2: Polymerase Chain Reaction and Detection of Products

IV. PCR Using Oligo(dT)-primed or Random Hexamer-primed cDNA from Section I or II

16. Program a thermal cycler with the following program:

No. of cycles	Denaturation	Annealing	Polymerization
1	3 min at 94°C		
15–35	30 sec at 94°C	30 sec at 55°C	30 sec at 72°C
1			5 min at 72°C
Hold at 4°C			

    For amplification products longer than 500 nucleotides, increase the polymerization time by 30 sec for each additional 500 nucleotides. The number of cycles will need to be in the linear range of amplification to make this assay "semiquantitative." See Discussion.

17. Mix the following:

H_2O	40.5 µL
AmpliTaq buffer (10X)	5 µL
dNTPs (each 10 mM)	1 µL
Gene-specific 5′ primer (15 pmol/µL)	1 µL
Gene-specific 3′ primer (15 pmol/µL)	1 µL
[^{32}P]dCTP (3000 mCi/mmol, 20 µCi/mL)	0.5 µL
AmpliTaq DNA polymerase (5 units/µL)	0.5 µL
cDNA from Sections I (Step 5) or II (Step 10)	0.5 µL

18. Place the reaction in the thermal cycler and run the program from Step 16.

19. Load 5 µL of each reaction on a 5%–7.5% native polyacrylamide gel (0.8:30 bis-acrylamide: acrylamide) in 1X TBE.

20. Fix the gel in 10% methanol/10% acetic acid and dry on a vacuum gel drier. Subject the gel to quantitation by phosphorimaging.

    If the gel is thin (<0.75 mm), gel fixation is optional. See Tips and Troubleshooting.

V. PCR Using cDNA Produced with Gene-specific Primers from Section III

Below we outline a manual hot-start procedure; see Discussion for additional advice.

21. Program a thermal cycler with the desired parameters for the amplification reaction. Use an annealing temperature that is 5°C less than the melting temperature (T_m) of the primers.

No. of cycles	Denaturation	Annealing	Polymerization
1	3 min at 94°C		
20–30	30 sec at 94°C	1 min at 45°C–58°C	30 sec at 72°C
1			5 min at 72°C
Hold at 4°C			

22. Set up the cDNA amplification reaction as follows:

AmpliTaq buffer (10×)	2 µL
Gene-specific primer (3′) used in Section III (Step 12)	25 pmol
Other gene-specific primer (5′) that used in Section III (Step 12)	50 pmol
cDNA from Step 15	5–10 µL
H_2O	to 20 µL

Typically, it is not necessary to add more dNTPs to the PCR because there are plenty in the RT reaction.

23. Place it in the thermal cycler with the cap open.
24. Begin the program. Once the thermal cycler has reached ~80°C, add 5 µL of the following mix:

AmpliTaq buffer (10×)	0.5 µL
AmpliTaq DNA polymerase (5 units/µL)	0.25 µL
H_2O	4.25 µL

This is called a manual "hot start" and helps to reduce nonspecific annealing of the primers. Make sure that the mix has been added to all samples before the annealing/extension cycles begin. See Discussion.

25. Load 5–10 µL of each amplification reaction on a 1.8% agarose gel and detect using standard procedures. See Tips and Troubleshooting.

TIPS AND TROUBLESHOOTING

1. It is critical that the two PCR primers be T_m-matched (Sambrook et al. 1989). When this procedure does not work, it is often due to poor primer design. It is worth spending some effort on primer design before using this protocol (see Discussion).
2. If the reactions are not visible by agarose gel electrophoresis in Sections IV or V, try blotting to a membrane and hybridizing with a specific ^{32}P-labeled probe. Also, try radiolabeling the PCR product(s) during the PCR. If using this method, analysis by native polyacrylamide gel electrophoresis will need to be performed as described in Section IV.
3. Check primers for structure or potential homodimer formation.

DISCUSSION

Advice on Primer Design

The success of an RT-PCR (or any PCR) depends critically on primer design. Below are some suggestions for designing RT-PCR primers (adapted from www1.qiagen.com/literature/render.aspx?id= 23570).

1. Typical primers for RT-PCR are 18–30 nucleotides long and have a G + C content of 40%–60%.
2. It may be helpful to access web-based tools such as Primer3 (Rosen and Skaletsky 2000) to aid in primer design (http://biotools.umassmed.edu/bioapps/primer3_www.cgi).
3. A simple formula that can be used to estimate the melting temperature (T_m) is

$$T_m = 2°C \times (A + T) + 4°C \times (G + C).$$

4. If at all possible, make an effort to design primer pairs with similar or identical T_m values.

5. Optimal annealing temperatures are empirical, i.e., the optimal annealing temperature for a given primer pair may be either above or below the calculated T_m. As an initial starting point, try an annealing temperature at 5°C below the T_m.
6. Make every attempt to avoid complementarity of the 3′ terminal two or three bases on the primer pairs so as to reduce the formation of primer dimers. Use available software programs for this.
7. Make sure to avoid base sequence mismatches between the very 3′ end of the primer and the template cDNA target sequence.
8. Make sure to avoid putting extended runs of three or more Gs or Cs at the 3′ end of the primer.
9. If possible, avoid a T at the extreme 3′ end of the primer because it has been observed that primers with a T at the 3′ end have a potential for mispriming due to mismatch tolerance.
10. Avoid any extended complementary sequences within a single primer and between the two primers of a primer pair.
11. For single-stranded DNA oligonucleotide primers, the spectrophotometric conversion for primers is as follows: 1 A_{260} unit = 33 μg/mL or 20 A_{260} units = 660 μg/mL.
12. Molar conversions:

Primer length (nucleotides)	pmol/μg	20 pmol
18	168	119 ng
20	152	132 ng
25	121	165 ng
30	101	198 ng

Hot-start PCR

Because primer annealing and mispriming of spurious cDNA or the formation of "primer dimers" can occur during the initial phases of a PCR in which the temperature is raised from ambient to 95°C, it is useful in RT-PCRs to perform a manual "hot-start" reaction in which the *Taq* (or other thermostable) DNA polymerase is added after the reaction tubes reach at least 80°C. Alternatively, it is possible to use a commercial polymerase, available from many companies. These polymerases are a modified form of the recombinant 94-kDa *Taq* DNA polymerase that is provided in an inactive state with no polymerase activity at ambient temperatures. HotStarTaq DNA Polymerase allows for a quick PCR setup because all reaction components can be combined at room temperature, and it prevents the formation of misprimed products and primer dimers at low temperatures. HotStarTaq DNA Polymerase is activated by a 15-min, 95°C incubation step that can easily be incorporated into existing thermal cycling programs. A hot-start protocol usually provides high PCR specificity and often increases the yield of the specific PCR product.

Determining Exponential Range in PCR

The exponential phase of amplification occurs during PCR when the PCR products are accumulating at a constant (exponential, the so-called "linear" phase) rate (see Fig. 3-20) and when the reaction components are still in excess over the product(s). To empirically determine the exponential range, typically a PCR master mix is prepared that includes 5–10 μL/Ci of [α-^{32}P]dCTP (10 mCi/mL; 400–800 Ci/mmol) in addition to the normal PCR components and cDNA template. The master mix is then split into 10 different aliquots that are then separately subjected to PCR. Samples are removed from the thermal cycler at the indicated cycle numbers and resolved by electrophoresis on a native 5% polyacrylamide gel in TBE buffer (see Fig. 3-21). The gel is then dried and the radiolabeled cDNA products are quantitated by phosphorimaging. The cycle number (*x* axis) is plotted versus the log of the ^{32}P signal (*y* axis) and a straight line is obtained for samples in the exponential (linear) phase of the amplification curve (see Fig. 3-20).

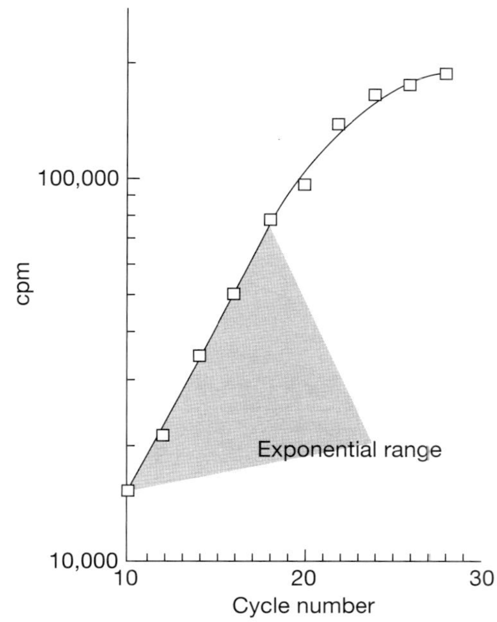

FIGURE 3-20. Plot of cycle number versus log of the phosphorimaging signal. The exponential (so-called linear) range of the curve is indicated. (Redrawn from www.ambion.com/ [www.ambion.com/techlib/append/pcr_linearrange.html/].)

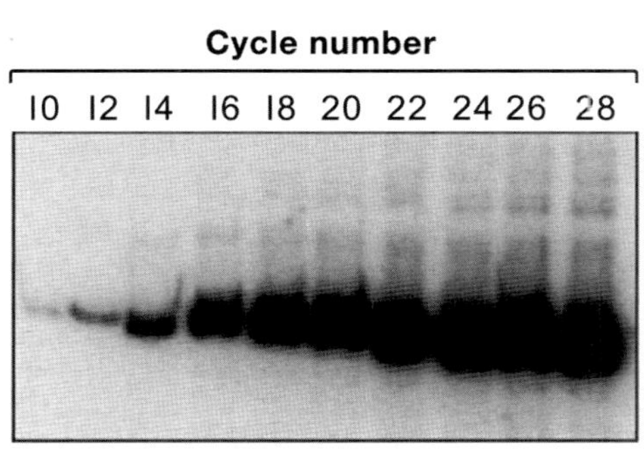

FIGURE 3-21. Phosphorimaging of PCRs at different cycle numbers. This gel using ^{32}P-labeled RT-PCRs with samples taken at different cycle numbers indicates the increase in product formed and the lack of linearity in the accumulation of cDNA product with increasing cycle number. (Reprinted from www.ambion.com/[www.ambion.com/techlib/append/pcr_linearrange.html/].)

SOLUTION AND REAGENT RECIPE

CAUTION: See the Cautions Appendix for appropriate handling of material marked with <!>.

First-strand Buffer (5x)

Reagent	Quantity (for 1 mL)	Final concentration
Tris-HCl (1 M, pH 8.3)	250 µL	0.25 M
KCl (1 M) <!>	375 µL	375 mM
$MgCl_2$ (1 M) <!>	15 µL	15 mM
DTT (0.5 M) <!>	200 µL	0.1 M
H_2O	160 µL	

Store for 6–12 mo at –20°C.

REFERENCES

Castle J, Garrett-Engele P, Armour CD, Duenwald SJ, Loerch PM, Meyer MR, Schadt EE, Stoughton R, Parrish ML, Shoemaker DD, et al. 2003. Optimization of oligonucleotide arrays and RNA amplification protocols for analysis of transcript structure and alternative splicing. *Genome Biol* **4:** R66.

Hammond L, Rudner DZ, Kanaar R, Rio DC. 1997. Mutations in the *hrp48* gene, which encodes a *Drosophila* heterogeneous nuclear ribonucleoprotein particle protein, cause lethality, developmental defects and affect P-element third-intron splicing in vivo. *Mol Cell Biol* **17:** 7260–7267.

Labourier E, Blanchette M, Feiger JW, Adam MP, Rio DC. 2002. The KH-type RNA-binding protein PSI is required for *Drosophla* viability, male fertility, and cellular mRNA processing. *Genes Dev* **16:** 72–84.

Rozen S, Skaletsky H. 2000. Primer3 on the WWW for general users and for biologist programmers. *Methods Mol Biol* **132:** 365–386.

Sambrook J, Fritsch EF, Maniatis T. 1989. *Molecular cloning: A laboratory manual*, 2nd ed. Cold Spring Harbor Laboratory Press, Cold Spring Harbor, NY.

READING LIST

Dieffenbach CW, Dveksler GS., eds. 2003. *PCR primer: A laboratory manual*, 2nd ed, pp. 163–219. Cold Spring Harbor Laboratory Press, Cold Spring Harbor, NY.

WWW RESOURCES

http://biotools.umassmed.edu/bioapps/primer3_www.cgi Primer3: WWW primer tool. Pick primers from a DNA sequence. University of Massachusetts Medical School, Worcester, MA.

www.ambion.com/techlib/append/per_linearrange.html/ Determing exponential range in PCR. Applied Biosystems.

Protocol 19

Basic Quantitative PCR Using Real-time Fluorescence Measurements

This protocol uses quantitative PCR to measure the number of DNA molecules containing a specific contiguous sequence in a sample of interest (e.g., genomic DNA or cDNA generated by reverse transcription as described in Protocol 18). The sample is subjected to PCR amplification, and theoretically, during each cycle, two new duplex DNA molecules are produced for each duplex DNA molecule present in the sample. The progress of the reaction during PCR is evaluated by measuring the fluorescence of dsDNA–dye complexes in real time. In the early cycles, DNA duplication is not detected because inadequate amounts of DNA are made. At a certain threshold cycle, DNA–dye complexes double each cycle for 8–10 cycles, until the DNA concentration becomes so high and the primer concentration so low that the reassociation of the product strands blocks efficient synthesis of new DNA and the reaction plateaus. There are two types of measurements: (1) the relative change of the target sequence compared to a reference sequence and (2) the determination of molecule number in the starting sample. The first requires a reference sequence, and the second requires a sample of the target sequence with known numbers of the molecules of sequence to generate a standard curve. By identifying the threshold cycle at which a sample first begins to accumulate DNA–dye complexes exponentially, an estimation of the numbers of starting molecules in the sample can be extrapolated.

MATERIALS

CAUTION: See the Cautions Appendix for appropriate handling of material marked with <!>.

Reagents

Master mix (2x) (e.g., Eurogentec qPCR MasterMix Plus for SYBR Green I <!>–Low ROX or Bio-Rad iQ SYBR Green Supermix)

The master mix has nucleoside triphosphates, dye, buffer, magnesium, and heat-stable DNA polymerase. Avoid multiple freeze–thaw cycles; it may be helpful to aliquot and freeze in useful volumes.

Sample of DNA (e.g., genomic DNA or cDNA) containing target sequence of interest

Equipment

Gel electrophoresis apparatus (*optional*; see Step 10)

Microtiter plate seals with proper optical properties for real-time PCR (e.g., Bio-Rad Optical Sealing Tape)

Microtiter plates (96 well, 0.2 mL/well) with proper optical properties for real-time PCR (e.g., Bio-Rad iQ Real-Time PCR Plates)

Primer3

Use the Primer3 web interface at http://frodo.wi.mit.edu/; alternatively, download the program from http://primer3.sourceforge.net/.

Real-time thermal cycler

A special thermal cycler equipped with cameras and optics for measuring fluorescence in microtiter wells is necessary for this procedure. We will not detail the specifics of any one machine. Numerous machines are available and each has its own quirks and software solutions. Check with the instrument manufacturer to ensure proper fit and optical properties of the microtiter plates and seals.

Spreadsheet program (see Step 4)

Standards derived from samples of known quantity (*optional*; see Step 5)

For example, use a purified plasmid carrying the target sequence. Because each plasmid molecule has two single-stranded template molecules and the molecular number of plasmid molecules in a sample can be determined from the mass and the molecular weight of the plasmid, the number of starting template molecules can be determined.

Text file of target sequence to be used for primer design

Alternatively, use published primer sequences if available (see Steps 1–3).

OVERVIEW

Measuring the amount of DNA (or cDNA derived from RNA) in a sample by qPCR requires some advanced planning. Primer pairs must be carefully designed. Samples including reconstructed test reactions with known numbers of molecules to generate a standard curve must be planned. Once planned, the plates must be loaded carefully with the precise components needed for each of the many reactions, and finally, the machine must be run and the data analyzed. Performed with attention, this method can provide accurate quantification of specific sequences in a complex mixture.

METHOD

Design the Primers

If published primer pairs for a target exist, it may be advisable to use the published sequences, especially if the goal is to compare results to the previous work. Otherwise, design primers as described here. A good primer pair will allow replication to proceed at its theoretical rate of two new DNA molecules for each starting DNA molecule per cycle.

1. Input a text file of the target sequence into the Primer3 program.
2. Adjust the primer selection parameters in Primer3 as follows:
   i. Select primers that are 17–22 nucleotides in length with a T_m between 56°C and 66°C (ideally, 60°C).

      Primers with a lower T_m may be necessary for some targets, but the risk of off-target products may increase.

   ii. Select primers that are located on opposite strands, complementary to the ends of the target sequence, oriented toward one another (such that their 3′ ends are closer than their 5′ ends) and ~120–200 bp apart.

      The target should be short (120–200 bp) to decrease the amount of incomplete products during the amplification.

   iii. If multiple targets with different primer sets are to be detected in the same microtiter plate, ensure more uniform reaction efficiencies by using primer sets that produce amplicons of approximately the same length and with a similar T_m.
3. Before attempting qPCR, test and optimize the primers using standard PCR to ensure that a single product of the expected size is obtained.

   For many real-time applications where the predicted T_m of the primer pair is ~60°C, a cycle program of 30 sec at 95°C, 30 sec at 55°C, and 1 min at 72°C will work. Because the reaction should be followed to completion, 40 cycles are recommended.

Plan the Reactions

4. Plan out the experiment on a spreadsheet that mirrors the 96-well plate to keep track of which wells have which reactions. Run the samples at least in triplicate.

   Complications do arise when running multiple DNA samples with multiple primer sets in triplicate on the same plate. The key to avoiding wasteful errors is organization and planning.

5. Plan to set up 15–18 reactions (5–6 dilutions, each in triplicate) for a standard curve if estimating the absolute numbers of molecules in the sample. Determine the approximate number of template molecules expected in the test samples; the standards should cover a range of 100-fold above and below these numbers (e.g., 0.01*X*, 0.1*X*, 1*X*, 10*X*, 100*X*, where *X* is the estimated number). A convenient standard can be made from a plasmid clone carrying the target sequence. If standardizing for a cDNA sample, note that the plasmid standard will give twice the number of initiating molecules on a molar basis than the cDNA because the cDNA is single stranded.

   As few as several thousand template molecules can be accurately measured per reaction. Below this number, quantification becomes unreliable because template molecules are lost to surface adsorption, and dilution errors become large when sorting few molecules. Although it may be theoretically possible to detect a single molecule by PCR, this is usually impractical. The Poisson distribution predicts that if the template is distributed to reactions at an average of one molecule per reaction, about 37% of reactions will have no molecules.

6. Become familiar with the real-time thermal cycler well before using it. Take time to set up a program that specifies the run program with optimized conditions from Step 3, melting curves, dyes used, which wells in the plate are unused, and which wells have standards for use in calculating a standard curve.

Run the Reactions

7. Turn on the real-time thermal cycler to warm it up.
8. Assemble the reactions on a microtiter plate. To each well, add the equivalent of the following:

Master mix (2x)	10 μL
Target DNA	4 μL
H_2O	2 μL
Primer 1 (5 pmol/μL)	2 μL
Primer 2 (5 pmol/μL)	2 μL
Total	20 μL

   Many vendors suggest 50-μL reactions, likely hoping that more master mix will be purchased than is necessary. It is possible that 20–25-μL reactions will work just fine if the cycler can be set for small volumes. To minimize pipetting error and avoid repeated pipetting of small volumes (<4 μL), make appropriate premixes when assembling the reactions. For example, if the same target DNA is being tested with many primer pairs, mix the DNA with the 2x master mix and H_2O, and add this premix to the primer pairs distributed on the plate. Or, if testing many different DNAs with one primer pair, mix the primers with the 2x master mix and H_2O, and add this premix to the DNA samples distributed on the plate.

9. After the plate reactions are assembled, seal the plate with microtiter plate seal and start the run using the program from Step 6.
10. Analyze the data and inspect the melting curve, making sure that the product formed in each well has a uniform melting transition at its characteristic temperatures.

    Wells that show a fluorescent product that has a heterodispersive melting profile have probably not amplified the desired target and have instead produced a complex mix of strange DNA. If desired, run the final product on a gel to confirm that amplification has been accurate. A sharp band the size of the expected target should appear. See Tips and Troubleshooting. For example of how to analyze the data, see Discussion.

TIPS AND TROUBLESHOOTING

1. If a heterogeneous, multiply peaked trace is observed in Step 10, the reaction may have produced primer dimers or multiple products instead of clean, single-peaked transitions.
2. If there is any question concerning the homogeneity of the product, the samples in question should be subjected to gel electrophoresis.

DISCUSSION

Depending on the real-time machine and its software capabilities, different presentations of data may be available. Most will calculate a threshold cycle (Ct) for each sample and will display a melting curve of the end product. Some will calculate a standard curve and apply it to an unknown, provided standards are identified with their expected molecule numbers.

Threshold Cycle

The Ct is identified by a curve-fitting equation that determines the cycle at which the reaction is detectably above background fluorescence and during which accumulation of fluorescence is progressing exponentially. The base of the exponential curve is also identified and used to estimate the replication efficiency, in the event that the theoretical twofold increase in fluorescence per cycle is not achieved.

Figure 3-22 shows the results of a quantitative PCR assay using reverse-transcribed yeast mRNA and primer pairs for two targets: *SEC65* and *RPL34B*. To a first approximation, these data show that

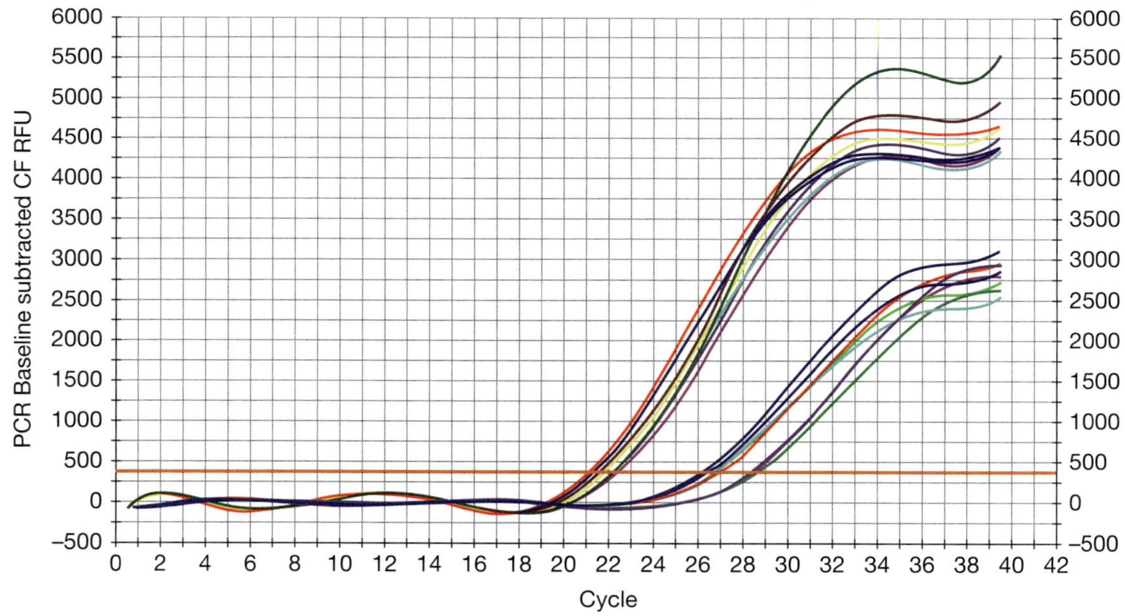

FIGURE 3-22. Reaction course and threshold cycle for a quantitative PCR with two different primer pairs using reverse-transcribed yeast mRNA. The two sets of primers measured the abundance two different yeast mRNAs, *SEC65* and *RPL34B* (left and right trace clusters, respectively), at 11 h of the meiotic gene expression program in synchronized wild-type and mutant (*mer3*) lines. The calculated threshold is presented as a horizontal orange line just below 500 fluorescence units (FU). The Ct for each reaction is the cycle at which the reaction crossed the threshold. The primer pairs were optimized and produced no fluorescence in the absence of template (not shown). The RNA sample was exhaustively treated with DNase before reverse transcription so that no contaminating DNA was present, as confirmed by reactions in which the RNA sample was used as a template.

TABLE 3-2. Cycle Threshold (Ct) Data for the Samples Described in Figure 3-22

Sample *RPL34B*	Ct	Sample *SEC65*	Ct
mer3_34-1	28.6	mer3_65-1	21.3
mer3_34-2	28.3	mer3_65-2	21.2
mer3_34-3	28.3	mer3_65-3	21.1
Ave ± S.D.	28.4 ± 0.2	Ave ± S.D.	21.2 ± 0.1
Wt_34-1	26.9	Wt_65-1	22.4
Wt_34-2	26.7	Wt_65-2	22.4
Wt_34-3	26.5	Wt_65-3	22.1
Ave ± S.D.	26.7 ± 0.2	Ave ± S.D.	22.3 ± 0.2

for many of the samples tested, the threshold cycle for *RPL34B* is about five cycles later than that for *SEC65*, indicating that the *SEC65* mRNA is about 32 times more abundant than the *RPL34B* mRNA in these samples. This estimate depends on whether the two primer sets amplify their targets with the same efficiency.

In many applications, comparisons are made between a target sequence of interest and a reference or control sequence (often, β-*actin* or *GAPDH* mRNA) whose abundance is expected to remain constant. In this case, *SEC65* mRNA was used as the reference sequence. The levels of *RPL34B* mRNA were compared to the levels of *SEC65* mRNA in mutant (*mer3*) and wild-type cells. The $2^{(-\Delta\Delta Ct)}$ method (Livak and Schmittgen 2001) can be used for such simple comparisons: The relative difference in the Ct is used to determine the difference in the abundance of an mRNA of interest (*RPL34B*) relative to a reference mRNA (*SEC65*) in two samples (here, a mutant [mer3] and a wild type). The real-time PCR machine software called the Ct data as shown in Table 3-2. Using these data,

$$\Delta\Delta Ct = (Ct_{RPL34B} - Ct_{SEC65})_{mer3} - (Ct_{RPL34B} - Ct_{SEC65})_{\text{wild type}},$$

$$\Delta\Delta Ct = (28.4 - 21.2) - (26.7 - 22.3),$$

$$\Delta\Delta Ct = 7.2 - 4.4 = 2.8.$$

Therefore, the difference in the level of *RPL34B* mRNA in the *mer3* mutant compared to the wild type was

$$\text{relative difference} = 2^{(-\Delta\Delta Ct)} = 2^{-2.8} = 0.14.$$

On the basis of these data, we would conclude that *mer3* mutant cells have 14% of the wild-type *RPL34B* mRNA level. Another way to say this is that *RPL34B* mRNA in *mer3* mutant cells is down sevenfold compared to wild type.

Melting Curve

The real-time PCR machine reads total fluorescence in the reaction well. An assumption that influences the accuracy of the results is that a single product is being amplified and that the fluorescence is due solely to target amplification. If true, the product DNA should melt with a simple profile indicative of a single molecular species, at a characteristic and reproducible melting point.

Figure 3-23 presents the melting profile of the products amplified in Figure 3.22. The *RPL34B*-derived products melted in a distinct transition at ~79°C, and the *SEC65*-derived products melted at ~83°C. These uniform melting curves indicate that, in both cases, a single species has been amplified.

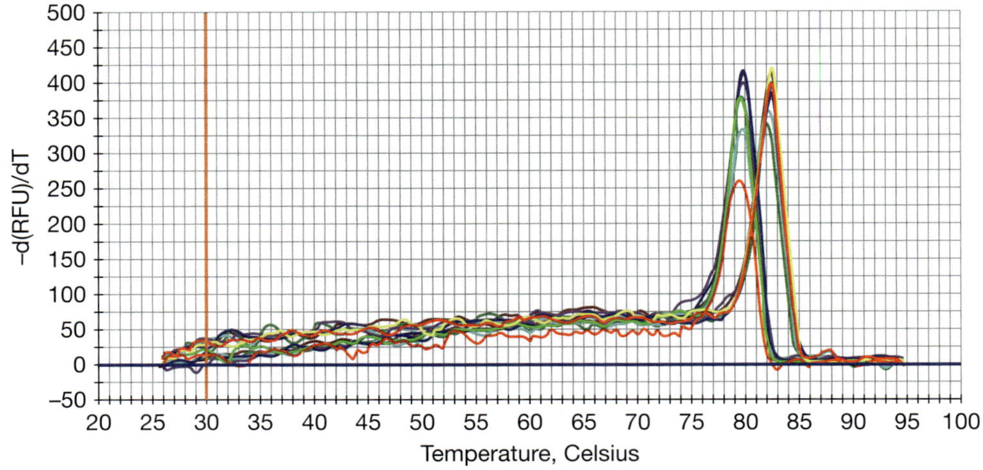

FIGURE 3-23. Melting curve for the samples described in Figure 3-22, displayed as the derivative of the change in fluorescence with the change in temperature.

REFERENCE

Livak KJ, Schmittgen TD. 2001. Analysis of relative gene expression data using real-time quantitative PCR and the $2^{-\Delta\Delta C_T}$ method. *Methods* **25:** 402–408.

Protocol 20

5′ Rapid Amplification of cDNA Ends

Characterizing the 5′ end of an RNA transcript is often the first step in defining a gene's promoter and identifying potential modifications (e.g., cytosine methylation and histone modifications) that may impact gene expression and chromatin structure. Rapid amplification of cDNA ends (RACE) remains the most commonly used technique for the characterization of individual mRNA ends. Here, we provide a general set of methods for 5′ RACE (Fig. 3-24).

MATERIALS

The recipes for the items marked with <R> are at the end of the protocol or in the Common Recipes Appendix.

CAUTION: See the Cautions Appendix for appropriate handling of materials marked with <!>.

Reagents

Alkaline phosphatase (10 units/μL) and alkaline phosphatase buffer (10x) <R>

Dithiothreitol (DTT) (100 mM) <!>

dNTPs (10 mM each)

Ethanol (100%) <!>

GlycoBlue (15 mg/mL) (*optional*; see Step 4)

PCR buffer (10x) <R>

Phenol:chloroform:isoamyl alcohol (PCA) (25:24:1) <!>

Primers (each at 10 μM) and RNA oligonucleotide anchor (20 pmol/μL)

At a minimum, one anchor primer (anchor primer 1) and two gene-specific primers (GSP-1 and GSP-2) are required for this procedure (see Fig. 3-25). In general, GSPs are used because they reduce the level of nonspecific products in subsequent PCRs. However, in some cases, we have observed that GSPs do not prime cDNA as well as random primers (6 to 9 mers), which may be used instead with modifications to the incubation temperatures (see Steps 18 and 20). For the optional nested PCR procedure (Steps 29–31), a third GSP (GSP-3) and a second anchor primer (anchor primer 2) are required. We recommend using anchor primers and RNA oligonucleotides that have been previously used for RACE and that are available in commercial kits; these can be synthesized by Integrated DNA Technologies or another oligonucleotide provider. The following are used in the FirstChoice RLM-RACE Kit (Ambion):

5′-GCUGAUGGCGAUGAAUGAACACUGCGUUUGCUGGCUUUGAUGAAA-3′ (RNA oligo)

5′-GCTGATGGCGATGAATGAACACTG-3′ (anchor primer 1)

5′-GAACACTGCGTTTGCTGGCTTTGATG-3′ (anchor primer 2)

Primer design can be performed with Primer3 (http://frodo.wi.mit.edu/) or Oligo Primer Analysis Software (http://www.oligo.net/). If possible, primer design should be performed so that ~10–20 nucleotides of known sequence are incorporated into the PCR product; this sequence can be used to confirm that the RACE product corresponds to the mRNA of interest.

QIAquick PCR Purification Kit (QIAGEN)

Reverse transcriptase buffer (5x) <R>

RNase H (5 units/μL) (*optional*; see Step 23)

This protocol was provided by Richard Davis, University of Colorado Health Sciences Center.

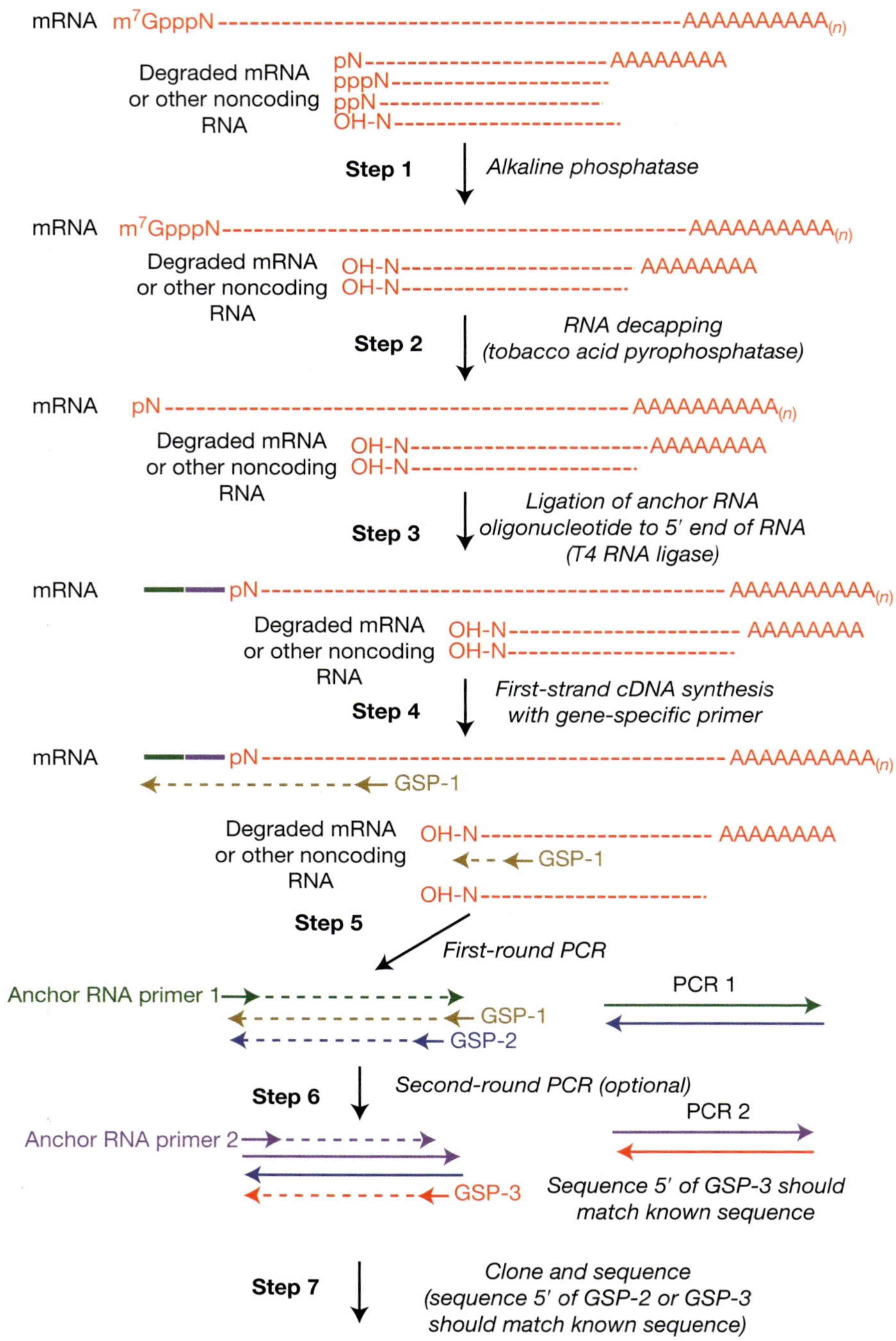

FIGURE 3-24. 5′ RACE. (Courtesy of Richard Davis, University of Colorado Health Sciences Center.)

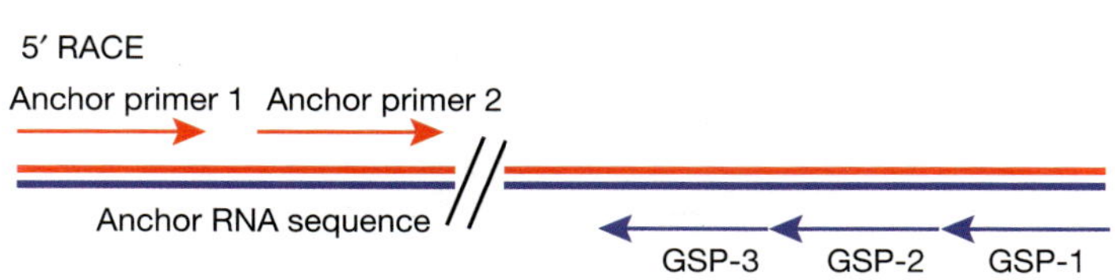

FIGURE 3-25. 5′ RACE anchor, anchor primers, and GSPs.

RNasin (40 units/μL)
Sodium acetate (3 M, pH 5.2)
SuperScript III (200 units/μL)
T4 RNA ligase (100 units/μL) and T4 RNA ligase buffer (10×) <R>
Taq DNA polymerase (2 units/μL)
TE buffer <R>
Tobacco acid pyrophosphatase (TAP) (10 units/μL) and TAP buffer (10×) <R>
Total RNA or polyadenylated RNA

Isolate total RNA from the cells, tissue, or organism of interest using TRIzol <!> (see Chapter 2, Protocol 2). If the mRNA of interest is very low in abundance, polyadenylated RNA can be isolated from the total RNA (see Chapter 2, Protocol 10) and used as the starting material. Treatment of RNA with DNase I (see Chapter 2, Protocol 11) is recommended for removing any contaminating genomic DNA from the RNA preparation.

Equipment

Bucket with ice
Dry ice <!>
Materials for agarose gel electrophoresis and detection (see Step 27 and Chapter 2, Protocol 12)
Microcentrifuge
Microcentrifuge tubes (0.5 or 1.5 mL)
Thermocycler
Water baths, incubators, or temperature blocks (37°C, 42°C [or 25°C; see Step 20], 55°C, and 65°C–70°C)

OVERVIEW

The theoretical advantage of the method described here and outlined in Figure 3-25 is that it should only generate PCR products from mRNA molecules with intact 5′ ends. In contrast to noncoding and partially degraded RNAs (pppN-, ppN-, pN-, or OH-N-RNA, where N represents the first transcribed nucleotide of the RNA), full-length eukaryotic mRNAs are posttranscriptionally capped with a guanosine that is methylated at the N7 position (m^7GpppN-RNA). In this protocol, a mixed population of RNA is treated with alkaline phosphatase. This treatment converts the 5′-phosphate groups of noncoding and partially degraded RNAs to 5′-hydroxyl groups (OH-N-RNA) but leaves the 5′ ends of capped eukaryotic mRNAs untouched. Subsequently, treatment of the RNA with TAP, an RNA-decapping enzyme, cleaves capped mRNAs to generate 5′-phosphate mRNAs (m^7GpppN → m^7Gpp + pN) but leaves RNAs with a 5′-hydroxyl group untouched. Following this step, an RNA oligonucleotide is ligated to RNAs with a 5′-phosphate group (oligo-pN) using T4 RNA ligase; the RNAs with a 5′-hydroxyl group are not substrates for the RNA ligation reaction. Following this series of reactions, only the mRNAs that were originally intact (and capped) will have an RNA oligonucleotide anchor sequence at the 5′ end of the mRNA. After reverse transcription to form first-strand cDNA, this anchor sequence is used as a priming site for PCR. When setting up a 5′ RACE experiment, consider including the following control reactions:

- Include a control in which RNA is processed through all steps except the decapping procedure (Steps 8–13). Total RNA that has not undergone the treatments described in Steps 1–24 can also be included as a negative control.
- In the reverse transcription procedure (Steps 17–24), it is recommended that a "no-RT" control reaction be performed. This will enable identification of DNA contamination of the RNA in subsequent PCRs.
- In all cases, use the primers alone, without template, as a negative control in the PCR.

- If possible, include a positive control for cDNA synthesis and PCR. The optimal control is a standard RT-PCR for a known region of the RNA for which 5′ RACE is being performed. Lack of a PCR product from this reaction indicates problems during either cDNA synthesis or PCR. The efficiency of 5′ (and 3′) RACE is typically significantly less than standard RT-PCR, and therefore this control provides a framework for the expected level of PCR product that can be obtained with two specific primers. Thus, if a PCR product for a known region of the mRNA cannot be obtained, the likelihood of getting a 5′ RACE product is greatly reduced. If a region of the RNA of interest cannot be used as a control, use a pair of primers corresponding to a region of another mRNA in the sample.

METHOD

Treat RNA with Alkaline Phosphatase

1. Assemble the following 20-µL reaction on ice in a 1.5-mL microcentrifuge tube:

RNA	10 µg of total RNA or 250 ng of polyadenylated RNA
RNasin (40 units/µL)	1 µL
DTT (100 mM)	1 µL
Alkaline phosphatase (10 units/µL)	1 µL
Alkaline phosphatase buffer (10x)	2 µL
H_2O	to 20 µL

   Larger amounts of RNA can be used for low abundance or difficult substrates. See Tips and Troubleshooting.

2. Incubate for 1 h at 37°C. See Tips and Troubleshooting.
3. Add 150 µL H_2O and 30 µL 3 M sodium acetate (pH 5.2) and then PCA-extract (see Chapter 2, Protocol 1).
4. Precipitate the RNA with 2.5 volumes of 100% ethanol (see Chapter 2, Protocol 3). Freeze for 15 min on dry ice.

   If desired, add 1 µL of 15 mg/mL GlycoBlue carrier. This provides a visible pellet after RNA precipitation and can help to prevent loss of the RNA.

5. Centrifuge in a microcentrifuge at maximum speed for 20 min at room temperature.
6. Discard the supernatant and remove any residual ethanol.
7. Dissolve the RNA in 15 µL of H_2O.

   It is advisable at this point to determine if the phosphatase treatment has caused any RNA degradation. The procedure can be stopped now; the sample should be stored frozen at –80°C.

Treat Dephosphorylated RNA with Tobacco Acid Pyrophosphatase

8. Assemble the following 20-µL reaction on ice:

RNA from Step 7	to 15 µL
TAP buffer (10x)	2 µL
TAP (10 units/µL)	1 µL
H_2O	to 20 µL

9. Incubate the reaction for 1 h at 37°C.
10. Dilute the reaction with 200 µL of TE buffer. Add 25 µL of 3 M sodium acetate (pH 5.2) and 2.5 volumes of 100% ethanol. Freeze for 15 minutes on dry ice.

11. Centrifuge in a microcentrifuge at maximum speed for 20 min at room temperature.
12. Discard the supernatant and remove any residual ethanol.
13. Dissolve the RNA in 15 µL of H_2O.

    The procedure can be stopped at this point; the sample should be stored frozen at –80°C.

Ligation of Anchor RNA Oligonucleotide to 5′ End of RNAs

14. Assemble the following 20-µL reaction on ice:

RNA from Step 13	to 15 µL
RNA oligonucleotide (20 pmol/µL)	1–2 µL (20–40 pmol)
T4 RNA ligase buffer (10×)	2 µL
T4 RNA ligase (100 units/µL)	1 µL
H_2O	to 20 µL

15. Incubate the reaction for 1 h at 37°C.
16. If desired, clean up the RNA as described in Steps 3–7.

    The procedure can be stopped at this point; the sample should be stored frozen at –80°C.

Reverse Transcription of RNA Using Gene-specific Primers or Random Hexamers

For 5′ RACE, a gene-specific primer (GSP) or random primers (6 to 9 mers) can be used to prime the RNA for cDNA synthesis. In general, using a GSP reduces the subsequent level of nonspecific PCR products and is the preferred method. However, in some cases, we have observed that GSPs do not prime cDNA as well as random primers or even oligo(dT). Consequently, we often use random primers and modify the temperatures.

17. Assemble the following reaction on ice:

5′-anchored RNA from Step 15 or 16	5–10 µg of total RNA or 125–250 ng of polyadenylated RNA
GSP-1 (10 µM)	2 pmol
H_2O	to 12 µL

    As an alternative to GSP-1, use 250–500 ng of random primers (6 to 9 mers).

18. Heat the sample for 5–10 min at 70°C and place on ice.
19. Add the following components on ice:

Reverse transcriptase buffer (5×)	4 µL
DTT (100 mM)	1 µL
RNasin (40 units/µL)	1 µL
dNTPs (10 mM each)	1 µL
SuperScript III (200 units/µL)	1 µL

20. Incubate the sample for 10 min at 42°C.

    If using random primers, incubate for 30 min at 25°C. The incubation for 10 min at 42°C is to ensure that gene-specific primers with lower melting temperatures can anneal. If using a gene-specific primer with a melting temperature greater than 55°C, omit this step.

21. Add 1 µL of 200 units/µL SuperScript III and incubate for 30 min at 55°C.
22. Heat-kill the reaction for 15 min at 70°C.
23. If desired, add 1 µL of 5 units/µL of RNase H and incubate for 20 min at 37°C.

    This step removes the RNA in the RNA:DNA duplex and in some cases can improve the efficiency of PCR.

24. Use a QIAquick PCR Purification Kit to remove the GSP-1. Elute the cDNA in 30 μL of EB buffer (supplied in kit).

    cDNAs of 100–125 nucleotides or smaller will be lost using this kit; therefore, if the RNA of interest is <100–125 nucleotides, an alternate method for removing the GSP-1 is required. The procedure can be stopped at this point; the sample should be stored frozen at –80°C.

PCR and Analytical Analysis of PCR Products

We have observed significant differences in the specificities and levels of products produced with different polymerases, reaction conditions, templates, and primers. Therefore, the optimal concentration of *Taq* polymerase, $MgCl_2$, and reaction buffer components, as well as other parameters of the PCR, may need to be determined empirically for each template-primer pair.

25. Assemble the following 50-μL reaction on ice in a PCR tube:

PCR buffer (10×)	5 μL
dNTPs (10 mM each)	1 μL
5′-anchored cDNA from Step 24	1–5 μL
Anchor primer 1 (10 μM)	1 μL
GSP-2 (10 μM)	1 μL
Taq DNA polymerase (2 units/μL)	1 μL
H_2O	to 50 μL

    The amount of cDNA is dependent on the abundance of the RNA and efficiency of previous steps.

26. Perform 25–30 PCR cycles using the annealing temperature that was recommended by the primer design software program.
27. Separate the PCR products on an agarose gel and visualize the products with ethidium bromide staining (see Chapter 2, Protocol 12). See Tips and Troubleshooting.
28. If a single PCR product or only a few PCR products are observed, gel-purify, clone, and sequence the products to confirm their identity.

    Note that if a mixture of products is observed, longer fragments will clone less efficiently; thus, it may prove judicious to separately gel-purify bands containing different sized fragments. Also, it may be necessary to sequence from both ends of the clone if the PCR product is long or if cloning is not directional.

Nested PCR (Optional)

Nested PCR can be used to provide support for the specificity of products observed in the first-round PCR or to reduce the complexity of the products observed. Note that although this additional PCR often leads to discrete products, heterogeneous mixtures of products may also occur. Empirical determination of the amount of input from the first-round PCR and the number of PCR cycles may be required.

29. Assemble the following reaction on ice:

PCR buffer (10×)	5 μL
dNTPs (10 mM each)	1 μL
QIAquick-purified PCR products or 1/10 of PCR product from Step 26	1 μL
Anchor primer 2 (10 μM)	1 μL
GSP-3 (10 μM)	1 μL
Taq DNA polymerase (2 units/μL)	1 μL
H_2O	to 50 μL

30. Perform 25–30 PCR cycles using the annealing temperature that was recommended by the primer design software program.
31. Repeat Steps 27–28 to analyze the PCR products.

TIPS AND TROUBLESHOOTING

1. Steps 1 and 2 can be performed with alkaline phosphatase from Roche at 50°C. This may improve the dephosphorylation of some RNAs.
2. In the event that no specific PCR products are evident in Step 27, perform nested PCR as described in Steps 29–31. In some cases, the nested PCR may lead to multiple products, heterogeneous mixtures, or, again, no products. When this occurs, it may be possible to identify 5′ RACE products from the first round of PCR by Southern blot analysis using the gene-specific primer 2.
3. If heterogeneous mixtures are observed in Step 27, this may be the result of too much cDNA input, genomic DNA contamination, or a primer issue. In some cases, performing PCR with a single primer can identify the problematic primer as it alone may contribute to generating the smear. Smears can often be observed without any discrete PCR product evident. This is likely the result of too much cDNA input or additional sites where the primers can anneal. Reducing the amount of input cDNA or redesigning the primers may resolve this problem.
4. If multiple products are observed in Step 27, this may be the result of nonspecific PCR products (i.e., multiple species of RNA were amplified with the primers). If this occurs, perform nested PCR as described in Steps 29–31. If multiple products or smears are again produced in the nested PCR, it may be possible to distinguish the 5′ RACE products by Southern blot analysis using the gene-specific primer 2. Alternatively, separate products from the first round of PCR on a gel, gel-purify DNA from the region of the gel corresponding to the expected product size, and then perform nested PCR as described in Steps 29–31 to enrich for the desired product with less background.

SOLUTION AND REAGENT RECIPES

CAUTION: See the Cautions Appendix for appropriate handling of materials marked with <!>.

Alkaline Phosphatase Buffer (10x)

Reagent	Quantity (for 1 mL)	Final concentration (10x)
Tris-HCl (1 M, pH 7.9)	50 µL	50 mM
NaCl (5 M)	20 µL	100 mM
$MgCl_2$ (1 M) <!>	10 µL	10 mM
DTT (1 M) <!>	1 µL	1 mM
H_2O	919 µL	

Store for up to 3 mo at −20°C.

Tobacco Acid Pyrophosphatase (TAP) Buffer (10x)

Reagent	Quantity (for 1 mL)	Final concentration (10x)
Sodium acetate (3 M, pH 6.0)	166.6 µL	500 mM
EDTA (0.5 M, pH 8.0) <!>	20 µL	10 mM
β-mercaptoethanol (100%) <!>	1 µL	0.1% (v/v)
Triton X-100 (10%) <!>	10 µL	0.1%
H_2O	802.4 µL	

Store for up to 3 mo at −20°C.

T4 RNA Ligase Buffer (10x)

Reagent	Quantity (for 1 mL)	Final concentration (10x)
Tris-HCl (1 M, pH 7.5)	500 µL	500 mM
$MgCl_2$ (1 M) <!>	100 µL	100 mM
DTT (1 M) <!>	100 µL	100 mM
ATP (100 mM)	100 µL	10 mM
Bovine serum albumin (BSA) (20 mg/mL)	30 µL	0.6 mg/mL
H_2O	170 µL	

Store for up to 3 mo at −20°C.

Reverse Transcriptase Buffer (5x)

Reagent	Quantity (for 1 mL)	Final concentration (10x)
Tris-HCl (1 M, pH 8.3)	250 µL	250 mM
KCl (2 M) <!>	187.5 µL	375 mM
$MgCl_2$ (1 M) <!>	15 µL	15 mM
H_2O	547.5 µL	

Store at −20°C.

PCR Buffer (10x)

Reagent	Quantity (for 1 mL)	Final concentration (10x)
Tris-HCl (1 M, pH 8.6)	100 µL	100 mM
KCl (2 M) <!>	250 µL	500 mM
$MgCl_2$ (1 M) <!>	15 µL	15 mM
H_2O	635 µL	

Store at −20°C.

READING LIST

Anonymous. 2005. Rapid amplification of 5′ complementary DNA ends (5′ RACE). *Nat Methods* **2:** 629–630.

Frohman MA, Dush MK, Martin GR. 1988. Rapid production of full-length cDNAs from rare transcripts: Amplification using a single gene-specific oligonucleotide primer. *Proc Natl Acad Sci* **85:** 8998–9002.

Fromont-Racine M, Bertrand E, Pictet R, Grange T. 1993. A highly sensitive method for mapping the 5′ termini of mRNAs. *Nucleic Acids Res* **21:** 1683–1684.

Scotto-Lavino E, Du G, Frohman MA. 2006. 5′ End cDNA amplification using classic RACE. *Nat Protoc* **1:** 2555–2562.

WWW RESOURCES

http://frodo.wi.mit.edu/ Primer3 (v.0.4.0) Pick primers from a DNA. Whitehead Institute for Biomedical Research, Cambridge.

http://www.oligo.net/ Oligo primer analysis software. Molecular Biology Insights, Inc.

Protocol 21

3′ Rapid Amplification of cDNA Ends

The 3′ RACE method described here is based on the original method by Frohman et al. (1988) (Fig. 3-26); see also Scotto-Lavino et al. (2006).

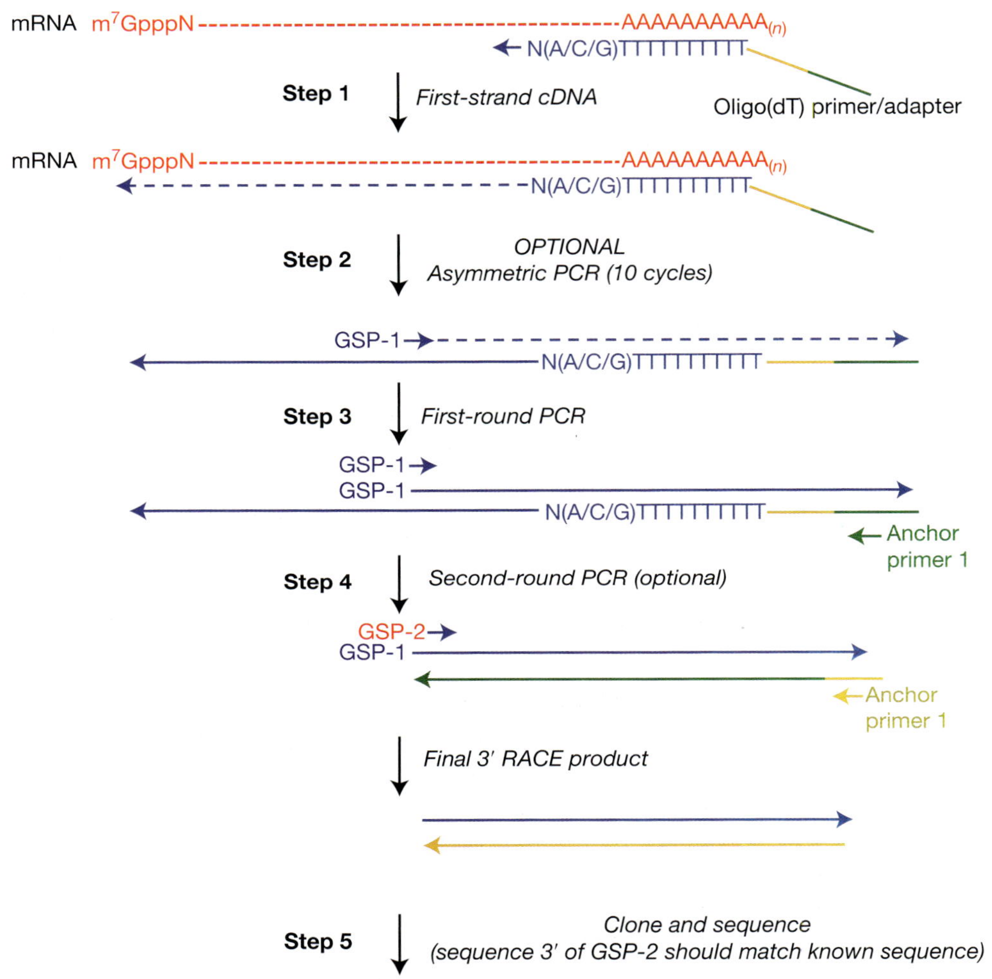

FIGURE 3-26. 3′ RACE. This illustration uses a primer that will anneal to the 5′ end of the mRNA poly(A) tail. Use of a primer that is solely T residues 3′ of the adapter region can also be used. For a detailed description, see the text. (Courtesy of Richard Davis, University of Colorado Health Sciences Center.)

MATERIALS

The recipes for the items marked with <R> are at the end of the protocol or in the Common Recipes Appendix.

CAUTION: See the Cautions Appendix for appropriate handling of material marked with <!>.

This protocol was provided by Richard Davis, University of Colorado Health Sciences Center.

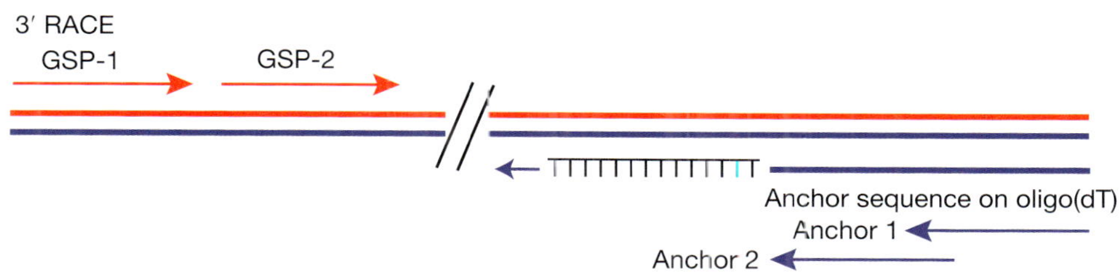

FIGURE 3-27. 3′ RACE anchor, anchor primers, and GSPs.

Reagents

Dithiothreitol (DTT) (100 mM) <!>
dNTPs (10 mM each)
PCR buffer (10x) <R>
Primers (each at 10 μM) and oligo(dT) adapter (15 pmol/μL)

The oligo(dT) adapter is a DNA oligonucleotide with 16–24 thymidine residues at the 3′ end and an anchor sequence containing a set of sequences for subsequent PCR priming at the 5′ end. Runs of thymidine residues do not have a high melting temperature. Therefore, to facilitate high specificity and matched PCR primers, the anchor sequence is used as the priming site for PCR. All mRNAs converted to cDNA within a given sample will have the anchor sequence. In addition to the oligo(dT) adapter, an anchor primer (anchor primer 1) and a gene-specific primer (GSP-1) are required (see Fig. 3-27). The GSP-1 primer is also required for the optional asymmetric PCR (Steps 9–10), but at a concentration of 2 μM instead of 10 μM. For the optional nested PCR procedure (Steps 15–17), a second gene-specific primer (GSP-2) and a second anchor primer (anchor primer 2) are required. We recommend using primers and oligo(dT) adapters that are available in kits; these can be synthesized by Integrated DNA Technologies or another oligonucleotide provider. The following are used in the FirstChoice RLM-RACE Kit (Ambion):

5′-GCGAGCACAGAATTAATACGACTCACTATAGGT$_{12}$VN-3′ (oligo[dT] adapter)

5′-GCGAGCACAGAATTAATACGACT-3′ (anchor primer 1)

5′-GAATTAATACGACTCACTATAGG-3′ (anchor primer 2)

Primer design can be performed with Primer3 (http://frodo.wi.mit.edu/) or Oligo Primer Analysis Software (http://www.oligo.net/).

QIAquick PCR Purification Kit (QIAGEN)
Reverse transcriptase buffer (5x) <R>
RNase H (5 units/μL) (*optional*; see Step 6)
RNasin (40 units/μL)
SuperScript III (200 units/μL)
Taq DNA polymerase (2 units/μL)
Total RNA or polyadenylated RNA

Total RNA is often used for 3′ RACE experiments. Isolate total RNA from the cells, tissue, or organism of interest using TRIzol <!> (see Chapter 2, Protocol 2). If the mRNA of interest is very low in abundance, polyadenylated RNA can be isolated from total RNA (see Chapter 2, Protocol 10) and used as the starting material. Treatment of RNA with DNase I (see Chapter 2, Protocol 11) is recommended to remove any contaminating genomic DNA from the RNA preparation.

Equipment

Bucket with ice
Materials for agarose gel electrophoresis and detection (see Step 13 and Chapter 2, Protocol 12)
Microcentrifuge tubes (0.5 or 1.5 mL)

PCR tubes
Thermal cycler
Water baths, incubators, or temperature blocks (37°C [*optional*; see Step 6], 42°C, 55°C, and 65°C–70°C)

OVERVIEW

First-strand cDNA is synthesized using an oligo(dT) adapter and reverse transcriptase (Steps 1–8). Subsequent PCRs are performed with a gene-specific primer (GSP-1) and anchor primer 1 (Fig. 3-27) (Steps 11–14). Asymmetric PCR is an optional procedure (Steps 9–10) that can increase the starting material for PCR and enhance product formation. Likewise, improved specificity and yield can be obtained by performing a second round of PCR using a nested gene-specific primer (GSP-2) and anchor primer 2 (Steps 15–17). When setting up a 3′ RACE experiment, consider including the following control reactions:

- In the reverse transcriptase procedure (Steps 1–8), it is recommended that a "no-RT" control reaction be performed. This will enable identification of DNA contamination of the RNA in subsequent PCRs.
- In all cases, use the primers alone, without template, as a negative control in the PCR.
- If possible, include a positive control for cDNA synthesis and PCR. The optimal control is a standard RT-PCR for a known region of the RNA for which 3′ RACE is being performed. Lack of a PCR product from this reaction indicates problems during either cDNA synthesis or PCR. The efficiency of 3′ (and 5′) RACE is typically significantly less than standard RT-PCR, and therefore this control provides a framework for the expected level of PCR product that can be obtained with two specific primers. Thus, if a PCR product for a known region of the mRNA cannot be obtained, the likelihood of getting a 3′ RACE product is greatly reduced. If a region of the RNA of interest cannot be used as a control, use a pair of primers corresponding to a region of another mRNA in the sample.

METHOD

Reverse Transcription of RNA

1. Assemble the following 20-μL reaction on ice in a 0.5-mL tube:

RNA	5–10 μg of total RNA or 125–250 ng of polyadenylated RNA
Oligo(dT) adapter (15 pmol/μL)	2 μL
H_2O	to 20 μL

Larger amounts of RNA can be used for low abundance or difficult substrates.

2. Heat for 5–10 min at 65°C–70°C.
3. Transfer the reaction to ice and add the following components:

Reverse transcriptase buffer (5×)	4 μL
DTT (100 mM)	1 μL
RNasin (40 units/μL)	1 μL
dNTPs (10 mM each)	1 μL
SuperScript III (200 units/μL)	1 μL

4. Incubate for 30 min at 42°C.

5. Add 1 µL of 200 units/µL SuperScript III and incubate for 30 min at 55°C.
6. If desired, add 1 µL of 5 units/µL RNase H and incubate for 20 min at 37°C.

   This optional step removes the RNA in the RNA:DNA duplex and in some cases can improve the efficiency of PCR.
7. Heat-kill the reaction for 15 min at 70°C.
8. Use a QIAquick PCR Purification Kit to remove the oligo(dT) adapter. Elute the cDNA in 30 µL of EB buffer (supplied in kit).

   This is a recommended step, but it is not always required. cDNA of 100–125 nucleotides or smaller will be lost during this procedure. Therefore, if the RNA of interest is <100–125 nucleotides, an alternate method for removing the oligonucleotide will be needed.

Asymmetric PCR (Optional)

9. Assemble the following 50-µL reaction on ice in a PCR tube:

PCR buffer (10x)	5 µL
dNTP (10 mM each)	1 µL
First-strand cDNA from Step 8	1–5 µL
GSP-1 (2 µM)	1 µL
Taq DNA polymerase (2 units/µL)	1 µL
H_2O	to 50 µL

   The amount of first-strand cDNA is dependent on the abundance of the RNA and efficiency of the previous steps.
10. Perform 10–15 PCR cycles using the annealing temperature that was recommended by the primer design software program. Continue with Step 11 using dilutions 1–10 and 1–40 of the asymmetric PCR product.

PCR and Analytical Analysis of PCR Products

We have observed significant differences in specificities and the levels of products produced with different polymerases, reaction conditions, templates, and primers. Therefore, the optimal concentration of *Taq* polymerase, $MgCl_2$, and reaction buffer components, as well as other parameters of the PCR, may need to be determined empirically for each template–primer pair.

11. Assemble the following reaction on ice:

PCR buffer (10x)	5 µL
dNTPs (10 mM each)	1 µL
First-strand cDNA from Step 8 or asymmetric PCR product from Step 10	1–5 µL
Anchor primer 1 (10 µM)	1 µL
GSP-1 (10 µM)	1 µL
Taq DNA polymerase (2 units/µL)	1 µL
H_2O	to 50 µL

   The amount of first-strand cDNA is dependent on the abundance of the RNA and efficiency of the previous steps. As an alternative to first-strand cDNA, use 25 µL of asymmetric PCR product from Step 10.
12. Perform 25–30 PCR cycles using the annealing temperature that was recommended by the primer design software program.
13. Separate the PCR products on an agarose gel and visualize the products with ethidium bromide staining. See Tips and Troubleshooting.

14. If a single PCR product or only a few PCR products are observed, gel-purify, clone, and sequence the products to confirm their identity (Chapter 2, Protocol 12).

    Note that if a mixture of products is observed, longer fragments will clone less efficiently; thus, it may prove judicious to separately gel-purify bands containing different sized fragments. Also, it may be necessary to sequence from both ends of the clone if the PCR product is long or if cloning is not directional.

Nested PCR (Optional)

Nested PCR can be used to provide support for the specificity of products observed in the first-round PCR or to reduce the complexity of the products observed. Note that although this additional PCR often leads to discrete products, smears of products may also occur. Empirical determination of the amount of input from the first-round PCR and the number of PCR cycles may be required.

15. Assemble the following reaction on ice in a PCR tube:

PCR buffer (10x)	5 µL
dNTPs (10 mM each)	1 µL
QIAquick-purified PCR products or dilution of PCR product from Step 8	1 µL
Anchor primer 2 (10 µM)	1 µL
GSP-2 (10 µM)	1 µL
Taq DNA polymerase (2 units/µL)	1 µL
H_2O	to 50 µL

16. Perform 25–30 PCR cycles using the annealing temperature that was recommended by the primer design software program.

17. Repeat Steps 13–14 to analyze the PCR products.

TIPS AND TROUBLESHOOTING

In situations where smears are observed for the PCR product, this may be the result of too much cDNA input, genomic DNA contamination, or a primer issue. In some cases, performing the PCR with a single primer can identify the problematic primer as it alone may contribute to generating the smear.

SOLUTION AND REAGENT RECIPES

Reverse Transcriptase Buffer (5x)

Reagent	Quantity (for 1 mL)	Final concentration (10x)
Tris-HCl (1 M, pH 8.3)	250 µL	250 mM
KCl (2 M) <!>	187.5 µL	375 mM
$MgCl_2$ (1 M) <!>	15 µL	15 mM
H_2O	547.5 µL	

Store at −20°C.

PCR Buffer (10×)

Reagent	Quantity (for 1 mL)	Final concentration (10×)
Tris-HCl (1 M, pH 8.6)	100 μL	100 mM
KCl (2 M) <!>	250 μL	500 mM
$MgCl_2$ (1 M) <!>	15 μL	15 mM
H_2O	635 μL	

Store at −20°C.

REFERENCES

Frohman MA, Dush MK, Martin GR. 1988. Rapid production of full-length cDNAs from rare transcripts: Amplification using a single gene-specific oligonucleotide primer. *Proc Natl Acad Sci* **85:** 8998–9002.

Scotto-Lavino E, Du G, Frohman MA. 2006. 3′ End cDNA amplification using classic RACE. *Nat Protoc* **1:** 2742–2745.

Protocol 22

RACE on Uncapped or Nonpolyadenylated RNAs

A variety of noncoding RNAs of interest may be uncapped and lack a poly(A) tail. RACE can also be performed on these RNAs as long as some minimal sequence information is known. For 5′ RACE on uncapped RNAs (pppN-, ppN-, pN-, or OH-N-RNA), the phosphates can be removed with alkaline phosphatase (as described in the 5′ RACE protocol), and the RNAs can then be phosphorylated using T4 polynucleotide kinase. This generates RNAs with a single 5′ monophosphate that is the substrate for the ligation of an RNA oligonuleotide to the RNA as an anchor (as described in the 5′ RACE protocol). Subsequent 5′ RACE steps are the same as those described for 5′ RACE (Protocol 20). For 3′ RACE on RNAs without a polyadenylated tail, the tail can be added enzymatically using poly(A) polymerase. Subsequent 3′ RACE steps are the same as those described above.

MATERIALS

The recipes for the items marked with <R> are at the end of the protocol or in the Common Recipes Appendix.

CAUTION: See the Cautions Appendix for appropriate handling of materials marked with <!>.

Reagents

Alkaline phosphatase (10 units/μL)
Alkaline phosphatase buffer (10x) <R>
ATP (10 mM)
DTT (100 mM) <!>
EDTA (0.5 M, pH 8) <!>
Ethanol (100%) <!>
$MgCl_2$ (25 mM) <!>
OptiKinase (10 units/μL)
OptiKinase buffer (10x) <R>
Poly(A) polymerase (10 units/μL)
Poly(A) polymerase buffer (5x) <R>
RNA (10 μg of total RNA or 250 ng of polyadenylated RNA)
RNase inhibitor

Equipment

Bucket with ice
Freezer
Incubator
Microcentrifuge
Microcentrifuge tubes (0.5 or 1.5 mL)
PCR tubes
Water baths or incubator (37°C)

OVERVIEW

This protocol is used to generate RNAs with either a 5′ monophosphate or 3′ poly(A) tail to be used as substrates in 5′ or 3′ RACE.

This protocol was provided by Richard Davis, University of Colorado Health Sciences Center.

METHOD

Dephosphorylation of RNA for 5′ RACE (see Protocol 20)

1. Assemble the following on ice in a 1.5-mL microcentrifuge tube:

RNA (10 µg of total RNA)	up to 15 µL
Alkaline phosphatase buffer (10×) (provided with enzyme)	2 µL
RNase inhibitor	1 µL
DTT (100 mM)	1 µL
Alkaline phosphatase	1 µL
H_2O	to a total volume of 20 µL

Incubate for 1 h at 37°C and purify RNA as described in Protocol 20, Steps 3–6. Resuspend RNA in 10–20 µL of H_2O.

Phosphorylation of Dephosphorylated RNAs for 5′ RACE

2. Assemble the following 50-µL reaction on ice in a PCR tube:

RNA (5 µg of RNA, dephosphorylated in Step 1)	up to 32 µL
OptiKinase buffer (10×) (provided with enzyme or see Recipes)	10 µL
RNase inhibitor	1 µL
100 mM DTT	1 µL
10 mM ATP	5 µL
OptiKinase (10 units/µL)	1 µL
H_2O	to a total volume of 50 µL

3. Incubate for 1 h at 37°C.
4. Purify by one of the methods described for RNA purification in the 5′ RACE protocol (Protocol 20).
5. Proceed to the 5′ RACE protocol starting at the ligation step (Step 14).

A-tailing of RNA for 3′ RACE

6. Assemble the following reaction on ice in a 1.5-mL microcentrifuge tube:

RNA (5–10 µg of RNA)	up to 27 µL
Poly(A) polymerase buffer (5x) (provided with enzyme or see <R>)	10 µL
RNase inhibitor (40 units/µL)	1 µL
DTT (100 mM)	1 µL
ATP (10 mM)	5 µL
$MgCl_2$ (25 mM)	5 µL
Poly(A) polymerase (2–3 units/µL)	1 µL
H_2O	to a total volume of 50 µL

7. Incubate for 1 h at 37°C.
8. Add EDTA to 5 mM.
9. Purify RNA by one of the methods described for RNA purification in the 5′ RACE protocol (Protocol 20). Proceed to Step 1 in the 3′ RACE protocol (Protocol 21).

TIPS AND TROUBLESHOOTING

The tail length under these conditions should be ~25–50 bases. In cases where the RNA of interest does not tail efficiently, increase the amount of poly(A) polymerase.

SOLUTION AND REAGENT RECIPES

CAUTION: See the Cautions Appendix for appropriate handling of materials marked with <!>.

Alkaline Phosphatase Buffer (10x)

Reagent	Quantity (for 1 mL)	Final concentration (10x)
Tris-HCl (1 M, pH 7.9)	50 µL	50 mM
NaCl (5 M)	20 µL	100 mM
$MgCl_2$ (1 M) <!>	10 µL	10 mM
DTT (1 M) <!>	1 µL	1 mM
H_2O	919 µL	

Store for up to 3 mo at −20°C.

Poly(A) Polymerase Buffer (5x)

Reagent	Quantity (for 1 mL)	Final concentration
Tris-HCl (1 M, pH 7.0)	100 µL	100 mM
KCl (2 M) <!>	125 µL	250 mM
$MnCl_2$ (1 M) <!>	3.5 µL	3.5 mM
EDTA (0.5 M, pH 8.0) <!>	2 µL	1 mM
Bovine serum albumin (BSA), acetylated (20 mg/mL)	25 µL	0.5 µg/µL
Glycerol (100%)	0.5 mL	50%
H_2O	244.5 µL	

Store for up to 1 yr at −20°C.

Protocol 23

RNA Sequencing by Primer Extension

Direct sequencing of RNA by primer extension is a fast and accurate method that is useful in a variety of situations. It is a valuable technique for determining the faithfulness of in vitro processing reactions (see Chapter 6), such as splicing or RNA editing. It is often used as an alternative to RT-PCR followed by cloning and DNA sequencing.

MATERIALS

The recipes for the items marked with <R> are at the end of the protocol or in the Common Recipes Appendix.

CAUTION: See the Cautions Appendix for appropriate handling of materials marked with <!>.

Reagents

Annealing buffer (2x) <R>
Bromophenol blue <!>
Dideoxynucleotides ddCTP, ddATP, ddTTP, and ddGTP (0.5 mM each)
Formamide/dye loading buffer <R> <!>
Kinased primer (see Protocol 1)
Methanol (10%) acetic acid (10%) <!>
Reverse transcriptase buffer (2.5x) <R>
RNA, preferably enriched or gel purified
TE buffer <R>

Equipment

Autoradiography or phosphorimaging equipment
Ice
Incubators or temperature blocks (42°C, 65°C, and 95°C)
Microcentrifuge
Microcentrifuge tubes (0.5 mL)
Phosphorimaging screen or X-ray film
Polyacrylamide gel electrophoresis apparatus

> The gel should be 0.4-mm thick and consist of 8% polyacrylamide (19:1 acrylamide:bis-acrylamide) <!> and 8 M urea <!> in 1× TBE. For details about denaturing polyacrylamide gel electrophoresis, see Chapter 2, Protocol 12.

Vacuum gel drier
Whatman 3MM paper

OVERVIEW

This method uses a DNA primer reverse transcriptase and chain-terminating nucleotide triphosphates (dideoxy) to determine the sequence of a target RNA directly.

METHOD

1. Mix the following reagents together on ice in a 0.5-mL microfuge tube:

Kinased primer (10 ng)	1–2 µL
RNA	0.1–1 µg
Annealing buffer (2x)	15 µL
Water	to 30 µL

2. Incubate for 2 min at 95°C and for 10 min 65°C, and then chill on ice.
3. Centrifuge in microcentrifuge to collect liquid at the bottom of the tube.
4. Make up reverse transcriptase buffer/enzyme mix as follows:

Reverse transcriptase mix (2.5x)	24 µL
H_2O	3 µL
AMV reverse transcriptase (1 unit/µL)	3 µL

5. Aliquot 5 µL of the reverse transcriptase buffer/enzyme mix to five microfuge tubes (C, T, A, G, B; B = no ddNTP).
6. Add 1 µL of 0.5 mM dideoxy C, T, A, G, or H_2O to the appropriate tube.
7. Add 4 µL of annealed RNA mix to each tube.
8. Incubate for 1 h at 42°C.
9. If the kinased primer has been purified and does not have free [γ^{32}-P]ATP, there is no need to extract and precipitate the reaction. Aliquots of these reactions can be run directly on gels. They should be diluted with TE buffer at least twofold and mixed with formamide dye/loading buffer.
10. Heat the samples for 2 min at 95°C.
11. Prerun the gel for 45 min to allow for better resolution of the shorter fragments. Load 5 µL of each reaction onto a prerun polyacrylamide gel (0.4-mm thick, containing 8% polyacrylamide [19:1 acrylamide:bis-acrylamide] and 8 M urea in 1x TBE). Run at 50 W with constant power until the bromophenol blue is near the bottom of the gel.
12. Soak the gel briefly (10–15 min) in 10% methanol/10% acetic acid at room temperature without agitation.
13. Transfer the gel to Whatman 3MM paper and dry with heat on a vacuum gel dryer.
14. Expose the gel to a phosphorimaging screen or X-ray film to visualize the ladder of bands.

    See Tips and Troubleshooting.

TIPS AND TROUBLESHOOTING

1. If sequence is not readable in Step 14 (i.e., if there is too much full-length extension product or stops that reflect increased early termination), first check the ddNTP:dNTP ratio in each reaction. These ratios are critical for obtaining the correct amount of chain termination. When preparing the solutions containing dNTPs and ddNTPs, measure carefully. Use fresh dNTP and ddNTP solutions that have been stored at –20°C or, preferably, at –80°C.

2. Note that these are suggested dideoxy ranges to try for most templates where extension sizes are 50–200 nucleotides. Some templates require adjustment of individual dideoxy concentrations and may require titrations.

3. If the chosen primer does not give good extension products, it often is desirable to try different primers.

SOLUTION AND REAGENT RECIPES

CAUTION: See the Cautions Appendix for appropriate handling of materials marked with <!>.

Annealing Buffer (2x)

Reagent	Quantity (for 1 mL)	Final concentration
KCl (2 M) <!>	250 µL	500 mM
Tris (1 M, pH 8.3) <!>	50 µL	50 mM
H_2O	500 µL	

Store indefinitely at –20°C.

Formamide/Dye Loading Buffer

Reagent	Quantity (for 10 mL)	Final concentration
Deionized formamide <!>	9.5 mL	95%
Bromophenol blue <!>	2.5 mg	0.025% (w/v)
Xylene cyanol FF <!>	2.5 mg	0.025% (w/v)
EDTA (0.5 m, pH 8.0) <!>	100 mL	5 mm
H_2O	400 mL	

Store in aliquots at –20°C.

Reverse Transcriptase Buffer (2.5x)

Reagent	Quantity (for 1 mL)	Final concentration
Tris (1 M, pH 8.3) <!>	50 µL	50 mM
DTT (1 M) <!>	50 µL	50 mM
$MgCl_2$ (1 M) <!>	20 µL	20 mM
dATP (10 mM)	40 µL	0.4 mM
dCTP (10 mM)	40 µL	0.4 mM
dGTP (10 mM)	40 µL	0.4 mM
dTTP (10 mM)	40 µL	0.4 mM
H_2O	720 µL	

Store indefinitely at –20°C.

Protocol 24

Direct Chemical Sequencing of End-labeled RNA

This protocol is a variation on the chemical method of DNA sequencing developed by Maxam and Gilbert (1977) and was described by Debra Peattie (1979, 1983) in the late 1970s. As with DNA sequencing, the method relies on the fact that each of the four bases in RNA is susceptible to chemical modification in a different way. In this protocol, end-labeled RNAs are subjected to base-specific chemical modification leading to chain cleavage next to the modified base. After cleavage, the RNAs are resolved on denaturing polyacrylamide gels. Guanosine residues are preferentially alkylated at the N-7 position by methylating agents such as dimethylsulfate. Accordingly, limited methylation of an RNA chain followed by reduction with sodium borohydride and cleavage with aniline yields fragments of RNA ending with guanosine residues. Diethylpyrocarbonate carbethoxylates the N-7 positions of adenosine and guanosine; this ultimately causes ring opening and allows aniline-induced strand cleavage. Because diethylpyrocarbonate reacts with adenosine faster than with guanosine, the adenosine residues can be inferred by comparing the guanosine-specific reaction. Hydrazine preferentially attacks uridine bases relative to cytosine (U >> C); therefore, aniline cleavage after hydrazinolysis identifies uridine residues. However, hydrazinolysis under different conditions (i.e., using anhydrous hydrazine in the presence of 3 M NaCl) suppresses modification of uridine, and cytosine bases are preferentially modified and can then again be cleaved with aniline. Although this combination of chemical reactions is effective for sequencing RNA, we have found that it can be quite touchy. It is imperative to follow the Peattie (1979, 1983) protocols exactly. The following is our translation of her protocols.

Note that the chemical reaction is specific to which base is attacked. Therefore, cleavage is random regarding the position of the base, and if the reaction is limited, it generates a nested set of radioactive fragments that are resolved according to length via polyacrylamide gel electrophoresis.

MATERIALS

The recipes for the items marked with <R> are at the end of the protocol or in the Common Recipes Appendix.

CAUTION: See the Cautions Appendix for appropriate handling of materials marked with <!>.

Reagents

Because the reactions require small volumes and it is necessary to prepare fresh solutions for many steps, we advise that chemicals be purchased in small quantities and/or aliquoted into small quantities and stored according to the manufacturer's instructions.

A > G reaction buffer <R>
Aniline/acetate buffer <R> <!>
Borohydride stop mix <R>
C > U reaction buffer <R>
Diethylpyrocarbonate (DEPC) <!>
Dimethylsulfate (DMS) <!>
Ethanol (80% and 100%), prechilled to –20°C <!>
Ferric chloride ($FeCl_3$) (2 M) <!>
G reaction buffer <R>

G reaction stop buffer <R>
Hydrazine (anhydrous) <!>
Loading dye <R>
NaOH (5 N) <!>
RNA of interest, 3′-end labeled with [^{32}P]pCp (see Chapter 4, Protocol 5) <!>
Sodium acetate (0.3 and 1.5 M)
Sodium acetate/EDTA solution <R> <!>
Sodium borohydrate ($NaBH_4$) (0.2 M; freshly prepared) <!>
Tris-HCl (1.0 M, pH 8.2)
tRNA (10 mg/mL)
U reaction buffer <R>

Equipment

Aluminum foil
Chemical fume hood
Dry ice <!>
Gel electrophoresis apparatus and high-voltage power supply for 6% denaturing polyacrylamide gel
Heat block (90°C)
Ice
Incubator or water bath (60°C) (with a nontransparent lid to keep reactions inside dark)
Microcentrifuge
Microcentrifuge tubes (1.5 mL)
SpeedVac or lyophilizer
Vortex mixer or pipette

OVERVIEW

Chemical sequencing is a "difficult" protocol. For the inexperienced, it might be appropriate to run each base-specific reaction on consecutive days. For G, follow Steps 1–23 and then proceed to Step 57. For A > G, follow Steps 24–35 and then continue to Step 57. For C > U, follow Steps 36–45 and proceed to Step 57. For U, follow Steps 46–56 and then Step 57. Then perform aniline cleavage (Step 57) reactions and run the gel on the following day. It is valuable to practice on an abundant RNA before attempting these procedures on a valuable experimental sample. Reaction times in Steps 40 and 50 are given as ranges and should be adjusted for individual RNAs. Keep in mind that this protocol is time-consuming because of lyophilization steps and multiple precipitations. In cases of drying pellets, times are not generalizable. Use the shortest time possible to achieve dryness depending on the equipment. Cutting corners is strongly discouraged. Please keep in mind that dimethylsulfate, hydrazine, and diethylpyrocarbonate are potentially quite toxic and should be handled and disposed of properly. It is important to use enough counts per minute of RNA sample in each reaction so that the ladders are readable. Remember that the modifications are only partial.

METHOD

Most of the chemical reagents are hazardous. Reactions are best performed in a chemical fume hood with a lab coat and gloves. Discard DMS and DEPC waste in a container with 500 mL of 5 N NaOH. Discard hydrazine waste in a container with 500 mL of 2 M $FeCl_3$.

Guanosine Reaction

1. Assemble the following reagents in a 1.5-mL microcentrifuge tube:

tRNA (10 mg/mL)	1 µL
Gel-purified, 3′-end-labeled RNA	5 µL (20,000–50,000 cpm)
G reaction buffer	300 µL

2. Chill for 1 min on ice.
3. Add 0.5 µL of DMS. Mix and centrifuge for 5 sec.

   In this and subsequent steps, mix by vortexing or pipetting up and down until well mixed. Perform centrifugations at room temperature at maximum speed.
4. Incubate for 30–60 sec at 90°C.
5. Chill for 1 min on ice.
6. Add 75 µL of G reaction stop buffer and mix.
7. Precipitate the RNA by adding 900 µL of cold 100% ethanol.
8. Freeze for 10 min on dry ice or until frozen.
9. Thaw and centrifuge at maximum speed for 10 min at room temperature.
10. Discard the supernatant. Resuspend the pellet in 200 µL of 0.3 M sodium acetate.
11. Reprecipitate by adding 600 µL of cold (–20°C) 100% ethanol.
12. Repeat Steps 8 and 9.
13. Pour off the ethanol so as not to disturb the pellet. Wash the pellet with 1 mL of cold (–20°C) 100% ethanol and discard the ethanol.
14. Dry in a SpeedVac with no heat to remove the ethanol until the dried pellet appears as a flake on the bottom of the tube.
15. Resuspend the pellet in 10 µL of 1 M Tris-HCl (pH 8.2).
16. Add 10 µL of fresh 0.2 M $NaBH_4$.
17. Incubate for 30 min in the dark on ice. (Cover top of ice bucket with aluminum foil.)
18. Add 200 µL of cold (4°C) borohydride stop mix and 5 µL of 10 mg/mL tRNA.
19. Precipitate the RNA by adding 600 µL of cold 100% ethanol.

    Samples can be stored in ethanol at –20°C until ready to process for aniline cleavage.
20. Microcentrifuge for 10 min as above.
21. Discard the supernatant.
22. Wash the pellet with 100% ethanol by adding 0.5 mL of cold 100% ethanol to tube, centrifuging for 2 min, and discarding the supernatant.
23. Dry the pellet in a SpeedVac. Proceed to Step 57 to treat with aniline.

Adenosine > Guanosine Reaction

24. In a 1.5-mL microcentrifuge tube, mix the following:

tRNA (10 mg/mL)	1 µL
Gel-purified, 3′-end-labeled RNA	5 µL (at least 10,000 cpm)
A > G reaction buffer	100 µL

25. Place for 1 min on ice.
26. Add 1 µL of DEPC.

27. Mix with a micropipette and centrifuge briefly (5 sec) at room temperature.
28. Incubate for 5–10 min at 90°C.
29. Chill for 1 min on ice and then add the following:

Sodium acetate (1.5 M)	50 μL
tRNA (10 mg/mL)	1 μL

30. Precipitate by adding 750 μL of cold 100% ethanol.
31. Freeze for 10 min on dry ice.
32. Thaw and microcentrifuge for 10 min.
33. Pour off and discard supernatant in 200 μL of 0.3 M sodium acetate and reprecipitate using 600 μL of cold (4°C) 100% ethanol.
34. Repeat Steps 31 and 32.
35. Wash pellet with 500 μL of 100% ethanol and dry in a SpeedVac. Proceed to Step 57 to treat with aniline.

    If desired, the reactions can be stored at –20°C in the final ethanol precipitation. Samples can be processed for and cleaved with aniline all at once.

Cytidine > Uridine Reaction

36. Mix the following in a 1.5-mL microcentrifuge tube:

tRNA (10 mg/mL)	1 μL
RNA (3′-end labeled)	20,000–50,000 cpm

37. Dry in a SpeedVac.
38. Add 10 μL of fresh 3.0 M NaCl/hydrazine on ice (C > U reaction buffer).
39. Mix or vortex for 5 sec and centrifuge quickly (5 sec).
40. Incubate for 20–30 min on dry ice. It may be necessary to optimize the time of incubation for individual RNAs.
41. Add 1 mL of cold 80% ethanol and 1 μL of 10 mg/mL tRNA to precipitate.
42. Freeze for 10 min on dry ice, thaw, and centrifuge in a microcentrifuge for 10 min.
43. Decant the supernatant. Resuspend and reprecipitate from 200 μL of 0.3 M sodium acetate with 500 μL of cold (4°C) 100% ethanol.
44. As before, the RNA can be stored in ethanol, and the cleavage reactions can be processed together.
45. Wash the pellet with 0.5 mL of cold (4°C) 100% ethanol and dry in a SpeedVac. Proceed to Step 57 to treat with aniline.

Uridine Reaction

46. In a 1.5-mL microcentrifuge tube, mix the following:

tRNA (10 mg/mL)	1 μL
Gel-purified, 3′-end-labeled RNA	20,000–50,000 cpm

47. Dry in a SpeedVac.

    This step ensures that excess H_2O is not added to hydrazine.

48. Add 10 μL of 50% hydrazine (U reaction buffer).
49. Mix and centrifuge quickly.

50. Incubate for 10–15 min on ice.
51. Add 200 μL of cold 0.3 M sodium acetate/0.1 mM EDTA solution.
52. Add 750 μL of cold 100% ethanol.
53. Freeze on dry ice for 10 min, thaw, and centrifuge in a microcentrifuge for 10 min.
54. Remove and discard the supernatant into 2 M $FeCl_3$. Resuspend in 200 μL of 0.3 M sodium acetate, and reprecipitate the RNA with 600 μL of cold (4°C) ethanol.
55. The ethanol precipitation step can be stopped and saved at –20°C to process aniline reactions.
56. Wash the pellet with 100% ethanol. Dry in a SpeedVac. Proceed to Step 57 to treat with aniline.

Aniline Treatment

57. Centrifuge the reprecipitated samples in a microcentrifuge for 10 min at room temperature. Wash pellets with 0.5 mL of cold (4°C) 100% ethanol. Dry the pellets in a SpeedVac.
58. Resuspend each pellet in 20 μL of aniline/acetate buffer.
59. Mix well by pipetting up and down and centrifuge for 5 sec.
60. Incubate for 20 min at 60°C *in the dark*. Cover the microcentrifuge tubes with aluminum foil and place a lid over the water bath to keep the samples in the dark.
61. Freeze the samples on dry ice and then lyophilize in a SpeedVac.

    Samples must be well lyophilized after the aniline treatment or they will not enter the gel properly. There should be no moisture in the tubes. The tRNA should appear as a white speck of powder. Be careful that it does not leap out of the tube because of glove static.

62. Resuspend each pellet in 20 μL of H_2O, freeze, and then dry in a SpeedVac. Repeat.
63. Resuspend each pellet in 2–5 μL of loading dye.
64. Heat the samples for 30 sec at 90°C to denature them.
65. Place the samples on ice and load half of the sample on a denaturing 6% acrylamide gel (see Chapter 2, Protocol 12). If maximum cpm of RNA was used, it is practical to load one-half of the sample in case it needs to be rerun. Load the samples so that bands in adjacent lanes can be compared (i.e., A > G next to G and U next to C > U) and band intensities can be used to infer the sequence.
66. Fix and dry the gel. Expose to a phosphorimager (Fig. 3-28). See Tips and Troubleshooting.

TIPS AND TROUBLESHOOTING

1. If there is irregular spacing of the bands on the sequencing gel, rerun the gel with 7 M urea and 40% formamide in 1x TBE.
2. If the RNA has a heterogeneous end, the sequencing ladder may have bands in multiple lanes. If this occurs, (1) repurify the RNA by rerunning the gel for a longer period of time or on a different percentage gel to optimize separation of the species, (2) change the TBE concentration to 0.5x or change the acrylamide:bis-acrylamide ratio so that it differs from the original purification gel (i.e., 30:1), (3) cut a tighter band (i.e., cut it in half horizontally to repurify on gel), and/or (4) purify on formamide gel.
3. If gel bands are fuzzy or grainy, it may be necessary to fix with acetic acid and methanol and then wash with H_2O to remove urea before drying.

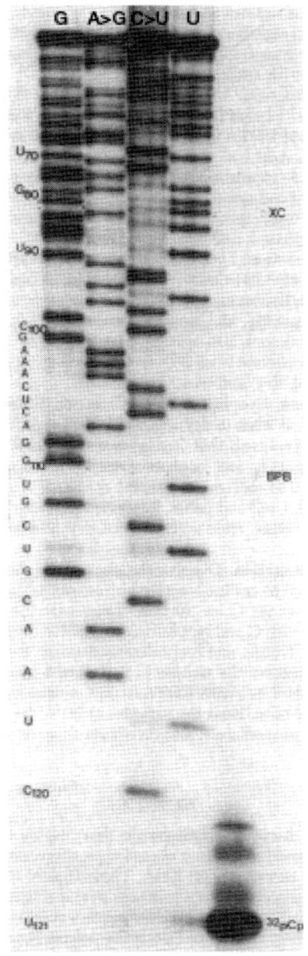

FIGURE 3-28. Example of direct chemical sequencing of RNA. As described in Peattie (1979), an end-labeled RNA was subjected to four separate chemical modification cleavage reactions diagnostic for the four bases present in RNA. Each lane of the gel displays cleavage products resulting form the four reactions. The derived sequence is shown at the left-hand side of the gel. (Reprinted, with permission, from Peattie 1979.)

SOLUTION AND REAGENT RECIPES

CAUTION: See the Cautions Appendix for appropriate handling of materials marked with <!>.

A > G Reaction Buffer

Reagent	Quantity (for 50 mL)	Final concentration
Sodium acetate (2.5 M)	2 mL	50 mM
EDTA (0.5 M, pH 8.0) <!>	0.2 mL	1 mM

Store at 4°C.

Aniline/Acetate Buffer (pH 4.5)

Reagent	Quantity (for 5.5 mL)	Final concentration
Aniline (11 M) <!>	1x (0.5 mL)	1 M
Glacial acetic acid (17 N) <!>	3x (1.5 mL)	to obtain a pH of 4.5
H_2O	7x (3.5 mL)	

Twenty µL of aniline/acetate buffer is needed per reaction. This final volume (5.5 mL) is convenient. Make fresh each time that cleavage reactions are performed.

Borohydride Stop Mix

Reagent	Quantity (for 10 mL)	Final concentration
Sodium acetate (3 M)	2 mL	0.6 M
Glacial acetic acid (17 M) <!>	0.35 mL	0.6 M
tRNA (10 mg/mL)	25 µL	25 µg/mL
H_2O	7.62 mL	

Store at 4°C. The pH should be 4.5.

G Reaction Buffer

Reagent	Quantity (for 50 mL)	Final concentration
Sodium cacodylate (0.1 M) <!>	25 mL	50 mM
EDTA (0.5 M, pH 8.0) <!>	0.1 mL	1 mM
H_2O	24.9 mL	

Adjust the pH to 5.5 with HCl <!>. Store at 4°C.

G Reaction Stop Buffer

Reagent	Quantity (for 50 mL)	Final concentration
Tris acetate (2 M)	25 mL	1 M
Sodium acetate trihydrate	10.2 g	1.5 M
2-Mercaptoethanol (14 M) <!>	3.6 mL	1 M
H_2O	to 50 mL	

Adjust the pH to 7.5 with acetic acid <!>. Store at 4°C.

U Reaction Buffer

Reagent	Quantity (for 1 mL)	Final concentration
Hydrazine (anhydrous) <!>	0.5 mL	50% (v/v)
H_2O	0.5 mL	

Prepare fresh just before use and chill on ice.

Loading Dye

Reagent	Quantity (for 10 mL)	Final concentration
Urea (ultrapure) <!>	4.8 g	8 M
Tris-HCl (1 M, pH 7.4)	200 µL	20 mM
EDTA (0.5 M, pH 8.0) <!>	20 µL	1 mM
Xylene cyanol (1%, w/v) <!>	0.5 mL	0.05% (w/v)
Bromophenol blue (1%, w/v) <!>	0.5 mL	0.05% (w/v)

Store at room temperature.

C > U Reaction Buffer

Reagent	Quantity (for 1 mL)	Final concentration
NaCl, baked at 110°C overnight and stored in a desiccator	0.175 g	3 M
Hydrazine (anhydrous) <!>	~945 µL	

Prepare fresh before use. The NaCl must be absolutely dry before combining with anhydrous hydrazine.

Sodium Acetate (0.3 M)/EDTA Solution (0.1 mM)

Reagent	Quantity (for 100 mL)	Final concentration
Sodium acetate (2.5 M)	12 mL	0.3 M
EDTA (0.1 M) <!>	0.1 mL	0.1 mM
H_2O	to 100 mL	

Store at 4°C.

REFERENCES

Maxam AM, Gilbert W. 1977. A new method for sequencing DNA. *Proc Natl Acad Sci* **74:** 560–564.

Peattie DA. 1979. Direct chemical method for sequencing RNA. *Proc Natl Acad Sci* **76:** 1760–1764.

Peattie DA. 1983. Direct chemical method for sequencing end-labeled ribonucleic acids. In *Methods of DNA and RNA sequencing* (ed. SM Weissman), pp. 261–304. Praeger, NY.

Protocol 25

Splinted Ligation Method to Detect Small RNAs

This protocol relies on the ability of T4 DNA ligase to covalently join the terminal 3′-hydroxyl group of an RNA molecule to the labeled 5′-phosphate group of a DNA chain in the presence of a DNA "bridge" oligonucleotide that is complementary to both (Fig. 3-29). Because DNA ligase will only join RNA and DNA molecules that are perfectly aligned on the bridge oligonucleotide (i.e., no gaps or mismatches at the junction), this technique is highly specific. We have found that ligation-mediated detection is significantly more sensitive than northern blotting for the detection of small RNAs (e.g., miRNAs and piRNAs) and can be performed much more quickly than blotting. The method is quantitative and can be used to compare levels of small RNA expression among different RNA samples (Maroney et al. 2007, 2008).

MATERIALS

The recipes for the items marked with <R> are at the end of the protocol or in the Common Recipes Appendix.

CAUTION: See the Cautions Appendix for appropriate handling of materials marked with <!>.

Reagents

Alkaline phosphatase (10 units/μL)

Annealing buffer (10x) <R>

Bridge (7.5 ng/μL) and ligation (100 ng/μL) oligonucleotides

Design the bridge oligonucleotide so that ~20 nucleotides of its 3′ end are perfectly complementary to the target RNA and its 5′ end is perfectly complementary to the 14-nucleotide ligation oligonucleotide (see Fig. 3-29). The ligation oligonucleotide (and its complement) can be composed of arbitrary sequence, but the sequence should not impart significant secondary structure to the bridge oligonucleotide as a whole. For best results, include a nonligatable end (e.g., dideoxy or carbon spacer) at both the 3′ end of the ligation oligonucleotide and the 3′ end of the bridge oligonucleotide. If the sequence in the ligation oligonucleotide and its complement in the bridge oligonucleotide are always the same, the same ligation oligonucleotide can always be used.

Denaturing acrylamide gel (6%; see Step 8) <!>

Formamide loading dye <R> <!>

[γ-^{32}P]ATP (6000 Ci/mmol, 150 μCi/μL) <!>

Kinase buffer (10x) <R>

Ligation buffer (2x) <R>

Alternative ligase buffers are available commercially. If choosing these buffers, adjust concentrations accordingly (e.g., 2.5 μL of 5x buffer per 12.5 μL of ligation reaction).

OptiKinase (10 units/μL; US Biochemicals)

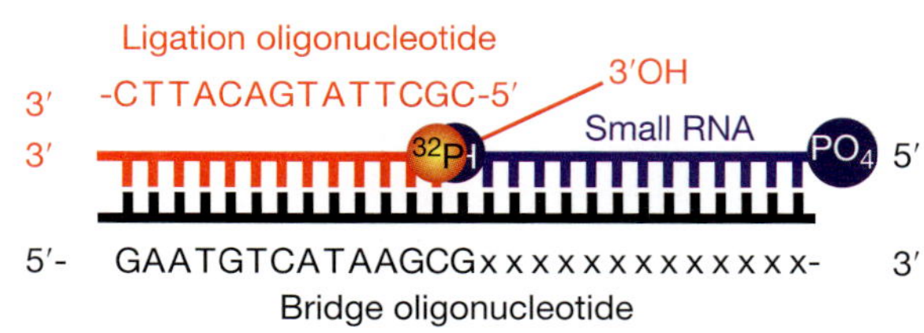

FIGURE 3-29. Bridge oligonucleotide sequence design. (Redrawn, with permission, from Maroney et al. 2008 [Macmillan].)

Reagents for phenol <!> extraction and ethanol <!> precipitation (for Step 1.iv)
T4 DNA ligase (10 units/μL)
Total RNA (1 μg/μL)

Equipment

Autoradiography or phosphorimaging equipment
Gel drier
Incubator or water bath (37°C)
Micro Bio-Spin 6 Column (Bio-Rad)
Microcentrifuge
Microcentrifuge tubes (1.5 mL)
PCR tubes
Power supply and electrophoresis apparatus for denaturing acrylamide gels (see Step 8)
Temperature block (95°C)
Thermal cycler

OVERVIEW

After isolation of total RNA or RNA enriched for small <30-nucleotide RNAs, this method can be used to detect specific small RNAs (e.g., miRNAs and piRNAs). The procedure involves covalent linking of a labeled DNA oligonucleotide to the RNA of interest using T4 DNA ligase and a splint or bridge oligonucleotide. Before beginning this procedure, it is important to design appropriate oligonucleotides. Following ligation, the small RNAs are detected using denaturing gel electrophoresis that separates the small RNA lengthened by the addition of the labeled DNA oligonucleotide from unreacted labeled DNA oligonucleotide.

METHOD

1. Label the ligation oligonucleotide.
   i. Assemble the following reaction in a 1.5-mL microcentrifuge tube:

Ligation oligonucleotide (100 ng/μL)	1 μL (20 pmol)
[γ-^{32}P]ATP (6000 Ci/mmol, 150 μCi/μL)	2 μL (300 μCi)
Kinase buffer (10×)	1 μL
OptiKinase (10 units/μL)	1 μL
H_2O	5 μL

   ii. Incubate the reaction for 20 min at 37°C.
   iii. Dilute with 50 μL of H_2O and run through a Micro Bio-Spin 6 Column.
   iv. Phenol-extract and then ethanol-precipitate with glycogen (see Chapter 2, Protocols 1 and 2).
   v. Resuspend the labeled ligation oligonucleotide in 75 μL of H_2O.
2. Prepare a mix containing the following:

Annealing buffer (10×)	5 μL
Labeled ligation oligonucleotide from Step 1.v	5 μL (1 pmol)
Bridge oligonucleotide (7.5 ng/μL)	1.5 μL (1 pmol)
H_2O	8.5 μL

   This volume is enough for 9–10 reactions.
3. Add 2.0 μL of the mix from Step 2 to 3 μL of total RNA (1 μg/μL) in a PCR tube. Place the tube in a thermal cycler. Denature for 1 min at 95°C and then anneal for 2 min at 65°C and for 10 min at 37°C.

If the RNA of interest (and bridge oligonucleotide) have secondary structure, the annealing conditions may need to be optimized.

4. While the annealing reaction is taking place, prepare the following ligation mix:

Ligation buffer (2X)	62 µL
T4 DNA ligase (10 units/µL)	2 µL
H_2O	11 µL

This mix is sufficient for 10 reactions.

5. Add 7.5 µL of ligation mix to the 5-µL annealing reaction and mix well by pipetting up and down. Incubate for 1 hr at room temperature.
6. Stop the reaction by adding 1 µL of 10 units/µL alkaline phosphatase and incubate for 15 min at 37°C. Samples can be stored for 2–4 wk at –20°C.
7. Remove 2.5 µL of the sample and add 5 µL of formamide loading dye.
8. Denature for 2 min at 95°C and place on ice.
9. Load samples on a denaturing 6% acrylamide gel that has been prerun for at least 45 min. After loading the samples, run the gel until the bromophenol blue dye has migrated 15 cm from the wells. Dry the gel and visualize the bands using autoradiography or phosphorimaging. See Tips and Troubleshooting.

TIPS AND TROUBLESHOOTING

1. If aberrant bands are detected in Step 9, these bands may represent fragments of degraded DNA in the RNA sample. To solve this problem, do not use DNase-treated RNA samples. Try incubating the ligation reactions at 30°C instead of at room temperature.
2. If no bands are detected in Step 9, excess bridge oligonucleotide may have been used. Carefully quantitate the oligonucleotides. The ligation oligonucleotide must be present in an equimolar ratio to or in excess of the bridge oligonucleotide. If the bridge oligonucleotide is in excess, the RNA of interest and the ligation oligonucleotide may have annealed to separate molecules and thus did not ligate.
3. If there is no signal in Step 9, the oligonucleotides may not have been designed appropriately. Make sure that the 3′ end of the RNA of interest is known to be accurate. To exclude technical problems, run a control reaction with reagents and RNAs of known quality using a bridge oligonucleotide to a known abundant RNA. Check for secondary structure in the bridge oligonucleotide and if secondary structure is present, optimize the annealing conditions and/or increase the ligation temperature to 30°C.

SOLUTION AND REAGENT RECIPES

CAUTION: See the Cautions Appendix for appropriate handling of materials marked with <!>.

Annealing Buffer (10X)

Reagent	Quantity (for 1 mL)	Final concentration (10X)
KCl (2 M) <!>	375 µL	750 mM
Tris-HCl (1 M, pH 7.5)	200 µL	200 mM
H_2O	425 µL	

Store indefinitely at –20°C.

Kinase Buffer (10x)

Reagent	Quantity (for 1 mL)	Final concentration (10x)
Tris-HCl (1 M, pH 7.5)	500 µL	500 mM
$MgCl_2$ (1 M) <!>	100 µL	100 mM
DTT (1 M) <!>	50 µL	50 mM
H_2O	350 µL	

Store for up to 6 mo at –20°C.

Ligation Buffer (2x)

Reagent	Quantity (1 mL)	Final concentration (2x)
Tris-HCl (1 M, pH 7.6)	132 µL	132 mM
$MgCl_2$ (1 M) <!>	20 µL	20 mM
DTT (1 M) <!>	2 µL	2 mM
ATP (100 mM)	10 µL	1 mM
Polyethylene glycol (PEG) (MW 6000) (20%, w/v)	750 µL	15% (w/v)
H_2O	86 µL	

Store for up to 6 mo at –20°C.

REFERENCES

Maroney PA, Chamnongpol S, Souret F, Nilsen TW. 2007. A rapid, quantitative assay for direct detection of microRNAs and other small RNAs using splinted ligation. *RNA* **13:** 930–936.

Maroney PA, Chamnongpol S, Souret F, Nilsen TW. 2008. Direct detection of small RNAs using splinted ligation. *Nature Protocols* **3:** 279–287.

4 Synthesis, Purification, Labeling, and Substitution of Transcripts Synthesized In Vitro

IN THE PRECEDING CHAPTER, we presented a variety of methods designed to detect, sequence, and, in some cases, quantify the levels of naturally occurring RNA molecules. It this chapter, we turn to in vitro synthesis of specific RNAs. A vast literature documents the importance of synthetic transcripts as experimental tools used to understand a wide variety of RNA-related questions. Specifically, synthetic transcripts have been essential for elucidating RNA-processing pathways, RNA-folding dynamics, RNA catalysis, RNA–protein interactions, and RNA structure.

Although there are literally hundreds of variations, depending on the specific experimental design (see below), all in vitro transcription reactions are basically the same, i.e., a DNA template corresponding to the sequence of interest is transcribed by a phage (T7, T3, or SP6) RNA polymerase; each phage polymerase recognizes a distinct promoter sequence. Moreover, in all in vitro transcription reactions, the same considerations apply, i.e., it is necessary to synthesize full-length transcripts and it is essential to minimize RNA degradation.

In this chapter, we first discuss the basics of in vitro transcription and then present strategies for optimizing yield for specific applications (e.g., are milligram or microgram quantities of RNA needed?), labeling strategies (e.g., 5′ or 3′ end, body, or site specific), and the introduction of modified nucleotides (5′ capping, internal modified bases [either random or site specific], and methods for purifying and quantifying synthetic transcripts).

PROTOCOLS

1 In Vitro Transcription of RNA: Synthesis, Labeling, and Substitution, 210
2 High-yield Synthesis of RNA Using T7 RNA Polymerase and Plasmid DNA or Oligonucleotide Templates, 216
3 Determining the Yield of RNA Synthesized In Vitro, 220
4 Gel Purification of RNA, 222
5 3′-End Labeling of RNA with [^{32}P]pCp and T4 RNA Ligase 1, 225
6 3′-End Labeling of RNA with Yeast Poly(A) Polymerase and 3′-Deoxyadenosine 5′-[α-^{32}P]Triphosphate (Cordycepin 5′-[α-^{32}P]Triphosphate), 228
7 5′-End Labeling of RNA with [γ-^{32}P]ATP and T4 Polynucleotide Kinase, 231
8 Site-specific Labeling and Substitution of RNA, 234

IN VITRO TRANSCRIPTION: THE BASICS

As noted above, all in vitro transcription reactions require a DNA template. Because most in vitro transcriptions are "run off," i.e., they are terminated by the end of the template, templates must have a free 3′ end. Accordingly, appropriate templates can be plasmids cut with a specific restriction endonuclease, purified polymerase chain reaction (PCR) products, or in some cases annealed oligonucleotides. It is essential that all templates be free of ribonuclease (we recommend phenol:chloroform:isoamyl alcohol [PCA] extraction of plasmid or PCR fragments) and carefully quantified. One major cause of failed in vitro transcriptions is inaccurate quantification of templates; another is templates contaminated with RNases.

The second essential component of in vitro transcription reactions is a phage RNA polymerase. Although SP6 was the first polymerase used, it is now much more common to use either T7 or T3 polymerases. Each polymerase is commercially available, but specific polymerases can also be purified in the investigator's laboratory (see Appendix 3, Protocol 3). Again, it is especially important to ensure that "homemade" polymerase preparations be free of contaminating RNases.

The third component common to all in vitro transcription reactions are ribonucleoside triphosphates (rNTPs), the substrates of the polymerases. rNTPs are available from several commercial sources, but it is essential to know whether or not the NTPs are supplied with or without magnesium. Each NTP chelates an equimolar amount of Mg^{++}; therefore, all transcription reactions must contain an excess of Mg^{++} over the nucleotide concentrations. This is a third and remarkably often overlooked cause of failed transcription reactions.

YIELD OF IN VITRO–TRANSCRIBED RNAS

When designing an experiment that uses synthetic transcripts, it is important to consider how much RNA will be needed. For applications such as structure determination, it is necessary to synthesize milligram amounts of RNA. However, most applications do not require large quantities of RNA, and microgram quantities are sufficient. This is particularly true when using labeled transcripts. Because the half-life of ^{32}P is 14.3 d, most labeled transcripts are only useful for 2–3 wk, and any high-specific-activity RNA "left over" after this time frame is essentially useless because it has already been damaged by radiolysis.

Four parameters should be considered when estimating expected yield of an in vitro transcript: (1) amount of template, (2) amount of RNA polymerase, (3) concentration of nucleotides, and (4) length of the transcript to be synthesized. In general, these parameters can be reduced to two: 2–5 μg of template and 1 unit of polymerase per microliter of transcription reaction (50 μL) are "standard" for most applications. Under these conditions, one can reasonably expect to obtain 10-fold amounts of RNA relative to the template with standard (0.5 mM) concentrations of rNTPs. This yield of full-length transcripts might be reduced by as much as one-half when long molecules ≥800 nucleotides are synthesized. The reduction in yield is due to premature termination by the polymerase and the greater vulnerability of long transcripts to any RNases that may be present.

This leaves one parameter, rNTP concentration, as the major determinant of yield. A good rule of thumb is that the amount of transcript synthesized will be reduced proportionately to the reduction in concentration of the *least* concentrated rNTP in the transcription reaction; i.e., if one rNTP is present at 25 μM, yield will be reduced 20-fold, 50 μM 10-fold, 100 μM 5-fold, and so on. As detailed below, exact quantitation of transcription yield is preferably obtained by determining incorporation of labeled rNTPs.

LABELING OF IN VITRO–SYNTHESIZED RNAS

Several strategies for labeling transcripts synthesized in vitro include body labeling, end labeling (either 5′ or 3′), and site-specific labeling. Each approach is appropriate for multiple applications and we discuss each in turn.

Body Labeling

Body labeling, as the name implies, involves incorporation of radiolabel throughout the transcript by random insertion of labeled rNTPs. Labeling is generally with [α-^{32}P]guanosine triphosphate (GTP) because the concentration of GTP is lowered when capped transcripts are synthesized (see below). However, uridine 5′-[α-^{32}P]triphosphate (UTP) and the other rNTPs can also be used, depending on the specific application (see below). Body-labeled transcripts can be synthesized with a wide range of specific activities. We describe three here.

High-specific-activity transcripts (50,000–100,000 cpm/ng) are useful for nuclease protection assays, as hybridization probes, and for highly sensitive RNA–protein interaction studies. Because unlabeled GTP must be lowered dramatically (25 μM) to achieve this specific activity, the yield of full-length transcripts will be quite low. Moreover, high-specific-activity transcripts are only useful for 1–2 wk after synthesis because they are subject to radiolytic decay.

Body-labeled transcripts of medium specific activity (10,000–15,000 cpm/ng) are useful for most in vitro processing reactions (e.g., splicing and polyadenylation) and are also useful for RNA–protein interaction studies. They can be synthesized capped (see below) and are not affected by radiolytic decay to the same extent as high-specific-activity transcripts. Accordingly, they can be used for up to 1 mo after synthesis, but it is important to remember that during storage, the specific activity will be reduced, i.e., 10,000 cpm/ng RNAs will be 2500 cpm/ng after 1 mo.

The third level of specific activity is low or trace labeled. Trace-labeled RNAs are useful for subsequent end labeling or site-specific labeling because labelings introduce high-specific-activity ^{32}P-labeled nucleotides that "mask" the trace radioactivity. We recommend that all in vitro transcripts (except of course those that require higher specific activity) be trace labeled. Labeling allows one to visualize the integrity of transcripts, facilitates gel purification, and perhaps most importantly provides a convenient and accurate way to precisely measure yield from any transcription reaction. Before moving on to other labeling strategies, we briefly discuss how yield is determined by incorporation or radioactivity.

DETERMINATION OF TRANSCRIPT YIELD BY MONITORING INCORPORATION OF LABEL

As discussed in Chapter 2, several methods are available for assessing quantity and quality of RNA purified from cells or tissues. These methods are in general not appropriate for analyzing transcripts synthesized in vitro. By far, the best way to quantitate yield of in vitro–synthesized RNA is simply to measure the amount of radioactivity incorporated into full-length transcripts. The following example is applicable to all in vitro transcription situations. Consider a 50-μL transcription reaction that contains a 1 mM solution of each rNTP. Here, there are 50 nanomoles of each nucleotide present or 200 nanomoles of total RNA (or 66 μg). Adding radiolabled nucleotide (the mass is negligible and irrelevant) and determining the input cpm by scintillation counting (i.e., spot 1 μL of the transcription reaction on a filter) provide the cpm of 1/50th of the total reaction. After transcription and purification of the synthesized product, the cpm contained in 1/50th of the resuspended RNA is determined (i.e., if resuspended in 50 μL, determine the cpm in 1 μL). Dividing this number by the cpm of input provides the yield and the specific activity. For example, if 40% of the label was incorporated into the product, 80 nanomoles of nucleotides were incorporated into RNA (or 26 μg). The actual quantity of your transcript in moles depends on the length of the transcript; i.e., a 20-mer will be 4 nanomoles and a 100-mer will be 0.8 nanomole. Specific activity can be expressed in cpm per nanomole or picomole or in cpm per nanogram.

SYNTHESIS OF RNA CONTAINING MODIFIED NUCLEOTIDES

For several applications, including affinity purification protocols, cross-linking protocols, and modification–interference protocols, it is desirable to synthesize RNAs containing internally modified nucleotides. The type and extent of modification are dictated by the specific experiment; e.g., for affinity selection, RNAs containing biotin are synthesized; for thiophosphate interference studies, RNAs substituted with α-thiophosphates are synthesized. Just as in the labeling strategies discussed above, introduction of modified nucleotides can be done throughout the RNA (analogous to body labeling) or through site-specific approaches. Body substitution is performed during transcription via the incorporation of modified NTPs; the degree of substitution is determined by the ratio of modified NTP to its unmodified counterpart (see Protocol 1).

Site-specific substitution is conceptually identical to site-specific labeling. With current technology, the best way to introduce site-specific modifications is to purchase an RNA oligonucleotide carrying the desired modification and to incorporate it into the RNA of interest using three-part or consecutive two-part ligations (see Protocol 3).

5′- and 3′-End Labeling

Often, it is desirable to generate 5′- or 3′-end-labeled RNAs. Such RNAs are useful for modification–interference approaches, RNA–protein interaction studies, chemical or enzymatic structure determination, footprinting, or sequencing. In addition, 5′-end-labeled RNAs are useful in RNase protection experiments, and RNAs labeled at either end can be used as hybridization probes.

The two most common ways to generate 5′-end-labeled RNAs are by direct labeling and kinasing with [γ-^{32}P]ATP and T4 polynucleotide kinase. For direct labeling, simply adding [γ-^{32}P]GTP to the transcription reaction produces 5′-end-labeled RNA because the γ-phosphate is only incorporated with the first nucleotide in the transcript; internal incorporation of G leads to the release of the labeled phosphate as PP_i. The advantage of this approach is that it is easy and does not involve any additional steps beyond the transcription reaction itself. The disadvantage is that specific activity obtained is much lower than with the kinase approach because the labeled GTP has to compete with unlabeled GTP in the transcription reaction.

To label 5′ ends using kinase, it is first necessary to dephosphorylate the 5′ end of the transcript to be labeled using phosphatase (Chapter 3, Protocol 6). After removal of 5′-phosphate groups, 5′ ends can be labeled to high specific activity using [γ-^{32}P]ATP and T4 polynucleotide kinase (Chapter 3, Protocol 1). This is a straightforward procedure, but it requires prior purification of the RNA of interest.

There are also two ways to label the 3′ ends of RNA molecules. The first and most common method uses T4 RNA ligase to covalently link [^{32}P]cytidine 3′,5′-bis(phosphate) ([^{32}P]pCp) to the 3′ end (Protocol 4). This procedure requires prior purification of the RNA of interest and a free 3′-hydroxyl end. Advantages to 3′-end labeling in this way include the high specific activity that can be obtained and the fact that the 3′-phosphate terminus protects the RNA from some 3′-to-5′ exonucleases. The second approach to 3′-end labeling uses yeast poly(A) polymerase and [α-^{32}P]cordycepin (3′-deoxyadenosine triphosphate) to add a single radiolabeled residue to the 3′ end (Protocol 6). In principle, this is a straightforward method, but it is not as popular as pCp labeling perhaps because of historical reasons (i.e., pCp labeling was described first).

Site-specific Labeling

Site-specific labeling provides an extremely powerful means of probing RNA–protein interactions, RNA–RNA interactions, or the importance of specific functional groups in an RNA. Three approaches can be used to generate RNAs containing a specific internally labeled phosphate; these are distinguished by the way the RNA of interest is prepared. In general, all site-specific approaches rely on the ability of T4 DNA ligase to covalently join two (or more) RNA strands when they are aligned on a complementary oligodeoxynucleotide (Moore and Sharp 1992).

In the first approach, the entire transcript of interest is synthesized. It is then digested with RNase H in the presence of a complementary chimeric 2′-*O*-methyl (2′OMe)/DNA oligonucleotide of the sequence 5′-MMMDDDDMMMMMMM-3′, where M is a 2′OMe ribonucleotide and D is a deoxynucleotide (see Yu et al. 1997). RNase H specifically cleaves the RNA in the oligonucleotide–RNA hybrid at the phosphodiester bond opposite the 5′-most deoxynucleotide, creating 3′-hydroxyl and 5′-phosphate termini. After dephosphorylation of the 5′ end of the 3′ fragment with phosphatase, this terminus is labeled to high specific activity with [γ-^{32}P]ATP. The fragments of the original RNA are then resealed with T4 DNA ligase in the presence of a splint (or bridge) oligonucleotide. Advantages of this approach include the fact that any phosphodiester bond can be labeled regardless of sequence context.

The second approach involves the independent synthesis of the 5′ and 3′ fragments. The 3′ fragment is labeled at its 5′ end after dephosphorylation, and the two fragments are then joined by ligation (Fig. 4-1). Although conceptually straightforward, the synthesis of two in vitro–transcribed RNAs can sometimes be problematic. Phage RNA polymerases often add untemplated nucleotides to the 3′ ends of transcripts, and any transcripts that contain such nucleotides cannot be ligated because they will not anneal perfectly with the splint oligonucleotide. A second potential problem is

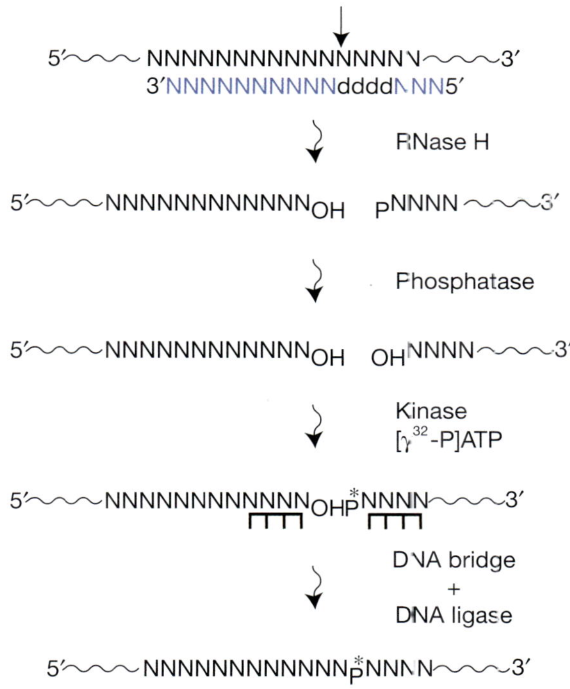

FIGURE 4-1. Site-specific labeling strategy. The site of cleavage by RNase H is indicated by the straight arrow. (Blue) 2′OMe residues, (lowercase letter "d"s) deoxy residues of the oligonucleotide. (*) Marker. Following phosphatase kinase and ligation steps, a full-length RNA with a site-specific label is produced. (Redrawn, with permission, from Maroney et al. 2000.)

the synthesis of the 3′ fragment. Because phage RNA polymerases greatly prefer to initiate with a purine (usually G), yields of transcripts initiating with a pyrimidine are quite low. This consideration places constraints on the specific phosphodiesters that can be labeled. For example, placing a label in a run of pyrimidines is extremely difficult using two-part synthesis.

The third approach, three-part ligation, circumvents the constraints described above. In this approach, an RNA oligonucleotide of the desired sequence is purchased or synthesized chemically and then labeled at its 5′ end. The 5′ and 3′ fragments are synthesized by in vitro transcription and the three RNAs are ligated simultaneously or in steps. Although this method is a bit cumbersome for generating site-specifically labeled RNAs, it does bypass the enzymatic dephosphorylation step and can be combined with the site-specific RNase H technique such that all fragments have appropriate ligatable termini.

The protocols described in this chapter should provide all of the tools necessary to synthesize, label, and create substituted RNAs. Such RNAs have myriad uses as described in Chapter 5.

REFERENCES

Maroney PA, Romfo C, Nilsen TW. 2000. Nuclease protection of RNAs containing site-specific labels: A rapid method for mapping RNA-protein interactions. *RNA* **6:** 1905–1909.

Moore MJ, Sharp PA. 1992. Site-specific modification of pre-mRNA: The 2′-hydroxyl groups at the splice sites. *Science* **256:** 992–997.

Yu YT, Shu MD, Steitz JA. 1997. A new method for detecting sites of 2′-O-methylation in RNA molecules. *RNA* **3:** 324–331.

Protocol 1

In Vitro Transcription of RNA: Synthesis, Labeling, and Substitution

As discussed in the introduction to this chapter, all in vitro transcription reactions involve the use of common components, a template, phage RNA polymerase, and rNTPs. The concentration of rNTPs is a crucial variable that is manipulated to determine the specific activities of labeled RNAs. This protocol describes methods used to synthesize RNAs of low specific activity (trace labeled), medium specific activity, and high specific activity. Also described is how the same principles that determine the level of incorporation of labeled rNTPs are applicable to the synthesis of transcripts containing modified nucleotides including 5′ cap and internal (body) modifications such as biotinylated uridine, 4-thiouridine, and phosphorothiotated nucleotides.

MATERIALS

The recipes for the items marked with <R> are at the end of the protocol or in the Common Recipes Appendix.

CAUTION: See the Cautions Appendix for appropriate handling of materials marked with <!>.

Reagents

5′ Cap analog (10 mM)

[α-^{32}P]GTP (800 Ci/mmol; 10 μCi/μL) or [α-^{32}P]UTP (800 Ci/mmol; 10 μCi/ μL) <!>

We recommend labeling capped transcripts with [α-^{32}P]GTP because the concentration of GTP is already lowered for capping to be more efficient.

ATP, GTP, UTP, and CTP stocks (10 mM each, ~pH 7.0)

Chloroform:isoamyl alcohol (24:1) <!>

Dithiothreitol (DTT) (1 M) <!>

DNase I (10 units/μL)

EDTA (0.5 M) <!>

Ethanol (100%) <!>

GlycoBlue (15 μg/μL)

Phenol:chloroform:isoamyl alcohol (PCA) <!>

Proteinase K (optional; see Tips and Troubleshooting)

RNasin (RNase inhibitor) (40 units/μL)

SDS extraction buffer <R>

Sodium acetate (3 M, pH 5.2)

T7 RNA polymerase (50 units/μL)

Template DNA, linearized plasmid or gel-purified PCR product

Transcription buffer (5x) <R>

Equipment

Gel-purification equipment (as in Protocol 3)

Dry ice <!>

Micro Bio-Spin Columns with Bio-Gel P-6 in Tris buffer (Bio-Rad)

Microcentrifuge

Microcentrifuge tubes (1.5 mL, polypropylene, capped)

OVERVIEW

The methods below are designed to prepare RNAs by in vitro transcription. Before setting up an in vitro transcription reaction, it is necessary to have a clear idea of how much RNA is needed and at what specific activity. In addition, is the RNA to be capped or substituted? We recommend that the entire set of protocols be read before designing the optimal in vitro transcription reaction; in particular, for reactions for capping and substitution of in vitro–synthesized transcripts and random incorporation of modified nucleotides during in vitro transcription.

METHOD

1. Assemble the following reaction(s) at room temperature. Choose i, ii, or iii depending on the intended specific activity.
   i. For unlabeled or trace-labeled RNAs, combine the following in the order listed:

Reagent	Quantity (for 50 μL)	Final concentration
H_2O	to 50 μL	
Template DNA (linearized plasmid or PCR product)	2 μg	0.04 μg/μL
ATP (10 mM)	2.5 μL	0.5 mM
CTP (10 mM)	2.5 μL	0.5 mM
UTP (10 mM)	2.5 μL	0.5 mM
GTP (10 mM)	2.5 μL	0.5 mM
Transcription buffer (5x)	10 μL	1x
RNasin (40 units/μL)	1 μL	0.8 unit/μL
DTT (1 M)	1 μL	20 mM
T3 or T7 RNA polymerase (50 units/μL)	1 μL	1 unit/μL

   To monitor the nonradioactive transcription reaction, remove 10 μL of the reaction and transfer to a separate tube. Add 1 μL of [α-^{32}P]GTP or UTP (800 Ci/mmol; 10 μCi/μL) and process in parallel with unlabeled GTP. For trace labeled, add 0.2–0.5 μCi of GTP or UTP (800 Ci/mM). Remove and set aside 1 μL from any labeled reaction to use for input quantitation (see Protocol 7).

   This reaction should yield ~10–20 μg of RNA at ~100 cpm/ng.

   ii. For high-specific-activity RNA, combine the following in the order listed:

Reagent	Quantity (for 50 μL)	Final concentration
H_2O	to 50 μL	
Template DNA (linearized plasmid or PCR product)	2 μg	0.04 μg/μL
ATP (10 mM)	2.5 μL	0.5 mM
CTP (10 mM)	2.5 μL	0.5 mM
UTP (10 mM)	2.5 μL	0.5 mM
GTP (0.5 mM)	2.5 μL	25 μM
Transcription buffer (5x)	10 μL	1x
RNasin (40 units/μL)	1 μL	0.8 unit/μL
DTT (1 M)	1 μL	20 mM

(Continued on next page)

Reagent	Quantity (for 50 μL)	Final concentration
*[α-^{32}P]GTP 800 Ci/mM	10 μCi	2 μCi/μL
T3 or T7 RNA polymerase (depends on template promoter) (50 units/μL)	1 μL	1 unit/μL

*If labeling with [α-^{32}P]UTP, substitute 2.5 μL of 10 mM GTP (instead of 0.5 mM) and 2.5 μL of 0.5 mM UTP for 10 mM UTP and add 10 μL of [α-^{32}P]UTP in the above reaction instead of [^{32}P]GTP. Remove and set aside 1 μL to use for input quantitation (see Protocol 7).

This reaction will yield ~3 μg of RNA at ~50,000–100,000 cpm/ng.

iii. For moderate-specific-activity RNA, combine the following in the order listed:

Reagent	Quantity (for 50 μL)	Final concentration
H_2O	to 50 μL	
Template DNA (linearized plasmid or PCR product)	2–4 μg	0/04–0.08 μg/μL
ATP (10 mM)	2.5 μL	0.5 mM
CTP (10 mM)	2.5 μL	0.5 mM
UTP (10 mM)	2.5 μL	0.5 mM
GTP (1 mM)	5 μL	0.1 mM
Transcription buffer (5x)	10 μL	1x
RNasin (40 units/μL)	1 μL	0.8 unit/μL
DTT (1 M)	1 μL	20 mM
T3 or T7 RNA polymerase (depends on template promoter) (50 units/μL)	1–4 μL	~1 unit/μL
[α-^{32}P]GTP (800 Ci/mmol; 10 Ci/μL)	5 μL	0.01 μCi/μL

Adjust the amount of [α-^{32}P]GTP to get the desired specific activity. This reaction will yield ~10 μg of RNA at ~4000–8000 cpm/ng.

2. Incubate the reaction for 1–2 h at 37°C.

3. Add 1 μL of 10 units/μL (for each 50 μL transcription reaction) DNase I and incubate for 15 min at 37°C.

4. Add EDTA to 10 mM.

5. For labeled reactions, remove unincorporated nucleotides with a Micro Bio-Spin Column and then dilute to 200 μL with SDS extraction buffer. For unlabeled or trace-labeled reactions, dilute with an equal volume of SDS extraction buffer. Add 1/10 volume of 3 M sodium acetate (pH 5.2) to both labeled and unlabled reactions.

6. Extract with PCA (see Chapter 2, Protocol 1).

7. Add 2 μL of 15 μg/μL GlycoBlue or glycogen to serve as carrier or to aid in resuspending the pellet.

8. Add 2.5 volumes of ethanol. Freeze on dry ice or in dry ice/ethanol bath (~15 min). Thaw. Centrifuge at maximum speed in a microcentrifuge for 10 min. Carefully pipette off the supernatant and dispose by an appropriate waste method.

9. Resuspend the pellets in 2 mM EDTA (2–5 µL for each 50 µL of transcription reaction for gel purification). If RNA is to be used directly, resuspend in H_2O. See Tips and Troubleshooting.

10. Gel-purify the RNA according to the procedure described in Protocol 4 of this chapter.

    Check a small aliquot of longer transcripts (>600–700 nucleotides) on the gel for intactness. If the RNA is a single band with no trailing of radioactivity below, the transcripts can be used without gel purification. See Tips and Troubleshooting.

Capping and Substitution of In vitro–synthesized Transcripts

The presence of a cap at the 5′ end of in vitro–synthesized transcripts can be useful in many situations, e.g., by preventing 5′ to 3′ nuclease activity or by promoting translation. "Caps" come in several forms: ^{7}mGpppG, ApppG, and GpppG. ^{7}mGpppG serves as a recognition factor for the translation initiation factor eIF-4E and thus is important when making RNAs that will be used for translation. The other analogs are primarily used to prevent nucleolytic attack. Capping during transcription involves the initiation of RNA chains with the cap (obviously, cap cannot be incorporated internally). The extent of capping is determined by competition between cap and GTP, therefore cap must be present in excess of GTP. We recommend that the concentration of cap be 10-fold that of GTP; under such conditions, ~90% of the RNAs will be capped. Because of constraints on the concentrations of cap that can be achieved (1 mM is practical), it is necessary to reduce the concentration of GTP to 100 µM to achieve efficient capping. A typical synthesis reaction for capping follows:

11. For capping; mix the following reagents in the order listed:

Reagent	Quantity for 50 µL	Final concentration
Template DNA (linearized plasmid or PCR product)	2–4 µg	0.04–0.08 µg/µL
ATP (10 mM)	2.5 µL	0.5 mM
CTP (10 mM)	2.5 µL	0.5 mM
UTP (10 mM)	2.5 µL	0.5 mM
GTP (1 mM)	5 µL	0.1 mM
Transcription buffer (5x)	10 µL	1x
RNasin (40 units/mL)	1 µL	0.8 unit/µL
DTT (1 M)	1 µL	20 mM
*Cap analog (see above) (40 mM)	1.25 µL	1 mM
T3 or T7 polymerase (50 units/µL)	1 µL	1 unit/µL
[α-^{32}P]GTP (800 Ci/mmol; 10 µCi/µL)	5 µL	1 µCi/µL
H_2O	to 50 µL	

*Substitute any cap analog, dinucleotide (NpG), or GMP for ^{7}MGpppG at a final concentration of 1 mM.

12. Incubate for 1–2 h at 37°C.

13. Continue with Step 3 of Protocol 1 (p. 212).

Random Incorporation of Modified Nucleotides during In Vitro Transcription

Many modified nucleotides can be incorporated into in vitro–synthesized RNA, including, but not limited to, biotinylated UTP, 4-thio-UTP, bromo-UTP, iodo-UTP, and many adenosine analogs (with α-thiophosphates). Biotinylated RNAs are used in affinity selection experiments; 4-thio-, 5-bromo-, and 5-iodo-U-substituted RNAs are useful for cross-linking studies; and α-thio-based analogs are primarily used for modification–interference experiments (see Chapter 5, Protocols 7, 8, and 10). When designing in vitro transcription reactions that incorporate modified nucleotides, it is important to consider the level of incorporation desired. For example, in modification–interference approaches, transcripts should contain one or even fewer modifications per transcript, whereas in biotin applications, heavier substitution (several per transcript) is desirable. The degree of substitution is determined by competition between the unmodified and modified nucleotide just as in capping (see above). Because some modified nucleotides are incorporated less efficiently than their unmodified counterparts, the degree of substitution may not correspond exactly to the ratio of modified to unmodified nucleotides present in the transcription reaction. Thus, it is sometimes necessary to titrate the level of the modified nucleotide to achieve the desired level of substitution. Oversubstitution with biotin can sometimes cause RNA to partition with the organic phase because of the hydrophobic spacer. Transcripts heavily substituted with biotinylated nucleotides should not be extracted with phenol. The following is a "typical" in vitro transcription reaction designed to obtain a capped, moderate specific activity, biotin- or 4-thio-U-substituted RNA; the modified uridine is incorporated at approximately every fourth uridine residue.

14. For substitution, mix the following reagents in the order listed:

Reagent	Quantity for 50 μL	Final concentration
Template DNA (linearized plasmid or PCR product)	2–4 μg	0.04–0.08 μg/μL
ATP (10 mM)	2.5 μL	0.5 mM
CTP (10 mM)	2.5 μL	0.5 mM
UTP (5 mM)	3.75 μL	0.375 mM
Biotin-UTP or 4-thio-UTP (1 mM)	6.25 μL	0.125 mM
GTP (1 mM)	5 μL	0.1 mM
Transcription buffer (5x)	10 μL	1x
RNasin (40 units/μL)	1 μL	0.8 unit/μL
DTT (1 M)	1 μL	20 mM
M7GpppG (40 mM)	1.25 μL	1 mM
T3 or T7 polymerase (50 units/μL)	1 μL	1 unit/μL
[α-^{32}P]GTP (800 Ci/mmol; 10 μCi/μL)	5 μL	1 μCi/μL
H_2O	to 50 μL	

15. Incubate for 1–2 h at 37°C.

16. Continue with Step 3 of Protocol 1 (p. 212).

TIPS AND TROUBLESHOOTING

1. As noted in the chapter introduction, there are hundreds of possible in vitro transcription reactions. The only significant variable is rNTP concentration, which is manipulated to achieve different levels of radioactive nucleotide incorporation (different specific activities), the efficiency of capping, and the degree of substitution. Careful attention to detail in setting up the transcription reaction will permit the synthesis of any number of "designer RNAs."

2. In addition to "conventional" cap analogs, which can be incorporated in two ways to produce mixtures of differently capped mRNAs, Ambion sells an antireverse cap analog (ARCA), which ensures that the cap is in the right orientation.

3. It is possible to initiate transcription with dinucleotides, i.e., UpG or CpG, or even mononucleosides (i.e., G). Transcripts initiated in this way can be 5′-end labeled without prior phosphatase treatment (see Protocol 7).
4. Capping is usually done in conjunction with radiolabeling and/or with other internal modifications (see below).
5. If the RNA does not completely resuspend in Step 9, try resuspending in a larger volume or diluting and reprecipitating with ethanol.
6. If the yield of RNA is low in Step 10, rerun the reaction, making sure that the mixture remains in solution while assembling the reaction. The reactions are assembled at room temperature in Step 1 to avoid precipitation. If the reactions containing high concentrations of cap and nucleotides form a precipitate, heat them for 5 min at 37°C and briefly vortex before adding the enzyme.
7. If the RNA is degraded in Step 10, make sure that the template was nuclease-free. If necessary, after plasmid linearization, dilute 100 µL of digestion with 200 µL of SDS extraction buffer. Add proteinase K (100 µg/mL), digest for 15 min at 37°C, add 1/10 volume 3 M NaOAc (pH 5.2) and proteinase K, phenol-extract, and ethanol-precipitate. Make sure that all solutions are RNase-free.
8. Very efficient transcriptions may produce a PP_i precipitate in large amounts and interfere with solubilty. This problem can be overcome by addition of pyrophosphatase (see Large-scale Synthesis protocol, Steps 4 and 9).

SOLUTION AND REAGENT RECIPES

CAUTION: See the Cautions Appendix for appropriate handling of materials marked with <!>.

Transcription Buffer (5x)

Reagent	Quantity (for 10 mL)	Final concentration
Tris-HCl (1 M, pH 8.0)	2 mL	200 mM
$MgCl_2$ (1 M) <!>	400 µL	40 mM
NaCl (5 M)	0.5 mL	250 mM
Spermidine (100 mM) <!>	1 mL	10 mM
H_2O	6.1 mL	

Store for up to 1 yr at –20°C.

SDS Extraction Buffer

Reagent	Quantity (for 500 mL)	Final concentration
Tris-HCl (1 M, pH 7.5)	10 mL	20 mM
EDTA (0.5 M, pH 8.0) <!>	1 mL	1 mM
SDS (10% w/v) <!>	25 mL	0.5% (w/v)
H_2O	464 mL	

Store indefinitely at room temperature.

Protocol 2

High-yield Synthesis of RNA Using T7 RNA Polymerase and Plasmid DNA or Oligonucleotide Templates

When a large amount (milligram quantities) of RNA is desired (e.g., for making RNAs for RNA interference studies or for structural analyses), it is advantageous to use optimized transcription conditions. The method described here uses high millimolar concentrations of rNTPs, pyrophosphatase, and a different buffer than that described in Protocol 1. These conditions prevent product (pyrophosphate) inhibition of polymerase, which permits the accumulation of large quantities of RNA.

MATERIALS

The recipes for the items marked with <R> are at the end of the protocol or in the Common Recipes Appendix.

CAUTION: See the Cautions Appendix for appropriate handling of materials marked with <!>.

Reagents

ATP, GTP, UTP, and CTP stocks (100 mM each, pH ~7.0),
Dithiothreitol (DTT) (0.5 M) <!>
High-yield transcription buffer (10x) <R>
Inorganic pyrophosphatase (IPP) in storage buffer (0.5 unit/μL) <R>
$MgCl_2$ (1 M) <!>
RNasin RNase inhibitor (40 units/μL; Promega)
RQ1 RNase-free DNase (1 unit/μL; Promega)
Sodium acetate (3 M, pH 5.2) <R>
T7 RNA polymerase (commercial or homemade; see Appendix 3)
Template DNA, either 0.5 mg/mL cleaved plasmid DNA, gel-purified PCR product (0.1 mg/μL), with sequence of interest adjacent to a T7 promoter, or 20 μM annealed synthetic oligonucleotides

For hybridizing oligonucleotides to prepare templates 10–80 nucleotides in length, make a 20-μM mix of T7 top strand (17-mer) and the oligonucleotide sequence of interest in TE buffer containing 0.1 M NaCl. Heat the mixture for 2 min to 90°C and then leave it at room temperature until it cools to <37°C.

Equipment

Air chamber incubator or room at 37°C
Desalting column
Dry ice <!>
Ice
Heat block (90°C)
Microcentrifuge tubes
Preparative gel (optional; see Step 4)

OVERVIEW

Before beginning this protocol, generate template DNA either by restriction enzyme cleavage of plasmid DNA, by gel purification of a PCR product, or by annealing two synthetic DNA oligonucleotides. When setting up the reaction, add the components in the order listed; the RNA polymerase enzyme is added last. Note that the 10x high-yield transcription buffer contains spermidine, which can form an insoluble complex with the template DNA if the reaction is set up on ice. Therefore, it is critical that the components for the reaction (which *are kept on ice*) be added to a reaction tube at room temperature. After the reaction components are mixed, the tubes are incubated at 37°C. The transcripts are isolated from a desalting column or by preparative gel electrophoresis.

METHOD

Preparation of a Standard Unlabeled 100-μL Reaction

1. Combine the following in a microcentrifuge tube. Keep the components for the reaction on ice, but add them to a reaction tube at room temperature.

High-yield transcription buffer (10x)	10 μL
H_2O	up to 44 μL
DTT (0.5 M)	2 μL
*$MgCl_2$ (1 M)	2.6 μL
ATP (100 mM)	7.5 μL
CTP (100 mM)	7.5 μL
UTP (100 mM)	7.5 μL
GTP (100 mM)	7.5 μL
RNasin RNase inhibitor (40 units/μL)	2 μL
IPP (0.5 unit/μL)	2 μL
Template DNA (20 μM), annealed, gel-purified oligonucleotides or 0.5 mg/mL cleaved plasmid DNA or 0.1 mg/mL gel-purified PCR fragment	5 μL
T7 RNA polymerase (500 units/μL)	2 μL
Total	100 μL

*This $MgCl_2$ is added if the 100 mM NTP stocks do not contain magnesium. $MgCl_2$ concentrations should be 6 mM greater than the total NTP concentration.

2. Incubate the reaction in an air chamber incubator or room for 8 h to overnight at 37°C. Do not use a temperature block.
3. Add 5 μL of 1 unit/μL RQ1 RNase-free DNase and incubate for 30 min at 37°C.
4. Purify by a desalting gel filtration column and then a preparative gel if necessary. See Tips and Troubleshooting.

   Alternatively, do a larger reaction and remove 10 μL and add 1 μL of [α-^{32}P]UTP or [α-^{32}P]GTP and purify in parallel (see Protocol 3). To help quantite, add α-^{32}P nucleotides as in trace labeling. The remainder can be purified by a desalting column. If 20% of the nucleotides are incorporated, you should have made 200 μg per 100-μL reaction.

TIPS AND TROUBLESHOOTING

1. The most common problems with this procedure are inaccurate quantification of the template DNA and annealing the synthetic oligonucleotide templates. Usually, it is better to quantitate DNA on a gel rather than use OD_{260}.
2. The yields of synthetic RNAs from oligonucleotide templates are often much lower per template molecule than those with cleaved plasmid DNA templates. If this is the case and a partial duplex

oligonucleotide template is being used, such as that described in Milligan et al. (1987), try using a full-length top-strand oligonucleotide instead of just a top-strand promoter oligonucleotide.

3. If products are shorter than expected, either the transcription reaction produced truncated RNAs (which can occur when the enzymes pause at A/T-rich stretches in the template) or RNase contamination was introduced at some point during the procedure. Often, if this is the case, it is better to prepare all new reagents and begin again, rather than individually test each component separately. Create an RNase-free environment and a workflow to ensure that you and your labmates are free of RNases; this will go a long way to eliminate frustration caused by degraded RNA preps.
4. The reaction can be scaled up many-fold if necessary.

SOLUTION AND REAGENT RECIPES

CAUTION: See the Cautions Appendix for appropriate handling of materials marked with <!>.

High-yield Transcription Buffer (10x)

Reagent	Quantity (for 1 mL)	Final concentration (10x)
HEPES-KOH (2 M, pH 7.5)	0.5 mL	1 M
$MgCl_2$ (1 M) <!>	0.1 mL	0.1 M
Spermidine-HCl (1 M) <!>	20 µL	0.02 M
DTT (2 M) <!>	0.2 mL	0.4 M
Bovine serum albumin (BSA), RNase-free (20 mg/mL)	50 µL	1 mg/mL
H_2O	30 µL	

Store for up to 1 wk at −20°C.

Inorganic Pyrophosphatase (IPP) Stock and Storage Buffer

Reagent	Quantity (for 10 mL)	Final concentration
HEPES-KOH (1 M, pH 7.5)	500 µL	50 mM
DTT (2 M) <!>	5 µL	0.5 mM
Glycerol, autoclaved	5 mL	50%
H_2O	4.45 mL	

Resuspend the IPP (Sigma I1891; 1000 units/mg, 100 µg/vial) in the solution above at a concentration of 0.5 unit/µL. Store it for up to 1 yr at −20°C.

Sodium Acetate (0.3 M, pH 5.2)

For 1 L, add 40.83 g of NaOAc • $3H_2O$. Adjust the pH to 5.2 with glacial acetic acid <!> and bring the final volume to 1 L with H_2O. Filter and store at room temperature.

REFERENCE

Milligan JF, Groebe DR, Witherell GW, Uhlenbeck OC. 1987. Oligoribonucleotide synthesis using T7 RNA polymerase and synthetic DNA templates. *Nucleic Acids Res* **15:** 8783–8798.

READING LIST

Cunningham PR, Ofengand J. 1990. Use of inorganic pyrophosphatase to improve the yield of in vitro transcription reactions catalyzed by T7 RNA polymerase. *Biotechniques* **9:** 713–714.

Gurevich VV, Pokrovskaya ID, Obukohova TA, Zozylya SA. 1991. Preparative in vitro mRNA synthesis using SP6 and T7 RNA polymerases. *Anal Biochem* **195:** 207–213.

Pokrovskaya ID, Gurevich VV. 1994. In vitro transcription: Preparative RNA yields in analytical scale reactions. *Anal Biochem* **220:** 420–423.

Sambrook J, Russell DW. 2001. *Molecular Cloning: A laboratory manual*, 3rd ed. Cold Spring Harbor Laboratory Press, Cold Spring Harbor, NY.

Protocol 3

Determining the Yield of RNA Synthesized In Vitro

A simple way to determine the amount of RNA made by in vitro transcription is to include a trace amount of radioactive nucleotide and calculate the amount of nucleotides incorporated into the RNA, as described here. This protocol can also be used to quantitate unlabeled transcripts, assuming that the unlabeled RNA is equivalent to an aliquot of labeled RNA synthesized in parallel. Alternatively, if the transcription reactions were performed with no radioactive tracers, the RNA can be quantitated with a spectrophotometer (see Chapter 2, Protocol 13). Radioactive RNA should be analyzed by gel electrophoresis on an appropriate gel to check quality. The quality of large unlabeled transcripts can be checked by running them through an agarose gel and staining them with SYBR Gold or ethidium bromide.

MATERIALS

Reagents

RNA sample (1 µL) from transcription reaction (Protocol 1)

Equipment

Micro Bio-Spin Columns with Bio-Gel P-6 in Tris buffer (Bio-Rad)
Microcentrifuge tubes
Scintillation vial and counter
Spectrophotometer

OVERVIEW

This protocol describes a simple way to quantitate RNA and determine its specific activity.

METHOD

1. When setting up the labeled transcription reactions (Protocol 1), remember to transfer 1 µL of the reaction to a microcentrifuge tube containing 10 µL of H_2O. Keep the tube in a safe place until after the RNA has been purified.

   This sample will be used as the input during the calculation in Step 5.

2. After the desired synthesis reaction has been performed, remove the unincorporated nucleotides by gel purification (Chapter 2, Protocol 12).

   If preferred, remove the nucleotides with a Micro Bio-Spin Column. The RNA is then precipitated with ethanol (see Chapter 2, Protocol 3) and resuspended in H_2O.

3. After RNA purification, transfer an appropriate quantity of purified RNA to a microcentrifuge tube.

   For example, if the reaction volume was 50 µL and 1 µL of input was saved in Step 1, transfer 1/50 of the purified RNA to the microcentrifuge tube.

4. Place each microcentrifuge tube directly into a scintillation vial and count in a scintillation counter.
5. Calculate the percentage of nucleotides incorporated:

$$\frac{\text{cpm purified RNA} \times 100}{\text{cpm input}} = \text{\% of NTPs incorporated.}$$

   If the transcription reaction was performed with 50 µM GTP (the limiting nucleotide), the contribution of the [α-^{32}P]GTP to the concentration of nucleotides is insignificant.

6. Determine the total amount of NTPs in the reaction. (If one of the nucleotides is at a lower concentration, assume that the other nucleotides are at the same concentration as the limiting nucleotide.)

   For example, if the NTPs are at a final concentration of 50 µM in a 50-µL reaction, then there are 2.5 nmol of each of the four NTPs in the reaction. Therefore, there is a total of 10 nmol of NTPs (2.5 nmol x 4) in the reaction.

   If 10% of the radioactive nucleotides were incorporated (as determined in Step 5), 0.1 x 10 nmol = 1 nmol of RNA was synthesized in the 50-µL reaction described in the example in Step 6.

Protocol 4

Gel Purification of RNA

For many applications, including size selection of RNAs and purification of in vitro transcription products, it is necessary to purify RNAs on a denaturing gel. This procedure describes how to purify transcripts that have been synthesized in vitro (e.g., from Protocols 1 and 2). It can also be used to isolate small RNAs (Chapter 8, Protocol 6). In general, RNA purification by denaturing gel electrophoresis is practical only when the size of the desired RNA is 600 nucleotides or less.

MATERIALS

The recipes for the items marked with <R> are at the end of the protocol or in the Common Recipes Appendix.

CAUTION: See the Cautions Appendix for appropriate handling of materials marked with <!>.

Reagents

Bromophenol blue dye <!>
Denaturing polyacrylamide gel <!>
EDTA (2 mM and 0.5 M, pH 8.0) <!>
Ethanol (100%) <!>
Formamide gel-loading buffer <R> <!>
Gel elution buffer <R>
GlycoBlue (15 μg/μL) or tRNA carrier
Phenol:chloroform (1:1) <!>
RNA from transcription reaction (see Protocols 1 and 2)
SDS (10%) <!>
Sodium acetate (3 M)
TBE running buffer <R> (see Appendix 1)

Equipment

Dry ice <!>
Electrophoresis apparatus for denaturing polyacrylamide gel and power supply
Fluorescent markers (luminescent paint) also available commercially from art supply stores
Hand-held UV light (254 nm) <!>
Ice
Microcentrifuge
Microcentrifuge tubes
Plastic wrap
Scalpel or single-edge razor blade
Spin columns (e.g., Bio-Spin 6 columns in Tris buffer; Bio-Rad)
Temperature block (95°C)
X-ray film and developer
X-ray intensifying screen

OVERVIEW

This procedure is reliable for extracting RNAs from denaturing polyacrylamide gels. It is useful for lableled RNAs or unlabeled RNAs when sufficient mass is present. RNA that has been synthesized or labeled in vitro can be first run through a spin column to remove excess unincorporated radioactivity and then precipitated with ethanol. Start the protocol with RNA prepared by in vitro transcription using any of the protocols in this chapter or other labeled cellular RNAs that need to be gel-purifed for further anlysis.

METHOD

1. After PCA extraction and precipitation of the RNA with ethanol, resuspend RNA from a 50-µL transcription or 5′- or 3′-end-labeling reaction in 2.5 µL of 2 mM EDTA (pH 8), and add 5 µL of formamide gel-loading buffer.
2. Prepare a denaturing polyacrylamide gel as described in Chapter 2, Protocol 12. Set up the gel in the gel box, add running TBE buffer to the upper and lower reservoirs, and prerun the gel for 15–45 min at a maximum of 1500 V/45 mA.

   If the RNA transcript is greater than 100 nucleotides, do shorter preruns (15–20 min).
3. Heat the sample from Step 1 for 1 min at 95°C and then place it on ice.
4. Load the sample at 7.5 µL/lane. Use one lane for the RNA from one 50-µL transcription reaction (see Protocol 1). If total cellular RNA is labeled, load it at 5–10 µg per lane. These amounts will not overload the lanes. (Refer to thick and thin gels in lanes section of Protocol 12 in Chapter 2 to decide which gel to run.) Run the gel at a maximum of 1500 V/45 mA until the bromophenol blue dye is about halfway down the gel.

   If the sample is run too far, unincorporated radiolabel will be present in the lower buffer chamber, increasing the chances of unwanted contamination. This will not affect the RNA preparation, but it makes the cleanup process more time-consuming.

   If the RNA is not radiolabeled, proceed to Step 5; if the RNA is radiolabeled, proceed to Step 6.
5. If the RNA is not radiolabeled, visualize the band by UV shadowing.
   i. Remove one glass plate and cover the gel with plastic wrap.
   ii. Cut a wide area of the gel expected to contain the band of RNA and transfer it (plastic wrap side down) to a clean X-ray intensifying screen.
   iii. Shine the UV light on the gel piece.

      The exact location of the RNA band will appear purple in the gel under UV light.
   iv. Excise the band with a razor blade and then proceed to Step 7.
6. If the RNA is radiolabeled, visualize the band as follows:
   i. Remove one glass plate and cover the gel with plastic wrap. Place fluorescent ink markers on the corners of the gel and expose the gel to X-ray film.

      High-specific-activity transcripts should require 10–30 sec; trace-labeled transcripts may take 5–10 min.
   ii. Identify the location of the radiolabeled RNA on the gel, using the fluorescent ink markers on the gel to line up the gel with the marks on the film.
   iii. Use a razor blade to excise the pieces of the acrylamide gel that contain the RNA and then continue to Step 7.
7. Place the gel slices in a microcentrifuge tube with 400 µL of gel elution buffer (up to three slices from the same RNA in one tube).

8. Freeze the tube containing the buffer and gel pieces on dry ice for 15 min and then allow the RNA to diffuse from the gel by leaving the tube overnight at room temperature.
9. Centrifuge the tube at maximum speed for 10 min at room temperature, and draw off and save the supernatant in a clean microfuge tube.
   This supernatant should contain ≥70% of the RNA.
10. Extract the RNA with 1–2 volumes of 1:1 phenol:chloroform, then with 1 volume of chloroform. Precipitate the RNA with 2 volumes of ethanol.
11. Resuspend in 5–10 µL of H_2O and quantitate by incorporation of label or by A_{260} if necessary. Dilute to 500 ng/µL with H_2O.
    RNA should be stored short term at –20°C or indefinitely at –80°C.

SOLUTION AND REAGENT RECIPES

CAUTION: See the Cautions Appendix for appropriate handling of materials marked with <!>.

Formamide/Dye Loading Buffer

Reagent	Quantity (for 10 mL)	Final concentration
Deionized formamide <!>	9.5 mL	95%
Bromophenol blue <!>	2.5 mg	0.025% (w/v)
Xylene cyanol FF <!>	2.5 mg	0.025% (w/v)
EDTA (0.5 m, pH 8.0) <!>	100 mL	5 mm
H_2O	400 mL	

Store in aliquots at –20°C.

Gel Elution Buffer

Reagent	Quantity (for 500 mL)	Final concentration
Tris-HCl (1 M, pH 7.5)	10 mL	20 mM
Sodium acetate (2.5 M, prepared without pH adjustment)	50 mL	0.25 M
EDTA (0.5 M, pH 8.0) <!>	1 mL	1 mM
SDS (10%, w/v) <!>	12.5 mL	0.25%
H_2O	426.5 mL	

If the SDS precipitates, warm to 37°C until the precipitate disappears. Store indefinitely at room temperature.

Protocol 5

3′-End Labeling of RNA with [^{32}P]pCp and T4 RNA Ligase 1

This protocol is used to radiolabel the 3′ ends of RNAs, either synthesized by in vitro transcription or purified from cells or tissues, by ligation of [^{32}P]cytidine 3′,5′-bis(phosphate) (pCp). [^{32}P]pCp can be obtained commercially or be prepared in the laboratory using polynucleotide kinase to phosphorylate Cp with [γ-^{32}P]ATP. "Homemade" pCp is considerably cheaper and has a higher final concentration than pCp obtained from commercial sources. The labeling protocol uses T4 RNA ligase 1, which covalently joins pCp to the free 3′ hydroxyl of RNA. 3′-End-labeled RNAs are useful for in vitro processing reactions such as splicing (Chapter 6, Protocol 2), footprinting (Chapter 5, Protocol 6), chemical sequencing, gel-shift analyses (Chapter 5, Protocol 2), and chemical probing (Chapter 5, Protocol 4). Ligation of unlabeled pCp to 3′ ends can also inhibit nuclease activity because of the 3′ phosphate.

MATERIALS

The recipes for the items marked with <R> are at the end of the protocol or in the Common Recipes Appendix.

CAUTION: See the Cautions Appendix for appropriate handling of materials marked with <!>.

Reagents

ATP (10 mM)
Dimethylsulfoxide (DMSO) (100%) <!>
Dithiothreitol (DTT) (0.1 M) <!>
EDTA (0.5 M, pH 8.0) <!>
Ethanol (100%, prechilled to –20°C) <!>
$MgCl_2$ (1 M) <!>
Phenol:chloroform:isoamyl alcohol (PCA) <!>
[5′-^{32}P]pCp (cytidine 3′,5′-bis[phosphate]) (3000 Ci/mmol; 10 mCi/mL) <!>
    PerkinElmer NEG019A has a concentration of 3.3 µM or [5′-^{32}P]pCp prepared in one's own laboratory (see below).
RNA ligase buffer (10x) <R>
RNA sample of interest (in vitro synthesized or in vivo purified)
Sodium acetate (NaOAc; 3 M, pH 5.2)
T4 RNA ligase 1 (900 units/mL) (US Biochemicals)
TE buffer <R>

Equipment

Dry ice <!>/ethanol <!> bath
Microcentrifuge
Microcentrifuge tubes
Sephadex G-25 gel filtration spin columns
SpeedVac

OVERVIEW

This procedure is used to radiolabel the 3′ ends of in vitro–synthesized or in vivo–purified RNA molecules that have free 3′-hydroxyl termini. For best labeling, labeled pCp should be at least equimolar or higher to available 3′-hydroxyl ends. The reaction takes overnight at low temperature and requires the use of [5′-^{32}P]pCp and T4 RNA ligase 1. At the end of the procedure, the reaction is desalted by gel filtration to remove any unincorporated [5′-^{32}P]pCp.

METHOD

1. Set up the following labeling reaction in a microcentrifuge tube:

RNA ligase buffer (10×)	3 µL
ATP (10 mM)	3 µL
DTT (0.1 M)	1 µL
DMSO	3 µL
RNA sample of interest	*x* µL (~1 µM final concentration of 3′ ends of single-stranded RNA)
[5′-^{32}P]pCp (3000 Ci/mmol; 10 mCi/mL)	9.1 µL (1 µM final concentration)
T4 RNA ligase 1 (9 units/µL)	3 µL
H_2O	to 30 µL

2. Incubate the reaction overnight at 4°C–5°C.
3. Stop the reaction by adding 2 µL of 0.5 M EDTA (pH 8.0).
4. Remove unused [^{32}P]pCp from the reaction using a gel filtration spin column.
5. PCA-extract and ethanol-precipitate the RNA as follows.
   i. Dilute to 200 µL with TE buffer and add sodium acetate to 0.3 M. Phenol-extract with an equal volume of phenol.
   ii. Add 3 volumes of cold 100% ethanol.
   iii. Chill for at least 10 min in a dry ice/ethanol bath.
   iv. Centrifuge at maximum speed in a microcentrifuge for at least 10 min.
   v. Discard the supernatant. Add 100 µL of cold 100% ethanol to the pellet and repeat Steps 5.iii–iv.
   vi. Discard the supernatant. Briefly dry the RNA pellet in a SpeedVac.
6. Resuspend the RNA pellet in 10 µL of H_2O and store for up to 2 wk at –20°C or –80°C.

TIPS AND TROUBLESHOOTING

Important note: Make sure any ammonium (NH_4^+) ions are removed (e.g., after precipitation with ammonium acetate), because any free NH_4^+ ions will inhibit T4 RNA ligase. RNAs with secondary structure at their 3′ ends are labeled inefficiently.

SOLUTION AND REAGENT RECIPE

CAUTION: See the Cautions Appendix for appropriate handling of materials marked with <!>.

RNA Ligase Buffer (10x)

Reagent	Quantity (for 1.0 mL)	Final concentration (10x)
Tris-HCl (1 M, pH 8.0)	500 µL	0.5 M
$MgCl_2$ (1 M) <!>	150 µL	0.15 M
DTT (0.5 M) <!>	66 µL	33 mM
H_2O	284 µL	

Store for up to 6 mo at –20°C.

Homemade pCp

Reagent	Quantity
OHcP (1 mM stock)	1 µL
OptiKinase buffer	4 µL
[γ-^{32}P]ATP (150 µCi/µL; 6000 Ci/mM)	33 mL
OptiKinase	2 µL

Incubate for 1 h at 37°C. Treat for 10 min at 65°C and store at –20°C.

READING LIST

England TE, Uhlenbeck OC. 1978. 3′-Terminal labelling of RNA with T4 RNA ligase. *Nature* **275:** 560–561.

Protocol 6

3′-End Labeling of RNA with Yeast Poly(A) Polymerase and 3′-Deoxyadenosine 5′-[α-^{32}P]Triphosphate (Cordycepin 5′-[α-^{32}P]Triphosphate)

This procedure can be used to radiolabel the 3′ ends of RNA molecules using yeast poly(A) polymerase and 3′-deoxyadenosine 5′-[α-^{32}P]triphosphate (cordycepin 5′-[α-^{32}P]triphosphate), a chain-terminating nucleotide. Poly(A) polymerase catalyzes the addition of adenosine residues to the 3′ ends of RNA molecules (Sippel 1973; Edmonds 1982). The reaction requires Mn^{++} or Mg^{++}, ATP and RNA molecules with 3′-hydroxyl termini. In this protocol, ^{32}P-labeled cordycepin-5′-triphosphate (3′-dATP) is substituted for ATP; this results in the addition of a single radiolabeled nucleotide to the 3′ end of each RNA molecule (Martin and Keller 1998). Other modified nucleotides (e.g., aminohexyl ATP) can also be used to tail the 3′ end of RNA molecules using yeast poly(A) polymerase (Martin and Keller 1998).

MATERIALS

The recipes for the items marked with <R> are at the end of the protocol or in the Common Recipes Appendix.

CAUTION: See the Cautions Appendix for appropriate handling of materials marked with <!>.

Reagents

Chloroform <!>

3′-Deoxyadenosine 5′-triphosphate (cordycepin 5′-[α-^{32}P]triphosphate) (5000 Ci/mmol; 10 mCi/mL) <!>

PerkinElmer NEG026 has a concentration of 2 μM.

Ethanol (100%, prechilled to –20°C) <!>

Phenol-SEVAG <!>

Poly(A) polymerase buffer (5x) <R>

RNA sample of interest (in vitro synthesized or in vivo purified)

SDS extraction buffer <R>

Sodium acetate (3 M, pH 5.2)

TE buffer <R>

Yeast poly(A) polymerase (600 units/μL; US Biochemicals)

Equipment

Dry ice <!>/ethanol <!> bath

Incubator or water bath

Microcentrifuge

Microcentrifuge tubes

Sephadex G-25 gel filtration spin columns

SpeedVac

OVERVIEW

This protocol labels in vitro–synthesized or in vivo–purified RNA molecules that have free 3′-hydroxyl termini. The reaction takes 10 min at 37°C and requires the use of 3′-deoxyadenosine 5′-

[α-^{32}P]triphosphate (cordycepin 5′-[α-^{32}P]triphosphate) and yeast poly(A) polymerase. At the end of the procedure, the reaction is desalted by gel filtration to remove any unincorporated nucleotides.

METHOD

1. Set up the labeling reaction in a microcentrifuge tube:

Poly(A) polymerase buffer (5x)	2 µL
3′-Deoxyadenosine 5′-[α-^{32}P]triphosphate (cordycepin 5′-[α-^{32}P]triphosphate) (5000 Ci/mmol; 10 mCi/mL)	2 µL (0.4 µM final concentration)
RNA sample of interest	*x* µL (0.2 µM final concentration)
H_2O	to 9 µL
Yeast poly(A) polymerase (600 units/µL)	1 µL
Total	10 µL

2. Incubate the reaction for 30 min at 37°C.
3. Add 40 µL of TE buffer and remove unincorporated nucleotides using a gel filtration spin column.
4. Dilute to 200 µL with SDS extraction buffer plus 0.3 M sodium acetate. SDS must be removed for any subsequent enzymatic reaction.
5. Precipitate the RNA as follows:
   i. Add 0.1 volume of 3 M sodium acetate (pH 5.2) and 3 volumes cold 100% ethanol.
   ii. Chill for at least 10 min in a dry ice/ethanol bath.
   iii. Centrifuge at maximum speed in a microcentrifuge for at least 10 min at 4°C.
   iv. Discard the supernatant. Wash the pellet by adding 100 µL of cold 100% ethanol and chilling and centrifuging again, as described in Steps 5.ii and 5.iii.
   v. Discard the supernatant. Air-dry the RNA pellet or briefly dry the pellet in a SpeedVac.
6. Resuspend the pellet in H_2O and store at –80°C.

SOLUTION AND REAGENT RECIPES

CAUTION: See the Cautions Appendix for appropriate handling of materials marked with <!>.

Poly(A) Polymerase Buffer (5x)

Reagent	Quantity (for 1.0 mL)	Final concentration (5x)
Tris-HCl (1 M, pH 8.0)	100 µL	0.1 M
$MnCl_2$ (0.3 M) <!>	10 µL	3 mM
EDTA (0.1 M) <!>	1 µL	0.1 mM
DTT (0.5 M) <!>	2 µL	1 mM
Bovine serum albumin (BSA), acetylated (50 mg/mL)	10 µL	500 µg/mL
H_2O	877 µL	

Store indefinitely at –20°C.

SDS Extraction Buffer

Reagent	Quantity (for 500 mL)	Final concentration
Tris-HCl (1 M, pH 7.5)	10 mL	20 mM
EDTA (0.5 M, pH 8.0) <!>	1 mL	1 mM
SDS (10% w/v) <!>	25 mL	0.5% (w/v)
H_2O	464 mL	

Store indefinitely at room temperature.

REFERENCES

Edmonds M. 1982. In *The enzymes*, 3rd ed. (ed. PD Boyer), vol. 15, pp. 217–244. Academic, NY.

Martin G, Keller W. 1998. Tailing and 3′-end labeling of RNA with yeast poly(A) polymerase and various nucleotides. *RNA* **4:** 226–230.

Sippel AE. 1973. Purification and characterization of adenosine triphosphate: Ribonucleic acid adenyltransferase from *Escherichia coli*. *Eur J Biochem* **37:** 31–40.

Tu CP, Cohen SN. 1980. 3′-End labeling with [α-^{32}P]cordycepin-5′-triphosphate. *Gene* **10:** 177–183.

Protocol 7

5′-End Labeling of RNA with [γ-^{32}P]ATP and T4 Polynucleotide Kinase

This procedure uses T4 polynucleotide kinase to catalyze the transfer of a radiolabeled, terminal (γ) phosphate of ATP to the 5′-hydroxyl terminus of a DNA or RNA molecule. The reaction is very efficient and hence is a general method for phosphorylating polynucleotides or oligonucleotides; it can be used to radiolabel the 5′ ends of RNA molecules for protein footprinting (Chapter 5, Protocol 6), chemical probing or protection (Chapter 5, Protocols 4 and 5), chemical sequencing, and sequencing by specific RNase digestions.

MATERIALS

The recipes for the items marked with <R> are at the end of the protocol or in the Common Recipes Appendix.

CAUTION: See the Cautions Appendix for appropriate handling of materials marked with <!>.

Reagents

EDTA (0.5 M, pH 8.0) <!>
Ethanol (100%) <!>
[γ-^{32}P]ATP (>7000 Ci/mmol; 167 mCi/mL) <!>
$MgCl_2$ (1 M) <!>
Phenol <!>
RNA sample of interest (~20 pmol of 5′ ends)
Sodium acetate (3 M, pH 5.2)
T4 kinase buffer (10x) <R>
T4 polynucleotide kinase (10 units/μL) (e.g., US Biochemicals OptiKinase)
TE buffer <R>

Equipment

Dry ice <!>/ethanol <!> bath
Incubator or water bath (37°C)
Microcentrifuge
Microcentrifuge tubes
Pasteur pipette
Sephadex G-25 gel filtration spin columns
SpeedVac

OVERVIEW

This protocol uses in vitro–synthesized or synthetic RNA that has free 5′-hydroxyl termini. If the RNA was prepared by in vitro transcription (and has a 5′ phosphate or a 5′ triphosphate), it must be treated with calf intestine alkaline phosphatase (CIP) to remove the 5′ phosphate(s) (see Chapter 3, Protocol 6) before rephosphorylation with T4 polynucleotide kinase.

METHOD

1. Set up a kinase reaction with RNA (~20 pmol of 5′ ends) in the following order in a microfuge tube:

H_2O	15 µL
T4 kinase buffer (10x)	2 µL
[γ-^{32}P]ATP (>7000 Ci/mmol; 167 mCi/mL)	2 µL (~100 pmol)
T4 polynucleotide kinase (10 units/µL)	2 µL

2. Incubate the reaction for 1 h at 37°C.
3. Stop the reaction by adding 2 µL of 0.5 M EDTA (pH 8.0).
4. Remove unincorporated radioactivity using a gel filtration spin column.
5. Precipitate the RNA with ethanol.
   i. Dilute the RNA to 200 µL with SDS extraction buffer (see Appendix) and add sodium acetate to a final concentration of 0.3 M.
   ii. Add 3 volumes of cold 100% ethanol.
   iii. Chill in a dry ice/ethanol bath for at least 10 min.
   iv. Centrifuge at maximum speed in a microcentrifuge for 10 min.
   v. Aspirate the supernatant with a pasteur pipette. Add 100 µL of cold 100% ethanol to the pellet and repeat Steps 6.iii–iv.
   vi. Discard the supernatant. Air-dry the RNA pellet or briefly dry the pellet in a SpeedVac.
6. Resuspend the RNA pellet in 10 µL of H_2O. Store for up to 2 wk at –20°C or –80°C.

TIPS AND TROUBLESHOOTING

It is important to make sure that any ammonium (NH_4^+) ions are removed (if RNA has been precipitated with ammonium acetate), because any free NH_4^+ ions will inhibit T4 polynucleotide kinase.

SOLUTION AND REAGENT RECIPE

CAUTION: See the Cautions Appendix for appropriate handling of materials marked with <!>.

T4 Kinase Buffer (10x)

Reagent	Quantity (for 1 mL)	Final concentration (10x)
Tris-HCl (1 M, pH 7.5)	500 µL	0.5 M
$MgCl_2$ (1 M) <!>	100 µL	0.1 M
DTT (0.5 M) <!>	100 µL	50 mM
H_2O	300 µL	

Store for up to 6 mo at –20°C.

READING LIST

Chaconas G, van de Sande JH. 1980. 5′-^{32}P labeling of RNA and DNA restriction fragments. *Methods Enzymol* **65:** 75–88.

Conway L, Wickens M. 1989. Modification interference analysis of reactions using RNA substrates. *Methods Enzymol* **189:** 369–379.

Harrison B, Zimmerman SB. 1986. T4 polynucleotide kinase: Macromolecular crowding increases the efficiency of reaction at DNA termini. *Anal Biochem* **158:** 307–315.

Richardson CC. 1981. Bacteriophage T4 polynucleotide kinase. In *The enzymes* (ed. PD Boyer), vol. 14A, pp. 299–314. Academic, NY.

Protocol 8

Site-specific Labeling and Substitution of RNA

In this protocol, in vitro–transcribed RNA is specifically cleaved in a functionally important region, radiolabeled, and religated to produce the original RNA with an internal radiolabel. The introduced label can then be used to analyze the RNA in subsequent procedures such as cross-linking (Chapter 5, Protocol 10) and nuclease protection analyses (Chapter 5, Protocols 5 and 6).

MATERIALS

The recipes for the items marked with <R> are at the end of the protocol or in the Common Recipes Appendix.

CAUTION: See the Cautions Appendix for appropriate handling of materials marked with <!>.

Reagents

Annealing buffer (6x) <R>
Calf intestine alkaline phosphatase (CIAP) (New England BioLabs) (10 units/μL)
Cap analog (GpppG) (40 mM)
Capped RNA with low specific activity (prepared according to Protocol 1)
Chimeric oligonucleotide (100 ng/μL)

Chimeric oligonucleotides are typically 14-mer 2′-*O*-methyl RNA oligonucleotides that contain four consecutive deoxynucleotides beginning at the fourth base from the 5′ end. These deoxynucleotides determine the cut site: RNase H cleaves the RNA at the position that corresponds to the 5′ side of the first deoxy residue of the oligonucleotide (see Fig. 4.1).

DNase I, RNase-free (10 units/μL) (Roche)
EDTA (0.5 M, pH 8.0; 2 mM, pH 8.0) <!>
EGTA (0.1 M, pH 8.0)
Ethanol (100% and 70%) <!>
Formamide loading buffer <R>
Gel elution buffer <R>
GlycoBlue (10 μg/μL)
Ligase (5 units/μL)
OptiKinase buffer (10x) <R>
Phosphatase buffer (10x) <R>
Rapid ligation mix (2x) <R>
RNase H buffer (5x) <R>
RNase H (Invitrogen) (2 units/μL)
RNase inhibitor (40 units/μL)
T4 DNA
TE buffer <R>

Equipment

Denaturing polyacrylamide gel apparatus
Dry ice <!>
Heat block (95°C)
Micro Bio-Spin Columns with Bio-Gel P-6 in Tris buffer (Bio-Rad)

Micro Bio-Spin Columns with Bio-Gel P-30 in Tris buffer (Bio-Rad)
Microcentrifuge
Water bath (37°C) or thermal cycler (55°C)

OVERVIEW

This method has many steps that must be checked before beginning the protocol. RNase H cleavage by chimeric oligonucleotides should be tested on a labeled substrate to make sure that the oligonucleotides direct complete and specific cleavage. All fragments should be carefully quantitated so that equimolar amounts of the ligation components are used. These substrates must be capped so that the only site labeled is the internal site of interest.

METHOD

RNA Transcription and Purification

1. Prepare capped RNA with low specific activity according to Protocol 1, using a total reaction volume of 200 µL. Add 4 µL of 10 units/µL DNase I during the appropriate step. When setting up the reaction, remove 1 µL to count as input for calculation of yield in Protocol 7.
2. Purify the RNA as described in Protocol 4. Quantify the RNA as described in Protocol 3.

Cleavage of RNA with a Chimeric Oligonucleotide

3. Mix the following in a microcentrifuge tube:

Trace-labeled RNA from Step 2	2–5 µg
Chimeric oligonucleotide	500 ng
H_2O	to 14 µL (total volume)

4. Heat for 2 min at 95°C and then place on ice.
5. Add

RNase H buffer (5x)	4 µL
RNase H (2 units/µL) (Invitrogen)	2 µL

6. Incubate for 1 h at 37°C.
7. Add

Gel elution buffer without SDS	400 µL
EDTA (0.5 M, pH 8.0)	10 µL
GlycoBlue 15 µg/µL	1–2 µL

8. PCA-extract and precipitate with 2 volumes of 100% ethanol (see Chapter 2, Protocols 1 and 3).
9. Resuspend the cut RNA in 75 µL of TE buffer and pass it through a Micro Bio-Spin Column with Bio-Gel P-30.

   This step removes the chimeric oligonucleotide without an additional gel-purification step, but only works with capped RNA. If using uncapped RNA or if cleavage is inefficient, gel-purify the 5′ and 3′ pieces after RNase H digestion.

10. Add 325 µL of gel elution buffer without SDS and add 1 µL of 15 µg/µL GlycoBlue. Precipitate the RNA with 2.5 volumes of ethanol, freeze-dry on ice for 15 min, thaw, and microcentrifuge at maximum speed for 10 min. Pour off the supernatant and resuspend the cut RNA in 42 µL of H_2O. Store at –20°C or –80°C until needed.

Dephosphorylation of Cleaved RNA

11. To the cut RNA (42 µL) from Step 10, add the following to produce a 50-µL reaction:

Phosphatase buffer (10x)	5 µL
RNasin (40 units/µL)	1 µL
CIAP (10 units/µL)	2 µL

12. Incubate for 30 min at 37°C.
13. Add

Gel elution buffer with SDS	300 µL
EGTA (0.1 M, pH 8.0)	8 µL

14. PCA-extract the RNA and precipitate it with 2 volumes of ethanol. Freeze on dry ice for 15 min, thaw, and microfuge at maximum speed for 10 min. Pour off the supernatant and wash the pellet with 70% ethanol.
15. Resuspend the RNA at 300 ng/µL in H_2O.

    RNA can be stored at –20°C until needed.

Phosphorylation of RNA with [γ-^{32}P]ATP

16. Mix the following reagents in the order listed in a microcentrifuge tube:

H_2O	8 µL
RNA (300 ng/µL, cleaved and CIAP-treated from Step 15)	1 µL
OptiKinase buffer (10x)	1.5 µL
[γ-^{32}P]ATP (6000 Ci/mmol; 150 µCi/µL)	3 µL
OptiKinase (10 units/µL)	1.5 µL

17. Incubate for 30 min at 37°C.
18. Add 35 µL of TE and pass through a Micro Bio-Spin Column with Bio-Gel P-6. Set aside on ice 2.5 µL of flowthrough.

    This will be run on the gel in Step 28 and serve as a reference marker.
19. Add 250 µL of gel elution buffer with SDS, 1 µL of 10 µg/µL GlycoBlue, PCA-extract, and precipitate with 2 volumes of ethanol. Wash the RNA pellet with 500 µL of 70% ethanol.
20. Resuspend the radiolabeled RNA in 4 µL of H_2O. Proceed to Step 21.

    Radiolabeled RNA can be stored for a few days at –20°C in ethanol if necessary. However, we recommend proceeding to Step 21 immediately.

Ligation of RNA using a Bridge Oligonucleotide

21. Mix the following reagents in the following order in a microcentrifuge tube:

Radiolabeled RNA from Step 20	4 µL
Bridge oligonucleotide (resuspended to be equimolar to the RNA half-molecule to be ligated)	1 µL (no more than 1 M equivalent)
Annealing buffer (6x)	1 µL
Total	6 µL

    It is important not to use too much bridge oligonucleotide. Excess bridge oligonucleotide may cause each bridge oligonucleotide to hybridize to only one of the RNA half-molecules (either the 5′ half-molecule or the 3′ half-molecule). For ligation to occur, both the 5′ and 3′ half-molecules need to be base-paired to the same bridge oligoculeotide.

22. Heat the mixture for 1 min at 95°C and then for 3 min at 55°C, and place on ice.
23. Add the following reagents in the following order to the mixture from Step 22:

Rapid ligation mix (2x)	10 µL
RNasin (40 units/µL)	1 µL
Ligase (5 units/µL)	2 µL
H_2O	1 µL

24. Incubate for 1 h at room temperature.
25. Add the following to the mixture from Step 24:

Gel elution buffer with SDS	380 µL
GlycoBlue (10 µg/µL)	2 µL

26. PCA-extract and precipitate with 2 volumes of 100% ethanol.
27. Resuspend in 2.5 µL of 2 mM EDTA and add 5 µL of formamide dye.
28. Add 5 µL of formamide dye to the nonligated sample from Step 18. Heat both samples for 2 min at 95°C, place on ice, and load on a denaturing polyacrylamide gel. Cut out the ligated band and elute the RNA from the gel slice as described in Protocol 4.

TIPS AND TROUBLESHOOTING

1. If the starting RNA is uncapped, the 5′ and 3′ pieces of the RNase H cleavage reaction must be gel-purified. Otherwise, the 5′ fragment will be dephosphorylated and labeled along with the 3′ fragment.
2. If extra bands appear, make sure that the chimeric oligonucleotide cuts only once at the desired position.
3. If efficiency is low, check phosphatase and labeling steps.
4. If ligation efficiency is low, make sure fragments and the bridge oligonculeotide are quantitated correctly.

SOLUTION AND REAGENT RECIPES

CAUTION: See the Cautions Appendix for appropriate handling of materials marked with <!>.

Annealing Buffer (6x)

Reagent	Quantity (for 1 mL)	Final concentration (6x)
Tris-HCl (1 M, pH 7.4)	60 µL	60 mM
KCl (2 M) <!>	150 µL	300 mM
H_2O	0.79 mL	

Store indefinitely at –20°C.

Gel Elution Buffer (with SDS)

Reagent	Quantity (for 500 mL)	Final concentration
Sodium acetate (2.5 M, unpH'd)	50 mL	250 mM
Tris-HCl (1 M, pH 7.5)	10 mL	20 mM
EDTA (0.5 M, pH 8.0) <!>	1 mL	1 mM
SDS (10%, w/v) <!>	12.5 mL	0.25% (w/v)
H_2O	to 500 mL	

Store indefinitely at room temperature.

Gel Elution Buffer (no SDS)

Reagent	Quantity (for 500 mL)	Final concentration
Tris-HCl (1 M, pH 7.5)	10 mL	20 mM
Sodium acetate (2.5 M, unpH'ed)	50 mL	0.25 M
EDTA (0.5 M, pH 8.0) <!>	1 mL	1 mM
H_2O	to 500 mL	

Store indefinitely at 4°C.

OptiKinase Buffer (10x)

Reagent	Quantity (for 1 mL)	Final concentration (10x)
Tris-HCl (1 M, pH 7.5)	500 µL	0.5 M
$MgCl_2$ (1 M) <!>	100 µL	100 mM
DTT (1 M) <!>	50 µL	50 mM
H_2O	350 µL	

Store for up to 6 mo at –20°C.

Phosphatase Buffer (10x)

Reagent	Quantity (for 1 mL)	Final concentration (10x)
Tris-HCl (1 M, pH 7.9)	0.5 mL	0.5 M
$MgCl_2$ (1 M) <!>	0.1 mL	100 mM
NaCl (5 M)	0.2 mL	1 M
DTT (1 M) <!>	10 µL	10 mM
H_2O	0.19 mL	

Store for up to 6 mo at –20°C.

Rapid Ligation Mix (2x)

Reagent	Quantity (for 1 mL)	Final concentration (2x)
Tris-HCl (1 M, pH 7.5)	132 µL	132 mM
$MgCl_2$ (1 M) <!>	20 µL	20 mM
DTT (1 M) <!>	2 µL	2 mM
ATP (100 mM)	20 µL	2 mM
Polyethylene glycol (PEG) (50%, w/v)	300 µL	15% (w/v)
H_2O	526 µL	

Store for up to 6 mo at –20°C. Note that rapid ligation buffers are available from many commercial suppliers.

RNase H Buffer (5x)

Reagent	Quantity (for 10 mL)	Final concentration (5x)
Tris-HCl (1 M, pH 7.5)	1 mL	100 mM
KCl (2 M) <!>	2.5 mL	500 mM
$MgCl_2$ (1 M) <!>	0.5 mL	50 mM
DTT (1 M) <!>	0.5 mL	50 mM
Sucrose	2.5 g	25 (w/v)
H_2O	to 10 mL	

If this buffer is used infrequently, prepare a stock without DTT, aliquot into 1-mL aliquots, and add 50 µL of 1 M DTT to each aliquot before use. Store working aliquot with DTT for up to 1 mo at –20°C.

READING LIST

Maroney PA, Romfo C, Nilsen TW. 2000. Nuclease protection of RNAs containing site-specific labels: A rapid method for mapping RNA-protein interactions. *RNA* **6:** 1905–1909.

Moore MJ, Sharp PA. 1992. Site-specific modification of pre-mRNA: The 2′-hydroxyl groups at the splice sites. *Science* **256:** 992–997.

Yu YT, Shu MD, Steitz JA. 1997. A new method for detecting sites of 2′-*O*-methylation in RNA molecules. *RNA* **3:** 324–331.

5 Detecting and Defining RNA–Protein and RNA–RNA Interactions

PROTOCOLS

1 Filter-binding Assay for Analysis of RNA–Protein Interactions, 254
2 Native Polyacrylamide-gel-electrophoresis-binding Assay for RNA–Protein Complexes: Gel Mobility Shift Assay or Electrophoretic Mobility Shift Assay, 257
3 SELEX to Identify Protein-binding Sites on RNA, 262
4 RNA Structure Determination Using Chemical and Nuclease Digestion Methods, 269
5 Mapping Sites of RNA–Protein Interactions Using Chemical Methods, 276
6 Mapping Sites of RNA–Protein Interactions Using Nuclease-based Methods: RNase Footprinting and Reverse Footprinting, 284
7 Modification Interference and Nucleotide Analog Interference Mapping, 291
8 Boundary Analysis, 298
9 Ultraviolet Photochemical Cross-linking to Detect RNA-binding Proteins, 301
10 Photocross-linking Using Uridine Analogs, 304
11 Detecting RNA–RNA Interactions Using Psoralen Derivatives, 307
12 Cross-linking with Formaldehyde, 311
13 Basic Affinity Selection Methods, 314

IN THE PRECEDING CHAPTERS, WE described methods for isolating cellular RNAs and for synthesizing and labeling RNAs in vitro. This chapter presents a variety of methods that are useful for determining whether specific proteins bind to specific RNAs and, if they do, how to identify their binding sites. In addition, we detail methods that are designed to detect RNA–protein interactions in situations in which the binding protein has not yet been identified as well as methods to detect RNA–RNA interactions both in vitro and within cells. This array of techniques is extremely important for understanding many aspects of RNA biology because all RNAs within the cell do not function as "naked" nucleic acids but rather as ribonucleic acid–protein complexes (RNPs). Therefore, elucidating the composition of RNPs is essential to deciphering their biochemical role(s).

We first describe rather general methods for detecting specific RNA–protein interactions and then present a series of techniques that can be used to determine protein-binding sites. We then detail how various cross-linking methodologies can be used to identify interacting partners, and finally, we discuss how a variety of affinity purification schemes are helpful in characterizing the components of RNPs. Many of the methods described in this chapter make use of reagents prepared using methods detailed in earlier chapters (e.g., labeled synthetic RNAs and gels). Thus, it is essential to become familiar with those methods before attempting experiments using the techniques described in this chapter.

GENERAL METHODS FOR DETECTING RNA–PROTEIN INTERACTIONS

Filter-binding Assays

One of the oldest and simplest (and still very useful) methods for detecting RNA–protein interactions is nitrocellulose filter binding (Protocol 1). This method relies on the fact that almost all proteins (there are rare exceptions) bind to nitrocellulose with high affinity, whereas nucleic acids do not bind. Therefore, if a mixture of RNA and protein is passed through a nitrocellulose filter, the protein will be retained and the RNA will pass through. But if the protein is capable of binding the RNA of interest, the RNA will be retained on the filter as well. Filter-binding assays are conceptually quite straightforward: The protein of interest is mixed with a specific labeled (end or body labeled) synthetic RNA under conditions of choice, and whether RNA is bound to the filter is measured. If no RNA is bound, it can be deduced that it does not contain a binding site for the protein; if RNA is bound, it must contain a binding site(s) for the proteins. In the case that RNA does contain a binding site(s) for the protein, it is then possible, by manipulating the concentrations of protein and/or RNA, to determine a variety of parameters, including affinity constants (on and off rates) and the number of binding sites. It is also possible to determine the nature of the binding site(s) using this method by performing systematic mutagenesis of the RNA (although preferable methods for determining binding sites are outlined below). It is important to realize that filter-binding assays are only useful if a very pure (preferably recombinant) preparation of RNA-binding protein or RNP is being used. If this technique is attempted using crude mixtures of proteins, it is a near certainty that the presence of "nonspecific" RNA-binding proteins will obscure any meaningful binding. However, filter binding can be used with a purified protein to select, with high sensitivity, specific RNAs from a complex mixture (e.g., 3′-end-labeled cellular RNAs or RNAs from a randomized pool; see Protocol 3).

Gel Mobility Shift Assays

Gel mobility shift (or electrophoretic mobility shift assay, EMSA) assays (Protocol 2) are extremely popular and powerful techniques for the detection of RNA–protein interactions. In principle, like filter-binding assays, the methodology is straightforward. The technique relies on the fact that naked RNA has a certain mobility on nondenaturing gels. If the RNA is bound by protein, the mobility of the RNA is altered (reduced), and therefore the binding of protein results in a characteristic upward shift of the RNA, as monitored using radiolabled RNA. For reasons that are not completely understood, most RNA–protein complexes—especially those that result from high-affinity interactions—do not dissociate during the prolonged times required for electrophoretic separation. Because high-affinity interactions are more stable, it is often possible to identify specific interactions over a "background" of weak interactions. Accordingly, gel mobility shift analyses can be performed using complex mixtures of proteins such as cell extracts.

These assays are quite versatile and can be used for a wide variety of purposes. For example, they can be valuable for determining if a protein of interest is present in a particular preparation of proteins (e.g., cytoplasmic or nuclear extract). Here, a labeled synthetic RNA would be incubated with the protein preparation. Following nondenaturing gel electrophoresis, the labeled RNA is visualized by phosphorimaging or autoradiography, and the presence of shifted (or not) material is observed. If a gel shift is observed, it is then necessary to determine if the shift resulted from the binding of a protein that has a specific binding site in RNA. To distinguish between specific and nonspecific binding, competition experiments are carried out. If the shift is abolished in the presence of an excess of specific competitor (unlabeled RNA identical to the labeled RNA), but not by an excess of nonspecific competitor (unlabeled unrelated RNA or a mutant version of the labeled RNA), it can be concluded that the observed shift resulted from specific binding. If specific binding is observed and one suspects that a specific protein is responsible and an antibody is available for that protein, it is possible to perform what is called supershift analysis. Here, the antibody is included in the initial binding incubation before electrophoresis. Binding of the antibody to a complex of interest with the RNA results in a further reduction in mobility of the complex, i.e., a supershift. Techniques designed to characterize specific binding proteins without prior knowledge of their identify are described below.

The example outlined above illustrates only one simple use of the gel mobility shift technique. Other more complex questions can also be addressed. EMSA approaches have been extremely valuable in deciphering the ordered pathway of assembly of large (megadalton) complexes such as the spliceosome. In this case, assembly can be monitored in the presence or absence of an energy source (ATP) or a specific RNP (small nuclear ribonucleoprotein, snRNP) or auxiliary factor(s). We do not detail all of the potential applications of EMSA approaches because they are innumerable. Suffice to say that EMSA is often the method of choice whenever an RNA–protein interaction (or more complex assembly) can be monitored in vitro.

Although conceptually straightforward, it is important to note that gel composition and percentage as well as electrophoresis conditions can have profound effects on the results obtained. Finally, it is worth noting that gel mobility shift assays can (like filter binding) be used to quantitate various kinetic parameters of RNA–protein interactions. Of course, pure proteins (or RNPs) is required for such applications.

SELEX

SELEX (systematic evolution of ligands by exponential enrichment) techniques encompass a variety of methods that are extremely useful for determining preferred RNA-binding sites for specific RNA-binding proteins and for determining preferred RNA sequences that promote or inhibit specific reactions (see Protocol 3). All SELEX, or randomization–selection, techniques rely on the same principles. A large pool of RNAs containing randomized (with respect to sequence) sequences is synthesized and allowed to bind to a specific protein or participate in a specific reaction. Those RNAs that bind productively or participate are then recovered, converted to DNA, and used to generate a new pool of transcripts. This new pool, which is enriched for productive sequences, is then used for a second round of binding or is allowed to participate in the same reaction. This iterative selection progressively enriches for production sequences until a maximum is reached, i.e., ≥90% of the pool binds to a specific protein or participates in a specific reaction. This final pool is then converted to DNA, and the nucleotide sequence of a large number (≥100) of clones is determined. This analysis reveals the consensus sequence or sequences of RNA elements that bind to a specific protein or participate in (or inhibit) a specific reaction.

It is often the case that suboptimal (but functional) sequence elements can be determined if the cloning and sequencing are carried out at earlier "rounds" of selection. In general, the number of iterative selection rounds required is determined by the size of the starting pool (e.g., RNA containing seven randomized positions contains only ~16,000 individual molecules, whereas an RNA randomized at 10 positions has a pool size of ~10^6 individual molecules) and the stringency of selection. To ensure adequate coverage, it is important that ~10 times the theoretical pool size be synthesized for each reaction. It is essential that the investigator know and keep in mind the number of distinct molecules required for adequate representation of the pool and take steps to ensure that each molecule of distinct sequence is present in the starting pool. It is amazing and disconcerting that this crucial consideration is often not taken into account in the design of SELEX experiments.

RNA STRUCTURE PROBING

Before outlining approaches designed to determine sites of RNA–protein interactions, we briefly discuss methods for determining RNA structures (i.e., single- and double-strand regions). Such structural information is often useful when assessing the potential for certain RNAs to interact with proteins or when determining whether RNAs that are dissimilar in sequence can form the same structure. We have found that empirical structure determinations are more reliable than structure predictions generated in silico.

Only basic methods for determining secondary structure are considered here (see Protocol 4). More advanced methods used to elucidate tertiary interactions are not discussed. There are two types of approaches—chemical and enzymatic—for structure determination. Both are reliable; however, we prefer enzymatic methods because they are simpler and do not involve the use of haz-

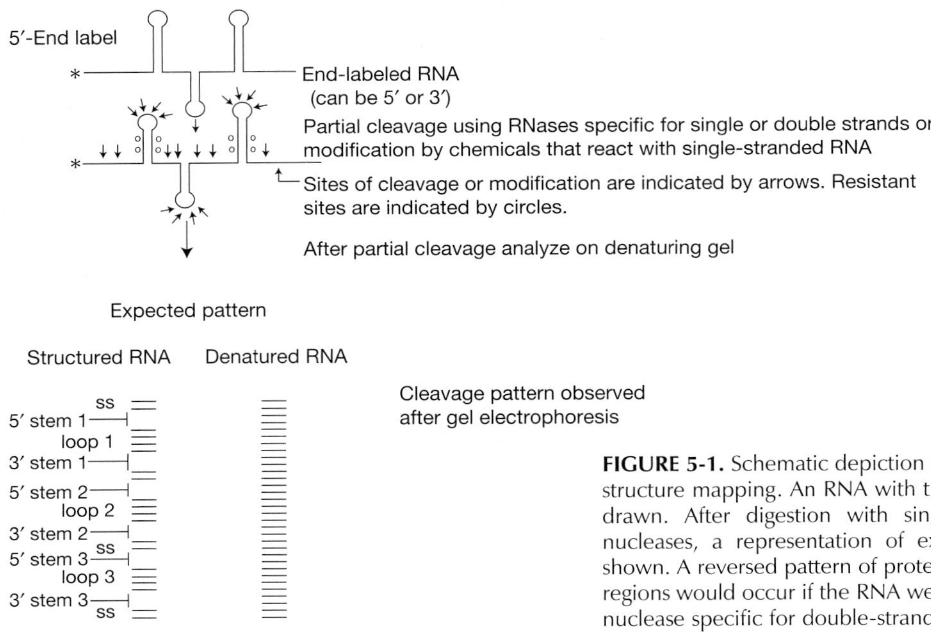

FIGURE 5-1. Schematic depiction of RNA secondary structure mapping. An RNA with three stem-loops is drawn. After digestion with single-strand-specific nucleases, a representation of expected results is shown. A reversed pattern of protected and sensitive regions would occur if the RNA were digested with a nuclease specific for double-strand regions.

ardous chemicals (Fig. 5-1). Chemical methods rely on the fact that certain reagents interact with RNA bases when they are single stranded but do not react when the bases are involved in Watson–Crick base pairs. For example, dimethylsulfoxide (DMS) methylates the N1 position of adenosine, the N7 position of guanine, and the N3 position of cytosine only when these bases are in single-strand regions. Modifications of adenosine and cytosine create blocks to reverse transcriptase; accordingly, these modifications are detected as stops to primer extension. Modification of guanine does not create reverse transcriptase stops, but these modifications can be detected by cleavage of the modified RNA after borohydride reduction and aniline cleavage (see protocols for direct RNA sequencing). Other chemicals that are commonly used for chemical determination of RNA structure are kethoxal (Protocol 5), which modifies guanine at the N1 and N2 positions, and 1-cyclohexyl-3-(2-morpholinoethyl)carbodiimide metho-*p*-toluene sulfonate (CMCT) (Protocol 5), which reacts with uridine at the N3 position. Kethoxal and CMCT modifications can be detected by blocks to reverse transcriptase. All of these reagents modify single-stranded RNA; therefore, double-strand regions are inferred by lack of modification.

In contrast to chemical methods, enzymatic methods use end-labeled RNAs, and sites of cleavage are determined directly be electrophoresis on denaturing gels. Moreover, because there are enzymes that can cleave only when the RNA is single stranded (e.g., RNase T1 and RNase A) or double stranded (e.g., V_1), it is possible to do parallel analyses; this is another reason that we prefer enzymatic to chemical approaches.

MAPPING SITES OF RNA–PROTEIN INTERACTIONS

Direct Methods: Footprinting and Reverse Footprinting

When a specific RNA–protein interaction has been identified (e.g., by EMSA or filter binding), it is often desirable to determine the binding site of the protein on the RNA. We describe several methods to achieve this goal, the first of which is known as footprinting (see Protocol 6). As illustrated in Figure 5-2, footprinting is based on the fact that a protein bound to an RNA protects the RNA from attack by chemicals or nucleases. The portion of the RNA that is not bound by protein is accessible and therefore vulnerable to attack. For footprinting to be successful, it is necessary that a substantial (≥80%) fraction of the target RNA be bound. If less RNA is bound, the unbound RNA, even in the protein-binding site, will be digested (or modified), thus obscuring the signal.

A "typical" footprinting experiment using nuclease digestion is described here; however, the same principles apply whether nucleases or chemicals are used to attack the accessible RNA. The procedure begins with the preparation of an end-labeled (5′ or 3′) synthetic RNA that contains a binding site(s) for the protein of interest; the presence of a binding site has previously been demonstrated via filter binding or gel shift. After the protein has been allowed to bind to the RNA, the RNA is then fragmented by mild nuclease (or chemical) digestion. It is important to determine the nuclease concentration that yields a pattern of fragments that covers the entirety of the RNA (≤1 per molecule). Overdigestion will yield only small fragments and defeat the purpose of the experiment. Titrations are performed on naked end-labeled RNA.

If the experiment is carried out properly, a comparison of the pattern of fragments generated from naked RNA with that obtained from digestion of the RNA with protein bound should reveal the binding site of the protein. The binding site will be evident by the absence of fragments in the protein-bound sample resulting from protection of the RNA by the protein; the protected region is known as the protein's "footprint."

Reverse Footprinting

As the name implies, reverse footprinting is a mirror image of a conventional footprint (Fig. 5-2B). The "footprint" is revealed by the presence, not the absence, of bands. In this approach, site-specifi-

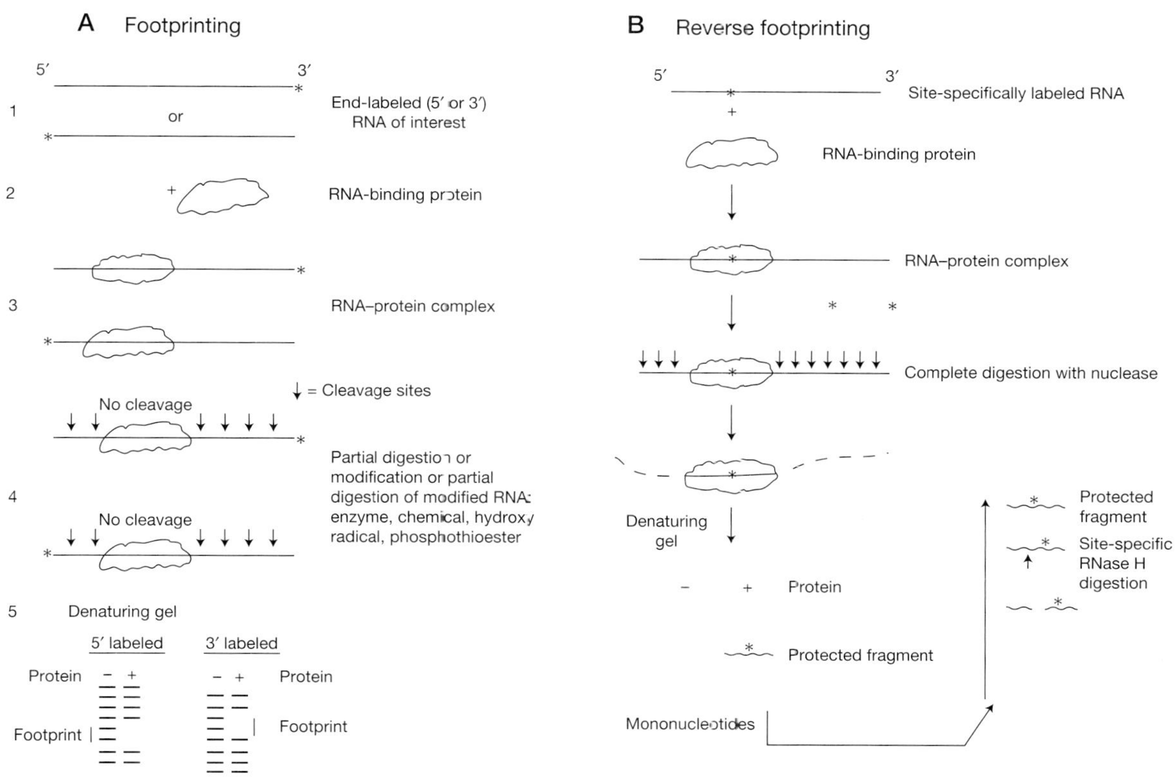

FIGURE 5-2. (*A*) Schematic representation of a "typical" footprinting experiment. For a detailed description, see the accompanying text. (*B*) Schematic representation of a "typical" reverse footprinting experiment. For a detailed description, see the accompanying text.

cally labeled RNAs are used rather than end-labeled RNA. If a protein binds to the region that contains the site-specific label, it will protect the label and adjoining sequences from nuclease digestion; the protected region is thus identified after extensive fragmentation of the labeled RNA. Reverse footprinting has a distinct advantage over conventional footprinting in that only a small fraction of the input RNA needs to be protected to visualize resistant bands. We have found this technique to be very valuable in elucidating sites of protein–RNA interactions as well as snRNP–mRNA interactions.

Determining Sites of RNA–Protein Interaction by Boundary Analysis

Another often overlooked but powerful way to determine sites of RNA–protein interaction is known as boundary analysis (Protocol 8; Fig. 5-3). In this technique, end-labeled RNAs are partially fragmented by alkaline hydrolysis to generate a ladder of bands. This pool of fragments is then allowed to bind to the protein of interest. Bound fragments (e.g., selected by affinity or antibody binding) are then displayed on a gel. The point at which binding is lost for 5′-end-labeled RNA defines the 3′ boundary of the binding site, and the point at which binding is lost for 3′-end-labeled RNAs defines the 5′ boundary of the binding site.

Modification Interference and Nucleotide Analog Interference Mapping

Methods collectively known as modification interference are exceptionally powerful approaches used to identify functionally important chemical groups in the phosphodiester backbone or nucleobases of an RNA (see Protocol 7). In principle, modification interference can be used whenever it

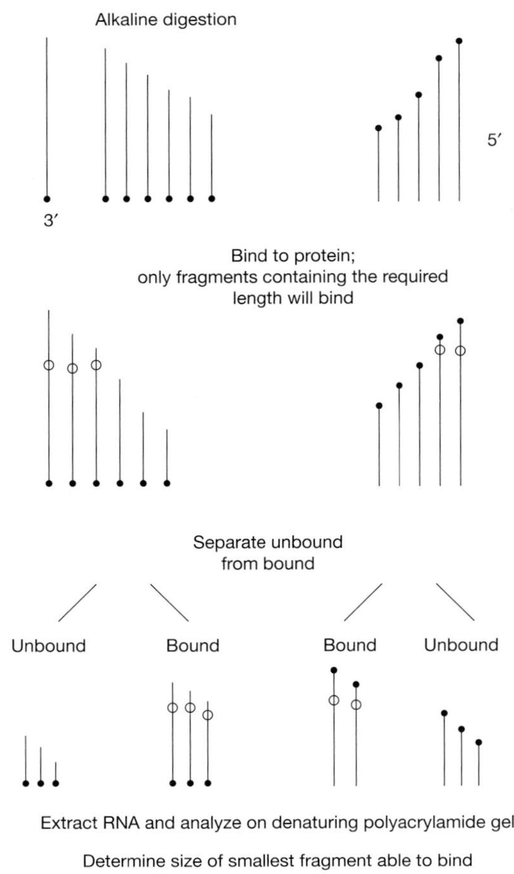

FIGURE 5-3. Schematic representation of a "typical" boundary experiment. For a detailed description, see the accompanying text.

is possible to separate productive RNAs (e.g., those bound by protein or that successfully participate in a processing reaction) from unproductive RNAs (e.g., those not bound by protein or unable to participate in a processing reaction). The basic idea is to create a pool of end-labeled RNAs wherein each RNA molecule is modified at a different position along the chain (Fig. 5-4). This is achieved by modifying or synthesizing an RNA such that the frequency of modification or incorporation is less than one per molecule. The pool of RNAs is then allowed to participate in the reaction of interest. For example, consider a simple RNA–protein interaction in which the protein binds to a specific RNA sequence. Also assume that productive binding requires recognition of the phosphodiester backbone within that sequence. If a pool of RNAs containing random substitutions (thiophosphates) within the backbone is allowed to bind to the protein, some RNAs (with thiophosphates outside of the critical region) will bind. RNAs containing substitutions that prevent recognition by the protein will not bind. The unbound RNAs when cleaved with iodoethanol will yield a sequencing ladder in which all positions of modification in the original pool will be identified. Conversely, when the bound RNAs are analyzed similarly, bands corresponding to crucial sites with thiophosphate substitution will be absent. A straightforward comparison of the ladders generated from bound versus unbound RNA will reveal sites that are important for binding. Note that in this analysis, every possible position is examined in one experiment. This is the power of modification interference.

This example considered modification of the phosphate backbone with thiophosphates. The same principles apply to any situation where a specific modification can be introduced and the sites of modification identified. For example, RNAs can be modified chemically with DMS, diethyl pyro-

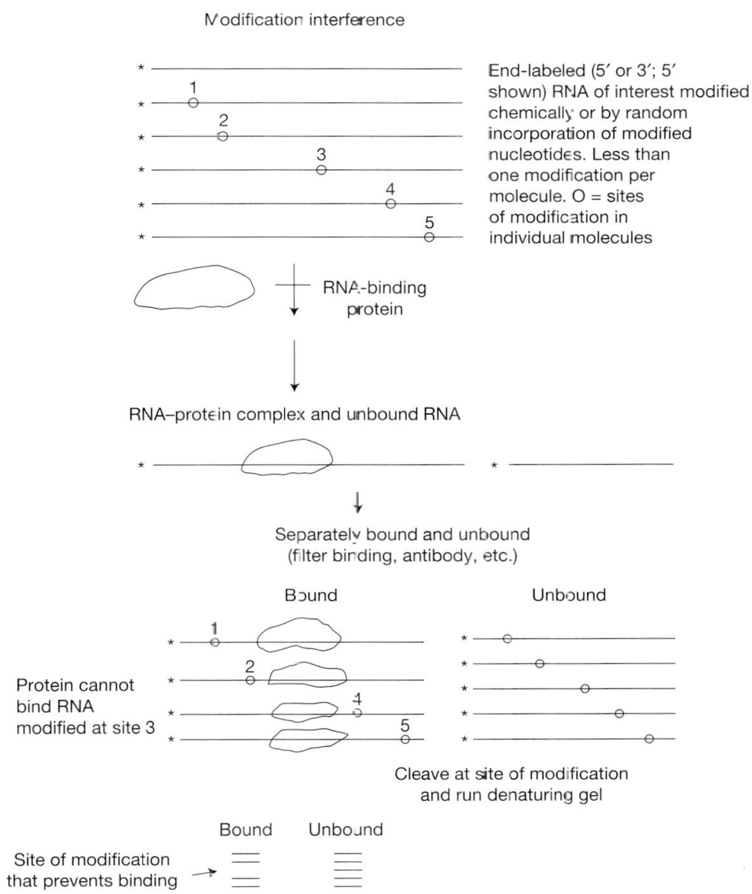

FIGURE 5-4. Schematic representation of a "typical" modification interference experiment. For a detailed description, see the accompanying text.

carbonate (DEPC), CMCT, or kethoxal as described above to identify functional groups required within the bases of an RNA for a reaction in which it is possible to separate productive from non-productive molecules.

Nucleotide Analog Interference Mapping

As discussed above, many modifications can be used for interference analysis. However, many other modifications cannot be identified (e.g., by chemical cleavage or blockage of reverse transcriptase). If sites of modification cannot be identified, it is impossible to carry out modification interference. To overcome this problem and to allow the analysis of more modifications, nucleotide analog interference mapping (NAIM) was invented by Strobel and colleagues (Strobel and Shetty 1997). This technique is based on the ability of thiophosphate substitution to be identified by iodoethanol cleavage. Thus, if α-thionucleotides containing a modified base are incorporated into an RNA chain, the sites containing the modified base can be identified by cleavage with iodoethanol. This technique is useful whenever the thiophosphate substitution on its own does not prevent or inhibit a specific reaction. To perform NAIM, it is first necessary to perform a thiophosphate interference analysis. Any (perhaps all) positions that are not affected by thiophosphate substation can then be analyzed by NAIM. As described in Protocol 7, several useful base analogs are now available for this powerful approach.

Cross-linking Approaches

Cross-linking techniques comprise a wide variety of methods designed to form covalent attachment of proteins to RNAs, proteins to proteins, and RNAs to RNAs (Protocols 9–12). Before considering any cross-linking approach, it is critical to understand which cross-linking agent (or treatment) best suits the needs of the particular question. It is also important to know the limitations and advantages of each technique, as well as potential artifacts and how to interpret the data generated. In our experience, cross-linking studies are perhaps the most overinterpreted of all analyses. Before outlining the most common cross-linking approaches and the advantages and disadvantages of each, we discuss a few considerations that apply to all cross-linking studies.

The primary goals of any cross-linking experiment are to determine sites of RNA–protein interactions, the identity of protein–protein interacting partners, or the identity of RNA–RNA interacting partners. Two parameters of each approach that must be considered are (1) the mechanism(s) of cross-linking using a particular approach and (2) the anticipated efficiency of cross-linking. Each cross-linking technique has intrinsic limitations that are imposed by the mechanisms by which a specific agent or treatment works. For example, direct photocross-linking with short-wavelength ultraviolet (UV) (254-nm) light requires a specific juxtaposition of the protein of interest with its RNA-binding site. Therefore, many specific and high-affinity interactions are not cross-linkable. Similarly, formaldehyde, which is widely considered to be a "promiscuous" cross-linker, requires that ε-amino groups of lysine be in close proximity to amine groups of RNA bases or lysine residues of another protein. These considerations lead to the first fundamental rule of thumb in all cross-linking analyses: Negative results (lack of cross-linking) are almost never interpretable. The only exception to this principle is when during the course of a reaction (e.g., splicing), a cross-link is observed at one stage of the reaction but is absent at another. In such cases, it can be inferred that a change in the physical organization of the participants in the reaction has occurred.

A second (oftentimes ignored) consideration that applies to all cross-linking approaches is that the efficiency at (or extent to) which a specific cross-linking event is observed does not necessarily provide any quantitative information regarding affinity, occupancy, or any other kinetic parameter. For example, in HITS-CLIP (high-throughput sequencing of RNA isolated by cross-linking immunoprecipitation) studies (see Chapter 8), the number of reads of a specific RNA in an RNA–protein complex is often interpreted to be an indication of occupancy of a specific site, when in fact the observed frequency could (and probably does) reflect the favorable geometry of a specific interaction.

A third consideration when interpreting cross-linking experiments is the fact that these techniques can "trap" transitory or nonspecific interactions. For example, consider a case in which an abundant RNA-binding protein with weak affinity for a specific RNA sequence interacts with that sequence in a transitory way. Because cross-linking introduces covalent linkages, that "nonspecific" interaction can be "fixed."

Finally, it must be remembered that all cross-linking approaches provide a limited "snapshot" of a specific cellular state at a specific time. Unless experiments are designed specifically to provide time- or condition-dependent data, no inference can be made regarding the dynamics of a specific detected interaction.

UV Cross-linking Using RNAs Substituted with Photoactivatable Nucleosides

Certain analogs of RNA bases—4-thiouridine, 5-iodouridine, 5-bromouridine, and 6-thioguanosine—are called photoactivatable because they absorb and become activated at specific wavelengths (>304 nm) of UV light. They are significantly more efficient than unsubstituted RNA for cross-linking purposes. Recently, it has been shown that 4-thiouridine is useful both in vivo and in vitro. These substituted RNAs are particularly valuable when used for site-specific cross-linking. Their advantages, in addition to enhanced efficiency, are the same as those for short-wavelength UV cross-linking, and 4-thiouridine and 6-thioguanosine are effective for detecting RNA–RNA interactions. Use of these derivatives also shares some of the same disadvantages as those for short-wavelength UV cross-linking. In addition, the use of these derivatives obviously introduces base composition effects, and the substitutions themselves have the potential to interfere with specific RNA–protein interactions (although this possibility has not been extensively investigated).

Formaldehyde

Once an extremely popular cross-linking agent, formaldehyde fell out of favor for a time, possibly because of the tendency to make insoluble aggregates. However, because of its versatility, the use of formaldehyde is undergoing a resurgence (see Protocol 12). Its advantages and disadvantages are summarized in Table 5.1.

Psoralen

Cross-linking with psoralen derivatives is the preferred method to identify RNA–RNA interactions, particularly those that are mediated by base pairing (see Protocol 11). Psoralens are bifunctional tricyclic compounds that function by intercalating into double-strand regions of nucleic acids. Upon irradiation with long-wavelength (365-nm) UV light, covalent linkages are formed between pyrimidines located in adjacent positions on opposite strands (i.e., an AU on one strand and a UA on the other). Particularly useful is the water-soluble psoralen derivative 4′-aminomethyl trioxsalen (AMT). AMT has been used successfully to elucidate RNA–RNA interactions in the spliceosome as well as to identify double-strand regions in RNA. It is the agent of choice whenever two RNAs (intermolecular interactions) or double-strand regions of a single RNA (intramolecular interactions) are suspected to interact via base pairing.

Affinity Approaches

Affinity-based techniques comprise a broad spectrum of approaches that have wide applicability to the analysis of RNA–protein, protein–protein, and RNA–RNA interactions (see Protocol 13). In principle, all of these approaches have a common design: The RNA or protein of interest is "tagged" in a way such that it can specifically be retrieved from a complex (or simple) mixture. Following retrieval, interacting partners, either proteins or RNAs, can be identified by a variety of methods (commonly mass spectrometry for proteins, sequencing for RNAs). We discuss a number of poten-

TABLE 5-1. Advantages and Disadvantages of Various Cross-linking Methodologies

Direct cross-linking with UV light

Advantages

- Captures a wide variety of RNA–protein interactions
- Can be used effectively in vitro or in vivo
- Can be used for "label transfer" to cross-linked proteins using body-labeled or site-specifically labeled RNA
- Broadly applicable
- Highly specific; only zero-length cross-links observed (i.e., protein must be in intimate contact with RNA of interest)
- Not highly sensitive to base composition

Disadvantages

- Low efficiency
- Highly dependent on specific geometry of RNA–protein interaction
- Poor for detecting RNA–RNA interactions
- Not useful for detecting protein–protein interactions
- Not reversible

Formaldehyde

Advantages

- Highly efficient at inducing protein–RNA and protein–protein cross-links. Because of this efficiency, it is possible to obtain simultaneous RNA–protein–protein cross-links, thus enabling a broader view of RNP composition.
- Formaldehyde cross-links are readily reversible by treatment with heat or acid
- Useful both in vitro and in vivo
- Small size of formaldehyde makes most interactions accessible to the agent

Disadvantages

- Dependent on the geometry of specific interaction
- Easy to "overcross-link" and thus trap fortuitous interactions

Psoralens

Advantages

- Can be used in vitro or in vivo
- Effective for cross-linking double-stranded RNAs
- Cross-links readily reversible by irradiation with short (254-nm)-wavelength UV light

Disadvantages

- Less effective at detecting RNA–RNA interactions not mediated by base pairing
- Poor for cross-linking protein to RNA

tial approaches involving the use of tagged RNAs or proteins and then discuss potential problems (often overlooked) that can confound these approaches.

Tagged RNAs

The following four methods are commonly used to tag RNAs for subsequent affinity purification. The first is direct incorporation of biotinylated uridine. Tagging RNA with biotin involves the synthesis of transcripts using substitution with biotinylated uridine 5′ triphosphate (UTP) (see Chapter 4, Protocol 1). After incubation with a cellular extract, the biotinylated RNA is retrieved by its affinity to immobilized streptavidin. Although the interaction of biotin to streptavidin is remarkably strong (10^{-15} M), recovery of biotinylated RNA is often disappointingly low (≤20%) and nonspecific background can be quite high (see below).

The next method is indirect tagging with biotin, which we have found can be much more efficient than direct tagging. In this approach, a chimeric 2′-*O*-methyl (2′OMe) RNA–DNA oligonucleotide is designed to have partial complementarity to the RNA of interest. The 2′OMe part of the oligonucleotide contains the hybridizing region (to avoid digestion of the RNA by RNase H), whereas the DNA part of the oligonucleotide contains multiple biotinylated residues. The oligonucleotide is preannealed to the RNA of interest and then retrieved after incubation again by streptavidin affinity. Because the DNA portion of the oligonucleotide can be cleaved with DNase I, DNase digestion is used to elute the RNA from the beads. We have found that this type of elution greatly reduces background inherent in direct biotinylation approaches.

Direct tagging using aptamers is the third approach. Here, an aptamer (a short RNA sequence that has been selected to bind to a ligand) is synthesized as part of the RNA of interest. If the aptamer folds properly in the context of the RNA to which it is appended, the RNA containing the aptamer can be directly retrieved via its affinity to the ligand. An avidin-binding RNA aptamer has been used successfully to select proteins bound to specific RNAs. Potential problems with this approach include "nonspecific" protein binding to the aptamer sequence and improper folding of the aptamer when appended to an RNA.

The fourth method is indirect tagging of RNAs using fusion proteins. This approach takes advantage of the fact that certain stem-loop sequences present in bacteriophage genomes (e.g., MS2, R17, and λN) serve as high-affinity binding sites for small phage proteins. Fortuitously, these proteins still function in RNA binding when they are fused to other proteins. The property of these proteins to function as fusion proteins provides the basis for "tethering" experiments. For affinity approaches, the phage proteins can be fused to proteins that can bind to affinity resins. For example, phage proteins have been fused to maltose-binding protein (MBP) or glutathione S-transferase (GST). If an RNA is designed to have λN binding sites and is prebound by λN–GST fusion protein, it can be retrieved by affinity to a reduced glutathione resin. Subsequent elution with reduced glutathione captures the RNA and any associated proteins and/or RNAs. There are literally dozens of variations on this common theme, but they all follow the same common principles of design and execution.

Tagged Proteins

Fusion proteins containing affinity tags have become exceptionally useful tools with a wide variety of applications in the analyses of RNA–protein and protein–protein interactions. Some uses for such proteins in indirect tagging of RNA are described above. This section outlines many of the tags that are in common use.

Epitope Tags. In epitope tagging, a recognition element for a specific antibody is appended (either amino or carboxyl terminally) to the protein of interest. Commonly used tags of this sort are FLAG, V5, and Myc. Because epitopes are normally short peptide sequences, these types of tags are unlikely to interfere with a protein's function (but see Common Potential Problems, below). In addition, the ability to elute material bound by antibody affinity by competition with the peptide ligand greatly reduces background (but see Common Potential Problems, below). In general, epitope tags are very useful for expression studies (by western blotting), localization studies, and affinity purifications. The latter two applications can be occasionally confounded if the protein of interest is involved in an interaction that blocks recognition of the epitope.

Functional Tags. A second type of tag—functional tags—does not rely on antibody-mediated affinity but rather on the ability of the tag to interact with a specific affinity matrix. Such tags include oligohistidine (six or more consecutive histidine residues appended to either terminus of the protein), GST, MBP, and various tandem affinity purification (TAP) tags. Oligohistidine has the advantage of being short and therefore unlikely to interfere with a protein's function (but see below). GST, MBP, and many TAP tags are all large and bulky, and therefore one should be very cautious when using them; however, they all have the advantage of binding to specific affinity resins. TAP tags are particularly valuable because sequential affinity matrices are used, therefore reducing background. We have found that a tandem tag composed of protein G (for the first round of affinity) separated from avidin-binding protein (for the second round of affinity purification) by a protease cleavage site (tobacco etch virus [TEV] protease) is particularly useful. However, it should be noted that "loosely" associated proteins (i.e., those bound with low affinity to the RNA) can be lost during purification, and it is often advisable to incorporate a cross-linking step (see above) into these types of approaches.

Common (and Often Overlooked) Potential Problems with Affinity Approaches

- **Background.** No matter what affinity approach is chosen, it is remarkable how much background (i.e., nonspecific proteins) is present after purification. In our experience, we have observed several background proteins persisting through four biochemical steps followed by an affinity purification. The presence of background can be explained by the fact that the protein of interest represents only a minute fraction of the total protein in a cell or extract. Thus, it is always essential to carry out parallel purifications using a tagged irrelevant protein. Moreover, the presence of highly abundant cellular proteins (e.g., ribosomal proteins) in any purified sample should be viewed with extreme caution.

- **The tag causes loss or alteration of function of the protein of interest.** This is a pernicious problem that is often overlooked. Introduction of "foreign" sequence into the protein of interest (even epitopes and His residues) can potentially inactivate or otherwise debilitate the protein of interest. In this case, the results of any purification are meaningless. Whenever possible, it is absolutely required to ensure that the protein of interest retains its native activity when tagged. Doing so can be quite difficult, especially in mammalian cells where genetic complementation is often impossible. Moreover, in most cases, antibodies to the endogenous protein are simply not available. Nevertheless, this is an important concern that cannot be ignored.
- **In vivo (and in vitro) approaches using tags involving overexpression.** To determine the proteins or RNAs with which a tagged protein interacts, it is obvious that the tagged protein be introduced into a cell (or extract). Often large (nonphysiological) levels of the tagged protein may be present. In such cases, it is possible (if not likely) that the introduced protein may mislocalize or interact with unnatural partners. Other than adjusting the level of expression, this problem is quite difficult to overcome or circumvent, but it must be considered when interpreting any results.

SUMMARY

This chapter covers a broad array of methods that can be used to detect RNA–protein, RNA–RNA, and protein–protein interactions. All are useful when applied appropriately. It is essential to know the advantages and disadvantages of each method before choosing the approaches that best suit a particular question. Often, it is desirable to use as many methods as possible to answer a specific question. A convergence of results provides high confidence that the answers obtained are both reliable and reproducible.

REFERENCE

Strobel SA, Shetty K. 1997. Defining the chemical groups essential for Tetrahymena group I intron function by nucleotide analog interference mapping. *Proc Natl Acad Sci* **94:** 2903–2908.

READING LIST

Bouvet P. 2001. Determination of nucleic acid recognition sequences by SELEX. *Methods Mol Biol* **148:** 603–610.

Chanfreau G, Jacquier A. 1994. Catalytic site components common to both splicing steps of a group II intron. *Science* **266:** 1383–1387.

Christiansen J, Garrett R. 1988. Enzymatic and chemical probing of ribosomal RNA-protein interactions. *Methods Enzymol* **164:** 456–468.

Clarke PA. 1999. RNA footprinting and modification interference analysis. *Methods Mol Biol* **118:** 73–91.

Das R, Reed R. 1999. Resolution of the mammalian E complex and the ATP-dependent spliceosomal complexes on native agarose mini-gels. *RNA* **5:** 1504–1508.

Dixon WJ, Hayes JJ, Levin JR, Weidner MF, Dombroski BA, Tullius TD. 1991. Hydroxyl radical footprinting. *Methods Enzymol* **208:** 380–413.

Eckstein F. 1985. Nucleoside phosphorothioates. *Annu Rev Biochem* **54:** 367–402.

Ehresmann C, Baudin F, Mougel M, Romby P, Ebel J-P, Ehresmann B. 1987. Probing the structure of RNAs in solution. *Nucleic Acids Res* **15:** 9109–9128.

Fabrizio P, Abelson J. 1992. Thiophosphates in yeast U6 snRNA specifically affect pre-mRNA splicing in vitro. *Nucleic Acids Res* **20:** 3659–3664.

Gish G, Eckstein F. 1988. DNA and RNA sequence determination based on phosphorothioate chemistry. *Science* **240:** 1520–1522.

Hall KB, Kranz JK. 1999. Nitrocellulose filter binding for determination of dissociation constants. *Methods Mol Biol* **118:** 105–114.

Hannon GJ, Maroney PA, Yu YT, Hannon GE, Nilsen TW. 1992. Interaction of U6 snRNA with a sequence required for function of the nematode SL RNA in *trans*-splicing. *Science* **258:** 1775–1780.

Hartz D, McPheeters DS, Traut R, Gold L. 1988. Extension inhibition analysis of translation initiation complexes. *Methods Enzymol* **164:** 419–425.

Hausner T-P, Giglio LM, Weiner AM. 1990. Evidence for base-pairing between mammalian U2 and U6 small nuclear ribonucleo-protein particles. *Genes Dev* **4:** 2146–2156.

Kenan DJ, Keene JD. 1999. In vitro selection of aptamers from RNA libraries. *Methods Mol Biol* **118:** 217–231.

Kent OA, MacMillan AM. 2002. Early organization of pre-mRNA during spliceosome assembly. *Nat Struct Biol* **9:** 576–581.

Konarska MM. 1989. Analysis of splicing complexes and small nuclear ribonucleoprotein particles by native gel electrophoresis. *Methods Enzymol* **180:** 442–453.

Lipson SE, Hearst JE. 1988. Psoralen cross-linking of ribosomal RNA. *Methods Enzymol* **164:** 330–341.

Maroney PA, Romfo C, Nilsen TW. 2000. Nuclease protection of RNAs containing site-specific labels: A rapid method for mapping RNA-protein interactions. *RNA* **6:** 1905–1909.

Moore MJ, Sharp PA. 1992. Site-specific modification of pre-mRNA: The 2′-hydroxyl groups at the splice sites. *Science* **256:** 992–997.

Parker KA, Steitz JA. 1989. Determination of RNA-protein and RNA-ribonucleoprotein interactions by nuclease probing. *Methods Enzymol* **180:** 454–468.

Peattie DA, Gilbert W. 1980. Chemical probes for higher-order structure in RNA. *Proc Natl Acad Sci* **77:** 4679–4682.

Reed R, Chiara MD. 1999. Identification of RNA-protein contacts within functional ribonucleoprotein complexes by RNA site-specific labeling and UV crosslinking. *Methods* **18:** 3–12

Ringquist S, Gold L. 1998. Toeprinting assays. Mapping by blocks to reverse transcriptase primer extension. *Methods Mol Biol* **77:** 283–295.

Sachs MS, Wang Z, Gaba A, Fang P, Belk J, Ganesan R, Amrani N, Jacobson A. 2002. Toeprint analysis of the positioning of translation apparatus components at initiation and termination codons of fungal mRNAs. *Methods* **26:** 105–114.

Setzer DR. 1999. Measuring equilibrium and kinetic constants using gel retardation assays. *Methods Mol Biol* **118:** 115–128.

Sontheimer EJ, Steitz JA. 1993. The U5 and U6 small nuclear RNAs as active site components of the spliceosome. *Science* **262:** 1989–1996.

Stern S, Moazed D, Noller HF. 1988. Structural analysis of RNA using chemical and enzymatic probing monitored by primer extension. *Methods Enzymol* **164:** 481–489.

Stockley PG. 2001. Filter-binding assays. *Methods Mol Biol* **148:** 1–11.

Strobel SA, Shetty K. 1997. Defining the chemical groups essential for Tetrahymena group I intron function by nucleotide analog inteference mapping. *Proc Natl Acad Sci* **94:** 2903–2908.

Tijerina P, Mohr S, Russell R. 2007. DMS footprinting of structured RNAs and RNA-protein complexes. *Nat Protoc* **2:** 2608–2623.

Wassarman DA, Steitz JA. 1992. Interactions of small nuclear RNA's with precursor messenger RNA during in vitro splicing. *Science* **257:** 1918–1925.

Wollenzien PL. 1988. Isolation and identification of RNA cross-links. *Methods Enzymol* **164:** 319–329.

Wong I, Lohman TM. 1993. A double-filter method for nitrocellulose-filter binding: Application to protein-nucleic acid interactions. *Proc Natl Acad Sci* **90:** 5428–5432.

Yu YT, Maroney PA, Darzynkiwicz E, Nilsen TW. 1995. U6 snRNA function in nuclear pre-mRNA splicing: A phosphorothioate interference analysis of the U6 phosphate backbone. *RNA* **1:** 46–54.

Ziehler WA, Engelke DR. 2001. Probing RNA structure with chemical reagents and enzymes. *Curr Protoc Nucleic Acid Chem*, Chapter 6, Unit 6.1.

Protocol 1

Filter-binding Assay for Analysis of RNA–Protein Interactions

The ability of proteins to be retained on nitrocellulose filters while free radiolabeled nucleic acids flow through provides the basis of filter-binding assays. These assays are rapid and sensitive and can be used to determine a variety of constants as well as on and off rates. The method can also be used to determine effects of salt concentrations or detergents on known binding interactions.

MATERIALS

The recipes for the items marked with <R> are at the end of the protocol or in the Common Recipes Appendix.

CAUTION: See the Cautions Appendix for appropriate handling of material marked with <!>.

Reagents

Master binding mix <R>
Protein sample
RNA, 5′ or 3′ end or uniformly ^{32}P labeled <!>
Wash buffer <R>

Equipment

"Dot-blot" apparatus (96 well; e.g., BioRad Bio-Dot 170-6545)
Eppendorf tubes (1.5 mL) or 96-well plate
GE Healthcare Hybond-ECL (RPN303D) (pure nitrocellulose membrane; retains bound RNA)
GE Healthcare Hybond-N+ (RPN303B) (positively charged nylon membrane; retains free RNA)
Incubator
Nitrocellulose filters (2.5-cm diameter, 0.45-µm pore size; Millipore HAWP02400 or Schleicher & Schuell BA85)
Paper towels
Petri plate (15 cm)
Phosphorimager
Pipetman (P200)
Plastic wrap
Scintillation counter
Tuffryn membrane (Pall Corporation, Tuffryn Membrane Filter 142 mm, 0.45 µM, P/N 66228) (optional)
Vacuum filtration manifold (e.g., Hoefer or Millipore)

OVERVIEW

To prepare for this protocol, a purified protein (or chromatographic fractions) of interest and 5′-end or uniformly ^{32}P-labeled RNA are needed. The protein sample is serially diluted to several concentrations, and the protein is then mixed with a fixed amount of radioactive RNA and allowed to bind under desired conditions for 30–60 min. The binding reactions are then applied to a 96-well dot-blot apparatus with low vacuum to trap the complexes on membranes. Alternatively, manifolds or single chimneys can be used. After washing and drying, the membranes are exposed to phosphorimaging screens for quantitation. Here, we present two methods. The first uses of three membranes. The top membrane traps aggregates, the middle (nitrocellulose) binds proteins and RNA–protein complexes, and the bottom membrane, which is charged, collects free RNA. The second method uses single, larger filters (that can be counted in a scintillation counter) and a filter manifold.

METHOD

Method 1: Using Three Membranes

1. Cut each of the required membranes to fit the filter-binding apparatus. Make sure that the filters cover all of the wells and that the apparatus can still be closed tightly.
2. Wet the membranes in H_2O and then soak them for at least 30 min in the same buffer and salt conditions used in the final concentration of the binding assay (wash buffer).
3. Prepare 1 mL of a binding master mix.

   The conditions of the buffer should be the same as those of the protein assay. See Tips and Troubleshooting.
4. Prepare 5–10 threefold serial dilutions of the protein stock in the same buffer in which the protein is stored.

   Final concentration range should be from ~100-fold less to ~100-fold more than the molar concentration of the RNA.
5. Aliquot 34 µL of binding mix into 1.5-mL Eppendorf tubes or the wells of a 96-well plate.
6. Mix 16 µL of each protein sample with the binding mix in the tubes or plate. Incubate for 20–60 min at the appropriate temperature.

   Generally, these parameters will need to be determined empirically, but ice, room temperature, and 37°C for 20, 40, and 60 min are common.
7. Place the membranes in the dot-blot apparatus, from bottom to top: Tuffryn (no. 3, optional), Hybond-ECL (no. 2), and Hybond-N (no. 1). Turn the vacuum on low.

   For the alternate method using a filter manifold, skip to Step 12.
8. Use a P200 Pipetman to apply all of each binding assay (50 µL) into separate holes on the apparatus. Make sure that all of the liquid passes through the membranes.
9. Once the liquid has passed through the membranes, wash each well with 100 µL of wash buffer.
10. When the top membrane is dry, turn off the vacuum, disassemble the apparatus, remove the membranes, and let them air-dry for ~5 min on paper towels.
11. Cover the membranes with plastic wrap. Expose the three membranes to a phosphorimaging screen overnight.

Alternative Method Using a Filter Manifold

12. Wet and soak the nitrocellulose filters (2.5-cm diameter, 0.45-µm pore size) in a dish or large (15-cm) Petri plate for at least 30 min.

    Extra unused filters can be soaked overnight.
13. Place the nitrocellulose filters on a vacuum filtration manifold and prewet the filters with 5 mL of washing buffer at low vacuum pressure. See also Tips and Troubleshooting.
14. Use a Pipetman P200 to apply the binding reaction from Step 6 to the filter. Allow the fluid to go through under very low house vacuum. Collect the unbound RNA in tubes placed in a rack in the bottom chamber.
15. Immediately wash the filters with 5 mL of wash buffer.
16. Air-dry and count the filters in a scintillation counter.

TIPS AND TROUBLESHOOTING

1. For an uncharacterized RNA–protein complex, a range of monovalent cation concentrations (e.g., 20–200 mM NaCl/NaOAc or KCl/KOAc) with or without 0.5–10 mM $MgCl_2$/$Mg(OAc)_2$ may need to be tested to find the optimum ionic strength to stabilize the protein–RNA complex of interest.
2. If nonspecific binding is a problem, it may be helpful to include nonspecific RNA competitors (such as yeast tRNA or total RNA) or small amounts of heparin (10 µg/mL) in the binding reaction.
3. For single filters, a simple apparatus with a chimney clamped to a stoppered glass frit in a side-armed flask may be used.

SOLUTION AND REAGENT RECIPES

CAUTION: See the Cautions Appendix for appropriate handling of materials marked with <!>.

Master Binding Mix

Reagent	Quantity (for 1 mL)	Final concentration
Tris (1 M, pH 7.6) <!>	20 µL	20 mM
NaCl or KCl (1 M) <!>	20–200 µL	20–200 mM
Dithiothreitol (DTT) (1 M) <!>	1 µL	1 mM
BSA (20 mg/mL)	2.5 µL	50 µg/mL
Glycerol (50%, w/v)	200 µL	10%
H_2O	to 1 mL	

Store for 6 mo at –20°C. Add $MgCl_2$ <!> or magnesium acetate to 1–5 mM if needed.

Wash Buffer

Reagent	Quantity (for 100 mL)	Final concentration
Tris (1 M, pH 7.6) <!>	2 mL	20 mM
NaCl or KCl (1 M) <!>	2–20 mL	20–200 mM
H_2O	to 100 mL	

Store indefinitely at 4°C. Add 1 mM DTT <!> just before use. Adjust salt concentrations as needed and add detergent as needed to wash buffer.

READING LIST

Hall KB, Kranz JK. 1999. Nitrocellulose filter binding for determination of dissociation constants. *Methods Mol Biol* **118:** 105–114.

Ross W, Gourse RL. 2009. Analysis of RNA polymerase-promoter complex formation. *Methods* **47:** 13–24.

Stockley PG. 2001. Filter-binding assays. *Methods Mol Biol* **148:** 1–11.

Wong I, Lohman TM. 1993. A double-filter method for nitrocellulose-filter binding: Application to protein-nucleic acid interactions. *Proc Natl Acad Sci* **90:** 5428–5432.

Protocol 2

Native Polyacrylamide-gel-electrophoresis-binding Assay for RNA–Protein Complexes: Gel Mobility Shift Assay or Electrophoretic Mobility Shift Assay

Gel-shift assays are powerful methods to detect specific RNA–protein interactions. The principle of these assays is that RNA–protein complexes have slowed mobility relative to RNA alone during gel electrophoresis, and thus binding is monitored as an upward shift on a gel. These assays can be used to investigate a wide range of RNA–protein interactions from single protein-binding events to assembly of large complexes such as the spliceosome. Gel-shift assays can also be useful for determining kinetic parameters (such as affinity constants) for RNA–protein interactions.

MATERIALS

The recipes for the items marked with <R> are at the end of the protocol or in the Common Recipes Appendix.

CAUTION: See the Cautions Appendix for appropriate handling of materials marked with <!>.

Reagents

Acrylamide (40% stock) <!>
Ammonium persulfate (10%, w/v) in ddH$_2$O <!>
Binding buffer <R>
Bisacrylamide (2% stock) <!>
Bovine serum albumin (BSA; 1 µg/mL)
Bromophenol blue (1%) <!>
Glycerol (5%)
Glycine (1 M, electrophoresis grade)
Heparin (50 mg/mL, w/v) in ddH$_2$O (Sigma-Aldrich H-3393) <!>
Loading buffer <R>
Protein of interest, purified
RNA of interest, uniformly labeled or purified (see Step 1)
Splicing mix <R>
TEMED (*N,N,N′,N′*-tetramethylethylenediamine) <!>
Total RNA or tRNA (10 mg/mL from yeast or *Escherichia coli*)
Tris (pH 7.6) <!>
Tris base (1 M) (Trizma base; Sigma-Aldrich T-6066) <!>
  Do not pH.
Tris/borate/EDTA buffer (TBE) <R> <!>
Tris glycine buffer (10x) <R>
Xylene cyanol (1%) <!>

Equipment

Electrophoresis equipment
  This should include vertical gel, comb (0.75-cm-wide wells), spacers (0.8- or 1.5-mm thick), and high-voltage power supply.

Gel dryer
Ice
Incubator
Microcentrifuge tube (1.5 mL)
Phosphorimager equipment or X-ray film and intensifying screen
Pipette
Plates
Vacuum gel dryer
Whatman 3MM paper

OVERVIEW

It is important to note that there are a large number of potential variations to gel-shift assays. Thus, the buffers, times, temperatures, nonspecific competitors, and even gel conditions may vary depending on the complexes being studied. Below are example protocols for different types of RNA–protein interactions. Although these describe specific conditions, they are readily adaptable to any analogous situation.

Simple Interactions Such as U1 RNA with Purified U1A Protein

1. Label the RNA of interest, either uniformly with [α-^{32}P]NTP by in vitro transcription or with [γ-^{32}P]ATP or [^{32}P]cytidine 3′,5′-bis(phosphate) (^{32}pCp) for an oligoribonucleotide or a 100–200-nucleotide purified RNA.
2. Incubate the RNA (~50,000 cpm) with purified protein under suitable binding conditions such as the binding buffer listed in Reagents supplemented with 10 μg/mL BSA. For example, mix together on ice:

Binding buffer	20 μL
RNA (50,000 cpm)	1 μL
Purified protein (serial dilutions from excess to less than equimolar)	4 μL

   Incubate the mixture for 30–60 min at room temperature (30°C or 37°C).
3. After incubation, dilute the sample with 4 μL of loading buffer (without heparin) and load 10–20 μL of the mixture onto a native polyacrylamide gel. For a single protein (~50,000 MW) and a single RNA (~100 nucleotides long), start with a 0.8-mm-thick 6% acrylamide:bis-acrylamide (30:1) gel made with Tris glycine or TBE. Prerun the gel for 30 min.
4. Electrophorese the gel at 300 V for 3 h (or 13 V/cm), making sure that it does not heat up. If the interactions are suspected to be unstable, run the gel in a cold room.
5. After the gel is run, use a sheet of Whatman 3MM paper to pick up the gel. Dry the gel with vacuum at 80°C. Expose the gel to a phosphorimager or X-ray film.

Longer RNA Forming a Small Complex Such as U1 snRNA and Its Associated Proteins (U1 snRNP)

1. Prepare a labeled RNA target by in vitro transcription and gel-purify it to make sure that the substrate will be a single band on a gel.
2. Incubate ~50,000 cpm RNA with purified proteins or cell extract in binding buffer for 20–60 min at the preferred temperature (4°C–37°C). For example, mix together on ice:

Binding buffer	20 μL
RNA (50,000 cpm)	1 μL
Purified proteins or cell extract	1–4 μL

Incubate for 30–60 min at room temperature (30°C or 37°C).

3. Following incubation, dilute the reaction to 100 μL with 500 mM NaCl, 10 mM Tris (pH 7.6), 5% glycerol, and 200 μg/mL heparin (in cases where cellular extract is used). Incubate for 10 min at room temperature.
4. Remove 1 μL of the reaction and add it to 30 μL of loading buffer. Make sure that the glycerol concentration is at least 5% so that the sample does not float away.
5. Load this sample onto a 1.5-mm-thick, 20-cm-long 0.5x TBE, 6% acrylamide:bis-acrylamide (60:1) gel. Electrophorese at 25 mA until the xylene cyanol is about two-thirds down the gel.
6. Disassemble the gel, pick it up with Whatman 3MM paper, and dry it at 80°C. Expose the gel to a phosphorimager or X-ray film.

Large Multicomponent RNA–Protein Complexes, Such as the Spliceosome

1. Prepare a native gel:

   3.75%–4.2% polyacrylamide
   Acrylamide:bis-acrylamide (80:1)

2. Pour the solution into 17 x 14.7 x 0.15-cm plates, insert a comb with 0.75-cm-wide wells, and allow the gel to polymerize (this usually takes at least 30 min).

   The gel will be very jelly-like. Narrower wells, such as those in a 20-well comb, can also be used.

3. Mix the following components in a 1.5-mL microfuge tube by pipetting up and down. The following conditions are the same as those used for in vitro splicing (see Chapter 6), except that the polyvinyl alcohol or polyethylene glycol is omitted.

Nuclear extract (15–20 mg/mL) or extract dialysis buffer (see Chapter 6, Protocol 1 or 2)	5 μL
RNA (usually 25,000–50,000 cpm)	1 μL
Splicing mix (4x)	5 μL
H_2O	9 μL
Total	20 μL

   Incubate for 30–60 min at 20°C–30°C.

4. Stop the reaction by placing the tubes on ice. Add heparin (titrate a range between 5 and 500 μg/mL) to reduce nonspecific binding.
5. Add 3 μL of 1% xylene cyanol and bromophenol blue to the reactions. Mix by pipetting up and down.

   Alternatively, the dye can be added to a free lane on the gel to monitor electrophoresis and prevent possible interference with protein binding.

6. Prerun the gel for 20 min at no more than 6 W.
7. Load 5 μL of the binding reaction mixture from Step 5 onto the gel.
8. Run the gel at no more than 6 W at 120–210 V maximum at room temperature until the dye is 1–2 cm from the bottom (for a 70-nucleotide RNA).
9. Transfer the gel onto Whatman 3MM filter paper and dry it for 30–60 min at 80°C on a vacuum gel dryer.
10. Expose the gel to X-ray film with an intensifying screen at –80°C.

TIPS AND TROUBLESHOOTING

1. It is often useful to include controls, such as inhibitors of the interaction and both specific and nonspecific RNA competitors.
2. Try varying the pH or the gel and buffers. Vary the ratio of Tris to glycine until the optimal pH is found.
3. The percentage of gel and ratio of acrylamide:bis-acrylamide may need to be varied to optimize separation of the bound and free RNA.
4. Heparin and high salt are included to resolve what could otherwise appear as a radioactive smear into discrete complexes on the gel. However, heparin also likely strips some proteins from the complexes. Total RNA or tRNA might also be added to improve resolution.
5. Because of the poor stacking capabilities of the gel, with a single polyacrylamide concentration and buffer pH, it is important not to overload the wells. For small wells (20-well comb), use a maximum volume of 5 µL/well.
6. For large protein–RNA assemblies, such as pre-mRNA splicing complexes, it may be useful to try native agarose gels instead of polyacrylamide gels.

SOLUTION AND REAGENT RECIPES

CAUTION: See the Cautions Appendix for appropriate handling of materials marked with <!>.

Binding Buffer

Reagent	Quantity (for 10 mL)	Final concentration
Tris (1 M pH 8.0) <!>	0.4 mL	40 mM
KCl (2 M) <!>	0.15 mL	30 mM
$MgCl_2$ (1 M) <!>	10 µL	1 mM
NP40 (10%, w/v) <!>	10 µL	0.01% (w/v)
DTT (1 M) <!>	10 µL	1 mM
H_2O or extract	9.47 mL	

If the proteins or extract used for binding are not in glycerol, add 5% to the binding buffer. Store indefinitely at –20°C.

Loading Buffer

Reagent	Quantity (for 10 mL)	Final concentration
KCl (2 M) <!>	0.3 mL	60 mM
Tris (1 M, pH 7.6) <!>	0.1 mL	10 mM
Glycerol (50%, w/v)	1 mL	5% (w/v)
Heparin (50 mg/mL) <!>	40 µL	200 µg/mL
Xylene cyanol (0.1%, w/v) <!>	1 mL	0.01% (w/v)
Bromophenol blue (0.1%, w/v) <!>	1 mL	0.01% (w/v)
H_2O	7.1 mL	

Store at 4°C. The amount of heparin may need to be adjusted or omitted.

Splicing Mix (4x)

Reagent	Quantity (for 100 µL)	Final concentration (4x)
Creatine phosphate (0.2 M)	10 µL	20 mM
HEPES/KOH (1 M, pH 7.6) <!>	8 µL	80 mM
$MgCl_2$ (100 mM) <!>	12 µL	12 mM
ATP (100 mM)	8 µL	8 mM
DTT (100 mM) <!>	40 µL	4 mM
H_2O	22 µL	

Prepare fresh for each use. Keep on ice.

Tris Glycine Buffer (10x)

Reagent	Quantity (for 100 µL)	Final concentration
Tris base (1 M) <!>	500 mL	0.5 M
Glycine (1 M)	500 mL	0.5 M

Store at 4°C. The pH should be ~8.8. See also Tips and Troubleshooting.

READING LIST

Bowers HA, Maroney PA, Fairman ME, Kastner B, Lührmann R, Nilsen TW, Jankowsky E. 2006. Discriminatory RNP remodeling by the DEAD-box protein DED1. *RNA* **12:** 903–912.

Das R, Reed R. 1999. Resolution of the mammalian E complex and the ATP-dependent spliceosomal complexes on native agarose mini-gels. *RNA* **5:** 1504–1508.

Goodrich JA, Kugel JF. 2007. *Binding and kinetics for molecular biologists*. Cold Spring Harbor Laboratory Press, Cold Spring Harbor, N.Y.

Konarska MM. 1989. Analysis of splicing complexes and small nuclear ribonucleoprotein particles by native gel electrophoresis. *Methods Enzymol* **180:** 442–453.

Setzer DR. 1999. Measuring equilibrium and kinetic constants using gel retardation assays. *Methods Mol Biol* **118:** 115–128.

Protocol 3

SELEX to Identify Protein-binding Sites on RNA

SELEX is a very powerful method for determining the binding site of a protein on RNA. It relies on the ability of an RNA-binding protein to select high-affinity RNA ligands from a randomized pool of RNAs. SELEX experiments are carried out over several "rounds," with each round resulting in increased enrichment of RNAs capable of binding to the protein. There are many variations of SELEX strategies, but they all rely on the ability to separate bound RNA from unbound RNA. The method presented here uses tagged proteins; however, it is also possible to use other separation techniques (e.g., filter binding or antibodies).

MATERIALS

The recipes for the items marked with <R> are at the end of the protocol or in the Common Recipes Appendix.

CAUTION: See the Cautions Appendix for appropriate handling of materials marked with <!>.

Reagents

ATP, CTP, GTP, UTP (100 mM each)
BamHI (200 μL)
DNase I (RNase-free; 2 units/μL)
dNTP mix (10 mM each of dATP, dCTP, dGTP, dTTP)
Ethanol (70%, 100%) <!>
Klenow buffer <R>
Klenow enzyme
NT2 buffer <R>
PCR buffer (10x) <R>
Phenol:chloroform:isoamyl alcohol (PCA) (25:24:1) <!>
Primers
  "Forward" primer sequence: 5′-CGC GAA TTC TAA TAC GAC TCA CTA TAG GGG CCA CCA ACG ACA TT-3′
  EcoRI restriction site is nucleotides 4–9, T7 promoter is nucleotides 10–29.
  "Reverse" primer sequence: 5′-CCC GAC ACC GCG GGA TCC ATG GGC ACT ATT TAT ATC AA-3′
  BamHI restriction site is nucleotides 13–18.
  "Library" primer sequence: 5′-TTACAGCAACCACCG GG GAT CCA TGG GCA CTA TTT ATA TCA AC $(N)_{25}$ AAT GTC GTT GGT GGC CC-3′
Proteinase K (10 mg/mL) <!>
Proteinase K buffer (2x) <R>
Protein of interest (such as GST or histidine) fused to a tag for purification
Reverse transcriptase (M-MuLV 200 units/μL)
Reverse transcriptase buffer (10x) <R>
RNase inhibitor (40 units/μL)
SDS extraction buffer <R>
Sepharose beads for purification of complexes
Sodium acetate (3 M, pH 5.2)

Protocol courtesy of James L. Manley, Biological Sciences, Columbia University, New York, New York.

T7 RNA polymerase
T7 transcription buffer (10×) <R>
Taq DNA polymerase (1 unit/µL)
TBE polyacrylamide (19:1 acrylamide:bis-acrylamide) gel
TE buffer (pH 8.0) <R>
Urea (8 M), 15% polyacrylamide TBE gel
Vector for cloning and bacteria for transformation (see Step 46)
Yeast tRNA (10 mg/mL)

Equipment

Beaker with 50 mL of H_2O at 75°C
Conical centrifuge tube (15 mL)
Dry ice <!>
Ice
Microcentrifuge
Microcentrifuge tubes (1.5 mL)
PCR tube
Polypropylene tubes (12 mL)
Shaker
Spectrophotometer
Thermal cycler
Tube rocker
UV light source (254 nm, handheld) <!>
Vortex mixer
Water baths (37°C, 42°C, 55°C, and 75°C)
X-ray intensifying screen for UV shadowing

OVERVIEW

Before beginning this protocol, it is necessary to understand the experimental flow. Each "round" of SELEX consists of the same sequence of steps. First, it is necessary to construct DNA templates for synthesizing the randomized RNA pool. The pool is then synthesized, labeled, and allowed to bind to the protein of interest. Bound RNAs are then converted back into DNA templates by reverse-transcriptase–polymerase chain reaction (RT-PCR). This process is reiterated until the bulk (≥90%) of the pool is capable of binding to the protein. At this point, the "final" pool is converted back to double-stranded DNA, cloned, and sequenced. The following primers need to be purchased for this protocol: (1) a "forward" primer, containing T7 promoter and a restriction enzyme site for cloning, (2) a "reverse" primer, containing a restriction enzyme site for cloning and hybridization sequence for reverse transcription, and (3) a "library" primer, containing a region of randomized sequence as well as sequences that anneal to the "forward" and "reverse" primers (see Reagent list and Fig. 5-5).

METHOD

Convert Library of Oligonucleotides into Double-stranded DNA Transcription Templates

1. Anneal the "forward" and "library" primers by mixing the following in a 1.5-mL tube:

Klenow buffer (5×)	200 µL
100 µM "forward" primer	85 µL
100 µM "library" primer	85 µL
H_2O	590 µL

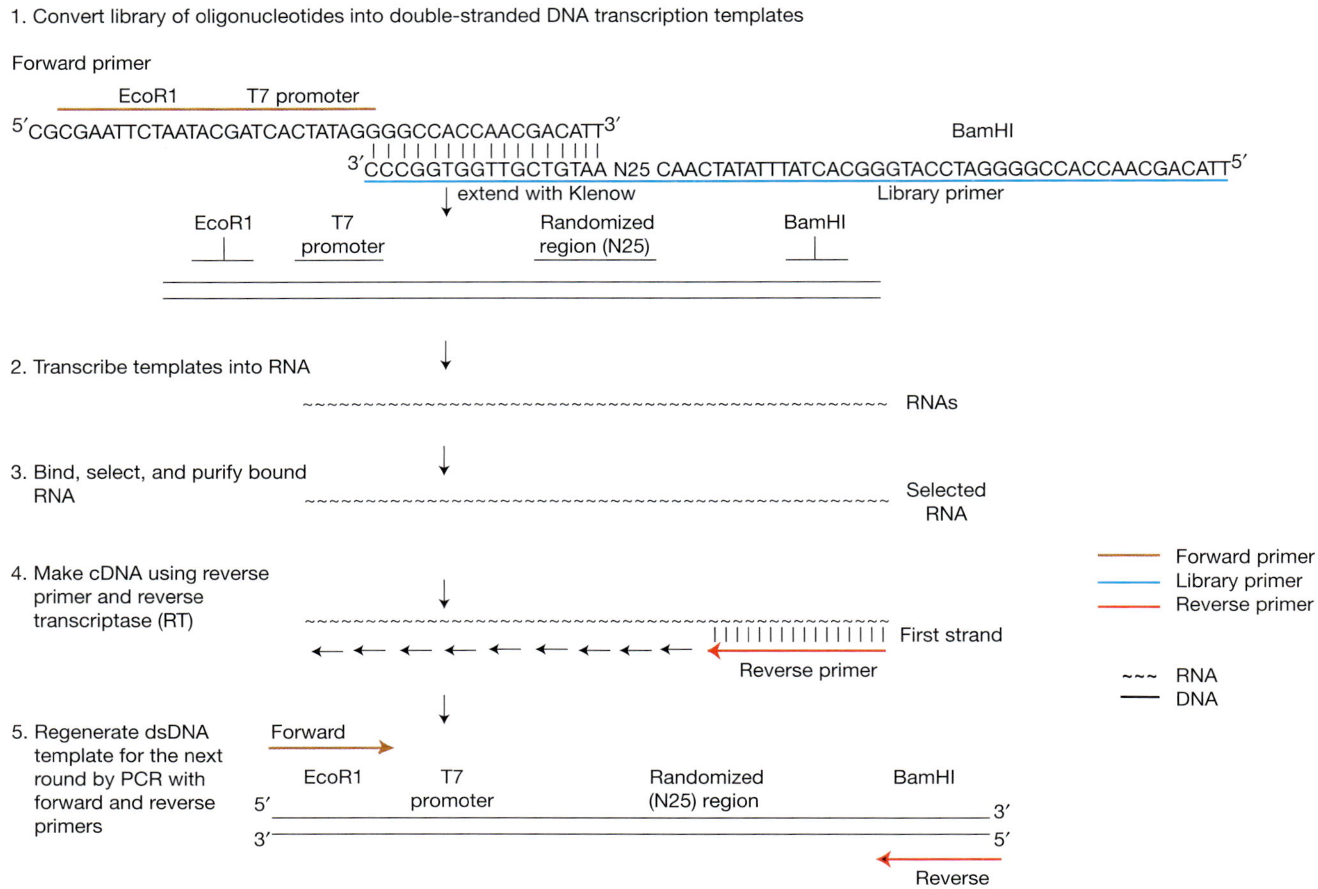

FIGURE 5-5. Schematic representation of a "round" of SELEX. For a detailed description, see accompanying text.

2. Heat the mixture for 15 min to 75°C and then transfer it to a beaker containing 50 mL of H_2O at 75°C. Allow the tube to cool slowly to room temperature. Centrifuge briefly for 5 sec at room temperature.
3. Extend the annealed primers by adding 20 µL of dNTP mix (10 mM each of dATP, dCTP, dGTP, and dTTP) and 20 µL of Klenow enzyme. Incubate for 1 h at 37°C.
4. Add 100 µL of 3 M sodium acetate (pH 5.2).
5. Divide the sample into four 275-µL aliquots and add 600 µL of 100% ethanol to each.
6. Freeze on dry ice for 10 min and then centrifuge at ≥12,000*g* for 15 min at room temperature.
7. Decant the supernatant and wash the pellet with 70% ethanol. Centrifuge at ≥12000*g* for 5 min at room temperature. Decant the supernatant and air-dry the pellet.
8. Resuspend the double-stranded DNA library from Step 7 and gel-purify using a 15% nondenaturing TBE polyacrylamide (19:1 acrylamide:bis-acrylamide) gel. Electrophorese until the xylene cyanol dye is close to the bottom of a 15-cm-long gel.
9. Visualize the nucleic acids by UV shadowing.
10. Excise the band containing the full-length double-stranded library. Crush the excised gel band and elute the DNA using 1–2 mL of 0.5 M sodium acetate overnight with shaking.
11. Collect supernatant, precipitate the DNA with ethanol, and wash the precipitate with 70% ethanol. Repeat the elution once to increase recovery if desired.

12. Dissolve the DNA in ~100 µL of TE buffer (pH 8.0).
13. Determine the OD_{260} and convert the value to molar concentration.

    For example, for 50 µg/mL per OD_{260}, the 95-nucleotide product has a molecular weight of 62,440 g/mol.
14. Dilute the DNA to 1.7 nmol/100 µL and divide it into 100-µL aliquots.

Transcribe the Double-stranded DNA Template for First Round

15. Set up the transcription reaction in a 15-mL conical centrifuge tube as follows:

1.7-nmol double-stranded DNA library (from Step 14)	100 µL
10x T7 transcription buffer	1 mL
100 mM ATP	100 µL
100 mM CTP	100 µL
100 mM GTP	100 µL
100 mM UTP	100 µL
T7 RNA polymerase	1 mL
H_2O	7.5 mL

16. Mix the reagents gently by inverting and then incubate for 2 h at 37°C.

    The solution should become cloudy.
17. Add 100 µL of RNase-free DNase I to the transcription reaction and incubate for a further 15 min at 37°C.
18. Split into 2.5-mL aliquots in 12-mL polypropylene tubes and precipitate the RNA with ethanol.
19. Purify the full-length RNA transcripts on an 8 M urea, 15% polyacrylamide gel.
20. Visualize the nucleic acids by UV shadowing.
21. Excise the band containing full-length RNA transcripts. Elute RNA overnight as in Step 10. Collect the supernatant, precipitate with ethanol, and wash the precipitate with 70% ethanol.
22. Dissolve the RNA in ~100 µL of TE buffer (pH 8.0) or in SDS extraction buffer.
23. Determine the OD_{260} and convert the value to molar concentration.

    For example, for 40 µg/mL per OD_{260}, the 69-nucleotide, single-stranded RNA has a molecular weight of 23,630 g/mol.
24. Dilute the RNA to 17 nmol/100 µL and store in 100-µL aliquots at –80°C.

    Each aliquot on average contains 10 RNA copies of each of the 1 × 10^{15} DNA templates.

Protein–RNA Binding and Selection

25. Select RNA from Step 24 using a matrix bound to the protein of interest (e.g., glutathione–Sepharose beads with a GST-tagged protein or Ni-NTA agarose with a histidine-tagged protein).

    Immunoprecipitation, filter binding, or a mobility shift assay can also be used to separate protein–RNA complexes from free RNA.
26. Prepare sufficient beads for 2 mg of each protein of interest and an equal volume of beads to preclear the RNA library. Wash the beads in NT2 buffer.
27. Bind 1–10 µg of protein to Sepharose or agarose beads for at least 2 h at 4°C with rocking.
28. Pellet the beads by centrifugation at 8000*g* for 10 sec at room temperature. Then, wash the beads three times with 1 mL of NT2 buffer each time.
29. Preclear a 17-nmol aliquot of the RNA library from Step 24 by binding it to beads for 30 min at 4°C. Pellet the beads by centrifugation at 8000*g* for 10 sec at room temperature. Keep the supernatant, which contains the precleared RNA library.

30. Bind the precleared RNA library from Step 29 to the proteins from Step 28 by incubating for at least 2 h at 4°C with rocking.
31. Wash the beads five times with 1 mL of NT2 buffer each time. After the last wash, leave 100 µL of NT2 buffer on the beads.
32. Add 100 µL of H_2O, 200 µL of 2x proteinase K buffer, and 5 µL of proteinase K. Incubate for 15 min at 37°C.
33. Add 400 µL of PCA (25:24:1), vortex for 30 sec, and then centrifuge at 12,000*g* for 1 min at room temperature.
34. Transfer aqueous layer, which contains the RNA, to a clean tube. Add 2 µL of 10 mg/mL of yeast tRNA, 40 µL of 3 M sodium acetate (pH 5.2), and 1 mL of 100% ethanol. Mix thoroughly, freeze on dry ice for 10 min, and then centrifuge at ≥12,000*g* for 15 min at room temperature.
35. Decant the supernatant, wash the RNA pellet with 70% ethanol, and centrifuge at ≥12,000*g* for 5 min at room temperature. Decant the supernatant and air-dry the pellet.

Reverse Transcription PCR

36. Resuspend the RNA pellet from Step 35 in 13 µL of H_2O. Add the following:

"Reverse" primer	3 µL
Reverse transcriptase buffer (10x)	2 µL
dNTP mix	2 µL
Reverse transcriptase	0.5 µL

Incubate the reaction for 5 min at 55°C and then for 1 h at 42°C.

37. Transfer 6 µL from the reverse transcription reaction into a PCR tube and add the following:

H_2O	75.5 µL
Taq buffer (10x)	10 µL
dNTP mix	2 µL
"Forward" primer	3 µL
"Reverse" primer	3 µL
Taq polymerase	0.5 µL

38. Denature for 5 min at 94°C. Then, amplify for 25 cycles of 30 sec at 94°C, 30 sec at 50°C, and 30 sec at 72°C followed by a final extension of 10 min at 72°C.
39. Add 1 µL of BamHI to the reaction and incubate at 37°C to regenerate the 3′ ends of original library.
40. Extract with an equal volume of PCA. Precipitate the double-stranded DNA template with ethanol, air-dry the pellet, and resuspend it in 15 µL of H_2O.

Transcribe Double-stranded DNA for Subsequent Rounds

41. Combine the following:

T7 transcription buffer (10x)	2 µL
NTP mix (2.5 mM each ATP, CTP, UTP, GTP)	8 µL
Double-stranded DNA template (from RT-PCR of previous round [Step 40])	3 µL
RNase inhibitor (40 units/mL)	1 µL
H_2O	4 µL
T7 RNA polymerase	2 µL

Incubate for 1 h at 37°C.

42. Add 1.5 µL of RNase-free DNase I to the reaction and incubate for an additional 10 min at 37°C.

43. Add 78.5 µL of H_2O and 10 µL of 3 M sodium acetate (pH 5.2). Extract the reaction with an equal volume of PCA, precipitate with ethanol, air-dry the RNA pellet, and then resuspend RNA in 20 µL of H_2O.

44. Use one-quarter (5 µL) of the RNA from Step 43 to repeat the cycle for the binding to protein–RNA-binding step (Steps 25–35). Repeat the RT-PCR (Steps 36–40). Usually 4–10 cycles of enrichment are required.

45. Check for protein–RNA binding by electrophoretic mobility shift assay using radiolabeled RNA (either by incorporation of radioactive nucleotide during transcription or by end-labeling RNA transcripts) and the purified protein of interest (see Protocol 2).

Determine the Sequence Recognized by the Protein

46. After sufficient enrichment has been achieved (≥90% of the transcribed RNA is bound by the protein), clone the double-stranded DNA templates into a suitable vector and transform bacteria.

47. Pick a large number (50–100) of transformants and determine the nucleotide sequence of the inserts. Consensus sequences for binding sites should emerge from this analysis.

SOLUTION AND REAGENT RECIPES

CAUTION: See the Cautions Appendix for appropriate handling of materials marked with <!>.

Klenow Buffer (5x)

Reagent	Quantity (for 1 mL)	Final concentration
NaCl (5 M)	50 µL	250 mM
Tris-HCl (1 M, pH 7.9)	50 µL	50 mM
DTT (1 M) <!>	5 µL	5 mM
H_2O	895 µL	

Store for 6 mo at –20°C.

NT2 Buffer

Reagent	Quantity (for 100 mL)	Final concentration
Tris-HCl (1 M, pH 7.4)	5 mL	50 mM
NaCl (5 M)	3 mL	150 mM
$MgCl_2$ (1 M) <!>	0.1 mL	1 mM
Nonidet P-40 (NP-40; 10%)	0.5 mL	0.05%
H_2O	91.4 mL	

Store for 6 mo at 4°C.

PCR Buffer (10x)

Reagent	Quantity (for 1 mL)	Final concentration
Tris-HCl (1 M, pH 8.6)	100 μL	100 mM
KCl (2 M) <!>	250 μL	500 mM
$MgCl_2$ (1 M) <!>	15 μL	15 mM
H_2O	635 μL	

Store for 6 mo at –20°C.

Proteinase K Buffer (2x)

Reagent	Quantity (for 50 mL)	Final concentration
SDS (10%) <!>	2.5 mL	0.5%
EDTA (0.5 M, pH 8.0) <!>	0.4 mL	4 mM
H_2O	47.1 mL	

Store indefinitely at room temperature.

Reverse Transcription Buffer (10x)

Reagent	Quantity (for 1 mL)	Final concentration
Tris-HCl (1 M, pH 8.3)	500 μL	500 mM
KCl (2 M) <!>	200 μL	400 mM
$MgCl_2$ (1 M) <!>	60 μL	60 mM
DTT (1 M) <!>	10 μL	10 mM
H_2O	230 μL	

Store for 6 mo at –20°C.

T7 Transcription Buffer (10x)

Reagent	Quantity (for 1 mL)	Final concentration
Tris-HCl (1 M, pH 8.0)	400 μL	400 mM
$MgCl_2$ (1 M) <!>	200 μL	200 mM
DTT (1 M) <!>	50 μL	50 mM
H_2O	350 μL	

Store for 6 mo at –20°C.

READING LIST

Bouvet P. 2001. Determination of nucleic acid recognition sequences by SELEX. *Methods Mol Biol* **148:** 603–610.

Kenan DJ, Keene JD. 1999. In vitro selection of aptamers from RNA libraries. *Methods Mol Biol* **118:** 217–231.

Protocol 4

RNA Structure Determination Using Chemical and Nuclease Digestion Methods

As discussed in the introduction to this chapter, a large number of methods can be used to determine RNA structure. These methods can be divided into two main classes: chemical probing (Method 1) and nuclease probing (Method 2). Detailed methods for examples of each are provided, as are abbreviated methods for variations (e.g., using different chemicals or different nucleases). Following structure determination methods, footprinting techniques are given, again arranged first with chemical approaches followed by nuclease-based methods.

Method 1: RNA Structure Determination Using Chemical Methods

MATERIALS

The recipes for the items marked with <R> are at the end of the protocol or in the Common Recipes Appendix.

CAUTION: See the Cautions Appendix for appropriate handling of materials marked with <!>.

Reagents

5′-End-labeled oligonucleotide primers prepared using T4 polynucleotide kinase (see Chapter 3, Protocol 1)
Aniline (1 M) <R>
Annealing buffer (2x) <R>
Aurintricarboxylic acid (ATA; 5 mg/mL in 50 mM Tris, pH 7.5) (Sigma-Aldrich) <!>
CMCT in H_2O (cyclohexyl-3-(2-morpholinoethyl)-carbodiimide metho-*p*-toluene sulfonate) (25 mg/mL) <!>
Diethylpyrocarbonate (DEPC; Sigma-Aldrich) <!>
Dimethylsulfate (DMS) (diluted 1:6 in 100% ethanol) <!>
Dithiothreitol (DTT; 0.2 M) <!>
EDTA (0.2 M, pH 8.0) <!>
Ethanol (100%) <!>
Loading dye <R>
Modification buffers <R>
- A native
- A denaturing
- B native
- B denaturing

Reverse transcriptase (M-MuLV, 200 units/µL or SuperScript II, 200 units/µL)
Reverse transcription buffer (5x) <R>
RNases
- T1 (0.25 unit/µL)
- 1 (12.5 units/µL)
- V5 (6.25 x 10^{-3} units/µL)

RNA to be probed (~0.2–1 µg)

This protocol was supplied by Mark Caprara, RNA Center, Case Western Reserve University.

Sodium acetate (NaOAc; 3 M, pH 5.2)
Sodium borohydride (0.2 M) <!>
Sodium hydroxide (1 N) <!>
SuperScript II (200 units/mL)
T1 sequencing buffer <R>
TBE buffer <R>
Tris-HCl (1 M, pH 8.3)
tRNA carrier (*E. coli* 10 µg/µL)

Equipment

Denaturing polyacrylamide gel electrophoresis system and power supply to run an 8% gel
Dry ice <!>
Gel dryer
Heat block (95°C)
Ice
Microcentrifuge
Phosphorimager or autoradiography materials
SpeedVac
Vortex mixer
Water baths (set to temperatures to accommodate specific reactions)

OVERVIEW

The conditions described in this method provide a starting point for concentrations of reagents and times of incubation. These should be titrated on denatured RNA (unfolded) to find the optimum modification conditions for individual substrates, usually ≤1 per molecule. After optimizing conditions for each reagent, assemble two reactions, one in "native" buffer (folded RNA) and one in "denaturing" buffer (unfolded RNA). The sites of modification are monitored by primer extension using a 5′-end-labeled oligonucleotide (see Chapter 3, Protocol 14). If RNAs are longer than 100 nucleotides, more than one primer may be needed (one every 100 nucleotides). The primer extension ladders from two samples are compared to each other and a residue is scored as protected if the band intensity of the "native" is less than 50% of the intensity of the "denatured." DMS and DEPC should be used in a chemical fume hood and must be disposed of properly (see also chemical sequencing, Chapter 3, Protocol 24).

METHOD

I. Modification with DMS

DMS methylates the N1 position of unpaired adenines and the N3 position of unpaired cytosines. The reactivity of these positions to this reagent is a good indicator for whether the bases are paired. DMS can also modify the N7 position (in the major groove) of guanines, and its reactivity reflects either noncanonical base-pairing interactions or coordination with metal ions.

1. Set up two reactions containing 25–100 nM RNA, one in 100 µL of modification buffer A (native) and one in 100 µL of modification buffer A (denaturing).
2. Preincubate the reactions for 10 min at 37°C.
3. Add 1 µL of a 1:6 dilution of DMS in 100% ethanol to each reaction, mix by vortexing, and incubate for 4 min at 37°C (native) or 1 min at 55°C (denaturing).

4. Stop the reactions by adding 0.3 mL of 200 mM EDTA (pH 8) and 10 μg of tRNA carrier. Add NaOAc to 0.3 M and precipitate the RNA with 2.5 volumes of 100% ethanol.
5. Freeze on dry ice, thaw, and centrifuge in a microcentrifuge at maximum speed for 10 min at room temperature.
6. Remove and properly discard the supernatant. Save the pellet and continue with Step 7 for visualization of modified guanine. Proceed to Section V for visualization of modified adenosines or cytosines.

   Modified guanines are sensitive to aniline-induced cleavage after reduction of methylated guanine by sodium borohydride. The sites of cleavage are then detected by primer extension in Section V.

7. To reduce the modified guanines, resuspend the RNA pellet in 10 μL of 1 M Tris-HCl (pH 8.3). Add 10 μL of 0.2 M sodium borohydride, mix by vortexing, and incubate in the dark for 30 min on ice. Precipitate the RNA with ethanol twice (see Chapter 2, Protocol 3). After the second precipitation, dry the pellets in a SpeedVac and proceed to Section IV (Aniline Cleavage).

II. CMCT

CMCT reacts with the N3 of unpaired uridines and to a lesser extent with the N1 of unpaired guanines. The sites of modification are detected directly by stops in reverse transcription.

1. Set up two reactions containing 25–100 nM RNA, one in 100 μL of modification buffer B (native) and one in 100 μL of modification buffer B (denaturing).
2. Preincubate the reactions for 10 min at 37°C.
3. Add 1 μL of 25 mg/mL CMCT to each reaction, mix by vortexing, and incubate for 4 min at 37°C (native) or for 1 min at 55°C (denaturing).
4. Stop the reactions by adding 0.3 mL of 200 mM EDTA and 10 μg of tRNA carrier. Add sodium acetate to 0.3 M and precipitate the RNA with 2.5 volumes of ethanol.
5. Save the RNA from Step 4 (–20°C in 100% ethanol) or proceed to detection by primer extension (Section V).

III. DEPC

DEPC carbethoxylates the N7 of adenines. It is sensitive to base stacking and therefore adenines that are base paired will not react. Modification of the adenine does not cause a block to primer extension and the RNA must be cleaved with aniline. In this case, reduction by sodium borohydride is not necessary.

1. Set up two reactions containing 25–100 nM RNA, one in 100 μL of modification buffer A (native) and one in 100 μL of modification buffer A (denaturing).
2. Preincubate the reactions for 10 min at 37°C.
3. Add 2 μL of undiluted DEPC to each reaction, mix by vortexing, and incubate for 4 min at 37°C (native) or 1 min at 55°C (denaturing).

   Additional vigorous mixing with a Pipetman is suggested because DEPC is not readily insoluble in H_2O.

4. Stop the reactions by adding 0.3 mL of 0.2 EDTA (pH 8.0) and 10 μg of tRNA carrier. Add sodium acetate to 0.3 M and precipitate with 2.5 volumes of 100% ethanol.
5. Freeze on dry ice, thaw, and then microcentrifuge at maximum speed for 10 min at room temperature. Pour off the ethanol and resuspend the RNA pellets in 0.2 mL of 0.2 M EDTA.

6. Add sodium acetate to 0.3 M and reprecipitate with 2.5 volumes of 100% ethanol. Dry the pellets in a SpeedVac and proceed to Section IV (Aniline Cleavage).

IV. Aniline Cleavage

1. Resuspend pellets from Section I or III in 20 μL of 1 M aniline (pH 4.5) and incubate in the dark for 15 min at 60°C.
2. Dilute with 200 μL of H_2O and add NaOAc to 0.3 M. Precipitate with ethanol twice.
3. Continue with Section V.

V. Detection of Modifications by Primer Extension and Blocks to Reverse Transcription

1. Prepare one or more 5′-end-labeled oligonucleotide primers using T4 polynucleotide kinase as described in Chapter 3, Protocol 1.
2. Resuspend the RNA pellets in 10 μL of H_2O.
3. Set up the following mixture for each sample:

Kinased oligonucleotide	1–4 μL	5 pmol
Modified RNA	1–4 μL	
Annealing buffer (2x)	8 μL	1x
H_2O	to 16 μL	

   Heat for 30 sec at 95°C, then for 15 min at 55°C.
4. Place the mixture on ice and add 4 μL of 5x reverse transcriptase buffer and 1 μL of 200 units/μL SuperScript II. Incubate for 10 min at 50°C.
5. Stop the reactions by adding EDTA to a final concentration of 50 mM. Add an equal volume of loading dye.
6. Analyze the products on 8% denaturing polyacrylamide gels with TBE as running buffer (see Chapter 2, Protocol 12). Dry the gels and quantitate the bands with a phosphorimager.
7. After subtracting the background, the band intensities are normalized to the amount of full-length RNA in each lane. Protections or enhancements in the presence of $MgCl_2$ are assigned on the basis of the change in band intensity relative to the denatured RNA sample. A position can be considered protected or enhanced if the band intensity is less or greater than twofold of that in the denatured RNA sample, respectively. To identify primer extension products, compare a sequencing ladder made with the same 5′-end-labeled primers to the sequence (with Sequenase or reverse transcriptase) of a plasmid containing a cDNA copy of the RNA being analyzed.

TIPS AND TROUBLESHOOTING

1. Avoid buffers containing amines (such as Tris) because many of the reagents used here modify primary amines.
2. The concentration of the reagents should be adjusted so that they modify an RNA substrate only once. A good rule of thumb is to ensure that the total amount of modified or cleaved material in a sample is no greater than 20% of the total amount of RNA.
3. Tracking dyes can distort bands that comigrate and can affect quantitation. If this is a problem, load the tracking dyes in parallel lanes.

4. Gels should be prerun for 30–45 min to prevent hyperfocusing and distortion of the unused primer and shorter products (see Chapter 2, Protocol 12).

Method 2: RNA Structure Determination Using Nuclease Digestion Methods

A number of ribonucleases (RNases) have been used to map RNA structure, but sadly, many of these are no longer available. However, three commonly available RNA endonucleases can provide a wealth of structural information. RNase T1 cleaves 3′ to an unpaired guanine, RNase I cleaves 3′ to all bases with a slight preference for pyrimidines, and RNase V1 cleaves double-strand regions with a preference to either end of a helix. RNase V1 will also cleave single-strand stacked regions. RNase V1 cleavage requires $MgCl_2$, so a "denaturing" control (no $MgCl_2$) cannot be performed with this enzyme.

Ribonuclease probing is performed similarly to chemical probing, but end-labeled RNA is used. Cleavages are detected directly after electrophoresis on denaturing polyacrylamide gels. The cleavage reactions are initiated by an RNase or H_2O (for mock-treated controls) and terminated with aurintricarboxylic acid (a potent ribonuclease inhibitor).

MATERIALS

See the Materials list for Method 1.

METHOD

1. Dilute end-labeled RNA with reaction buffer A (native or denaturing) to obtain 25,000 cpm in 25 μL for each reaction.
2. Preincubate the reaction for 10 min at 37°C.
3. Add RNase enzyme (as follows) to one reaction at a time (1 μL to 25 μL). Final concentrations will be 0.01 unit/μL RNase T1, 0.5 unit/μL RNase I, or 2.5×10^{-4} units/μL RNase V1. Because times are short, perform one reaction at a time.
4. Incubate the reactions for 5 sec at 37°C, quickly add 3 μL of ATA, and mix by pipetting up and down.
5. Add an equal volume of loading dye. Analyze on an 8% denaturing polyacrylamide gel with markers described in Steps 6 and 7.

   Markers for nuclease probing are generated from the end-labeled RNA by partial RNase T1 digestion or base hydrolysis.
6. For T1 digestion, incubate the RNA (100,000 cpm) in 10 μL of T1 sequencing buffer for 10 min at 50°C. Stop the reaction with 2 μL of ATA and 12 μL of Loading dye.
7. To generate an alkaline ladder, use 100,000 cpm end-labeled RNA in 9 μL of 2 mM EDTA (pH 8.0) containing 20 μg of *E. coli* tRNA. Add 1 μL of 1 N NaOH and incubate for 20 min at 30°C. Stop the reaction by adding 2 μL of 1 M Tris-HCl (pH 8) and freeze on dry ice until solid. Add an equal volume of Loading dye and thaw to room temperature. Load equal cpm of markers or sample on a denaturing polyacrylamide gel. Also see Appendix for an alternate procedure to prepare alkaline ladders.

SOLUTION AND REAGENT RECIPES

CAUTION: See the Cautions Appendix for appropriate handling of materials marked with <!>.

Aniline (1 M)

Reagent	Quantity (for 10 µL)	Final concentration
Aniline (11 M)	1 µL	1 M
Glacial acetic acid (17 M) <!>	7 µL	pH 4.5
H_2O	3 µL	100 mM

Make fresh for each use. Scale up amounts according to numbers of reactions.

Annealing Buffer (2x)

Reagent	Quantity (for 10 mL)	Final concentration
Tris (1 M, pH 8.3) <!>	0.2 mL	20 mM
KCl (2 M) <!>	0.1 mL	80 mM
H_2O	9.4 mL	

Store indefinitely at 4°C.

Urea Loading Dye

Reagent	Quantity (for 10 µL)	Final concentration
Urea (ultrapure) <!>	5.4 g	9 M
TBE (see Appendix)*		
Xylene cyanol <!>/	0.05% w/v	
bromophenol blue <!>		
(1%, w/v) (optional)	500 µL	0.5%

*Add enough to reach the 10-mL mark in a 15-mL conical tube containing urea. Heat at 50°C until dissolved. Check before using to make sure urea is still in solution.

Store at room temperature and make sure urea is in solution before using.

Modification Reaction Buffer A (Native)

Reagent	Quantity (for 100 mL)	Final concentration
HEPES/KOH (1 M, pH 7.5) <!>	1 mL	10 mM
KCl (2 M) <!>	5 mL	100 mM
$MgCl_2$ (1 M) <!>	0.5 mL	5 mM
H_2O	93.5 mL	

Store indefinitely at 4°C. Adjust $MgCl_2$ concentration between 5 and 25 mM for individual substrates.

Modification Reaction Buffer A (Denaturing)

Reagent	Quantity (for 100 mL)	Final concentration
HEPES/KOH (1 M, pH 7.5) <!>	1 mL	10 mM
EDTA (0.5 M, pH 8) <!>	0.5 mL	1 mM
H_2O	98.5 mL	

Store indefinitely room temperature.

Modification Reaction Buffer B (Native)

Reagent	Quantity (for 100 mL)	Final concentration
Potassium borate (0.2 M, pH 7.9)	5 mL	10 mM
KCl (2 M) <!>	5 mL	100 mM
$MgCl_2$ (1 M) <!>	0.5 mL	5 mM
H_2O	89.5 mL	

Store indefinitely at 4°C. Adjust $MgCl_2$ concentration between 5 and 25 mM for individual substrates.

Modification Buffer B (Denaturing)

Reagent	Quantity (for 100 mL)	Final concentration
Potassium borate (0.2 M, pH 7.9)	5 mL	10 mM
EDTA (0.2 M, pH 8.0) <!>	0.5 mL	1 mM
H_2O	94.5 mL	

Store indefinitely at 4°C.

Reverse Transcription Buffer (5x)

Reagent	Quantity (for 1 mL)	Final concentration
$MgCl_2$ (1 M) <!>	30 μL	30 mM
DTT (1 M) <!>	100 μL	100 mM
dNTPs (25 mM each)	100 μL	2.5 mM
H_2O	770 μL	

Store for 6 mo at –20°C.

T1 Sequencing Buffer

Reagent	Quantity (for 1 mL)	Final concentration
Sodium acetate (3 M, pH 5.3)	6.6 μL	20 mM
Urea (Molecular Biology grade) <!>	0.42 g	7 M
EDTA (0.5 M, pH 8.0) <!>	2 μL	1 mM
H_2O	adjust to 1 mL	

Store for 6 mo at room temperature.

Protocol 5

Mapping Sites of RNA–Protein Interactions Using Chemical Methods

The preceding section described methods for determining the structure of naked RNAs via chemical and enzymatic approaches. Here, we present analogous methods for determining the site(s) of RNA–protein interactions. Collectively, these approaches are known as footprinting and, as above, we first describe techniques based on chemical modifications, followed by nuclease-based techniques.

CHEMICAL FOOTPRINTING

Fe(II)-EDTA

Fe(II)-EDTA in the presence of a reducing agent generates hydroxyl radicals that indiscriminately cleave the ribose in regions of the ribose-phosphate backbone if it is exposed to solvent. Therefore, it has been a useful reagent to monitor sites of RNA–protein interactions. Because the reactive species is smaller and less specific than RNases, this method, if feasible, is often the method of choice. Fe(II)-EDTA can also be used to monitor higher-order (tertiary) structures of naked RNAs.

MATERIALS

The recipes for the items marked with <R> are at the end of the protocol or in the Common Recipes Appendix.

CAUTION: See the Cautions Appendix for appropriate handling of materials marked with <!>.

Reagents for Section I–IV. (Consult the specific protocols to determined the reagents needed.)

Acetic acid (10%)/methanol (10%) <!>
Affinity reagents for purification of RNA/protein complexes
Ammonium iron (II) sulfate (50 mM)
Aniline (1 M) <R> (see p. 274)
Annealing buffer (2x) <R> (see p. 274)
Ascorbic acid (0.25 M) (Sigma-Aldrich 26,855-0)
β-Mercaptoethanol (14 M) <!>
Binding buffer, specific for each protein such as Modification Buffer A or B
Chloroform:isoamy alcohol (24:1) <!>
CMCT in H_2O (cyclohexyl-3-(2 morpholinoethyl) carbodiimide metho-*p*-toluene sulfonate) (25 mg/mL) <!>
DEPC (diethylpyrocarbonate; Sigma-Aldrich) <!>
Dithiothreitol (DTT; 0.2 M) <!>
DMS (dimethylsulfate) <!>
    Diluted 1:6 in 100% ethanol.
dNTPs (10 mM)
EDTA (0.2 M, pH 8.0) <!>
Ethanol (100%) <!>

Extract or protein for binding reactions
Formamide loading dye <R>
GlycoBlue (10 µg/µL)
Heparin (50 mg/mL, w/v) in ddH_2O (Sigma-Aldrich H-3393) <!>
HEPES (0.2 M, pH 7.6 with KOH)
H_2O_2 (2.5%, 30%; hydrogen peroxide) (Fisher H325-100) <!> (see p. 289)
Kethoxal <!>
$MgCl_2$ (1 M) <!>
Modification buffers <R>
  A native (see p. 274)
  B native (see p. 275)
$(NH_4)_2Fe(SO_4)_2 \cdot 6H_2O$ (50 mM; Baker 2054-1)
Phenol <!>
Potassium borate (0.2 M, pH 7.9)
Potassium chloride (2 M) <!>
Proteinase K (10 mg/mL) <!>
Reverse transcriptase (M-MuLV 200 units/µL or SuperScript II 200 units/µL)
Reverse transcription buffer (5x) <R> (see p. 275)
RNA, ^{32}P 5′- or 3′-end labeled and gel purified <!>
RNA to be probed (~0.2–1 µg)
SDS extraction buffer <R>
Sodium acetate (3 M, pH 5.2)
Sodium borohydride (0.2 M) <!>
T1 sequencing buffer <R> (see p. 275)
TBE buffer <R>
TEMED (*N,N,N′,N′*-tetramethylethylenediamine) <!>
Thiourea (0.1 M; Sigma-Aldrich T-7875) <!>
Tris base (1 M; Trizma base, Sigma-Aldrich T-6066) <!>
    Do not pH.
tRNA, yeast or *E. coli* (10 mg/mL)
Urea loading dye <R>
Xylene cyanol (1%) <!>

Equipment

Chemical fume hood
Denaturing polyacrylamide gel electrophoresis system and power supply to run an 8% gel
Dry ice <!>
Gel dryer
Heat block (95°C)
Ice
Microcentrifuge
Microcentrifuge tubes (1.5 mL)
Phosphorimager or autoradiography materials
Pipettor
SpeedVac
Vortex mixer
Water baths or heat blocks (binding temperature and 95°C)
Whatman 3MM paper

OVERVIEW

Before attempting to perform hydroxyl radical footprinting, determine the optimum binding parameters for the protein and RNA using gel shifts or filter binding (see Prococols 1 and 2). Glycerol and Tris must be absent or greatly reduced for hydroxyl radical footprinting. Either 5′- or 3′-end-labeled RNA can be used (see Chapter 4), but it is often advisable to analyze 5′- and 3′-end-labeled RNA in parallel.

METHOD

I. Fe(II)-EDTA

1. Allow the protein to bind to labeled RNA for the appropriate amount of time in 25 μL of binding buffer in a 1.5-mL microcentrifuge tube. At the same time, prepare a control reaction with no protein, substituting protein dilution buffer.

   Binding must be close to 100% or the footprint will not be clear (purify the complex if necessary).

2. Initiate the cleavage reaction by adding 1 μL each of the following reagents to the side of the 1.5-mL microfuge tube containing the reaction:

Ammonium iron (II) sulfate	50 mM
EDTA <!>	100 mM
Ascorbic acid	250 mM
H_2O_2 (2.5%) <R>	

   Centrifuge in a microcentrifuge for 2–5 sec to collect the reagents in the bottom of the tube with the reaction mix.

3. Allow the cleavage reaction to proceed for ~5 min at room temperature.
4. Stop the reaction by adding 10 μL of 0.1 M thiourea.
5. Phenol-extract the samples and ethanol-precipitate with 10 μg of GlycoBlue (see Chapter 2, Protocols 1 and 3).
6. Resuspend the RNA pellet in 200 μL of 2 mM EDTA and then add 0.3 M NaOAc (pH 5.2) and 600 μL of 100% cold (–20°C) ethanol. Freeze on dry ice, thaw, and centrifuge in a microcentrifuge for 15 min at room temperature. Carefully remove the ethanol supernatant and discard.
7. Resuspend the RNA in 5–10 μL of H_2O. Denature in urea loading dye and analyze the cleavage products on the appropriate denaturing polyacrylamide gel. Depending on the size of the RNA, it may be necessary to use two different gel percentages for best resolution. Alternatively, the footprint can be made with unlabeled RNA and detected by primer extension blocks to reverse transcriptase.
8. Subject the gel to phosphorimaging or autoradiography. Compare the no protein plus Fe^{++}-EDTA lanes to the samples bound to protein (see Fig. 5.6).

PROBING WITH OTHER CHEMICALS

Probing with other chemicals is conceptually the same as that with Fe(II)-EDTA.

Allow the protein to bind RNA for the appropriate amount of time in 25 μL of binding buffer in a 1.5-mL microcentrifuge tube. At the same time, prepare a control reaction with no protein, substituting protein dilution buffer. Binding must be 100% or the footprint will not be clear (purify the complex if necessary).

The sites of modification are detected as was described for structure mapping above. In all cases, modification conditions need to be optimized on RNA without protein. The modifications should be only ~1 per molecule as assayed by primer extension. Dideoxy sequencing reactions performed on the same substrate provide good markers.

Reagent-specific alterations to the binding conditions or handling of the substrate are described on the following page.

II. DMS

1. Dilute DMS 1:6 with 95% ethanol immediately before use.
2. Use modification reaction buffer A (native) for binding. Adjust KCl and $MgCl_2$ to be compatible with the reaction.
3. Set up a 50-μL binding reaction and incubate under binding conditions.
4. Add 1–5 μL of DMS (see Reagents list) and incubate for 30 min to 2 h at 4°C.
5. Stop the reaction with 200 μL of binding buffer containing 20 μg of tRNA followed by precipitation with ethanol.

III. Kethoxal

1. For binding, use modification reaction buffer A (native) as described above in Step 2.
2. After binding, chill on ice.
3. Add 1–5 μL of kethoxal.
4. Incubate for 30 min to 2 h at 4°C.
5. Stop the reaction by dilution with reaction buffer containing 20 μg of carrier tRNA. Add 25 mM potassium borate (pH 7.0) to stabilize the kethoxal adduct and 0.3 M NaOAc. Precipitate with ethanol.
6. After precipitation, use 25 mM potassium borate (pH 7.0) to resuspend kethoxal-treated RNA.

IV. CMCT

1. Substitute potassium borate buffer for HEPES. Use modification reaction buffer B.
2. Allow the substrate to bind.
3. Add 1–5 μL of CMCT.
4. Incubate for 30 min to 2 h at 4°C.
5. Chill on ice and then add 200 μL of binding buffer containing 20 μg of tRNA carrier.
6. Add NaOAc to 0.3 M. Precipitate with ethanol.
7. After ethanol precipitation, resuspend the samples in 200 μL of H_2O or 25 mM potassium borate (for kethoxal method). Add NaOAc to 0.3 M, PCA-extract, and precipitate using ethanol.

    Analyze the samples by primer extension on a denaturing polyacrylamide gel (see Protocol 4, Method 1, Section V). Remember to keep the kethoxal samples in potassium borate.

TIPS AND TROUBLESHOOTING (FOR SECTIONS I–IV)

1. It may be necessary to modify the binding buffer. Find a condition that permits binding of the protein to the RNA and is also compatible with hydroxyl-radical (or other chemical reagent) cleavage of the RNA. Use buffers such as HEPES instead of Tris and omit glycerol and bovine serum albumin (BSA). The glycerol concentration in the final binding reaction buffer should be no more than 0.5%.
2. The H_2O_2 dilution should always be made fresh in H_2O from a 30% stock. The Fe(II)-EDTA and ascorbate solutions can be frozen in 1-mL aliquots at –20°C.
3. It will be necessary to optimize the cleavage times and temperature. Cleavage can be conducted on ice instead of at room temperature, but it will take longer.
4. Add more Fe(II)-EDTA to increase cleavage. However, do not increase the amount of H_2O_2 because this may affect binding by the protein.

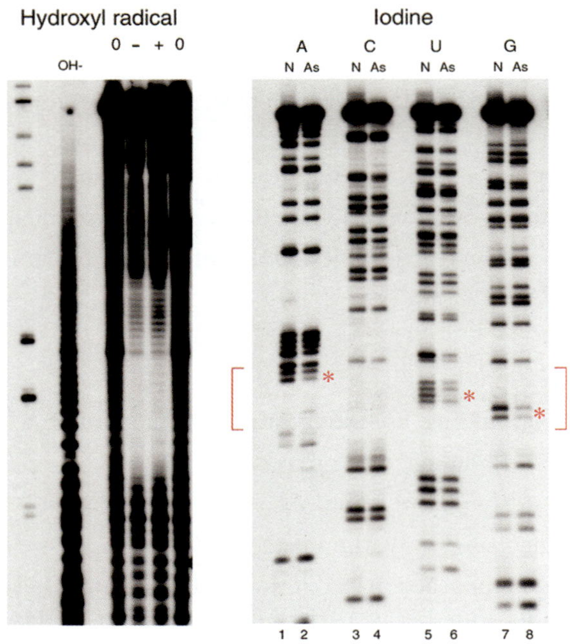

FIGURE 5-6. Examples of hydroxyl-radical and iodine footprinting. The gel on the left shows results obtained when end-labeled nematode SL RNP either without associated proteins (lanes *0*) or bound by Sm proteins (lane –) or Sm proteins plus SL RNP-specific proteins (lane +) was exposed to hydroxyl radicals. A clear footprint is present in the – and + lanes. The gel on the right shows iodine footprinting of the same SL RNP. Here, end-labeled SL RNAs randomly substituted with phosphorothioated nucleotides were digested with iodoethanol, either without incubation (lanes *N*) or after incubation with whole-cell extract (lanes *As*). Asterisks show specific protected bases.

V. Footprinting with Phosphorothioate-substituted RNA

This method relies on the ability of iodoethanol to cleave the backbone of RNA when a phosphodiester bond contains sulfur in place of a nonbridging oxygen. Because iodine is a small molecule, it provides information similar to that provided by hydroxyl-radical cleavage regarding solvent accessibility of the ribose-phosphate backbone. In this case, the accessible phosphodiester bond is cleaved. A potential advantage of this process over hydroxyl radical cleavage is that one can prepare RNAs that contain only selective thiol substitutions such that the resulting cleavage patterns contain fewer bands, making quantification "easier" and the results cleaner. For example, a population of RNAs that only contains nonbridging thiol substitutions 5′ to each adenine can be prepared by including αS ATP in the transcription reaction. To survey all positions, a series of RNAs is made by transcription in the presence of either αS ATP, αS CTP, αS UTP, or αS GTP. In general, the concentrations of NTP and αS NTP should be adjusted to achieve a ratio of 10:1. Each of the selectively substituted RNAs is probed separately.

MATERIALS

The recipes for the items marked with <R> are at the end of the protocol or in the Common Recipes Appendix.

CAUTION: See the Cautions Appendix for appropriate handling of materials marked with <!>.

Reagents

αS NTPs
ATP, CTP, UTP, GTP (10 mM)
β-Mercaptoethanol <!>

Dithiothreitol (DTT; 1 M, 2 mM) <!>
Ethanol <!>
[γ-^{32}P]ATP <!>
Formamide loading dye <R>
Gel-purification materials
Glycogen
Iodoethanol (0.5 mM) <R> (see p. 289)
Linear DNA (1 µg/mL)
m7GpppG
Phenol:chloroform <!>
[α-^{32}P]pCp <!>
[α-^{32}P]rRNT (0.5 mCi/µL) <!>
RNA polymerase, T7 or T3
RNasin
SDS buffer <R>
Transcription buffer (5x) <R>

Equipment

Denaturing gel equipment (8% denaturing gel)
Gel-purification or spin-column equipment
Tubes

OVERVIEW

RNAs appropriately substituted with thiophosphates (≤1 per molecule) are synthesized and checked by cleavage with iodoethanol. Once conditions are established, the transcriptions can be done on a larger scale and the substituted RNAs stored at –80°C. Prepare end-labeled RNA as needed.

METHOD

1. Set up transcription reactions including one thiophosphate nucleotide triphosphate to obtain ≤1 substitution per molecule (~10:1 to 20:1 NTP:αS NTP). The correct ratio must be determined empirically.
2. Start with one master transcription reaction as follows:

Reagent	Quantity	Final concentration
Transcription buffer (5x)	50 µL	1x
ATP (10 mM)	12.5 µL	0.5 mM
CTP (10 mM)	12.5 µL	0.5 mM
UTP (10 mM)	12.5 µL	0.5 mM
GTP (10 mM)	12.5 µL	0.5 mM
DTT (1 M)	5 µL	20 mM
RNasin	5 µL	0.8 unit/µL
Linear DNA (1 µg/µL)	10 µL	0.04 unit/µL
[^{32}P]rNTP (0.5 µCi/µL)	2 µL	0.004 µCi/µL
T7 or T3 RNA polymerase	10 µL	
H_2O to 225 µL	93 µL	

3. Transfer 45 µL of the transcription mix to each of five tubes and add αS NTPs or H_2O to yield the desired ratio in each tube labeled C, U, A, G, and no thio. For example, add 1.5 µL of 1 mM αS CTP to the tube designated C. Incubate for 2 h at 37°C.

To make capped transcripts, reduce GTP in all reactions to 0.2 mM and include 2 mM m7GpppG in all of the reactions. Add the appropriate amount of αS GTP to the substituted G reaction to reach the appropriate ratio for one substitution per molecule in the presence of reduced GTP.

Add a small amount of ^{32}P nucleotide to allow quantitation and gel purification of transcripts (see Chapter 4, Protocol 1).

4. Remove and save 1 µL of the reaction to count input-labeled nucleotide.
5. Dilute transcription mix reactions from Step 3 with SDS buffer containing 2 mM DTT.
6. Extract with phenol:chloroform (see Chapter 2, Protocol 1).
7. Precipitate with ethanol (see Chapter 2, Protocol 3).
8. Resuspend the RNA pellet in 2.5 mM EDTA and 2 mM DTT.
9. Gel-purify the transcripts.
10. Ethanol-precipitate with glycogen.
11. Resuspend the RNA pellet in H_2O and quantitate the RNA by calculating the percentage of ^{32}P nucleotide incorporated or by A_{260}. Remove 200 ng of each transcript and end label with [^{32}P]pCp or, if uncapped, with [γ-^{32}P]ATP after dephosphorylation.
12. Remove unincorporated label by gel purification or spin column. Check the level of substitution by cleavage with iodine.
13. Store unused RNA stocks at –80°C in H_2O containing 2 mM DTT. Ethanol-precipitate to remove the DTT before use.

Cleavage with Iodethanol

14. Resuspend the RNA from Step 13 in 9 µL of H_2O. The RNA must be free of reducing agents such as DTT.
15. Add 1 µL of 10 mM iodine (made in ethanol). After 5 min at room temperature, add 1 µL of β-mercaptoethanol to stop the reaction.
16. Add an equal volume of denaturing gel-loading dyes and analyze the cleavage patterns (i.e., read the RNA sequence) on an 8% denaturing gel. If the RNA ladders look good, the substrate is now ready to be used in modification interference (see Protocol 7) or iodine footprinting reactions.

Footprinting with I_2

17. Set up five identical binding reactions with ~50,000 cpm of each of the four thiosubstituted RNAs and an unsubstituted substrate in 10–25 µL (see Chemical Footprinting above). In this case, omit DTT or β-mercaptoethanol from buffers or protein dilutions. In addition, set up the same reactions with reaction buffer and no protein.
18. If binding is <80%, separate bound from unbound RNA (by affinity purification).
19. Make a 10 mM stock of iodine in 100% ethanol. Add I_2 to the binding reactions to achieve a final concentration of 1 mM.
20. Incubate the reactions for 15 min at room temperature.
21. Stop the reaction with β-mercaptoethanol.
22. Dilute the reactions with 200 µL SDS extraction buffer. Add 0.3 M NaOAc and extract with phenol:chloroform:isoamyl alcohol. If protein concentrations are high, digest with proteinase K for 15 min at 37°C before extraction with PCA.
23. Ethanol-precipitate with 20 µg GlycoBlue.

24. Resuspend the RNA in 10 µL of H_2O. Analyze 2.5 µL on a denaturing polyacrylamide gel.
25. Dry the gel and expose it with a phosphorimager.

    The absence of cleavage products in the lanes containing protein relative to cleavage products observed with the RNA alone indicates those nucleotides that are protected by the protein.

SOLUTION AND REAGENT RECIPES

CAUTION: See the Cautions Appendix for appropriate handling of materials marked with <!>.

Hydrogen Peroxide (2.5%, w/v)

Reagent	Quantity (for 10 mL)	Final concentration
H_2O_2 (30%) <!>	0.25 mL	2.5% (v/v)
H_2O	9.75 mL	

Prepare fresh for each use. Store 30% stock tightly capped at 4°C.

Protocol 6

Mapping Sites of RNA–Protein Interactions Using Nuclease-based Methods: RNase Footprinting and Reverse Footprinting

As described in the introduction to this chapter, binding of a protein to an RNA protects the protein from nuclease digestion. Here, we describe methods for using end-labeled RNA and nuclease digestion for conventional footprint analysis, followed by reverse footprinting wherein the protein protects a site-specifically labeled RNA. A conventional footprint is visualized by the absence of digestion products whereas a reverse footprint is detected as the presence of protected bands following nuclease digestion.

Method 1: RNase Footprinting

Ribonuclease footprinting is performed similarly to chemical footprinting except that end-labeled RNAs are used. Cleavages are detected directly after electrophoresis on denaturing polyacrylamide gels. The cleavage reactions are initiated by an RNase or H_2O (for mock-treated samples).

MATERIALS

The recipes for the items marked with <R> are at the end of the protocol or in the Common Recipes Appendix.

CAUTION: See the Cautions Appendix for appropriate handling of materials marked with <!>.

Reagents

Acetic acid (10%)/methanol (10%) <!>
Affinity reagents for purification of RNA–protein complexes
[α-^{32}P]NTP (800 Ci/mmol; 10 mCi/mL) <!>
ATA (aurintricarboxylic acid; Sigma-Aldrich; 5 mg/mL in 50 mM Tris; pH 7.5) <!>
Binding buffer, specific for each protein <R>
Chloroform:isoamy alcohol (24:1) <!>
Denaturing gel-loading dyes <R>
EDTA (0.2 M, pH 8.0) <!>
EGTA (100 mM, pH 8.0)
Ethanol (100%) <!>
Extract or protein for binding reactions
Formamide loading dye <R>
$MgCl_2$ (1 M) <!>
Nucleases
  micrococcal (60 units/µL) <R>
  RNase H (3 units/µL)
Phenol <!>
Proteinase K (10 mg/mL) <!>
Ribonuclease 1 (Ambion)
Ribonculease T1 (1 mg/mL in TE buffer; pH 7.5; Sankyo)
Ribonuclease VI

RNA, ^{32}P 5′- or 3′-end labeled and gel purified <!>
RNase H buffer (5x) <R>
RNase I dilution buffer <R>
RNase V5 (6.25×10^{-3} units/μL)
RNA site-specifically labeled with ^{32}P <!>
SDS extraction buffer <R>
Sodium acetate (3 M, pH 5.2)
TBE buffer <R>
TEMED (*N,N,N′,N′*-tetramethylethylenediamine) <!>
Tribasic sodium phosphate (100 mM)
tRNA, yeast or *E. coli* (10 mg/mL)
Urea loading dye <R>
Xylene cyanol (1%) <!>

Equipment

Chemical fume hood
Denaturing polyacrylamide gel electrophoresis system and power supply to run an 8% gel
Dry ice <!>
Gel dryer
Gel-purification or spin-column equipment
Heat block (95°C)
Ice
Microcentrifuge
Microcentrifuge tubes (1.5 mL)
Pipettor
SpeedVac
Vortex mixer
Water baths or heat blocks (binding temperature and 95°C)
Whatman 3MM paper

OVERVIEW

RNase I cleaves after every nucleotide of RNA, works well under many assay conditions, and can be inactivated by SDS. It is very important to establish appropriate digestion conditions such that overdigestion is avoided. In this regard, it will be necessary to titrate the amount of enzyme under a range of time and temperature conditions. Addition of "carrier" RNA to the digestions will help to keep the reactions reproducible and easier to control. Begin by finding conditions necessary to digest the substrate ≤1 per molecule. The optimal concentration will show unreacted substrate and a ladder of bands below it.

METHOD 1

1. Prepare an end-labeled RNA (5′ or 3′). Remove unincorporated nucleotides by gel purification or spin column. Resuspend the RNA in H_2O and adjust to ~25,000 cpm/μL.
2. Mix 25,000–50,0000 cpm end-labeled RNA for each digestion with 10 μg of *E. coli* tRNA. Incubate the reactions with a series of dilutions of RNase I in 10 μL of RNA-binding buffer. For example, use 0, 0.3, 1.0, 3.0, and 0.9 unit for 5 min at 4°C and room temperature.
3. Stop the reactions by adding an equal volume of denaturing gel-loading dyes and check the extent of cleavage on an 8% denaturing polyacrylamide gel. Chose the amount of enzyme that

gives a pattern of bands representing cleavage after each nucleotide with some (~20%) of the transcript remaining uncleaved.

4. Using the same amount of substrate as in Step 2, perform binding reactions with increasing amounts of protein. Make sure that 100% of the RNA is bound in some reactions (see note in Step 8). Include a no-protein control.
5. After binding is complete, add 1 µg of tRNA and digest under the predetermined RNase I conditions. It may be necessary to try two to three enzyme concentrations (one above and one below the optimum).
6. Add 0.2 mL of SDS extraction buffer to stop the reactions.
7. Recover the RNAs by phenol extraction and ethanol precipitation (add 20 µg GlycoBlue). Alternatively, add 2 volumes of gel-loading dye and analyze ~10,000 cpm on a denaturing polyacrylamide gel. Freeze the remainder at –80°C.
8. Analyze footprints on a gel as in Step 3 (see Fig. 5-7). Carefully pour the bottom buffer out of the gel apparatus to avoid nuclease contamination. Clean gel plates with a solution of 5% SDS.

   If binding is not complete, the footprint will not be clear. If this is the case, affinity-purify the complex. The nuclease cleavages can be performed while the complex is bound to beads. If cleavages are performed in the presence of beads, it may be necessary to increase the time of digestion. Stop reactions by addition of 0.2 mL of SDS extraction buffer, remove beads by centrifuging in a microcentrifuge for 5 sec, and continue with RNA recovery by phenol extraction as in Step 7.

Footprinting with Other Nucleases

Other enzymes with different cleavage specificities or with preference for double-stranded (such as RNase VI) versus single-stranded (such as RNase I) RNA are also helpful. Once digestion conditions are found, the RNA–protein complex can also be probed with combinations of enzymes (such as

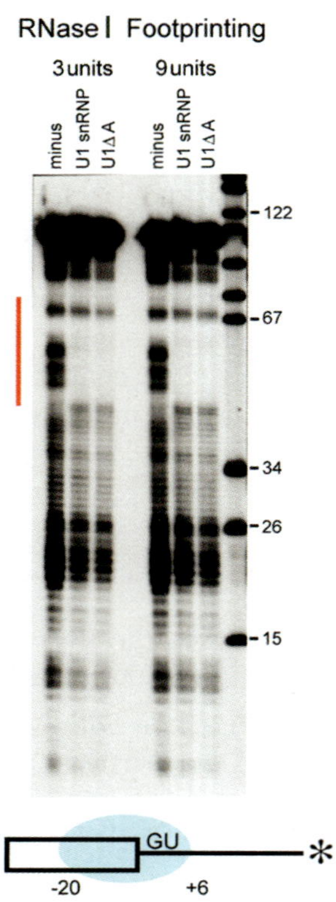

FIGURE 5-7. Example of nuclease footprinting. An end-labeled molecule containing a 5′ splice site (shown schematically below the gel) was digested with RNase I either in the absence (lanes *minus*) or presence of U1 snRNP either containing (lanes *U1 snRNP*) or lacking (lanes *U1ΔA*) U1-specific protein U1A. A clear footprint is observed that covers the 5′ splice site region. (Nilsen lab, unpubl.)

RNase A or T1). In particular, RNase VI is preferable when analyzing structured RNA and RNase A is preferable when using pyrimidine-rich RNAs. RNase T1 is useful for G-rich RNAs. Often, a combination of all three nucleases is useful.

1. Dilute end-labeled RNA with H_2O to obtain 25,000 cpm/µL for each reaction.
2. Titrate the enzymes as described above (in Method 1: RNase I Footprinting) until suitable cleavage patterns are observed on denaturing polyacrylamide gels.

   Look for a uniform ladder with a "core" of undigested RNA at the top. Once the optimal times and enzyme concentrations are found, they should remain consistent and reproducible.
3. Bind 25,000 cpm substrate to protein in 25 µL of binding buffer. Make sure that binding is complete or the footprint will not be clear (see note in Step 8, above). Add 10 µg of carrier tRNA to each sample.
4. Add enzyme to one reaction at a time (1 µL to 25 µL). Final concentrations will be 0.01–0.1 µg/µL RNase T1, 2.5×10^{-4} units/µL RNase V1, or 0.001 µg/mL RNase A.
5. Incubate the reactions for 0.5–2 min at room temperature and then add 3 µL of ATA. Mix by pipetting up and down.
6. Dilute the reactions with 200 µL of SDS extraction buffer, add 0.3 M NaOAc, extract with phenol, and precipitate with 20 µg GlycoBlue and ethanol.
7. Resuspend the RNA in 5 µL of H_2O and then add 10 µL of formamide loading dye. Load ~5 µL onto a denaturing 8%–10% polyacrylamide gel.
8. Dry the gel and visualize the bands by phosphorimager or autoradiography.

Markers

Sequencing markers for nuclease probing are generated from the end-labeled RNA by partial T1 digestion or base hydrolysis.

1. For T1 markers, incubate the RNA in 10 µL of T1 sequencing buffer (see Appendix) for 10 min at 50°C. Stop the reaction with 2 µL of ATA and 12 µL of loading buffer.
2. To generate an alkaline ladder, use 100,000 cpm end-labeled RNA in 9 µL of 2 mM EDTA (pH 8.0) containing 20 µg of *E. coli* tRNA. Add 1 µL of 1 N NaOH and incubate for 20 min at 30°C. Stop the reaction by adding 2 µL of 1 M Tris-HCl (pH 8) and freeze on dry ice until solid. Add an equal volume of loading buffer and thaw to room temperature. Load equal cpm of markers or sample onto a polyacrylamide gel.

TIPS AND TROUBLESHOOTING

1. If binding is not complete, the protein–RNA complex can be bound and selected by antibody or tagged proteins selected with an affinity reagent. After affinity purifications, the nuclease digestions are performed on beads.
2. Nuclease VI requires magnesium for cleavage, so remember to include it in the protein binding buffer.
3. After addition of ATA, if protein concentration is low and buffer does not interfere with gels, the samples can be analyzed without purification. Add an equal volume of urea loading dye and load 5 µL onto a gel. This is not advisable for samples containing RNase A.
4. Remember to clean gel plates thoroughly with SDS before running the gel.

Method 2: Reverse Footprinting

Reverse footprinting or nuclease protection of site-specifically labeled RNAs is an extremely powerful method for determining sites of RNA–protein interactions. It has a distinct advantage over conventional footprinting in that only a small fraction of the labeled RNA needs to be bound (as described above, ≥80% of input RNA must be bound to perform conventional footprinting). Moreover, because a specific site is being analyzed, nonspecific binding to other proteins of the input RNA does not interfere with the analysis.

MATERIALS

See the Materials list for Method 1.

METHOD

1. Prepare a substrate with a single label at the site for protein binding or complex formation (see Chapter 4).
2. Begin by titrating the amount of enzyme necessary to digest the substrate to completion as judged from electrophoresis. This provides a starting point. Next, titrate the amount of enzyme necessary to digest the substrate to completion in the presence of "carrier" RNAs present in extract.

   Titration is done by prediluting the extract before addition of the substrate to prevent the reaction from occurring. Nuclease-resistant fragments can then be compared to this background. The optimal conditions will be the least amount of enzyme required to eliminate background (i.e., digest all unprotected RNA) without disrupting the RNA–protein interaction.
3. The following steps are an example of RNA protection from digestion with micrococcal nuclease:
   i. Set up a binding reaction with site-specifically labeled substrate using either purified protein or cell extract. Incubate 50,000 cpm substrate in the extract under assay conditions in a final volume of 10 μL.
   ii. Quench the reaction on ice and dilute to 50 μL with assay buffer plus 2 mM $CaCl_2$.
   iii. Digest with 300 units of micrococcal nuclease (5 μL of 60 units/μL stock) for 10 min on ice.

      The digestion requires more time and enzyme on ice, but lower temperatures are likely to result in more stable complexes.
   iv. Add 4 mM EGTA to terminate the reaction.
   v. Dilute with SDS extraction buffer. Recover RNA after proteinase K digestion by extraction with phenol:chloroform:isoamyl alcohol (25:24:1) and precipitation with ethanol (see Chapter 2, Protocols 1 and 3). Add 10 μg of GlycoBlue as carrier.
   vi. Analyze the fragments on an 8% denaturing polyacrylamide gel and compare sizes to an alkaline ladder.

      This gel will resolve a fairly large range of sizes (~5–75 nucleotides). To determine the exact 5′ and 3′ boundaries of the protected fragments, site-specific digestion with RNase H and the chimeric 2′-*O*-methyl (2′OMe)/DNA oligonucleotide used to construct the substrate is used.
   vii. Elute the band of interest from the gel and perform an RNase H cleavage reaction such as that described in Chapter 4, Protocol 8. Recover the labeled fragment and determine its size compared to an alkaline ladder on a denaturing polyacrylamide gel (see Protocol 8).
4. RNase H leaves 5′ phosphates during digestion so that the label will be on the 3′ digestion product (see Fig. 5-8). The size of the 5′ product will be inferred by subtraction.

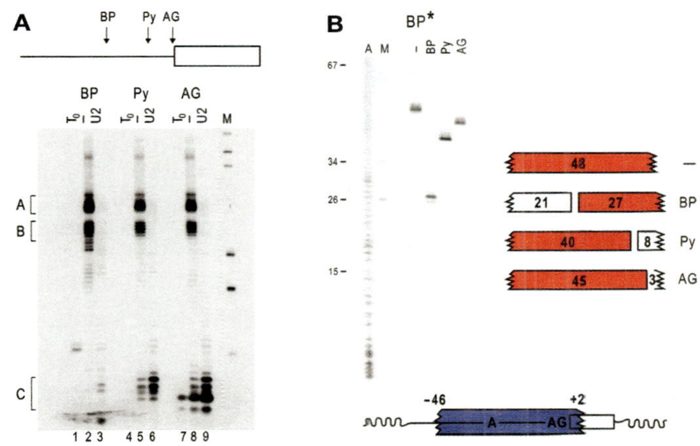

FIGURE 5-8. (*A*) An example of reverse footprinting. Site-specific labels at the branchpoint (BP), polypyrimide tract (PY), or 3′ splice site (AG) were introduced into 3′ half-transcripts that were then incubated with nuclear extract. Lanes T_0 indicate reactions that were diluted and digested with micrococcal nuclease before incubation. Lanes – were incubated and then digested with micrococcal nuclease. Lanes U2 were incubated in the presence of a 2′OMe oligonucleotide that prevents U2 snRNP from associating with the branchpoint. Families of bands that are designated A, B, and C represent nuclease-resistant fragments after digestion. For details, see Maroney et al. 2000. (Reprinted, with permission, from Maroney et al. 2000.) (*B*) Example of using site-specific RNase H digestion to map the sites of nuclease-resistant fragments generated by reverse footprinting. A specific 48-nucleotide-resistant fragment generated by an experiment such as that in Fig. 5-2 was excised from a gel and subjected to site-specific digestion with RNase H. In this case, the fragment resulted from protection of a specific label in the branchpoint sequence. Because the RNA is uniquely labeled, only one fragment is observed. When cut with RNase H with three different chimeric oligonucleotides, the resultant fragments reveal that protection encompasses the 3′ splice site and extends 46 nucleotides upstream (shown schematically at the bottom). (Reprinted, with permission, from Maroney et al. 2000.)

SOLUTION AND REAGENT RECIPES

CAUTION: See the Cautions Appendix for appropriate handling of materials marked with <!>.

Hydrogen Peroxide (2.5%, w/v)

Reagent	Quantity (for 10 mL)	Final concentration
H_2O_2 (30%) <!>	0.25 mL	2.5% (v/v)
H_2O	9.75 mL	

Prepare fresh for each use. Store 30% stock tightly capped at 4°C.

Iodoethanol

Weigh a few crystals of I_2 <!> and transfer to a 50-mL conical tube. Add 100% ethanol to make a 50 mM solution. Make a 10 mM solution by diluting 2 mL with 8 mL of ethanol <!>. Prepare fresh each time.

Micrococcal Nuclease

Carefully open the cap to keep crystals from escaping. Resuspend at 60 units/μL by adding 250 μL of 5 mM Tris <!> (pH 8.0) to a bottle containing 15,000 units of micrococcal nuclease. Store for 6 mo at –20°C.

RNase I Dilution Buffer

Reagent	Quantity (for 10 mL)	Final concentration
Tris (1 M, pH 8.0) <!>	0.2 mL	20 mM
NaCl (5 M)	0.2 mL	100 mM
EDTA (0.5 M, pH 8.0) <!>	20 µL	1 mM
Glycerol (50%, w/v)	4 mL	20% (w/v)

Store indefinitely at 4°C.

RNase H Buffer (5x)

Reagent	Quantity (for 10 mL)	Final concentration (5x)
Tris-HCl (1 M, pH 7.5)	1 mL	100 mM
KCl (2 M) <!>	2.5 mL	500 mM
$MgCl_2$ (1 M) <!>	0.5 mL	50 mM
DTT (1 M) <!>	0.5 mL	50 mM
Sucrose	2.5 g	25% (w/v)
H_2O	to 10 mL	

If this buffer is used infrequently, prepare a stock without DTT and aliquot into 1-mL tubes.

SDS Extraction Buffer

Reagent	Quantity (for 100 mL)	Final concentration
Tris (1 M, pH 7.5) <!>	2 mL	20 mM
EDTA (0.5 M, pH 8.0) <!>	0.2 mL	1 mM
SDS (10%, w/v) <!>	5 mL	0.5% (w/v)
H_2O	92.8 mL	

Urea Loading Dye

Reagent	Quantity (for 10 mL)	Final concentration
Urea <!>	4.8 g	8 M
Tris-HCl (1 M, pH 8.0)	200 µL	20 mM
EDTA (0.25 M) <!>	40 µL	1 mM
Bromphenol blue/xylene cyanol (1%, w/v) <!>	0.5 mL	0.05% (w/v)
ddH_2O	to 10 mL	

READING LIST

Clarke PA. 1999. RNA footprinting and modification interference analysis. *Methods Mol Biol* **118:** 73–91.

Dixon WJ, Hayes JJ, Levin JR, Weidner MF, Dombroski BA, Tullius TD. 1991. Hydroxyl radical footprinting. *Methods Enzymol* **208:** 380–413.

Jain SS, Tullius TD. 2008. Footprinting protein-DNA complexes using the hydroxyl radical. *Nat Protoc* **3:** 1092–1100.

Maroney PA, Romfo C, Nilsen TW. 2000. Nuclease protection of RNAs containing site-specific labels: A rapid method for mapping RNA-protein interactions. *RNA* **6:** 1905–1909.

Protocol 7

Modification Interference and Nucleotide Analog Interference Mapping

Modification interference and nucleotide analog interference mapping (NAIM) are extremely powerful methods for surveying an entire RNA molecule for functionally important bases and or functional groups on those bases. For protein binding or participation in an enzymatic reaction (e.g., splicing, polyadenylation, and modification), both approaches absolutely require the ability to separate functional from nonfunctional molecules. For example, when analyzing an enzymatic reaction, it is critical to be able to separate precursor and product (or intermediate). This is often achieved by straightforward gel electrophoretic analysis. In the case of protein binding, it is necessary to be able to separate bound protein from unbound RNA. This can be achieved in a number of ways including gel shift (EMSA), filter binding, or a variety of affinity approaches such as immunoprecipitation or the use of tagged proteins. Use of these techniques does not require that a large fraction of RNA be bound or reacted, and, as a result, they are quite sensitive. We describe one variation of chemical modification interference (i.e., modifying the RNA with a chemical before binding to a protein or participating in a reaction). This technique can easily be extrapolated to other chemicals. We then describe simple phosphorothioate interference and how this can be extended to NAIM.

Method 1: Chemical Modification Interference

MATERIALS

The recipes for the items marked with <R> are at the end of the protocol or in the Common Recipes Appendix.

CAUTION: See the Cautions Appendix for appropriate handling of materials marked with <!>.

Reagents

Acetic acid (10%)/methanol (10%) <!>
Affinity reagents for purification of RNA–protein complexes
A + G reaction buffer <R>
[α-^{32}P]NTP (800 Ci/mmol; 10 mCi/mL) <!>
αS nucleotide triphosphates (Trilink)
Aniline (1 M) <R>
β-mercaptoethanol (14 M; Sigma-Aldrich) <!>
Binding buffer, specific for each protein
Chloroform:isoamy alcohol (24:1) <!>
DEPC <!>
Dithiothreitol (DTT; 0.2 M) <!>
EDTA (0.2 M, pH 8.0) <!>
Ethanol (100%, 70%) <!>
Extract or protein for binding reactions
Formamide loading dye <R>
[γ-^{32}P]ATP or ^{32}pCp <!>
HEPES (0.2 M, pH 7.6 with KOH <!>)
Hydrazine (50%) <!> (Sigma-Aldrich; U reaction buffer)
Iodoethanol <R> <!>
$MgCl_2$ (1 M) <!>
Phenol <!>

RNA, ^{32}P 5′- or 3′-end labeled and gel purified <!>
RNA, thiophosphate substituted
SDS extraction buffer <R>
Sodium acetate (NaOAc; 3 M, pH 5.2)
T1 sequencing buffer <R> (see p. 275)
TBE
TEMED (*N,N,N′,N′*-tetramethylethylenediamine) <!>
tRNA, yeast or *E. coli* (10 mg/mL)
Urea loading dye <R> (see p. 290)
Xylene cyanol (1%) <!>

Equipment

Chemical fume hood
Denaturing polyacrylamide gel electrophoresis system and power supply to run an 8% gel
Dry ice <!>
Gel dryer
Heat block (60°C and 95°C)
Ice
Microcentrifuge
Microcentrifuge tubes (1.5 mL)
Phosphorimager or X-ray film
Pipettor
SpeedVac
Spin column
Vacuum gel dryer
Vortex mixer
Water baths or heat blocks (binding temperature and 95°C)
Whatman 3MM paper

OVERVIEW

Before beginning these procedures, it is essential to read the entire method from start to finish. It is necessary to establish conditions in which the end-labeled RNA of interest is modified at a frequency of ≤1 per molecule and that cleavage conditions be optimized to generate a ladder wherein each modified base can be visualized on a denaturing gel.

METHOD

1. Prepare RNA labeled at 5′ or 3′ ends (see Chapter 4, Protocols 5–7). Gel-purify or remove unincorporated label with a spin column and precipitate the RNA with ethanol (see Chapter 2, Protocol 3).

   DEPC A + G (purine) modifications are made as described in Chapter 3, Protocol 24.

2. Resuspend ~200,000 cpm of the labeled RNA in 200 µL of A + G reaction buffer containing 10 µg of yeast tRNA and transfer to a 1.5-mL microcentrifuge tube. Chill the tube for 5 min on ice.
3. In a chemical fume hood, add 2.3 µL of DEPC.

   CAUTION: DEPC is very toxic, so be sure to perform these steps in a chemical fume hood.

4. Vortex the reaction for 5 sec and then incubate for 2 min at 90°C.

   Time is critical, but it is also specific to given substrates. Longer times (e.g., 15 min) with less DEPC (0.5 µL) may be useful to try if the reaction is difficult to optimize. It will likely be necessary to try a range of times and temperatures for each different substrate.

5. Stop the reaction by adding NaOAc to 0.3 M and precipitate the RNA with ethanol.
6. Centrifuge in a microcentrifuge at full speed for 10 min at room temperature to pellet the RNA.
7. In a chemical fume hood, remove the supernatant and discard it in an appropriate manner.
8. Resuspend the RNA pellet in 200 µL of H_2O and add 0.3 M NaOAc. Reprecipitate the RNA with ethanol.
9. Centrifuge in a microcentrifuge for 10 min, pour off the ethanol, and dry the pellet in a SpeedVac. Continue with Step 10 if necessary or proceed to aniline cleavage (Step 16).
10. Gel-purify the chemically modified RNA (see Protocol 4).

    This is necessary only if there is degradation or fragmentation during the modification. After gel purification, precipitate the RNA with ethanol twice and dry the pellet with a SpeedVac before aniline cleavage.

11. Check the extent of modification by cleaving with aniline (Step 16). Analyze on an 8% polyacrylamide gel (see below).

Uridine Reaction

12. Use a SpeedVac to dry 200,000 cpm end-labeled RNA from Step 1 with 10 µg of yeast tRNA in a 1.5-mL microcentrifuge tube.
13. Chill the tube for 5 min on ice.
14. Add 50% hydrazine and resuspend by vortexing.
15. Incubate for 10 min on ice and then continue with Step 5, as above (A + G modification).

Aniline Cleavage at the Modified Nucleotides

16. Dissolve RNA pellets from Step 9 or 10 in 20 µL of 1 M aniline adjusted to pH 4.5 with glacial acetic acid.
17. Incubate in the dark for 20 min at 60°C. After incubation, dry the RNA three times in a SpeedVac, resuspending the pellet with 100 µL of H_2O between each time.

    Complete removal of aniline will result in better-loading gels.

Resolving the Cleavage Products

18. Separate the RNA fragments on a denaturing polyacrylamide gel (8%–12% acrylamide:bis-acrylamide).
19. Fix the gel by soaking for 10–15 min in 10% acetic acid/10% methanol and dry on a heated vacuum gel dryer.
20. Subject the gel to phosphorimaging or autoradiography.
21. If modification conditions are correct, continue with protein binding. If not, vary the time and temperature of the modifications until suitable ladders are observed.

RNA–Protein Binding and Separating Free RNA from RNA–Protein Complexes

22. For interference reactions, resuspend the modified RNA from Steps 9 or 10 in H_2O (at ~50,000 cpm/µL) and incubate 1 µL of the modified RNA with protein or extract under predetermined binding conditions.
23. Purify the RNA product or protein–RNA complex away from unreacted substrate using a gel or affinity method.
24. Recover the affinity-purified products and unreacted substrate. Dilute with SDS extraction buffer. Add NaOAc to 0.3 M and then extract with phenol and precipitate with ethanol. For

products and intermediates of a processing reaction, elute bands from the gel with SDS extraction buffer and then precipitate the RNA with ethanol twice before continuing with aniline cleavage (previous section) and gel electrophoresis.

The digestion patterns of free RNA will be compared to the digestion patterns of bound RNA for each chemical modification. It is generally useful to quantitate some of the cleavage bands using phosphorimaging.

TIPS AND TROUBLESHOOTING

1. This experiment works best if the RNA is modified at most only once per molecule. If anything, err on the side of undermodification. It is wise to first perform a trial modification and cleavage, titrating the amount of modifying reagent and/or time of modification, and choose conditions where there is uncleaved RNA remaining.
2. It is critical to use freshly diluted, recently purchased, high-quality chemicals (see also Chapter 3, Protocol 24).

Method 2: Nucleotide Analog Interference Mapping

Because NAIM requires a knowledge of which, if any, phosphorothioate modifications interfere with binding or a productive reaction, we first describe phosphorothioate interference analysis.

MATERIALS

See Materials list for Method 1.

METHOD

Start with RNA substituted with each of the four αS nucleotides and one control unsubstituted as described in Protocol 5. Verify by digestion with I_2 that substitution is ≤1 per molecule and that a nice sequencing ladder is formed with uncleaved RNA at the top of the gel. There will be five binding reactions for each substrate corresponding to RNA substituted with αS C, U, A, or G, or no thio. Keep the reactions separate and perform five separate purifications.

Synthesis of Substrates

1. Transcribe thiophosphate-substituted RNA as described in chemical footprinting (Protocol 5) with iodine. Prepare 5′- or 3′-end-labeled substrate with [γ-^{32}P]ATP or [^{32}P]cytidine 3′,5′-bis (phosphate) (^{32}pCp) and remove unincorporated label by spin column or gel purification.

   When using 3′-end-labeled substrate, make sure that the 3′ end is uniform or the RNA sequencing ladder will not be clear.
2. Verify by cleavage with iodine that the RNA is properly substituted (≤1 per molecule) and that a nice sequencing ladder is formed.

Binding and Purification

3. Set up the binding reactions or RNA-processing reactions, such as polyadenylation or splicing, in suitable reaction buffers. Include 1 mM DTT in reactions to prevent oxidation of the thio groups in the RNA.

4. After the reactions are complete, separate the bound and unbound substrates by affinity methods or gel shift and cleaved or spliced products or intermediates from precursor by purification on denaturing polyacrylamide gels. In cases where extracts or high concentrations of proteins are used, dilute the samples with SDS extraction buffer, digest with proteinase K, and extract with phenol:chloroform:isoamyl alcohol, followed by ethanol precipitation and gel purification. After recovery of purified RNA, perform cleavage reactions with iodine (Steps 5 and 6).

Cleavage

5. Resuspend the RNA in 2.5 μL of H_2O. Add 2.5 μL of 200 μM iodine prepared in ethanol (see above).
6. Heat the reaction for 1 min at 95°C. Chill on ice.
7. Add 10 μL of formamide loading dye.
8. Display the sequencing ladders on a 6%–8% denaturing polyacrylamide gel, as described in Protocol 5, chemical footprinting with I_2.
9. Compare the lanes of unreacted and reacted substrate in each of the substituted nucleotides. A band missing in the product lane but not in the precursor lane indicates a phosphate that was required for the reaction (Fig. 5-9, see asterisks in the figure).

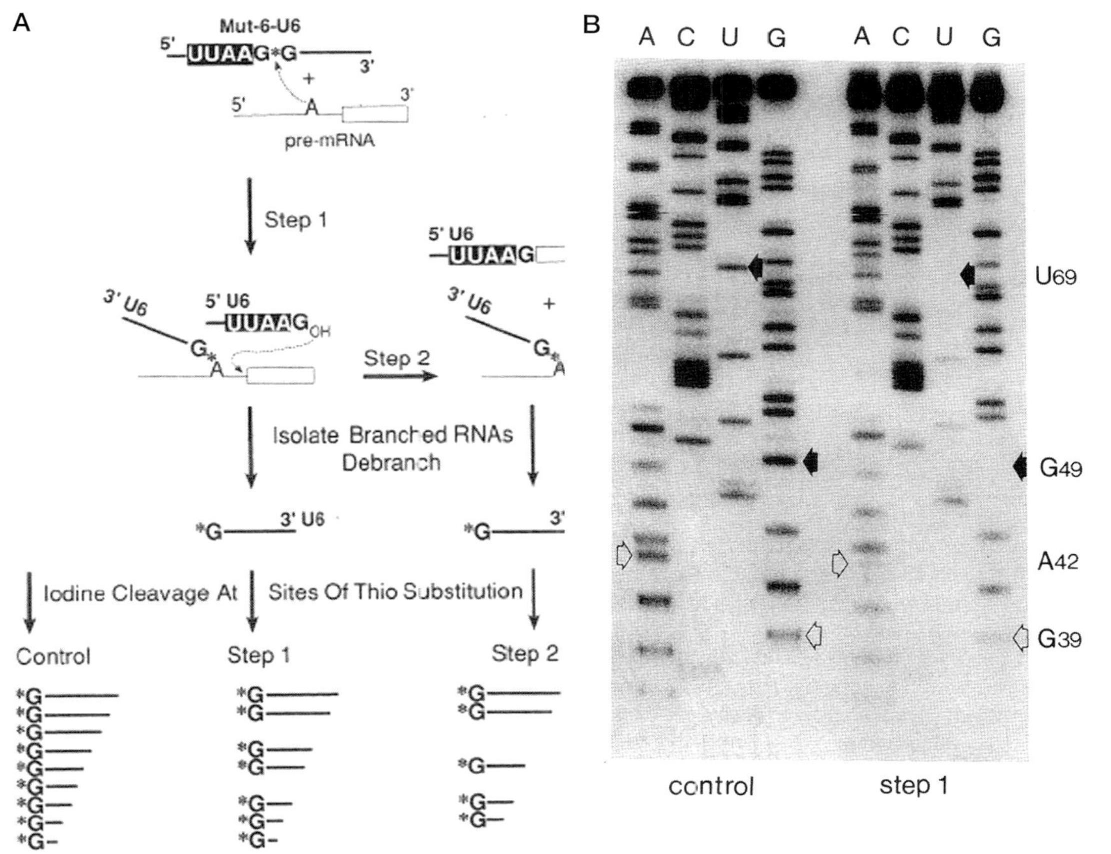

FIGURE 5-9. An example of phosphorothioate interference analysis. This experiment was designed to examine phosphates required for U6 snRNA function in pre-mRNA splicing. (*A*) Experimental strategy, (*B*) results of the experiment. (Dark arrows) Phosphates present in control (unreacted) RNA but absent in U6 snRNA, which has undergone the first step of splicing. (Open arrows) Position of phosphates that reduce but do not eliminate splicing. For details, see Yu et al. (1995).

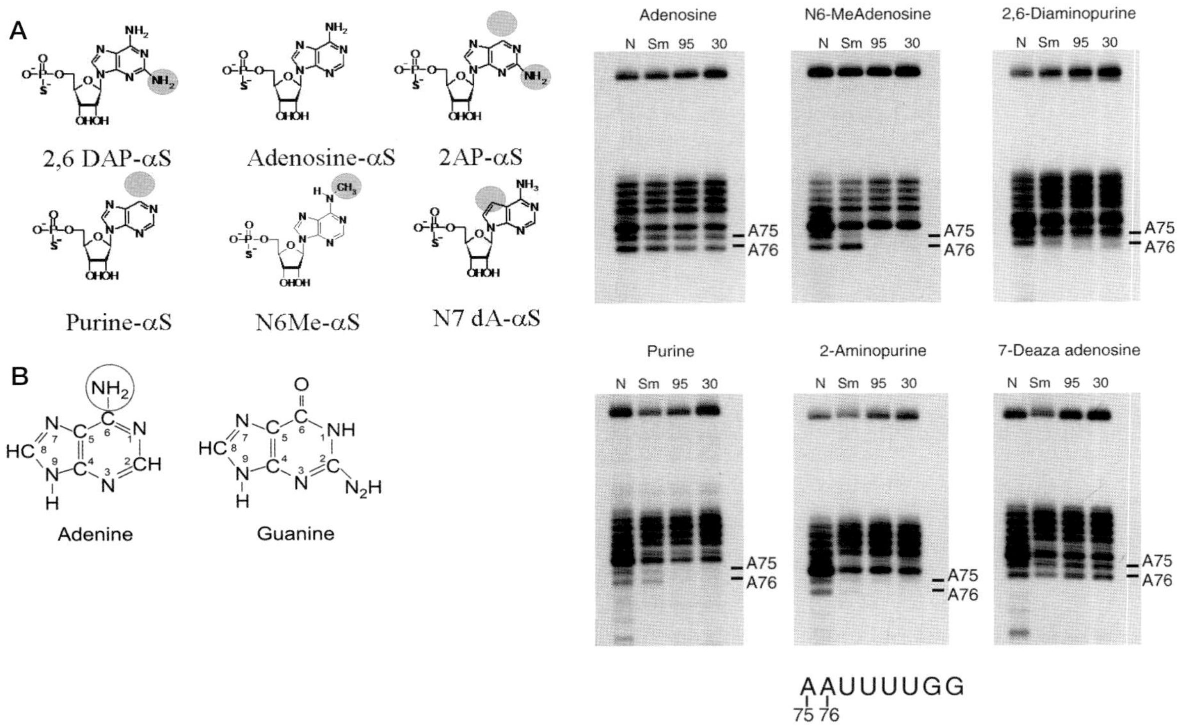

FIGURE 5-10. (*A*) Some αthio adenosine analogs available for NAIM. Each modification that differs from adenosine is highlighted with a circle. (*B*) Example of NAIM. A nematode SL RNA was substituted (≤1 per molecule) with the indicated adenosine analogs. After assembly into an SL RNP containing Sm proteins (Sm) or Sm proteins plus SL RNP-specific proteins (95 or 30), complexes were immunoprecipitated and cleaved with iodoethanol. Inspection of the gel shows that N^6 methylation of A75 inhibits both Sm assembly and the association of specific proteins, whereas the same modification at A76 does not inhibit Sm assembly but completely inhibits the association of specific proteins. Similar striking results are observed with the other adenosine analogs (with the exception of 7-deaza adenosine). (Nilsen lab, unpubl.)

Base Modifications

10. After eliminating a requirement for unsubstituted phosphates in a binding or processing reaction, perform a similar analysis to detect modified nucleotides.
11. To analyze modifications of adenosine, substitute αthio ATP with other αS analogs such as αS D-aminopurine triphosphate, αS 2-aminopurine, αS N6 Me A, αS 7, or αS GTP with αS 6-thioGTP (see Protocol 5, Footprinting with I_2). In addition, include αS nucleotide with unmodified adenosine. Compare the cleavage patterns of the modified A lanes to those unmodified on a sequencing gel (see Fig. 5-10).

TIPS AND TROUBLESHOOTING

1. Include 1 mM DTT during extraction and use fresh phenol.
2. Inefficient cleavage can result from the presence of reducing agents such as DTT or β-mercaptoethanol. If this is suspected, wash the pellet with 70% ethanol, remove all liquid with a pulled glass pipette, and resuspend in H_2O or iodine.

SOLUTION AND REAGENT RECIPES

CAUTION: See the Cautions Appendix for appropriate handling of materials marked with <!>.

A + G Reaction Buffer

Reagent	Quantity (for 100 mL)	Final concentration
NaOAc (1 M, pH 4.5)	5 mL	50 mM
EDTA (0.5 M, pH 8.0) <!>	0.2 mL	1 mM
H_2O	94.8 mL	

Adjust pH to 4.5 if necessary before adding all of the H_2O.

50% Hydrazine (U Reaction Buffer)

Reagent	Quantity (for 1 mL)	Final concentration
Hydrazine (anhydrous) <!>	0.5 mL	50% (v/v)
H_2O	0.5 mL	

Prepare fresh just before use and chill on ice. Handle with care in a chemical fume hood.

Iodoethanol

Weigh a few crystals of iodine <!> and transfer to a 50-mL conical tube. Add 100% ethanol <!> to make a 50 mM solution. Make a 1 mM solution by diluting 1 mL of this solution with 49 mL of ethanol. Prepare fresh each time.

SDS Extraction Buffer (100 mL)

Reagent	Quantity (for 100 mL)	Final concentration
Tris (1 M, pH 7.5) <!>	2 mL	20 mM
EDTA (0.5 M, pH 8.0) <!>	0.2 mL	1 mM
SDS (10%, w/v) <!>	5 mL	0.5% (w/v)
H_2O	92.8 mL	

REFERENCE

Yu Y-T, Maroney PA, Darzynkiewicz E, Nilsen TW. 1995. U6 snRNA function in nuclear pre-mRNA splicing: A phosphorothioate interference analysis of the U6 phosphate backbone. *RNA* **1:** 46–54.

Protocol 8

Boundary Analysis

A powerful and often overlooked method for determining the sequences within an RNA required for a specific RNA–protein interaction is called boundary analysis. In this approach, end-labeled RNAs (5′-end labeled and 3′-end labeled) are fragmented randomly (usually by limited alkaline hydrolysis, but nuclease fragmentation can also be used) such that a ladder of fragments covering the whole molecule of interest is produced. This mixture of fragments is then allowed to bind to the protein (or other molecule) of interest. Binding will be observed to all fragments that contain a complete binding site. If we imagine that a protein-binding site is present between nucleotides 20 and 30 of a 100-base 5′-end-labeled RNA, binding will be observed to all fragments that are longer than 30 nucleotides; any fragments shorter than 30 nucleotides will not be bound. This result gives the 3′ boundary of binding. Conversely, if the same RNA is 3′-end labeled, all fragments will not be bound, thus giving the 5′ boundary of the binding site (Fig. 5-11; see also Fig. 5-3).

MATERIALS

The recipe for the item marked with <R> is at the end of the protocol or in the Common Recipes Appendix.

CAUTION: See the Cautions Appendix for appropriate handling of materials marked with <!>.

Reagents

Acetic acid (10%)/methanol (10%) <!>
Affinity reagents for purification of RNA–protein complexes (see Protocol 13)
Binding buffer, specific for each protein
Chloroform:isoamyl alcohol (24:1) <!>
EDTA (0.2 M, pH 8.0) <!>
Ethanol (100%) <!>
Extract or protein for binding reactions
Formamide loading dye <!>
GlycoBlue
$MgCl_2$ (1 M) <!>
Phenol <!>
Proteinase K (10 mg/mL) <!>
RNA, ^{32}P 5′- or 3′-end labeled and gel purified <!>
SDS extraction buffer <R>
Sodium acetate (NaOAc; 3 M, pH 5.2)
Tribasic sodium phosphate (100 mM) <!>
Tris (1 M, pH 7.5) <!>
tRNA, yeast or *E. coli* (10 mg/mL)
Xylene cyanol (1%) <!>

Equipment

Denaturing polyacrylamide gel electrophoresis system and power supply to run an 8% gel
Dry ice <!>
Gel dryer

Ice
Microcentrifuge
Microcentrifuge tubes (1.5 mL)
Phosphorimager or autoradiography materials
Pipettor
SpeedVac
Vortex mixer
Water baths or heat blocks (binding temperature, 65°C and 95°C)
Whatman 3MM paper

OVERVIEW

This method is used to determine the minimal sequence required for protein binding to a site. Both 5′- and 3′-end-labeled RNAs are needed as well as an affinity reagent such as an antibody or tagged protein to separate bound RNA from free RNA. This example shows binding requirements for U1 snRNP (small nuclear RNP) to substrate with a 5′ splice site. RNA fragments bound to U1 snRNA are copurified along with U1 snRNA using a U1 snRNP-specific antibody.

METHOD

1. Prepare end-labeled RNA (see Chapter 4, Protocols 5 and 6).
2. Prepare an alkaline ladder from both 5′- and 3′-end-labeled RNA (see Appendix).
   i. Start with four aliquots of 5′- or 3′-end-labeled RNA (50,000 cpm each with 10 µg of tRNA in 10 µL of H_2O) to establish the best conditions for the RNA being used.
   ii. Add 10 µL of ~100 mM tribasic sodium phosphate.
   iii. Incubate one sample for 1 min at 95°C and the others for 2, 5, or 10 min at 65°C.
   iv. Neutralize with 25 µL of 1 M Tris (pH 7.5) and dilute with SDS extraction buffer. Add 20 µg of GlycoBlue and 0.3 M NaOAc and precipitate with ethanol.
   v. Check on an 8% polyacrylamide gel prerun for 45 min.
   vi. Dry the gel and visualize the bands by phosphorimager or autoradiography.
   vii. Chose the best conditions and prepare hydrolyzed substrate for reactions using 2×10^6 to 5×10^6 cpm. There should be a uniform distribution of bands below a core of intact RNA at the top (see Fig. 5-11).
3. Set up four binding reactions with 100,000 cpm of each end-labeled RNA in 25 µL of the following:
   i. 5′-end-labeled RNA control, no added protein
   ii. 5′-end-labeled RNA with protein
   iii. 3′-end-labeled RNA control, no added protein
   iv. 3′-end-labeled RNA with protein
4. Incubate the reactions for 30 min at 0°C or room temperature.
5. Prepare the affinity reagent (such as antibody coupled to beads; see Protocol 13).
6. Bind RNA from Step 3 samples ii and iii (with protein) to beads for 2 h at 4°C.
7. After binding, save the supernatant (as unbound), wash the beads, and elute the bound complex with SDS extraction buffer.

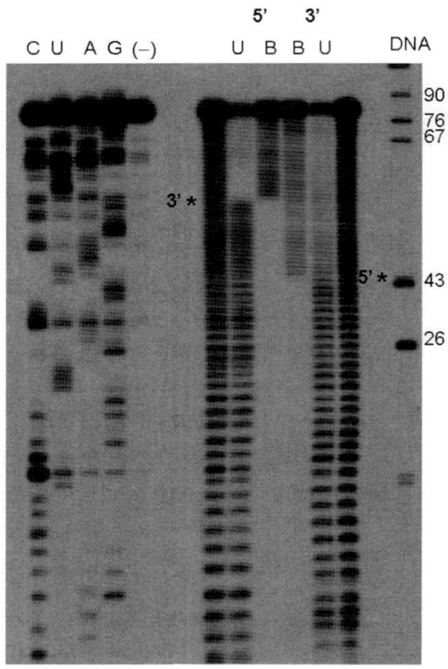

FIGURE 5-11. Example of a boundary experiment. In this case, a 5′- or 3′-labeled RNA containing a 5′ splice site was fragmented by alkali and the fragments allowed to bind to U1 snRNP. There are clear size "cut-off" boundaries between bound (B) and unbound (U) fragments. By determining the sizes of bound and unbound fragments relative to a sequencing ladder (*left*), it is possible to deduce the binding site. The boundary for 3′-end-labeled RNA is the 5′ end of the binding site; the boundary for the 5′-end-labeled RNA is the 3′ end of the binding site. (Nilsen lab, unpubl.)

8. Recover RNA by extraction with phenol and precipitation with ethanol. Recover RNA from samples i and iii by precipitation with ethanol. Add GlycoBlue (20 µg) to each sample for precipitation.
9. Resuspend RNA in 10 µL of H_2O.
10. Analyze 2.5 µL of RNA on an 8% polyacrylamide gel (the example shown in Fig. 5-11 is a 100-nucleotide RNA run on a 10% gel) with an RNA sequencing ladder in parallel.

    Run an RNA sequencing ladder such as iodine-cleaved thiophosphate-substituted RNA as well as an alkaline ladder in parallel to determine the size of bound and unbound fragments.
11. Compare the bands with the markers. Use the 3′-end-labeled substrate to determine the 5′ end of the "boundary." Use the 5′-end-labeled substrate to determine the 3′ end of the "boundary."

Protocol 9

Ultraviolet Photochemical Cross-linking to Detect RNA-binding Proteins

RNA is photoreactive upon exposure to UV light in the 250–270-nm range. Upon UV treatment, RNA bases absorb energy to generate free radicals that can covalently attach to nearby amino acid residues in RNA-bound proteins. This method has been used extensively as an analytical tool in efforts to identify and characterize RNA-binding sites for proteins bound to RNA in crude extracts or with purified proteins by using "label transfer" of ^{32}P-labeled RNA to the protein of interest (Reed and Chiara 1999; Luo and Reed 2003). UV cross-linking is also used to cross-link proteins to RNA in vivo (in intact cells). This approach is described in detail in Chapter 8 (see Protocol 8).

MATERIALS

The recipes for the items marked with <R> are at the end of the protocol or in the Common Recipes Appendix.

CAUTION: See the Cautions Appendix for appropriate handling of materials marked with <!>.

Reagents

Acetic acid (10%)/methanol (10%) <!>
[α-^{32}P]UTP (800 Ci/mmol; 10 mCi/mL; Perkin Elmer NEG007X) <!>
Bromophenol blue dye (powder) <!>
Coomassie blue protein gel stain <!> <R>
Protein of interest
RNase A (Worthington RAF LS005649) <!>
  Dissolve at 10 mg/mL in ddH_2O; store at –20°C.
SDS polyacrylamide gel
SDS–polyacrylamide gel electrophoresis (PAGE) protein sample buffer (2x) <R>

Equipment

Gel filtration spin chromatography or preparative denaturing urea–polyacrylamide gel electrophoresis equipment
Heating block
Ice
Metal block from dry bath incubator or square pipette can
Microcentrifuge tubes (1.5 mL)
Parafilm
Phosphorimager or X-ray film
Pipettor
Polyacrylamide gel equipment
Styrofoam box
UV lamp (254 nm, model UVG-54, Ultraviolet Products, Inc.) or Stratalinker <!>
Vacuum gel dryer
Whatman 3MM paper

OVERVIEW

Before beginning, it is necessary to prepare a labeled RNA of interest, either body labeled (Chapter 4, Protocol 1) or site-specifically labeled (Chapter 4, Protocol 8). It is also desirable to verify that the RNA of interest contains a binding site for the specific protein, i.e., by conducting gel-shift analysis.

METHOD

1. Set up an in vitro transcription reaction to incorporate [α-^{32}P]UTP (or another [α-^{32}P]NTP) into the transcript of interest using a standard phage RNA polymerase transcription reaction. Purify the RNA by gel-filtration spin chromatography or preparative denaturing urea–polyacrylamide gel electrophoresis.

   Detection of cross-links in a uniformly labeled RNA indicates an interaction some place in the RNA. Insertion of a site-specific label by splinted ligation will provide information about the location of the cross-link.

2. Incubate the labeled RNA under the appropriate conditions. Use assay conditions for the system of interest.

3. Cover the flat side of a chilled metal block (or fill a square pipette can lid with ice to chill) with Parafilm. Place the block on ice in a styrofoam box and then pipette the protein-binding reactions as drops on top of the Parafilm.

   An alternative is to have the samples in 0.5-mL microcentrifuge tubes in a flat rack on ice, with the tubes close together and open.

4. Irradiate the samples for 12 min using a handheld 254-nm UV lamp. Place the lamp directly over the sample drops with the lamp resting on the edges of the styrofoam box (~2 cm). Make sure the drops are directly exposed to the lamp bulb and are not beneath the plastic frame that surrounds the bulb on the lamp. In the case of the 0.5-mL sample tubes in a rack with the tubes evenly flat, place the lamp directly on top of the tubes.

   Cross-linking can also be done in a Stratalinker with 254-nm bulbs. In this case, the metal block is chilled in a 3-inch-high plastic box and placed in the cabinet. The samples are cross-linked using the time setting on the Stratalinker.

   Note that you may need to vary and empirically determine the irradiation time; see Tips and Troubleshooting below.

5. If using the drops-on-Parafilm method, after irradiation, transfer the drops to 0.5-mL microcentrifuge tubes.

6. Add 10 mg/mL RNase A to 1 mg/mL final concentration (i.e., 1/10 of the binding reaction volume) and incubate for 30 min at 30°C.

   The RNase may need to be titrated and/or other RNases used to obtain adequate digestion of the non-cross-linked RNA. In addition, note that too much RNase may decrease or abolish the cross-linking signal. As with all procedures that use RNase A, exteme caution is required.

7. Add 2x SDS–PAGE sample buffer (~15 µL) to the RNase-treated binding reactions. Boil the mixture in a boiling water bath for 5 min.

8. Load one-third volume (~10 µL) of the samples (RNase-treated UV-cross-linked RNA-binding reactions in SDS sample buffer) onto an SDS polyacrylamide gel (typically 7.5%–12% gel) and run the gel as usual for a normal protein separation.

9. When the bromophenol blue dye runs just off the bottom of the gel, stain the gel with Coomassie blue protein gel stain and destain to return to the normal size of the gel. Alternatively, fix the gel with 10% acetic acid/10% methanol for 15 min at room temperature.

10. Dry the gel onto Whatman 3MM paper using a heated vacuum gel dryer and expose the dried gel to a phosphorimaging screen or X-ray film overnight.

TIPS AND TROUBLESHOOTING

1. It is typical to load one-third to one-half of a 25 µL reaction (with at most 60 µg of extract) into one lane (4.5 mm wide) of a SDS polyacrylamide minigel.
2. It may be necessary to test different times of irradiation with 254-nm UV light to optimize the cross-linking of any given protein.
3. Be sure to load SDS–PAGE markers to determine the size of the cross-linked protein–RNA adduct.
4. Consider loading both a high- and low-percentage gel to avoid missing small proteins.
5. Be sure to use RNase A in a contained area.

SOLUTION AND REAGENT RECIPES

CAUTION: See the Cautions Appendix for appropriate handling of materials marked with <!>.

SDS–PAGE Protein Sample Buffer (2x)

Reagent	Quantity (for 50 mL)	Final concentration (2x)
Tris (1 M, pH 6.8) <!>	4.0 mL	80 mM
SDS (20%) <!>	5 mL	2.0%
Glycerol (10%, w/v)	5 mL	10.0%
Bromophenol blue (0.1%) <!>	300 µL	0.0006%
DTT (1 M) <!>	5 mL	0.1 M
H_2O	30.7 mL	

Omit DTT for longer-term storage. Store indefinitely at room temperature. Add DTT to 0.1 M on aliquots for daily use.

Coomassie Blue Stain

Reagent	Quantity (for 1 L)	Final concentration
Methanol <!>	500 mL	50% (v/v)
Glacial acetic acid <!>	100 mL	10% (v/v)
Coomassie blue <!>	1.25 g	0.125% (w/v)
H_2O	400 mL	

Store at room temperature. Add the acid last.

REFERENCES AND READING LIST

Luo MJ, Reed R. 2003. Identification of RNA binding proteins by UV cross-linking. *Curr Protoc Mol Biol* Chapter 27: Unit 27.2.

Reed R, Chiara MD. 1999. Identification of RNA-protein contacts within functional ribonucleoprotein complexes by RNA site-specific labeling and UV crosslinking. *Methods* **18:** 3–12.

Protocol 10

Photocross-linking Using Uridine Analogs

Straight UV cross-linking using short-wavelength UV light is inherently very inefficient. Therefore, it is often desirable to increase the efficiency of cross-linking. The uridine analogs 4-thiouridine and halogenated uridine (5-bromouridine or 5-iodouridine) are much more photoreactive than uridine. Accordingly, RNAs substituted with these bases are quite effective for cross-linking applications—both RNA–protein and RNA–RNA (see the introduction to this chapter). Although fully substituted RNAs can be useful (e.g., in determining whether any proteins [or RNAs] interact with the RNA of interest), it is often preferable to use site-specifically substituted RNA. Such RNAs provide valuable information regarding the specific sites of protein–RNA (and RNA–RNA) interactions. For any cross-linking application using substituted RNA, it is essential to remember that the uridine analogs are activated at a wavelength greater than 304 nm (312 nm) than is used for short-wavelength (254 nm) cross-linking.

MATERIALS

The recipes for the items marked with <R> are at the end of the protocol or in the Common Recipes Appendix.

CAUTION: See the Cautions Appendix for appropriate handling of materials marked with <!>.

Reagents

Acetic acid (10%)/methanol (10%) <!>
[α-^{32}P]ATP (800 Ci/mmol; 10 mCi/mL; Perkin Elmer NEG0003X) <!>
Buffer A
Coomassie blue protein gel stain <!> <R>
Cross-linking reaction buffer (specific for assay conditions)
Phage RNA polymerase transcription reaction materials (see Chapter 4)
Protein of interest
Reaction buffer (see Step 2 note) (specific for assay conditions)
RNase A (Worthington RAF LS005649) <!>
Dissolve at 10 mg/mL in ddH_2O; store at –20°C.
SDS–PAGE protein sample buffer (2x) <R>
4-Thio-UTP (1 µM stock; TriLink BioTechnologies N-1025)

Equipment

Flat rack
Gel purification equipment or gel filtration spin column
Ice
Microcentrifuge tubes (1.5 mL)
Phosphorimager or X-ray film
SDS protein gel equipment
UV lamp (302 nm) or Stratalinker with 302-nm bulbs <!>
Vacuum gel dryer
Whatman 3MM paper

OVERVIEW

Before beginning this type of analysis, it is necessary to decide whether to use uniformly substituted RNA or site-specifically substituted RNA. If one has foreknowledge of critical regions in the RNA of interest, it is advisable to use site- or region-specific substitution. The following protocol specifically describes the use of 4-thiouridine, but identical procedures apply to using 5-bromouridine or 5-iodouridine.

METHOD

1. Prepare an RNA substrate by in vitro transcription reaction, the same as for preparing internally labeled RNA (see Chapter 4, Protocol 1) but with the following changes:
   i. Add GTP, CTP, and 4-thio-UTP to a final concentration of 500 μM. See no. 2 in Tips and Troubleshooting.
   ii. Add cold (unlabeled) ATP to 50 μM and label the RNA with [α-^{32}P]ATP (2 μL = 20 μCi at 800 Ci/mmol; 10 mCi/mL).
   iii. Purify the RNA by gel purification or gel filtration spin column, depending on whether there are truncated products from the transcription reaction.
2. Set up the cross-linking reaction in a 0.5-mL microcentrifuge tube. This should be a 20-μL reaction with ~20,000 cpm of RNA (at a specific activity of ~10^6 cpm/pmol).

   The reaction buffer composition will depend on the protein or extract; use assay buffer or binding buffer compatible with your reaction.
3. Incubate the reaction for 30 min at room temperature or on ice to allow protein binding.
4. Place the reactions in a flat rack on ice with the 0.5-mL tubes uncapped. Irradiate the samples at 304 nm with a handheld UV lamp for 20 min by placing the lamp directly on top of the tubes. When working out conditions, also try a 365-nm light because the wavelengths of these lamps are not exact and one may be more efficient than the other.

   Alternatively, drops of the reaction mixture can be placed on a piece of Parafilm on an aluminum block kept on ice. Afer cross-linking is complete, transfer the drops to a 0.5-mL microcentrifuge tube for RNase digestion.
5. Add 2 μL of RNase A (10 mg/mL) to each tube and keep the tube for 20 min at room temperature.
6. Add 20 μL of 2x SDS sample buffer to the cross-linking reactions and boil for 3–5 min in a boiling water bath.
7. Load half of the sample (~25 μL) onto an SDS polyacrylamide protein gel. Stain the gel with Coomassie blue protein gel stain (containing acetic acid and methanol). Alternatively, fix the gel with 10% acetic acid/10% methanol for 15 min at room temperature.
8. Place the gel on Whatman 3MM paper and dry it using a heated vacuum gel dryer.
9. Expose the dried gel to a phosphorimaging screen or X-ray film overnight to determine the molecular weight of X-linked protein(s).

TIPS AND TROUBLESHOOTING

1. For site-specific substitution, see Chapter 4, Protocol 8 and proceed as above, Step 2.

2. The incorporation of 4-thio-UTP into synthetic RNA in vitro is routine using the phage T7, SP6, or T3 RNA polymerases. In some cases, however, the yield of RNA can be low. For long RNAs or those with stretches of U, it may be necessary to add some unsubstituted UTP to get full-length product.
3. As with all procedures that use RNase A, extreme caution is required.

SOLUTION AND REAGENT RECIPES

CAUTION: See the Cautions Appendix for appropriate handling of materials marked with <!>.

Coomassie Blue Stain

Reagent	Quantity (for 1 L)	Final concentration
Methanol <!>	500 mL	50% (v/v)
Glacial acetic acid <!>	100 mL	10% (v/v)
Coomassie blue <!>	1.25 g	0.125% (w/v)
H_2O	400 mL	

Store at room temperature. Add the acid last.

SDS–PAGE Protein Sample Buffer (2x)

Reagent	Quantity (for 50 mL)	Final concentration
Tris (1 M, pH 6.8) <!>	4.0 mL	80 mM
SDS (20%) <!>	5 mL	2.0%
Glycerol (10%, w/v)	5 mL	10.0%
Bromophenol blue (0.1%) <!>	300 μL	0.0006%
DTT (1 M) <!>	5 mL	0.1 M
H_2O	30.7 mL	

Omit DTT for longer-term storage. Store indefinitely at room temperature. Add DTT to 0.1 M on aliquots for daily use.

Protocol 11

Detecting RNA–RNA Interactions Using Psoralen Derivatives

Psoralens are tricyclic compounds that intercalate into double-stranded DNA or RNA and, upon photoactivation, covalently link pyrimidines on adjacent strands. More rarely, psoralen cross-links can be observed at the ends of helices (i.e., double-stranded–single-stranded boundaries). Although psoralens can in some instances cross-link protein to RNA, their primary application is to detect RNA–RNA base-pairing interactions. The most useful psoralen derivative is aminomethyltrioxalen (AMT), which is soluble in H_2O.

MATERIALS

The recipes for the items marked with <R> are at the end of the protocol or in the Common Recipes Appendix.

CAUTION: See the Cautions Appendix for appropriate handling of materials marked with <!>.

Reagents

Aminomethyltrioxalen (AMT; 1 mg/mL in H_2O) <!>
Cultured cells
Formamide loading dye <R>
Ethanol <!>
Phenol:chloroform:isoamyl alcohol (25:24:1) <!>
Phosphate-buffered saline (PBS)
Proteinase K (10 mg/mL) <!>
RNase H
RNase H buffer (5x) <R>
RNA transcribed and labeled with ^{32}P <!>
SDS extraction buffer <R>
Sodium acetate (pH 5.2)
TRIzol <!>

Equipment

Denaturing polyacrylamide gel electrophoresis equipment
Glass plate (2-mm thick, such as a gel plate)
Ice
Metal rack or square pipette can lid
Microcentrifuge
Microcentrifuge tubes (1.5 mL)
Parafilm
Petri dish or chilled metal block on ice
Phosphorimager or X-ray film
Pipettor
Styrofoam or 3-inch-high plastic box
Tissue culture dish (multiwell with lid)
UV lamp (365 nm) or Stratalinker with 365-nm bulbs

OVERVIEW

This method can be used to detect RNA–RNA interactions in tissue culture cells or in extracts. Labeled cross-linked RNAs are detectable by their reduced mobility in polyacrylamide gels. Unlabeled cross-linked RNAs can be detected by northern blots of similar gels. Cross-links can be reversed by exposure to short-wavelength (254 nm) UV light.

METHOD

Cross-linking in Cell Extracts

1. Prepare ^{32}P-labeled RNA by in vitro transcription either uniformly labeled with [α-^{32}P]nucleotide or end labeled using T4 polynucleotide kinase or RNA ligase and [γ-^{32}P]ATP or [^{32}P]cytidine 3′,5′-bis(phosphate) (pCp) (see Chapter 4, Protocol 4).
2. Incubate the RNA in cellular extract using appropriate reaction conditions (see Chapter 6, Protocol 1 for extract preparation). Include controls of unincubated reactions or reactions with conditions designed to be inhibitory.
3. Chill the samples on ice and then add psoralen (AMT to 40 μg/mL).
4. Spot the reaction mixes (10–50 μL) onto Parafilm in an open Petri dish or on a chilled metal block kept on ice.
5. Place a 2-mm glass plate between the UV source and the samples to block shortwave irradiation. If longer times are required, make sure that the samples remain ice cold. Irradiate the solution on ice with 365-nm UV light (handheld or in a Stratalinker with 365-nm bulbs) for times to be determined empirically (10–30 min).
6. After cross-linking, transfer the extract to a 1.5-mL microcentrifuge tube, dilute with 200 μL of SDS extraction buffer, and digest with 1 mg/mL proteinase K for 15 min at 42°C. Add sodium acetate (pH 5.2) to 0.3 M, extract with phenol:chloroform:isoamyl alcohol (25:24:1), and precipitate with ethanol (see Chapter 2, Protocols 1 and 3). Proceed to Step 12 for analysis of cross-linked RNA.

Cross-linking in Cells

7. For suspension cell cultures, centrifuge and resuspend cells in growth medium at a concentration of 2 x 10^7/mL and then aliquot to a multiwell tissue culture plate and add AMT to each well (see note). For monolayer cultures, remove medium from cell monolayers and replace it with medium containing the desired amount of AMT. Use just enough medium to cover the surface and prevent the cells from drying out.

   Titrate the amount of AMT by using 0, 5, 10, 20, and 50 μg/mL. Set one sample aside as a nonirradiated control. Add AMT to suspension cells from a 10x concentrated stock diluted in fresh media and mix by pipetting up and down.

8. Incubate the cells under growth conditions for 5 min and then chill on ice.
9. Remove the plate lids and place the plate on ice on a level metal block. Cover it with a 2-mm-thick glass plate and irradiate the cells for 15 min from a distance of 2.5 cm with a handheld 365-nm light or in a Stratalinker with 365-nm bulbs.
10. After cross-linking, transfer the suspension cells to a conical centrifuge tube and pellet the cells at 1000*g* for 5 min. If using adherent cells, leave on the plate until the cells are lysed.
11. Wash the cells with PBS and lyse the pellet of suspension cells or adherent cells with TRIzol using amounts suggested in Chapter 2 for making RNA from tissue culture cells (Protocol 5).

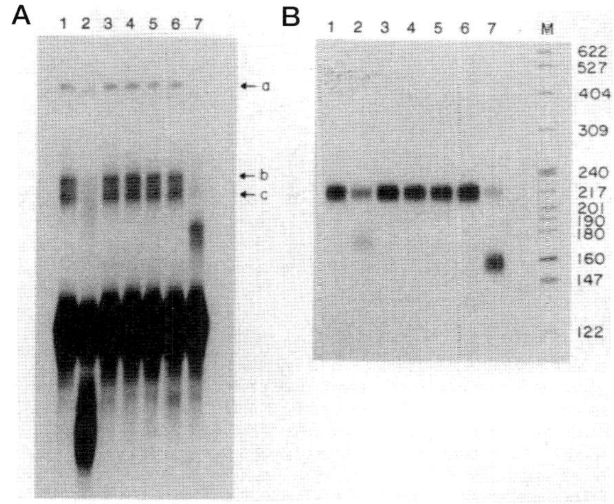

FIGURE 5-12. Example of psoralen cross-linking to identify interacting RNA. In this case, a radiolabeled nematode SL RNA was incubated in whole-cell extract and then cross-linked with psoralen. (*A*) Bands a, b, and c appeared to be cross-linked species because of their mobility. To determine which, if any, of the cross-linked species resulted from an interaction with another snRNP, the RNAs were digested with RNase H in the presence of a panel of oligodeoxynucleotides complementary to various RNAs. Lane *1* was undigested, and lane *2* was digested with an oligodeoxynucleotide complementary to the SL RNA. Although oligodeoxynucleotides complementary to U2, U4, U5, and U1 snRNAs (lanes *3–6*) did not elicit degradation, an oligodeoxynucleotide complementary to U6 snRNA did cause fragmentation in the presence of RNase H. (*B*) Digestion of a purified band (b) with RNase H as in *A*. These results, among others, showed that the SL RNA interacts with U6 snRNP via base pairing. (Reprinted, with permission, from Hannon et al. 1991.)

Extract the RNA using TRIzol (see Chapter 2, Protocol 2) and precipitate with ethanol (Chapter 2, Protocol 3).

12. Resuspend precipitated RNA in H_2O, add denaturing loading dye, and analyze on denaturing polyacrylamide gel (see Chapter 2, Protocol 12).

    The cross-linked RNAs will behave like structured RNA on higher-percentage gels (>6%) and will likely run slower than the sum of the sizes of the two RNAs. For example, two RNAs each 100 nucleotides long will migrate slower than a 200-nucleotide RNA in an 8% denaturing polyacrylamide gel. Compare cross-linked and non-cross-linked control RNAs.

13. For further analysis of cross-linked RNAs, identify the nonlabeled RNA cross-linked to the substrate by performing RNase H digestion with oligodeoxyribonucleotides complementary to candidate target RNAs. After digestion, the cross-linked species will show increased mobility (Fig. 5-12).

TIPS AND TROUBLESHOOTING

1. RNA from unlabeled reactions or from tissue culture cells can be detected by northern blot (see Hausner et al. 1990). Gels described in Step 12 are transferred to a hybridization membrane and probed with probes complementary to candidate RNAs.
2. Cross-links are reversible by exposure to shortwave UV light (254 nm) for 10–15 min at 4°C. Be careful not to irradiate too long or the RNA will be damaged.
3. Positions of cross-linking can be mapped by primer extension. Reverse transcription will "stop" at the cross-linked nucleotides (see Chapter 3, Protocol 15).

SOLUTION AND REAGENT RECIPE

CAUTION: See the Cautions Appendix for appropriate handling of materials marked with <!>.

SDS Extraction Buffer

Reagent	Quantity (for 100 mL)	Final concentration
Tris (1 M, pH 7.5) <!>	2 mL	20 mM
EDTA (0.5 M, pH 8.0) <!>	0.2 mL	1 mM
SDS (10%, w/v) <!>	5 mL	0.5%
H_2O	92.8 mL	

Store indefinitely at room temperature.

REFERENCES

Hannon GJ, Maroney PA, Yu YT, Hannon GE, Nilsen TW. 1992. Interaction of U6 snRNA with a sequence required for function of the nematode SL RNA in trans-splicing. *Science* **258:** 1775–1780.

Hausner T-P, Giglio LM, Weiner AM. 1990. Evidence for base-pairing between mammalian U2 and U6 small nuclear ribonucleoprotein particles. *Genes Dev* **4:** 2146–2156.

Protocol 12

Cross-linking with Formaldehyde

Formaldehyde is a powerful cross-linking agent that covalently attaches ε-amino groups of lysines to elicit protein–protein cross-links or the same groups to amines on nucleic acids (see the introduction to this chapter). It is important to remember that less is better when working with formaldehyde, and it is usually advisable to titrate down to the lowest concentration possible. Over-cross-linking can cause insoluble aggregates and make it impossible to isolate the desired molecule. It is important to design the experiment to use the minimum amount of formaldehyde necessary. Furthermore, always keep in mind that cross-linking is not a quantitative procedure, although one may get that impression from the literature. Finally, remember that formaldehyde is a highly toxic substance and treat it as such.

MATERIALS

The recipes for the items marked with <R> are at the end of the protocol or in the Common Recipes Appendix.

CAUTION: See the Cautions Appendix for appropriate handling of material marked with <!>.

Reagents

Buffer A <R>
Empigen (10%, v/v; Sigma-Aldrich) <R> <!>
Formaldehyde (10%, v/v) <R> <!>
Glycine (1.25 M) <R>
NaCl (5 M)
Phosphate-buffered saline (PBS)
SDS (10%) <!>
Suspension cells (2×10^7 to 1×10^8) or monolayer cell cultures removed from plates

Equipment

Cell scraper
Centrifuge for collecting cells or nuclei
Centrifuge with SS34 rotor or equivalent
Conical centrifuge tubes (50 or 15 mL)
Dounce homogenizer (Kontes) with type-B pestle
Ice
Phase-contrast microscope
Platform rocker (for mixing cells)
Sonicator
Vortex mixer

OVERVIEW

This method describes formaldehyde cross-linking of intact cells followed by sonication or fractionation into nuclear and cytoplasmic extracts. Cross-linked extracts are treated with high salt and SDS to disrupt non-cross-linked aggregates. The proteins or RNAs of interest are then purified by immunoprecipitation or affinity selection. Formaldehyde cross-links are reversible by heat.

METHOD

1. For suspension cells, centrifuge at 1000*g* for 5 min, wash with PBS, and continue with Step 2. For monolayer cells, wash the cells on the plate with PBS, cover cells with PBS, scrape the cells from the plate, and transfer them to a conical centrifuge tube (15 or 50 mL).
2. Centrifuge the cells at 1000*g* for 5 min and resuspend in three volumes (or at least 1 mL) of buffer A.
3. As with many other methods in this chapter, the cross-linking conditions must be determined empirically. Titrate formaldehyde concentrations between 0% and 1%, with temperatures at 4°C, room temperature, or 30°C and time from 10 min to 30 min.

   For example, the following conditions were used for S2 cells in our lab. Start with 1×10^6 to 5×10^6 cells/mL or ~1/10 mL packed cells/sample cell suspension. Transfer 1 mL of cells into each of five 15-mL conical tubes and add formaldehyde to final concentrations of 0.1%, 0.2%, 0.5%, and 1.0% from a 10% (v/v) stock.
4. Cross-link for 10 min on ice with occasional mixing or rocking. Quench the formaldehyde by adding one-fourth volume of 1.25 M glycine. Let stand for 10 min on ice.
5. Centrifuge the cells at 1000*g* for 5 min at 4°C. Discard the supernatant (appropriately in a chemical fume hood) and then resuspend the cells in three times the packed cell volume or at least 1 mL of buffer A.
6. For the whole-cell extracts, sonicate the cells with 10-sec pulses, keeping the cells cold between each pulse until they are broken. Monitor breakage with a phase-contrast microscope. Proceed to Step 10.
7. To make cytoplasmic and nuclear fractions, transfer the cells to a Dounce homogenizer. Break the cells with 10–20 (firm) strokes. Check the cells under a phase-contrast microscope to monitor cell breakage. Note the concentration of formaldehyde where at least 50% of the cells are broken.
8. Add Empigen to 1% and vortex briefly. The remainder of the cells should break immediately.
9. Centrifuge the nuclei at 2000 rpm for 5 min at 4°C. Save the supernatant for cytoplasmic extract. Recover the nuclei and resuspend them in buffer A. Sonicate as for whole-cell preparations (Step 6).
10. Add one-ninth volume of 5 M NaCl to the extract from Step 6 or the fraction from Step 9. Centrifuge in an SS-34 rotor or equivalent at 15,000 rpm for 6 min at 4°C. At this stage, the extract is ready for immunoprecipitation or even polysome analysis.
11. To reverse the formaldehyde cross-links, heat the sample for 1 h at 65°C.

TIPS AND TROUBLESHOOTING

1. Do not heat samples at 95°C before loading onto SDS polyacrylamide gels because cross-links are sensitive to heat. Instead, add SDS sample buffer and heat for 10 min at 37°C–42°C to break up aggregates.
2. Be sure to avoid amines in the buffer because formaldehyde will cross-link amines to the lysines of proteins.
3. Acid will reverse the cross-link.
4. Add β-mercaptoethanol (250 mM) during cross-link reversal to prevent protein oxidation and to accelerate the reversal.

SOLUTION AND REAGENT RECIPES

CAUTION: See the Cautions Appendix for appropriate handling of materials marked with <!>.

Buffer A

Reagent	Quantity (for 500 mL)	Final concentration
HEPES (1 M, pH 7.6)	10 mL	20 mM
KCl (2 M) <!>	2.5 mL	10 mM
$MgCl_2$ (1 M) <!>	0.75 mL	1.5 mM
H_2O	to 500 mL	

Store indefinitely at 4°C. For immediate use, make a fresh 50-mL aliquot with 1 mM DTT <!>. Add 50 mL of 1 M DTT to 50 mL of buffer A. It is important not to use Tris buffers in formaldehyde cross-linking.

Empigen (10%)

Reagent	Quantity (for 50 mL)	Final concentration
Empigen (100%, w/v) <!>	5 mL	10%
H_2O	45 mL	

Store indefinitely at room temperature.

Formaldehyde (10%)

Reagent	Quantity (for 10 mL)	Final concentration
Formaldehyde (37%, w/v) <!>	2.7 mL	10% (w/v)
Buffer A	7.3 mL	10% (v/v)

Prepare a fresh dilution for each use. Handle formaldehyde solution in a chemical fume hood and dispose of properly.

Protocol 13

Basic Affinity Selection Methods

A large number of affinity approaches are available to isolate RNAs associated with specific proteins or proteins associated with specific RNAs. Most, if not all, of these approaches involve the selective retention of tagged molecules on an affinity matrix composed of a ligand immobilized on an agarose or magnetic bead. Whenever possible, it is desirable to avoid elution methods that can strip nonspecific material from the affinity resin. Thus, methods that displace the desired material by competition are greatly preferred over denaturing (e.g., SDS) elution methods (Table 5-2). The presence of nonspecific contaminants is the most confounding problem in any affinity approach.

Method 1: Basic Selection with Beads

We describe here a "generic" method that can be used for any bead-based approach. It describes handling of both agarose and magnetic beads. In general, the basic selections remain the same, but buffer conditions for binding and washing are optimized for each purpose.

MATERIALS

The recipes for the items marked with <R> are at the end of the protocol or in the Common Recipes Appendix.

CAUTION: See the Cautions Appendix for appropriate handling of materials marked with <!>.

Reagents

Affinity reagent (magnetic beads or agarose beads coupled to the tag of interest)
Bind/wash buffer <R>
Elution buffer <R>
Ethanol <!>
GlycoBlue (10 μg/μL)
Phenol:chloroform:isoamyl alcohol <!>
RNA (in extract or purified)
Sodium acetate (NaOAc; 3 M, pH 5.2)

Equipment

Dry ice <!>
Ice
Magnet to hold 1.5-mL microcentrifuge tubes
Microcentrifuge tubes (1.5 mL)
Microcentrifuge with adjustable speed
Pipetman tips (with wide openings)
Pipettor
Rotating wheel or rocking device
Scintillation counter or Geiger counter
Water bath or incubator (4°C–30°C and 37°C)

TABLE 5-2. Elution Methods in Affinity Selection

Selection	Competitor for elution
Anti-TMG beads	7-methylguanosine
Nickel beads	Imidizole
Anti-Flag	Flag peptide
Anti-Myc	Myc peptide
GST fusion proteins	reduced glutathione
Aptamer binding	aptamer ligand

OVERVIEW

Before beginning any affinity purification, think the strategy through completely. If possible, use physical approaches (e.g., centrifugation and size chromatography) to achieve partial purification of the desired material. Dilute well-solubilized starting material is desirable; remember that the molecule(s) of interest is going to be concentrated by the procedure.

1. If using agarose beads, aliquot 30 µL of bead slurry for each sample (up to 10 samples) into a 1.5-mL microcentrifuge tube. Add 0.5 mL of bind/wash buffer and invert the tube to mix. If using magnetic beads, determine capacity of the beads in the manufacturer's protocol. Aliquot beads to a 1.5-µL centrifuge tube with 0.5 mL of bind/wash buffer. Follow the same steps as for agarose beads but substitute separation with a magnet for centrifugation.
2. Centrifuge at low speed (~2000 rpm) for 20 sec at room temperature to collect the beads at the bottom of the tube.
3. Carefully remove the supernatant without disturbing the beads and discard.
4. Add 0.5 mL of bind/wash buffer and repeat Steps 2 and 3 twice more.
5. Resuspend the agarose beads as a 25% slurry in bind/wash buffer and aliquot to a fresh tube for each sample *OR* resuspend the magnetic beads in 50 µL of bind/wash buffer per sample and aliquot to a fresh tube for each sample.
6. Dilute the sample to be selected with bind/wash buffer (final concentration will be 1x bind/wash buffer and RNA in 200 µL). Add the sample to the beads.
7. Mix on rotator or rocker for 1.5–2 h at 4°C.
8. Centrifuge the mixture at 2000 rpm for 20 sec at room temperature to collect the beads.
9. Remove the supernatant and save it as the unbound fraction.
10. Wash the beads as in Step 1 with 0.5 mL of ice-cold bind/wash buffer
11. Repeat Steps 1–3 three times.
12. For competitive elution, resuspend the beads in 100 µL of bind/wash buffer. Add competitor and mix on a rotator or rocker for 30 min. Separate the beads and repeat the elution step. Remove the beads and save the supernatant containing the eluted samples.
13. For recovery of RNA only, resuspend the beads in 200 µL of elution buffer. Add proteinase K to a final concentration of 100 µg/mL. Incubate for 15 min at 42°C. Centrifuge the mixture at 2000 rpm for 20 sec at room temperature to remove the beads. Carefully transfer the supernatant to a 1.5-mL microfuge tube and add NaOAc to 0.3 M. Extract the RNA with phenol:chloroform:isoamyl alcohol and precipitate it with ethanol and 20 µg of GlycoBlue (see Chapter 2, Protocols 1 and 3).
14. To remove protein or RNA + protein samples, strip the beads with SDS–PAGE 2x sample buffer (see Appendix). Transfer the supernatant with the eluted selected sample to a fresh tube.

TIPS AND TROUBLESHOOTING

1. When using beads, check their binding capacity. Use enough beads to bind completely, but not so many extra as to risk increasing nonspecific backgrounds.
2. Always carry out a control such as no antibody, no biotin, or no tag.
3. Determine the optimal amount of salt and detergent in wash buffers.
4. Use of different detergents, e.g., Empigen, Triton X-100, or NP-40, may reduce nonspecific contamination.
5. Elution by competitors of binding sites usually gives lower backgrounds, but elution is not always complete. In this case, use native conditions so that the competitor replaces the RNA or protein on the beads.

Method 2: Selection with a Biotinylated Oligonucleotide

To select an RNA, such as a substrate cross-linked to another RNA, we have found that using hybrid selection with a biotinylated oligonucleotide can be much more efficient than direct selection of biotinylated RNAs.

MATERIALS

CAUTION: See the Cautions Appendix for appropriate handling of material marked with <!>.

Reagents

Formamide for elution
Magnetic beads coupled to Streptavidin (M280)
Phenol extracted, ethanol-precipitated, and resuspended in H_2O
Purified RNA
SSC (20x)
Wash buffer 1 (1x SSC) (see SSC [20x] [Method 2], p. 319)
Wash buffer 2 (0.1x SSC) (see SSC [20x] [Method 2], p. 319)

Equipment

Dry ice <!>
Heat block (95°C)
Ice
Magnet to hold 1.5-mL microcentrifuge tubes
Microcentrifuge
Microcentrifuge tubes (1.5 mL)
Pipette with P1000 tip
Rotating wheel or rocking device
Water bath or incubator

OVERVIEW

There are several possible ways to proceed once the oligonucleotide is in hand; it can be preannealed to the RNA of interest or it can be bound to the bead for subsequent hybridization. We greatly prefer hybridizing in solution to a target RNA. Suppliers of magnetic beads (e.g., Dynabeads M-250

Streptavidin) have recommended bind and wash buffers but many different conditions can be used. The ones presented here provide an example. It is important to know the binding capacity of the beads and the amount of material to select.

METHOD

1. Resuspend the M280 Streptavidin beads by inverting the vial. Transfer an appropriate aliquot (usually 20–100 μL) to a 1.5-mL microcentrifuge tube containing 0.5 mL of bind/wash buffer. If more than one sample is being processed, beads can be washed together. Close the tube and invert it to mix.
2. Place the tube in the magnet for 1–2 min. The beads will collect on the side of the tube in contact with the magnet.
3. Remove supernatant with a P1000 pipette tip without disturbing the beads.
4. Repeat the wash steps twice more and then resuspend the beads in bind/wash buffer. Aliquot equal amounts of beads to each of the samples being selected.
5. Resuspend the beads in 200 μL of bind/wash buffer (1x SSC) or, if larger volumes are used, in twice their original volume.
6. To anneal the oligonucleotide to the RNA sample, mix together

RNA to be selected	
SSC (20x)	2.5 μL
Biotinylated oligonucleotide (100 μg/μL)	2.5 μL
H_2O	to 25 μL

   Heat the mixture at 95°C for 2 min, transfer to 65°C for 10 min, and then to 30°C for 10 min.
7. Add the hybridized mixture of RNA and oligonucletide to the tubes with beads and rotate or rock for 5–30 min at room temperature.
8. After binding is complete, separate the beads with the magnet and collect the unbound solution. Wash the beads twice more as in Steps 1 and 2 with wash buffer 1 and then twice more with wash buffer 2.
9. Resuspend the beads in formamide buffer (90% formamide, 10 mM EDTA) and heat for 2 min at 95°C to elute the hybridized RNA.
10. Collect beads with the magnet and recover the supernatant. The supernatant will contain the selected RNA and some of the oligonucleotide. Dilute with extraction buffer or H_2O. Add 10 μg GlycoBlue, sodium acetate to 0.3 M, and precipitate with ethanol.
11. Analyze selected RNA by primer extension or end labeling and gel electrophoresis.

Method 3: Selection of a Ribonucleoprotein Using a Complementary Biotinylated Oligonucleotide

Before beginning an experiment of this type, it is important to design the hybridizing oligonucleotide appropriately. The first consideration is the accessibility of the potential hybridizing site. To determine this, carry out pilot experiments with a panel of DNA oligonucleotides and RNase H. If the segment of RNA is accessible, it will be digested with RNase H in the presence of DNA oligonucleotide (see Chapter 3, Protocol 17). Once an accessible site has been identified, design a chimeric 2′-*O*-methyl (2′OMe) RNA DNA oligonucleotide. This oligonucleotide should have 20 2′OMe RNA residues complementary to the RNA of interest at the 5′ end of the oligonucleotide. The 3′ end of the oligonucleotide should contain at least 20 DNA residues with the three 3′-most residues biotinylated. Biotin at the 3′ end ensures that the oligonulceotide will have the biotin because it is synthesized first. This leaves the 5′ end for labeling with [γ-^{32}P]ATP and polynucleotide kinase to monitor for binding and elution with DNase I. Clearly, the oligonucleotide should be designed such that it will not contain strong secondary structure.

The advantages of this approach are twofold. First, if the oligonucleotide is designed correctly, it is possible to elute selected material through the use of DNase I, which greatly reduces nonspecific background. Second, multiple biotin groups can be clustered on the oligonucleotide, greatly enhancing binding.

MATERIALS

The recipes for the items marked with <R> are at the end of the protocol or in the Common Recipes Appendix.

CAUTION: See the Cautions Appendix for appropriate handling of material marked with <!>.

Reagents

Binding buffer (Method 3) <R>
DNase I buffer (Method 3) <R>
Magnetic beads coupled to streptavidin
Oligonucleotide complementary to the RNA of interest with biotin groups (see Overview for oligonucleotide design) and a control nonspecific oligonucleotide of the same design
Purified RNA
SSC (20x)
Wash buffer (Method 3) <R>

Equipment

Dry ice <!>
Heat block (95°C)
Ice
Magnet to hold 1.5-mL microcentrifuge tubes
Microcentrifuge
Microcentrifuge tubes (1.5 mL)
Pipette with P1000 tip
Rotating wheel or rocking device
Water bath or incubator

METHOD

1. Bind 500 μL of magnetic bead slurry to 100 μg of a nonspecific oligonucleotide to beads as described in Method 2 (Steps 1 and 2).
2. Preclear the extract by mixing with the nonspecific oligonucleotide bound to beads (Step 1) and rotating or rocking for 3 h at 4°C–30°C (depending on the stability of the complex and extract).
3. Meanwhile, prepare magnetic beads with a target-specific oligonucleotide and a separate set of beads with a nonspecific (control oligonucleotide). Add target or control oligonucleotide to beads and bind and wash as in Steps 1 and 2 in Method 2. Supplement each tube with ~100,000 cpm of 5′-end-labeled oligonucleotide to monitor binding and release.
4. Remove the beads from Step 2 with a magnet and transfer the supernatant (precleared extract) to the tubes from Step 3 using one-half for beads plus target oligonucleotide and one-half for beads plus control oligonucleotide.

5. Incubate with gentle mixing for 3 h at 4°C–30°C.
6. Collect beads as in Step 1. Save the supernatant at –80°C. It can be reused with other oligonucleotides.
7. Wash the beads three times with 1 mL of wash buffer, twice with 1 mL of binding buffer, and finally once with 300 μL of DNase I buffer. Remove liquid completely after the final wash.
8. Add 30 μL of DNase I buffer and 3 μL (6 units) of DNase I to the washed beads. Incubate for 20 min at 37°C. Remove and save the supernatant.
9. Repeat Step 7.
10. Determine the elution efficiency by counting in a scintillation counter or by monitoring with a Geiger counter. After counting, pool the samples if desirable.
11. Analyze recovered RNA or protein complexes by gel electorphoresis.

TIPS AND TROUBLESHOOTING

1. RNase H in extracts can be a problem when RNA is annealed to a DNA oligonucleotide. If this is the case, a chimeric oligonucleotide can be used. Design the oligonucleotide with 2′O methyl nucleotides for hybridization to the target followed by 5–10 DNA residues before the biotin group to provide a target for DNase I.
2. Elution with DNase may not be complete. Inclusion of a trace amount of ^{32}P 5′-end-labeled biotinylated oligonucleotide is very helpful in monitoring binding and elution. Make sure that the trace-labeled oligonucleotide is free of unincorporated label so that only the DNA is being monitored. Add ~100,000 cpm trace-labeled oligonucleotide to the selection oligonucleotide during the bead-binding step.
3. If volumes are large, the separation of beads with the magnet can be done sequentially until all of the sample has been separated. Alternatively, the sample can be centrifuged briefly and most of the supernatant removed before collecting the beads with the magnet.
4. Other detergents or wash buffers may be used. Check for compatibility with the beads being used.

SOLUTION AND REAGENT RECIPES

CAUTION: See the Cautions Appendix for appropriate handling of materials marked with <!>.

SSC (20x) (Method 2)

Reagent	Quantity (for 100 mL)	Final concentration
NaCl	17.5 g	3.0 M
Sodium citrate <!>	5 mL	0.3 M
H_2O	to 100 mL	

Use autoclaved H_2O and filter with a 0.2-μm filter. Store at room temperature.

Dilute 1:20 to make wash buffer 1 (1x SSC). Dilute 1:200 to make wash buffer 2 (0.1x SSC).

Bind/Wash Buffer (Method 1)

Reagent	Quantity (for 100 mL)	Final concentration
NaCl	3 mL	150 mM
Nonidet P-40 (NP-40; 10%, v/v) <!>	1 mL	0.1% (v/v)
Tris (1 M, pH 7.5) <!>	5 mL	50 mM
H_2O	91 mL	

Store at 4°C.

Binding Buffer (Method 3)

Reagent	Quantity (for 100 mL)	Final concentration
NaCl (5 M)	1 mL	50 mM
Tris (1 M, pH 7.5) <!>	2 mL	20 mM
EDTA (0.2 M, pH 8.0) <!>	0.1 mL	0.2 mM
Nonidet P-40 (NP-40; 10%, v/v) <!>	1 mL	0.1% (v/v)
H_2O	95.9 mL	

Store at 4°C.

DNase I Buffer (Method 3)

Reagent	Quantity (for 100 mL)	Final concentration
Tris (1 M, pH 7.5) <!>	100 µL	100 mM
$MgCl_2$ (1 M) <!>	100 µL	100 mM
H_2O	800 µL	

Store for 6 mo at –20°C.

Elution Buffer (Method 1)

Reagent	Quantity (for 100 mL)	Final concentration
NaCl (5 M)	3 mL	150 mM
Tris (1 M, pH 7.5) <!>	5 mL	50 mM
EDTA <!>	1 mL	5 mM
SDS (10%) <!>	10 mL	1%
H_2O	81 mL	

Store indefinitely at room temperature.

Wash Buffer (Method 3)

Reagent	Quantity (for 100 mL)	Final concentration
NaCl (5 M)	10 mL	500 mM
Tris (1 M, pH 7.5) <!>	2 mL	20 mM
EDTA (0.2 M, pH 8.0) <!>	0.1 mL	0.2 mM
Nonidet P-40 (NP-40; 10%, v/v) <!>	1 mL	0.1% (v/v)
H_2O	86.9 mL	

Store indefinitely at 4°C.

6 Analysis of RNA-processing Reactions Using Cell-free Systems

PROTOCOLS

1 Preparation of Nuclear Extracts from HeLa Cells, 325
2 Analysis of Pre-mRNA Splicing Using HeLa Cell Nuclear Extracts, 330
3 Preparation of *Drosophila* Kc Cell Nuclear Extracts for In Vitro Splicing, 334
4 In Vitro Splicing Reactions in *Drosophila* Kc Nuclear Extracts, 339
5 Preparation and Analysis of Cell-free Splicing Extracts from *Saccharomyces cerevisiae*, 343
6 Analysis of Splicing In Vitro Using Extracts of *Saccharomyces cerevisiae*, 348
7 Analysis of Splicing Complexes on Native Gels, 352

THE BIOCHEMICAL DISSECTION of complex RNA-processing reactions such as pre-mRNA splicing and 3′-end formation has been made possible through the development and use of cell-free systems that faithfully and efficiently carry out those reactions. These systems enable the use of a myriad of techniques, such as depletion–add back and site-specific cross-linking, that are impossible in intact cells. Moreover, in vitro analysis allows detection of processing (and assembly) intermediates, determination of energy requirements, and monitoring of kinetics. Indeed, most of our current knowledge regarding the mechanisms of and factors involved in pre-mRNA splicing and 3′-end formation derives from studies performed in cell-free extracts.

This chapter presents methods for preparing extracts from HeLa cells, *Drosophila* cells, and budding yeast that efficiently catalyze pre-mRNA splicing. We also present methods for using these extracts to monitor splicing. These procedures should prove to be useful for those who wish to analyze splicing in vitro. They also serve as guides for the preparation and use of cell-free systems for the analysis of any processing reaction.

GUIDELINES FOR THE PREPARATION OF ACTIVE CELL-FREE SYSTEMS

Making active cell-free extracts is more an art than a science. There are always students or postdocs with "green" thumbs and those who are all thumbs. Some people have a talent for preparing active extracts, and others, no matter how hard they try or how closely they follow directions, prepare nothing but dead extracts. The reasons behind this disparity are mysterious, but below we provide a few essential guidelines that should give everyone a fighting chance.

1. Always use actively growing, healthy cells. It is crucial that cells be in "log" phase when harvested for extract preparation. Cells that are over or under split will invariably yield poor extracts. For HeLa cells, we recommend that cells be 4×10^5 to 5×10^5 mL (if in suspension) or ~70% confluent when grown in monolayer.
2. Always use enough material. A common mistake is to attempt to prepare extracts from too few cells. Concentration is critical for producing active extracts, and dilute extracts are usually inactive. We recommend that 1 mL (~10^9) of packed animal cells be the minimum amount used.
3. Be prepared. It is absolutely essential to have all buffers premade and prechilled, all rotors chilled, and dialysis tubing prepared before beginning these protocols.

4. Speed is of the essence. By this, we do not mean haste, but it is critical that each step be timed precisely and that each transition be carried out smoothly. When the cells have pelleted is not the time to make or receive a phone call, eat lunch, or look up the next step. It is also helpful to perform a dry run to become familiar with all of the steps. This also ensures that all equipment and solutions are on hand and ready to go.
5. Always monitor the breakage of cells using a microscope. Perhaps the touchiest part of making an active extract is homogenization of the starting material. We have found through many years of experience that the mildest form of homogenization is always the best. For animal cells, the preferred method is Dounce homogenization of hypotonically swelled cells. Although quite simple in concept, there is an art to this technique. Smooth strokes that do not create bubbles are required. Moreover, it is necessary to make sure to use the right Dounce homogenizer. Despite protestations of the manufacturers, all Dounces are not created equal. We have found that Wheaton Dounces are too "loose" for good homogenization, whereas Kontes with B pestles are normally quite good. A "good" Dounce homogenizer will give ~90% cell breakage with 10 smooth strokes and negligible nuclear breakage. A Dounce that is too "tight" will break nuclei, whereas a Dounce that is too "loose" will not disrupt cells.

We recommend buying a few Dounce homogenizers and experimenting with different pestles before making extracts. Always monitor breakage under a phase-contrast microscope. We suspect that many failed extracts result from over or under breakage of the starting material.

GUIDELINES FOR TESTING AND OPTIMIZING EXTRACTS

It is now more than 25 years since the first active splicing extracts were described, and it has become "routine" to prepare such extracts. Accordingly, to check the quality of the extract, it is desirable to prepare a known splicing substrate and assay it for splicing. If splicing activity is low or nonexistent, it is best to discard the extract and start over. It is highly unlikely that a dead or dying extract can be resurrected (see below). If the extract is active for splicing, it is likely to be active for most if not all nuclear RNA-processing reactions.

To assay new activities, let us imagine that we have discovered a novel RNA-processing activity and wish to develop a cell-free system to analyze it. If the reaction occurs in the nucleus of a well-studied cell type (e.g., HeLa, *Drosophila*, or yeast), great! Start by making a cell-free system that is active for splicing. Then, design an assay to monitor the newly discovered reaction. We highly recommend "direct" assays, i.e., using radiolabled substrates that will produce readily identifiable products. If the desired activity is present, nothing more needs to be done. However, if the activity is not present, we recommend the manipulation of four variables in an attempt to observe the desired activity. In this regard, monovalent salt concentration can be critical in early steps; i.e., extraction. If extracts are inactive using "standard" monovalent conditions, try preparing extracts using lower or higher initial concentrations. Some reactions have quite defined optima for monovalent and divalent salts. In addition, the inclusion of molecular crowding agents such as polyethyleneglycol (PEG) can have profound effects on some processing reactions. Finally, inclusion of nonspecific RNAs such as poly(A), rRNA, or tRNA can "soak up" RNA-binding proteins that might be inhibiting the reaction of interest.

To address as many of these potential factors at once, we suggest setting up a matrix of reactions testing three or four concentrations of Mg^{++}, three or four concentrations of PEG, and three or four concentrations of nonspecific RNA. At this point, the least likely confounding variable is monovalent salt, so this is held constant for the starting matrix. A three-variable, three-concentration matrix involves setting up 81 independent reactions. This number may seem large, but these reactions can be done over several days.

If any reaction displays activity, focus on those conditions and optimize each variable; also include monovalent salt as a fourth variable at this point. If no activity is observed, it is likely that a key component is missing. In such instances, it is necessary to try different types of extracts, such as whole-cell extracts. It may also be necessary to try phosphatase, kinase, or proteinase inhibitors, because some proteins are exquisitely sensitive to proteolysis and some reactions require that a key component be properly phosphorylated or dephosphorylated.

This systematic approach may seem tedious; however, we have found that it has yielded active cell-free systems. The reward is well worth the work involved.

DEVELOPING CELL-FREE SYSTEMS FROM POORLY STUDIED ORGANISMS

Some processing reactions are only found in "obscure" organisms. For example, spliced leader (SL) *trans*-splicing is only found in a subset of lower eukaryotes, and certain types of RNA editing are only observed in organelles. Often, an understanding of these unusual reactions provides valuable insight not only into the diversity of RNA processing in general, but also into the evolution and mechanism of processing reactions in higher eukaryotes. Developing cell-free systems to study these reactions is somewhat more daunting than using well-studied systems. For such analysis, we recommend adhering to the following guidelines. First, choose the right organism for analysis. The "right" organism should be readily available and in sufficient quantity to allow many trials. Ideally, the organism should be cultureable, i.e., maintained and grown in defined conditions. However, this is not a prerequisite. For example, *Ascaris*, a large parasitic nematode, is not cultureable, but it has proven to be ideal for the analysis of nematode *trans*-splicing.

Once a suitable organism has been identified, model the extract preparation method after successful techniques used in other more well-studied creatures. However, be ready to adapt to the particular features of the organism. For example, it proved impossible to prepare nuclei from *Ascaris*, but whole-cell extracts were quite active. For making whole-cell extracts, we recommend breaking the cells with a metal Dounce (Kontes). The metal Dounce has a tight-fitting pestle and completely disrupts the nuclei. When beginning studies with a new organism, start simply, i.e., assay for an activity that is robust. RNA polymerase III activity is usually quite straightforward to assay. If that activity is present in the extract, move on to more complex assays. In that regard, we recommend that the same variables described above be manipulated when assaying for any activity. A particularly striking example comes again from studies in the *Ascaris* system. Extracts were completely inactive in the absence of PEG, but they demonstrated excellent activity when 3% PEG was included.

SUMMARY

The above guidelines provide a rather crude compass for preparing and assaying cell-free systems. A detailed road map is impossible to provide because every experimental scenario is likely to be distinct. When embarking on such approaches, we have found that "common sense" is the best tool in one's arsenal. Common sense plus patience and, above all, a systematic approach provide the best chance for success.

READING LIST

The following describe a variety of published extract preparation methods that serve as a basis for all of the protocols described in this chapter.

Colgan DF, Manley JL. 1997. Mechanism and regulation of mRNA polyadenylation. *Genes Dev* 11: 2755–2766.

Dignam JD, Lebovitz RM, Roeder RG. 1983. Accurate transcription initiation by RNA polymerase II in a soluble extract from isolated mammalian nuclei. *Nucleic Acids Res* **11:** 1475–1489.

Hannon GJ, Maroney PA, Denker JA, Nilsen TW. 1990. *Trans* splicing of nematode pre-messenger RNA in vitro. *Cell* **61:** 1247–1255.

Kataoka N, Dreyfuss G. 2008. Preparation of efficient splicing extracts from whole cells, nuclei, and cytoplasmic fractions. *Methods Mol Biol* **488:** 357–365.

Krainer AR., Maniatis T, Ruskin B, Green MR. 1984. Normal and mutant human β-globin pre-mRNAs are faithfully and efficiently spliced in vitro. *Cell* **36:** 993–1005.

Lin R-J, Newman AJ, Cheng S-C, Abelson J. 1985. Yeast mRNA splicing in vitro. *J Biol Chem* **260:** 14780–14792.

Mayeda A, Krainer AR. 1999. Mammalian in vitro splicing assays. *Methods Mol Biol* **118:** 315–321.

Moore CL, Sharp PA. 1985. Accurate cleavage and polyadenylation of exogenous RNA substrate. *Cell* **41:** 845–855.

Rio DC 1988. Accurate and efficient pre-mRNA splicing in *Drosophila* cell-free extracts. *Proc Natl Acad Sci* **85:** 2904–2908.

Seraphin B, Rosbash M. 1989. Identification of functional U1 snRNA-pre-mRNA complexes committed to spliceosome assembly and splicing. *Cell* **59:** 349–358.

Protocol 1

Preparation of Nuclear Extracts from HeLa cells

HeLa cells are the archetypal tissue culture cell line for the preparation of mammalian cell-free systems. These cells, derived from a cervical carcinoma, have been maintained in culture since the late 1940s. They grow with a doubling time of ~24 h and can be cultured in medium containing fetal bovine serum (in which they grow somewhat faster) or medium containing horse serum (which is considerably less expensive than fetal bovine serum). Extracts prepared from these cells have been used to determine the mechanism of splicing and of polyadenylation, and such extracts have been characterized extensively. HeLa cells are usually the cell of choice to initiate cell-free analysis of nearly any aspect of mammalian gene expression. In some instances, e.g., analysis of a tissue-specific alternative splicing, it is necessary to use a different cell type. We have found that the protocol described here can be used successfully to prepare active nuclear extracts from a wide variety of tissue culture cells including *Drosophila* S2 cells.

MATERIALS

The recipes for the items marked with <R> are at the end of the protocol or in the Common Recipes Appendix.

CAUTION: See the Cautions Appendix for appropriate handling of materials marked with <!>.

Reagents

Buffer A (hypotonic) <R>
Buffer B (low salt) <R>
Buffer C (high salt) <R>
Buffer D (dialysis) <R>
Dithiothreitol (DTT; 1 M) <!>
EDTA (0.5 M, pH 8.0) <!>
Gycerol
HeLa cells (10^9) (at least 1 mL of packed cells)
KCl (3 M) <!>
$MgCl_2$ (1 M) <!>
Phosphate-buffered saline (PBS; chilled at 4°C)
S100 buffer (10x) <R>
Tris-HCl (1 M, pH 7.9)

Equipment

Aspirator
Cell culture plates
Cell scraper
Centrifuge for collecting cells (low speed)
Centrifuge tubes (polypropylene) rated for 10,000*g* (15 or 50 mL)
Centrifuge with SS-34 or equivalent rotor
Conical centrifuge tubes (15 or 50 mL) and conical bottles (250 mL)

This protocol is modified from Dignam et al. (1983) by the Maniatis laboratory and communicated to us by Xiang-Dong Fu, University of California, San Diego.

Dialysis tubing (10,000 MW cutoff)
Dounce homogenizer (7 or 15 mL; Kontes) with type-B pestle
Dry ice <!>
Freezer (–80°C)
Ice
Magnetic stir plate or platform rocker
Phase-contrast microscope
Pipettes (5 or 10 mL) and pipetting device
Polycarbonate centrifuge tubes (for cytoplasmic S100)
Polypropylene centrifuge tubes, graduated
Ti-70 rotor
Vortex mixer

OVERVIEW

Before beginning this protocol, all buffers should be prepared and chilled and centrifuges should be available and cooled to 4°C. Careful planning is essential for success. Once initiated, this protocol must be carried through to completion. Any delays are likely to yield inactive extracts. Finally, it is crucial that cells be healthy and washed well. We have found that the optimum density of cells in suspension is 5×10^5/mL. For adherent cells, confluency should not be ≥70%–80%. In addition, make sure that the cells are not infected with mycoplasmas. Above all, speed is of the essence, but any mistakes can and will have serious consequences to the preparation. Do not multitask. Be there, on time, waiting for centrifuge spins or incubations to end.

METHOD

Prepare all manipulations on ice with cold buffers.

Preparation of Nuclear Extract

1. For harvesting suspension or nonadherent cells:
   i. Centrifuge the cells plus medium at 1000*g* for 5 min in 250-mL centrifuge bottles.
   ii. Remove all of medium, being careful not to disturb the pellet.
   iii. Wash the pellet three times with 100 mL of ice-cold PBS to remove any serum (which contains RNases).
   iv. For the final wash, use 10 mL of PBS to pool and transfer the cells to a 15-mL conical tube.
   v. Aspirate the PBS and immediately proceed to Step 3.
      If the cell pellet is <0.3 mL, add buffer A to 1 mL. Never try to process <1 mL.
2. For harvesting adherent cells. Start with 10–20 10-cm dishes.
   i. Wash the cells on the plate three times (10 mL/10-cm dish) with cold PBS.
      A few rocks back and forth should suffice.
   ii. Add 10 mL of PBS and then use a cell scraper to scrape the cells from the plate.
   iii. Transfer the cells to a 50- or 250-mL conical centrifuge tube as appropriate.
   iv. Centrifuge the cells at 1000*g* for 5 min. Resuspend in 10 mL of 4°C PBS and pool into one tube.
   v. Centrifuge as in Step 2.iv.
   vi. Aspirate the PBS.

3. Resuspend the packed cells in three times the packed cell volume of hypotonic buffer (buffer A) by pipetting up and down and vortexing briefly. Allow the cells to swell for 10 min on ice.
4. Homogenize the cells by 10 up-and-down strokes in a glass Dounce homogenizer (type-B pestle). Check for cell breakage with a microscope (cell breakage should be >80%–90%).
5. Transfer the cells to a 15-mL conical tube and centrifuge in a low-speed centrifuge (such as Beckman J6B) at 2000 rpm for 15 min at 4°C.

   Use graduated polypropylene centrifuge tubes for ease in measuring the packed nuclear volume after the centrifugation.
6. Remove the supernatant carefully and measure its volume. Add 0.11 volume of 10x S100 buffer to the supernatant and centrifuge at 3000 rpm (100,000*g*) for 1 h at 4°C in a polycarbonate tube in a Ti-70 rotor. The resulting supernatant is the cytoplasmic (S100) fraction. Dialyze as described in Step 11.
7. Measure the nuclear pellet volume and then add to it one-half times the packed nuclear volume of the low-salt extraction buffer (buffer B).

   This step gently resuspends the nuclei and eliminates clumps in the extraction to follow. This can be done by pipetting up and down with a 5–10-mL pipette. For volumes >5 mL, one or two gentle strokes in a 15-mL Dounce homogenizer can be done if necessary.
8. While gently stirring on ice, or with frequent mixing, add one-half times the packed nuclear volume (from Step 5) of the high-salt extraction buffer (buffer C). This must be added dropwise with frequent mixing or the nuclei will lyse. After addition is complete, mix the nuclei with two to three gentle strokes in a glass Dounce homogenizer (type-B pestle).

   The final salt concentration will be 0.3 M KCl. If desirable, it is possible to go to slightly higher salt concentrations (0.35 M KCl). However, the nuclei may be fragile under these conditions and lyse easily.
9. Transfer the nuclei to a 40-mL Oak Ridge tube and extract for 30 min by very gently stirring on a magnetic stirrer using a flea (very small magnetic stir bar) in a tube placed in an ice bath or gently tilting the tube on a rocker placed in a 4°C cold room.
10. After extraction, remove the stir bar and centrifuge at 10,000 rpm for 30 min at 4°C. The resulting supernatant is the nuclear extract. Carefully remove the extract to a clean tube to measure the volume.
11. Dialyze both the cytoplasmic (S100) (from Step 6) and the nuclear (from Step 10) extract against 100 volumes of dialysis buffer (buffer D) for 2 h at 4°C.
12. Change the buffer and dialyze for a further 2 h. Overnight dialysis is not recommended.
13. Remove extract from the dialysis tubing and centrifuge in an SS-34 or equivalent rotor at 10,000 rpm for 15–20 min at 4°C.
14. Make aliquots, freeze on dry ice, and store at –80°C.

    The extract can be refrozen and later thawed on ice but not repeatedly. Prepare aliquots according to the amounts likely to be used per experiment.

TIPS AND TROUBLESHOOTING

1. Make sure that the level of liquid in the homogenizer is below the bottom tube portion before the pestle is used.
2. Dounce homogenizers are available in 40-mL and even 100-mL sizes. If large-scale extraction preparations are being performed, the 40-mL size can be used, but it may not be as effective for breaking cells. The 40-mL size can be used for gently mixing nuclei for extraction; however, sequential use of a 15-mL Dounce is preferable for cell breakage.

SOLUTION AND REAGENT RECIPES

CAUTION: See the Cautions Appendix for appropriate handling of materials marked with <!>.

Buffer A (Hypotonic)

Reagent	Quantity (for 500 mL)	Final concentration
Tris (1 M, pH 7.9) <!>	5 mL	10 mM
$MgCl_2$ (1 M) <!>	0.75 mL	1.5 mM
KCl (3 M) <!>	1.6 mL	10 mM
H_2O	to 500 mL	

Add 0.5 mM DTT <!> to 10–20 mL. Store the remainder for 1 yr at 4°C.

Buffer B (Low Salt)

Reagent	Quantity (for 100 mL)	Final concentration
Tris-HCl (1 M, pH 7.9)	2 mL	20 mM
Glycerol (50 mL, ultrapure)	50 mL	25% (v/v)
$MgCl_2$ (1 M) <!>	0.15 mL	1.5 mM
EDTA (0.5 M, pH 8.0) <!>	0.04 mL	0.2 mM
KCl (3 M) <!>	0.66 mL	20 mM
H_2O	to 100 mL	

Store at 4°C.

Buffer C (High Salt)

Reagent	Quantity (for 100 mL)	Final concentration
Tris-HCl (1 M, pH 7.9)	2 mL	20 mM
Glycerol (ultrapure, 50%, v/v)	50 mL	25% (v/v)
$MgCl_2$ (1 M) <!>	0.15 mL	1.5 mM
EDTA (0.5 M, pH 8.0) <!>	0.04 mL	0.2 mM
KCl (3 M) <!>	40 mL	1.2 mM
H_2O	to 100 mL	

Store at 4°C.

S100 Buffer (10x)

Reagent	Quantity (for 100 mL)	Final concentration
Tris-HCl (1 M, pH 7.9)	30 mL	0.3 M
KCl (3 M) <!>	46.6 mL	1.4 M
$MgCl_2$ (1 M) <!>	3 mL	0.03 M
H_2O	to 100 mL	

Store at 4°C.

Buffer D (Dialysis)

Reagent	Quantity (for 2 L)	Final concentration
Tris-HCl (1 M, pH 7.9)	40 mL	20 mM
Glycerol (100%, v/v)	400 mL	20% (v/v)
KCl (3 M) <!>	66.6 mL	100 mM
EDTA (0.5 M, pH 8.0) <!>	0.8 mL	0.2 mM
DTT (1 M) <!>	2 mL	1 mM
H_2O	to 2 L	

Carefully pour glycerol into the bottom of a 2-L measuring cylinder so that it does not run down the sides. To store buffers for later use, make them without DTT and add fresh DTT to the portions used. Prepare buffers the day before so that they will be well chilled.

Protocol 2

Analysis of Pre-mRNA Splicing Using HeLa Cell Nuclear Extracts

HeLa cell nuclear extracts have proven to be exceptionally valuable for the analysis of a variety of RNA-processing reactions, including pre-mRNA splicing. This protocol is specifically designed for analysis of splicing, but similar protocols can be used to address almost any process that can be assayed. Before asking any specific question, make sure that the extract is active. We suggest using a splicing substrate that is known to be efficiently spliced in HeLa nuclear extract (e.g., adenovirus major late or MINX RNA). If these substrates are spliced ≥50% in the extract, proceed with specific questions. If these substrates are spliced poorly or not at all, review the steps in the splicing protocol, and even if everything was done correctly, a new extract must be prepared. An extract that is "dead" cannot be rescued. In addition, we recommend that all splicing substrates be capped and of sufficient specific activity (see Chapter 4).

MATERIALS

The recipes for the items marked with <R> are at the end of the protocol or in the Common Recipes Appendix.

CAUTION: See the Cautions Appendix for appropriate handling of materials marked with <!>.

Reagents

ATP (100 mM)
Creatine phosphate (0.5 M)
Denaturing polyacrylamide gel reagents
Dithiothreitol (DTT; 1 M) <!>
EDTA (0.5 M, 2 mM, pH 8.0) <!>
Ethanol (100%) <!>
GlycoBlue (10 μg/mL)
KCl (3 M) <!>
$MgCl_2$ (1 M) <!>
NaOAc (3 M, pH 5.2)
Phenol:chloroform:isoamyl alcohol (25:24:1 w/v) <!>
Polyethylene glycol (PEG; 20%, w/v) CarboWax 8000 (Fisher) <R>
Proteinase inhibitors (aprotinin, leupeptin) <R> <!>
Proteinase K solution (10 mg/mL) <!>
Radiolabeled DNA markers <!>
RNase inhibitor (40 units/μL)
RNA (labeled with 25 cpm; ~15,000–20,000 cpm/ng) <!>
RNA splicing substrate (known to be efficiently spliced in HeLa nuclear extract)
SDS (10%, w/v) <!>
SDS extraction buffer <R>
Splicing master mix (4x) <R>
Tris (1 M, pH 7.9) <!>
Tris/borate/EDTA (TBE) buffer <R> <!>

Equipment

Dry ice <!>
Electrophoresis equipment, including vertical gel, comb spacers, and high-voltage power supply (see Chapter 2, Protocol 12)
Gel dryer
Heat block (95°C)
Ice
Microcentrifuge
Microcentrifuge tubes (1.5 mL)
Phosphorimager equipment or X-ray film and intensifying screen
Vortex mixer
Water bath (30°C)

OVERVIEW

This protocol is used to determine the splicing behavior of pre-mRNAs in extracts that are capable of carrying out splicing. ^{32}P-labeled RNA is incubated under splicing conditions for various times, and the resulting products are analyzed on denaturing polyacrylamide gels.

METHOD

1. Prepare a denaturing polyacrylamide gel (see Chapter 2, Protocol 12)
2. Set the up splicing reactions. Make a splicing master mix fresh for each experiment.
   i. Mix the following in a 1.5-mL microcentrifuge tube:

Splicing extract	5 µL
RNasin	1 µL, 40 units/µL
Labeled RNA (25 cpm) (~15–20,000 cpm/ng)	1 µL
Splicing mix (4x)	2.5 µL

   ii. Mix briefly by vortexing and/or flicking the tube.
   iii. Incubate for desired time (10–90 min) at 30°C.
3. Process the splicing reactions for the polyacrylamide gel.
   i. Add to the reaction from Step 2:

SDS extraction buffer	200 µL
Proteinase K	5 µL, 10 mg/mL

   ii. Incubate for 15 min at 42°C.
   iii. Add 0.3 M NaOAc. Then, PCA-extract with phenol and precipitate with ethanol with 20 µg of GlycoBlue (see Chapter 2).
   iv. Resuspend the precipitate in 10 µL of 2 mM EDTA, aliquot 2.5 µL for the gel, and add 5 µL of RNA-loading buffer.
4. Run a denaturing polyacrylamide gel:
   i. Set up the gel in the gel box, add running buffer to the upper and lower reservoirs, and prerun the gel for 15–45 min at 1500 V, maximum 45 mA.

      Shorter preruns are allowed when the RNA fragments are all above 100 bases in length.

   ii. Heat the samples for 1 min at 95°C and then place on ice.

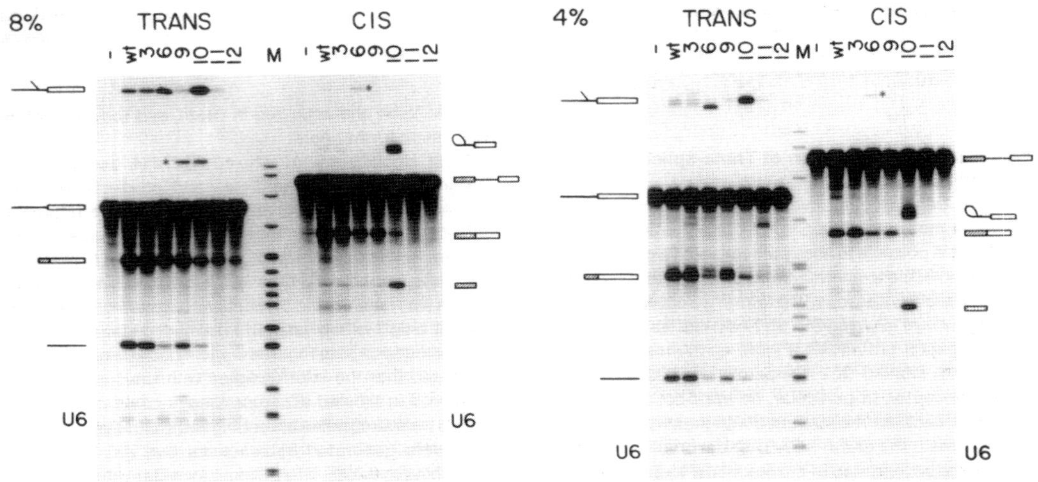

FIGURE 6-1. An example of a splicing reaction analyzed on high (7%) and low (4%) acrylamide gels. Note the striking difference in the mobility of spliced intermediates on the two gels, particularly in the CIS lanes 10. (Reprinted, with permission, from Yu et al. 1993 [© Elsevier]).

iii. Load the samples and run the gel at 1500 V/45 mA until the desired separation is reached for the expected products and intermediates from the reaction.

iv. Dry the gel and expose on a phosphorimager.

Splicing intermediates such as lariat exon 2 (product of Step 1) or other branched molecules (such as from *trans*-splicing) will have abnormal mobility on gels. This altered mobility will be more dramatic on higher-percentage gels. It may be necessary to run more than one gel to resolve all of the products and intermediates (Fig. 6-1).

TIPS AND TROUBLESHOOTING

1. If splicing efficiency is low or nonexistent, it may be necessary to prepare a new extract (see introduction to this chapter).
2. If nuclease activity is a problem (i.e., the substrate is degraded), make sure that the substrate is capped. In addition, ligation of unlabeled pCp to the 3′ end of the substrate can stabilize the substrate due to the inhibition of 3′-to-5′ exonucleases.
3. It is important to make sure that the serum is completely removed during washes; use at least two large-volume washes before pooling cells.

SOLUTION AND REAGENT RECIPES

CAUTION: See the Cautions Appendix for appropriate handling of materials marked with <!>.

Splicing Master Mix (4x)

Reagent	Quantity (for 62.2 μL)	Final concentration (1x splicing reaction)
Creatine phosphate (0.5 M)	10 μL	20 mM
ATP (100 mM)	5 μL	2 mM
KCl (3 M) <!>	2.5 μL	30 mM*
$MgCl_2$ (0.5 M) <!>	2.1 μL	4.2 mM
DTT (1 M) <!>	1 μL	4 mM
Tris (1 M, pH 8.0) <!>	1 μL	16 mM
Aprotinin and leupeptin <!>	1 μL	
Polyethyleneglycol (PEG 8000; 20%, w/v)	37.5 μL	3%
H_2O	2.1 μL	

Prepare fresh and keep on ice.

*The final KCl concentration is usually between 50 and 80 mM. The extract also contains KCl, which contributes to the final concentration. Adjust according to the concentration in dialysis buffer and the amount of extract used in the reaction.

Aprotinin and Leupeptin

Reagent	Quantity (for 40 μL)
Aprotinin (10 mg/mL) <!>	1 μL
Leupeptin (2 mg/mL) <!>	1 μL
H_2O	38 μL

Prepare fresh from concentrated stock. Use for adding to 4x splicing master mix.

PEG (20%, w/v)

For 40 mL, weigh out 8 g of PEG in a 50-mL conical tube. Add RNase-free H_2O to the 40-mL mark on the tube. Close the lid tightly and heat briefly at 65°C to dissolve. Store for 4 mo at room temperature.

SDS Extraction Buffer

Reagent	Quantity (for 100 mL)	Final concentration
Tris (1 M, pH 7.5) <!>	2 mL	20 mM
EDTA (0.5 M, pH 8.0) <!>	0.2 mL	1 mM
SDS (10%, w/v) <!>	5 mL	0.5% (w/v)
H_2O	92.8 mL	

Protocol 3

Preparation of *Drosophila* Kc Cell Nuclear Extracts for In Vitro Splicing

This protocol describes how to prepare nuclear extracts from cultured *Drosophila* cells. It uses the Kc cell line, which grows without fetal bovine serum and so is suitable for large-scale (4–32 L) suspension cultures. The procedure is similar to that used to prepare transcription-competent nuclear extracts from HeLa cells (see Protocol 1). These extracts are active for splicing of pre-mRNAs in vitro (see Protocol 4).

MATERIALS

The recipes for the items marked with <R> are at the end of the protocol or in the Common Recipes Appendix.

CAUTION: See the Cautions Appendix for appropriate handling of materials marked with <!>.

Reagents

Bradford assay materials
Buffer A <R>
Buffer B <R>
Buffer C <R>
Buffer D <R>
Dithiothreitol (DTT; 0.5 M) <!>
Drosophila Kc cells grown and maintained in 1x D22 medium
Phenylmethylsulfonyl fluoride (PMSF; 10 mg/mL) <R>
Phosphate-buffered saline (PBS) + $MgCl_2$ (1 g/L) <!> <R>

Equipment

Centrifuges
- Beckman fixed-angle 45Ti rotor (for S-100)
- Low-speed centrifuge for 1-L bottles (Sorvall RC3)
- Superspeed centrifuge for 50-mL polycarbonate tubes (Sorvall RC5C)
- Tabletop refrigerated centrifuge for 50-mL Corning tubes (Beckman Accuspin GS-6R)
- SS-34 fixed-angle rotor (Sorvall)

Conical tubes (50 mL; one for each 4 L of cells) (Corning)
Dialysis tubing (10,000–12,000 MW cutoff)
Dounce homogenizers
- Small 15-mL (Kontes) with type-A (loose) pestle, large 40 mL (Kontes) with type-B (tight) pestle

Eppendorf tubes (50 0.5-mL tubes for 100-μL aliquots)
Flasks (25 mL and 125 mL for 4–8-L prep or 500 mL for 16-L prep) + stir bar
Graduated cylinders (50 and 100 for 4–8-L prep, and 500 mL for 16-L prep)
Liquid nitrogen <!>
Microscope
Pipettes (two, wide bore; 10 mL)
Polycarbonate centrifuge tubes, round bottomed (three or eight for 16-L prep)
Polycarbonate screw-cap 50-mL Oak Ridge tubes
Teflon paddle flasks (Kontes Cytostir 882910-3000) for cell culture

OVERVIEW

These nuclear extracts are made according to the original protocol of Dignam et al. (1983) with some modifications. The suspension cells are harvested by centrifugation and washed with PBS + $MgCl_2$ buffer to remove media components. The cells are resuspended in a low-salt (hypotonic) buffer that makes the cells swell in size and easy to homogenize. Usually, nuclear extract preparations are performed on 4–32 L of cells at ~3.5×10^6 cells/mL. The preparation described below is for 4 L. The state of the cells is crucial. All steps are performed on ice in the cold room.

METHOD

1. Prepare and chill all equipment in the cold room 1 day ahead of the procedure.
2. Grow and maintain the cells in 1x D22 medium in 3-L Teflon paddle flasks. Grow them to confluency and then split them 1:5 (or less).

   The cells should reach confluency again in 3 days.

   If not using a shaking incubator, use Corning T-flasks. It is important to split the cells very carefully when they are growing in Corning T-flasks; specifically, they should be maintained at a fairly high density.
3. Harvest cells in two 1-L bottles with repeated spins in the Sorvall RC3 swinging bucket centrifuge at 3000 rpm (~2500*g*) for 10 min at 4°C.
4. Resuspend the cell pellets gently with a wide-bore (10-mL) pipette in a total of 30–40 mL (slightly less) of cold PBS containing $MgCl_2$ (1 g/L). Keep the 1-L bottles on ice and use a cold (4°C) pipette.
5. Transfer the cells to a 50-mL Corning conical tube that has been precooled on ice.
6. Pellet the cells in 50-mL Corning tubes in a Beckman Accuspin tabletop centrifuge or equivalent at 2300 rpm (1200*g*) for 10 min at 4°C.
7. Estimate the volume of packed cells (packed cell volume = PCV). Use a 10-mL wide-bore pipette to resuspend the pelleted cells gently in 5 PCV of buffer A. Then, incubate the resuspended cells for 10 min on ice.
8. Pellet the cells again in the Beckman Accuspin tabletop centrifuge or equivalent at 2300 rpm (1200*g*) for 10 min at 4°C.
9. Resuspend the pelleted cells in 2 PCV of buffer A with a 10-mL wide-bore pipette and transfer to a Dounce homogenizer on ice.
10. Break the cells by homogenizing in the Dounce homogenizer with a type-B pestle (tight) with 10–15 strokes (a firm downstroke and a slow, gentle upstroke). Check breakage with a microscope by spotting a drop of cell suspension on a microscope slide before and after douncing.

    After 10–15 strokes of the pestle, expect ~90% cell breakage. For this step, the PCV equals the PCV observed after Step 7, i.e., before the cells have swollen in buffer A.
11. Transfer the cell homogenate by pouring from the Dounce homogenizer into 50-mL round-bottomed, polycarbonate, clear, screw-capped Oak Ridge tubes.
12. Centrifuge the tubes in the Beckman Accuspin tabletop centrifuge or equivalent at 2300 rpm (1200*g*) for 10 min at 4°C to gently pellet the nuclei. Proceed to Step 15 with the nuclei.
13. Carefully pipette off most of the supernatant, being careful not to pipette off the nuclei, which are only loosely packed at this point.

    The low-speed centrifugation in Step 11 removes lysozomes, which contain nuclease and protease activities.

14. For the preparation of cytoplasmic S-100 fraction,
    i. Transfer the cytoplasmic supernatant fraction above from Step 12 to a 45Ti polycarbonate centrifuge tube.
    ii. Add 0.11 volume of buffer B to the cytoplasmic supernatant.
    iii. Centrifuge at 35,500 rpm in the Beckman 45Ti rotor for 1 h at 4°C.
    iv. Remove the floating lipid layer at the surface, if possible.
    v. Dialyze 5–8 h against 2x 50 volumes of buffer D with three changes of buffer.
    vi. Aliquot and freeze at –80°C.
15. Proceed by packing the nuclei by centrifugation in the Sorvall SS-34 rotor at 14,500 rpm (12,000*g*) for 20 min at 4°C.
16. Resuspend the nuclei in ~2.5 mL of buffer C.
17. Dislodge the nuclear pellet with a cold stainless steel spatula and transfer it with a 10-mL wide-bore pipette to a 15-mL Dounce homogenizer. Use the Dounce homogenizer to resuspend the nuclei with the type-A pestle (loose) using 5–10 strokes until the nuclei are fully resuspended and there are no clumps from the original pellet.
18. Transfer the nuclei with a plastic pipette to a 50-mL round-bottomed screw-cap Oak Ridge polycarbonate tube. Rotate the tube gently on a wheel for 45 min in the cold room.
19. Centrifuge the nuclei in a Sorvall SS-34 rotor at 15,000 rpm (1200*g*) for 30 min at 4°C.
20. Carefully remove the supernatant and transfer it to large-size 10,000–12,000 MW cutoff dialysis tubing.
21. Dialyze the nuclear extract for ~5 h versus 500 mL of buffer D (at least 50 volumes), with two changes to fresh buffer D after 1 h and another change after another 2 h. Do not let dialysis proceed overnight, because the activity will be markedly reduced. Do not be alarmed if a visible white precipitate forms.
22. Use a plastic pipette to transfer the liquid containing the nuclear extract to a fresh 50-mL round-bottomed screw-cap Oak Ridge polycarbonate tube.
23. Centrifuge the sample in the Sorvall SS-34 rotor at 12,000 rpm for 10 min at 4°C to remove any precipitate.
24. Prepare 50- to 100-μL aliquots in 0.5-mL Eppendorf tubes on ice and then snap-freeze in liquid nitrogen. Transfer to a –80°C freezer.

TIPS AND TROUBLESHOOTING

1. It is critical that the cells be growing well to obtain active splicing extracts. It is imperative to count the cells daily and keep them in the 2×10^6 to 6×10^6 density range.
2. Aeration of the cultures is critical for exponential growth. Typically, 3-L Teflon paddle flasks (e.g., Kontes cytostir) are used with 2 L of medium to ensure good growth. If the cells stop dividing or start clumping, discard them and start again with T-flasks and cells placed freshly into suspension.
3. It is necessary to monitor the cell lysis, because often the cells will begin lysing when placed in hypotonic buffer A. If this occurs, reduce the number of Dounce homogenizer strokes because the nuclei will be broken by extensive homogenization.

SOLUTION AND REAGENT RECIPES

CAUTION: See the Cautions Appendix for appropriate handling of materials marked with <!>.

We make all of our buffers without DTT <!> and PMSF <!>. We store the buffers in the cold room and then add DTT (0.5 mM final) and PMSF (0.5 mM final) immediately before use. PMSF is the only protease inhibitor we include. All buffers can be stored for up to 6 mo at 4°C.

Buffer A

Reagent	Quantity (for 250 L)	Quantity (for 2 L)	Final concentration
HEPES-KOH (1 M, pH 7.6)	2.5 mL	20 mL	10 mM
$MgCl_2$ (1 M) <!>	375 µL	3 mL	1.5 mM
KCl (3 M) <!>	833 µL	6.664 mL	10 mM
DTT (0.5 M) <!>	250 µL	2 mL	0.5 mM

Buffer B

Reagents	Quantity (for 250 mL)	Final concentration
HEPES-KOH (1 M, pH 7.6) <!>	75 mL	0.3 M
KCl (3 M) <!>	117 mL	1.4 M
$MgCl_2$ (1 M) <!>	7.5 mL	0.03 M
ddH_2O	to 250 mL	

Buffer C

Reagents	Quantity (for 100 mL)	Final concentration
HEPES-KOH (1 M, pH 7.6) <!>	2 mL	20 mM
Glycerol	25 mL	25%
NaCl (5 M)	8.4 mL	0.42 M
$MgCl_2$ (1 M) <!>	150 µL	1.5 mM
EDTA (0.25 M) <!>	80 µL	0.2 mM
DTT (0.5 M) <!>	100 µL	0.5 mM
PMSF (10 mg/mL) <!>	1 mL	0.5 mM
ddH_2O	to 100 mL	

Buffer D

Reagent	Quantity (for 1 L)	Quantity (for 4 L)	Final concentration
HEPES-KOH (1 mM, pH 7.6) <!>	20 mL	80 mL	20 mM
Glycerol (100%)	200 mL	800 mL	20%
KCl (3 M) <!>	33.3 mL	133.2 mL	0.1 M
EDTA (0.25 M) <!>	800 µL	3.2 mL	0.2 mM
DTT (0.5 M) <!>	1 mL	4 mL	0.5 mM
PMSF (10 mg/mL) <!>	10 mL	40 mL	0.5 mM
ddH_2O	to 1 mL	to 4 L	

Put into wide-mouth Erlenmeyer flasks.

*PMSF (10 mg/mL in 95% Ethanol)**

Reagent	Quantity (for a 4–8-L preparation)	Quantity (for a 16–32-L preparation)
PBS + $MgCl_2$ (1 g/L) <!>	200 mL	500 mL
Buffer A	75 mL	1 L
Buffer C	5 mL	50 mL
Buffer D (in 3x 500-mL flasks + stir bars)	1.5 L	3 L
DTT (0.5 M) <!>	1.6 mL	10 mL
PMSF* (10 mg/mL) <!>	16 mL	100 mL
ddH_2O	to 2 L	to 4 L

*Add slowly with stirring to each buffer right before use.

Protocol 4

In Vitro Splicing Reactions in *Drosophila* Kc Nuclear Extracts

This protocol describes how to generate and analyze the products and intermediates in the pre-mRNA splicing reaction using ^{32}P-labeled, GpppG-capped, synthetic pre-mRNAs prepared by in vitro transcription and *Drosophila* cell culture nuclear extracts. The pre-mRNA is generally less than 500 nucleotides long and contains a single intron. Visualization of splicing reactions will reveal the pre-mRNA, the spliced mRNA, and the intermediates generated by the first step of splicing.

MATERIALS

The recipes for the items marked with <R> are at the end of the protocol or in the Common Recipes Appendix.

CAUTION: See the Cautions Appendix for appropriate handling of materials marked with <!>.

Reagents

ATP (100 mM)
Chloroform:isoamyl alcohol (24:1 = SEVAG) <!>
Creatine phosphate (0.2 M), sodium salt, dissolved in sterile ddH$_2$O (CalBiochem 2380)
Denaturing polyacrylamide gel reagents
Ethanol (100%, 70%, –20°C) <!>
HEPES-KOH (1 M, pH 7.6) <!>
Kc cell *Drosophila* nuclear extract (10–12 mg/mL total protein) (see Protocol 3)
Kc splicing mix (4x) <R>
MgCl$_2$ (150 mM) <!>
Phenol <!>
PK buffer <R> (optional, see Step 3)
Polyvinyl alcohol (10%) dissolved in sterile ddH$_2$O (Sigma-Aldrich P-8136)
Proteinase K (10 mg/mL; Roche) in ddH$_2$O <!>
    Store frozen at –20°C.
RNasin (40 units/µL; Promega)
Stop buffer <R>
Urea/dyes <R> <!>

Equipment

Denaturing polyacrylamide gel equipment
Dry ice/ethanol bath <!>
Geiger counter, handheld monitor
Microcentrifuge
Phosphorimaging screen or intensifying screen
Shaker (Eppendorf 5432)
SpeedVac centrifugal concentrator (Savant)
Water bath or temperature block (20°C)
X-ray film

OVERVIEW

In this protocol, pre-mRNA substrate is incubated in nuclear extract under splicing conditions for 1–2 h. The products of the reaction are purified by phenol:chloroform extraction and precipitation with ethanol and then loaded directly onto a denaturing urea–acrylamide gel. For inefficient reactions, a more sensitive detection method, such as RNase protection, primer extension, or RT-PCR (reverse transcriptase–polymerase chain reaction), may be required.

METHOD

1. For each splicing reaction, mix the following in a 1.5-mL microcentrifuge tube:

Kc nuclear extract (10–12 µg protein/µL) (3–12 µL is usually optimal)	3–10 µL
Pre-mRNA substrate (50,000–150,000 cpm at ~10^6 cpm/pmol)	2 µL
Kc splice mix	6.25 µL
H_2O	to 25 µL in final reaction
10% polyvinyl alcohol (2.5% final)	6.25 µL

2. Incubate the reactions at 20°C.

   Splicing intermediates will appear after 20–30 min.

   We usually incubate for 45–120 min. It is best to do a time course to determine the point at which maximum splicing is observed.

3. Stop the reactions by adding 25 µL of 2x PK buffer.
   i. Add 5 µL of 10 mg/mL proteinase K (Roche) and digest for 30 min at room temperature.
   ii. Add 150 µL of stop buffer.
4. Extract the RNA with 200 µL of phenol-SEVAG (1:1, v/v). Centrifuge at maximum speed in a microcentrifuge for at least 5 min at room temperature to separate the phases.
5. Remove 170 µL of the aqueous phase, which contains the RNA, taking care not to disturb the interface.
6. Precipitate the RNA with 440 µL of cold 100% ethanol (–20°C) for at least 10 min in a dry ice/ethanol bath or on dry ice.
7. Pellet the RNA by centrifuging in a microcentrifuge at maximum speed for at least 10 min.
8. Wash the RNA pellet with 500 µL of cold 70% ethanol (–20°C).
9. Centrifuge in a microcentrifuge at maximum speed for 5 min at room temperature.
10. Remove the ethanol, taking care not to disturb the pellet.
11. Dry the pellet briefly in a SpeedVac and then resuspend it in 12 µL of urea/dyes for at least 5 min with shaking.
12. Load 4 µL per well on a denaturing acrylamide gel. Run at 25 W for 1–2 h.

    Loading more than one-third of the RNA per lane may cause the sample to smear.

13. After electrophoresis, dry the gel and expose X-ray film at –80°C with an intensifying screen or a phosphorimaging screen at room temperature.

    A 4–12-h exposure is usually sufficient.

TIPS AND TROUBLESHOOTING

1. The optimal concentration of $MgCl_2$ usually varies between 1 mM and 5 mM greater than the ATP concentration for most introns. Therefore, it is useful to titrate Mg^{++} for each specific substrate.
2. Negative control reactions lacking ATP and creatine phosphate should be performed. Additional controls without extract may also be desired.
3. The best way to visualize the intermediates and products is to use a gel of a high enough percentage of acrylamide such that the branched species, due to their unusual structure, migrate above the pre-mRNA. Thus, the ATP-dependent appearance of distinct bands migrating above the pre-mRNA, as well as bands running at the expected positions for the linear mRNA and 5′ exon, is a simple and believable reflection of an accurate splicing event.

SOLUTION AND REAGENT RECIPES

CAUTION: See the Cautions Appendix for appropriate handling of materials marked with <!>.

Stop Buffer

Reagent	Quantity (for 100 mL)	Final concentration
NaCl (5 M)	4 mL	0.2 M
EDTA (0.25 M, pH 8.0) <!>	8 mL	20 mM
SDS (10%, w/v) <!>	10 mL	1% (w/v)
ddH_2O	to 100 mL	

Store indefinitely at room temperature.

Kc Splicing Mix (4x)

Reagent	Quantity (for 62.5 µL)	Final concentration in splicing reactions
ATP (100 mM)	7.5 µL	3 mM
Creatine phosphate (0.2 M)	6.25 µL	5 mM
$MgCl_2$ (0.5 M) <!>	2 µL	4 mM
HEPES-KOH (1 M, pH 7.6) <!>	5 µL	20 mM
RNasin (40 units/µL)	2.5 µL	0.4 unit/µL
H_2O	39.25 µL	

PK Buffer (2x)

Reagent	Quantity (for 100 mL)	Final concentration
Tris-HCl (1 M, pH 7.5)	20 mL	0.2 M
EDTA (0.5 M, pH 8.0) <!>	10 mL	25 mM
NaCl (5 M)	6 mL	0.3 M
SDS (10%, w/v) <!>	20 mL	2%
H_2O	to 100 mL	

Store indefinitely at room temperature.

Urea/Dye Loading Buffer

Reagent	Quantity (for 10 mL)	Final concentration
Urea <!>	4.8 g	8 M
Tris-HCl (1 M, pH 8.0)	200 µL	20 mM
EDTA (0.25 M, pH 8.0) <!>	40 µL	1 mM
Bromophenol blue (0.5%) <!>	1 mL	0.05%
Xylene cyanol (0.5%) <!>	1 mL	0.05%
H_2O	to 10 mL	

Protocol 5

Preparation and Analysis of Cell-free Splicing Extracts from *Saccharomyces cerevisiae*

Much of our understanding of the mechanism of splicing comes from analysis of cell extracts able to carry out splicing complex formation and splicing reactions in vitro using exogenously added synthetic model pre-mRNA transcripts. This protocol describes the preparation and testing of whole-cell extracts of the budding yeast *Saccharomyces cerevisiae*, first prepared by Lin et al. (1985). Related protocols are given for carrying out splicing reactions and for analyzing yeast-splicing complexes on native gels, using modifications of a method from Seraphin and Rosbash (1989). Extracts prepared in this way allow the dissection of biochemical steps in the splicing reaction and determination of the macromolecules involved, cofactor requirements, and substrate features necessary for successful splicing.

MATERIALS

The recipes for the items marked with <R> are at the end of the protocol or in the Common Recipes Appendix.

CAUTION: See the Cautions Appendix for appropriate handling of materials marked with <!>.

Reagents

Buffer A <R>
Buffer D <R>
Dithiothreitol (DTT; 30 mM, 3 mM) <!>
H_2O (sterile)
KCl (2 M) <!>
SDS (10%) <!>
Sorbitol buffer <R>
Yeast medium (yeast extract/peptone/dextrose [YEPD]) <R>
Yeast strain starter culture
Zymolase (20 mg/mL) <!>
Zymolase buffer <R>

Equipment

Beaker with stir bar
Dialysis tubing (10,000 MW cutoff)
Dounce homogenizer (metal) with tight-fitting pestle
Fernbach flasks (2.5 L)
Glass rod
Ice
Liquid nitrogen <!>
Microcentrifuge
Phase-contrast microscope
Pipette
Polypropylene tubes (screw capped, 50 µL)
Spectrophotometer

SS-34 rotor, centrifuge tubes, and centrifuge or equivalent
Stir plate
Ultracentrifuge, 60Ti rotor, Oak Ridge centrifuge tubes, and/or equivalent

OVERVIEW

It is best to set aside a whole day for preparation of splicing extracts. At least 1 L of yeast culture in late log phase is needed to start, although larger volumes of culture can be extracted. There can be extract-to-extract variation; thus, any scientific conclusions to be drawn should be supported by reproducible results using multiple preparations of extract. Advanced preparation is key. Rotors, beakers, pipettes, and buffers should be at 4°C and spectrophotometers and centrifuges reserved. Time is of the essence, but rushing produces errors. One liter should produce ~3 mL of extract at ~25 mg/mL protein, an amount sufficient for several hundred 10-μL splicing reactions.

METHOD

1. Use a fresh starter culture of the desired yeast strain to seed a desired volume of YEPD medium in Fernbach flasks (1 L of medium per 2.5-L Fernbach flask) at 30°C so that the culture has an A_{600} of 3–5 to start (one A_{600} unit usually represents 2×10^7 yeast cells).

   This may require having measured the doubling time of the strain and backcalculating by the number of doubling times to the anticipated start time. For example, if the strain doubles every 90 min and the 1-L culture is seeded at 5 p.m., anticipating a 9 a.m. start the next morning, this is 16 h or 960 min, meaning ~10 doublings. Because 2^{10} is ~10^3, start at $A_{600} = 0.005$. Once the A_{600} has reached 3, it is worth beginning the preparation.

2. Centrifuge the cells at 3000 rpm in an appropriate rotor (e.g., GSA or GS-3) for 5 min at room temperature.
3. Resuspend the pellets in a total of 250 mL of sterile H_2O and centrifuge as in Step 2 again in one bottle.
4. Completely resuspend the cells in 15 mL of Sorbitol buffer + 30 mM DTT and incubate for 15 min at 23°C without shaking.
5. Pellet the cells at 3000 rpm for 5 min at 23°C and then resuspend them in 7.5 mL of Sorbitol buffer + 3 mM DTT. Remove 50 μL to ice as a prelysis comparison sample.
6. Add 100 μL of 20 mg/mL Zymolase (dissolved in Zymolase buffer) to the cells and gently shake (50–60 rpm) for 35 min at 30°C.
7. After 35 min, dilute 10 μL of cells in 1 mL of 10% SDS and read the A_{800}. Do the same for 10 μL of the prelysis sample.

   The A_{800} of the Zymolyase-treated cells should be <20% that of the untreated cells. If the treated cells are not completely spheroplasted, add another 50 μL of Zymolase and continue incubating for another 10 min at 30°C.

8. Chill everything and keep everything very cold from here on.
9. Centrifuge the Zymolyase-treated cells at 3000 rpm for 5 min at 4°C.
10. Gently resuspend the cells in 15 mL of Sorbitol buffer + 3 mM DTT. The cells will be sticky, so use a pipette or glass rod to resuspend them.
11. Centrifuge once more at 3000 rpm for 5 min at 4°C.
12. Resuspend the cells in 8 mL of buffer A.
13. Transfer the cells to a prechilled metal Dounce homogenizer with a tight-fitting pestle and break the cells with 8–10 hard strokes. Reserve 50 μL for a pre-Dounce comparison sample.

Check cells for breakage under a phase microscope, comparing them with the un-Dounced cells. More than 80% of the cells should be broken.

14. Transfer the homogenate to a prechilled beaker with a stir bar. Use a cold pipette to measure the volume of the transferred homogenate as accurately as possible. Expect perhaps 12 mL of homogenate from 1 L of cells.
15. Stir the homogenate on ice over a stir plate for 5 min in the cold room. Then, add cold 2 M KCl dropwise during a ~5-min period to the homogenate until the KCl concentration is 0.2 M (i.e., add one-ninth volume of 2 M KCl).
16. Stir gently for 30 min on ice.
17. Centrifuge the extract in an SS-34 or equivalent rotor at 17,000 rpm for 30 min at 4°C in prechilled 50-mL screw-capped Oak Ridge polycarbonate tubes.
18. Pipette the supernatant, which contains the extract, into prechilled Oak Ridge tubes of the correct size for the 60Ti rotor, avoiding the pellet and the oily material floating on the top.
19. Centrifuge the supernatant in the 60Ti rotor at 37,000 rpm for 60 min at 4°C.

    This will separate the extract into several visible phases. An opalescent ribosome pellet will be at the very bottom of the tube; lipoprotein scum will be at the top. The liquid near the bottom will be clear, but in the middle of the tube, it will be a pale yellow phase, which contains the splicing extract.

20. Avoiding the lipoprotein scum at the top and the bottom clear phase, remove as much as one-third or so (~3 mL total) of the pale yellow liquid from the middle of the tube.
21. Dialyze the splicing extract twice for 1.5 h each time against 1 L of buffer D (2 L total, change after 1.5 h).

    The volume will reduce during this time to ~60% of original extract and a precipitate may appear.

22. Remove the precipitate by centrifugation in an SS-34 rotor at 10,000*g* for 15 min at 4°C.
23. Aliquot extract into convenient volumes and freeze in liquid N_2. Store at –70°C.

TIPS AND TROUBLESHOOTING

1. Be sure to keep everything cold.
2. Make sure the cells lyse well.
3. Do not be greedy when attempting to recover the pale yellow phase.

DISCUSSION

Extracts should contain ~25 mg/mL protein and may vary in their activity, but they should be able to convert a substantial fraction of 10 fmol of pre-mRNA to splicing intermediates and products in a 5–10-µL reaction composed of 40% crude extract in ~20 min. This rate is pathetic compared to the rate of splicing in vivo (a single yeast cell splices ~150 transcripts per second). In a good extract, something more than 50% of the substrate molecules that react in the first chemical step go on to the second step. Extract-to-extract variation in the efficiency of second step relative to first step is often noticeable and is a sign of variable differential extraction or recovery of second-step factors. Note that unlike a simple single protein enzyme, the spliceosome is a very complex, heterogeneous crew of protein and RNA factors. It is actually amazing that one whole-cell extraction procedure recovers enough of each of these varied activities that the entire process proceeds at all.

SOLUTION AND REAGENT RECIPES

CAUTION: See the Cautions Appendix for appropriate handling of materials marked with <!>.

Buffer A

Reagent	Quantity (for 100 mL)	Final concentration
KCl (3 M) <!>	0.33 mL	10 mM
HEPES (1 M, pH 7.9 at 25°C)	1 mL	10 mM, pH 7.0 at 4°C
$MgCl_2$ (1 M) <!>	0.15 mL	1.5 mM
DTT (1 M) <!>	0.05 mL	0.5 mM
PMSF (50 mM) <!>	1 mL	0.5 mM
Leupeptin (1 mg/mL) <!>	0.1 mL	1 µg/mL
Pepstatin (1 mg/mL)	0.1 mL	1 µg/mL
H_2O	to 100 mL	

Add protease inhibitors just before use. Store at 4°C. Do not keep buffer to which protease inhibitors have been added for longer than 1 wk.

Buffer D

Reagent	Quantity (for 2 L)	Final concentration
HEPES (1 M, pH 7.9 at 25°C)	20 mL	20 mM
KCl (3 M) <!>	33.3 mL	50 mM
EDTA (0.5 M, pH 8.0) <!>	0.8 mL	0.2 mM
Glycerol (100%, v/v)	400 mL	20% (v/v)
H_2O	to 2 L	

Store at 4°C.

Sorbitol Buffer

Reagent	Quantity (for 100 mL)	Final concentration
Sorbitol	18.2 g	1 M
Tris-HCl (1 M, pH 7.8)	5 mL	50 mM
$MgCl_2$ (1 M) <!>	1 mL	10 mM
H_2O	to 100 mL	

Sterile-filter and store at 4°C.

Yeast Medium

Yeast medium (1 L)	Final concentration
Yeast extract	10 g
Bacto-peptone	20 g
D-glucose	20 g

Autoclave and store at room temperature.

Zymolase Buffer

Reagent	Quantity (for 100 mL)	Final concentration
Potassium phosphate (1 M, pH 7.2)	2 mL	20 mM
Glucose	5 g	5% (w/v)
H_2O	to 100 mL	

Protocol 6

Analysis of Splicing In Vitro Using Extracts of *Saccharomyces cerevisiae*

In vitro splicing studies are a powerful means of investigating the requirements and mechanisms of action of the many components of the splicing apparatus. The ability to add and subtract components, purify activities, and reconstitute activity, as well as to expose the apparatus to chemical probes of various types, allows a far more mechanistically detailed view of the process to emerge than is available from genetic or in vivo studies alone. Two kinds of activities are assayed during in vitro splicing. The first concerns the chemical conversion of the substrate pre-mRNA into splicing intermediates and products and is usually visualized using a labeled substrate followed by separation on a denaturing gel. The second concerns the assembly of noncovalent complexes between the substrate and the myriad components of the splicing apparatus. This is also visualized using a labeled substrate, but the separation of complexes is achieved using native gel electrophoresis or gradient sedimentation. In this protocol, we describe the splicing reaction and its preparation for analysis by denaturing gels (see Chapter 2, Protocol 12) and native splicing complex gels (see Protocol 7).

MATERIALS

The recipes for the items marked with <R> are at the end of the protocol or in the Common Recipes Appendix.

CAUTION: See the Cautions Appendix for appropriate handling of materials marked with <!>.

Reagents

Acid phenol <!>
Acrylamide (6%), 7 M denaturing gel reagents
Denaturing stop buffer <R>
D-glucose (1 mM)
Ethanol <!>
Formamide dye <R>
KPO_4 (0.6 M, pH 7.0)
$MgCl_2$ (125 mM) <!>
Native gel-loading buffer <R>
Native quench buffer <R>
^{32}P (10,000–20,000 cpm)
Polyethylene glycol, PEG 8000 (50%, w/v; use highest quality)
Radiolabeled pre-mRNA substrate (~40 nM)
  Prepare as described in Chapter 4, Protocol 1.
rATP (20 mM)
Spermidine (100 mM) <!>
Splicing buffer <R>
Yeast hexokinase (150–300 units/mg, dissolve at 10 units/μL) (Sigma-Aldrich H-4502)
Yeast-splicing extract
  Prepare as described in Protocol 5.

Equipment

Autoradiography or phosphorimaging equipment
Ice
Tubes
Water bath (~23°C)

OVERVIEW

Setting up and running splicing assays require organization and attention. If the samples are to be evaluated on denaturing gels (Chapter 2, Protocol 12), they can be extracted, processed, and stored until the gel can be run. However, if splicing complexes are to be evaluated, the gel must be poured, cured, mounted, and prerun before the splicing reactions are initiated. In essence, a reaction composed of 40% splicing extract, splicing buffer including $MgCl_2$, spermidine, PEG, and ATP is incubated with 4 nM labeled substrate (10,000–20,000 cpm, ^{32}P) for usually <30 min at 15°C–30°C and then stopped and extracted (denaturing gel analysis) or quenched and immediately run on a gel (native complex gel analysis). We also provide conditions for depletion of ATP, a critical cofactor that is hydrolyzed during numerous key steps in spliceosome assembly and splicing progression.

METHOD

Splicing Assay

1. Prepare fresh 5x splicing buffer and keep on ice.
2. Set up splicing reactions on ice. The design of the experiment may dictate whether and how a master mix can be made up. Each 10-µL reaction will contain the following reagents assembled on ice:

Splicing buffer (5x)(freshly prepared)	2 µL
rATP (20 mM)	1 µL
Yeast pre-mRNA substrate (40 nM)	1 µL
dH_2O	2 µL
Yeast-splicing extract	4 µL

   For reactions in which rATP is to be left out, see below. If a time course is being run, it is preferable to make a large reaction and remove aliquots rather than make several separate reactions.

3. At time zero, place the reaction tube in a water bath of the desired temperature, usually 23°C. The reaction is finished in 20 min.

   Short incubation times at near-ambient temperatures are best accomplished using an ice bucket with tap water of the desired temperature. The large volume of water in the ice bucket will not change the temperature significantly in 20 min.

4. Below are two ways to stop and process the reaction:
   - If the reaction is to be analyzed only for substrate reaction
     i. Add 200 µL of denaturing stop buffer, mix, and incubate for 10 min at 65°C.
     ii. Then, add 200 µL of acid phenol, extract, and precipitate with 2.5 volumes ethanol.
     iii. Resuspend in formamide dye and run on a 6% acrylamide, 7 M denaturing gel. Autoradiograph or collect data with a phosphorimager.
   - If the reaction is to be analyzed for splicing complex formation
     i. Add 10 µL of Native quench buffer, mix gently, and incubate for 10 min on ice.
     ii. Then, add 5 µL of Native gel-loading buffer (see Protocol 7) and load on a native gel.

ATP Depletion

Pre-mRNA splicing is an ATP-requiring reaction. Removal of ATP, as described here, at specific stages of the splicing reaction will arrest spliceosome assembly at specific points.

Splicing extracts can contain variable amounts of residual ATP, which can confound the analysis of ATP-dependent steps in the spliceosome assembly pathway. This modification allows the depletion of residual ATP, resulting in a truly ATP-minus extract. ATP can then be added when desired. The method relies on the enzyme hexokinase, which uses ATP to phosphorylate glucose. Because yeast-splicing extracts can have widely varying hexokinase activity, we recommend buying it from Sigma-Aldrich and adding it.

1. Premix the splicing buffer and the splicing extract and add (per 6 μL of extract plus buffer) 1 μL of 1 mM D-glucose and 1 μL of 10 units/μL yeast hexokinase.
2. Incubate the mixture for 15 min at 23°C and then place on ice.
3. Add 1 μL of 40 nM pre-mRNA substrate and 1 μL of either H_2O (–ATP) or 20 mM rATP (+ATP).
4. Incubate at 23°C to initiate splicing.
5. Stop the reactions as described in Step 4 of the splicing assay for either denaturing or native gel analysis.

TIPS AND TROUBLESHOOTING

1. Take care to be accurate when measuring viscous solutions such as PEG and glycerol.
2. Do not overload the splicing reaction with substrate because the extract has a limited ability to carry out splicing, and excess substrate will obscure the products of the reaction.
3. Remember that it is the physical concentration of the substrate that matters and not its radioactive content.
4. Make splicing buffer fresh to avoid precipitation of KPO_4 upon storage.

DISCUSSION

A typical splicing reaction will be finished in ~20 min. A fraction of the substrate will have been converted to intermediates (the free 5′ exon and the lariat intron-3′ exon or "lariat intermediate"), and a fraction of the material converted to intermediates will have been converted to products (the spliced exons and the lariat intron or "lariat product"). The two lariats will migrate on denaturing gels more slowly than expected from their length because the loop of the lariat retards migration. In some extracts and upon longer incubation, other reactions can occur. For example, if spliceosome disassembly is efficient, a "debranched" lariat representing the linearized intron may appear. A commonly used actin pre-mRNA substrate produces spliced exons that become fuzzy and move up the gel due to polyadenylation of the spliced RNA in the extract. Far more substrate will assemble into early-stage splicing complexes than will undergo the chemical reaction.

SOLUTION AND REAGENT RECIPES

CAUTION: See the Cautions Appendix for appropriate handling of materials marked with <!>.

Denaturing Stop Buffer

Reagent	Quantity (for 10 mL)	Final concentration
NaOAc (3 M)	1 mL	0.3 M
SDS (10%) <!>	0.2 mL	0.2%
EDTA (0.5 mM) <!>	20 μL	1 mM
Proteinase K (10 mg/mL) <!>	10 μL	10 μg/mL
H_2O	to 10 mL	

Native Quench Buffer

Reagent	Quantity (for 10 mL)	Final concentration
HEPES-KOH (1 M, pH 7.4) <!>	0.5 mL	50 mM
MgOAc (1 M)	20 μL	2 mM
Total yeast or other RNA 10 mg/mL)	20 μL	2 μg/mL
H_2O	to 10 mL	

Splicing Buffer (5x)

Reagent	Quantity (for 100 μL)	Final concentration
PEG 8000 (50%)	30 μL	15%
KPO_4 (0.6 M, pH 7.0)	50 μL	36 mM
Spermidine (100 mM) <!>	10 μL	10 mM
$MgCl_2$ (125 mM) <!>	10 μL	12.5 mM

Protocol 7

Analysis of Splicing Complexes on Native Gels

Splicing requires a complex set of ATP-dependent macromolecular associations that lead to the rearrangement of just a few covalent bonds in the pre-mRNA substrate. Seeing only the covalent bonds breaking and forming is seeing only a very small part of the process. Analysis of native splicing complexes into which the radiolabeled substrate has been assembled, but not necessarily completely reacted, has provided a good understanding of the process. This protocol describes a gel method originated by Seraphin and Rosbash (1989) for detecting and analyzing splicing complexes formed in vitro on radiolabeled pre-mRNA substrate. Complexes formed in the splicing reaction are quenched by dilution and addition of an excess of RNA, which is thought to strip nonspecifically bound proteins from the labeled substrate RNA. After loading on a low-percentage, low-cross-linking ratio composite agarose–acrylamide gel (in 10% glycerol), labeled bands are detected. These can be extracted and shown to contain small nuclear RNAs (snRNAs) and partly reacted pre-mRNA.

MATERIALS

The recipes for the items marked with <R> are at the end of the protocol or in the Common Recipes Appendix.

CAUTION: See the Cautions Appendix for appropriate handling of materials marked with <!>.

Reagents

Agarose
Ammonium persulfate (10%), freshly prepared (1 mL) <!>
Ethanol (70%) <!>
Glycerol (50%)
Native gel-loading buffer <R>
Native gel solutions:
  40% (w/v) monoacrylamide, ultrapure, ready to use (AMRESCO) <!>
  2% (w/v) bis-acrylamide, ultrapure, ready to use (AMRESCO) <!>
Native quench buffer or other RNA <R>
TBE running buffer (0.5x, 1x, 10x) <R>
TEMED (*N*, *N*, *N′*, *N′*-tetramethylethylene diamine) <!>

Equipment

Autoradiography or phosphorimaging equipment
Cold room
Erlenmeyer flask (200 mL)
Gel box for electrophoresis
Gel dryer
Microwave oven
Native gel plates (unsilanized; small plate, 20 x 20 cm)
Native gel spacers and 1.5-mm comb (serrated side spacers recommended)
Power supply
Whatman 3MM paper

METHOD

1. Wash the glass gel plates vigorously with soap and deionized H_2O. Rinse with H_2O and then rinse and wipe dry with 70% ethanol.
2. Assemble the unsilanized plates with spacers and as many tight clamps as possible.

   These gels are fragile, and if gel detaches from the side spacers, it may slide into the lower buffer chamber.
3. Set the plates horizontally on a raised box (an empty tip rack is useful) on a good-sized piece of lab mat.
4. Make 35 mL of a 1% agarose (*not* low melt) solution in 1x TBE. Dissolve the agarose by controlled boiling in a microwave oven and then place at 55°C.
5. Make 35 mL of a 6% acrylamide solution with an 80:1 mono:bis ratio (40% acrylamide:0.5% bis). Place at 55°C.
6. Prewarm a 250-mL Erlenmeyer flask to 55°C. Add the agarose from Step 4 to the acrylamide solution from Step 5 and mix well. Avoid introducing oxygen.
7. Add 125 µL of 10% ammonium persulfate and 100 µL of TEMED to the mixture from Step 6.
8. Mix and pour between the gel plates by tipping the plates to an ~45° angle and placing the mouth of the flask against the larger plate. As the plates fill, move them back to a horizontal position and insert the comb.
9. Allow the gel to polymerize for 2 h. Then, place it in the cold room to cure for 30 min.
10. In the cold room, carefully remove the comb first and then the bottom spacer without disturbing the attachment of the gel to the side spacers.
11. Mount the gel in a vertical gel box, again without detaching the gel from the side spacers or either plate.
12. Fill the gel box with 0.5x TBE running buffer and prerun at 90–100 V for 60 min in the cold room.
13. After performing in vitro splicing (Protocol 6), stop the 10-µL splicing reaction by mixing gently with 10 µL of native quench buffer. Incubate for 10 min on ice.
14. Add 5 µL of native gel-loading buffer to the stopped splicing reaction and load the sample onto the gel.
15. Run the gel at 90–100 V for 16 h in the cold room.
16. Remove the gel to Whatman 3MM paper and dry on a gel dryer.
17. Image by autoradiography or phosphorimaging.

TIPS AND TROUBLESHOOTING

1. It may take some experience to become skilled at handling these gels. Modifications include increasing the cross-linking ratio to 60:1 or increasing the agarose concentration slightly. In both cases, migration may be affected.
2. The RNA used for the quench can be a bit variable. Each preparation must be optimized to determine the amount of RNA that provides the best pattern. Too little RNA results in large and smeary complexes, whereas too much may overstrip the complexes.

3. If northern blotting to detect snRNAs is going to be done with a gel, then the quench RNA should not be from yeast.
4. Larger substrates may require longer electrophoresis times.

DISCUSSION

Splicing complexes as visualized on these and other similar gels are observed reproducibly, but their precise composition is unknown. It seems likely that the complexes observed are not exactly identical in composition to the functional complexes that existed in the reaction before quench, but they are likely derived from them. This is inferred because some complexes that form are dependent on certain components for their assembly but no longer seem to contain them. This is true for the prespliceosome complex in yeast, which requires U1 snRNP for its formation but seems to be absent from the complex observed in native gels. Thus, care must be taken in analysis of these complexes because they may only partly represent the native spliceosome.

SOLUTION AND REAGENT RECIPES

CAUTION: See the Cautions Appendix for appropriate handling of materials marked with <!>.

Native Gel-loading Buffer

Reagent	Quantity (for 10 mL)	Final concentration
TBE (10x)	2.5 mL	2.5x
Glycerol (100%)	5 mL	50%
Bromophenol blue (1%) <!>	1 mL	0.1%
H_2O	1.5 mL	

Store at room temperature.

Native Quench Buffer

Reagent	Quantity (for 10 mL)	Final concentration
HEPES-KOH (1 M, pH 7.4) <!>	0.5 mL	50 mM
MgOAc (100 mM)	0.2 mL	2 mM
Yeast total RNA (1 mg/mL)	20 µL	2 µg/mL
H_2O	9.3 mL	

Store at –20°C.

7 RNA Interference

PROTOCOLS

1 dsRNA-induced RNAi in *Drosophila* Cells by Soaking, 363
2 dsRNA-induced RNAi in *Drosophila* Cells by Transfection, 366
3 dsRNA-induced RNAi in *C. elegans* by Feeding with dsRNA-expressing *E. coli*, 370
4 RNAi in *C. elegans* by Injection of dsRNA, 374
5 RNAi in Cultured Mammalian Cells Using Synthetic siRNAs, 380
6 Construction of Transgenic *Drosophila* Expressing shRNAs in the miR-1 Backbone, 387
7 Creating an miR30-based shRNA Vector, 395
8 Packaging shRNA Retroviruses and Host Cell Infection, 399
9 Infection of Mammalian Cells with Retroviral shRNAs, 403
10 Creating Transgenic shRNA Mice by Recombinase-mediated Cassette Exchange, 407

RNA INTERFERENCE (RNAI) enables sequence-specific, experimentally induced silencing of virtually any gene by tapping into innate regulatory mechanisms that are conserved among virtually all eukaryotes. RNAi was discovered initially in *Caenorhabditis elegans* more than a decade ago by investigators who were interrogating gene function by using antisense RNA to inhibit expression (Fire et al. 1998). Essentially by accident, they uncovered a biological pathway which enabled genetic manipulations that are not only more predictable, but also much more potent. Since that time, the application of RNAi has come to pervade biology, changing the technical landscape in a manner perhaps last seen with the introduction of the polymerase chain reaction (PCR).

The biochemical pathways that comprise the RNAi mechanism emerged more than 3 billion years ago, roughly coinciding with the appearance of eukaryotes (Hannon 2002). Although the components of RNAi are found in every eukaryotic lineage examined to date, the biological functions of RNAi-related pathways are extremely diverse. One common thread that runs through plant, animal, and fungal kingdoms is a role for RNAi in the response to pathogenic and parasitic nucleic acids (Girard and Hannon 2008), providing clues to the ancestral function of the pathway. The biological impacts of RNAi now encompass gene regulation, control of chromosome structure and function, and transmission of epigenetic information from parent to progeny (White and Allshire 2008; Bartel 2009; Malone and Hannon 2009).

As a tool for manipulating gene expression, RNAi has been deployed in an astounding array of experimental systems, including both the commonplace and the exotic. It is impossible to gather and present the breadth of experimental protocols applicable to all of these models here. Instead, this chapter provides the basic information needed to carry out RNAi experiments in three commonly used experimental organisms: mammals, *C. elegans*, and *Drosophila*. Both in vivo and cell culture (where applicable) approaches are described. Although the precise protocols presented and experiments described are, of necessity, limited in scope, the basic principles provided in this chapter can be extended for use in virtually any organism where methodologies for gene delivery have been developed.

We thank Chris Hammel (CSHL) and Rene Ketting (Hubrecht Laboratories) for help in preparation of protocols for *C. elegans* RNAi. Rui Zhou, Jian Quan Ni, Norbert Perrimon, the *Drosophila* RNAi Screening Center (DRSC), and the Transgenic RNAi Project (TRiP, Harvard) provided many protocols for RNAi in *Drosophila*. Prem Premsrirut, Young Park, and Scott Lowe (CSHL) developed and provided the RMCE protocol for shRNA mice. Amy Valentine, Ken Chang, and Krista Marran provided protocols for RNAi in mammalian cells. Jim Duffy (CSHL) drew the figures for this chapter.

RNAi: THE BASICS

At its core, RNAi is a response to double-stranded RNA (dsRNA) (Fire et al. 1998; Hannon 2002). This unusual structure, often reflective of the presence of viral replication intermediates or high-copy genetic elements, is recognized and parsed into small RNAs that are a defining feature of the RNAi family of pathways (Fig. 7-1). In the classical response, the dsRNA processing step is carried out by an RNase III family enzyme called Dicer (Bernstein et al. 2001; Jinek and Doudna 2009). This produces species, called small interfering RNAs (siRNAs), with a double-stranded portion, often ~19–20 nucleotides in length, flanked by two-nucleotide 3′ overhangs (Elbashir et al. 2001b). Classical siRNAs bear 5′-phosphate and 3′-hydroxyl termini, although more recently discovered species, that may be produced by alterative biogenesis pathways, vary in their terminal structures (Siomi and Siomi 2009).

siRNAs become incorporated into an effector complex called RISC (RNA-induced silencing complex), which has an Argonaute family protein as its core component (Joshua-Tor 2006).

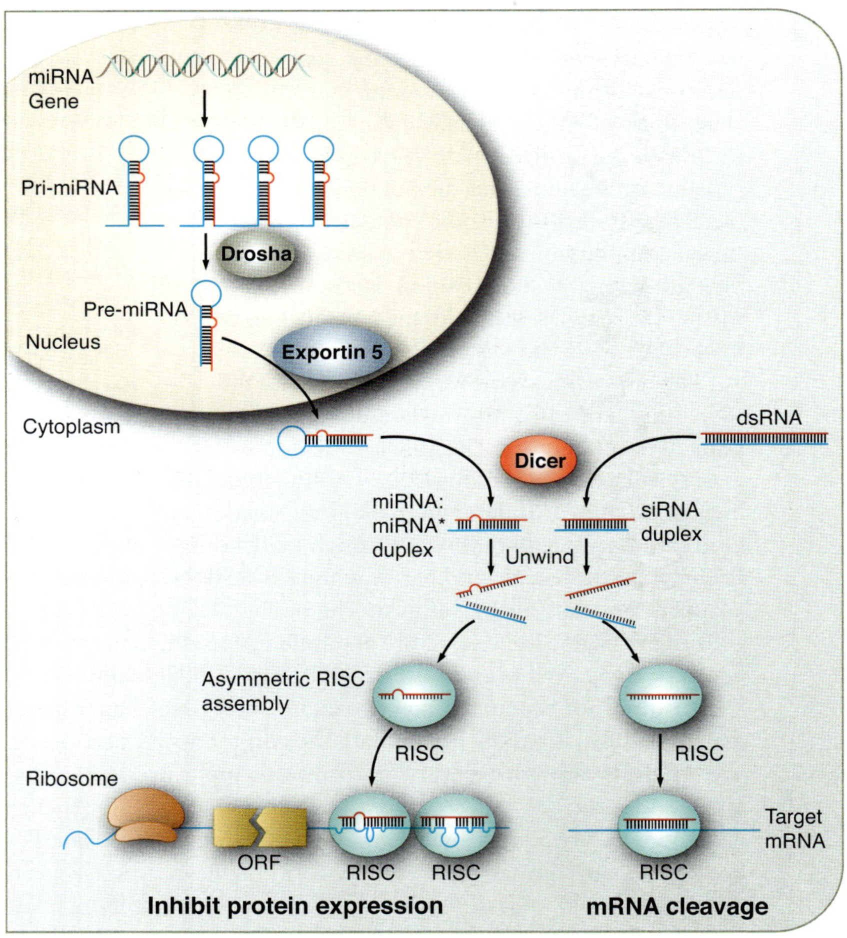

FIGURE 7-1. Simplified summary of the RNAi pathway. RISC, the effector complex of RNAi, can receive small RNAs based on a number of different triggering strategies, which are described in this chapter. Long dsRNAs can be synthesized exogenously and introduced into cells or animals. (Many of the complexities of the *C. elegans* pathway are suppressed in this simplified summary, but see excellent reviews on this topic by Mello and Conte [2004].) Dicer cleaves this substrate to generate an siRNA-like intermediate, which is loaded into RISC. miRNAs are generally derived from RNA polymerase II transcription units and processed through a two-step biogenesis mechanism. These can be mimicked with a remodeled stem-loop structure to redirect the response against any desired gene. Here, the generally imperfect base-pairing interactions between the small RNA and its target are changed to create perfect complementarity and favor target cleavage. siRNAs, an intermediate in the pathway, can also be supplied directly and loaded into RISC.

Argonaute proteins are highly conserved enzymes with several signature features (Hock and Meister 2008; Nowotny and Yang 2009). They contain a PAZ domain, a variant of the oligonucleotide-binding fold, that interacts with the 3′ overhang of the siRNA. The siRNA 5′ end engages a binding pocket in the mid domain that contributes specificity to the pathway by favoring selected terminal structures (Frank et al. 2010).

RISC uses the sequence of its bound small RNA to search for complementary substrates, which it engages via Watson–Crick base pairing. For this to occur, RISC must convert its initially double-stranded guide to a single-stranded form. The mechanism by which this occurs remains to be fully elucidated, but at least some studies indicate that removal of what is termed the "passenger strand" from the siRNA during RISC assembly occurs through a catalytic cycle identical to that which occurs during silencing of a target RNA (Matranga et al. 2005).

Recognition of silencing targets by RISC involves pairing with the so-called "guide strand" of the siRNA. This brings the target within a basic binding groove and positions the phosphodiester bond joining the bases that pair with nucleotides 10 and 11 of the guide within an RNase-H-like active site (Martinez and Tuschl 2004; Schwarz et al. 2004; Song et al. 2004). Here, a metal-assisted reaction cleaves the target, leaving 5′-phosphate and 3′-hydroxyl termini. The substrate is then released, freeing RISC to act on additional targets as a multiple turnover enzyme (Haley and Zamore 2004).

Many studies have examined the degree of complementarity needed to elicit an effective silencing response (Siomi and Siomi 2009). This is important especially for designing specific silencing triggers, a problem that is discussed in more detail below. For most purposes, the trigger should be designed to be fully complementary to the target, with pairing across the cleavage site being particularly important for permitting enzymatic cleavage.

In animals, one type of small RNA guide does not normally silence by directing enzymatic cleavage of its targets (Bartel 2009). microRNAs (miRNAs) are produced from endogenous transcripts, called primary or pri-miRNAs. These contain regions that fold into short, duplexed structures that are initially processed within the nucleus by Drosha, another conserved RNase III family nuclease (Fig. 7-1). This liberates a short hairpin structure, known as a precursor or pre-miRNA, that is exported to the cytoplasm. Further processing by Dicer yields something akin to an siRNA that becomes loaded into RISC. Here, the guide is known as the miR strand and the passenger as miR*. The mechanisms for liberating the "star" strand to form active RISC must differ from those used for siRNAs because many pre-miRNAs are imperfectly paired across the potential cleavage site and therefore cannot be cut by RISC.

miRNAs recognize their targets via interactions that depend disproportionately on a region spanning nucleotides 2–8 of the miR. This "seed" region is also likely important for initial target recognition by siRNAs, particularly in organisms where the miRNA and siRNA pathways are mediated by identical protein components (e.g., in mammals) (Bartel 2009).

THE DESIGN OF RNAi EXPERIMENTS: GENERAL CONSIDERATIONS AND STRATEGIES

In a typical RNAi experiment, an artificial silencing trigger is introduced such that the RNAi pathway is directed toward a target that it would not normally recognize. This is most often an endogenous protein-coding gene, although some noncoding RNAs can also be silenced effectively. The exact nature of the artificial trigger varies, depending on the experimental organism in which it will be used. However, the strategies most commonly used to provoke RNAi can be divided into two groups: those that use synthetic RNA triggers and those that use encoded triggers.

Synthetic Triggers

Synthetic triggers are delivered to cells or organisms in the form of RNA. Their simplest manifestation is a long dsRNA, ~500 bp in length, produced by in vitro transcription (see Chapter 4). These are effective in organisms or cells that lack a robust protein kinase R (PKR) response, which might confound the otherwise sequence-specific effects of RNAi (Sledz and Williams 2004). PKR and related responses are responsible for a largely nonselective inhibition of translation upon exposure to dsRNA. Such responses are particularly evident in most mammalian cell types, but they are largely absent in invertebrates such as *Drosophila* and *C. elegans*.

In the presence of PKR responses, RNAi is often triggered by synthetic species that are sufficiently short to evade the nonspecific pathway. Typically, the 30-bp length is used as a rule of thumb for predicting what species will provoke a PKR response, although in reality PKR is activated in a fashion that is graded with respect to length, depends on the subcellular localization of the dsRNA, and is effected by the precise structure adopted by the introduced element (Sadler and Williams 2007). Given these uncertainties, the most common synthetic RNAi trigger in mammalian cells is the siRNA.

Synthetic siRNAs mimic the natural products of Dicer and comprise duplexes with ~19–20 paired bases and two-nucleotide 3′ overhangs (Elbashir et al. 2001a,b). These can be manufactured in a number of ways. The simplest is to contract the production of synthetic oligoribonucleotides using one of many available manufacturers (e.g., Integrated DNA Technologies, Dharmacon RNAi Technologies, or QIAGEN). Chemically synthesized RNAs most commonly arrive annealed and ready for use. These do not bear 5′-phosphate groups important for incorporation into RISC; however, cellular kinases correct this efficiently, making it unnecessary to enzymatically phosphorylate synthetic siRNAs before use (Weitzer and Martinez 2007). Pools of siRNAs can also be produced from long dsRNAs by exposing these to recombinant Dicer enzyme in vitro (e.g., using enzymes available from New England BioLabs or Gelantis) (Myers et al. 2003). In this case, it is critical that long dsRNA precursors be removed before introduction into cells with intact PKR responses.

Encoded Triggers

Encoded triggers take the form of templates for the transcription of long or short dsRNAs by cellular polymerases, most often RNA polymerase II or III. Although there can be greater complexity in their design and construction, they offer some advantages over synthetic triggers. Because transcription templates can be stably integrated into the genome, the effects of encoded triggers can persist indefinitely, as compared to the more limited duration of most experiments with synthetic RNAs. Moreover, their expression can be restricted to specific tissues or provoked in response to inducing signals that activate conditional promoters (e.g., tetracycline-regulated expression). The forms of encoded triggers divide much as synthetic triggers do. Long dsRNAs (~200–500 bp) can be expressed as extended inverted repeats. The nearest equivalent of an siRNA is an artificial miRNA or short hairpin RNA (shRNA), where a natural miRNA backbone (also called a scaffold) is remodeled to produce a different small RNA or a small inverted repeat (<30 nucleotides) is simply expressed.

Basic Experimental Design Factors

The basic design of an RNAi experiment must take a number of factors into consideration. First and foremost is the palette of methods suitable for provoking a response in the organism in which the study will be performed (Protocols 1–5).

In *C. elegans*, long dsRNAs are overwhelmingly the trigger of choice. *C. elegans* has the unusual property that it can be fed either the dsRNA itself or *Escherichia coli*–expressing dsRNA to provoke a potent and systemic silencing response (Timmons and Fire 1998). Only isolated experiments have used encoded triggers, and their application is so specialized that they do not merit detailed description in this general chapter.

In *Drosophila*, RNAi can be used either in cultured cells or the intact organism. In cells, RNAi can be triggered by synthetic long dsRNAs, siRNAs, or encoded triggers, which are either transcribed long (~200–500 bp) hairpin loops or artificial miRNAs. Overall, the use of enzymatically synthesized long dsRNAs is most prevalent, and this method has been applied in many fly cell lines to study the functions of individual genes or for genome-wide scans using dsRNAs corresponding to each gene in the *Drosophila* genome (Clemens et al. 2000; Hammond et al. 2000; Kiger et al. 2003). One contributor to the popularity of this approach is that many fly cell lines spontaneously take up dsRNAs from media, obviating the need for assisted uptake methods such as transfection (Clemens et al. 2000). In the fly, RNAi must be performed by transgenesis, because, unlike worms, flies cannot be effectively fed dsRNA to induce a systemic silencing response. In this case, benefit is derived from the long history of producing *Drosophila* lines with Gal4 "drivers," wherein tissue-specific promoters cause the expression of this sequence-specific activator. Gal4-responsive promoters can therefore be used for expression of either long dsRNAs or shRNAs to inhibit gene function in selected cell types.

In mammals, cell lines are commonly transfected with siRNAs or engineered to carry genomic integrants of shRNAs that can be driven from constitutive or inducible promoters. Finally, any mammal for which transgenic technologies exist can be modified to express an shRNA in an inducible or cell-type-specific manner.

Once technological limitations are appreciated, the choice silencing modality must be governed by the characteristics of the target gene and the expected phenotype. The functions of proteins with short half-lives can effectively be studied in cultured cells using synthetic triggers; however, studies of long-lived proteins that might require several cell divisions for effective elimination often benefit from an encoded approach. Phenotypes that are reflected by changes in gene expression can easily be monitored in short-term assays, whereas more complex or multicomponent phenotypes, such as neoplastic transformation, almost always require time spans that dictate use of encoded RNAi.

The goal of presenting the following protocols is not to anticipate every possible experimental need but instead to provide standard methods that will suffice for the most common applications. By varying the protocols below, strategies may be tailored as needed in the subject organism or even adapted to a new experimental model.

RNAi WITH LONG dsRNAs

Trigger Selection and Generation of Transcription Templates

The first step in using long dsRNAs in an RNAi experiment is to synthesize a template for RNA synthesis. Because many different siRNAs will be derived from the long dsRNA, less attention is paid to the detailed design of these molecules. Templates are almost always produced by PCR to create a linear double-stranded fragment with phage polymerase promoter sequences at both ends (see Fig. 7-2 and Chapter 4). Amplicons can be derived from cDNA clones or genomic DNA; however, in the

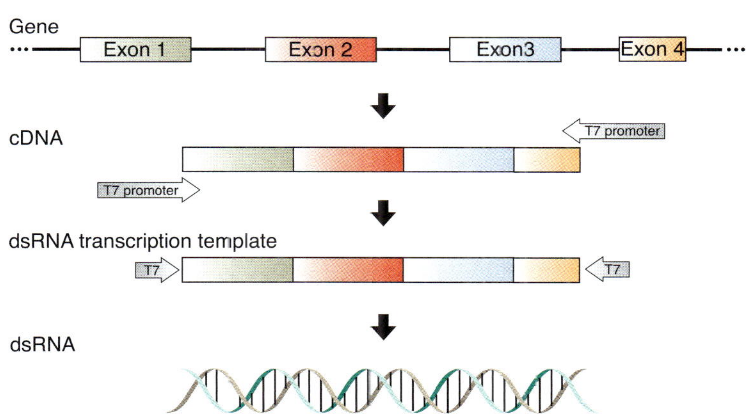

FIGURE 7-2. Synthesis of long dsRNAs as RNAi triggers. Templates for the synthesis of long dsRNAs can be made in several ways, but it is important to bear in mind that most studies indicate that only sequences in the mature transcript can be targeted. Here, an example is given where a fragment of a cDNA, spanning four exons, is amplified with custom primers, which append T7 RNA polymerase promoters to both ends of a transcription template. Simply using the resulting fragment in an in vitro transcription reaction will produce dsRNA suitable for use in *Drosophila* cells or in *C. elegans* RNAi experiments.

latter case, it is important to bear in mind that RNAi must target the mature mRNA. Targeting of intronic sequences is rarely effective; therefore, maximizing exonic content is key. Although few rules have emerged for long dsRNA design, experiments in *Drosophila* have indicated that triggers should not have regions of complete identity that are ≥19 nucleotides in length to any gene other than the desired target (Kulkarni et al. 2006; Ma et al. 2006). SnapDragon (found at www.flyrnai.org) is a web-based tool that can aid in primer design for studies in *Drosophila*; however, no similar tool exists for *C. elegans*. Although sequences as short as 100–150 nucleotides have been shown to function in both organisms, typical long dsRNA triggers range from ~300 to 600 nucleotides. The same phage polymerase promoter is typically used at each end of the transcription template (see Fig. 7-2). The most common arrangement is to use sequences that direct transcription by T7 RNA polymerase (TAATACGACTCACTATA**G**G; the bold G indicates the first base of the transcript); however, either T3 (AATTAACCCTCACTAAA**G**G) or SP6 (ATTTAGGTGACACTATA**G**A) RNA polymerase promoters can be substituted with good results.

The precise design of primers and PCR conditions differ according to the template, and many convenient online tools exist to facilitate design (e.g., Primer3, found at http://frodo.wi.mit.edu/primer3/). Most often, chimeric oligonucleotides consisting of the phage polymerase promoter sequence and a gene-specific sequence are used directly for amplification of a template (Fig. 7-2). However, in some cases, particularly when the amplicon is being prepared directly from genomic DNA or pooled reverse-transcribed cDNAs, this proves to be unacceptably inefficient. An alternative is to first amplify the template only with gene-specific primers and then only later append the appropriate promoter sequences in a second round of PCR. Sufficient template can be obtained, provided that PCRs are efficient, in a single 25–100-µL reaction. The product can optionally be purified by gel electrophoresis or column chromatography using any number of commercially available resins (e.g., QIAGEN QIAquick PCR Purification Kit).

It is essential that templates be generated under RNase-free conditions and that they be suspended after purification in nuclease-free H_2O or TE buffer. It is generally not necessary to use proofreading polymerases to generate transcription templates.

dsRNA Synthesis

dsRNA synthesis should be carried out as described in Chapter 4, Protocol 2 (High-yield Synthesis of RNA Using T7 RNA Polymerase and Plasmid DNA or Oligonucleotide Templates). The only twist on the standard protocol is that two strands are transcribed rather than one. Once synthesis and purification are completed, it is not necessary to perform further steps to hybridize the separately transcribed strands. This will have occurred during the organic extraction and clean-up steps associated with the protocol. The use of kits is often less than ideal, but there are several high-yield RNA transcription kits on the market (e.g., the Ambion MEGAscript T7 Kit, no. 1334). If this kit is used in lieu of the provided protocol, follow the manufacturers' instructions, carrying out transcription for 16 h. A PCR product can be used directly from the thermocycler without purification (although it is a good idea to validate the template by gel electrophoresis), adding 8 µL of the PCR per 20-µL transcription reaction. There are a number of ways to verify that sufficient dsRNA product has been generated using either strategy. The simplest is to run 5 µL of a 1:500 dilution of the transcription reaction directly on a nondenaturing 1% agarose gel in TAE or TBE buffer, taking care that this is done in a clean gel box under RNase-free conditions. Ethidium bromide or SYBR Green staining should suffice to evaluate whether a predominantly full-length product was produced. The product can also be analyzed using a Bioanalyzer, which would provide both size and concentration information.

The dsRNA product should be purified by phenol:chloroform:isoamyl alcohol extraction and ethanol precipitation (see Chapter 2) or using a commercial kit, such as QIAGEN RNeasy (no. 74104). Using the former method will maximize yield. If the latter, or a similar kit approach, is used, be careful not to overload the capacity of columns; this often requires dividing a single 20-µL reaction among multiple purifications. If RNA synthesis for many templates is being done, such as for a focused or full-genome RNAi screen (see below), manifolds from Millipore (Montage PCR-96 plate, no. MANU03050) have proven to be reliable for high-throughput purification.

To calculate RNA concentrations, measure the optical density (OD) 260 of a 1:50–1:100 dilution of the purified material. A rough estimate can be obtained by multiplying the OD_{260} by the dilution factor and then by the extinction coefficient of 45. The expected yield per 20 µL is 80–200 µg. dsRNAs can be stored frozen at either –20°C or –80°C.

REFERENCES

Bartel DP. 2009. MicroRNAs: Target recognition and regulatory functions. *Cell* **136:** 215–233.

Bernstein E, Caudy AA, Hammond SM, Hannon GJ. 2001. Role for a bidentate ribonuclease in the initiation step of RNA interference. *Nature* **409:** 363–366.

Clemens JC, Worby CA, Simonson-Leff N, Muda M, Maehama T, Hemmings BA, Dixon JE. 2000. Use of double-stranded RNA interference in *Drosophila* cell lines to dissect signal transduction pathways. *Proc Natl Acad Sci* **97:** 6499–6503.

Elbashir SM, Harborth J, Lendeckel W, Yalcin A, Weber K, Tuschl T. 2001a. Duplexes of 21-nucleotide RNAs mediate RNA interference in cultured mammalian cells. *Nature* **411:** 494–498.

Elbashir SM, Lendeckel W, Tuschl T. 2001b. RNA interference is mediated by 21- and 22-nucleotide RNAs. *Genes Dev* **15:** 188–200.

Fire A, Xu S, Montgomery MK, Kostas SA, Driver SE, Mello CC. 1998. Potent and specific genetic interference by double-stranded RNA in *Caenorhabditis elegans*. *Nature* **391:** 806–811.

Frank F, Sonenberg N, Nagar B. 2010. Structural basis for 5′-nucleotide base-specific recognition of guide RNA by human AGO2. *Nature* **465:** 818–822.

Haley B, Zamore PD. 2004. Kinetic analysis of the RNAi enzyme complex. *Nat Struct Mol Biol* **11:** 599–606.

Hammond SM, Bernstein E, Beach D, Hannon GJ. 2000. An RNA-directed nuclease mediates post-transcriptional gene silencing in *Drosophila* cells. *Nature* **404:** 293–296.

Hannon GJ. 2002. RNA interference. *Nature* **418:** 244–251.

Hock J, Meister G. 2008. The Argonaute protein family. *Genome Biol* **9:** 210.

Jinek M, Doudna JA. 2009. A three-dimensional view of the molecular machinery of RNA interference. *Nature* **457:** 405–412.

Joshua-Tor L. 2006. The Argonautes. *Cold Spring Harbor Symp Quant Biol* **71:** 67–72.

Kiger AA, Baum B, Jones S, Jones MR, Coulson A, Echeverri C, Perrimon N. 2003. A functional genomic analysis of cell morphology using RNA interference. *J Biol* **2:** 27.

Kulkarni MM, Booker M, Silver SJ, Friedman A, Hong P, Perrimon N, Mathey-Prevot B. 2006. Evidence of off-target effects associated with long dsRNAs in *Drosophila melanogaster* cell-based assays. *Nat Methods* **3:** 833–838.

Ma Y, Creanga A, Lum L, Beachy PA. 2006. Prevalence of off-target effects in *Drosophila* RNA interference screens. *Nature* **443:** 359–363.

Malone CD, Hannon GJ. 2009. Small RNAs as guardians of the genome. *Cell* **136:** 656–668.

Martinez J, Tuschl T. 2004. RISC is a 5′ phosphomonoester-producing RNA endonuclease. *Genes Dev* **18:** 975–980.

Matranga C, Tomari Y, Shin C, Bartel DP, Zamore PD. 2005. Passenger-strand cleavage facilitates assembly of siRNA into Ago2-containing RNAi enzyme complexes. *Cell* **123:** 607–620.

Mello CC, Conte D Jr. 2004. Revealing the world of RNA interference. *Nature* **431:** 338–342.

Myers JW, Jones JT, Meyer T, Ferrell JE Jr. 2003. Recombinant Dicer efficiently converts large dsRNAs into siRNAs suitable for gene silencing. *Nat Biotechnol* **21:** 324–328.

Nowotny M, Yang W. 2009. Structural and functional modules in RNA interference. *Curr Opin Struct Biol* **19:** 286–293.

Sadler AJ, Williams BR. 2007. Structure and function of the protein kinase R. *Curr Top Microbiol Immunol* **316:** 253–292.

Schwarz DS, Tomari Y, Zamore PD. 2004. The RNA-induced silencing complex is a Mg^{2+}-dependent endonuclease. *Curr Biol* **14:** 787–791.

Siomi H, Siomi MC. 2009. On the road to reading the RNA-interference code. *Nature* **457:** 396–404.

Sledz CA, Williams BR. 2004. RNA interference and double-stranded-RNA-activated pathways. *Biochem Soc Trans* **32:** 952–956.

Song JJ, Smith SK, Hannon GJ, Joshua-Tor L. 2004. Crystal structure of Argonaute and its implications for RISC slicer activity. *Science* **305:** 1434–1437.

Timmons L, Fire A. 1998. Specific interference by ingested dsRNA. *Nature* **395:** 854.

Weitzer S, Martinez J. 2007. The human RNA kinase hClp1 is active on 3′ transfer RNA exons and short interfering RNAs. *Nature* **447:** 222–226.

White SA, Allshire RC. 2008. RNAi-mediated chromatin silencing in fission yeast. *Curr Top Microbiol Immunol* **320:** 157–183.

Protocol 1

dsRNA-induced RNAi in *Drosophila* Cells by Soaking

This protocol is designed for Schneider S2 cells in one well of a standard six-well tissue culture plate. The protocol can be scaled up or down easily for different plate formats simply by multiplying all volumes and cell numbers by the surface area of the desired plate divided by the surface area of the six-well plate (see Table 7-1). Cell lines other than S2 can also be used, although the ability of each line to passively take up dsRNA does vary. Therefore, the efficiency of passive uptake should be carefully verified for each line. If the soaking protocol does not work for a particular cell line, one can almost always introduce dsRNA by transfection (see Protocol 2).

MATERIALS

The recipes for the items marked with <R> are at the end of the protocol or in the Common Recipes Appendix.

CAUTION: See the Cautions Appendix for appropriate handling of material marked with <!>.

Reagents

Complete S2 medium <R>
Drosophila S2 cells in semiadherent culture or suspension culture
dsRNA at 0.5–10 µg/µL in nuclease-free H_2O or 10 mM Tris, pH 8/1 mM EDTA <!> (TE buffer)
PBS/EDTA <!>/EGTA <R>
Phosphate-buffered saline (PBS) <R>
Trypsin/EDTA solution <R> <!>
Schneider's *Drosophila* medium without serum (Invitrogen 11720-034)

Equipment

Centrifuge (low speed)
Culture incubator (25°C)
Hemocytometer for cell counting
Microscope
Pipette
Tissue culture plates (six well)

OVERVIEW

The length of incubation required in this protocol varies depending on the target gene and the effectiveness of the dsRNA. Loss of mRNA, measured by semiquantitative reverse transcription–PCR

TABLE 7-1. Scaling Up or Down for Different Plate Formats

Multiply all volumes and cell numbers by the surface area of the desired plate divided by the surface area of the six-well plate

Culture vessel	96 well	48 well	24 well	12 well	6 well	35 mm	60 mm	100 mm	150 mm	T25	T75
Surface area (cm^2)	0.3	0.7	2	4	10	10	20	60	140	25	75

(RT-PCR) or northern blotting, is generally evident after 2–3 days. The time required to observe loss of protein is more variable and depends on many factors, including the protein half-life. In the majority of cases, a knockdown of ≥80% or greater at both mRNA and protein levels is observed after 3–4 days, and this can persist until about day 5–6. If achieving the desired knockdown proves to be problematic, a different dsRNA should be tested (although it is always wise to test several initially). Alternatively, a second round of soaking can be done 4 d after the initial treatment using this protocol. This may not only increase knockdown (although it does not in all cases), but it also affords the possibility of examining phenotypes that require longer times to become evident.

METHOD

1. Collect S2 cells from semiadherent culture using trypsin/EDTA (or by vigorous pipetting in PBS/EDTA/EGTA) or harvest from suspension culture. Pellet cells by centrifuging at 800–1000*g* for 5 min at room temperature.
2. Resuspend the cells in culture medium lacking serum, count cells with a hemocytometer, and adjust to a concentration of 1 x 10^6 to 5 x 10^6 cells/mL.
3. Add 10–30 µg of dsRNA per tissue culture plate well.
4. Add 1.5 mL of cell suspension and mix by pipetting.
5. Incubate the mixture for 30–60 min at room temperature.
6. Add 3 mL of complete S2 medium containing 10% FBS to each well.
7. Culture the cells for 3–6 d at 25°C.

   S2 cells can also be cultured at room temperature.
8. Measure knockdown by quantitative PCR or western blotting and determine the phenotypic consequences of knockdown as desired.

TIPS AND TROUBLESHOOTING

See Tips and Troubleshooting following Protocol 2.

SOLUTION AND REAGENT RECIPES

CAUTION: See the Cautions Appendix for appropriate handling of material marked with <!>.

Phosphate-buffered Saline (PBS)

Reagent	Quantity (for 1 L)	Final concentration
NaCl	8 g	137 mM
KCl <!>	0.2 g	2.7 mM
Na_2HPO_4	1.44 g	10 mM
KH_2PO_4	0.24 g	2 mM

Dissolve in 800 mL of Milli-Q H_2O. Adjust the pH to 7.4 and then add Milli-Q H_2O to 1 L. Autoclave or filter to sterilize. Store at room temperature. Alternatively, buy or make 10x PBS.

PBS/EDTA/EGTA

Reagent	Quantity (for 500 mL)	Final concentration
PBS (10x)	50 mL	1x
EDTA (0.5 M, pH 8.0) <!>	1 mL	1 mM
EGTA (0.1 M, pH 8.0)	5 mL	1 mM
H_2O	to 500 mL	

Complete S2 Medium

Schneider's complete medium (Invitrogen catalog no. 11720-034) including
Fetal bovine serum (10%)
Antibiotic/antimycotic solution (1x) (e.g., Invitrogen catalog no. 15240-062)

Trypsin/EDTA

0.25% Trypsin and 1 mM EDTA <!> in PBS.

Protocol 2

dsRNA-induced RNAi in *Drosophila* Cells by Transfection

For many *Drosophila* cell lines and cell types, passive dsRNA uptake is inefficient. In addition, silencing responses more complete than those seen with soaking can often be obtained in S2 cells using a variety of transfection protocols, perhaps because higher levels of intracellular dsRNA are achieved. In this protocol, S2 cells are transfected using QIAGEN's Effectene reagent, which has proven to be reliable for many investigators. Particularly when working with a new cell line, it is recommended that many different transfection reagents and protocols be tried in an effort to optimize knockdown efficiency. It would not be uncommon to scan a dozen different commercial reagents, including charged and neutral lipids, nanoparticles, and several different electroporation protocols (e.g., standard electroporation and Amaxa nucleofection) when developing a means to efficiently introduce dsRNA into a previously uncharacterized cell type.

MATERIALS

The recipes for the items marked with <R> are at the end of the protocol or in the Common Recipes Appendix.

CAUTION: See the Cautions Appendix for appropriate handling of material marked with <!>.

Reagents

Complete S2 culture medium <R>
Drosophila S2 cells
dsRNA at 0.5–10 µg/µL in nuclease-free H_2O or TE buffer
Transfection reagent
Effectene Transfection Reagent (QIAGEN) is used in this protocol.
Trypsin/EDTA <R> <!>

Equipment

Centrifuge (low speed)
Culture incubator (25°C)
Eppendorf tubes
Hemocytometer for cell counting
Microscope
Pipette
Tissue culture plates (six well)
Vortex mixer

OVERVIEW

This protocol presents a method for six-well plates, but it can be scaled as needed based on the ratio of the surface area of the desired vessel to that of a well of a six-well plate (see Table 7-1). In this case, a plasmid DNA has also been included in the transfection mix. This is typical in many RNAi experiments and can be used to provide additional functionality in a way that cannot be achieved in a soaking experiment. The plasmid DNA can encode, for example, a reporter of the activity of a pathway or specific transcription factor, or a marker that allows visualization of some cellular behavior or structure. It is also useful to include a plasmid that encodes a fluorescent protein simply to monitor transfection efficiency.

This method can be modified to accommodate any transfection protocol simply by following the manufacturer's instructions. In general, transfection methods require less dsRNA than does soaking and give somewhat better knockdown.

METHOD

1. Collect S2 cells from adherent culture using trypsin/EDTA or harvest from a suspension culture and pellet cells by centrifuging at 800–1000*g* for 5 min at room temperature.
2. Use a hemocytomoter to resuspend and count the cells. Dilute cells in fresh S2 culture medium (typically to a density of 1×10^6 cells/mL). Dispense 2 mL of diluted cells into each well of a six-well plate.

   At this point, cells are generally left from several hours to overnight to adhere to the plate. If this is the case, exchange the medium before transfection. Cells can also be used immediately with some (e.g., Effectene) but not all (e.g., calcium phosphate coprecipitation) transfection methods.
3. Prepare the nucleic acid mix in an Eppendorf tube according to the manufacturer's instructions. For each plate well, the mix consists of optionally DNA (as a transfection or functional marker), dsRNA (total nucleic acid comprising ~2 µg), and EC buffer, a component of the Effectene Kit. Mix the total volume (should be ~200 µL) by inversion or gentle vortexing.
4. Add Enhancer from the Effectine Kit to the mix so that there is 16 µL of Enhancer per well. Mix well by inversion and then incubate for 5 min at room temperature.
5. Prepare the transfection solution by adding Effectene to the mix so that there is 8 µL of Effectene per well. Mix by pipetting up and down 5–10 times or vortexing on high for 10 sec. Allow the solution to stand for 10 min at room temperature.
6. Dispense the mix dropwise into individual wells of six-well plates containing the cells from Step 2.
7. Incubate the plates in a humidified chamber in a tissue culture incubator for 2–6 d at 25°C.
8. Measure knockdown by quantitative PCR or western blotting and determine the phenotypic consequences of knockdown as desired.

TIPS AND TROUBLESHOOTING

1. Although media, PBS, and trypsin/EDTA can be made from scratch, it is often convenient to purchase tissue-culture–qualified reagents, which are sold at their final concentration or as stock solutions. Particularly for S2 cells, which adhere weakly to plates, cells can often be removed without the use of trypsin by incubation with PBS containing 1 mM EDTA and 1 mM EGTA. S2 cells can also be maintained in suspension culture.
2. Cells that are healthy and growing exponentially give the best results. It may be necessary to test different lots of fetal bovine serum to find one that works well with S2 or other fly cells. Doubling times of 18–24 h are normal for most lines. Note especially that many S2 lines are persistently virus infected (e.g., with flock house virus [FHV]). Care must be taken to obtain and work with noninfected cells.
3. Although each dsRNA is sampled to produce many different siRNAs, dsRNAs derived from different portions of target genes differ in their silencing efficiency. Therefore, it is generally a good idea to test several different dsRNAs in parallel for each gene to be silenced.

4. In some circumstances, for example, when close paralogs must be distinguished of if rescue experiments are planned (see below), it is desirable to use a single synthetic siRNA rather than a long dsRNA. In this case, use a transfection protocol similar to the one outlined above but substitute a synthetic siRNA for the dsRNA. Choosing an siRNA sequence is a complex topic and is detailed below in the discussion of RNAi in mammalian cultured cells. Presently, there is no reason to suspect that the characteristics of effective siRNAs differ in vertebrates and invertebrates (although new evidence may emerge), and the same design procedures should be followed in both cases.
5. It is important wherever possible to measure knockdown by quantifying changes in protein levels. Although it is easier to measure RNA knockdown, RNA and protein are not always equally affected nor does their decline follow the same time course.
6. In transfection protocols, it is typical to cotransfect a plasmid encoding green fluorescent protein (GFP) with and without a GFP dsRNA. It must be recognized, however, that it is often easier to reduce the levels of a cotransfected marker than an endogenous gene. Many cell lines are sensitive to the knockdown of *thread*, an endogenous inhibitor of apoptosis orthologous to mammalian inhibitor of apoptosis (IAP), and die in the presence of a *thread* dsRNA. This is also an effective control for endogenous knockdown in many situations.

SOLUTION AND REAGENT RECIPES

CAUTION: See the Cautions Appendix for appropriate handling of material marked with <!>.

Complete S2 Medium

Schneider's complete medium (Invitrogen catalog no. 11720-034) including
Fetal bovine serum (10%)
Antibiotic/antimycotic solution (1×) (e.g., Invitrogen catalog no. 15240-062)

Trypsin/EDTA

0.25% Trypsin and 1 mM EDTA <!> in PBS.

RNAi IN *C. ELEGANS*

RNAi can be induced in *C. elegans* via a number of methodologies. Initial studies used injection of dsRNA directly into the germline of animals, a procedure that was sufficient to induce a systemic response in injected animals and induce silencing in progeny for a limited number of generations (Fire et al. 1998). The biology of these processes is complex and has been the topic of several excellent reviews (Mello and Conte 2004; Grishok 2005). Following the realization that systemic responses could be provoked by introducing RNA into the intestine, protocols were developed that enabled RNAi by ingestion by feeding worms with enzymatically synthesized dsRNAs (see above) or even with *Escherichia coli* that had been engineered to express complementary RNA strands from a plasmid vector (Timmons et al. 2001). Due to the relative ease and high-throughput nature of ingestion-mediated RNAi, the feeding of engineered *E. coli* to wild-type and mutant animals has developed into the most productive and common methodology used to probe the functions of *C. elegans* genes.

The optimal RNAi protocol is determined by a multitude of factors, including the nature of the depleted gene, the particular phenotype that will be scored, the number of genes that will be screened in a given experiment, and the technical capabilities of the laboratory and experimenter. Most *C. elegans* researchers choose among three common methodologies. Protocol 3 includes two variations of RNAi by feeding: one in which L1 larvae are fed dsRNA-expressing *E. coli* in liquid culture (a protocol commonly used to score effects of RNAi on postembryonic larval development) and a second method involving feeding of late larval- or adult-staged animals on standard solid medium culture plates. These generalized protocols should provide a sufficient substrate for further adaptation to fit the needs of individual experiments.

REFERENCES

Fire A, Xu S, Montgomery MK, Kostas SA, Driver SE, Mello CC. 1998. Potent and specific genetic interference by double-stranded RNA in *Caenorhabditis elegans*. *Nature* **391:** 806–811.

Grishok A. 2005. RNAi mechanisms in *Caenorhabditis elegans*. *FEBS Lett* **579:** 5932–5939.

Mello CC, Conte D Jr. 2004. Revealing the world of RNA interference. *Nature* **431:** 338–342.

Timmons L, Court DL, Fire A. 2001. Ingestion of bacterially expressed dsRNAs can produce specific and potent genetic interference in *Caenorhabditis elegans*. *Gene* **263:** 103–112.

Protocol 3

dsRNA-induced RNAi in *C. elegans* by Feeding with dsRNA-expressing *E. coli*

The first step toward silencing a particular gene by ingestion-mediated RNAi is to construct or obtain an *E. coli* strain expressing dsRNA targeting the gene(s) of interest. Large collections of strains corresponding to nearly all identified genes in the *C. elegans* genome have been produced (Fraser et al. 2000; Kamath et al. 2003; Rual et al. 2004). These can be obtained as individual bacterial stocks or purchased as whole-genome RNAi libraries (e.g., see http://www.geneservice.co.uk/products/rnai/Celegans.jsp and www.openbiosystems.com). Construction of custom strains is also straightforward and recommended, because multiple feeding constructs can be tested for each gene of interest.

This approach begins identically to that described above for use of enzymatically synthesized long dsRNA with the selection of a target region of ~500 bp in length, corresponding to the mature mRNA. The PCR product should be generated without the appended phage polymerase promoters but with restriction sites that facilitate incorporation into purpose-built plasmids for the creation of feeding strains (see Fig. 7-2; replace polymerase promoters with appropriate restriction sites). As an example, the polylinker region of one such plasmid, L4440 (pPD129.36; http://www.addgene.org/pgvec1?cmd=findpl&identifier=1654), is shown (Fig. 7-3). This expression vector is based on a pUC backbone in which opposing T7 RNA polymerase promoters surround an extensive polylinker into which a PCR fragment can be inserted.

To create a feeding strain, the plasmid should be transformed into an *E. coli* strain that expresses an inducible T7 RNA polymerase. Although several strains for this purpose exist, an HT115 derivative is the most extensively used parental strain. This carries an isopropyl-β-D-thiogalactoside (IPTG)-inducible copy of the phage gene 1 encoding T7 polymerase and lacks a functional dsRNA-specific endonuclease (RNase III) encoded by the bacterial *rnc* gene (*rnc14::DTn10*) (Timmons et al. 2001).

MATERIALS

The recipes for the items marked with <R> are at the end of the protocol or in the Common Recipes Appendix.

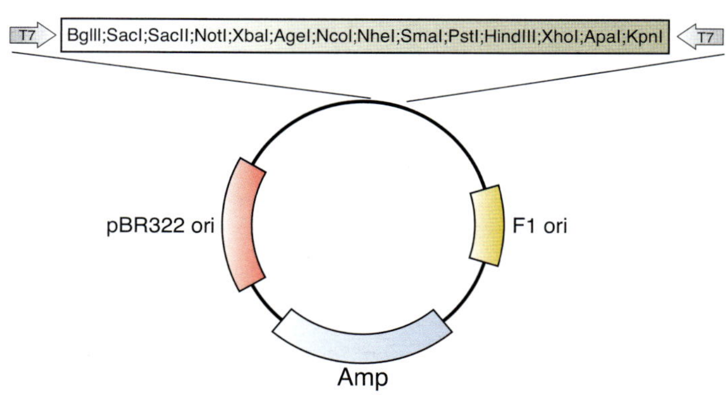

FIGURE 7-3. Map of pL4440, a vector commonly used for the construction of *E. coli* strains for the induction of RNAi by feeding in *C. elegans*. Inserts that will correspond to the dsRNA that will trigger RNAi are cloned into the polylinker shown. The T7 promoters are symmetric; thus, the orientation of the insert is unimportant.

Reagents

C. elegans of the appropriate genotype or strain
Feeding strains (*E. coli* expressing the appropriate dsRNA)
IPTG (isopropyl-β-D-thiogalactoside) <R> <!>
Luria broth (LB) medium + ampicillin (50 μg/mL or 100 μg/mL, as specified) <R>
M9 medium <R>
M9+ medium <R>
Nematode growth medium (NGM)/IPTG plates <R>

Equipment

Bacterial shaker (37°C)
Deep-well blocks (96 well)
Flat-bottomed plates (96 well)
Incubator (22°C) (optional)

OVERVIEW

Feeding of *C. elegans* with dsRNA-expressing *E. coli* in liquid and on plates often produces similar results with regard to penetrance and expressivity, but the procedures differ in experimental design and limitations. The 96-well RNAi format offers high throughput and can be scaled for RNAi screens of extensive gene sets, but this also bears a cost. For example, subtle and challenging phenotypes are difficult to score by microscopy in larval suspension and animals develop more slowly in liquid culture. In contrast, NGM plate RNAi allows animals to reproduce under more standardized growth conditions and also facilitates observations by microscopy and subsequent handling of animals.

METHODS

RNAi of L1 Larvae in 96-well Liquid Cultures

1. Fill 96-well deep-well blocks with 500 μL of LB medium containing 50 μg/mL ampicillin per well.
2. Inoculate each well with a bacterial colony (feeding strain) capable of expressing the desired dsRNA. In separate wells, inoculate control strains that either do not express a biologically relevant dsRNA (e.g., dsRNA against GFP L4417 [pPD128.110]) or express a dsRNA that will give a defined developmental phenotype.

   One convenient control is *unc-22* RNAi (pLT61.1), which produces a visible "twitching" or uncoordinated (unc) phenotype associated with defective muscle development. Control plasmids can be obtained from http://www.geneservice.co.uk/products/rnai/Celegans.jsp.
3. Grow the cultures overnight at 37°C.
4. Induce expression of specific dsRNA by adding 0.25 mg/mL of IPTG to each well.
5. Incubate the bacterial cultures for a further 4 h at 37°C.
6. In a fresh 96-well plate, inoculate 10–20 *C. elegans* L1 larvae in 50 μL of M9+ medium per well of a flat-bottomed 96-well tissue culture plate.
7. After the bacterial cultures have cooled to room temperature (20°C–23°C), transfer 100 μL of induced bacterial suspension to each well of the plate containing the larval animals.
8. Shake the plates at 150–200 rpm for 3–5 d at room temperature.
9. Score the animals for developmental phenotypes and/or gene knockdown.

RNAi on Plates

As described above, bacterial food sources that cause the expression of visible phenotypes (e.g., unc-22 RNAi [pLT61.1]) should be included to verify proper induction of dsRNA in the bacterial culture and to ensure that animals grow properly and respond by silencing target genes on the NGM + IPTG (1–3 mM) RNAi plates.

1. Culture the feeding and control bacterial strains overnight in LB + 100 µg/mL ampicillin in a shaker at 37°C.
2. Pipette 200 µL of each culture onto NGM + IPTG (1–3 mM) 60-mm plates. Allow the bacterial cultures to be absorbed into the NGM medium.
3. Grow the plates containing cultures overnight at room temperature. This also provides time for expression of specific dsRNAs.
4. Transfer 4–10 worms at the appropriate developmental stage (L3, L4, or adult, depending on the phenotype to be scored) to the culture plate containing the induced bacteria.

   For typical pilot experiments, three to seven late to L4/adult-stage animals are transferred to a single 60-mm plate and RNAi phenotypes are scored in their progeny.
5. Incubate for 1–4 d at 20°C–22°C and then score the relevant phenotypes.

   Depending on the specific phenotype to be observed and the effective time required to deplete target mRNA, one might wish to remove the original adult worms from the plate and only score their progeny, which typically reach adulthood in 48–52 h at 20°C. Removal of parental animals will allow sufficient bacterial food to remain on experimental plates for *C. elegans* progeny to be assayed.

SOLUTION AND REAGENT RECIPES

CAUTION: See the Cautions Appendix for appropriate handling of materials marked with <!>.

M9 Stock (1x)

KH_2PO_4	3.0 g
Na_2HPO_4	6.0 g
NaCl	5.0 g
$MgSO_4$ (1 M) <!>	1.0 mL
dH_2O	to 1 L

M9+ Medium

1x M9 stock plus:	
Cholesterol	10 µg/mL
Ampicillin <!>	50 µg/mL
Tetracycline <!>	12 µg/mL
IPTG <!>	200 µg/mL
Fungizone	0.1 µg/mL

NGM Plates

Bacto-peptone	2.5 g
NaCl	3 g
Bacto-agar	18 g
dH_2O	to 975 mL

Autoclave. After cooling, add

$CaCl_2$ (1 M) <!>	1 mL
$MgSO_4$ (1 M) <!>	1 mL
Cholesterol (5 mg/mL) (dissolved in 100% ethanol <!>)	1 mL
$NaPO_4$ (1 M, pH 6) <!>	25 mL

NGM RNAi Plates

NGM plates plus

Ampicillin <!>	50 µg/mL
IPTG <!>	1–3 mM

IPTG

Dissolve IPTG to 10 mg/mL in dimethylsulfoxide (DMSO) <!>.

Luria Broth (LB)

Add the following to 800 mL of H_2O:

Bacto-tryptone	10 g
Yeast extract	5 g
NaCl	10 g

Adjust the pH to 7.5 with NaOH and then adjust the volume to 1 L with dH_2O. Sterilize by autoclaving.

REFERENCES

Fraser AG, Kamath RS, Zipperlen P, Martinez-Campos M, Sohrmann M, Ahringer J. 2000. Functional genomic analysis of *C. elegans* chromosome I by systematic RNA interference. *Nature* **408:** 325–330.

Kamath RS, Fraser AG, Dong Y, Poulin G, Durbin R, Gotta M, Kanapin A, Le Bot N, Moreno S, Sohrmann M, et al. 2003. Systematic functional analysis of the *Caenorhabditis elegans* genome using RNAi. *Nature* **421:** 231–237.

Rual JF, Ceron J, Koreth J, Hao T, Nicot AS, Hirozane-Kishikawa T, Vandenhaute J, Orkin SH, Hill DE, van den Heuvel S, et al. 2004. Toward improving *Caenorhabditis elegans* phenome mapping with an ORFeome-based RNAi library. *Genome Res* **14:** 2162–2168.

Timmons L, Court DL, Fire A. 2001. Ingestion of bacterially expressed dsRNAs can produce specific and potent genetic interference in *Caenorhabditis elegans*. *Gene* **263:** 103–112.

Protocol 4

RNAi in *C. elegans* by Injection of dsRNA

Although injection of dsRNA into the somatic or germline tissues of animals requires both specific equipment and technical skills, the ability of *C. elegans* to amplify the initial dsRNA trigger and to transmit the RNAi activity to other somatic tissues and to the progeny of injected animals is one of the main advantages to using *C. elegans* as a model system. The direct injection of dsRNA into parental animals is the most reliable method for RNAi and also presents the least experiment-to-experiment and animal-to-animal variability.

MATERIALS

The recipes for the items marked with <R> are at the end of the protocol or in the Common Recipes Appendix.

Reagents

Agarose gel (1%)
Appropriate dsRNA at 0.25–1 μg/μL
C. elegans of the appropriate strain or genotype
E. coli strain OP50
M9 medium <R>
Nematode growth medium (NGM) plates <R>

Equipment

Microinjection setup (e.g., Inverted Zeiss AxioObserver, Narashige three-axis micromanipulator [MMO-202ND], Narashige Motor Drive Manipulator [MM-89], and Narashige oil-type microinjector [IM-9B])
Spectrophotometer

OVERVIEW

Although relatively straightforward, successful injection of *C. elegans* does require skill that can be acquired by practice. It is critical as one learns injection to include a number of positive controls with easily discernible phenotypes.

METHOD

Preparation of dsRNA

1. Prepare the dsRNA precisely as described above in Protocol 3 for soaking or transfection into *Drosophila* cells.
2. Quantify the yield of the in vitro transcription reaction with a spectrophotometer.
3. Dilute the sample to a working concentration of 0.25–1.0 μg/μL. For quality control purposes, resolve ~0.5–2.0 μg of the dsRNA sample on a 1% agarose gel to verify correct length of the transcriptional product.

Injection of dsRNA into Animals

4. Directly inject the dsRNA solution from Step 3 into early adult animals (containing less than five to seven retained, fertilized embryos). For this, the precise procedure depends on the microinjector apparatus, so follow the instructions from the manufacturer. As a rule of thumb, deliver a volume of the dsRNA injection mixture (30–50 pL) to the gonad of each worm, with injection being monitored by the visible movement of the syncytial germline cytoplasm.

   Avoid overloading the germline because this can cause a significant amount of lethality regardless of dsRNA content or concentration. Although an efficient RNAi response will occur if dsRNA is injected into almost any somatic tissue, direct microinjection of the dsRNA into the adult gonad leads to a higher efficiency of RNAi in F_1 progeny and presumably can more fully deplete maternally deposited gene products in developing embryos.

5. Recover the injected animal from the injection pad by resuspension in M9 medium and gently place it onto a fresh NGM plate containing standard *E. coli* OP50 as a food source.
6. After 24 h, transfer the animals to fresh NGM + OP50 plates.

   This procedure will leave unaffected embryos, laid before the RNAi response was fully mounted, on the original recovery OP50 plate and allow only RNAi-depleted F_1 progeny to be scored on the second plate.

   Typically, a fully potent effect of RNAi is seen ~24 h after microinjection. This longer incubation is particularly useful when targeting germline-expressed transcripts, which show a high degree of maternal contribution.

7. Score the progeny for developmental phenotypes and/or gene knockdown.

TIPS AND TROUBLESHOOTING

1. Be aware that the potency of each unique dsRNA-expressing vector can vary widely in its effectiveness in depleting the expression of a given target. In addition to differences in the penetrance or expressivity of phenotypes elicited by different RNAi constructs, the efficiency of RNAi-mediated depletion in different *C. elegans* tissues also varies widely. In particular, neuronally expressed mRNAs are especially refractory to RNAi-mediated depletion (Kennedy et al. 2004), although this limitation can be overcome (see below).
2. Both the efficacy and penetrance of RNAi induced by feeding dsRNA-expressing bacteria can be improved upon by performing RNAi in certain mutant backgrounds. Two often-used, RNAi-sensitized strains are NL2099, which harbors a specific deletion of an RNA-dependent RNA polymerase (*rrf-3[pk2426]*), and GR1379, which has a mutation in a cytoplasmic nuclease (*eri-1 [mg366]*). Both strains can be obtained from the *C. elegans* Genetics Center (http://www.cbs.umn.edu/CGC/index.html). These strains do have enhanced responses to experimentally triggered RNAi; however, the mutations that they contain are not neutral. Animals of both genotypes display lower brood sizes, defects in endogenous RNAi-based regulatory circuits, and a temperature-sensitive sterility phenotype when reared above 23°C (Simmer et al. 2002; Kennedy et al. 2004).
3. Specific genetic backgrounds can also be used to enhance RNAi induced by feeding in particular tissues. For instance, the NL2098 strain harbors a deletion in an additional RNA-dependent RNA polymerase *rrf-1* (*rrf-1[pk1417]*) that is required for the somatic amplification of the dsRNA trigger and therefore allows the experimental depletion via feeding to occur primarily in the germlines of parental animals (Sijen et al. 2001).
4. Multiple mRNAs can be targeted simultaneously using any of the protocols presented. In the case of ingestion-mediated RNAi, the liquid cultures of each bacterial strain can be mixed after induction (in 96-well format) or before spreading on NGM + RNAi plates. For injection, equal

molar quantities of each dsRNA can be combined into the injection mixture. It must be noted that the overall efficiency of each dsRNA is usually lowered when several mRNAs are targeted in an individual animal.

5. One peculiarity of the genomic organization of *C. elegans* is that as many as 25% of individual genes exist as parts of transcriptional operons that are *trans*-spliced before transport from the nucleus to the cytoplasm. One caveat implicit to this transcriptional organization is that RNAi of individual target transcripts can potentially also occur within the nucleus, and an RNAi trigger can therefore also effect the expression of other genes derived from the same operon (Bosher et al. 1999). A definitive conclusion associated with an individual RNAi clone is best verified with traditional loss-of-function alleles in the corresponding gene.
6. Recently, two important phenomena have been identified in individual *C. elegans* strain backgrounds that directly affect the process of RNAi and therefore should be considered when comparing the effects of RNAi on separately derived mutant animals. The first is a natural polymorphism in the *mut-16* gene and can be traced in origin to one of several wild-type isolates originally isolated at the Medical Research Council. Unfortunately, this polymorphism has unknowingly been propagated in a variety of laboratories. This *mut-16* loss-of-function allele (*mut-16[mg461]*) primarily affects RNAi in somatic tissues and is considered to be a weak RNAi-defective strain. The presence of the specific *mut-16(mg461)* allele can be detected by PCR with the following primers using single-worm genotyping protocols: primer 1, CCCGCCGATACAGAAACTAA; primer 2, AATATTCGATCGGCAAGCAG. Wild-type *mut-16* PCR products are 824 bp, whereas animals containing *mut-16(mg461)* produce a product of 373 bp. In addition to polymorphisms in the *mut-16* gene, recent reports indicate that a prevalent spontaneous deletion in *eri-6/7* occurs with high frequency and is facilitated by the direct, inverted repeats that are present at the *eri-6/7* genomic locus. The presence of *eri-6/7* mutations can be determined empirically by feeding query *C. elegans* strains *E. coli* producing *lin-29* or *nhr-23* dsRNAs. RNAi of *lin-29* or *nhr-23* does not produce overt or visible phenotypes in wild-type animals but causes cuticle defects as well as an Egl (egg-laying defective) phenotype in *eri-6/7*–containing strains. Importantly, the spontaneous deletions in the *eri-6/7* locus cause animals to more efficiently deplete a targeted transcript via RNAi and therefore may lead to incorrect or improper interpretations of RNAi data in different strains.

SOLUTION AND REAGENT RECIPES

CAUTION: See the Cautions Appendix for appropriate handling of materials marked with <!>.

M9 Stock (1×)

KH_2PO_4	3.0 g
Na_2HPO_4	6.0 g
NaCl	5.0 g
$MgSO_4$ (1 M) <!>	1.0 mL
dH_2O	to 1 L

NGM Plates

Bacto-peptone	2.5 g
NaCl	3 g
Bacto-agar	18 g
dH_2O	to 975 mL

Autoclave. After cooling, add

$CaCl_2$ (1 M) <!>	1 mL
$MgSO_4$ (1 M) <!>	1 mL
Cholesterol (5 mg/mL) (dissolved in 100% ethanol <!>)	1 mL
$NaPO_4$ (1 M, pH 6) <!>	25 mL

READING LIST

Boutros M, Kiger AA, Armknecht S, Kerr K, Hild M, Koch B, Haas SA, Paro R, Perrimon N. 2004. Genome-wide RNAi analysis of growth and viability in *Drosophila* cells. *Science* **303:** 832–835.

Clemens JC, Worby CA, Simonson-Leff N, Muda M, Maehama T, Hemmings BA, Dixon JE. 2000. Use of double-stranded RNA interference in *Drosophila* cell lines to dissect signal transduction pathways. *Proc Natl Acad Sci* **97:** 6499–6503.

Doumanis J, Wada K, Kino Y, Moore AW, Nukina N. 2009. RNAi screening in *Drosophila* cells identifies new modifiers of mutant Huntingtin aggregation. *PLoS One* **4:** e7275.

Fire A, Xu S, Montgomery MK, Kostas SA, Driver SE, Mello CC. 1998. Potent and specific genetic interference by double-stranded RNA in *Caenorhabditis elegans*. *Nature* **391:** 806–811.

Piano F, Schetter AJ, Mangone M, Stein L, Kemphues KJ. 2000. RNAi analysis of genes expressed in the ovary of *Caenorhabditis elegans*. *Curr Biol* **10:** 1619–1622.

Timmons L, Court DL, Fire A. 2001. Ingestion of bacterially expressed dsRNAs can produce specific and potent genetic interference in *Caenorhabditis elegans*. *Gene* **263:** 103–112.

RNAi WITH SYNTHETIC siRNAs

The ability to knock down genes in mammalian cells using RNAi has been nothing short of revolutionary. A key innovation came with the realization that one could deliver to mammalian cells synthetic mimics of the products of Dicer cleavage and that these would program RISC without triggering nonspecific suppression of gene expression via protein kinase R (PKR) (Elbashir et al. 2001). Creating an siRNA is a simple matter of annealing two chemically synthesized oligoribonucleotides to generate a molecule with 19 paired bases and two overhanging nucleotides on the 3′ end of each strand. A number of vendors offer this configuration as a standard product, and the strands almost invariably come preannealed into a functional molecule. Some vendors have developed proprietary variations on the standard siRNA configuration that are reputed to have improved properties. These include enhanced efficiency (e.g., dsiRNAs, IDT; stealth siRNA, Invitrogen), restriction of activity to only one of the two strands (e.g., Dharmacon, On-Target; see below), or the ability to be taken up by mammalian cells passively, without the need for a transfection reagent (e.g., Accel, Dharmacon).

Many manufacturers offer predesigned and, in some cases, prevalidated siRNAs to certain genes or gene sets. Although these siRNAs eliminate one preparation phase of the siRNA-silencing experiment, it is important to understand the restrictions that accompany the use of stock tools. For example, some companies charge an additional fee to reveal sequences, structures, and chemical modifications (or refuse outright) and others place some restrictions on use. Many stock siRNAs were designed with proprietary algorithms that may or may not reflect the latest understanding of siRNA design rules. Therefore, one must always weigh the risks and benefits of using a predesigned tool.

Choosing a custom siRNA requires more effort but also affords flexibility. Sequences can be chosen to target specific alternatively spliced isoforms or even specific mutant or variant alleles. In all cases, the sequence of the siRNA itself is of critical importance to two properties: potency and specificity.

siRNAs are composed of two strands termed the guide and the passenger. The guide strand is complementary to the target RNA. The passenger strand has the same sequence as the target. The first goal is therefore to create a sequence that leads to predominant guide-strand incorporation into RISC, maximizing the concentration of active species and minimizing the possibility that RISC loaded with the passenger strand could recognize undesired targets (off-target effects). In fact, On-Target from Dharmacon modifies the passenger strand to prevent its incorporation into active RISC.

The design of effective siRNAs also relies both on the understanding of RNAi biochemistry and on large-scale studies of effective and ineffective siRNAs. The former has led to the realization that the siRNA strand with the less stable 5′ end has a greater probability of being incorporated as the guide strand into RISC. Thus, so-called "thermodynamic asymmetry" became an important consideration in effective small RNA design. Improvements were subsequently made by a simple brute-force measurement of the silencing potential of more than 2000 siRNAs. Feeding these data into an artificial neural network, Biopredsi (Huesken et al. 2005), which is no longer available online, or a linear regression model (dsir; Vert et al. 2006; http://biodev.extra.cea.fr/DSIR/DSIR.html) led to what may be the most effective algorithms at the time of the preparation of this chapter. However, RNAi is a fast-moving field, and it is critical that investigators keep abreast of new developments that will undoubtedly occur in the future. Among these, one would predict the development of algorithms that take into account features not only of the small RNA but also of its target.

Once siRNAs have been designed and purchased, the same considerations apply for introducing these into mammalian cells as do for *Drosophila* cell lines. Each mammalian line is most efficiently transfected with a different reagent or protocol. The best approach is to follow the guidance of the literature where possible and to test the panoply of available transfection methods where necessary. If using reagents suitable for the transfection of DNA and RNA, transfection of a plasmid that encodes GFP is a convenient way to test and calibrate efficiencies. As a rule of thumb, small RNAs will give greater effective transfection than the GFP plasmid, so even if 50% of cells become GFP positive, it is still possible to achieve 80%–90% knockdown with an effective siRNA. A number of reagents have been developed specifically for small RNA delivery. Although most will also deliver plasmids, one can calibrate small RNA transfection using fluorescently labeled siRNAs that can be obtained from a number of commercial vendors (e.g., siGLO Red, Dharmacon).

REFERENCES

Bosher JM, Dufourcq P, Sookhareea S, Labouesse M. 1999. RNA interference can target pre-mRNA: Consequences for gene expression in a *Caenorhabditis elegans* operon. *Genetics* **153:** 1245–1256.

Elbashir SM, Harborth J, Lendeckel W, Yalcin A, Weber K, Tuschl T. 2001. Duplexes of 21-nucleotide RNAs mediate RNA interference in cultured mammalian cells. *Nature* **411:** 494–498.

Huesken D, Lange J, Mickanin C, Weiler J, Asselbergs F, Warner J, Meloon B, Engel S, Rosenberg A, Cohen D, et al. 2005. Design of a genome-wide siRNA library using an artificial neural network. *Nat Biotechnol* **23:** 995–1001.

Kennedy S, Wang D, Ruvkun G. 2004. A conserved siRNA-degrading RNase negatively regulates RNA interference in *C. elegans*. *Nature* **427:** 645–649.

Sijen T, Fleenor J, Simmer F, Thijssen KL, Parrish S, Timmons L, Plasterk RH, Fire A. 2001. On the role of RNA amplification in dsRNA-triggered gene silencing. *Cell* **107:** 465–476.

Simmer F, Tijsterman M, Parrish S, Koushika SP, Nonet ML, Fire A, Ahringer J, Plasterk RH. 2002. Loss of the putative RNA-directed RNA polymerase RRF-3 makes *C. elegans* hypersensitive to RNAi. *Curr Biol* **12:** 1317–1319.

Vert JP, Foveau N, Lajaunie C, Vandenbrouck Y. 2006. An accurate and interpretable model for siRNA efficacy prediction. *BMC Bioinformatics* **7:** 520.

Protocol 5

RNAi in Cultured Mammalian Cells Using Synthetic siRNAs

This method is an example in which Lipofectamine 2000 is used for siRNA delivery into HEK293 cells. This lipid reagent has proven to be effective for many different cultured mammalian cell lines. Alternative reagents should be used according to the manufacturer's instructions with the following protocol serving as a guide.

MATERIALS

Reagents

Appropriate cell culture medium (e.g., Dulbecco's modified Eagle's medium [DMEM])
Chemically synthesized siRNA
Cultured mammalian cells (e.g., HEK293 for the example protocol)
Lipofectamine 2000
Opti-MEM I Reduced Serum Medium (31985-062, GIBCO-Invitrogen)
siGLO Red Transfection Indicator (D-001630-02-20, Thermo Scientific, Dharmacon) for preparation of negative control siRNA
Transfection reagent (e.g., Trans-IT TKO, Lipofectamine 2000)

Equipment

Centrifuge (low speed)
Hemocytometer for cell counting
Microscope
Platform rocker
Tissue culture incubator (37°C)
Tissue culture plates

OVERVIEW

This procedure is optimized to transfect siRNA into mammalian cells in a 24-well format. All amounts and volumes are given on a per-well basis. For other formats, scale the reagent amounts based on the surface area of the culture vessel according to Table 7-1.

METHOD

1. At 12–24 h before transfection, plate the cells in 500 μL of cell culture medium without antibiotics so that they will be 30%–50% confluent at the time of transfection.

   The precise number of cells depends on the doubling time and cell size and must be determined somewhat empirically for each cell type. About 5×10^4 to 10×10^4 cells serves as a good starting point.

2. For each transfection sample, prepare oligomer–Lipofectamine 2000 complexes as follows:

i. Resuspend the siRNA in H_2O to a concentration of 100 µM. Dilute this to 20 µM in Opti-MEM I Reduced Serum Medium. Add 3 µL of the 20 µM dilution to 50 µL of Opti-MEM I Reduced Serum Medium to have a final RNA concentration of 100 nM when added to the cells (total volume will be ~600 µL). Mix gently.

ii. Prepare a suitable negative control siRNA.

This can vary by experiment, but a convenient control is siGLO Red Transfection Indicator. This permits both processing for biological assays and confirmation of transfection. A third control that is often desirable is a mock transfection with no small RNA added.

iii. Before use, mix Lipofectamine 2000 gently and then dilute 1 µL in 50 µL of Opti-MEM I Reduced Serum Medium. Mix gently and incubate for 5 min at room temperature. It is critical to proceed to Step iv within 25 min.

iv. Combine the diluted siRNA with the diluted Lipofectamine 2000, mix gently, and incubate for 20 min at room temperature. The solution may appear cloudy.

3. Add the oligomer–Lipofectamine 2000 complexes (from Step 2.iv) drop by drop to each of the wells containing cells and medium (from Step 1). Mix gently by rocking the plate back and forth.

4. Incubate the cells in the presence of 5% CO_2 for 24–96 h at 37°C.

5. *Optional:* After 4–6 h, remove the transfection mixture and add the culture medium in which the cells are normally grown (DMEM/10% FCS for 293 cells).

Most cells will tolerate culture in Opti-MEM for the duration of the assay; however, in many cases, medium exchange may be desirable.

Assessing Knockdown

6. As with experiments in *Drosophila* cells, it is best to assess knockdown at the protein level by western blotting.

This is even more important in mammalian cells because only a fraction of RISCs are capable of carrying out small RNA-directed RNA cleavage, whereas the remainder suppress expression by interfering with protein synthesis. Thus, checking knockdown only by measuring RNA levels (e.g., by quantitative PCR or northern blotting) can overestimate knockdown for long-lived proteins or underestimate knockdown due to contributions of noncatalytic RISCs.

TIPS AND TROUBLESHOOTING

1. The most common reason for a failed siRNA experiment is the use of a small RNA that is not effective against the target. For this reason, it is critical to test several different siRNAs against any given target gene. Although design algorithms continuously improve, their predictive power is not yet sufficient to guarantee that every siRNA will potently knock down its target.

2. Occasionally, chemically synthesized siRNAs will arrive in poor condition or degrade from improper handling. Particularly if an experiment fails with previously validated tools, it is important to check the integrity of the siRNA dilution by gel electrophoresis or using a bioanalyzer.

3. It is critically important either that transfection be efficient or that transfected cells somehow be distinguished from untransfected cells in scoring knockdowns and their associated phenotypes. This can be accomplished through the use of a cotransfected GFP expression plasmid, by expression of ectopic cell surface markers that could facilitate selection on magnetic beads, or by FACS (fluorescence-activated cell sorting).

IMPORTANT CONSIDERATIONS FOR THE USE OF SYNTHETIC RNAi TRIGGERS

When considering the biological consequences of a gene knockdown, it is critical to consider that the specificity of any given siRNA reagent is an unknown. Even if a target is knocked down effectively, other genes may also be silenced (so-called off-target effects) (Jackson et al. 2003). Off-target effects can lead to misinterpretation of results and to incorrect biological inferences being drawn. A few rules of thumb should be considered as good practice in the use of RNAi methodologies. First, it is essential to use multiple, independent RNAi triggers, either long dsRNAs or siRNAs, and to assess both the phenotypic effects and the degree of knockdown caused by each trigger. The phenotypic effect should not be produced by negative controls and should correlate with the degree of silencing seen with the targeted agents. Ideally, any phenotype caused by silencing should be rescued by expression of an RNAi-resistant version of the target gene. This last suggestion is, of course, much more straightforward if siRNAs are used to elicit silencing, because the binding site of the siRNA can be altered by mutation or by eliminating those sequences from the rescue construct (e.g., if the original target site was in a 3′- or 5′-untranslated region [UTR]). If the binding site cannot be eliminated, it is a good idea to introduce multiple mutations into the resistant target, focusing on alterations in the seed region and in the RISC cleavage site (opposite guide nucleotides 10–11).

Ultimately, the specificity of the RNAi process is intrinsically limited. Endogenous triggers of RNAi and their associated RISC components have evolved to recognize and regulate targets with remarkably little complementarity (Bartel 2009). Only a seed match is sufficient in some cases to reduce expression of a complementary target by an miRNA. As we come to understand the biochemistry of RNAi in more detail, rules may emerge that minimize undesired side effects. However, at present, the best practice is to draw conclusions on the basis of multiple independent RNAi alleles that all behave in a manner consistent with the hypothesis being tested.

READING LIST

Bartel DP. 2009. microRNAs: Target recognition and regulatory function. *Cell* **136:** 215–233.

Elbashir SM, Harborth J, Lendeckel W, Yalcin A, Weber K, Tuschl T. 2001. Duplexes of 21-nucleotide RNAs mediate RNA interference in cultured mammalian cells. *Nature* **411:** 494–498.

Tiemann K, Rossi JJ. 2009. RNAi-based therapeutics: Current status, challenges and prospects. *EMBO Mol Med* **1:** 142–151.

Vert JP, Foveau N, Lajaunie C, Vandenbrouck Y. 2006. An accurate and interpretable model for siRNA efficacy prediction. *BMC Bioinformatics* **7:** 520.

RNAi WITH ENCODED TRIGGERS

Chemically or enzymatically synthesized triggers are undoubtedly potent and convenient; however, their effects are generally transient. This limitation exists because most organisms, unlike *C. elegans*, have no means of amplifying the silencing signal produced by a dsRNA trigger. Therefore, effects are diluted by cell division or limited by RISC half-life. Long-term effects in flies and mammals must be achieved by integrating into the genome a transgene from which an RNAi trigger is transcribed by a cellular RNA polymerase, generally RNA polymerase II or III. Not only can essentially permanent silencing be achieved, but control can also be exerted over the level of trigger expression, with a resulting variation in the degree to which the target is silenced. Knockdown can also be controlled in a temporal and cell-type-dependent fashion through the use of well-established transgenic methodologies and well-tested promoters.

The many advantages of encoded triggers are balanced by several drawbacks. Expression from integrated transgenes does not generally yield as high an intracellular concentration of trigger as can be achieved by introduction of RNA. Therefore, better-designed, more potent triggers are often necessary to achieve the same degree of effect.

Constructs containing inverted repeats can be unstable, both in bacteria and in mammalian cells. Triggers are often introduced into mammalian cells through the use of viral vectors, which require specific biosafety precautions. The generation of transgenic animals, either flies or mammals, requires both specialized skills and equipment. Overall, the effort of generating stable RNAi alleles is usually well spent, because they permit sophisticated analyses of gene function in vivo for an investment in time and resources that would be generally impossible to match using more conventional genetic methods.

Encoded Triggers in *Drosophila*

Both long inverted repeats and shRNAs can be used as silencing triggers for transgenic RNAi in *Drosophila* (Fortier and Belote 2000; Kennerdell and Carthew 2000; Lam and Thummel 2000; Martinek and Young 2000; Dietzl et al. 2007; Haley et al. 2008). Long inverted repeats mimic endogenous loci that give rise to endo-siRNAs in many different tissue types (Czech et al. 2008; Ghildiyal et al. 2008; Okamura et al. 2008). Target sequences are chosen much as they are for RNAi in *Drosophila* cells using long dsRNAs. Inverted repeats are typically 400–600 bp in length. For use, the two arms of the inverted repeat are cloned serially into a plasmid suitable for expression in the fly (Figs. 7-4 and 7-5). Creating and maintaining such structures in *E. coli* is challenging, but it is greatly aided by using strains specifically designed for problematic sequences (e.g., STBL2, 3, and 4, Invitrogen). In general, most problems are encountered during the original cloning, requiring careful checking of plasmid structures during construction. However, once cloned, the inverted repeats are relatively stable, provided that they are propagated appropriately.

shRNAs are essentially miRNA mimics. Basically, two general configurations are used in flies and mammals (Fig. 7-4). In the simplest form, a short inverted repeat is created with a perfect stem of 19–29 nucleotides and a variable length loop (usually six to eight bases) (Brummelkamp et al. 2002; Paddison et al. 2002). The entire structure can be created by annealing two synthetic oligonucleotides such that appropriate overhangs are created for direct cloning into an expression vector (Fig. 7-5). Often, this type of shRNA is transcribed from an RNA polymerase III promoter such that initiation and termination define the 5′ and 3′ ends of the guide and passenger strands of the mature small RNA. Dicer, at least in the case of the longer triggers, recognizes and processes primary transcripts to yield an siRNA-like molecule that is incorporated into RISC. For the shorter stems, it is unclear how maturation occurs. It is also worth noting that the 5′-triphosphate terminus on the RNA-polymerase-III–generated versions is not the configuration that is normally recognized by RISC and that this type of RNA can sometimes generate stress or antiviral responses, e.g., via RIG-I (Kim et al. 2004; Kenworthy et al. 2009).

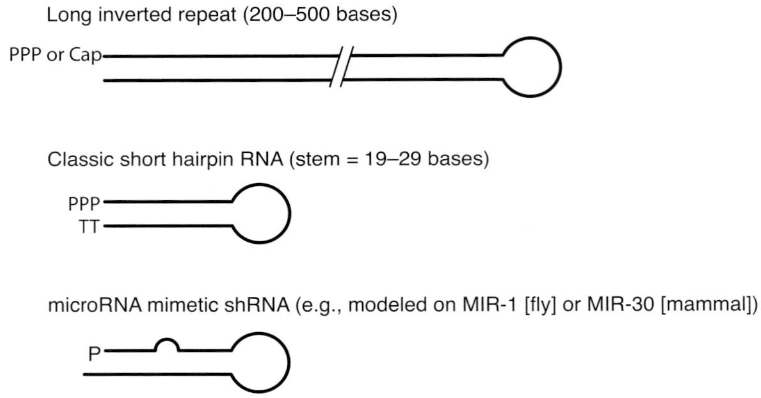

FIGURE 7-4. Hairpin strategies for triggering RNAi. Three basic strategies are used for encoded RNAi. The first uses a long inverted repeat that can be transcribed by RNA polymerase II or III. This results in a structure that is capped or initiates with a triphosphate. Dicer presumably initiates processing at a random location within the double-strand region and produces a sampling of siRNAs. Classic short hairpins are simple inverted repeat structures. These are almost always transcribed using RNA polymerase III promoters to give structures with 19–29-nucleotide stems, 5′ triphosphates, and one or more Ts at the 3′ end of the transcript. The processing pathway for these is not well studied, but at least in cases of longer stems does involve Dicer cleavage. miRNA mimetic shRNAs are carried in a specific miRNA-derived scaffold and are processed to intermediate pre-miRNAs (shown) by Drosha. These have stems of generally >21 bases and bear a 5′-phosphate group. This structure is cleaved specifically by Dicer, which appears to recognize and cut a precise distance from the overhung end.

shRNAs modeled more closely on endogenous miRNAs appear to offer advantages over the simpler design, including more efficient incorporation into RISC and more flexibility in the design of expression vectors (Zeng et al. 2002; Dickins et al. 2005; Stegmeier et al. 2005; Dickins et al. 2007). A number of endogenous miRNA backbones have been remodeled to produce experimentally designed small RNAs, with miR-1 in flies (see Protocol 6) (Haley et al. 2008) and miR-30 in mammals (see Protocol 7) (Zeng et al. 2002) being those in widest use. Construction proceeds much as with an inverted repeat shRNA, often by synthesizing and annealing two oligonucleotides that fully span the guide and passenger strands. In this case, the guide and passenger are separated by the loop derived from the miRNA. Some additional miRNA-derived flanking sequence is also included, because in this case, the artificial sequence must be engrafted into a larger fragment of the primary miRNA transcript (Fig. 7-5). Thus, expression vector designs incorporate restriction sites in a manner that minimizes predicted effects on the folding of the primary transcript, whose structure is critical for efficient and accurate processing (Han et al. 2006). Just as do endogenous miRNAs, miR-like shRNAs mature via a two-step biogenesis pathway. The primary transcript, virtually always transcribed by RNA polymerase II, is first cleaved by Drosha in the nucleus to produce a pre-shRNA. This is exported to the cytoplasm, where it is cleaved by Dicer and incorporated into RISC. The passenger strand is then lost to form an active complex.

Natural miRNAs are tightly regulated. Virtually all show defined, tissue-specific expression patterns. Many are also subject to posttranscriptional controls that influence the efficiency of their maturation by Drosha and Dicer. Thus far, miR-1 and miR-30 appear to be constitutively processed in all cell types examined, but this does not ensure that there will not be some cell type in which their expression is repressed. It is not difficult to create new shRNA scaffolds by remodeling other endogenous miRNAs (e.g., for use in an exotic cell type or model organism). However, empirical experience has shown that not all pri-miRNAs tolerate restructuring, so several must be tested. It is also a good guideline to choose an miRNA that is broadly and highly expressed as a backbone for creating artificial structures.

Several large-scale collections of *Drosophila* strains have been built in which a substantial fraction of fly genes have been targeted by long inverted repeat RNAs (Dietzl et al. 2007; Ni et al. 2009).

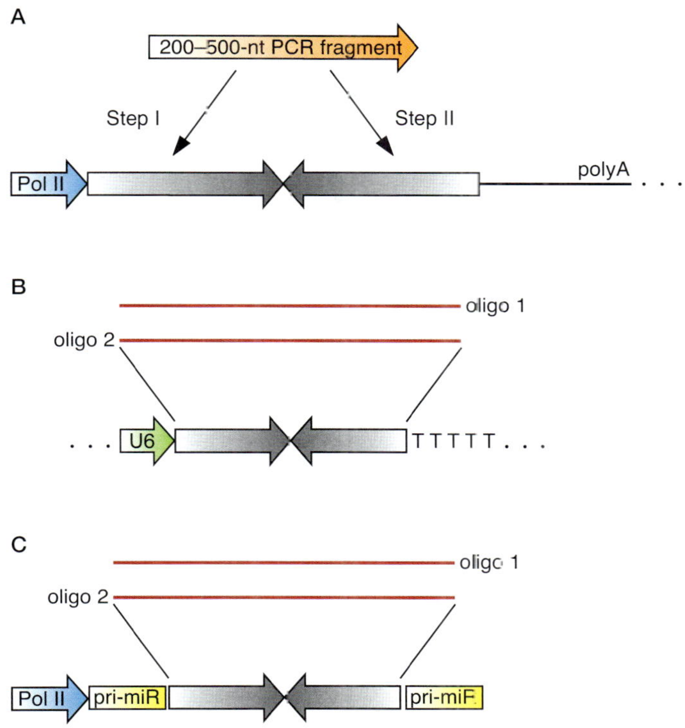

FIGURE 7-5. Cloning strategies for encoded triggers. (*A*) Cloning long inverted repeats is generally a two-step process. A PCR product, generally designed in the manner one would use for creating a template for synthesis of dsRNA, is inserted into a vector downstream from an RNA polymerase II promoter. A second copy of the fragment is then inserted adjacent to the first in the opposite orientation. Often, a marker or an intron is inserted between the two halves of the hairpin to stabilize the structure in *E. coli*. (*B*) "Classic" shRNAs are often constructed simply by annealing two synthetic oligonucleotides such that the double-stranded product has the appropriate overhangs for cloning into the recipient vector. In this case, transcription by RNA polymerase III initiates at the first base of the inverted repeat and termination at a stretch of Ts adds an overhang of several (usually two to four) bases. (*C*) Cloning of miRNA mimetic shRNAs is similar, in that inserts can be most easily made by the annealing of two oligonucleotides. The product is then cloned into a vector that embeds the inverted repeat within a fragment of the primary transcript of an miRNA (pri-miR, not shown to scale). Virtually any strong RNA polymerase II promoter can be used to drive shRNA expression, including inducible and tissue-specific promoters. Often, a functional spacer is placed between the promoter and the beginning of the pri-miR cassette, which both reports indirectly the expression level of the shRNA and seems (for reasons unknown) to enhance shRNA processing.

However, given the flexibility, specificity, and potency of shRNAs, these are becoming both more commonly used and more highly recommended. Although the miR-1 backbone is the standard choice for fly shRNA construction at the time of the preparation of this chapter, it must be appreciated that this is a fast-moving area and the user is cautioned to investigate the most current literature available on this topic.

REFERENCES

Bartel DP. 2009. microRNAs: Target recognition and regulatory functions. *Cell* **136:** 215–233.

Brummelkamp TR, Bernards R, Agami R. 2002. A system for stable expression of short interfering RNAs in mammalian cells. *Science* **296:** 550–553.

Czech B, Malone CD, Zhou R, Stark A, Schlingeheyde C, Dus M, Perrimon N, Kellis M, Wohlschlegel JA, Sachidanandam R, et al. 2008. An endogenous small interfering RNA pathway in *Drosophila*. *Nature* **453:** 798–802.

Dickins RA, Hemann MT, Zilfou JT, Simpson DR, Ibarra I, Hannon GJ, Lowe SW. 2005. Probing tumor phenotypes using stable and regulated synthetic microRNA precursors. *Nat Genet* **37:** 1289–1295.

Dickins RA, McJunkin K, Hernando E, Premsrirut PK, Krizhanovsky V, Burgess DJ, Kim SY, Cordon-Cardo C, Zender L, Hannon GJ, et al. 2007. Tissue-specific and reversible RNA interference in transgenic mice. *Nat Genet* **39:** 914–921.

Dietzl G, Chen D, Schnorrer F, Su KC, Barinova Y, Fellner M, Gasser B, Kinsey K, Oppel S, Scheiblauer S, et al. 2007. A genome-wide transgenic RNAi library for conditional gene inactivation in *Drosophila*. *Nature* **448:** 151–156.

Fortier E, Belote JM. 2000. Temperature-dependent gene silencing by an expressed inverted repeat in *Drosophila*. *Genesis* **26:** 240–244.

Ghildiyal M, Seitz H, Horwich MD, Li C, Du T, Lee S, Xu J, Kittler EL, Zapp ML, Weng Z, et al. 2008. Endogenous siRNAs derived from transposons and mRNAs in *Drosophila* somatic cells. *Science* **320:** 1077–1081.

Haley B, Hendrix D, Trang V, Levine M. 2008. A simplified miRNA-based gene silencing method for *Drosophila melanogaster*. *Dev Biol* **321:** 482–490.

Han J, Lee Y, Yeom KH, Nam JW, Heo I, Rhee JK, Sohn SY, Cho Y, Zhang BT, Kim VN. 2006. Molecular basis for the recognition of primary microRNAs by the Drosha-DGCR8 complex. *Cell* **125:** 887–901.

Jackson AL, Bartz SR, Schelter J, Kobayashi SV, Burchard J, Mao M, Li B, Cavet G, Linsley PS. 2003. Expression profiling reveals off-target gene regulation by RNAi. *Nat Biotechnol* **21:** 635–637.

Kennerdell JR, Carthew RW. 2000. Heritable gene silencing in *Drosophila* using double-stranded RNA. *Nat Biotechnol* **18:** 896–898.

Kenworthy R, Lambert D, Yang F, Wang N, Chen Z, Zhu H, Zhu F, Liu C, Li K, Tang H. 2009. Short-hairpin RNAs delivered by lentiviral vector transduction trigger RIG-I-mediated IFN activation. *Nucleic Acids Res* **37:** 6587–6599.

Kim DH, Longo M, Han Y, Lundberg P, Cantin E, Rossi JJ. 2004. Interferon induction by siRNAs and ssRNAs synthesized by phage polymerase. *Nat Biotechnol* **22:** 321–325.

Lam G, Thummel CS. 2000. Inducible expression of double-stranded RNA directs specific genetic interference in *Drosophila*. *Curr Biol* **10:** 957–963.

Martinek S, Young MW. 2000. Specific genetic interference with behavioral rhythms in *Drosophila* by expression of inverted repeats. *Genetics* **156:** 1717–1725.

Ni JQ, Liu LP, Binari R, Hardy R, Shim HS, Cavallaro A, Booker M, Pfeiffer BD, Markstein M, Wang H, et al. 2009. A *Drosophila* resource of transgenic RNAi lines for neurogenetics. *Genetics* **182:** 1089–1100.

Okamura K, Chung WJ, Ruby JG, Guo H, Bartel DP, Lai EC. 2008. The *Drosophila* hairpin RNA pathway generates endogenous short interfering RNAs. *Nature* **453:** 803–806.

Paddison PJ, Caudy AA, Bernstein E, Hannon GJ, Conklin DS. 2002. Short hairpin RNAs (shRNAs) induce sequence-specific silencing in mammalian cells. *Genes Dev* **16:** 948–958.

Stegmeier F, Hu G, Rickles RJ, Hannon GJ, Elledge SJ. 2005. A lentiviral microRNA-based system for single-copy polymerase II-regulated RNA interference in mammalian cells. *Proc Natl Acad Sci* **102:** 13212–13217.

Zeng Y, Wagner EJ, Cullen BR. 2002. Both natural and designed micro RNAs can inhibit the expression of cognate mRNAs when expressed in human cells. *Mol Cell* **9:** 1327–1333.

Protocol 6

Construction of Transgenic *Drosophila* Expressing shRNAs in the miR-1 Backbone

This protocol applies to the use of miRNA-based shRNAs, but most of the vectors, principles of experimental design, and protocols are also applicable to long inverted repeat transgenes, which are not extensively discussed here.

For expression in flies, the miR-1 cassette is generally cloned downstream from a basal promoter linked to multiple copies of upstream activator sequences (UASs) for the Gal4 transcriptional activator. These are combined with specific Gal4 enhancer-trap transgenes (also known as Gal4 drivers) in which the expression of Gal4 is under the control of various tissue- and developmental-stage-specific enhancer sequences. In this way, siRNAs can be generated from long- or short-hairpin-based silencing triggers, and endogenous genes can be silenced under spatial and temporal control (Brand and Perrimon 1993). The versatility of the modular Gal4/UAS system and the large collection of available Gal4 driver lines offer significant advantages for using *Drosophila* as a model system to interrogate gene function via RNAi.

Two approaches can be used to generate UAS–shRNA lines. The first is P-element-mediated transposition that leads to random insertion of the UAS–hairpin construct into the genome. A second approach is the insertion of transgenes via site-specific recombination into genomic loci known to be advantageous for expression (Groth et al. 2004; Markstein et al. 2008).

These two approaches use different vector backbones. Most P-element-based transgenic RNAi vectors are modified from the original pUAST vector (Brand and Perrimon 1993), which typically carries a mini-*white* gene as selection marker and an expression cassette consisting of multiple GAL4-responsive UAS elements, the basal hsp70 promoter, an intronic sequence, and a polyadenylation signal sandwiched between the minimal P-element sequences, which are required for genomic integration. Sequences are inserted into cloning sites that are placed downstream from the basal hsp70 promoter. For shRNAs, inserts for cloning can be prepared simply by annealing synthetic oligonucleotides (Fig. 7-5 and below). However, it should be noted that the same vectors can also be used for expression of long inverted repeat hairpins, necessitating a two-step cloning in which serial insertion of PCR fragments creates the inverted repeat. The vectors used in the site-specific integration system are referred to as the "VALIUM series" (Vermilion-AttB-Loxp-Intron-UAS-MCS) (Ni et al. 2008, 2009). These vectors contain *vermilion* as a selectable marker and an *attB* sequence to allow ϕC31-targeted integration at genomic *attP*-landing sites (Bischof et al. 2007). On the basis of the finding that UAS sites promote transcription in an additive fashion, the VALIUM vectors contain two pentamers of UAS sequences, one of which can be removed using the Cre/*loxP* system. VALIUM vectors use the *hsp70* TATA promoter, which has been shown to be effective in UAS vectors.

A number of parameters have been thoroughly tested to achieve optimal RNAi potency, including the placement of insulator sequences to increase hairpin expression, the identity, number, and location of intronic sequences (to reduce difficulties in hairpin propagation in bacteria), the identity of the polyadenylation signals, and the orientation of the hairpin. The latest vector as of the preparation of this chapter, VALIUM20 (Fig. 7-6), has been demonstrated to be highly effective for RNAi knockdown, but the experimenter is cautioned to stay informed regarding the almost continuous vector development in the field.

The targeted method to generate shRNA lines has many practical advantages over P-element-based methods. The frequency of recovering transformants using the integrase method, either following coinjection of integrase mRNA or by injecting into the nanos-integrase strain, is almost

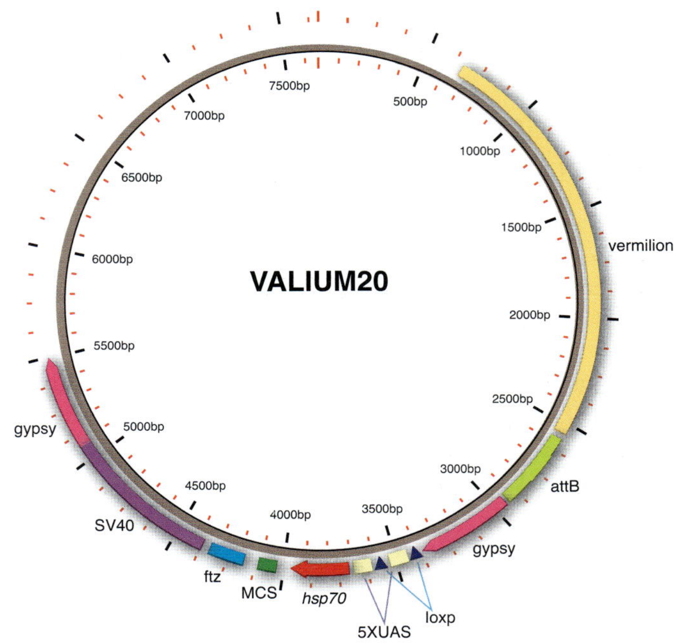

FIGURE 7-6. Map of VALIUM20, an example of an optimized vector for creating transgenic flies carrying shRNA inserts. It uses the ϕC31 recombinase for site-specific integration into any of a number of different, well-characterized landing sites in the *Drosophila* genome. The shRNA is driven by a minimal *hsp70* promoter linked to up to 10 UASs for activator binding. This permits tissue-specific expression using any of the many Gal4 driver lines that are available. The vector carries a *vermillion* marker to track integration and has been optimized (Ni et al., in prep.) by the placement of introns and insulator sequences to promote effective shRNA expression. In this case, miR-1-based inserts are cloned into NheI and EcoR1 sites in the vector polylinker, as described in the accompanying protocols.

fivefold higher than in conventional P-element transformation. Establishment of the lines is greatly facilitated because no mapping of the transformants to a specific chromosome (for the purpose of future crosses) is needed. Moreover, unlike P-element-based methods, insertions into the *attP*-landing sites are usually homozygous viable. Because the potency of transgenic RNAi depends on the level of expression of the UAS-driven constructs, the VALIUM vectors with their modular number of UAS copies allow the generation of a phenotypic series. From the original 10XUAS construct, a 5XUAS derivative can be recovered, and because most *attP*-containing chromosomes can be made homozygous, it is also possible to generate both 15XUAS and 20XUAS combinations (Ni et al. 2008). The ability to generate a phenotypic series from 5XUAS to 20XUAS may prove to be useful, in particular, when variation of the gene expression dosage is important for phenotypic studies of genes with pleiotropic impacts.

MATERIALS

The recipes for the items marked with <R> are at the end of the protocol or in the Common Recipes Appendix.

CAUTION: See the Cautions Appendix for appropriate handling of material marked with <!>.

Reagents

Annealing buffer <R>
ddH_2O
DNA ligase (Roche 10716359001)

Drosophila fly stock

The genetic background for making transgenics is y^1, *sc*, v^1, *nanos integrase* (y^+); *attP2* (y^+). A preferred *attP*-docking site is *attP2* located on the third chromosome at 68A4.

Injection buffer (1×) <R>

Ligation buffer (10×; Roche) <R>

Luria broth (LB) + ampicillin (50 mg/mL) <!>

Oligonucleotides (see Step 2)

pVALIUM20 vector

QIAprep Spin Miniprep Kit (QIAGEN 27106)

shRNA (see Step 1)

Equipment

Bacterial incubator (37°C)

Bacterial shaker (37°C)

Benchtop centrifuge

METHOD

Design of *Drosophila* shRNAs

1. The principles of shRNA construction are precisely the same as those discussed above for the design of synthetic siRNAs. The key is to choose a target site for which the corresponding small RNA is incorporated into RISC and for which that enzyme in turn efficiently recognizes and cleaves its target.

   In *Drosophila*, there are a few twists on the process, discussed below for mammals, that must be appreciated. Flies have two Argonaute proteins, Ago1 and Ago2, that are dedicated to the miRNA and endo-siRNA pathways, respectively. Although shRNAs are miRNA mimetics, it is important that they be funneled into the endo-siRNA pathway for maximum potency (Czech et al. 2008; Ghildiyal et al. 2008; Haley et al. 2008; Okamura et al. 2008). Although the Dicers, their obligate cofactors, and Argonautes differ for each of the two small RNA fates, they are not completely separate, and it is possible to direct an artificial miRNA selectively into Ago2 based on insights into rules for small RNA sorting that have emerged in the past 2 years (Czech et al. 2009; Okamura et al. 2009; Ghildiyal et al. 2010).

   Algorithms developed for mammalian siRNA design (see above) seem to work effectively also for *Drosophila* cells, demonstrating conservation of many of the essential features of effective silencing triggers. Loading of those RNAs into Ago2 can be directed by maintaining a perfectly paired stem instead of an miRNA-like stem, which almost invariably contains one or more bulged residues. Although siRNAs are designed as 21-nucleotide species, the small RNAs that emerge from the miR-1 backbone will be 22 nucleotides in length. The design principles of the construct compensate for this difference and result in a nucleotide being appended to the 3′ end of the guide strand, which is not thought to impact substrate specificity.

Oligonucleotide Design

2. The design of oligonucleotides for ligation into a miR-1 vector is straightforward.

   i. Begin with the 21-nucleotide siRNA prediction that can form perfect complementary base pairing with the target mRNA and add tagttatattcaagcata to the 5′ end. Infer the complement of the original 21-nucleotide siRNA prediction and add this to the 5′ end of the composite sequence. The resulting sequence will be the backbone of the "top" oligonucleotide.

ii. Generate the complementary sequence of the "top" oligonucleotide to create the backbone of the "bottom" oligonucleotide.

iii. Add ctagcagt and gcg to the 5′ and 3′ ends, respectively, of the "top" oligonucleotide.

iv. Add aattcgc and actg sequences to the 5′ and 3′ ends, respectively, of the "bottom" oligonucleotide.

v. Order the oligonucleotides from a reputable purveyor. There is no need for any special purification.

Annealing of the oligonucleotide pair will create a duplexed molecule with overhangs compatible with NheI and EcoRI sites for cloning into doubly cut VALIUM vectors, as described in detail below. Overhangs compatible with alternative vectors can be created simply by modifying the sequences appended in the preceding protocol.

Cloning miR-1-based shRNAs for Expression in Flies

3. Anneal as described below.

i. Dissolve the oligonucleotides in ddH_2O to make stock solutions (20 mM).

ii. Add 10 µL of top-strand and 10 µL of bottom-strand oligonucleotides to 100 µL of 1.2x annealing buffer and mix well.

iii. Heat the mixture for 5 min at 95°C and then slowly cool to room temperature.

The resulting DNA fragment features overhangs suitable for ligation with NheI- and EcoRI-digested vectors.

4. Clone as described below.

i. Directly ligate the DNA fragment from Step 3 with the pVALIUM20 vector that has been linearized with NheI and EcoRI:

Annealed DNA fragment	6 mL
Linearized vector (40 ng/µL)	2 mL
Ligation buffer (10x) (Roche)	2 mL
DNA ligase (Roche 10716359001)	1 mL
ddH_2O	to 20 mL

Mix well and incubate for 1 h at 16°C.

ii. Once ligations are transformed and individual colonies are isolated, screen the colonies for the presence of inserts by PCR using the following primer pairs: ACCAGCAACCAAG TAAATCAAC (forward) and TAATCGTGTGTGATGCCTACC (reverse).

The correct colony should yield an ~300-bp PCR product.

Alternatively, colonies can be screened by digestion with NheI and EcoRI.

iii. Sequence-verify the constructs with the following primer: ACCAGCAACCAAGTAAAT CAAC.

5. Purify the DNA for injection.

The correct clones can be expanded and DNA can be purified for injection in a number of ways. It is best to follow the recommendations of the injection facility (either internal or a commercial vendor). Virtually any high-quality DNA preparation can be used. Purified material should be suspended in 1x injection buffer and supplied to the facility at their requested concentration.

6. Create transgenic *Drosophila* carrying shRNA inserts.

Injection of hairpin-containing constructs can be performed either in-house following well-established protocols or by using commercial services (e.g., Genetic Services, Inc. [GSI] or BestGene).

SOLUTION AND REAGENT RECIPES

CAUTION: See the Cautions Appendix for appropriate handling of materials marked with <!>.

Annealing Buffer

Tris-HCl (12 mM, pH 7.5)
NaCl (0.12 M)
EDTA (1.2 mM) <!>

Injection Buffer (1x)

NaH_2PO_4 (0.1 mM, pH 6.8)
KCl (5 mM) <!>

Ligation Buffer (10x)

Reagent	Quantity (for 1 mL)	Final concentration
Tris-HCl (1 M, pH 7.5)	0.5 mL	500 mM
$MgCl_2$ (1 M) <!>	0.1 mL	100 mM
DTT (1 M) <!>	0.1 mL	100 mM
ATP (100 mM)	0.1 mL	10 mM
H_2O	0.2 mL	

Store for 6 mo at –20°C.

TIPS AND TROUBLESHOOTING

1. Although most transgenic lines are homozygous viable, a small percentage of the lines cannot be made homozygous. The underlying cause is still not clear, but it is conceivable that some basal level of transcription of the minimal promoter-driven short hairpin transgene can lead to the silencing of an essential target gene and thus contribute to the observed lethality.
2. Typically, a hairpin targeting the *white* gene, which encodes a transporter required for proper pigmentation in the fly eyes and is dispensable for viability, serves as a negative control for an in vivo RNAi experiment. A short hairpin transgene targeting a heterologous gene, such as *GFP* or *firefly luciferase*, can also serve as a negative control.
3. Sometimes the target gene is essential and severe knockdown can lead to lethality. An allelic series can be created by setting up the cross between the short hairpin transgenic RNAi lines and Gal4 driver lines at different temperatures (ranging from 18°C to 29°C). Higher temperatures generally lead to more efficient knockdowns. In addition, various ubiquitous Gal4 drivers of different strengths can be sampled.

READING LIST

Dietzl G, Chen D, Schnorrer F, Su KC, Barinova Y, Fellner M, Gasser B, Kinsey K, Oppel S, Scheiblauer S, et al. 2007. A genome-wide transgenic RNAi library for conditional gene inactivation in *Drosophila*. *Nature* **448:** 151–156.

Haley B, Hendrix D, Trang V, Levine M. 2008. A simplified miRNA-based gene silencing method for *Drosophila melanogaster*. *Dev Biol* **321:** 482–490.

Saj A, Arziman Z, Stempfle D, van Belle W, Sauder U, Horn T, Durrenberger M, Paro R, Boutros M, Merdes G. 2010. A combined ex vivo and in vivo RNAi screen for notch regulators in *Drosophila* reveals an extensive notch interaction network. *Dev Cell* **18:** 862–876.

Schnorrer F, Schonbauer C, Langer CC, Dietzl G, Novatchkova M, Schernhuber K, Fellner M, Azaryan A, Radolf M, Stark A, et al. 2010. Systematic genetic analysis of muscle morphogenesis and function in *Drosophila*. *Nature* **464:** 287–291.

ENCODED TRIGGERS IN MAMMALIAN CELL LINES

Encoded RNAi triggers in mammalian cells are essentially restricted to shRNAs—either the simple inverted repeat design or the miRNA mimic (Brummelkamp et al. 2002; Paddison et al. 2002; Zeng et al. 2002). Although both formats have their proponents, the latter are recommended both for their potency and for the ease of driving their expression from a variety of different promoters. The use of miRNA-based designs is assumed in the following discussions, but many of the points raised also apply to the inverted repeat format.

shRNAs can be used in transient transfections, similar to the procedure described for siRNAs, but this sacrifices many of their advantages. Instead, to exploit the full range of their possibilities, shRNAs are usually integrated into the host cell genome as a stable transgene. In this format, there are many options for expressing the shRNA from constitutive or regulated promoters and for monitoring shRNA expression.

Design of Mammalian shRNA Expression Vectors

shRNA expression vectors consist of several modular elements. The core of their functionality lies in the shRNA cassette. For mammals, the miR-30a backbone is often used. On the basis of early studies of ectopic miRNA expression constructs, ~200–300 bases of the primary miRNA transcript are generally included in the vector (Zhou et al. 2007). The miRNA hairpin itself is placed at the center of the fragment, and the natural miRNA stem is replaced with the desired sequence. The most critical aspects of the design are to maintain the overall structure of the RNA, so that Drosha and Dicer cleavage sites are efficiently recognized, and to place the artificial sequence within the stem such that the desired 5′ end of the guide strand is predictably formed. For miR-30 cassettes, this involves creating a perfectly duplexed stem with the 5′ end of the guide at the position that normally forms the 5′ end of the miR strand of miR-30. The proper processing of this modified miR-30 design has been rigorously verified with many different shRNAs, initially by rapid amplification of cDNA ends (RACE) and later by small RNA sequencing (Silva et al. 2005). If a new miRNA backbone is to be used, it is imperative that its processing be well understood and that the RNAs produced from many remodeled designs be determined.

The second key element of the shRNA cassette is the promoter that drives the expression of the primary transcript. miRNA mimetics are expressed well from RNA polymerase II promoters that are used to drive protein-encoding transgenes, including strong viral promoters (e.g., cytomegalovirus [CMV], Moloney murine leukemia virus [M-MuLV], or murine stem cell virus [MSCV] long terminal repeat [LTR] promoters), cellular promoters (e.g., eukaryotic initiation factor-1a [EIF-1a]), and inducible chimeric promoters (e.g., tet-CMV). The use of tissue-specific promoters is also possible. When choosing a promoter, there are several issues to consider. First, the promoter must be highly transcribed in the cell line in which knockdown is desired. Even strong viral promoters such as CMV are not well expressed in some cell types, and many commonly used promoters are selectively silenced in stem cells. The tet-CMV promoter seems to be active in most cell types, provided an appropriate activator (tet-VP16) is present at sufficient levels. The literature contains guidance for many commonly used cell lines, in the form of reports wherein shRNAs have been successfully expressed. If no prior information is available, it is a good idea to test several promoters comparatively, both for small RNA expression and target knockdown.

Mature miRNA sequences arise from nearly every possible area within annotated primary transcripts, including coding and noncoding exons and introns (Bartel 2009). In the artificial setting, the miRNA backbone can be separated from the promoter by a spacer that can provide additional functionality (Stegmeier et al. 2005). A common design is to place the shRNA within the 3′UTR of an expression cassette that encodes a fluorescent marker protein. This bifunctional transcript can produce either the small RNA or the encoded protein, but expression levels of the two are operationally linked. In this manner, the relative level of marker expression serves as a proxy for expression of the small RNA. This is particularly useful in situations in which induction of expression is to be monitored (e.g., from an inducible promoter) or in which an investigator wants to purify a subset of cells (e.g., by FACS) with the highest shRNA expression levels.

READING LIST

Bischof J, Maeda RK, Hediger M, Karch F, Basler K. 2007. An optimized transgenesis system for *Drosophila* using germ-line-specific ϕC31 integrases. *Proc Natl Acad Sci* **104:** 3312–3317.

Brand AH, Perrimon N. 1993. Targeted gene expression as a means of altering cell fates and generating dominant phenotypes. *Development* **118:** 401–415.

Brummelkamp TR, Bernards R, Agami R. 2002. A system for stable expression of short interfering RNAs in mammalian cells. *Science* **296:** 550–553.

Czech B, Malone CD, Zhou R, Stark A, Schlingeheyde C, Dus M, Perrimon N, Kellis M, Wohlschlegel JA, Sachidanandam R, et al. 2008. An endogenous small interfering RNA pathway in *Drosophila. Nature* **453:** 798–802.

Czech B, Zhou R, Erlich Y, Brennecke J, Binari R, Villalta C, Gordon A, Perrimon N, Hannon GJ. 2009. Hierarchical rules for Argonaute loading in *Drosophila. Mol Cell* **36:** 445–456.

Ghildiyal M, Seitz H, Horwich MD, Li C, Du T, Lee S, Xu J, Kittler EL, Zapp ML, Weng Z, et al. 2008. Endogenous siRNAs derived from transposons and mRNAs in *Drosophila* somatic cells. *Science* **320:** 1077–1081.

Ghildiyal M, Xu J, Seitz H, Weng Z, Zamore PD. 2010. Sorting of *Drosophila* small silencing RNAs partitions microRNA* strands into the RNA interference pathway. *RNA* **16:** 43–56.

Groth AC, Fish M, Nusse R, Calos MP. 2004. Construction of transgenic *Drosophila* by using the site-specific integrase from phage ϕC31. *Genetics* **166:** 1775–1782.

Haley B, Hendrix D, Trang V, Levine M. 2008. A simplified miRNA-based gene silencing method for *Drosophila melanogaster. Dev Biol* **321:** 482–490.

Markstein M, Pitsouli C, Villalta C, Celniker SE, Perrimon N. 2008. Exploiting position effects and the gypsy retrovirus insulator to engineer precisely expressed transgenes. *Nat Genet* **40:** 476–483.

Ni JQ, Liu LP, Binari R, Hardy R, Shim HS, Cavallaro A, Booker M, Pfeiffer BD, Markstein M, Wang H, et al. 2009. A *Drosophila* resource of transgenic RNAi lines for neurogenetics. *Genetics* **182:** 1089–1100.

Ni JQ, Markstein M, Binari R, Pfeiffer B, Liu LP, Villalta C, Booker M, Perkins L, Perrimon N. 2008. Vector and parameters for targeted transgenic RNA interference in *Drosophila melanogaster. Nat Methods* **5:** 49–51.

Okamura K, Chung WJ, Ruby JG, Guo H, Bartel DP, Lai EC. 2008. The *Drosophila* hairpin RNA pathway generates endogenous short interfering RNAs. *Nature* **453:** 803–806.

Okamura K, Liu N, Lai EC. 2009. Distinct mechanisms for microRNA strand selection by *Drosophila* Argonautes. *Mol Cell* **36:** 431–444.

Paddison PJ, Caudy AA, Bernstein E, Hannon GJ, Conklin DS. 2002. Short hairpin RNAs (shRNAs) induce sequence-specific silencing in mammalian cells. *Genes Dev* **16:** 948–958.

Silva JM, Li MZ, Chang K, Ge W, Golding MC, Rickles RJ, Siolas D, Hu G, Paddison PJ, Schlabach MR, et al. 2005. Second-generation shRNA libraries covering the mouse and human genomes. *Nat Genet* **37:** 1281–1288.

Stegmeier F, Hu G, Rickles RJ, Hannon GJ, Elledge SJ. 2005. A lentiviral microRNA-based system for single-copy polymerase II-regulated RNA interference in mammalian cells. *Proc Natl Acad Sci* **102:** 13212–13217.

Zeng Y, Wagner EJ, Cullen BR. 2002. Both natural and designed micro RNAs can inhibit the expression of cognate mRNAs when expressed in human cells. *Mol Cell* **9:** 1327–1333.

Zhou B, Wang S, Mayr C, Bartel DP, Lodish HF. 2007. miR-150, a microRNA expressed in mature B and T cells, blocks early B cell development when expressed prematurely. *Proc Natl Acad Sci* **104:** 7080–7085.

Protocol 7

Creating an miR30-based shRNA Vector

Generating expression constructs for artificial miRNAs is relatively straightforward. This protocol describes the creation of miR-30-based shRNA cassettes that are compatible with a number of standard vector systems (Fig. 7-7). The principles outlined below can also be easily applied to other miRNA scaffolds or to simple snapback shRNAs. Again, it is important to note that one must understand the processing of the artificial scaffold and be able to predict precisely the small RNAs that will be generated. Otherwise, no design principles can be effectively applied and the probability that any individual shRNA clone will work effectively will be greatly reduced.

MATERIALS

The recipes for the items marked with <R> are at the end of the protocol or in the Common Recipes Appendix.

Reagents

Annealing buffer <R>
Cells (competent)
ddH_2O or EB buffer (QIAGEN)
Ligation buffer <R>
siRNA antisense guide sequence (see Step 1)

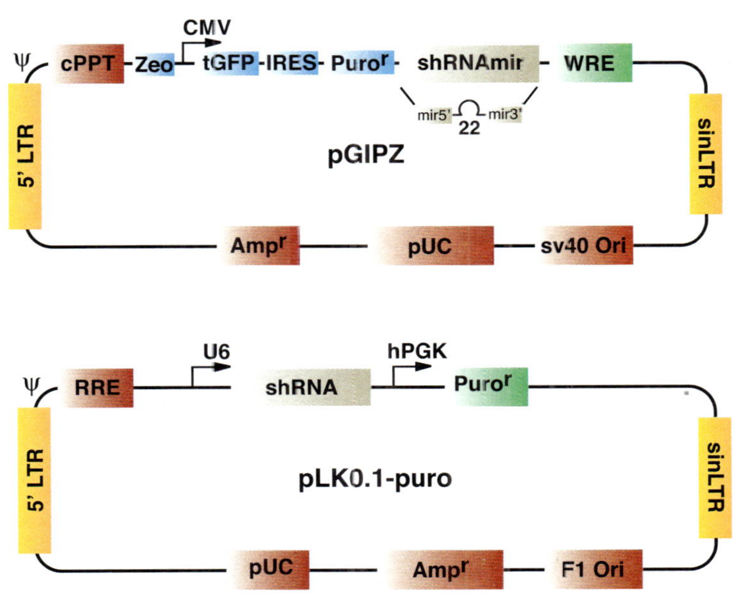

FIGURE 7-7. Examples of mammalian shRNA vectors. Two of the most prevalently used mammalian shRNA expression vectors are shown as examples. These carry the genome-wide shRNA collections and individual clones that are available from Sigma-Genosys (pKL0.1-puro) and Open Biosystems (pGIPZ), but they are also used extensively for cloning custom shRNAs. One (pGIPZ) carries miRNA mimetic shRNAs, whereas the other (pLKO.1-puro) typically carries simple snapback hairpins. Variations on both are available, but a key difference is that one uses RNA polymerase II promoters for shRNA expression, offering the potential for inducible and tissue-specific expression. The other uses an RNA polymerase III promoter that may be less subject to cell-type-specific variation in its expression.

Suitable vector in which XhoI and EcoRI sites permit engraftment into an miR-30 pri-miRNA context
Synthetic oligonucleotides (designed as described in Step 3)
T4 DNA ligase (New England BioLabs M0202)
T4 DNA ligation buffer (New England BioLabs B0202)

Equipment

Bacterial incubator
PCR machine
PCR tubes
Water bath

METHOD

1. Convert the predicted 21-nucleotide siRNA guide sequence to a 22-nucleotide guide sequence. This step is critical because miR-30-based vectors generate 22-nucleotide rather than 21-nucleotide guide and passenger strands.
   i. Begin with a predicted (e.g., predicted by DSIR, Biopred, or another algorithm, or found in the literature) 21-nucleotide antisense guide sequence.
   ii. Reverse-complement this sequence to find the target site on the mRNA and verify that the predicted guide targets the mRNA of interest.
   iii. Extend the target site by appending the base immediately 5′ on the mRNA to generate the 22-nucleotide target.
   iv. Reverse-complement this to produce the guide strand of the shRNA.
2. The miR-30-based shRNAs retain a bulge at the base of the shRNA stem that is predicted to exist in the bona fide miR-30 primary transcript. This impacts the design of the sense strand. Beginning with the target site derived above, alter the sense strand as follows. If the first base is A or T, convert it to C. If the first base is C or G, convert it to A. This generates the appropriate bulge in the final, folded RNA.
3. The oligonucleotides for annealing and cloning are designed by adding flanking sequence derived from the miR-30 pri-miRNA and sequences compatible with restriction sites. For miR-30-based vectors that use XhoI and EcoR1 as cloning sites, create sequences of oligonucleotides according to the following formulas.
   i. ***Top strand:*** Insert the gene-specific 22-nucleotide sense and 22-nucleotide antisense sequences into the following pattern and chemically synthesize the resulting 110-base oligonucleotide.

      5′-[TCGA]GAAGGTATATTGCTGTTGACAGTGAGCG + sense 22 mer + TAGTGAAGC CACAGATGTA (mir30 loop) + antisense 22 mer + TGCCTACTGCCTCGG-3′
   ii. ***Bottom strand:*** Remove the bracketed TCGA, which formed the overhang compatible with XhoI. Reverse-complement the sequence and append AATT at its 5′ end to create an overhang compatible with EcoR1. Synthesize the 110-base oligonucleotide. The annealed oligonucleotides will have the following structure (X denotes a variable position):

      5′-TCGAGAAGGTATATTGCTGTTGACAGTGAGCGXXXXXXXXXXXXXXXXXXXXXXTA GTGAAGCCACAGATGTAXXXXXXXXXXXXXXXXXXXXXXTGCCTACTGCCTCGG-3′

      3′-CTTCCATATAACGACAACTGTCACTCGCXXXXXXXXXXXXXXXXXXXXXXATCACTT CGGTGTCTACATXXXXXXXXXXXXXXXXXXXXXXACGGATGACGGAGCCTTAA-5′

iii. Order the designed oligonucleotides from an appropriate supplier. Once the ordered oligonucleotides arrive, anneal and ligate them into the vector as described in the following steps.

4. Begin by reconstituting each olignucleotide in H_2O to identical concentrations (0.5 µg/mL is convenient).
5. Set up the annealing reaction by mixing the following reagents in a PCR tube:

Top strand	20 µL
Bottom strand	20 µL
Annealing buffer (5x)	10 µL

6. Execute the following program in a PCR machine:

   5 min at 95°C
   10 min at 80°C
   5–7 h ramp from 80°C to 4°C (–0.5°C every 2.5 min)

7. Dilute 10 µL of annealed oligonucleotides 1:10 with ddH_2O or EB buffer. Store the undiluted and diluted shRNA inserts at –20°C.
8. Ligate the diluted shRNA insert into a suitable vector that has been digested with EcoR1 and Xho1 and purified (e.g., pSM2, pGIPZ, pTRIPZ, Open Biosystems) as follows:

XhoI/EcoRI-cut miR-30 vector (50–200 ng)	2 µL
Diluted annealed insert	2 µL
H_2O	12 µL
T4 DNA ligation buffer (10x)	2 µL
ATP (10 mM)	1 µL
T4 DNA ligase	1 µL

   Ligate for 2 h at 20°C.

9. Use 3–5 mL of the ligation mixture from Step 8 to transform 50–100 µL of competent cells.
10. Pick three to five colonies, prepare plasmid DNA, and check by restriction digest and sequencing.

SOLUTION AND REAGENT RECIPES

CAUTION: See the Cautions Appendix for appropriate handling of materials marked with <!>.

Annealing Buffer (5x)

Potassium acetate (500 mM)
HEPES-KOH (150 mM, pH 7.4) <!>
Magnesium acetate (10 mM)

Ligation Buffer (10x)

Reagent	Quantity (for 1 mL)	Final concentration
Tris-HCl (1 M, pH 7.5)	0.5 mL	500 mM
$MgCl_2$ (1 M) <!>	0.1 mL	100 mM
DTT (1 M) <!>	0.1 mL	100 mM
ATP (100 mM)	0.1 mL	10 mM
H_2O	0.2 mL	

Store for 6 mo at –20°C.

STRATEGIES FOR INTEGRATION OF SHRNA EXPRESSION CASSETTES INTO THE GENOME

To achieve long-term stable knockdown, shRNA expression cassettes must be integrated into the host genome. There are a number of ways in which this can be accomplished. The easiest, but perhaps least efficient, method is simply to transfect cells with the construct and select for stable integration either using a cotransfected selectable marker or by including a selectable marker on the shRNA plasmid. Individual colonies or pools of colonies are then scored for shRNA expression or knockdown. A much more efficient, reliable, and therefore popular method is to exploit the ability of recombinant retroviruses to stably integrate into the host genome.

Many retroviral backbones are available, and the most commonly used segregate into two classes. M-MuLV and its close relatives (e.g., MSCV) infect only dividing cells. Lentiviruses can infect both dividing and nondividing cells. Both use a fairly conventional plasmid backbone into which the *cis*-acting elements of the virus have been inserted, enabling it to be used as a vehicle for gene delivery. Long terminal repeats (LTRs) define the outer boundaries of the sequences that will be packaged into the virus. These provide promoters, poly(A) signals, and sequences required for successful reverse transcription and integration. Adjacent to the 5′ LTR is a packaging signal, a noncoding portion of the virus that forms a secondary structure that specifies efficient incorporation into the viral particle. The vectors also must include the primer-binding site for reverse transcription. All of the coding capacity of the virus is removed and replaced by desired functional elements, including the shRNA expression cassette and often a selectable marker gene.

Protocol 8

Packaging shRNA Retroviruses and Host Cell Infection

Once an shRNA is inserted into the backbone, the viral plasmid is transfected into a packaging cell line, which provides the *trans*-acting factors necessary for virus production. These include minimally capsid proteins and reverse transcriptase, but they also can include other regulatory factors (e.g., tat for some lentiviral vectors). It is critical to choose the correct packaging cell system for the viral backbone to be used. The packaging cell also defines the host range of the virus, depending on the envelope protein that it expresses. Ecotropic viruses are limited to rodent hosts, whereas amphotropic viruses have a broader host range that also includes humans. Often, investigators will express a nonretroviral envelope, e.g., vesicular stomatitus virus (VSV) G protein, to enhance virus stability and host range and to enable viruses to be concentrated following production.

Although viruses carrying shRNAs are packaged almost identically to viruses carrying protein-encoding genes, one twist is worth noting. Because shRNAs are efficiently cleaved by the host RNAi biogenesis machinery, this can reduce the level of viral genomic RNAs and consequently viral titers. Therefore, titers can be enhanced by cotransfecting the viral plasmid with an siRNA targeting a core miRNA biogenesis component, DGCR-8/Pasha (an effective sequence is given in the protocol). siRNAs against Drosha can also be used.

MATERIALS

The recipes for the items marked with <R> are at the end of the protocol or in the Common Recipes Appendix.

CAUTION: See the Cautions Appendix for appropriate handling of materials marked with <!>.

Reagents

$CaCl_2$ (2 M, 62.5 µL)

Calcium phosphate transfection solutions (or a commercial transfection reagent) (2 M $CaCl_2$, 2x HEPES-buffered saline [HBS])

Chloroquine (100 mM) <!>

Complete tissue culture medium (as recommended for each packaging line)

Custom oligonucleotide DGCR-8/Pasha siRNA (0.3 µg/µL; from QIAGEN or other vendor)

DMEM (Dulbecco's modified Eagle's medium), 10% FBS (fetal bovine serum), 1x penicillin/streptomycin

DNA to be transfected (27.5 µg)

HBS solution (2x) <R>

H_2O (nuclease-free, sterile)

NIH-3T3 or HEK-293 cells (to measure virus titer)

Packaging cells (Phoenix [Grignani et al. 1998] or a commercially available cell line)

Purified shRNA retroviral vector DNA

siRNA sequence: CGGGTGGATCATGACATTCCA (dXdY overhang, all annealed, no modification)

Sodium butyrate solution (1000x) (optional, see Step 9) <R>

Equipment

Automatic pipettor
Cell culture incubator (30°C, 37°C)
Falcon tubes (15 mL)
Pasteur pipette
Snap-cap tube (2–3 mL)
Tissue culture pipettes (disposable or autoclaved)
Tissue culture plates (10 cm)

OVERVIEW

The degree of safety precautions required is determined by the type of virus, its host range, and the character of the insert it carries (e.g., an shRNA targeting a human tumor suppressor requires more caution than an shRNA targeting a mouse gene). For example, an MSCV-based vector packaged with an ecotropic envelope requires a BL-2 setup, whereas lentiviruses with broad host range require BL-2+ to BL-3. Investigators must contact local biosafety officers when setting up virus production to ensure that all appropriate safety precautions are adequately followed.

METHOD

1. At 1–2 days before transfection, plate packaging cells in a 10-cm tissue culture dish so that on the day of transfection, they will be ~60% confluent and actively dividing. These cells double about every 24 h, so this will serve as a guide.
2. Harvest the cells and aliquot 6 million cells each into 15-mL Falcon tubes, bringing the final volume to 9 mL with complete tissue culture medium. Add 2.5 µL of 100 mM chloroquine to each tube and invert to mix.
3. For each sample, combine DNA to be transfected (27.5 µg total; see Tips and Troubleshooting below), DGCR-8/Pasha siRNA (16.7 µL of 0.3 µg/µL), and 62.5 µL of 2 M $CaCl_2$. Adjust the volume to 500 µL with nuclease-free, sterile water in a small sterile snap-cap tube (~2–3 mL). Place 500 µL of 2x HBS in a similar tube
4. Attach a sterile pasteur pipette to an automatic pipettor and gently bubble the 2x HBS solution.
5. While bubbling the 2x HBS, add the DNA/calcium solution from Step 3 drop by drop.
6. After finishing the addition, continue bubbling for 30–60 sec.
7. Cap the tubes and allow them to sit for 15–20 min. The mixture will be cloudy.
8. Add 1 mL of the transfection solution from Step 7 to the tubes containing the cells (from Step 2) and invert to mix. Immediately plate the cells on a 10-cm tissue culture plate.
9. Incubate the cells in a 5% CO_2 atmosphere for 14–18 h at 37°C. Then exchange the medium. *Optional:* Add 10 µL of 1000x sodium butyrate solution to the medium to enhance viral titers.
10. Harvest the virus for infection 36–72 h (typically 48 h) posttransfection. Filter the viral supernatants through a 0.45-µm filter to ensure that no packaging cells are transferred to the target culture during the infection process.

    Titers for freshly harvested virus from well-transfected cells should be ~1×10^6 infectious particles per milliliter when measured on NIH-3T3 or HEK-293 cells. Titers on other cell types may be lower, depending on many factors, including viral receptor expression.

TIPS AND TROUBLESHOOTING

1. The DNA solution to be applied to the cells depends largely on the experiment. It can be composed of only the purified shRNA retroviral vector DNA or can have other components. If an envelope protein that is not expressed in the packaging cells is to be used, a vector encoding this protein must be added to the DNA mixture. For example, if viruses are to be pseudotyped with VSV glycoprotein (VSV-G), the mixture would contain 25 µg of viral vector DNA and 2.5 µg of a plasmid in which the VSV-G-coding sequence is driven by the CMV promoter. Some investigators also like to cotransfect a plasmid that encodes packaging functions (*gag-pol* and sometimes *env* genes). This generally increases viral titers, but it can also have undesirable side effects. When packaging lines are constructed, they are designed to minimize the potential for RNA recombination to generate replication-competent viral genomes and are rigorously tested for the absence of intact virus. Cotransfection of a packaging plasmid can lead to the production of replication-competent virus, which can be potentially hazardous and have confounding effects in some experiments. The best practice is to use an established packaging line and avoid cotransfection of helper packaging plasmids. However, if their use is desired, the typical mixture is 15 µg of viral DNA, 10 µg of helper plasmid, and 2.5 µg of envelope plasmid.
2. Retroviruses, particularly those lacking a VSV-G envelope protein, are relatively unstable at 37°C. Many reports suggest that incubating transfected packaging cells at 30°C can increase viral titers. In this case, cells should be moved to a 30°C/5% CO_2 incubator after medium exchange.
3. Virus production is vigorous generally throughout the 36–72-h time frame. Because of their instability, retroviruses tend to reach an equilibrium titer, with viral decay equaling production. Therefore, one can harvest virus every 12 h from well-transfected packaging cells. Virus can be stored for up to 1 wk at 4°C with little loss of titer. Virus can be frozen for later use with some loss in titer, simply by placing collected culture media at –80°C. If viruses are pseudotyped with VSV-G, one should expect about a twofold to fourfold reduction in titer for each freeze–thaw cycle. Losses are slightly greater with ecotropic and amphotropic envelope proteins.
4. Lentiviruses, pseudotyped with VSV-G, can be concentrated to increase titers. Many protocols for achieving from a 10-fold to 100-fold increase in titers can be found in the literature.
5. Human cells can be made susceptible to infection with viruses bearing the ecotropic envelope protein by enforced expression of the ecotropic receptor (Wang et al. 1991). This can be quite convenient and can significantly reduce risks associated with recombinant retroviruses while allowing work in human cells.
6. Although calcium phosphate transfection solutions can be made in the laboratory, the pH is extremely critical. Therefore, this must be optimized empirically by making and testing many 2x HBS solutions with their pH surrounding the optimal 6.9–7.2 in increments of ~0.05 units. For very heavy users, this is practical, but well-tested calcium phosphate transfection solutions can be purchased commercially (e.g., Invitrogen K2780-01), and this offers the best approach for many more casual users. Even frozen, these transfection solutions have a usable life of ~6 mo. One can also transfect packaging cells with commercially available lipid reagents (e.g., Lipofectamine 2000) simply by following the manufacturer's instructions. High transfection rates are critical for obtaining good titers, so this should be monitored carefully with appropriate reporter plasmids, e.g., those encoding GFP, or by following a fluorescent marker on the shRNA vector.

SOLUTION AND REAGENT RECIPES

CAUTION: See the Cautions Appendix for appropriate handling of material marked with <!>.

HBS Solution (2x)

Reagent	Quantity (for 1000 mL)	Final concentration
NaCl	16.4 g	280 mM
HEPES	11.92 g	50 mM
Na_2HPO_4 (anhydrous)	0.21 g	1.5 mM
Dextrose	2.0 g	12 mM
KCl <!>	0.74 g	10 mM

Adjust to 1000 mL with H_2O. Split into 4 x 250 mL. Adjust pH to 6.96, 7.00, 7.04, and 7.08 with 0.5 M NaOH and test each empirically.

Sodium Butyrate (1000x)

Adjust 1.1 g of sodium butyrate (MW 110.1) to 10 mL with H_2O. Filter through 0.20-µm syringe filter.

READING LIST

Grignani F, Kinsella T, Mencarelli A, Valtieri M, Riganelli D, Lanfrancone L, Peschle C, Nolan GP, Pelicci PG. 1998. High-efficiency gene transfer and selection of human hematopoietic progenitor cells with a hybrid EBV/retroviral vector expressing the green fluorescence protein. *Cancer Res* **58:** 14–19.

Wang H, Kavanaugh MP, North RA, Kabat D. 1991. Cell-surface receptor for ecotropic murine retroviruses is a basic amino-acid transporter. *Nature* **352:** 729–731.

Protocol 9

Infection of Mammalian Cells with Retroviral shRNAs

Viral infection is a quite simple approach to stably introducing transgenes into the genome. The critical aspects are that the virus and the target cell should be appropriately matched. For example, a virus bearing an ecotropic envelope protein will not infect a human cell line unless the appropriate receptor has been purposefully expressed. VSV-G pseudotyped viruses have the greatest host range. Nondividing cells can only be infected with lentiviruses, but the additional safety precautions necessary for the use of these tools should dissuade their application to routinely cultured cell lines.

MATERIALS

The recipe for the item marked with <R> is at the end of the protocol or in the Common Recipes Appendix.

CAUTION: See the Cautions Appendix for appropriate handling of material marked with <!>.

Reagents

Cell type to be infected
Culture medium
Polybrene solution <R>
Virus for infection <!> as prepared in Protocol 8

Equipment

Centrifuge (low speed) and adapters that can hold multiwell plates or desired culture plates
Platform rocker
Tissue culture incubators (37°C, 32°C)
Tissue culture plates (10 cm) and plasticware

OVERVIEW

Although viral infection is a simple matter of adding culture supernatants from virus-producing cells to the cells targeted for infection, a few tricks can increase infection rates, and these are outlined below. There can be difficulties if the target cells grow in a specialized medium different from that used for the packaging cells. Here, the investigator should experiment with exchanging the medium on the packaging cells with that used for the target cell type 12 h before virus harvest. Alternatively, the target cells might remain unaffected by a brief exposure to virus-containing media while still being infected efficiently. Most variations on the protocols for virus production and infection will not make the difference between success and failure but will instead impact the overall efficiency of gene transfer.

METHOD

1. The method is given for 10-cm plates. Perform infection in any standard tissue culture vessel, scaling all amounts based on surface area (see Table 7-1 for guidance).

2. In general, plate cells at least 1 d before infection so that they are 60%–70% confluent at the time of initial virus exposure.

   The optimal density can vary for different cell lines and should be determined empirically.

   Plating cells too soon before virus infection can reduce titers. This may be related to the use of trypsin to recover cells in adherent culture, which may proteolyze cell surface receptors used by the virus.

3. Exchange culture medium 3–4 h before infection.
4. For infection, remove the medium and then add a combination of viral supernatant and fresh medium for a total volume of 10 mL.

   The ratio depends on the desired infection rate and the viral titer. For most cells to have a single integrant, aim for an infection rate of ~30%.

5. Add Polybrene to a final concentration of 8 µg/mL.

   Polybrene aids in infection. However, some cells are sensitive to Polybrene exposure. In such cases, this step can be omitted or the level of Polybrene can be reduced to 4–6 µg/mL.

6. *Optional step:* Load the tissue culture plates into a low-speed centrifuge and centrifuge at 800*g* for 1 h at room temperature. Be sure to turn braking off to avoid spilling the medium.

   This step will significantly increase infection for some cell types but is not recommended for use with viruses that can infect human cells because of the risks of contamination of the centrifuge and the exposure of investigators to potentially hazardous viruses.

7. Transfer the cells to a tissue culture incubator. If viruses are pseudotyped with VSV-G, incubate the cells immediately at 37°C. However, for viruses bearing ecotropic or amphotropic envelope proteins, incubate for the first 12 h at 32°C for increased infection.
8. After 12–24 h, make sure to exchange medium.

   This is not necessary if the viral supernatant corresponded to only a small percentage of the medium used to infect cells.

9. Assess infection rates, for example, by counting GFP-positive cells 48–96 h postinfection.

TIPS AND TROUBLESHOOTING

1. Cells need not be split on a particular schedule before infection. However, cells to be infected with M-MuLV- or MSCV-based vectors should be growing exponentially, and cells to be infected with lentiviral vectors should be healthy, even if they are nonproliferative.
2. Virus should have been titered on a standard cell line, for example, NIH-3T3 or HEK-293, both of which are highly infectable. Infection rates on any new cell line must be determined somewhat empirically, but once done, they can generally be predicted by reference to a standard titration. In general, infection rates of ~30% will result in one viral integration per cell, on average. If this is not necessary or desired, the amount of virus can generally be increased to give nearly 100% infection.

SOLUTION AND REAGENT RECIPE

Polybrene Solution

Suspend 8 mg of hexadimethrine bromide in 1 mL of H_2O for an 8 mg/mL stock. Filter through a 0.2-µm syringe filter.

READING LIST

Brummelkamp TR, Bernards R, Agami R. 2002. A system for stable expression of short interfering RNAs in mammalian cells. *Science* **296:** 550–553.

Dickins RA, Hemann MT, Zilfou JT, Simpson DR, Ibarra I, Hannon GJ, Lowe SW. 2005. Probing tumor phenotypes using stable and regulated synthetic microRNA precursors. *Nat Genet* **37:** 1289–1295.

Paddison PJ, Caudy AA, Bernstein E, Hannon GJ, Conklin DS. 2002. Short hairpin RNAs (shRNAs) induce sequence-specific silencing in mammalian cells. *Genes Dev* **16:** 948–958.

Stegmeier F, Hu G, Rickles RJ, Hannon GJ, Elledge SJ. 2005. A lentiviral microRNA-based system for single-copy polymerase II-regulated RNA interference in mammalian cells. *Proc Natl Acad Sci* **102:** 13212–13217.

Zeng Y, Wagner EJ, Cullen BR. 2002. Both natural and designed micro RNAs can inhibit the expression of cognate mRNAs when expressed in human cells. *Mol Cell* **9:** 1327–1333.

ENCODED RNAI IN MICE

In theory, precisely those principles that enable transgenic RNAi in cell lines can be used to create transgenic animals, which express shRNAs in a regulated or tissue-specific fashion. Unless clonal lines are derived, silencing in cultured cells always reflects an average of the entire population in which each integration site is differentially active. Moreover, each cell within the population can receive more than one copy of the transgene, depending on how the construct was introduced. In animals, the situation is much more complex. The activities of promoters that are commonly used for shRNA expression in cell culture can vary enormously in different tissues. Founder lines also typically vary in transgene expression due to the effects of their single integration sites. Combined, these factors make RNAi in animals somewhat more challenging than RNAi in cultured cells. Given the investment required to create genetically engineered animals, special care must be taken at every step to maximize the probability of success. However, the investment is almost invariably worthwhile, given the value of an animal in which a gene of interest can be reversibly silenced in virtually any desired cell type.

There are many ways to produce mice carrying shRNA transgenes. The most common is also the simplest, specifically pronuclear injection of constructs essentially identical to those used in cultured cells. Several reports have described shRNA-expressing mice built in precisely this fashion (Fedoriw et al. 2004; Rao and Wilkinson 2006; Dickins et al. 2007). Protocols for pronuclear injection are not presented, because creation of transgenic mice in this way is invariably accomplished by providing DNA to a core facility where trained personnel perform DNA injection and embryo implantation. Although straightforward, this procedure has significant disadvantages that must be overcome by detailed characterization of many individual founder lines. Founder mice, produced by pronuclear injection, are notorious for their variation in transgene expression levels due to differences in the chromatin structure surrounding the random sites into which transgenes are inserted. Therefore, many founders must be screened to identify one that is suitable for use. Because the chromatin structure of each integration site will vary by tissue, it is also a good idea to test shRNA expression and knockdown directly in cell types of interest and to realize that the behavior of each transgene might not be predictable in other tissues.

To maximize the probability of producing a useful animal, one can take advantage of RMCE (recombinase-mediated cassette exchange) (Protocol 10) (Beard et al. 2006). RMCE permits insertion of the shRNA (or any other) transgene into a well-characterized locus that gives reproducible and predictable expression in each founder and enhances the probability of potent expression in many cell types. This procedure is more involved and complex than simple pronuclear injection. Thus, the choice between the two approaches should be dictated by long-term plans of each laboratory. If there is a likelihood that only shRNA-expressing animal will be needed to test the function of only one gene product, the investment in setting up the RMCE system is probably not justified. However, if even a few shRNA mice are envisioned, for example, to probe the functions of several genes, the effort of setting up the processes outlined below are well worthwhile. Note that when creating a transgenic mouse through either route, one should take care to use the most potent shRNA possible. As a rule of thumb, the sequence chosen should provide more than 90% knockdown when introduced into cultured cells at single copy (e.g., upon retroviral infection at a multiplicity of ≤0.3).

REFERENCES

Beard C, Hochedlinger K, Plath K, Wutz A, Jaenisch R. 2006. Efficient method to generate single-copy transgenic mice by site-specific integration in embryonic stem cells. *Genesis* **44:** 23–28.

Dickins RA, McJunkin K, Hernando E, Premsrirut PK, Krizhanovsky V, Burgess DJ, Kim SY, Cordon-Cardo C, Zender L, Hannon GJ, et al. 2007. Tissue-specific and reversible RNA interference in transgenic mice. *Nat Genet* **39:** 914–921.

Fedoriw AM, Stein P, Svoboda P, Schultz RM, Bartolomei MS. 2004. Transgenic RNAi reveals essential function for CTCF in H19 gene imprinting. *Science* **303:** 238–240.

Rao MK, Wilkinson MF. 2006. Tissue-specific and cell type-specific RNA interference in vivo. *Nat Protoc* **1:** 1494–1501.

Protocol 10

Creating Transgenic shRNA Mice by Recombinase-mediated Cassette Exchange

This is a long protocol, involving many individual steps and was developed by Premsrirut and colleagues (in prep.). It is presumed that the investigator will have already created an appropriate miR-30-based shRNA (or used another scaffold) (Protocols 7 and 8) and carefully tested the properties of the construct. It also assumes that the laboratory is relatively proficient in embryonic stem (ES) cell culture. Careful culture of ES cells is essential for maintenance of pluripotency and therefore for the creation of mice with a germline derived from the engineered cells. The ES cells recommended in the following protocol are F_1 hybrids and are more robust than many ES cells, derived from pure backgrounds, that are often used for the creation of knockout mice by homologous recombination. For use in this protocol, shRNA inserts must be cloned into the Flp-In vector, shown in Figure 7-8, or a similarly constructed derivative. This can be accomplished precisely as described in Protocol 7. DNA for RMCE should be of high quality, prepared using a large-scale preparation protocol such as CsCl purification or isolated using a QIAGEN-tip 100 or 500 (QIAGEN 12143 or 12165).

MATERIALS

The recipes for the items marked with <R> are at the end of the protocol or in the Common Recipes Appendix.

CAUTION: See the Cautions Appendix for appropriate handling of materials marked with <!>.

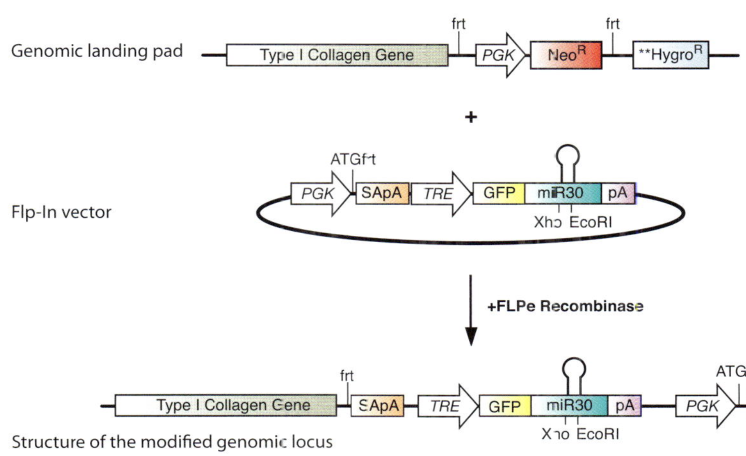

FIGURE 7-8. Creating an RNAi mouse by RCME. (*Upper panel*) Representation of the genomic locus in KH2 ES cells that will receive the integrated shRNA vector. Key features are two frt sites that flank a neomycin/G418-resistance gene and a hygromycin-resistance gene that lacks both a promoter and a translation initiation site. (*Middle panel*) The integration vector, modified from Beard et al. (2006), in which a tetracycline-responsive promoter drives a miR-30-based shRNA cassette. This can be inserted into the landing pad via the action of FLP recombinase, supplied by the cotransfected FLP-e plasmid. Once integrated (*lower panel*), the G418-resistance cassette is lost, replaced by the tetracycline-responsive shRNA cassette. Additionally, the previously inactive hygromycin-resistance gene receives both a PGK promoter and an initiating ATG to permit its expression.

Reagents

- β-Mercaptoethanol (100×) <!> <R>
- DNA for electroporation
- Doxycyline solution
- EDTA <!>
- Embryonic stem (ES) cells
- ES Cell Nucleofector Kit (Lonza DHPH1001)
- ES-cell-tested fetal bovine serum (FBS; Stem Cell Technologies 6952)
- Feeder cells

  Use mouse embryo fibroblasts (immortalized and irradiated, prepared as in optional protocol) or DR4 cells. Note that the feeders must be resistant to the antibiotic used for selection of modified ES cell clones, in this case hygromycin and neomycin. For the protocol presented here, DR4 cells, which are hygromycin, puromycin, neomycin, and 6-thioguanine resistant, are used. These can be obtained in a ready-to-use form from Open Biosystems (MES3948).
- Feeder medium <R>
- Flp-e plasmid
- Flp-In vector
- Freezing medium <R>
- Gelatin solution (0.1%)
- Genticin (Invitrogen 10131-035)
- GPS (glutamine/penicillin/streptomycin) solution <R>
- Hybrimax dimethysulfoxide (DMSO; Sigma-Aldrich D2650)
- Hygromycin solution <!>
- Knockout Dulbecco's modified Eagle's medium (DMEM; Invitrogen 10829-018)
- L-Glutamine (200 mM; Invitrogen 15400-054)
- LIF ESGro (Millipore ESG1107)
- M15 medium <R>
- M15 medium + hygromycin <!> <R>
- miR-30-based shRNA (from Protocol 7)
- Neomycin solution <!>
- Penicillin G (ICN 194537)
- Phosphate-buffered saline (PBS)
- QIAGEN Gentra Puregene DNA Kit
- Streptomycin sulfate (Invitrogen 11860-038) <!>
- Trypsin-EDTA (0.25%; Invitrogen 25200-056) <!>
- Trypsin-EDTA (0.5%; Invitrogen 15400-054) <!>

Equipment

- Centrifuge (low speed)
- CO_2 incubator
- Electroporation cuvettes
- Electroporator or Amaxa nucleofector (single or 96-well model; Lonza)
- Falcon tube (15 mL)
- Fluoresence microscope, FACS, or benchtop Guava instrument (Millipore)
- γ-Irradiation source <!>
- Gene Pulser (Bio-Rad)
- Hemocytometer or Coulter counter
- Liquid nitrogen storage tank <!>
- Microscopes for viewing and picking ES cell clones
- P20 pipette (Ranin)

Plates (12 well, 24 well, 96 well, round bottomed)
Rocking platform
Tissue culture incubator
Tissue culture plates (6–10 cm)
Tubes (50 mL, 15 mL)
Water bath

OVERVIEW

This protocol is broken into many substeps, each of which must be executed precisely to ensure the ultimate successful creation of a transgenic animal.

METHOD

Preparation of Gelatin-coated ES Culture Plates

These steps should be done immediately before use.

1. Add 0.1% of gelatin solution directly to tissue culture plates at least 1 h before use. Use 6 mL for a 10-cm plate, 2 mL for a 6-cm plate, 1.5 mL per well for a six-well plate, and 0.5 mL per well for 24-well plate.
2. Rock the plates to spread the solution evenly over the surface. Allow the plates to sit for 5–30 min at room temperature.
3. Immediately before use, aspirate the gelatin solution and then add tissue culture medium.
4. If the plates are not to be used immediately, keep them with the gelatin solution for up to several hours at room temperature.

Plating of Irradiated Feeder Cells

5. Culture ES cells on a layer of irradiated feeder cells.

   These feeder cells are generally created from immortalized mouse embryo fibroblasts (MEFs). Feeders ready for use can be purchased commercially or created by each laboratory by treatment in a γ irradiator or with a DNA-damaging agent. For an optional protocol for preparing feeder cells, see Preparation of Feeder Cells, below.

Preparation of Feeder Cells

Note that it is wise to prepare large batches of feeder cells and bank them for future use.

1. Grow primary or immortalized MEF derivatives in feeder medium to confluency in a 15-cm plate.
2. Add trypsin (3 mL) to the cells and incubate them for 3–5 min at 37°C.
3. Add 10 mL of M15 medium and mix the cells by pipetting them up and down more than five times.
4. Transfer the cells in medium to a 50-mL tube and expose them to 4300 rad of γ irradiation.
5. Centrifuge the cells at 1000 rpm for 5 min at room temperature.
6. Resupend the irradiated cells at a concentration of 5×10^6 cells/mL in freezing medium. Count the cells using a hemocytometer or Coulter counter.

6. Thaw a vial of DR4 feeder cells containing ~7 × 10^6 to 8 × 10^6 cells.

   About 0.9 × 10^6 to 1 × 10^6 feeders will be needed for a 10-cm gelatin-coated plate. This amount can be scaled as needed for other culture vessels (according to their surface area).

7. Add 5 mL of feeder medium to the thawed cells and recover the feeder cells by centrifuging at 800–1000*g* for 5 min at room temperature.
8. Aspirate the medium and suspend the cells in 80 mL of fresh feeder medium.
9. Aspirate the medium from a gelatin-coated 10-cm plate (see Preparation of Gelatin-coated ES Culture Plates, Step 3, above).
10. Add 10 mL of the cell suspension to the gelatin-coated plate.

Culture of ES Cells Before Electroporation

KH2 ES cells are an F_1 hybrid line that has been engineered with an integration site at the *Col1a* locus (Beard et al. 2006). This site contains a Flp-recognition target (FRT) site and a hygromycin-selectable marker that lacks both a promoter and a translation initiation site. Cotransfection of KH2 cells with the shRNA homing cassette and a plasmid encoding Flp-e recombinase facilitates integration specifically into that site. This not only places the shRNA at a defined chromosomal locus but also reconstitutes the hygromycin marker, allowing positive selection for proper integration (see Fig. 7-8).

11. Thaw the KH2 ES cells from archival vials in liquid nitrogen.
12. Add the thawed cells to 5 mL of ES cell medium and recover them by centrifugation at 800–1000*g* for 5 min at room temperature.
13. Plate the cells onto a 6–10-cm plate coated with DR4 feeder cells.
14. Culture the cells for at least 2 d before electroporation.

    Because ES cells divide very quickly, care must be taken to maintain them in a healthy, proliferative state. Cells can, of course, be carried for a longer period, particularly if many constructs are being introduced and large numbers of cells are needed.

15. Exchange the medium 2 h before electroporation.
16. Trypsinize to recover the ES cells. Plate the mixture of ES cells and feeder cells on a gelatin-coated plate for 30 min before electroporation. The feeders will attach to the plate, whereas the ES cells will not.
17. Recover the medium containing the ES cells and depleted of feeder cells. Determine the concentration of ES cells using a hemocytometer or Coulter counter.

Introduction of Transgenes into KH2 ES Cells

Plasmids are generally introduced into ES cells by electroporation. The following section contains a standard method that can be used for integration of RMCE cassettes. Although this will work well, nucleofection using an Amaxa nucleofection instrument (Lonza) is more efficient and more amenable for scaling up. This instrument allows 96 constructs to be easily introduced in one session into KH2 ES cells. If using the Amaxa Nucleofector, follow the manufacturer's instructions (using the ES Cell Nucleofector Kit, Lonza DHPH1001), but use 6 × 10^4 cells, 5 µg of the Flp-In vector, and 2.5 µg of the Flp-e plasmid (Beard et al. 2006). The following procedure uses the Bio-Rad Gene Pulser, but other electroporation devices can also be adapted to perform similarly.

18. Prepare one 6-cm plate per construct for ES cell culture (with gelatin and feeder cells).
19. For each electroporation, mix 50 mg of Flp-In vector (containing the shRNA of interest) with 25 µg of Flp-e plasmid in a 15-mL tube.

20. Harvest healthy KH2 cells for electroporation in 1 mL of trypsin.

    Generally, one 10-cm plate will yield 2×10^7 cells, which is sufficient for three to four different electroporations.

21. Suspend cells with 10 mL of complete M15 medium.

    Usually, ~5 min at 37°C in trypsin will be sufficient to detach the ES cells before suspension.

22. Transfer the cells to a 15-mL tube and collect the cells by centrifugation at 800–1000*g* for 5 min at room temperature.
23. Suspend the cells thoroughly in 10 mL of PBS and recover them again by centrifugation at 800–1000*g* for 5 min at room temperature.
24. Suspend the cells in PBS at ~5×10^6 cells per 900 µL.
25. Add 900 µL of KH2 cells to each tube containing DNA and mix by gently pipetting.
26. Transfer the cells to electroporation cuvettes (400-mm gap) at room temperature.
27. Pulse the cells twice at 125 µF (capacitance) and 400 V. If using one pulse, use 500 µF and 240 V. Two pulses are recommended. To prevent arcing, be sure to thoroughly wipe the cuvette before placing it in the electroporator. The time constant should be between 1.0 and 1.7, and ideally 1.6.
28. Allow cells to remain in the electroporation cuvette for 15 min at room temperature.
29. Change the medium in the prepared 6-cm plates from Step 18, replacing it with 3 mL of complete M15 medium.
30. Transfer the cells to the 6-cm plate.
31. After 2 d, proceed with hygromycin selection.

Selection of ES Cell Clones Containing Properly Integrated shRNA Cassettes

Once DNA constructs have been introduced into cells, those clones that arise from cells with the proper modification must be selected and isolated as clonal lines. After electroporation, cells will have been plated on feeders in nonselective M15 medium, and they should be left this way for 24 h.

32. At 24 h postelectroporation or nucleofection, exchange medium for M15 + hygromycin (140 µg/mL). Be sure to exchange this medium daily during selection.
33. Culture cells in M15 + hygromycin medium for 10–12 d, at which point the majority of ES cells will have died and a few discrete colonies should be easily visible.

    Although the number of colonies will depend on the efficiency with which DNA was introduced, dozens to hundreds are expected for a 10-cm plate.

34. Most hygromycin-resistant colonies contain a correctly integrated cassette, and thus only a few (in this case, four to six are recommended) need to be tested further for their integrity. Collect colonies while viewing under a microscope, aspirating them with a P20 pipette (e.g., Ranin) set to 7 µL.
35. Transfer each of the picked colonies, which are aggregates, to a well of a 96-well (round-bottomed) plate containing 25 µL of PBS/trypsin/EDTA.
36. Incubate the cells at 37°C until they are dispersed.
37. Transfer the dispersed cells, representing each colony, to a well of a 24-well plate prepared with gelatin and feeder cells.
38. Culture the cells in M15 + hygromycin medium, with daily medium changes, until the cells become confluent.
39. Once confluent, recover the cells by treatment with trypsin to a total volume of 200 µL.

40. Add 300 μL of M15 medium and aliquot the cells as follows:
    i. For each clone, place 5 μL of cells into each of two wells of a 96-well plate prepared with gelatin and feeders.
    ii. This will be used to test for neomycin resistance (see Testing Neomycin Sensitivity, below). Two wells of a prepared 24-well plate should receive 30 μL of cells for testing shRNA expression (see Testing for shRNA Expression, below). Both plates should also contain at least two wells of plated parental KH2 cells.
    iii. Transfer the remaining cells to a prepared 6-cm plate, from which they should be expanded and archived by freezing in liquid nitrogen.

Testing Neomycin Sensitivity

The proper insertion of the shRNA-resistance cassette replaces a neomycin-resistance gene. Therefore, although parental KH2 cells are neomycin resistant, their properly engineered progeny should be sensitive.

41. On the day following plating, add neomycin to a final concentration of 140 μg/mL to one of the two wells representing each clone and to one of the two KH2 wells.
42. Incubate the plates for 4–5 d at 37°C.
43. Score neomycin sensitivity visually. Cells with the correct integrations should be killed by neomycin treatment.

Testing for shRNA Expression

KH2 cells have a tetracycline *trans*-activator (tet-on) integrated at the *ROSA-26* locus, and the shRNA in Flp-In is under the control of the tet-CMV minimal promoter. Therefore, the shRNA can be induced by an activator, such as tetracycline or doxycycline. The shRNA resides within the 3′ UTR of GFP, and as a result, the ability of doxycycline to induce the linked GFP reporter serves as a proxy for the ability to induce expression of the shRNA.

44. At the time of plating, or up to 1 d thereafter, add doxycycline to a final concentration of 1 μg/mL to one of the two wells containing each clone and to one of the two KH2 parental wells.
45. After 2 d of exposure to the inducing agent, score for GFP expression.

    This can be done in any of several ways. The simplest is to view the cells on an appropriate fluorescence microscope. However, it is difficult to be quantitative using this approach. A better method is to analyze the cells for GFP fluorescence by FACS or using a benchtop instrument such as a Guava flow cytometer (Millipore).

Validation of Clones and Preparation of Frozen Stocks

Clones that have strong GFP induction and that display neomycin sensitivity should be tested by Southern blotting (see Validation of Clones by Southern Blotting, below) for proper integration and can be tested by northern blotting or quantitative PCR for induction of the shRNA (although this is optional).

46. About 4–5 d after plating, clones should reach ~80% confluency on the 6-cm dishes. These clones are now ready for storage and further validation. Begin by recovering cells using 600 μL of trypsin/EDTA with an incubation time of 5–10 min at 37°C.
47. Add 3 mL of M15 medium and transfer the cells to a 15-mL Falcon tube.

48. Remove 100 µL of cells and plate them in a gelatin-coated 12-well plate for isolation of genomic DNA.
49. Recover the cells for archiving by centrifugation at 800–1000*g* for 5 min at room temperature.
50. Aspirate the medium and suspend the cells in 1 mL of ice-cold freezing medium.
51. Divide the cells among three vials, label the vials, and place them overnight at –80°C before transfer to liquid nitrogen storage the next day.

Validation of Clones by Southern Blotting

This approach allows correct and single-copy transgene integration to be verified.

52. Purify genomic DNAs using the QIAGEN Gentra Puregene DNA Kit or through any other genomic DNA preparation method suitable for Southern blotting. To verify correct integration, digest genomic DNAs with SpeI.

    Using a ColA1 region probe (see below), a band of 6 kb, corresponding to the wild-type allele, and a band of 4 kb, corresponding to the correctly modified allele, should be detected. Genomic DNA digested with EcoRI can be used to verify single-site integration, using a fragment of GFP as a probe (see below). Positive clones with a single-copy integration produce a band of 4 kb. Bands of other sizes indicate multiple integrations.

    ColA1 probe:

    GACCTGAGGCCTGGGAGGTGTTGCCCATGGATCCTGGGAGGTTCATGAGCCCT
    CAAAGGGATTCAAGCAGCAGAAACAGGGACTGAGACATGGAGGACCAGGTTGCTC
    CTTGGTCACACATGGAGGGGTAGCTGGCCTCTCCCCACCTCTAGCCCAAAGAAAC
    CATAATTTAGTATAGAAGGGGCCTTCTAATGCTGGGGTGTCACAAGGAGGATGA
    GAATTTCTTGGGTCACCCTTGATCTCACCAGAGAGAGGTGCCCAGCCCTCAAG
    GAACTCTCCCCGGGGGATCTTAGAGAGCTGTGGTGGAACTTCCGGGGTGTCACCA
    GAAAGGACAGGACCCCACATCACAGAGGTGCCCGGGACGATAGCATGCTGGCTCG
    GTACAGGGAGCCCGTGCAATTGTTCCCCCCTTCTTCCTGCCCCCTTCTTTCCTTTTC
    CTTTATCCCCAAAACTTTTTGCTTTTTCTTCTTCCTTTTCCCTCTCTCCCTCTGTC
    CATCTGAGAACCTGAGGCCCTAGAGGGATGGTAACTAATTGTCCCCCACATCTCAGA
    GAATGGGGACGGTGACATAGTACAAGGCCCGAGAATCCAGCAGGCAGAAGTCTGGGGT
    GAACCAGACAGTAGACATAAGGACACAGATGACCCCTTTCACTCGGGGAGACAGGTCT
    CAAATTTGAAAGGAAGGAACAGGCCATTTGTACCTTCCCTCTCTTGTGCTACCAGTC
    TACTGTGGCACCCATAGGACACGTTCATATGGTCACGGAGTATCCAAGACTATTAATGATG
    GCCTTTAATAATTTATTATTACACTTTGCATGAGGACAACGTGAAAGGAAAATTTAT

    GFP probe (entire GFP-coding sequence):

    ATGGTGAGCAAGGGCGAGGAGCTGTTCACCGGGGTGGTGCCCATCCTGGTCGAGCTG
    GACGGCGACGTAAACGGCCACAAGTTCAGCGTGTCCGGCGAGGGCGAGGGCGATGC
    CACCTACGGCAAGCTGACCCTGAAGTTCATCTGCACCACCGGCAAGCTGCCCGTGCC
    CTGGCCCACCCTCGTGACCACCCTGACCTACGGCGTGCAGTGCTTCAGCCGCTACCCC
    GACCACATGAAGCAGCACGACTTCTTCAAGTCCGCCATGCCCGAAGGC
    TACGTCCAGGAGCGCACCATCTTCTTCAAGGACGACGGCAACTACAAGACCCGCGCC
    GAGGTGAAGTTCGAGGGCGACACCCTGGTGAACCGCATCGAGCTGAAGGGCATC
    GACTTCAAGGAGGACGGCAACATCCTGGGGCACAAGCTGGAGTACAACTACAACAGCCA
    CAACGTCTATATCATGGCCGACAAGCAGAAGAACGGCATCAAGGTGAACTTCAAGATC
    CGCCACAACATCGAGGACGGCAGCGTGCAGCTCGCCGACCACTACCAGCAGAACACCC
    CCATCGGCGACGGCCCCGTGCTGCTGCCCGACAACCACTACCTGAGCACCCAGTCCGC
    CCTGAGCAAAGACCCCAACGAGAAGCGCGATCACATGGTCCTGCTGGAGTTCGTGAC
    CGCCGCCGGGATCACTCTCGGCATGGACGAGCTGTACAAGTAATGA

Creation of Transgenic Mice

53. At this stage, use the cells to analyze the consequences of suppressing gene expression in the ES cells themselves or in derivatives of the ES cells, differentiated in vitro. However, the ultimate

goal is to produce a mouse in which expression can be suppressed in tissues. This is generally accomplished by delivery of the ES cells to a core facility or company that will create the mice. Virtually all institutions have facilities capable of generating mice by blastocyst injection or by production of aggregation chimeras. In this case, the investigator must screen chimeric founders for those that transmit the shRNA allele through the germline. However, KH2 are robust F_1 hybrid ES cells that can be used in tetraploid complementation to create mice wholly derived from the engineered cells. For institutions with this capability, this is by far the preferred route toward lines of shRNA-expressing animals.

TIPS AND TROUBLESHOOTING

1. ES cells must be maintained in good condition for proper gene transfer and for maintaining potency. Split ES cells in routine culture once the cells are 70%–80% confluent. Always pass them by dividing 1:2–1:3. Always exchange medium on the cells a few hours before splitting, transfecting, or freezing. ES cells tend to turn their medium yellow over time. This is a sign that they are doing well in culture. In general, the plating density should be 4 million cells per 10-cm plate.
2. Positive clones become visible usually from 9 to 10 d posthygromycin treatment. Pick the clones at days 10–12. Do not pick clones if they become visible only after 14 d of hygromycin treatment. They usually turn out to be incorrect.
3. Store freezing medium for no longer than 1 wk after it is made.
4. Be sure to use high-quality DNA for electroporation. It is suggested that the DNA be precipitated with ethanol twice and pellets washed well with 70% ethanol. Dissolve DNA at least to a concentration of above 1.5 µg/µL. Run the DNA on a gel to make sure that there is no genomic DNA or RNA contamination. Store plasmids frozen at –20°C.
5. This is a very long procedure. Therefore, it is very important to sequence-verify the entire targeting cassette before transfection. It is also a good idea to resequence the genomically integrated shRNA sequence before making mice.
6. Plates can be prepared with feeders and kept in the CO_2 incubator for up to 2–3 wk. Change the medium on feeders at least once a week.
7. Most importantly, change the medium daily on ES cells.

SOLUTION AND REAGENT RECIPES

CAUTION: See the Cautions Appendix for appropriate handling of materials marked with <!>.

GPS Solution (Glutamine/Penicillin/Streptomycin)

Penicillin	500 mg
Streptomycin <!>	300 mg

Dissolve the powders to 100 mL of 100x glutamine solution. Filter through a 0.2-µm filter unit. Aliquot (6 mL is a convenient size) and freeze at –20°C.

β-Mercaptoethanol (100x)

Add 75 mL of β-mercaptoethanol <!> to 500 mL of PBS. Filter the mixture and aliquot 50 mL each to individual tubes. Store for up to several weeks at 4°C.

M15 Medium

Knockout DMEM	500 mL
ES-cell-tested FBS	90 mL
GPS solution	6 mL
β-Mercaptoethanol <!>	6 mL
Leukemia inhibitory factor (LIF)	60 mL

Store for up to several weeks at 4°C.

M15 Medium + Hygromycin

Hygromycin stock (140 mg/mL) <!>	1.68 mL
M15 medium	600 mL

Store for up to several weeks at 4°C.

Feeder Medium

Knockout DMEM	500 mL
ES-cell-tested FBS	60 mL
GPS solution	6 mL

Store for up to several weeks at 4°C.

Freezing Medium

M15 medium	6 mL
DMSO <!>	1 mL
FBS	3 mL

This can be kept for up to 1 wk at 4°C.

DISCUSSION

RNAi Screens

RNAi screens simply extend the concepts and approaches outlined in this chapter from the analysis of individual genes to the analysis of many genes. Screens can be focused on small numbers of structurally or functionally related genes or can extend to the entire genome. For all three of the example organisms discussed in this chapter, whole-genome collections of RNAi-inducing constructs are available in one or more formats.

Irrespective of the organism or approach, the absolute key to all successful RNAi screens is the design of the assay. It must be robust and provide sufficient signal to noise that positive signals can be detected in a multiplexed format, where control over experimental conditions is often not as precise as in studies of single genes. Statistical rigor is also critically important, because variation in measurements or even in the biology of a particular pathway will almost always lead to at least some cohort of false positives.

The topic of RNAi screening and the various ways in which it can be applied could easily consume an entire manual. The intention here is to simply provide an overall outline that can be merged with the protocols given above and further reading in the literature to enable investigators to take advantage of this challenging aspect of RNAi technology. In practically every situation, the availability of liquid-handling systems greatly facilitates screening in well-by-well formats. However, it is possible with the cooperative effort of several individuals to carry out screens using repeating multichannel pipettes and manual tools for transfer of liquids (e.g., bacterial cultures) among 96-well plates. It is critical that investigators account in their planning for the several independent, biological replicates that must be performed to ensure accuracy in each screen.

RNAi Screens in C. elegans

In *C. elegans*, RNAi screening is almost invariably carried out by feeding with *E. coli* that express dsRNA. Genome-wide collections are available from several sources (http://www.geneservice.co.uk/products/rnai/Celegans.jsp; www.openbiosystems.com). The easiest method for triggering the response is to feed larvae in liquid culture in 96-well plates; however, on-plate protocols can also be used with a bit more effort. Here, the key is simply having a phenotypic or reporter assay that can easily be screened, for example, visually, without the need for an enormous effort to be devoted to each sample. One may also take advantage of automated microscopy and its associated software, should such equipment be available, for identifying candidate dsRNAs that affect a phenotype of choice. Although screening through 40,000–60,000 individual worm cultures manually may seem to be a daunting task, in reality, with a good and obvious phenotype, a few investigators working together can accomplish this in only a few, admittedly long, days.

RNAi Screens in Drosophila

In *Drosophila* cell culture, screening is most often accomplished by the introduction of dsRNAs, either by soaking or with the aid of a transfection reagent. Excellent guidelines for RNAi screens in cultured fly cells are provided by the *Drosophila* RNAi Screening Center (DRSC, www.flyrnai.org). There is also an existing collection of flies in which the expression of dsRNAs is driven from a Gal-4-inducible promoter. This is maintained at the Institute of Molecular Pathology in Vienna (www.vdrc.at). This enables temporal and cell-type-specific control of suppression via crossing to specific Gal4 drivers and up to genome-wide screens for specific phenotypes in animals. A second collection of flies, harboring artificial miRNAs under Gal4 control, is currently under construction (N. Perrimon, G. Hannon, pers. comm.).

RNAi Screens in Mammalian Cells

Mammalian cells offer two approaches to large-scale RNAi screens. The first is the use of chemically synthesized siRNAs. For this strategy, individual wells receive either single siRNAs or pools in which several siRNAs are directed against the same gene. Following transfection, phenotypes are generally assayed after short-term culture by tracking a convenient output, such as a luciferase or GFP reporter or using a viability test such as the 3-(4,5-dimethylthiazol-2-yl)-2,5-diphenyltetrazolium bromide (MTT) reduction assay. Analysis of the outcome can be greatly aided by using a multi-well plate reader, or detailed phenotypes can be examined by automated microscopy. Many companies sell genome-wide collections of siRNAs that are suitable for screening (e.g., Dharmacon, QIAGEN). shRNAs offer a complementary strategy. Here, collections are presently available corresponding to the human, mouse, and rat genomes. These can be used in well-by-well formats just as siRNAs by transfection of expression constructs or by infection with large collections of viruses corresponding to individual shRNAs (Luo et al. 2008; Barbie et al. 2009).

shRNAs can also be screened in pooled formats, something not possible for siRNAs, where individual cells respond to single shRNA integrants in a manner that can be scored (Stegmeier et al. 2005; Silva et al. 2008). This method requires virtually no specialized experience or expensive equipment and can be carried out by any investigator with basic cell culture skills and access to microarray hybridization facilties or next-generation sequencing. Phenotypes scored can range from those that result in a positive selective advantage, for example, survival, transformation, or the activation of a reporter that allows purification by FACS. Negative selections can also be used, for example, by assessing loss of growth potential or viability under specific conditions.

On the whole, RNAi screens are no longer mainly limited by technology but by the creativity that can be applied to the clever design of a screen that will highlight genes involved in a particular biological process. It must be noted that in every large-scale RNAi study, all of the caveats presented above for single-gene studies must be considered. Therefore, the screen itself must simply be viewed as a starting point in the generation of a list of candidate genes involved in a particular process. Each candidate must be validated to ensure that effects are on target before any substantial investment is made in detailed follow-up studies. Given the range of tools that are now accessible for flies, worms, and mammals, and the improvements to these tools that are constantly being made, RNAi screening is likely to remain a productive genetic strategy long into the future.

REFERENCES

Barbie DA, Tamayo P, Boehm JS, Kim SY, Moody SE, Dunn IF, Schinzel AC, Sandy P, Meylan E, Scholl C, et al. 2009. Systematic RNA interference reveals that oncogenic KRAS-driven cancers require TBK1. *Nature* **462:** 108–112.

Beard C, Hochedlinger K, Plath K, Wutz A, Jaenisch R. 2006. Efficient method to generate single-copy transgenic mice by site-specific integration in embryonic stem cells. *Genesis* **44:** 23–28.

Luo B, Cheung HW, Subramanian A, Sharifnia T, Okamoto M, Yang X, Hinkle G, Boehm JS, Beroukhim R, Weir BA, et al. 2008. Highly parallel identification of essential genes in cancer cells. *Proc Natl Acad Sci* **105:** 20380–20385.

Silva JM, Marran K, Parker JS, Silva J, Golding M, Schlabach MR, Elledge SJ, Hannon GJ, Chang K. 2008. Profiling essential genes in human mammary cells by multiplex RNAi screening. *Science* **319:** 617–620.

Stegmeier F, Hu G, Rickles RJ, Hannon GJ, Elledge SJ. 2005. A lentiviral microRNA-based system for single-copy polymerase II-regulated RNA interference in mammalian cells. *Proc Natl Acad Sci* **102:** 13212–13217.

READING LIST

Berns K, Hijmans EM, Mullenders J, Brummelkamp TR, Velds A, Heimerikx M, Kerkhoven RM, Madiredjo M, Nijkamp W, Weigelt B, et al. 2004. A large-scale RNAi screen in human cells identifies new components of the p53 pathway. *Nature* **428:** 431–437.

Dickins RA, McJunkin K, Hernando E, Premsrirut PK, Krizhanovsky V, Burgess DJ, Kim SY, Cordon-Cardo C, Zender L, Hannon GJ, et al. 2007. Tissue-specific and reversible RNA interference in transgenic mice. *Nat Genet* **39:** 914–921.

Gazin C, Wajapeyee N, Gobeil S, Virbasius CM, Green MR. 2007. An elaborate pathway required for Ras-mediated epigenetic silencing. *Nature* **449:** 1073–1077.

Ramadan N, Flockhart I, Booker M, Perrimon N, Mathey-Prevot B. 2007. Design and implementation of high-throughput RNAi screens in cultured *Drosophila* cells. *Nat Protoc* **2:** 2245–2264.

Sasaguri H, Mitani T, Anzai M, Kubodera T, Saito Y, Yamada H, Mizusawa H, Yokota T. 2009. Silencing efficiency differs among tissues and endogenous microRNA pathway is preserved in short hairpin RNA transgenic mice. *FEBS Lett* **583:** 213–218.

Schlabach MR, Luo J, Solimini NL, Hu G, Xu Q, Li MZ, Zhao Z, Smogorzewska A, Sowa ME, Ang XL, et al. 2008. Cancer proliferation gene discovery through functional genomics. *Science* **319:** 620–624.

Silva JM, Li MZ, Chang K, Ge W, Golding MC, Rickles RJ, Siolas D, Hu G, Paddison PJ, Schlabach MR, et al. 2005. Second-generation shRNA libraries covering the mouse and human genomes. *Nat Genet* **37:** 1281–1288.

Stern P, Astrof S, Erkeland SJ, Schustak J, Sharp PA, Hynes RO. 2008. A system for Cre-regulated RNA interference in vivo. *Proc Natl Acad Sci* **105:** 13895–13900.

Whitehurst AW, Bodemann BO, Cardenas J, Ferguson D, Girard L, Peyton M, Minna JD, Michnoff C, Hao W, Roth MG, et al. 2007. Synthetic lethal screen identification of chemosensitizer loci in cancer cells. *Nature* **446:** 815–819.

Zipperlen P, Fraser AG, Kamath RS, Martinez-Campos M, Ahringer J. 2001. Roles for 147 embryonic lethal genes on *C. elegans* chromosome I identified by RNA interference and video microscopy. *EMBO J* **20:** 3984–3992.

WEB RESOURCES

BestGene, Inc.
www.thebestgene.com
info@thebestgene.com

C. elegans *Genetics Center*
    www.cbs.umn.edu/CGC/index.html
C. elegans *RNAi feeding libraries*
    www.geneservice.co.uk/products/rnai/Celegans.jsp
    www.openbiosystems.com
Control plasmids, C. elegans
    www.geneservice.co.uk/products/rnai/Celegans.jsp
Drosophila *RNAi Screening Center (DRSC)*
    www.flyrnai.org
dsir; linear regression model (Vert et al. 2006)
    biodev.extra.cea.fr/DSIR/DSIR.html
Genetic Services, Inc.
    www.geneticservices.com
    info@geneticservices.com
Institute of Molecular Pathology in Vienna (expression of dsRNAs is driven from a Gal-4-inducible promoter)
    www.vdrc.at
L4440 (pPD129.36)
    www.addgene.org/pgvec1?cmd=findpl&identifier=1654
Primer3
    http://frodo.wi.mit.edu/primer3
SnapDragon (primer design for studies in Drosophila*)*
    www.flyrnai.org
Whole-genome RNAi libraries
    www.geneservice.co.uk/products/rnai/Celegans.jsp
    www.openbiosystems.com

8 Genomic Approaches

PROTOCOLS

1 Preparation of Fluorescent-dye-labeled cDNA from RNA for Microarray Hybridization, 435
2 Microarray Slide Hybridization, 440
3 Scanning Microarray Slides, 446
4 Tips on Hybridizing, Washing, and Scanning Affymetrix Microarrays, 451
5 Fragmentation of Whole-transcriptome RNA Using *E. coli* RNase III, 457
6 Preparation of Small RNA Libraries for High-throughput Sequencing, 460
7 Large-scale Immunopurification of RNP Complexes from *Drosophila* Nucleoplasmic Extracts for Tiling Microarrays (RNA Immunoprecipitation Microarray; RIP-Chip), 471
8 CLIP (Cross-linking and Immunoprecipitation) Identification of RNAs Bound by a Specific Protein, 476
9 Preparation of mRNA-Seq Libraries from Poly(A)$^+$ mRNA for Illumina Transcriptome High-throughput Sequencing, 491
10 Methods for Processing Microarray Data, 496
11 Methods for Processing High-throughput Sequencing Data, 503

As TECHNICAL ADVANCEMENTS in the study of genes have expanded our reach from individual genes to whole genomes, our ability to analyze RNA has also expanded from individual RNAs to whole transcriptomes. The development of these new RNA analysis methods has required miniaturization, parallelization, and automation, but analysis still relies on the fundamental principles of nucleic acid chemistry and enzymology that are exploited in the traditional methods (described in the previous chapters). These new methods, like the traditional ones, are designed to determine where each RNA starts and ends, which sequences are contained within the RNA, and, in some cases, the identity of single nucleotides at defined positions in the RNA. Combined with a broad spectrum of fractionation methods from gradient centrifugation to fluorescence-activated cell sorting, the composition of discrete RNA pools can be analyzed with respect to their protein-binding partners and their localization in a cell or tissue.

There are two main methods gaining wider use in the genome-scale analysis of RNA transcription, localization, function, and processing: microarrays and high-throughput RNA (actually cDNA) sequencing (HTS). Each method is designed to estimate the amount and genomic origins of the RNA molecules in a given experimental sample. In the case of microarrays, the sequences in the sample are identified through hybridization capture by their DNA complementary strands on the surface of the array (Pease et al. 1994; Schena et al. 1995; DeRisi et al. 1996; Shalon et al. 1996). The amount of a sample sequence captured is a reflection of its concentration in the original sample. In the case of HTS, many millions of cDNA fragments of the RNA molecules in the sample are sequenced at once on a surface using cycles of nucleotide incorporation linked to photon generation and imaging (Mitra et al. 2003; Ehn et al. 2004; Russom et al. 2005; Shendure et al. 2005; Guo et al. 2008). The "reads" are then mapped to the genome and tallied with respect to the genomic location from which they came. For both microarrays and HTS, there are numerous technical steps between the RNA sample and the analysis of the derived cDNA sequences that can introduce biases. In particular, fragmentation, ligation, and amplification steps pose some risk of bias. Usually, the effect of these biases can be minimized by performing replicate analyses and limiting amplification cycles.

Below, we outline and compare basic features of microarrays and HTS. In addition to the use of microarrays and HTS for analysis of gene expression and splicing profiles, these tools can be used to identify RNA transcripts and transcript regions using antibody immunopurification of RNA–protein complexes from cell extracts (RNA immunoprecipitation, or RIP) (Matunis et al. 1992; Keene et al. 2006). We also outline methods to prepare extracts for immunopurification, methods to perform in vivo RNA–protein cross-linking before immunoprecipitation (cross-linking immunoprecipitation, or CLIP) (Ule et al. 2003; Ule et al. 2005; Licatalosi et al. 2008), and methods used to analyze the RNA retrieved by these methods. Finally, all of these methods generate large data sets that require the use of bioinformatics tools for their analysis (Mortazavi et al. 2008; Pepke et al. 2009). We also provide some basic methods for analysis of microarray and HTS data.

MICROARRAYS

Microarrays provide the ability to measure the expression of every gene in the genome simultaneously using a single sample. A microarray designed to measure RNA levels from many genes at once is a large, miniaturized, parallel set of "reverse northerns," in essence, a tiny dot blot. In a typical northern blot, the RNA target is fixed to a surface, and a probe sequence is labeled and hybridized. The signal intensity and location on the surface of the captured probe indicates the amount and size of the RNA target in the sample. In a reverse northern ("dot blot" or "slot blot"), an unlabeled probe is fixed to a surface (such as nitrocellulose paper) and a labeled RNA target (or labeled cDNA derived from the RNA target) is hybridized from solution. An expression microarray is identical to this on a miniature scale, and multiplied thousands of times, using a glass surface instead of nitrocellulose and fluorescent rather than radioactive detection (Pease et al. 1994; Schena et al. 1995; DeRisi et al. 1996; Shalon et al. 1996).

The probe sequences for each gene or gene region are fixed at known positions at high density on the surface of the array. A labeled target mixture derived from the sample is hybridized to the array, and, after washing, the amount of label is detected. To a first approximation, the amount of target captured at the location of the probe represents the amount of that sequence in the sample. The main differences between a "dot blot" and an expression microarray concern the number and density of the probes on the surface, the volumes of hybridization fluid, the kind of label on the target (radioactive decay vs. fluoresence emission), and the method of signal capture (autoradiography or phosphorimaging vs. laser scanning and fluorescence detection). Because of the vast number of probes on a microarray (10^3–10^7), data must be analyzed by computational methods (see below).

A simple kind of microarray design carries one probe for each gene in a genome of interest and is able to detect and measure the overall level of mRNA derived from each gene expressed in a given sample. For higher-resolution analysis of transcription and RNA processing, more probes that target the entire span of a gene, in particular near the processing sites, are desirable. Given a large enough array, probes can be set at uniform intervals along the entire genome ("tiling arrays") to detect sequences from any possible genomic location that might be expressed. In addition to genome sequences, arrays can contain probes for nongenomic sequences created by RNA processing, for example, splice junctions, edited sites, or polyadenylation sites, to identify the products of RNA processing in a sample. Basically, any DNA sequence that can be synthesized either in situ on the surface of the array or placed by printing methods can be used as a probe.

At this time, there are a number of commercial providers of microarrays designed for a variety of studies in various model systems, and many universities have core facilities that will accept RNA samples and provide array data, occasionally with some analysis. Such services may be useful for general application; however, they may lack the flexibility to perform specialized kinds of experiments that do not meet exactly the services they provide. A number of companies offer custom array services in which the investigator may specify the design of the array and may also contract for hybridization, scanning, and analysis services. This can be convenient (and expensive), but it places some aspects of the process outside the control of the investigator. Investigators who opt to outsource their research to a company should ensure that their contract allows important experimental details on key elements of the work to be published. A company may have unrealistic expectations concerning their ability to simultaneously contract publicly supported research and maintain the proprietary nature of their protocols and materials. Investigators should be wary of agreements that do not allow publication of important scientific elements of the methods.

Two Experimental Strategies: Single-label One-Sample-per-Array versus Double-label Cohybridization of Two Samples

One of the technical challenges intrinsic to the DNA microarray approach is the nonuniform behavior of probes. Variation in probe hybridization characteristics and deposition on the surface of the array means that different probe spots on the array will have different abilities to capture signal, even though they may target the same gene or even contain the same sequence. There are several ways around these two problems. The first is to optimize probe hybridization characteristics by applying design criteria that make probes similar in their G + C content, melting temperatures, and length. The second is to optimize the construction of the array so that the spots are of extremely uniform size and have nearly identical amounts of DNA.

To the extent that these factors can be controlled, the intensity of fluorescence captured at a spot will reflect the amount of target in the sample, and these intensity values can be used to estimate changes in gene expression. Application of both theoretical and empirical probe design principles allows the first problem—variation in probe hybridization characteristics—to be approached. However, sometimes, the target to be probed is constrained in sequence (e.g., a microRNA [miRNA] or a splice junction), and few alternative sequences can be chosen. To approach the second problem—depositing a constant amount of probe at each position in the array—rigorous manufacturing conditions are applied to ensure uniform and efficient DNA synthesis at each of the many locations on the array for some platforms. For arrays where spot variation is hard to control (see below), this problem is sidestepped by cohybridizing two differently labeled samples for direct comparison by the ratio of the intensities of the two fluors (the "ratio-based" method). The ratio of intensities will be independent of spot size and hybridization characteristics because these will be the same for both labeled targets in the mixture. This allows relative (rather than absolute) RNA changes to be captured using arrays that are far cheaper and more flexible to produce.

Different Array Platforms

Microarrays can be constructed by several different means. Among the simplest are "pin-printed" arrays created by a robotic printer with a printing head consisting of a number of pins that are essentially quills, usually 32 or 48 in number, which are spaced 4 mm apart in a regular 4 x 8 (32, as shown in Fig. 8-1) or 4 x 12 (48) and can be adjusted to slightly touch the glass slide and deposit an ~80-µm-diameter spot of DNA. The pin arrangement mates with the standard 384-well microtiter plate dimensions, and the source DNAs (oligonucleotides or polymerase chain reaction [PCR] products) are carried in a set of such plates for printing. The printhead dips into the source and then prints the first set of 32 DNAs, 4 mm apart, with the 4-pin side of the array aligned with the width of the standard-sized specially treated microscope slide on which each array is printed. After the first set of DNAs is printed, the pins are washed and reloaded with the next 32 DNAs, which are printed ~160 µm offset from the first set, so that the edges of adjacent 80-µm spots are separated by ~80 µm (about one spot diameter, give or take) (Fig. 8-1). This is repeated until 25 spots are built into a row by each pin, which is now working nearly 4 mm from where it started. This creates the first row for each of the 32 sectors that will eventually appear as more rows are printed. The final array will be ~16 x 38 mm, covering the central area of the slide. Each of the 32 sectors will have 25 rows of 25 spots for a total of 20,000 spots (Fig. 8-1). Usually a large set of slides (150–200) is printed at the same time. Arrays of this type can be printed using a relatively inexpensive robot that can be assembled in-house from precision motors and other preassembled units; commercially built array printers are also available. These arrays will have variable spot diameters due to differences in the pins,

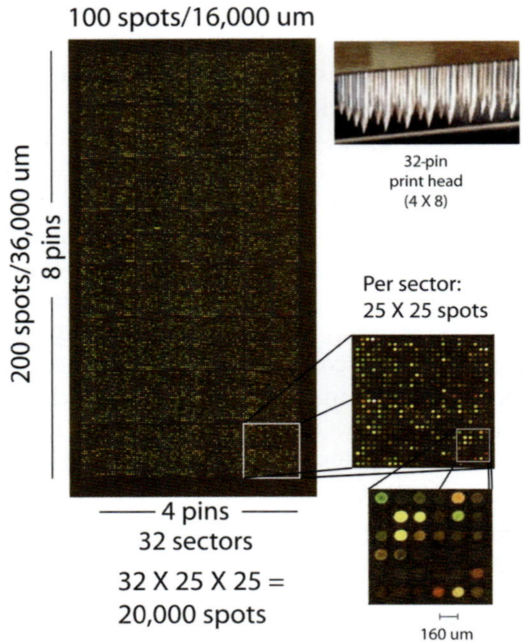

FIGURE 8-1. Pin-printed microarray slide and sector detail. (Courtesy of Ares lab, University of California, Santa Cruz.)

loading, humidity, and other factors and are thus usually best analyzed by cohybridization of the test target with a reference target using the ratio-based approach.

It is also possible to print microarrays using noncontact printing methods similar to "ink-jet" printing. In this case, rather than loading a set of quills with the oligonucleotide or PCR product to be printed as described above, piezoelectric printheads control the delivery of the DNA to the slide electronically, or thermally triggered bubble-jet printheads are used instead of contact printing. These have the advantage of better control of volume delivered and uniform spot size, as well as being more rapid and without the wear-related changes observed in contact printing. Nonetheless, in the end, the DNA is applied to the array from a source plate, and the analysis is the same whether contact or ink-jet printing is used.

Three other popular array formats are constructed by chemically synthesizing each DNA probe in situ, at a predetermined position on the slide, using phosphoramidite chemistry. In a format sold by Agilent, Inc., an ink-jet printer is used to deposit the appropriate phosphoramidite monomer to the spot during the standard DNA chemical synthesis steps of deprotecting, coupling the next base, capping unreacted ends, and deprotecting for the next step. By delivering only the appropriate phosphoramidite to each spot on the array during the coupling phase, the sequence at each ~80-µm spot is built up from 3′ to 5′, as for standard oligonucleotide synthesis. Usually, a 60-mer sequence is synthesized, and the accuracy of the noncontact ink-jet printing method with respect to volume makes the spots more uniform in size and DNA content than the typical pin-printed microarray.

Affymetrix, Inc. sells arrays made by in situ DNA synthesis as well, but they have pushed feature size down to 5 µm, achieving very high densities, ~5 million probes per array, which are usually 25 nucleotides in length using this format. The phosphoramidite strategy used by Affymetrix involves a light-activated deprotection step that controls which positions on the array will be reactive during a subsequent coupling step. The light activation is restricted by the use of physical masks that prevent deprotection of positions that will not require the next nucleotide. Upon light-activated deprotection, the next phosphoramidite monomer is allowed to react with the entire surface, but it is only coupled at positions that were exposed to light. After capping the oligonucleotides, a new mask is applied and a different set of positions is deprotected (depending on which position needs the next monomer). Thus, for each of the 25 nucleotides of the DNA being synthesized on the array, four masks are required, for a total of 100. Manufacturing the masks is a major expense; however, once made, a very large number of arrays can be made relatively cheaply. This means that Affymetrix arrays must be designed carefully and with the total numbers of arrays needed in mind, because the initial investment in masks is so high.

More recently, Nimblegen, Inc. has been selling arrays made using a similar light-dependent deprotection strategy, but without the use of masks. The Nimblegen method uses a tiny deformable mirror array that is electronically controllable to direct a laser to the sites of deprotection on the surface of the array (Fig. 8-2). Thus, the "mask" in this case is a virtual mask that is represented by the orientation of the mirrors which lead to the pattern of deprotection that allows coupling of the next nucleotide. This makes the arrays faster to build and cheaper when made one or a few at a time, allowing cycles of design testing and modification.

Nimblegen arrays can have probes that are more than 60 nucleotides long, but 60 is the standard length for most applications. These arrays have become popular because they can approach the high densities of Affymetrix arrays at lower cost for smaller numbers of arrays and with longer probes. They may be purchased, hybridized, and scanned using standard slide scanners such as those from Axon, Inc.; however, Nimblegen also promotes these services and will perform them for RNA researchers who prefer a "hands-off" approach to microarray experiments.

Microarrays allow a rapid, cost-effective, and reproducible method for detecting and measuring specific RNA sequences in a sample. But because they rely on hybridization, they are subject to the

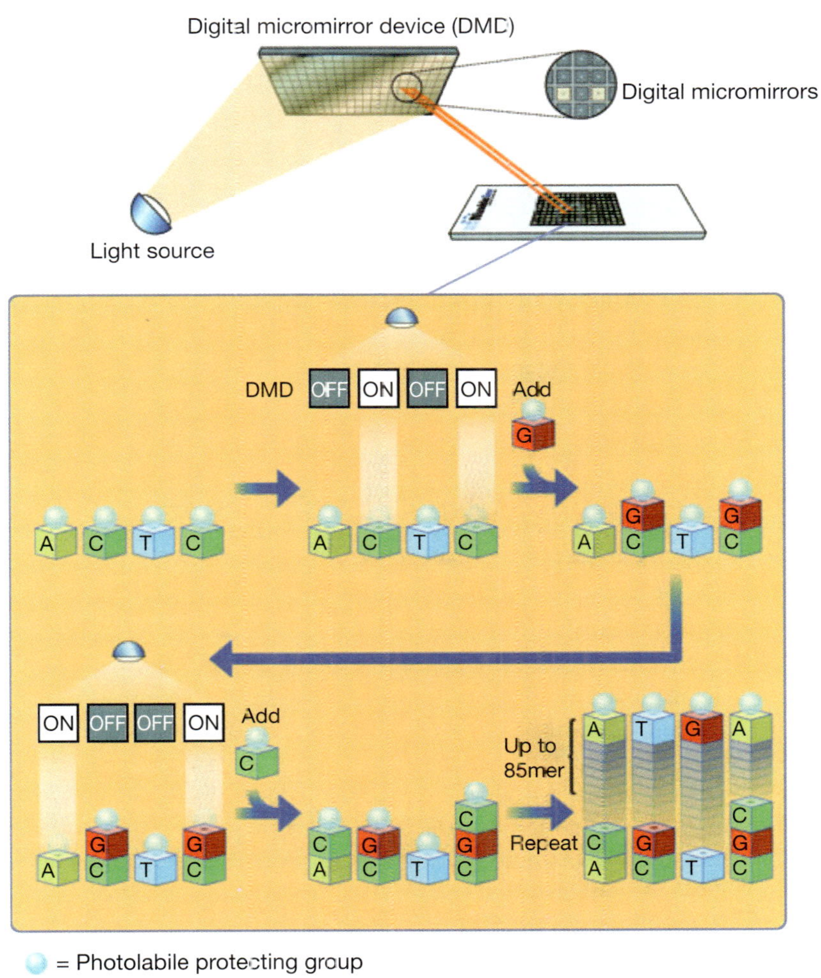

FIGURE 8-2. NimbleGen microarray light-activated deprotection strategy. (Photo courtesy of Roche NimbleGen. ©2010 Roche NimbleGen, Inc. [http://www.nimblegen.com/technology/manufacture.html].)

same problems as other hybridization methods. Hybridization capture of target is strongly dependent on the stringency of the hybridization and wash conditions, and hybrids between the probe and a sequence with one or several mismatched base pairs may not be distinguished from hybridization to a perfectly matching sequence. Because genes and gene families often contain multiple copies of closely related sequences, the opportunity exists for signal captured by a probe to come from more than one gene, complicating the analysis. Some of this can be overcome by careful probe design; however, in the case where a specific short sequence in the target must be sampled, the choice of probe sequence is limited. Despite these limitations, the cost effectiveness of microarray comparisons for many samples is currently much higher than that for HTS. Analysis is simpler, and the results, although perhaps not as comprehensive, allow rapid identification of the key trends in an experiment (Protocols 1–4). Several robust and well-tested platforms are available.

SEQUENCING OF RNAs BY HIGH-THROUGHPUT METHODS

Advances in genome sequencing technology have quickly been applied to the study of RNA. It has always been important to sequence RNA products (or their cDNA derivatives) to confirm their identity and structure. Identifying splice junctions requires sequencing across the joined exons of the spliced product, usually in a cloned cDNA or a purified reverse transcriptase (RT)–PCR product using standard (Sanger) DNA sequencing methods. Today's HTS methods provide short sequence reads for many millions of individual molecules in a sample, allowing the use of sequencing to measure the abundance of RNA molecules (Protocols 5 and 6). Once the sequence reads are obtained, their genomic origin is identified by matching their sequence to the genome, and the number of reads found to match a particular gene can be taken as a measure of expression level. Reads that cross splice junctions (if there are enough of them captured) can be taken as a measure of how often splicing occurs between two sites, and alternative splicing can thus be measured. Unlike a microarray signal, which is an analog measure of capture efficiency that is only partly related to concentration of the target in the sample, mapped sequence reads are a digital measure of the concentration.

HTS is expensive and still evolving in ways that should improve its performance and cost effectiveness. HTS reads tend to be short (but they are getting longer) and have different kinds of errors depending on the method. Thus, during mapping, choices must be made about whether a sequence read should count or be filtered out, given the likelihood of its match to the genome at a unique location. One could accept only perfect matches to absolutely unique regions of the genome and this might suffice for highly abundant RNAs. However, rare RNAs, and in particular, splice junction reads for rare RNAs, may be too infrequent to obtain accurate quantification of alternative splicing. As read length and accuracy improve, these problems should be minimized. In the meantime, sequence read files contain information on the base call and the confidence that the call is accurate for each position. Computational methods are applied to accept reads for which the high-confidence nucleotide positions match, with lesser penalties for mismatch at low-confidence positions, allowing more reads to be counted and improving quantification.

At the moment, there are three main HTS systems—454 or "pyrosequencing," SOLiD sequencing, and Illumina sequencing—that operate on similar principles (Mitra et al. 2003; Ehn et al. 2004; Russom et al. 2005; Shendure et al. 2005; Guo et al. 2008). Each method creates a library of seed molecules from the sample that are separated, amplified individually, and placed in an array of positions on a two-dimensional surface, with each reaction center composed of many copies of the original individual DNA molecule. Sequencing reactions are performed by flooding the array with reagents (DNA polymerase, nucleotides, etc.) in a specified reaction cycle, using light-emitting reactants that allow capture of an image of the array from the photons that arise at each cycle. The reactions are set up so that the identity of each nucleotide in the seed molecule copies can be determined by analysis of sequential images. The systems use different methods to amplify each individual seed molecule and place it on the array, and they use different light-emitting chemistries to follow the sequencing reaction. Because an array of many millions of reaction centers is imaged multiple times, the number of reads obtained in a single run can be quite huge. The amount of data to be processed, especially considering that the raw data are a large stack of images, makes the analysis for all three methods computationally intensive.

Illumina HTS

The simplest system to understand is the Illumina method. The RNA sample is used to make a seed molecule library containing special adapter oligonucleotide sequences at the ends, using ligation, PCR amplification, and size selection. The RNA-derived seed molecules with their adapters are attached randomly and at low concentration by their 3′ ends on a surface to which is also attached (at high concentrations by their 5′ ends) two oligonucleotides that can PCR-amplify seed DNA molecules containing the adapter sequences. The 5′ ends of attached sample-derived molecules are free to reach back toward the surface and anneal with an attached PCR primer, which is extended to produce a complement of the sample-derived molecule that, in the next cycle of PCR, can anneal by its (complementary) 5′ end to the other attached PCR oligonucleotide. After ~10 cycles of this "bridge amplification," ~1000 copies of the seed molecule accumulate in a tiny spot immediately surrounding the original seed molecule. Next, sequence data are obtained from all of the spots simultaneously using a set of four reversibly terminating deoxynucleotide triphosphates (dNTPs), each with a different fluorescent dye attached to the 3′ OH of the nucleotide (Fig. 8-3). In the reaction cycle, primer is annealed to the 5′ end of each seed molecule and extended with polymerase, leading to the incorporation of only one of the dye-labeled dNTPs, depending on the next base in the DNA template. The chain terminates after a single incorporation due to the blocked 3′ end of the incorporated nucleotide. Laser scanning and imaging of the surface create a color map in which each reaction center signals the identity of its incorporated base by the color of the dye. The dye-containing blocking group on the 3′ OH is removed, preparing the DNA end for the next cycle of nucleotide addition, which will reveal the identity of the next base for each of the reaction centers. This works well for 70–100 cycles until conditions degrade and reading errors become too high.

Pyrosequencing from Roche, Inc.

The pyrosequencing or "454" method also uses the concepts of creating a reaction center at a precise location and then divining the sequence by observing light from the reaction center, but the strategies are different. To create a reaction center, seed molecules from the sample fragment library are mixed with beads containing the attached amplification oligonucleotides in an emulsion of oil and aqueous buffer. The ratio of sample to beads is adjusted to create many aqueous droplets in the oil that contain only one bead and just one seed molecule (plus adequate polymerase and dNTPs) so that amplification and attachment of seed copies to the bead result in a bead covered with copies of one and only one molecule from the seed library. The beads are recovered from the emulsion and then scattered onto a "picotiter" plate, whose surface has a honeycomb of tiny wells into which one and only one bead can fit. These are the reaction centers, with each well containing a single bead carrying the products of (in theory) a single seed molecule. In practice, the ratio of seed molecules to beads is critical, and even under optimal conditions, many beads will have been reacted with no seed molecule, whereas others will have reacted with two or more. These reaction centers will be ignored in subsequent analysis because they will give either no sequence or multiple unreadable, superim-

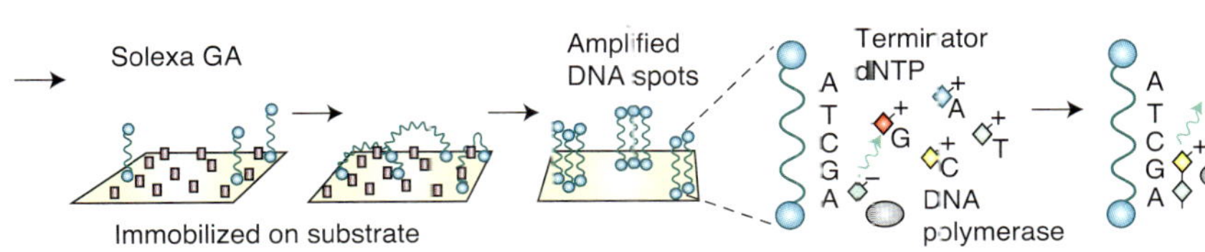

FIGURE 8-3. Illumina sequencing process. (Reprinted, in part, with permission, from MacLean et al. 2009 [©Macmillan].)

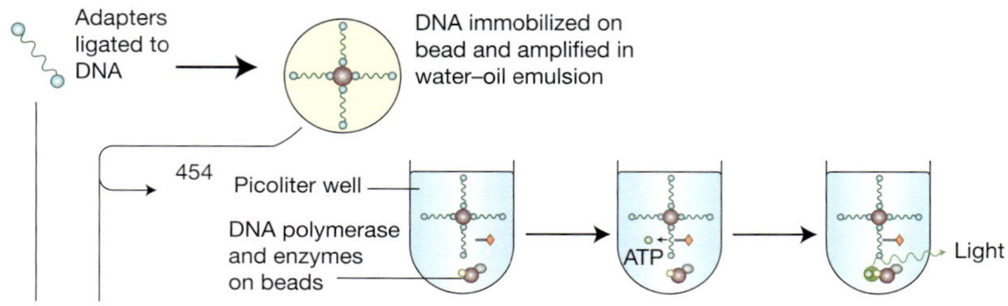

FIGURE 8-4. Pyrosequencing, or 454, method. (Adapted, with permission, from MacLean et al. 2009 [©Macmillan].)

posed sequences. The pyrosequencing reaction cycle does not rely on fluorescent dyes but rather on the generation of light by luciferase in response to the presence of adenosine triphosphate (ATP) (Fig. 8-4). ATP is generated by the pyrophosphate (PP_i) that is released as a product of dNTP incorporation. The amount of PP_i is detected and measured by its conversion to ATP using the reverse reaction catalyzed by ATP-sulfurylase, the enzyme that normally creates adenosine phosphosulfate (APS) from ATP and inorganic sulfate, releasing PP_i. In pyrosequencing, the enzyme runs in reverse, reacting PP_i with APS to generate sulfate and ATP, which then feeds photon production by luciferase, in proportion to the amount of nucleotide incorporated by the polymerase at the reaction center. Thus, the plate containing the reaction centers is flooded sequentially with each dNTP in turn, and the photons arising from each reaction center are captured to identify the base incorporated and how many bases of that type are incorporated. Note that because no chain terminators are used, if a template encodes the same base two or three times in a row, then two or three times as much PP_i is generated. This method currently allows longer reads than the Illumina method; however, long runs (>6) of the same nucleotide in the sequence cannot be read with high accuracy. Because the plate will be flooded repeatedly with different nucleotides, the reaction center wells containing the single large seed molecule bead are also filled with tiny beads carrying immobilized ATP-sulfurylase and luciferase, to ensure uniform and reproducible distribution of the enzymes necessary for the cascade from PP_i to photons. This high local concentration of enzymes also ensures that the diffusible limiting substrates in the chain reaction (PP_i and ATP) are consumed in the same location where they are generated.

SOLiD Sequencing from Applied Biosystems, Inc.

The third method is called SOLiD, and it uses a bead emulsion strategy for capturing and amplifying seed molecules from the library, with beads arrayed on the surface of a flow cell in a fashion similar to that of 454 sequencing. In SOLiD sequencing, however, a ligase rather than a polymerase is used to determine the sequence, and due to the requirement for proper matching of bases at the ligation site, two bases can be determined at a time. A set of 16 ligation substrates in four sets of four are dye-labeled at the 5′ end; they have a specific dinucleotide at their 3′ end and are randomized at all other positions (Fig. 8-5). A primer is annealed, and the first set of four ligation substrates is washed over the beads. If one of the dinucleotide substrates matches, it and its dye are ligated. After imaging, the second set of four is added and imaged, and so on, determining the first two nucleotides at each position at each reaction center. The ligation substrates are cleaved, removing the dye and all but the three randomized bases adjacent to the dinucleotide, and the next round of four-dinucleotide ligation, imaging, and washes is done, until seven such cycles have been performed. At this point, the sequence is determined at every two nucleotides, with three unknown positions in between (i.e., sequence is known at positions 1, 2, 6, 7, 11, 12, 16, 17, 21, 22, 26, 27, 31, and 32). Next, the DNA strand comprising the ligation products is denatured and removed, and a new primer is annealed, whose 5′ end is offset by one position relative to the first primer, and the ligation sequencing procedure is repeated. These cycles then determine the nucleotides at positions 0, 1, 5, 6, 10, 11, 15, 16, 20, 21, 25, 26, 30, and 31. Note that in the second round, positions 1, 6, 11, 16, 21, 26, and 31 have been read again.

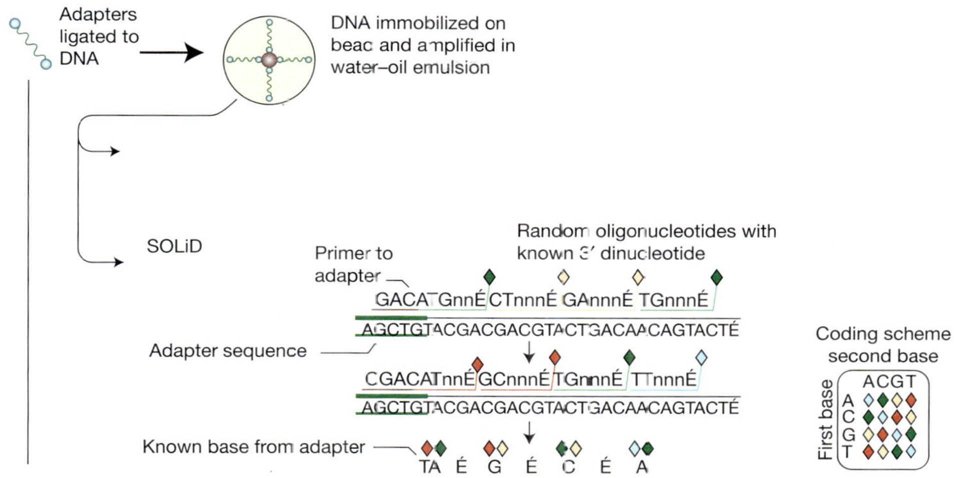

FIGURE 8-5. SOLiD sequencing method. (Adapted, with permission, from MacLean et al. 2009 [©Macmillan].)

Once the cycles are completed with this primer, a third primer is used, then a fourth and a fifth, each offset by one nucleotide each time. Ultimately, each nucleotide in the sequence is read twice, a feature that improves read accuracy.

Creating Seed Molecule Libraries for Sequencing

Despite the various approaches taken in the different HTS methods, all start with a library of seed molecules derived from the sample by ligation of adapters, cDNA synthesis, a second ligation step in some cases, PCR amplification, and size selection. These steps are necessary to properly capture the seed molecule in a form that allows bead or bridge amplification and analysis. Because the reactivity of all RNA fragments in the sample toward the primers or ligation substrates in these reactions may not be equal, the library construction steps can introduce bias into the population of seed molecules. At the same time, the addition of linkers affords an opportunity for two modifications that improve cost effectiveness and add information: "mate pair" or "paired end" read libraries and "bar coding" of samples. In libraries constructed for mate pair reads, the bead array can be denatured and reacted with a second sequencing primer to obtain sequence for the complementary strand at the other end of the seed molecule. Because seed molecules are size-fractionated during library construction, the paired end method provides pairs of sequence reads that are an approximately known distance apart. Such distance information can be very useful for genome assembly or, in the case of RNA analysis, splicing.

A second modification that can be incorporated during library construction is bar coding. In this case, the oligonucleotides used to generate the seed molecule libraries have extra bases that allow their origin to be determined in mixed runs. For example, if an investigator wants to run a time course and obtain sequence for samples at each time point, separate libraries can be constructed with different bar codes for each time point. To save on sequencing costs, the libraries can be mixed and sequenced together, and the reads can then be sorted out by their bar codes before analysis to separate molecules by their sample identity. Such multiplexing strategies reduce costs, but of course, they divide the total number of reads, so that coverage of the transcriptome may be reduced for each time point.

Many Applications of Sequencing to the Study of RNA

Sequencing of RNA may be used in a variety of applications. The simplest might be to understand which RNAs are expressed in a particular cell type or how expression changes after an experimental treatment. In this case, the number of reads obtained may define the limits of detection of rare mRNAs, as well as the confidence in the quantification of lower-abundance mRNAs and their spliced isoforms. A major component in RNA samples from cells is rRNA, and the presence of large amounts of rRNA in the initial sample can cause the library to be dominated by rRNA seed molecules. To avoid obtaining large numbers of rRNA reads at the expense of rare mRNAs, the RNA samples can be treated to remove rRNA using beads that carry oligonucleotides complementary to rRNAs, thus removing them from the starting material. (See Chapter 2, Protocols 6 and 9.) If the goal is to sequence small RNAs such as miRNAs, alternative library construction methods are used that enrich for these in the library. The companies that make high-throughput sequencers also sell a variety of kits and accessories for making libraries, although in principle it is possible for investigators to make their own libraries without kits (Protocols 5 and 6).

For many experiments of interest, making the library for sequencing will be initiated once a highly selected population of RNAs is obtained. For example, the CLIP method (see Protocol 10) has been adapted to interface with HTS so that small RNA fragments that have been cross-linked to a target protein and captured by immunoprecipitation of the target protein can be isolated, sequenced, and analyzed (Licatalosi et al. 2008; Sanford et al. 2009; Yeo et al. 2009; Zisoulis et al. 2010). In principle, any manipulation that provides a selected set of RNA molecules, such as gradient centrifugation, subcellular fractionation, microdissection, and even elution from microarrays, can be used to initiate library construction and sequencing. This approach is described in the next section.

IDENTIFYING RNAs ASSOCIATED WITH SPECIFIC PROTEINS

A major biological question that arises repeatedly during the study of RNA concerns which proteins in the cell bind an RNA of interest and which RNA sequences they bind. To answer this question, it is advantageous to use antibodies against RNA-binding proteins to immunoaffinity-purify endogenous RNA–protein complexes from cells and then identify the associated RNAs. This approach has been used to identify the cytoplasmic mRNAs that are bound to specific mRNA-binding proteins, the nuclear pre-mRNAs that bind to specific heterogeneous nuclear ribonucleoproteins (hnRNPs) and splicing factors, and the micro-ribonucleoproteins (miRNPs) that bind mRNA targets (Protocols 7 and 8).

Cross-linking or Not

There is some uncertainty regarding the generality of the phenomenon of dissociation and reassociation of RNA-binding proteins in vitro to RNAs that they normally do not bind in vivo, a process referred to as "redistribution" of RNA-binding proteins. This phenomenon has been documented for the HuR protein in mammalian cell extracts wherein rebinding of the protein to a nonphysiological target transcript in an extract mixing experiment occurred (Mili and Steitz 2004). In constrast, there are many reports of isolation of physiologically relevant target transcripts for both nuclear and cytoplasmic RNA-binding proteins without the use of cross-linking agents (Keene et al. 2006; Piacentini et al. 2009; Selth et al. 2009; Wang et al. 2010). The redistribution problem may be a major concern for certain "exchangeable" RNA-binding proteins and their binding sites, but it may not be problematic for others. In any case, the following discussion of cross-linking approaches includes cases where cross-linking was either used or not to find bona fide RNA targets for specific RNA-binding proteins.

UV Cross-linking

In the presence of short-wave ultraviolet (UV) light (254 nm), single-stranded RNA bases are activated to form free radicals that can covalently join with amino acid side chains on closely juxtaposed proteins. This method was used in the original identification of mammalian hnRNPs by Dreyfuss and colleagues following UV irradiation of cells and subsequent isolation of covalent RNA–protein adducts (Matunis et al. 1992; Burd and Dreyfuss 1994). More recently, in vivo UV cross-linking has been combined with RNA target identification for examining the binding sites for the mammalian Nova-1, ASF/SF2, and FOX2 splicing factors (Licatalosi et al. 2008; Sanford et al. 2009; Yeo et al. 2009). This method, called cross-linking immunoprecipitation (CLIP) (Protocol 10), was originally followed by molecular cloning methods to identify the cross-linked RNA sequence. More recently, high-throughput cDNA sequencing has eliminated the need for cloning the target RNA molecules (see Fig. 8-6). Briefly, the method involves UV cross-linking of intact cells, immunopurification of the protein of interest, fragmentation of the bound RNA (generally with T1 ribonuclease or micrococcal nuclease), radiolabeling the short RNAs, SDS-PAGE, and electrotransfer of immunopurified material to a membrane (like a "western" blot). This is followed by excision of the radioactive RNA–protein adducts from the membrane and protease treatment and isolation of the short RNA molecules that were covalently attached to the protein of interest. From here, RNA adapters are ligated on the RNA molecules and converted to cDNA that is either cloned into plasmid vectors for traditional sequencing or PCR-amplified for HTS (Fig. 8-6). Thus, the more recent modifications of the CLIP method can be used to map thousands of RNA fragments to a eukaryotic genome, affording a broad identification of the many RNAs that may bind to a specific protein.

One issue with UV cross-linking is that the yield of cross-linking is relatively low. It is very likely that some RNA sites are more efficiently cross-linked to a given protein by UV than other sites due to differences in protein–RNA juxtaposition and the precise sequence of the binding site. Moreover, there is differential photoactivation of the bases on the RNA and differences in reactivity of the amino acid side chains in the proteins. The most UV-activated base is uridine, which is more reactive than cytidine (rU > rC). The most reactive amino acid side chains are tyrosine, histidine, phenyl-

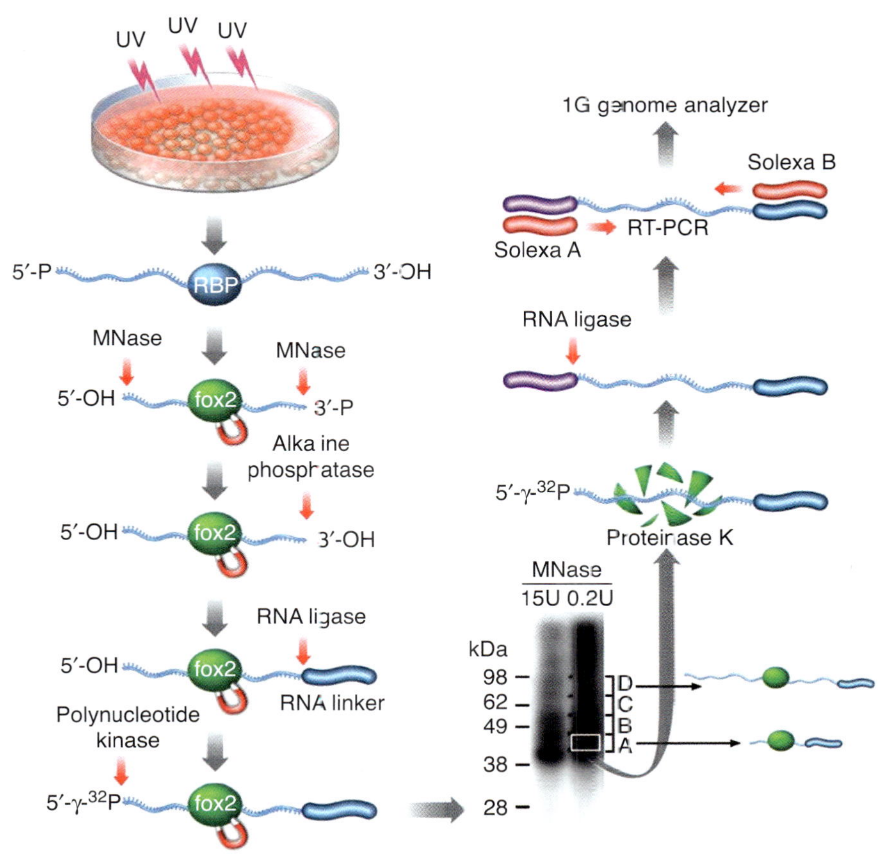

FIGURE 8-6. CLIP method of high-throughput sequencing. (Reprinted, with permission, from Yeo et al. 2009 [©Macmillan].)

alanine, leucine, and cysteine. Because UV-activated bases are essentially "zero-length" cross-linkers, the photoactivated base must be within a few angstroms of a reactive amino acid side chain in the protein of interest. Thus, one should consider UV RNA–protein cross-linking both inefficient and quite dependent on both the base composition of the RNA-binding site and the specific amino acid residues that are juxtaposed in the RNA–protein complex. This means that not every protein can cross-link to its RNA targets, and even for those proteins that can cross-link, not every RNA target will cross-link at the same efficiency.

Formaldehyde Cross-linking

An alternative to UV cross-linking is to use cell-permeable chemical cross-linking reagents that traverse the cell membrane and can covalently attach protein to RNA (or DNA) or protein to protein. By far, the most commonly used cross-linking agent is formaldehyde, which has been used extensively for DNA–protein cross-linking in chromatin immunopreciptation procedures (chromatin IP, ChIP, ChIP-Chip, ChIP-Seq) (see Protocol 10). Formaldehyde has been used to stabilize RNA–protein interactions in vivo, as well as small nuclear ribonucleoprotein (snRNP) binding to pre-mRNAs in yeast and mammalian cells. It is not clear whether formaldehyde is superior to UV as a stabilizing agent for in vivo RNA–protein complexes. Almost certainly, formaldehyde cross-linking can occur at more sites and with greater efficiency than UV cross-linking; however, its specificity remains suspect. Until more experiments are performed, the question concerning which method is better remains open.

No Cross-linking

Many proteins retain noncovalent interactions with their target RNAs in vitro in the absence of covalent cross-links. It is clear from the literature that it is possible to purify biologically relevant RNA–protein complexes without any prior covalent cross-linking. In numerous cases in yeast, *Drosophila*, and mammalian cells, mRNAs associated with specific RNA-binding proteins have been identified. In fact, in a series of plant RNA immunoprecipitation-microarray (RIP-Chip) (see Protocol 9) experiments, mRNAs associated with 11 different RNA-binding proteins were identified without the use of cross-linking (Schmitz-Linneweber et al. 2005; Barkan 2009). In both *Drosophila* and mammalian cells, miRNAs and mRNAs associated with specific members of Argonaute protein family members have been identified by immunopurification without prior cross-linking (Easow et al. 2007; Karginov et al. 2007; Hendrickson et al. 2008; Hong et al. 2009). So, one should consider whether a protein of interest can be retrieved in association with its RNA without cross-linking before moving on to the cross-linking methods.

Antibodies

An important technical concern in the execution of any immunoaffinity purification is the quality of the antibody. In some cases, the target RNA-binding protein can be engineered with an epitope tag to make use of a uniform source of antibody. This is most easily done in yeast, where homologous gene targeting can place an epitope-tagged version of the gene at its endogenous locus under control of its own promoter, and without competition from untagged protein for assembly into endogenous RNA–protein complexes. This requires more complicated manipulations for other organisms, but it can be done in both *Drosophila* and mammalian cells. It is also possible to transfect a copy of the epitope-tagged gene into mammalian cells, but precautions must be taken to ensure that the expression level of the exogenous gene is closely matched to the level of the endogenous protein; otherwise, nonphysiological targets may be bound. Nonetheless, in this scenario, the transfected tagged expressed protein will encounter competition with the endogenous protein for RNA–protein complex incorporation.

In contrast, preparation of a polyclonal antibody or perhaps autoimmune antisera to a protein of interest will be most useful to immunopurify a protein of interest. In this case, the best way to proceed is to express a large segment of the protein of interest (but lacking simple sequences, such as RS repeats or G-rich or Q-rich regions commonly found in RNA-binding proteins) as a recombinant protein in bacteria with a His6 tag. These proteins are generally easy to purify by nickel-chelate chromatography, and they can then be used to immunize rabbits for polyclonal reagent antibody production. For efficiency, and for specificity, it is essential that the immune antibodies be purified from the crude rabbit serum to improve the yield and reduce the background of contaminating proteins. Typically, these affinity-purified antibodies are bound to a solid support, such as protein A–Sepharose or protein A–magnetic beads, to retrieve the antibody-bound RNA–protein complexes. It is possible to conjugate the target protein to resin to purify the protein of interest by affinity purification.

Extracts

Once an antibody that will recognize the protein of interest is available, the next step is the preparation of cell extracts (Protocols 7–10). There are several reviews on this topic for mammalian, yeast, and plant cells. A few general points will be made here. Often, certain types of proteins, such as pre-mRNA splicing factors or translation factors, will be localized to particular cellular compartments (e.g., the nucleus or cytoplasm). In these cases, it might make sense to perform subcellular fractionation before preparing an extract. No matter how the cells are broken and fractionated, it is necessary to maintain physiological ionic strength (100–200 mM monovalent cation, typically K^+, because this ion predominates inside the cell) to stabilize the RNA–protein complexes. The magnesium ion (Mg^{++}) also has a stabilizing effect on RNA–protein complexes. Once the cells are broken and the debris is removed by centrifugation, the soluble extract can be used for immunopurification. For nuclear proteins, nucleoplasmic fractions can be prepared by sonication of the nuclei and step gradient sedimentation of chromatin. This procedure may or may not be advisable considering whether one is interested in capturing nascent RNA transcripts associated with transcribing RNA polymerase II because nascent transcripts will remain associated with the chromatin. One final consideration is how to wash the immobilized antibody-containing RNP complexes. Generally, harsh washing conditions, such as high salt and detergents, should be avoided because these treatments will dissociate the RNA–protein complexes. However, if cross-linking procedures were used, it may be possible to use harsher treatments to isolate and wash the RNA–protein complexes.

Extraction and Analysis of Coimmunopurified RNA

Once the immunopurified RNA–protein complexes have been bound and washed to remove unbound RNPs, RNA can be isolated from them by standard procedures, such a TRIzol extraction (see Chapter 2, Protocol 2) or hydrophobic interaction chromatography on silica gel columns (e.g., RNA isolation kits; see Chapter 2, Protocol 4). Once RNA is purified, the investigator has several options. The RNA can be converted into cDNA (generally by random hexamer priming) and cloned for sequencing of individual clones. Alternatively, the RNA can be converted to cDNA and labeled for microarray analysis, allowing the mapping of RNAs in the sample to the genome. Another option is to prepare cDNA libraries for HTS, which allows identification of RNAs for which microarray probes may not be available.

Although many of these procedures are extended, multistep operations, good planning, and attention to detail should give good results. However, the most important reagents are the antibodies. It may well be worth the investment to find or make the best possible antibody to the protein of interest before attempting to identify RNA.

ANALYSIS OF GENOMIC DATA

For both microarrays and HTS data, a signal is obtained: an analog fluorescent signal for microarrays or a digital count of reads for sequencing. These signals must be mapped to a reference genome as a first step in understanding the data. For microarrays, the probe sequence identifies the likely genomic origin of the signal (assuming no cross-hybridization), whereas for sequencing, the read is matched to the genome using sequence-matching algorithms. For arrays, the strength of the signal is judged by the amount of captured fluorescence, but for sequencing, one can simply count the number of reads that cover each genomic region (Fig. 8-7). Problems arise because closely related gene families exist in many genomes, and thus many reads may map to more than one gene. It is also important to note that in most cases, sequencing is currently done on an amplified library of sequences, rather than an unamplified population of molecules, making it possible that multiple reads arise from one seed molecule. This is another good reason for constructing replicate libraries for sequencing. Microarray data are subject to uncertainty of the extent of cross-hybridization, unless probe sequences capturing the signal are unique in the genome. Thus, for both sets of data, judicious filtering of signals is important. For sequences that do not exist in the genome, such as exon junctions created by splicing, special measures are necessary. For arrays, this may include the use of splice junction probes designed to capture spliced sequences; however, for sequencing, the splice junction reads will not match the genome and must be recovered separately. One way around this problem is to create a "transcriptome" database that contains all known splice junctions ever observed in the expressed sequence tag (EST) or mRNA sequence library, or any combination of bioinformatically predicted splice junctions, whether or not there is molecular evidence for them. The reads that fail to match the genome can be aligned to this transcriptome sequence set to recover and count the reads that cross the splice junctions. Computational approaches can also be taken during mapping in which a first round maps reads to the genome and a second round tests the unmapped reads for the ability to span two collections of mapped reads (transfrags, representing presumptive exons) by a putative splice junction.

Raw counts of sequencing reads, or raw intensities of microarray signals, are essentially meaningless in the absence of an understanding of at least three general features of the data: (1) attribution of signals to a unique location in the genome or transcriptome, (2) distribution of signal across the sampled sequence space (e.g., the genome, or the fraction of the genome represented on an array), and (3) relationships of the signals in the different samples to each other. As is generally the case with any method, quantifying relative changes in mRNA or splicing levels is easier than determining absolute numbers of molecules in the original sample, once unique reads or probes are identified. For example, RNA sequencing of a whole transcriptome leads to mRNA levels that are can be evaluated by counting "reads per kilobase pair per million reads." This number is a function of both the number of reads available for mapping and the number of unique positions within the gene or region to which reads can theoretically be mapped. As with array data, the intensity of the "read sig-

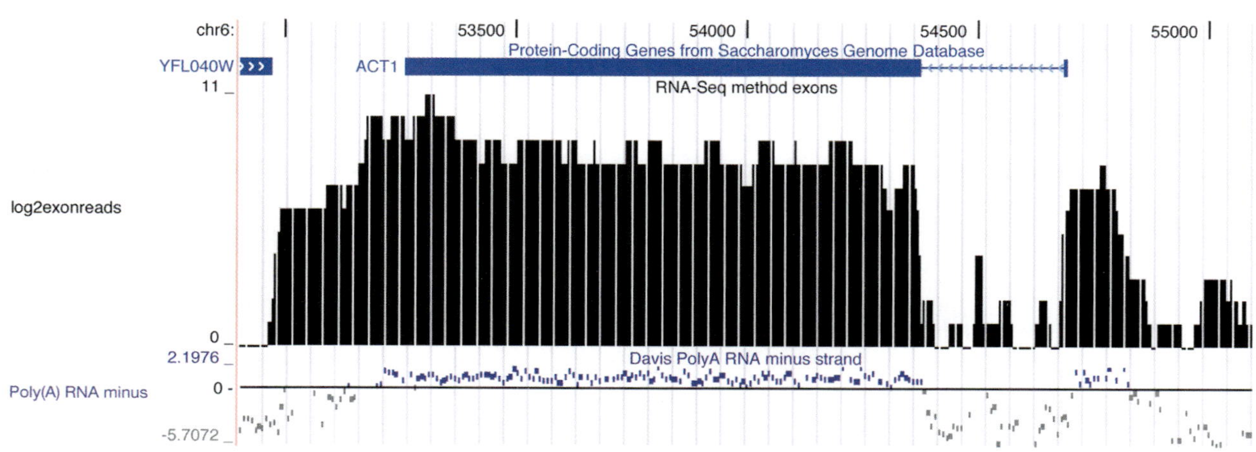

FIGURE 8-7. Comparison of expression of the yeast actin gene *(top)* using high-throughput RNA sequencing. *(Upper track)* RNA-Seq method exons (Nagalakshmi et al. 2008) and genomic tile micorarray analysis. *(Lower track)* Poly(A) RNA minus strand (David et al. 2006) in the UCSC Genome Browser (courtesy Ares lab, University of California, Santa Cruz).

nal" over a gene is generally taken as an indicator of mRNA level, although as described above, performance of both methods can be nonlinear with respect to RNA concentration in the original sample. For either sequencing or microarrays, it is possible to calibrate samples by addition of known amounts of standard RNAs not found in the natural sample and then measure these "spiked in" sequences in the library or on arrays.

Whether sequence data or microarray data are obtained, comparing replicate experiments to evaluate signal reproducibility is an essential first step. Comparing signals of like samples provides a basis for determining whether the measurements are reproducible enough for comparison to samples from other conditions. To compare distributions of distinct samples, it is necessary to "normalize" the distributions with each other. The term "normalize" has come to refer to a collection of operations that reshape and realign distributions that differ with respect to technical variations rather than biological ones; for example, some arrays are faint and some are bright. Some libraries provide deep coverage and some do not. Normalization provides a way to align distributions that differ primarily for technical reasons, so that their biological differences can be extracted. In Protocols 11 and 12, we provide an overview of methods and approaches to analyze RNA expression using these genome-wide techniques.

This chapter presents protocols for microarrays (Protocols 1–4), constructing libraries from RNA for HTS (Protocols 5 and 6), and capturing specific subclasses of RNA from cells for sequencing (Protocols 7–10). In many cases, the technical steps can be carried out by a core facility. Even so, the informed researcher will want a basic understanding of these processes. Although in some cases protocols refer to specific software or instrumentation, this reflects our personal experience rather than an endorsement for a particular commercial product. In our collective experience, different instruments and their software interfaces are not dramatically distinct in their fundamental processes, so we have written mostly about the generic aspects of these tools. Finally, once the data are captured, they must be tamed, and we provide suggestions for approaching this problem (Protocols 11 and 12).

REFERENCES

Barkan A. 2009. Genome-wide analysis of RNA-protein interactions in plants. *Methods Mol Biol* **553:** 13–37.

Burd CG, Dreyfuss G. 1994. RNA binding specificity of hnRNP A1: Significance of hnRNP A1 high-affinity binding sites in pre-mRNA splicing. *EMBO J* **13:** 1197–1204.

David L, Huber W, Granovskaia M, Toedling J, Palm CJ, Bofkin L, Jones T, Davis RW, Steinmetz LM. 2006. A high-resolution map of transcription in the yeast genome. *Proc Natl Acad Sci* **103:** 5320–5325.

DeRisi J, Penland L, Brown PO, Bittner ML, Meltzer PS, Ray M, Chen Y, Su YA, Tren JM. 1996. Use of a cDNA microarray to analyse gene expression patterns in human cancer. *Nat Genet* **14:** 457–460.

Easow G, Teleman AA, Cohen SM. 2007. Isolation of microRNA targets by miRNP immunopurification. *RNA* **13:** 1198–1204.

Ehn M, Nourizad N, Bergstrom K, Ahmadian A, Nyren P, Lundeberg J, Hober S. 2004. Toward pyrosequencing on surface-attached genetic material by use of DNA-binding luciferase fusion proteins. *Anal Biochem* **329:** 11–20.

Guo J, Xu N, Li Z, Zhang S, Wu J, Kim DH, Sano Marma M, Meng Q, Cao H, Li X, et al. 2008. Four-color DNA sequencing with 3′-*O*-modified nucleotide reversible terminators and chemically cleavable fluorescent dideoxynucleotides. *Proc Natl Acad Sci* **105:** 9145–9150.

Hendrickson DG, Hogan DJ, Herschlag D, Ferrell JE, Brown PO. 2008. Systematic identification of mRNAs recruited to argonaute 2 by specific microRNAs and corresponding changes in transcript abundance. *PLoS One* **3:** e2126. Erratum 2008. *PLoS One*.

Hong X, Hammell M, Ambros V, Cohen SM. 2009. Immunopurification of Ago1 miRNPs selects for a distinct class of microRNA targets. *Proc Natl Acad Sci* **106:** 15085–15090.

Karginov FV, Conaco C, Xuan Z, Schmidt BH, Parker JS, Mandel G, Hannon GJ. 2007. A biochemical approach to identifying microRNA targets. *Proc Natl Acad Sci* **104:** 19291–19296.

Keene JD, Komisarow JM, Friedersdorf MB. 2006. RIP-Chip: The isolation and identification of mRNAs, microRNAs and protein components of ribonucleoprotein complexes from cell extracts. *Nat Protoc* **1:** 302–307.

Licatalosi DD, Mele A, Fak JJ, Ule J, Kayikci M, Chi SW, Clark TA, Schweitzer AC, Blume JE, Wang X, et al. 2008. HITS-CLIP yields genome-wide insights into brain alternative RNA processing. *Nature* **456:** 464–469.

MacLean D, Jones JD, Studholme DJ. 2009. Application of "next-generation" sequencing technologies to microbial genetics. *Nat Rev Microbiol* **7:** 287–296.

Matunis MJ, Matunis EL, Dreyfuss G. 1992. Isolation of hnRNP complexes from *Drosophila melanogaster*. *J Cell Biol* **116:** 245–255.

Mili S, Steitz JA. 2004. Evidence for reassociation of RNA-binding proteins after cell lysis: Implications for the interpretation of immunoprecipitation analyses. *RNA* **10:** 1692–1694.

Mitra RD, Shendure J, Olejnik J, Edyta-Krzymanska O, Church GM. 2003. Fluorescent in situ sequencing on polymerase colonies. *Anal Biochem* **320:** 55–65.

Mortazavi A, Williams BA, McCue K, Schaeffer L, Wold B. 2008. Mapping and quantifying mammalian transcriptomes by RNA-Seq. *Nat Methods* **5:** 621–628.

Nagalakshmi U, Wang Z, Waern K, Shou C, Raha D, Gerstein M, Snyder M. 2008. The transcriptional landscape of the yeast genome defined by RNA sequencing. *Science* **320:** 1344–1349.

Pease AC, Solas D, Sullivan EJ, Cronin MT, Holmes CP, Fodor SP. 1994. Light-generated oligonucleotide arrays for rapid DNA sequence analysis. *Proc Natl Acad Sci* **91:** 5022–5026.

Pepke S, Wold B, Mortazavi A. 2009. Computation for ChIP-seq and RNA-seq studies. *Nat Methods* (suppl.) **6:** S22–S32.

Piacentini L, Fanti L, Negri R, Del Vescovo V, Fatica A, Altieri F, Pimpinelli S. 2009. Heterochromatin protein 1 (HP1a) positively regulates euchromatic gene expression through RNA transcript association and interaction with hnRNPs in *Drosophila*. *PLoS Genet* **5:** e1000670.

Russom A, Tooke N, Andersson H, Stemme G. 2005. Pyrosequencing in a microfluidic flow-through device. *Anal Chem* **77:** 7505–7511.

Sanford JR, Wang X, Mort M, Vanduyn N, Cooper DN, Mooney SD, Edenberg HJ, Liu Y. 2009. Splicing factor SFRS1 recognizes a functionally diverse landscape of RNA transcripts. *Genome Res* **19:** 381–394.

Schena M, Shalon D, Davis RW, Brown PO. 1995. Quantitative monitoring of gene expression patterns with a complementary DNA microarray. *Science* **270:** 467–470.

Schmitz-Linneweber C, Williams-Carrier R, Barkan A. 2005. RNA immunoprecipitation and microarray analysis show a chloroplast Pentatricopeptide repeat protein to be associated with the 5′ region of mRNAs whose translation it activates. *Plant Cell* **17:** 2791–2804.

Selth LA, Gilbert C, Svejstrup JQ. 2009. RNA immunoprecipitation to determine RNA-protein associations in vivo. *CSH Protoc* **2009:** pdb prot5234.

Shalon D, Smith SJ, Brown PO. 1996. A DNA microarray system for analyzing complex DNA samples using two-color fluorescent probe hybridization. *Genome Res* **6:** 639–645.

Shendure J, Porreca GJ, Reppas NB, Lin X, McCutcheon JP, Rosenbaum AM, Wang MD, Zhang K, Mitra RD, Church GM. 2005. Accurate multiplex polony sequencing of an evolved bacterial genome. *Science* **309:** 1728–1732.

Ule J, Jensen K, Mele A, Darnell RB. 2005. CLIP: A method for identifying protein-RNA interaction sites in living cells. *Methods* **37:** 376–386.

Ule J, Jensen KB, Ruggiu M, Mele A, Ule A, Darnell RB. 2003. CLIP identifies Nova-regulated RNA networks in the brain. *Science* **302:** 1212–1215.

Wang WX, Wilfred BR, Hu Y, Stromberg AJ, Nelson PT. 2010. Anti-Argonaute RIP-Chip shows that miRNA transfections alter global patterns of mRNA recruitment to microribonucleoprotein complexes. *RNA* **16:** 394–404.

Yeo GW, Coufal NG, Liang TY, Peng GE, Fu XD, Gage FH. 2009. An RNA code for the FOX2 splicing regulator revealed by mapping RNA-protein interactions in stem cells. *Nat Struct Mol Biol* **16:** 130–137.

Zisoulis DG, Lovci MT, Wilbert ML, Hutt KR, Liang TY, Pasquinelli AE, Yeo GW. 2010. Comprehensive discovery of endogenous Argonaute binding sites in *Caenorhabditis elegans*. *Nat Struct Mol Biol* **17:** 173–179.

WWW RESOURCE

http://www.nimblegen.com/technology/manufacture.html Roche NimbleGen, Inc., manufacturer of custom, high-density DNA arrays based on its proprietary Maskless Array Synthesizer (MAS) technology.

Protocol 1

Preparation of Fluorescent-dye-labeled cDNA from RNA for Microarray Hybridization

This protocol describes how to prepare fluorescent-dye-labeled cDNA for hybridization to microarrays. It consists of two steps. First, a mixture of anchored oligo(dT) and random hexamers is used to prime amine-modified cDNA synthesis by reverse transcriptase using a modified deoxynucleotide with a reactive amine group (aminoallyl-dUTP) and an RNA sample as a template. Second, the cDNA is purified and exchanged into bicarbonate buffer so that the amine groups in the cDNA react with dye *N*-hydroxysuccinimide (NHS) esters, covalently joining the dye to the cDNA. The dye-coupled cDNA is purified again, and the amount of dye incorporated per microgram of cDNA is determined. The resulting fluorescent dye-labeled cDNAs are ready for use in microarray hybridization (see Protocol 2).

MATERIALS

The recipes for the items marked with <R> are at the end of the protocol or in the Common Recipes Appendix.

CAUTION: See the Cautions Appendix for appropriate handling of materials marked with <!>.

Reagents

5-(3-Aminoallyl)-dUTP (50 mM)
Anchored oligo(dT) (dT24VN, HPLC pure; 4 µg/µL, prepared in H_2O)
Dithiothreitol (DTT; 0.1 M) <!>
DNA Clean & Concentrator-5 Kit (Zymo Research)

The kit contains DNA-binding buffer and DNA wash buffer, as well as filters and collection tubes. In addition, ethanol (95% or 100%) <!> is required to prepare the DNA wash buffer, but it is not supplied with the kit.

Ethanol (95%, 100%) <!>
First-strand buffer (5x; Invitrogen)
HEPES-K+ (1 M, pH 7.0)
NaOH mix <R> <!>
N-Hydroxysuccinimide (NHS) esters of fluorescent dyes (e.g., CyDye Post-labeling Reactive Dye Pack; GE Healthcare RPN5661)

One aliquot of dye contains ~40 nmol of reactive dye ester.

Nucleotide mix (50x) <R>
Random hexamers (3 µg/µL, prepared in H_2O)
RNase-free H_2O
RNaseOUT (40 units/µL; Invitrogen)
Samples of total RNA (20 µg per reaction; see Step 1)

The single most important determinant in success of this protocol is RNA integrity.

Sodium bicarbonate ($NaHCO_3$; 50 mM, pH 9.0) <R>
SuperScript III RT (reverse transcriptase; 200 units/µL; Invitrogen)
Tris-Cl (10 mM, pH 8.0)

Equipment

Heat blocks or water baths (48°C, 65°C, and 70°C)
Ice
Microcentrifuge
Microcentrifuge tubes (1.5 mL)
NanoDrop spectrophotometer
Vortex machine

METHOD

Reverse Transcription Reaction

1. Determine the number of samples needed for the set of arrays to be hybridized.

   Each array will have two samples (e.g., one labeled with Cy3 and one with Cy5). For additional considerations, see Discussion.

2. Thaw all reagents and RNA samples. Vortex and centrifuge to at least 100*g* for ~20 sec to collect droplets of liquid in the bottom of the tube ("quick-spin").

3. Mix 62.5 µL of 4 µg/µL anchored oligo(dT), 16.7 µL of 3 µg/µL random hexamers, and 20.8 µL of RNase-free H_2O.

   This primer mix will contain final concentrations of 2.5 µg/µL anchored oligo(dT) and 0.5 µg/µL random hexamers. It can be prepared in advance and stored indefinitely at –20°C.

4. Add 3 µL of primer mix to 20 µg of total RNA in a total volume of 23 µL. Anneal for 10 min at 70°C.

   Run individual samples separately. Do not scale up multiple reactions; instead, carry out additional reactions at this scale. If desired, pool and split the samples at critical points to maintain consistency. See Discussion.

5. Meanwhile, prepare the master mix for the reverse transcription reactions.

H_2O	0.96 µL
First-strand buffer (5x)	8 µL
DTT (0.1 M) <!>	4 µL
Nucleotide mix (50x)	0.8 µL
5-(3-aminoallyl)-dUTP (50 mM)	0.24 µL
RNaseOUT (40 units/µL)	1 µL
SuperScript III (200 units/µL)	1 µL
Total	16 µL

   The volumes shown are per reaction (1x). For eight reactions, make 9x master mix in the event of pipetting errors.

6. Cool the annealing reactions on ice and then quick-spin.
7. Add 16 µL of the master mix to each annealing reaction. Mix and quick-spin.
8. Incubate the reverse transcription reactions for 5 min at room temperature and then for 1 h at 48°C. Quick-spin the reactions.
9. Add another 1 µL of SuperScript III (200 units/µL) to each reaction. Mix and quick-spin.
10. Incubate the reactions for 1 h at 48°C. Quick-spin the reactions.
11. Hydrolyze the RNA by adding 4.4 µL of NaOH mix. Mix and quick spin.
12. Incubate for 20 min at 65°C.

13. Neutralize with 44 µL of 1 M HEPES-K+ (pH 7.0).

The cDNA can be stored for several days at –20°C before proceeding.

Purification of cDNA

In this part of the procedure, the DNA Clean & Concentrator-5 Kit (Zymo Research) is used to purify single-stranded cDNA. This part of the protocol includes slight modifications to the manufacturer's instructions.

14. Set up and label a filter/collection tube for each reaction.
15. Add 1 mL of Zymo DNA-binding buffer to each reaction and mix gently.
16. Load ~550 µL of the reaction into the filter cup.
17. Centrifuge at maximum speed in a microcentrifuge for 10 sec at room temperature.
18. Discard the filtrate. Load the remainder of the reaction into the filter cup.
19. Centrifuge at maximum speed in a microcentrifuge for 10 sec at room temperature.
20. Discard the filtrate. Return the filter to the tube.
21. Load 200 µL of Zymo wash buffer into the filter cup.

    Make sure that ethanol has been added to the wash buffer as recommended by the manufacturer.
22. Centrifuge at maximum speed in a microcentrifuge for 10 sec at room temperature.
23. Add another 200 µL of Zymo wash buffer into the filter cup.
24. Centrifuge at maximum speed in a microcentrifuge for 20 sec at room temperature.
25. Place the filter into a clean, labeled 1.5-mL tube.
26. Elute by pipetting 8 µL of 50 mM sodium bicarbonate (pH 9) directly onto the filter (make sure that the drop gets onto the filter). Wait for 5 min.
27. Centrifuge at maximum speed in a microcentrifuge for 20 sec at room temperature.
28. Repeat the elution procedure (Steps 26 and 27). There should be ~14–15 µL of total eluate.
29. Determine the yield of cDNA as follows:
    i. Initialize the NanoDrop under the "Nucleic Acid" setting.
    ii. Set the conversion factor to 33 µg/mL per A_{260} unit.
    iii. Blank (zero) the instrument with 50 mM sodium bicarbonate (pH 9).
    iv. Apply 1.3 µL of each eluate to the NanoDrop.
    v. Record the amount of cDNA. The cDNA can be stored for several days at –20°C before proceeding.

    There should be a total of ~2 µg of cDNA if starting with either 20 µg of total yeast RNA or 10–15 µg of total vertebrate cell RNA. Reactions that yield less than 1 µg of cDNA should be abandoned before Step 30; it is not worth wasting the expensive dye esters. See Tips and Troubleshooting.

Coupling Reaction and Purification

30. For the coupling reaction, add the entire eluate from the equivalent of one reaction above (or up to 2 µg of cDNA) to one dried-down aliquot of dye ester. Pipette carefully to resuspend. Incubate in the dark for 1 h at room temperature and flick to mix every 20 min.

    NHS esters of fluorescent dyes are extremely sensitive to hydrolysis and are thus often sold in dry aliquots. In this protocol, one aliquot of dye ester as indicated in the supply list contains ~40 nmol of reactive dye ester.

31. Set up and label a filter/collection tube for each reaction. Add 500 μL of Zymo DNA-binding buffer (in the DNA Clean & Concentrator-5 Kit) to each reaction and mix gently.
32. Load the entire reaction into the filter cup.
33. Follow Steps 19–25.
34. Elute by pipetting 8 μL of 10 mM Tris-Cl (pH 8.0) directly onto the filter (make sure that the drop gets onto the filter). Wait for 5 min.
35. Centrifuge at maximum speed in a microcentrifuge for 20 sec at room temperature.
36. Repeat the elution procedure (Steps 34 and 35). There should be ~14–15 μL of total eluate.
37. Determine the yield of dye-labeled cDNA as follows:
    i. Initialize the NanoDrop under the "Microarray" setting.
    ii. Set the conversion factor to 33 μg/mL per OD for single-stranded cDNA.
    iii. Blank (zero) the instrument with 10 mM Tris-Cl (pH 8.0).
    iv. Apply 1.3 μL of each eluate to the NanoDrop.
    v. Record the amount of cDNA and pmol of dye incorporated.

       The NanoDrop initialized under the "Microarray" settings will take a spectrum that includes both the nucleic acid region and the dye regions, allowing the determination of the number of pmol of dye incorporated per nanogram of cDNA. A number called the FOI (frequency of incorporation) is formally similar to "specific activity" and expresses the number of dye molecules incorporated per 1000 nucleotides of cDNA. This is determined by dividing the total pmol of dye incorporated times 324.5 ng/nmol (the average molecular weight of a nucleotide) by the total number of nanograms of cDNA. For example, 14 μL of a solution with 9 pmol/μL dye and 38 ng/μL cDNA will have ~0.5 μg of cDNA with an FOI of 77, meaning that 77 of every 1000 nucleotides are coupled to a dye molecule. The calculation is Total pmol of dye = 14 x 9 = 126 pmol. Total nanograms of cDNA = 38 x 14 = 532. Then, (126 x 324.5)/532 = 76.85 pmol dye per nmol of nucleotide (i.e., per 1000 bases). FOI values in excess of 100 are unnecessary and may inhibit hybridization, whereas those below 15 may not provide adequate signal after hybridization.

TIPS AND TROUBLESHOOTING

If no cDNA elutes from the column with sodium bicarbonate in Step 29.v, ensure that the ethanol has been added to the wash buffer in the DNA purification kit. If the ethanol is omitted, the DNA will not be retained on the column.

DISCUSSION

The goal of a microarray experiment determines which and how many samples are to be labeled. A typical experiment uses two arrays hybridized with the same samples, but with the dyes reversed. For example, in comparing a mutant strain to a wild-type strain, one slide would be hybridized with a mixture of mutant Cy5-cDNA and wild-type Cy3-cDNA, whereas the other slide would be hybridized with a mixture of mutant Cy3-cDNA and wild-type Cy5-cDNA. Data from this "dye-swapped pair" of slides are combined. The single cDNA synthesis reaction described in this protocol provides material for one dye-labeled sample for one array. Thus, to produce cDNA for one dye-swapped pair of slides, four reactions should be set up. There should be two reactions with mutant RNA and two reactions with wild-type RNA. One mutant reaction and one wild-type reaction should be coupled to a Cy3 ester; likewise, one reaction of each type should be coupled to a Cy5 ester. Many other NHS-dye esters, in addition to the commonly used Cy3 and Cy5 dyes, are available from different companies.

Consistency in the properties of the samples (e.g., concentration and extent of coupling) is important. For example, if comparing four different mutant strains to a wild-type strain, it would be best to use the same wild-type reference sample for all comparisons. To account for the variation in sample preparation that occurs in replicate wild-type reaction tubes, wild-type samples can be pooled after cDNA synthesis, their total concentration can be determined with one reading, and they can then be split into four reactions for coupling. After coupling, they can be pooled again; their concentration and FOI values can be determined with one reading, and they can then be split again for application to the arrays. Although this seems to be complicated, it allows for the production of a consistent wild-type sample that provides better microarray results.

SOLUTION AND REAGENT RECIPES

CAUTION: See the Cautions Appendix for appropriate handling of materials marked with <!>.

NaOH Mix

Reagent	Quantity (for 600 µL)	Final concentration (10x)
NaOH (10 M) <!>	60 µL	1 M
EDTA (500 mM) <!>	24 µL	20 mM
H_2O	516 µL	

Prepare fresh.

Nucleotide Mix (50x)

Reagent	Quantity (for 100 µL)	Final concentration
A (100 mM)	25 µL	25 mM
C (100 mM)	25 µL	25 mM
G (100 mM)	25 µL	25 mM
T (100 mM)	10 µL	10 mM
H_2O	15 µL	

Store indefinitely at –20°C.

Sodium Bicarbonate (1 M, pH 9.0)

For 100 mL, add 8.4 g of $NaHCO_3$ <!> to H_2O and adjust the pH to 9.0 with ~1 mL of NaOH <!>. Store indefinitely at room temperature.

WWW RESOURCE

http://cmgm.stanford.edu/pbrown/protocols/index.html P. Brown's Lab, Experimental Protocols, Stanford University School of Medicine, Stanford, California.

Protocol 2

Microarray Slide Hybridization

This protocol describes a blocking and hybridization protocol for microarray slides. The blocking step is particular to the chemistry of "CodeLink" slides, but it serves to remind us that almost every kind of microarray has a treatment step that occurs after printing but before hybridization. We recommend making sure of the precise treatment necessary for the particular chemistry used in the slides to be hybridized because the attachment chemistries differ significantly. Hybridization is similar to northern or Southern blots but on a much smaller scale.

MATERIALS

The recipe for the item marked with <R> is at the end of the protocol or in the Common Recipes Appendix.

CAUTION: See the Cautions Appendix for appropriate handling of material marked with <!>.

Reagents

Blocking buffer (50 mM ethanolamine, 100 mM Tris at pH 9, 0.1% SDS; add SDS to buffer just before use) <R>
HEPES (1 M, pH 7.0)
Hybridization solution (see Step 10)
Poly(A) (polyadenylic acid, potassium salt, 2 μg/μL; Sigma-Aldrich)
SDS (1%) <!>
SSC (20x)
Wash buffer 1 (4x SSC/0.1% SDS)
Wash buffer 2 (2x SSC/0.1% SDS)
Wash buffer 3 (0.2x SSC)
Wash buffer 4 (0.05x SSC)

Equipment

Canned air (e.g., "Dust-Off")
Centrifuge with microtiter plate rotor capable of carrying slide racks
CodeLink slides
Coplin jar
Coverslips (LifterSlip; Erie Scientific)
Make sure that they are the correct size to cover the array.
Desiccator
Forceps
Glass microscope slide (clean)
Heat block (95°C)
Humidification chamber
Microarray hybridization chambers (e.g., Die-Tech)
Microcentrifuge
Microtiter plate carrier rotor
Permanent marker or diamond pen
Pipettors

Shaking incubator
Slide racks and wash tubs
Squirt bottle
Water bath (63°C)
Whatman 3MM paper cut into ~0.5-cm squares (four per array)

OVERVIEW

CodeLink slides have a surface designed to capture amines, such as those incorporated onto oligonucleotides at the 5′ end. Upon printing, the slides are incubated in a humidification chamber to allow covalent attachment of the oligonucleotides. After this reaction, but before hybridization, the unreacted groups on the slide surface must be blocked so that they do not capture the fluorescent target cDNA (which also contains free amines) and contribute to background on the slide. This is done using ethanolamine treatment. Blocking the slide often makes the spots invisible to the naked eye because the salt from the printing buffer left at the spot is washed away during blocking (Fig. 8-8). Thus, a template slide must be made or each slide must be marked so that the location of the printed spots is not lost (see Step 1).

Once blocked, the slides can be hybridized. This is usually done with solutions that are very similar to northern and Southern hybridization buffers, except that the volumes are tiny and are applied under a coverslip only to the area of the slide where the spots have been printed. A special coverslip called a LifterSlip has thin strips along two opposite sides that hold the coverslip a set distance above the slide. The slide is placed in a metal chamber that encloses it in a small water-tight space, the coverslip is placed over the spots, the hybridization solution is delivered under the coverslip, and two buffer-soaked bits of filter paper are placed in the chamber as well to keep it humid. The chamber is closed and placed in a water bath for hybridization. At the end, slides are washed in Coplin jars through a wash buffer series very similar to those used for northern or Southern blots.

FIGURE 8-8. Unblocked (*left*) versus blocked (*right*) microarray slides. The spots on the unblocked slide are from evaporated printing buffer, which is washed away during blocking.

METHOD

Blocking the Slides

1. Mark the location of the array on the CodeLink slide with a permanent marker or diamond pen.

   If the slide has visible spots, but there are no permanent marks on the slide to indicate the boundary of the array, we recommend either marking the back of the slide (the side without the spots) a bit further out from each corner of the block of spots using a diamond pen so that the coverslip can be aligned after the spots have disappeared. Another method is to place the array slide on top of a clean glass microscope slide and use a permanent marker to mark the glass slide rather than the array. After the array spots have vanished, alignment of the array with the marked glass slide template will allow correct placement of the coverslip later. Because most experiments involve multiple sequential slides from the same print run, use a single template slide for each array in the experiment.

2. Preheat the Blocking buffer and Wash buffer 1 to 55°C for blocking the slides. Add SDS to the Blocking buffer just before use. Place the slides in plunger racks fit to containers that allow complete immersion of the slides.
3. Block the slides for ~45 min, with slow shaking, starting with 55°C Blocking buffer (can shake container at room temperature).
4. Rinse the slides twice with dH_2O at room temperature.
5. Block the slides for another 45 min at 55°C in Wash buffer 1.
6. Rinse the slides four times with dH_2O. Rinse vigorously to avoid streaks.
7. Place the slide rack on a microtiter plate carrier rotor with paper towels underneath. Centrifuge at 800 rpm for 4 min at room temperature. Carefully clean any droplets adhering to the edges with a Kimwipe before they splatter back onto the array surface. If there are visible streaks upon careful inspection, rinse again in dH_2O and centrifuge again.

   It may take a little fussing to be satisfied with the centrifuge time and speed depending on the centrifuge, but the idea is to have a pristine surface with as little residue of salt or SDS as possible.

8. The processed slides can be stored in a desiccator chamber for 1 mo at room temperature. For long-term storage (more than several months), leave the slides unblocked.

Hybridization of Microarray Slides

It is assumed that the investigator has planned out which samples should be mixed and hybridized with which other samples and has created fluorescent target using a method similar to Protocol 1.

9. Using the method described in Protocol 1, which yields ~14 µL of target per sample and should not require concentration of the target, mix the two target samples labeled with different dyes together. If a different protocol is used or the yields are very different, equalize the samples using the FOI measurement so that they have the same number of moles of coupled dye molecules. Mix samples to be cohybridized together in a total of ~28 µL (adjust to 28 µL with dH_2O if necessary).
10. Prepare the hybridization solution. The size of the array will determine the size of the LifterSlip to use, and this will determine the volume of hybridization solution to be prepared. A 20,000 spot array will fit under a 25 x 44-mm LifterSlip (25X44I), which will support ~73 µL of hybridization solution. Add hybridization mix reagents to the samples. Alternatively, make the hybridization solution as follows and add per mixed pair of samples (i.e., per slide):

HEPES (1 M, pH 7.0)	1.5 µL
Poly(A) (2 µg/µL; Sigma-Aldrich)	3.7 µL
SSC (20x)	14.6 µL
dH_2O	17.9 µL
Total volume	37.7 µL

Pipette gently up and down to mix. Add 7.3 µL of 1% SDS to the above mixture and pipette gently to mix.

From this point on, bubbles are unwanted, so avoid making bubbles.

11. Heat the hybridization mixes for 2 min at 95°C to denature the samples. Centrifuge at full speed in a microcentrifuge for 5 min and then cool to room temperature.

    This step pellets any precipitate that may cause background on the scan.

12. While the denaturation and centrifugation steps are proceeding, set up the arrays as follows. Have forceps handy in case the LifterSlips need to be shifted or straightened on the slides.

    i. Blow dust/particulates off the LifterSlips and processed CodeLink slides with Dust-Off or filtered air. Do not contact the printed surface of the array. If the LifterSlips are really dirty, wipe or wash them with H_2O, ethanol, and a Kimwipe. Wipe the back of the slide, but be sure not to touch the side with the spots.

    ii. Position the LifterSlips on the slides using a template slide or following the marks etched in the arrays slide (see Fig. 8-9A). Note that the template slide is on the bottom, next is the "CodeLink" slide with the array, and finally the LifterSlip is on top.

    iii. Place one slide in each microarray hybridization chamber.

    iv. Pipette up the hybridization solution from the microcentrifuge tube, avoiding any pelleted material from the denaturation and spin step.

    v. Slowly pipette the mixture under the LifterSlip. Start by placing the tip next to one corner and then slowly move across that edge as the sample is delivered, until the entire surface covered by the LifterSlip is filled (Fig. 8-9B). Avoid bubbles!

    vi. Load additional sample across the opposite edge of the LifterSlip as well, once the liquid front has reached it, again avoiding bubbles.

    vii. Stack two small (0.5 x 0.5 cm) squares of Whatman 3MM paper on the slide label and stack another two at the other end of the slide. Both stacks should be sufficiently far from the edge of the LifterSlip so that the hybridization fluid cannot touch it and be wicked away.

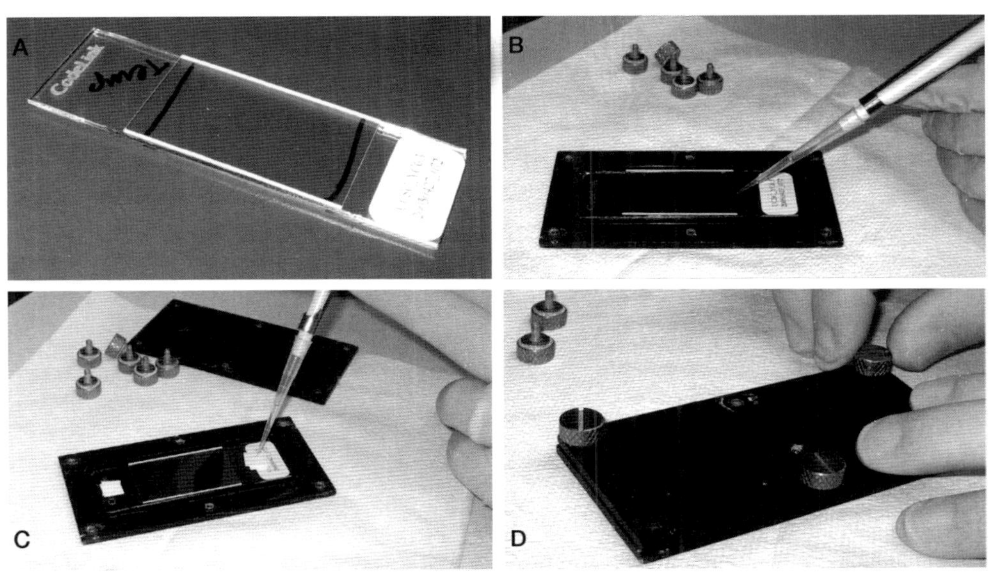

FIGURE 8-9. Hybridization of microarray slides. (*A*) Position the LifterSlip using a template slide underneath the array slide. (*B*) Pipette hybrization solution carefully, allowing it to wick under the LifterSlip. (*C*) Pipette 4x SSC onto squares of absorbent paper to maintain humidity in the chamber. (*D*) Close the chamber by hand-tightening the thumb screws.

viii. Pipette ~50 µL of 4x SSC onto each of the paper stacks to maintain humidity inside the chamber (Fig. 8-9C).

ix. Close the chambers, making sure that the gaskets are aligned so that the chamber is sealed, and hand-tighten the screws (Fig. 8-9D). Do not tighten each screw all the way down in turn. Instead, tighten each a bit first, and then go back and retighten each, little by little so that the gasket is evenly pressed around the groove in which it sits. This will prevent leaks. Do not overtighten.

13. To hybridize the slides, place the chambers in a Tupperware container and place the container in a water bath overnight (12–16 h) at 63°C.

    Avoid placing the chambers on top of heating elements under the water-bath pan because this can cause transient overheating of the metal slide chambers and thus the arrays.

14. Wash the arrays.

i. Prepare the wash buffers. Fill a small plastic squirt bottle with Wash buffer 1 and warm it to 63°C. In addition, warm Wash buffer 2. Fill a Coplin jar with Wash buffer 2. Prepare the slide racks and tubs. Fill one tub with warm Wash buffer 2 and insert an empty rack.

ii. Working as quickly as possible, take the chambers out of the water bath and blot dry. Make sure to remove the water from between the chamber plates before opening, so that water-bath fluid does not splash onto the arrays.

iii. Unscrew the holding screws and open the chamber. Carefully pry the slide out with forceps, lifting and grasping the slide on the end with the label.

iv. Hold the slide with the label end up over the Coplin jar. Use the squirt bottle to gently wash the LifterSlip straight down off the slide with a squirt or two of Wash buffer 1; avoid having the edges of the LifterSlip scrape the array area (Fig. 8-10). Rinse the front and back of the slide and then place the slide in the slide rack in the tub of prewarmed Wash buffer 2 so that the slide is completely covered. Repeat for any additional slides.

v. Shake the tub with the rack slowly for 5 min at room temperature.

vi. Pour out the Wash buffer and fill with with fresh Wash buffer 2.

vii. Transfer the slides individually to a new slide rack in Wash buffer 3 at room temperature. Try to minimize carryover of Wash buffer 2 into the tub of Wash buffer 3.

viii. Shake or dunk the slide rack for 1 min in Wash buffer 3 at room temperature.

ix. Transfer the slide rack to Wash buffer 4 and shake or dunk it for 1 min in Wash buffer 4.

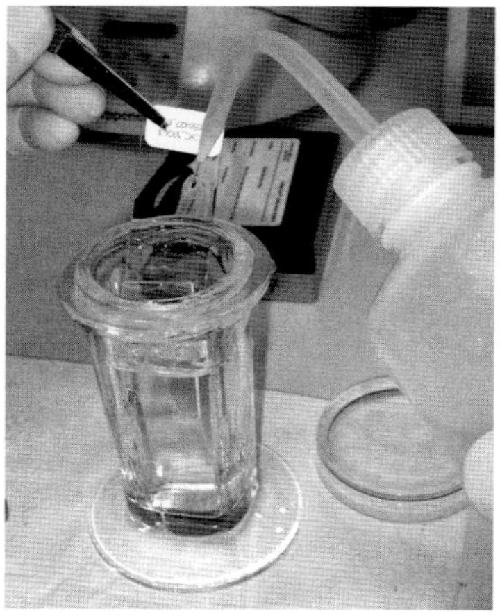

FIGURE 8-10. Rinse off the LifterSlip into a Coplin jar with a gentle stream of Wash buffer from a squirt bottle.

x. Immediately centrifuge the slides dry at 800 rpm for 4 min at room temperature.

xi. Check for streaks of buffer residue by carefully inspecting each slide. If there are streaks, wash the slides again with Wash buffer 4 in a fresh container and centrifuge again as in Step 14.x.

xii. Transfer slides to a slide box and protect them from light. Scan the slides as soon as possible.

Optional: Use Dyesaver (Genisphere) for longer-term storage before scanning. Dyes have different stabilities; in particular, Cy5 is more sensitive to oxidation and bleaching than Cy3.

TIPS AND TROUBLESHOOTING

1. Carefully separating the hybridization buffer from any precipitate formed during the denaturation step will keep tiny particulates from creating high background on the slide.
2. Once the slide is in the chamber and the LifterSlip is positioned, keep the slide chamber as close to horizontal and handle as carefully as possible to avoid shifting the position of the LifterSlip. Use caution when placing the chambers in the water bath.
3. Do not allow the Whatman 3MM paper squares to touch the residual hybridization fluid that may escape from under the LifterSlip. Doing so may cause the hybridization solution to wick out from under the LifterSlip.
4. Ensure that nothing scrapes against the printed surface of the slide. Do not allow buffer to dry onto the slide at any time. Immediately after rinsing and spinning to dry, inspect for streaks and if any are visible, wash again.

SOLUTION AND REAGENT RECIPE

CAUTION: See the Cautions Appendix for appropriate handling of materials marked with <!>.

Blocking Buffer

Reagent	Quantity (for 1000 mL)	Final concentration
Ethanolamine (16.7 M) <!>	3 mL	50 mM
Tris base (1 M, pH 9.0) <!>	100 mL	100 mM
SDS (10%, w/v) <!>	10 mL	0.1%
H_2O	887 mL	

Add SDS just before use. Store indefinitely at room temperature *without* SDS.

WWW RESOURCE

http://cmgm.stanford.edu/pbrown/protocols/index.html P. Brown's Lab Experimental Protocols, Stanford University School of Medicine, Stanford, California.

Protocol 3

Scanning Microarray Slides

This protocol describes some practical concerns surrounding the scanning of microarray slides that have been hybridized with fluorescent dyes. A laser scanner similar to instruments built by Axon, Inc. (now part of Molecular Devices, Inc., http://www.moleculardevices.com/) has two lasers, each set to excite a different fluor, with separate detectors to capture emission from each fluor. The laser passes over an address (position on the scanned surface) and photons emitted from each address are captured by the detectors. Two superimposed image files are written that carry intensities for each channel for each pixel of the image scan. This is the raw data. Image analysis software is used to identify and summarize the intensities of the pixels that make up each spot. After comparison to background pixels, the processed intensity levels representing the gene expression measurements are associated with the identity of each spot. The commonly used Axon scanner operates well with the Windows-based software package called GenePix also sold by Axon, Inc. However, publicly available software can also be used, e.g., "SCAN-ALIZE" at Mike Eisen's website, http://rana.lbl.gov/EisenSoftware.htm. Other scanning and analysis programs can be found at http://www.statsci.org/micrarra/image.html.

MATERIALS

Reagents

Canned Air (e.g., Dust-Off) or filtered house air

Microarrays (hybridized, washed, and dried from Protocol 2)

Equipment

Computer with spotfinding and image analysis software

Most of these programs are compatible with Windows systems only.

Dual laser-scanning microscope (microarray scanner)

Files

Array file

Reveals the spot location (block, row, column) on the array and the identity (sequence, genomic coordinates, PCR product) of the probe at each spot location

Image file

Generated by the scanning of the array. Has two superimposed images in it, one for each channel

Results file

Generated by the analysis of the data extracted from the image once the settings file is aligned

Settings file

Projects spot boundaries on the scanned image and can be edited so that each spot boundary fits its spot on the image

OVERVIEW

Hybridized, washed, and dried arrays are inspected for dust particles. If dust is observed, a stream of air can be applied to attempt to remove it. Slides are placed in the scanner (array side down in the Axon instrument). Low-resolution "prescans" of the array are used to adjust the laser power settings so that similar intensity profiles are obtained for the two dyes. Once this adjustment is made, the array is scanned at high resolution to produce the image files. The files are then analyzed by superimposing a template of the array carrying the spot identities and approximate positions onto the image. The template is then adjusted to align perfectly with the actual spots on the array image ("spot-finding"). Then, spots that do not conform to specified "good spot" characteristics are flagged and excluded from analysis. Such spots may be saturated with dye (too bright), not high enough above background, or may have other features that indicate that the signal is not reliable.

METHOD

Scan Slides

1. Turn on the scanner to warm up the lasers for at least 10 min before scanning.
2. Inspect and, if necessary, dust off the microarray.
3. Perform a prescan using the instrument settings in the scanner software. Modify the laser power (PMT) settings to avoid saturation, but try to maximize the number of spots in the linear range for both channels. Balance histogram of signals or spiked target-control spot signals so that the dyes are approximately equally detected. In the GenePix program, an "AutoPMT" function can do this automatically, and it performs well if both labeled samples hybridized about equally well.
4. Set scan resolution at 5 μm (10 μm is probably acceptable for arrays with larger [80–100 μm] spots) and two scans per line (this averages the counts from two independent scans of each line). Set the scan area to include all spots, and scan. Save the image file using an informative, systematic, and unambiguous file name.

Generate Template for Spotfinding

5. Use a file describing spot locations and identities for all arrays printed in the same run ("Array file" above, called a ".gal" file in GenePix) to generate a template for spotfinding ("Settings file" above, called a ".gps" file or "settings file" in GenePix).

   Most spotfinding programs provide many user-editable parameters that improve performance in the face of specific array irregularities that commonly occur during printing, such as variable spot size, variable pin-to-pin distances in the array head, and pin failure. Thus, unique characteristics of such arrays can be saved as part of the settings template, allowing more rapid and accurate spotfinding.

6. Open the image to be analyzed. Open a template "settings file" appropriate for the array being analyzed. Each image must have its own specific settings file; save the open settings file as a new file with a name like the image file name. From here on, save any changes (alt + S in GenePix) to the settings file associated with the image (and the template will not be overwritten).

Spotfinding

7. Align the template with the array spots on the image. To best extract the intensity data, make sure that the borders of each spot are accurately delineated. The goal here is to quickly edit the settings file so that the virtual spot in the settings file aligns well with the actual spot on the image file. The following instructions are for GenePix, but other programs have similar functions for semiautomated spotfinding.

i. In Block mode (B), select all blocks (ctrl + A; a "block" is the collection of spots printed by the same pin and appears as a distinct sector on the array). Drag them (and use up/down arrows for fine movement) to align with the image, looking at the top left corner block for reference.

ii. Because each block may not be squarely aligned with the others, align each separately. Using the > and < keys, toggle between each individual block. For each block, use the arrow keys to adjust the block more closely, so that all of the circles in the template are at least overlapping with the spots in the image. If there is severe skew or movement in a block (this may occur during print runs in which a pin fails and is replaced, or a restart is necessary after a power failure), further alignment will need to be done later. Save the settings periodically.

iii. Once all of the blocks are somewhat aligned, switch to Feature mode ("Feature" is another name for "spot") by hitting the F key. Select all features (spots) (ctrl + A) and then hit (alt + F5) to align all spots individually. Alternatively, after aligning each block of spots, hit F5 to align all spots in the block selected. In GenePix, the F5 key initiates a "spotfinding" program that automatically tries to align each template spot with the spot pixels in the underlying image using the parameters in the settings file. The program is pretty good at doing this, but it can sometimes get confused by spot artifacts or dust. Because this defines the data, check each spot again.

iv. Go through each block again (using < and >) and check for individual feature misalignments. In Feature mode (F), select individual or blocks of features to move (arrow keys), resize (ctrl + arrow keys), or realign (alt + F5). Once all of the spots are aligned, save the settings.

Create the Results File

8. Once the settings file is edited, click the Analyze button. This generates the results file that summarizes intensities, backgrounds, subtracted backgrounds, ratio of the medians, median of the ratios, and all sorts of other data for each spot. Some of these data come from bad spots that should be excluded from the final results and thus must be "flagged" as such. In GenePix, this can be done both automatically and by hand.

   i. To automatically find and flag bad spots, go to the Results tab and click on Flag Features.

   ii. Click on New to write a new query or open a previously saved query. The query is a logical filter that can be applied to data in the results file to identify features that match the query (in principle, data could be flagged as good, rather than bad). To see how many spots will be affected by the query, click Evaluate without flagging the spots. Select the flag to be applied (e.g., if the query is designed to identify "bad" spots, flag those satisfying the query as "Bad" by selecting that flag and clicking OK). This action edits the settings file and marks the bad spots.

   iii. Reopen Flag Features to run each subsequent query and apply additional flags.

   iv. Once all automatic flagging has been done, save the settings file again.

   v. To manually flag spots, go back to the image and toggle through the blocks (< and >) to check all spots again for those that should be flagged "Bad" but were missed by autoflagging (hitting the A key in Feature mode after selecting the spot will flag it as "Bad"). If any template spots are adjusted or moved, save the settings file and click Analyze again to regenerate a new results file.

9. Once all flagging is done, save the settings file one last time, click Analyze one last time, and finally save the finished results file. In GenePix software, the results file names have .gpr at the end.

   Most experiments require multiple arrays and the next level of analysis involves combining array results together using additional computer methods that operate on a set of such files from several arrays.

DISCUSSION

Below are examples of useful automatic flagging files that can be used in GenePix. The first flag identifies spots that probably have too little or no target hybridizing to them. It requires that the spot's median pixel intensity in the 635-nm fluor channel plus the median pixel intensity in the 532-nm channel be greater than 100 and that more than 50% of the pixels in the spot have intensities greater than 2 standard deviations above background in both channels. All spots that fail to meet all of these criteria are called "not found" and are not included in subsequent analysis. The second flag removes saturated spots, those that have more than 20% of the pixels saturated in either channel. GenePix instructions provide more information about writing flag scripts.

flagNF: (Flag Spots as Not Found)

[Flags] = [Clear] And
[Sum of Medians (635/532)] < 100 And
[% > B635+2SD2] < 50 And
[% > B532+2SD2] < 50

flagSATD: (Flag Spots as Bad)

[Flags] = [Clear] And
[F532 % Sat.] > 20 Or [F635 % Sat.] > 20

TIPS AND TROUBLESHOOTING

1. Before spending time on analysis, make sure that the microarray is clean and hybridization is robust with low and uniform background. If a sample is extremely precious, it may make sense to analyze marginal arrays, but otherwise including data from such arrays only increases noise. After scanning, problems with technique (if any) are readily apparent. Background comes in many forms including streaking of one color, differential sticking of precipitate that binds one of the dyes but not the other, scratches, fuzz from sweaters, and unbalanced dye incorporation into target. Some of these problems can be addressed in data extraction but others cannot, and it must be decided whether the hybridization is of sufficient quality to proceed with analysis.
2. Consistency is more important than perfection in aligning the spots. Median statistics are typically used to suppress the need for a strict and tight boundary for extraction, and spotfinding software has improved significantly. Thus, making slight adjustments that affect only a few pixels has more impact on investigator time than it does on the data quality.
3. Recognizing bad spots is easy with a little practice. Spots that bleed into other spots are obviously mixed results that should not be trusted. A spot that hybridizes both samples to similar extents over its whole area will be a nearly uniform color (see Fig. 8-11; unmarked spots in the figure are probably "good"). Spots in which only a few pixels give data (e.g., spot A) are not good spots and the data are likely not valid. Spot B may be okay, because most of the pixels seem to give the same results. Signal can be saturated (e.g., spot C) and thus, although qualitatively strong, quantification of the signal is compromised. Saturation is usually taken care of by automatic "flagging." Misshapen spots such as spot D are usually good, provided that they have not run into other spots. Spot E has two regions of dramatically different intensities and should be excluded from the analysis. Spot F may be of too low intensity to be analyzed; usually low-intensity spots are excluded automatically. One good test of spot quality could be whether the median of the ratios of each pixel's intensities is close to the ratio of the median intensities for the whole spot. This

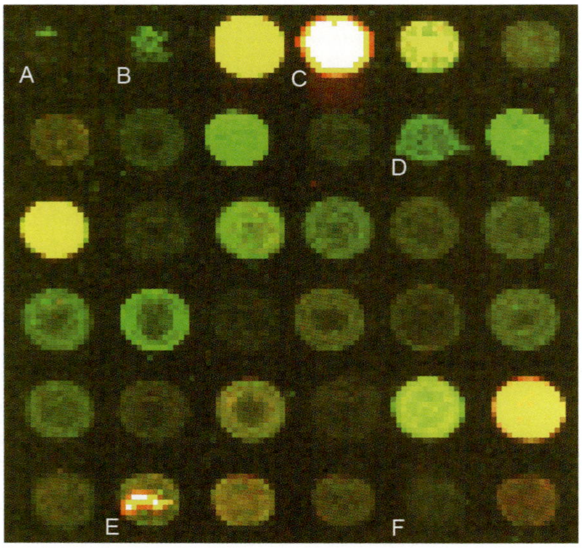

FIGURE 8-11. Good versus bad microarray spots. (*A*) Signal not coherent with a spot, flag as bad. (*B*) Small spot, most pixels give same result, probably good. (*C*) Saturated signal, quantification compromised, flag as bad. (*D*) Misshapen spot, but data are probably good. (*E*) Dramatically different pixel intensities within the spot, flag as bad. (*F*) Intensity too low, best to flag as undetected.

would say that signal from each individual pixel is quite similar to the signal from the whole spot. If not, there could be an artifact causing nonuniform binding of target over the spot (DNA–DNA hybridization would be expected to be uniformly distributed over the spot).

4. Excluding spots that are bad or close to background intensities is important for robust downstream analysis. Including data from a bad spot may cause a gene to be recognized as changing when it is not. If multiple arrays are combined, as is the usual case, it may not be practical or easy to go back and evaluate the spot quality for subtle or questionable gene changes found by the analysis. So it is best to eliminate anything that looks questionable during data extraction. Array analysis is intrinsically going to provide false-negative results, i.e., changes that occur but are not detected. These are far less damaging to the analysis than false positives, i.e., changes that do not occur but are called changes by the analysis. False positives will sometimes be discovered by validation experiments; however, it is not practical to independently confirm every statistically significant result. Thus, careful and rigorous data extraction will pay dividends down the line.

WEB RESOURCES

http://www.moleculardevices.com/ Molecular Devices, Inc. is a supplier of high-performance analytical systems, including instruments, software, and reagents, to accelerate and improve drug discovery and basic life science research.

http://rana.lbl.gov/EisenSoftware.htm Michael B. Eisen lab software page. Lawrence Berkeley National Lab, University of California, Berkeley.

http://www.statsci.org/micrarra/image.html StatSci.org, Microarray Data Analysis, Image Analysis Software. Institutions or companies that produce scanning or image analysis software for spotted DNA microarray slides.

Protocol 4

Tips on Hybridizing, Washing, and Scanning Affymetrix Microarrays

Starting in the late 1990s, Affymetrix, Inc. produced for sale a system for performing, hybridizing, washing, and scanning microarrays that was designed to be easy to operate and reproducible. The system used arrays packaged in a plastic cassette or chamber in which the prefabricated array was mounted and could be filled with fluid through resealable membrane ports either by hand or by an automated "Fluidics station" specially designed to handle the arrays. A special rotating hybridization oven and a specially designed scanner were also required. Primarily because of automation and standardization the Affymetrix system was and still remains popular. Here, we provide a skeleton protocol with the potential pitfalls identified. It is designed to augment the protocols provided by Affymetrix.

MATERIALS

Reagents

See Affymetrix protocols

Target sample(s) of interest

Depending on the type of array being used, this target might be composed of cDNA, double-stranded cDNA, or "cRNA" made by amplification of cDNA derived from the original sample RNA of interest. This protocol assumes that the investigator has converted the RNA samples of interest into the appropriate target preparation. Several different kits are available for these various target preparation methods.

Equipment

Affymetrix fluidics station
Affymetrix hybridization oven
Affymetrix microarray chips
Affymetrix scanner
Computer
Heat blocks (99°C, 65°C, and 50°C)
Microcentrifuge
Microcentrifuge tubes (1.5 mL)
Pipettors and tips
Racks for chips
Rotating oven (50°C)
Safe-Lock tubes
Tape or Tough-Spots

OVERVIEW

There are three main parts to this process. First, the chips must be prehybridized briefly (~10 min) and then the biotinylated target sample prepared from the isolated RNA of interest is combined in a hybridization fluid and loaded into the chip. After hybridization overnight, the target is removed and the chip is filled with a high-salt (nonstringent) buffer and loaded into the fluidics station, along with tubes of stain (streptavidin–phycoerythrin fusion protein, SAPE) and a biotinylated antibody against streptavidin (available from Affymetrix). The fluidics station runs a program to wash the array with stringent buffer at the appropriate temperature, then applies a first staining step in which the SAPE binds to biotin moieties on the hybridized target. After washing off unbound SAPE, this signal is amplified by then binding a biotinylated antistreptavidin antibody to the SAPE left after the first staining. In a third staining step, a second binding of fresh SAPE is used, which binds to biotins on the antibody. Finally, the array is washed again and ready for scanning. About 1 h is needed for the hybridization setup. The next day, ~1 h is needed for the wash setup; the fluidics station then runs for ~1.5 h to stain and wash. It takes 5–10 min to scan each chip.

METHOD

1. Gather reagents and set up equipment.
   i. Turn on heat blocks to 99°C, 50°C, and 65°C.
   ii. Turn on the Affymetrix rotating hybridization oven to 50°C.
   iii. Take chips to be hybridized out of 4°C to equilibrate to room temperature on the bench.
   iv. Take out –20°C reagents (BSA, HS DNA, 20x Hyb controls, B2 oligo) to thaw.
   v. Get 2x hybridization buffer from 4°C (50-mL tube covered with foil).
2. Make hybridization cocktail mix.
   i. Follow recipe in the appropriate Affymetrix Protocol, for 131 μL per chip.
   ii. Make a little extra to account for pipetting variability (i.e., if making cocktail for four chips, multiply all amounts by 4.2).
   iii. Vortex to mix and quick-spin down.
3. Add 131 μL of hybridization premix cocktail to each of the target sample tubes that have been prepared from RNA (~69 μL). Mix well and quick-spin down.

   There should be ~200 μL of 1x hybridization solution with the target sample.
4. Make enough 1x hybridization buffer to prehybridize all chips (200 μL per chip).
5. Prehybridize the chips.
   i. Take the chips out of their packets, being careful not to touch the glass surface, and place them on a a clean bench with the clean lab mat face down.
   ii. Carefully insert a pipette tip into the top septum (note that regular yellow tips, unbeveled, probably work best) to create an exit path for air in the chamber, which will be replaced by hybridization buffer.
   iii. Take up 200 μL of 1x hybridization buffer in a pipettor.
   iv. Holding the chip upright, pierce the bottom septum with the loaded pipette tip and slowly fill the chip while keeping the tip perpendicular to the chip surface, so as not to damage the septum. Wiggling or twisting the tip while in the septum may damage the septum.
   v. Pull the pipette tip out of the septum while the plunger is depressed at the *first stop*, not the "blow out" stop—all of the buffer should be out.
   vi. Remove the other pipette tip, pulling straight out again, so as not to damage the septum.
   vii. Rotate the chip manually while inspecting the glass side to ensure that the entire surface is wetted and that liquid is moving freely. The fluid can be seen in the chamber through the glass facing.

viii. When all of the chips are loaded with buffer, put them in racks (balance the racks by putting the chips in opposite positions on two racks; use an old empty chip to balance) and rotate the chips in the hybridization oven at 60 rpm for at least 10 min at 50°C.

6. Denature the samples (now in 1× hybridization mix) for 10 min at 99°C. Use Safe-Lock tubes or place a weight on top so that the tubes do not pop their tops at this higher than usual temperature. Transfer the tubes for 5 min at 50°C.
7. Spin the tubes at full speed in a microcentrifuge for 5 min to pellet denatured and aggregated material. Orient the tubes all in one direction to avoid the part where pelleted particulates will be.
8. Remove the prehybridized chips from the oven.
9. Remove the buffer from the chips, reversing the previous manipulations.
   i. Pierce the top septum with a pipette tip.
   ii. With the P200 set to 200 μL (and empty), depress the plunger to the first stop.
   iii. Holding the chip vertically, pierce the bottom septum with the pipette tip and slowly release the plunger to suck out the prehybridization buffer.
   iv. Pull the pipette tip out of the bottom septum and discard the fluid and the tip.
10. Fill the chips with denatured, pelleted target samples.
   i. When pipetting the sample, take care to position the tip at the bottom of the tube on the side away from the pellet. Get all of the sample (~200 μL).
   ii. Note which sample was applied to which chip. It is best to write the sample name on the chip label.
11. Apply small pieces of tape or "Tough-Spots" over both the septa on each chip in case of leakage.
12. Hybridize in a rotating oven at 60 rpm for 16 h at 50°C. Before the end of the hybridization, start preparing for the wash and scan.
   i. Ready the reagents for the stains—BSA from –20°C, SAPE, antibodies, and 2× Stain buffer (stored in a foil-wrapped tube) from 4°C. Vortex and quick-spin each before use. See the Affymetrix standard protocol for more information.
   ii. Turn on the fluidics station.
   iii. Prepare buffers for the fluidics station.
      a. Fill each bottle with the appropriate buffer or water, checking for particulates, and empty the waste bottle.
      b. Nonstringent buffer is stored at room temperature. Stringent buffer is stored in foil at 4°C. Match bottles up to the labeled intake lines on the right of the fluidics machine.
   iv. Log onto the computer and start up GCOS Affymetrix software that runs the fluidics station and the scanner.
   v. Open the fluidics station window.
13. Prime the fluidics station. The fluidics station has four "modules," each able to wash one chip.
   i. For each module to be used, select <no probe array> and run the appropriate priming program (e.g., <PRIME_450>). Click Run to start the priming protocol for each module.
   ii. Follow the instructions on the module windows to start the protocol. What to do when the window asks to:
      a. Remove vials: This means lift up the lever for the needles (no need to actually remove vial except when loading the stains).
      b. Load empty vials: This means gently push the lever back down.

c. Eject washblock: This means push down on the cartridge lever to open the module so that a chip can be inserted or retrieved.

d. Engage washblock: This means flip up the cartridge lever. *Do not do this too gently*, because a good amount of pressure is required to get the septa pierced appropriately so that the chip is engaged properly with the washblock.

When a protocol finishes, a notification will appear on the computer screen for each module.

14. For each chip, make an "experiment" entry.

    i. Use templates for sample and experiment to record the lot numbers for the chips and notes for the samples.

    ii. Choose the appropriate target sample type (cDNA, cRNA) and the proper array type. Select the appropriate analysis desired. This can be customized depending on the kind of analysis performed.

    iii. Typically, the sample name and experiment name will be the same (but use whatever system that is both informative and systematic).

    iv. Group samples together under project headings.

    v. After saving an entry, edit it to save as a new experiment.

15. After 16 h (or pick a consistent time to stop all hybridizations for a given project), remove the chips from the oven.

16. Remove the samples from the chip by removing the tape from the ports, piercing the upper port septum with a yellow 200-μL pipettor tip and pulling the sample from the lower port. Save the samples because they can be hybridized to a second chip if necessary. Store them at –20°C.

17. Fill each chip completely (~250 μL) with nonstringent wash buffer.

    i. Dial a 200-μL pipettor to ~135 μL and fill with nonstringent wash buffer. Empty into chip as described above. Reload the pipettor with buffer and insert into the bottom port septum again. Holding the chip upright, slowly fill the chip to ensure that there are no bubbles. The air space in the chip chamber should slowly disappear from the top center of the chip when the first stop on the plunger is reached.

    ii. With the plunger depressed, remove the pipette tip. Check that the chip surface is completely filled (no new bubbles appeared); remove the other pipette tip and check the chip surface again.

18. Allow the filled chips to sit at room temperature while the stains are prepared. Store hybridized chips (after replacing the hybridization fluid with nonstringent wash buffer) for a few hours at 4°C if necessary (e.g., eight chips hybridized but only four at a time on fluidics machine). Bring them to room temperature before putting them on the fluidics station.

19. Prepare the stain solutions.

    i. Make one master mix in a 15-mL tube. Make up to 10% additional mix to account for pipetting variation.

    ii. For each chip, aliquot 2 x 600 μL of the SAPE mix and 1 x 600 μL of the antibody mix into black 1.5-mL microcentrifuge tubes (this protects the samples from light).

    iii. Store tubes to be used later at 4°C along with chips waiting to be stained.

20. Start washing the chips.

    i. In the fluidics station window on the computer, select an experiment from the pull-down list for each module (these were previously entered above).

    ii. Run a protocol that is appropriate to the chip and target sample of interest.

iii. Click Run to start when ready and then follow the instructions on the fluidics station windows.

   a. Eject the washblock and *make sure to load the chip matching the experiment selected* for each module.

   b. When loading stain vials, *make sure to load the antibody mix in position 2.*

   c. To begin the washing protocol, first press the lever down to snap the needles into the stain vials. The washing takes ~1 h 15 min.

iv. When the protocol is almost done, follow the prompt to remove the vials and the chip. Before engaging the washblock again, check the chip for bubbles—it should be completely full.

   a. If bubbles are present (that cover any part of the scan area), place the chip back into the washblock and engage the washblock. The fluidics station will drain and fill the array and again prompt to eject cartridge.

   b. Repeat as necessary, but after several tries, manual filling of the chip cartridge with nonstringent buffer is required.

v. After vials and chip are removed, the module performs a 5-min purging procedure.

21. Cover septa with tape or "Tough-Spots" again to protect scanner from possible leaks.

22. Place the chips in a rack and put them in a drawer to protect them from the light until scanning. Store the chips at 4°C if it will be several hours before a scan.

23. Scan the chips. Turn on the scanner and let it warm up for 10 min.

   i. Open the scanner window in the software (still open from the wash protocol).

   ii. Select the experiment to be scanned from the pull-down menu.

   iii. Check the chip for streaks and dust. Blow off the dust carefully. If absolutely necessary, remove streaks by wiping from center to edge (glue at edges may smear even worse), using ethanol if desired.

   iv. Click the Start button. The sample door on the top of the scanner opens automatically.

   v. Load the chip (*make sure to match the chip to the experiment selected*) chamber window facing away, gently. It only fits one way so if there is a problem, do not force!

   vi. Click OK to close the door and start the scan.

   vii. The scan is automatically completed and the image is processed into a .cel file.

   viii. The sample door opens automatically at the end of the scan for chip removal. If it closes again before retrieval, simply click Load Sample in the scanner window to open the door.

24. Perform a quick quality check and initial analysis of images and .cel files.

   i. In the Expression assay tab, set scaling to user-defined (1), normalization to (1). This means no massaging of the data.

   ii. Right click on the .cel file and select Analyze. This generates a .chp file.

      If a modified analysis method was selected when first inputting and saving the experiment, this may already be done.

   iii. Right click on the .chp file and select Report. This generates a .rpt file, which can be printed or simply noted for comparison between chips:

      - Noise
      - Signals, 5′/3′ ratios for some test gene probes, Present/Absent calls for housekeeping genes and spiked controls (meaning if intensity is enough to determine the presence of mRNA for the given gene)

25. Shut down and clean up.

i. Place all fluidics station lines into the dH_2O bottle (fill again if necessary).
ii. For all modules used, select <no probe array> and Shutdown_450.
iii. Run the fluidics station shut down protocols, following the instructions on the computer.
iv. When completed, the fluidics station can be turned off. Dump waste, remaining buffers, and rinse bottles.
v. Turn scanner off (front button).
vi. Log off computer.

TIPS AND TROUBLESHOOTING

1. It essential to mark the chips with the name of the target. Although each array has a bar code, in practice, only core or heavy-using labs use bar-code readers to ensure that the right sample is associated with the right array.
2. It is essential to double check the naming of the computer files and the association of the files with the array and with the sample. The computer believes what it is told, and thus it may be difficult to figure out why a sample that is named "brain" looks exactly like it came from liver if in fact the file name is incorrect.
3. Protect the SAPE and the stained arrays from light as much as possible because SAPE will bleach and signal will be lost. Keep the arrays face down on a soft paper or cloth in a dark drawer or in the refrigerator when not actually in the scanner.

REFERENCE

Schinke-Braun M, Couget JA. 2007. Expression profiling using Affymetrix GeneChip probe arrays. *Methods Mol Biol* **366:** 13–40.

WEB RESOURCE

http://www.affymetrix.com/support/downloads/manuals/expression_analysis_technical_manual/

Protocol 5

Fragmentation of Whole-transcriptome RNA Using *E. coli* RNase III

Current high-throughput sequencing (HTS) methods are unable to read more than several tens of nucleotides in sequence reliably. To quantify the amount of a particular sequence in a sample of large RNAs (e.g., mRNAs), it is important to fragment the RNA into short pieces that can be ligated to oligonucleotides that allow PCR amplification and sequencing. The most desired end structure of RNA for such ligation steps is a 5′ phosphate and a 3′ OH. Thus, enzymes that leave these groups after cleavage are of particular utility because this activity avoids the need to dephosphorylate the 3′ end with phosphatases or phosphorylate the 5′ end with kinase before proceeding. Two enzymes are widely available to use: RNase III and P1 nuclease. Of these, RNase III is the more easily controlled. Although it primarily cuts duplex RNA, this specificity is salt and concentration dependent, and many RNAs that lack strong extended duplexes are nonetheless susceptible to RNase III cleavage at many spots (Dunn 1976). P1 nuclease is a very efficient enzyme with a preference for single-stranded regions. RNA fragmentation by RNase III does not seem to grossly affect the distribution of RNA sequencing reads in several studies and so has become a standard method for creating nominally representative pools of transcriptome sequences with 5′ phosphates and 3′ OH for library construction.

MATERIALS

The recipe for the item marked with <R> is at the end of the protocol or in the Common Recipes Appendix.

CAUTION: See the Cautions Appendix for appropriate handling of materials marked with <!>.

Reagents

Acid phenol <!>
EDTA (100 mM, pH 8.0) <!>
Ethanol (100%, 70%) <!>
RNA (0.5–1 µg of poly[A] RNA or 0.5–1 µg of rRNA-depleted total RNA)
RNase III buffer (10×) <R>
RNase III from *E. coli* (Ambion)
Sodium acetate (3 M, pH 5.2)

Equipment

Bioanalyzer (Agilent)
Ice
Incubator or heat block
Microcentrifuge
Microcentrifuge tubes
Microspectrophotometer
RNA 6000 Pico Chip Kit
Vortex machine
Water bath

OVERVIEW

Three important steps for preparing fragmented transcriptome RNA for sequencing library construction are (1) fragment the RNA with RNase III to the extent that ~60–100-nucleotide fragments are created, (2) purify the RNA from the RNase III reaction, and (3) analyze the digestion products for their suitability in library production.

METHOD

1. For each RNA sample, assemble the following reaction mixture on ice in a microcentrifuge tube:

mRNA	0.5–1.0 µg
H_2O	to 8 µL
RNase III buffer (10×)	1 µL
E. coli RNase III (stock is 1 unit/µL)	1 unit

   For the mRNA, use either 2× oligo(dT)-selected or rRNA-depleted (e.g., RiboMinus-treated) RNA (see Chapter 2, Protocol 10 or 6 and 9).

2. Gently mix the reaction and then "quick-centrifuge" at room temperature.
3. Incubate the samples for 5–10 min at 37°C.

   It may be advisable to take samples after different times of incubation to obtain the desired level of fragmentation.

4. Stop the reaction with 1.1 µL of 100 mM EDTA (pH 8.0).
5. Add 80 µL of H_2O, 10 µL of 3 M sodium acetate (pH 5.2), and 100 µL of acid phenol. Vortex to mix the phases.
6. Centrifuge at 12,000*g* for 10 min at room temperature. Pipette to remove the aqueous phase.
7. Add 250 µL of 100% ethanol and chill for at least 30 min at –20°C or –80°C. Then, centrifuge at 12,000*g* for 15 min at room temperature. Carefully remove the ethanol by pipetting without disturbing the pellet.
8. Rinse the pellet with ice cold 70% ethanol, "quick-centrifuge" at room temperature, and remove the ethanol. *Briefly* dry the pellet in a SpeedVac and resuspend it in 10 µL of H_2O.
9. Use a microspectrophotometer to assess the yield of RNA.

   Recovery should be at least 50%. Poor recovery may indicate overdigestion. If this is the case, perform the digestion step again but for a shorter time.

10. Dilute a small aliquot of the RNA to 5 ng/µL and analyze 1 µL on a Bioanalyzer using the RNA 6000 Pico Chip Kit following the manufacturer's protocol.

    The mean size of RNA fragments should be ~150 nucleotides with a modal value of ~60–100 nucleotides. Size fractionation will generally occur after linker ligation in most sequencing library protocols; thus, there is some flexibility and opportunity to optimize the size distribution at this step, depending on the exact sequencing protocol that follows.

TIPS AND TROUBLESHOOTING

Use of RNase III to generate sets of fragments for sequencing has a number of advantages, primarily that the RNA products have ends compatible for ligation to linkers. The enzyme is easy to control, and at the low salt used in the reaction (50 mM NaCl final), the specificity of the enzyme for double-stranded RNA is reduced, allowing more or less uniform fragmentation (Dunn 1976). If the RNA is overdigested, reducing the amount of enzyme, the time of incubation, or both may be necessary.

SOLUTION AND REAGENT RECIPE

CAUTION: See the Cautions Appendix for appropriate handling of material marked with <!>.

RNase III Buffer (10x)

Reagent	Quantity (for 10 mL)	Final concentration
NaCl (5 M)	1 mL	500 mM
Tris-Cl (1 M, pH 7.9)	1 mL	100 mM
$MgCl_2$ (1 M) <!>	1 mL	100 mM
DTT (1 M) <!>	0.1 mL	10 mM
dH_2O	6.9 mL	

Dilute 1:10 into the final RNase III reaction to make 1x buffer.

REFERENCE

Dunn JJ. 1976. RNase III cleavage of single-stranded RNA. Effect of ionic strength on the fidelity of cleavage. *J Biol Chem* **251:** 3807–3814.

Protocol 6

Preparation of Small RNA Libraries for High-throughput Sequencing

The following protocol details the process of small RNA cloning for sequencing on the Illumina/Solexa sequencing platform, but it can be easily modified for use on other next-generation platforms (e.g., SOLiD, 454). The general procedure and 3′ cloning linker were adapted from Lau et al. (2001). This procedure is designed to clone canonical small RNA molecules with 5′-monophosphate and 3′-hydroxyl termini. Modifications, such as the presence of a 2′-*O*-methyl group, can reduce efficiency, although not sufficiently to negate the utility of the approach. Other termini modifications, such as a 5′ triphosphate or a 3′ phosphate, can be altered by enzymatic treatment before cloning.

MATERIALS

The recipe for the item marked with <R> is at the end of the protocol or in the Common Recipes Appendix.

CAUTION: See the Cautions Appendix for appropriate handling of materials marked with <!>.

Reagents

Bovine serum albumin (BSA, 100x)
Choloroform <!>
Decade Marker (Ambion)
DEPC <!>–Milli-Q H_2O (see Step 24)
Dimethylsulfoxide (DMSO) <!>
Ethanol (100%, 70%) <!>
Ethidium bromide <!>
[γ-^{32}P]ATP <!> (Perkin Elmer)
Gel-loading buffer II (1x; Ambion)
Gel-loading dye (6x) (Fisher)
GeneRuler 50-bp DNA ladder (Crystalgen)
GlycoBlue (Ambion)
NaCl (0.4 M)
Nusieve GTG agarose (low melting, <1-kb DNA/RNA)
Oligonucleotides
3′ ligation
Modban: AMP-5′*p* = 5′-*p*CTGTAGGCACCATCAATdideoxyC-3′
miRNA cloning linker 1 (IDT)
5′ ligation
Solexa linker: 5′-rArCrArCrUrCrUrUrUrCrCrCrUrArCrArCrGrArCrGrC
rUrCrUrUrCrCrGrArUrC-3′
Custom oligonucleotide–HPLC purified
PCR
Sol_5_SBS3: 5′-AATGATACGGCGACCACCGAACACTCTTTCCCTACACGACG-3′
Custom oligonculeotide–HPLC purified
Sol_3_Modban: 5′-CAAGCAGAAGACGGCATACGATTGATGGTGCCTACAG-3′
Custom oligonculeotide–HPLC purified
PCR buffer (Roche)

Phenol (equilibrated, pH 8) <!>
Phenol:chloroform:isoamyl alcohol (PCA) <!>
PmeI (New England BioLabs)
Includes 10× buffer and BSA.
PNK buffer (Roche)
Polyacrylamide/urea gel (Sequagel, National Diagnostics)
Reverse transcriptase mix <R>
Reverse transcription

3′ RT primer:	5′-ATTGATGGTGCCTACAG-3′
(BanOne)	Custom oligonculeotide–HPLC purified

RNA oligonucleotides (19, 24, 28, and/or 33 nucleotide)
Seakem GTG agarose (low melting, <1-kb DNA/RNA)
SuperScript III RT (Invitrogen)
Includes 5× buffer, DTT, and dNTP mix.
SYBR Gold nucleic acid stain (Molecular Probes)
T4 polynucleotide kinase (New England Biolabs)
Includes 10× buffer.
T4 RNA ligase (Ambion)
Includes 10× buffer.
T4 RNA ligase 2 (truncated; New England Biolabs)
Includes 10× buffer.
TAE
Taq polymerase (Roche)
Includes 10× buffer + $MgCl_2$ and dNTP mix.
TBE (1×) (see Step 24)

Equipment

Chemical fume hood
Dialyzer tube (Novagen Dialyzer Midi, MWCO 3.5 kD, 71506-3) (optional; see Step 24)
Ethanol bath <!>
Add 100% ethanol to dry ice in an insulated container.
Gel box (see Step 24)
Gel electrophoresis equipment
Ice
Microcentrifuge tubes (low retention; Axygen MAXYMum Recovery microtubes)
Microcentrifuge tubes, micropore filter (Ultrafree-MC, 0.22 µm; Millipore)
Pestle
Phosphorimaging screen
Razor blade
Sepharose column (G-50; GE Healthcare Life Sciences)
Thermal cycler
UV light (long wave)

METHOD

Section I: Size Selection and Gel Purification of RNA Sample

1. Prepare a 15% vertical polyacrylamide/urea gel.

   The thickness of the gel depends on the amount of total RNA being run. As a guide, we use 1.0 mm for <20 µg and 1.5 mm for >20 µg.

2. Prepare the following 5′ oligonucleotide-labeling reaction for 19-, 24-, 28-, and 32-nucleotide RNA oliogonucleotides (depending on the small RNA sizes to be cloned) and incubate for 30 min at 37°C:

PNK buffer (10×)	2 μL
DNA oligonucleotide (10 μM)	1 μL
[γ-^{32}P]ATP	2–5 μL
T4 polynucleotide kinase	1 μL
H_2O	to 20 μL

Eliminate free ATP by running the samples through a G-50 Sepharose column.

3. Load 10 ng of ^{32}P-labeled Decade Marker as a size marker.

   These gels, samples, and supplies will contain radioactivity, so always dispose of materials properly and follow proper radiation safety procedures.

   RNA samples are sensitive to degradation at all steps of the procedure; use RNase-free reagents and supplies, when able to do so.

4. Add an equal amount of Gel-loading buffer II to the mixture. Heat the samples for 5 min at 95°C and then chill for 1 min on ice.

5. Spike the RNA samples with ^{32}P-labeled oligonucleotides (~10,000 counts/oligonucleotide per min).

   Oligonucleotides must contain an internal PmeI recognition sequence.

6. Load the sample(s) on a 15% polyacrylamide gel and run the gel at a constant 10W for 1–1.5 h (until the first dye front reaches the middle of the gel).

   In addition, 1–5 μg of each RNA sample can be loaded on a separate 1.0-mm gel, run, stained with SYBR Gold, and visualized under a UV lamp to verify RNA integrity.

7. Once the run has completed, pull one glass plate away from the gel and place it face up behind a clear radioactive shield. Then, blot a tiny amount of radioactivity onto small cut-out paper triangles and place them at three points on the face of the gel.

   These will serve as references when lining up the gel on the radioactive image where the bands should be cut out.

8. Expose a phosphor screen over the gel for 10–15 min (time will vary depending on radioactivity intensity and imager sensitivity) (Fig. 8-12). Print the gel image at actual size so that it can be placed under the glass plate and gel to use as a reference to cut out small RNAs of desired size in the following step.

9. Excise the precise band from the gel that corresponds to the desired size of small RNA (including radiolabeled oligonucleotide[s]) into a low-retention microcentrifuge tube (use these throughout the protocol to minimize sample loss).

   The selection of 20–28-nucleotide-sized small RNAs is shown in Figure 8-12.

10. Reexpose the gel under a phosphor screen to ensure that the desired small RNA fragments have been extracted from the gel.

11. Centrifuge the excised gel slice briefly in a centrifuge at maximum speed for 1 min at room temperature.

12. Carefully grind gel slices by hand using a pestle.

13. Centrifuge the microfuge tube for 1 min at room temperature to pellet the ground gel to the bottom of the tube.

14. Add 420 μL of 0.4 M NaCl to the ground gel slices.

15. Flash freeze the samples for 1 min in a bath of ethanol and dry ice.

16. Incubate overnight at room temperature with agitation (secure the samples on a vortex).

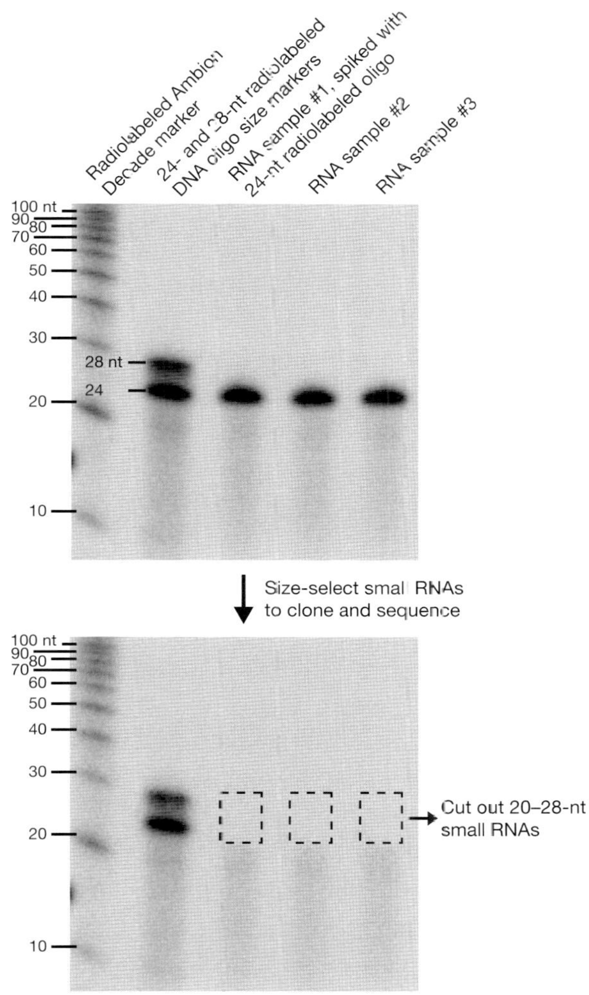

FIGURE 8-12. Size selection and gel extraction of small RNAs from total RNA. Radioactively labeled markers and oligonucleotides are resolved with unlabeled total RNA on a 12% polyacrylamide gel; 24- and 28-nucleotide markers were run in a separate lane to allow for more accurate size resolution. Total RNA samples were spiked with a radiolabeled 24-nucleotide oligonucleotide (although any combination of labeled oligonucleotides can be used), allowing one to visually follow the ligation process and also ensuring that the desired size fraction has been extracted from the gel. Here, 20–28-nucleotide small RNAs (including the radiolabeled 24-nucleotide oligonucleotide) have been cut from the gel. The gel was then reexposed, showing the removal of radiolabeled oligonucleotides (and underlying small RNAs).

17. Centrifuge the gel slice homogenate through a micropore filter at full speed for 1 min at room temperature.
18. Transfer the filtrate to a fresh microcentrifuge tube.
19. Add 20 µg of GlycoBlue, vortex to mix, and add 2.5 volumes of 100% ethanol.
20. Incubate the mixture for 3–6+ h at –20°C.
21. Centrifuge at full speed for 30 min at 4°C.
22. Wash with 70% ethanol and then remove *all* of the ethanol using a Pipetman with a fine, RNase-free tip.
23. Air-dry the pellet for <5 min and resuspend it in 13 µL of DEPC-Milli-Q H_2O.
24. Alternatively, the sample can be extrated from the gel using a gel box and dialyzer.
    i. Insert a gel slice into a dialyzer tube with 450 µL of H_2O.

ii. Position the dialyzer into a gel box, submerged in 1× TBE, such that the dialysis tubing has the current running perpendicularly through it.
iii. Run at 100 V for 18 min.
iv. Reverse the poles and run for 2 min to pull the sample off the dialysis tubing wall.
v. Pull out all of the water into a fresh microcentrifuge tube, mix in 20 µg of GlycoBlue, bring the solution to 0.4 M NaCl, mix, and add 2.5 volumes of 100% ethanol.
vi. Mix by inverting and incubate for 3–6+ h at –20°C.
vii. Centrifuge at full speed for 30 min at 4°C.
viii. Wash the pellet with 70% ethanol and then remove all of the ethanol.
ix. Air-dry the pellet for <5 min and resuspend it in 13 µL of DEPC-Milli-Q H_2O. Small RNAs of the desired size range have now been purified.

Section II: 3′ Linker Ligation

25. Set up the following reaction in a microcentrifuge tube and incubate for 1 h at 37°C.

Gel-purified RNA (in H_2O) (from Step 23 or 24)	13 µL
ATP-free T4 RNA ligase buffer (10×)	2 µL
DMSO	2 µL
Modban oligonucleotide (50 µM)	1 µL
T4 RNA ligase 2, truncated	2 µL
Total	20 µL

26. Add 20 µL of Gel-loading buffer II and then heat-inactivate for 5 min at 95°C. Proceed to Section III.

Section III: Gel Purification of 3′-ligated RNA Product

27. Heat samples for 5 min at 95°C and then chill on ice for 1 min.
28. Load 5 ng of ^{32}P-labeled Decade Marker as a size marker.
29. Prepare a 1.0-mm 15% vertical polyacrylamide/urea gel.
30. Load the sample(s) and run the gel at a constant 10W for 1.5–2 h (until the first dye front reaches the bottom of the gel).
31. Once the run has completed, pull one glass plate away from the gel and place it face up. Then, blot a tiny amount of radioactivity onto small cut-out paper triangles and place them at three points on the face of the gel. These will serve as references when lining up the gel on the radioactive image and align where the bands should be cut out.
32. Expose a phosphorimager screen over the gel for 30–45 min (time will vary) (Fig. 8-13). Print the gel image at actual size so that it can be placed under the glass plate and gel to use as a reference to cut out small RNAs of desired size in the following step.
33. Excise the precise band corresponding to the desired size of small RNA (including the ligated radiolabeled oligonucleotide[s]) into a microcentrifuge tube.

    Selection of 38–47-nucleotide-sized small RNAs is shown in Figure 8-13.
34. Centrifuge the gel slice briefly at maximum speed for 1 min at room temperature.
35. Carefully grind the gel slices using a pestle.
36. Centrifuge the microfuge tube for 1 min at room temperature to pellet the ground gel to the bottom of the tube.
37. Add 420 µL of 0.4 M NaCl to the ground gel slice.
38. Quickly freeze the samples for 1 min in a bath of ethanol and dry ice.

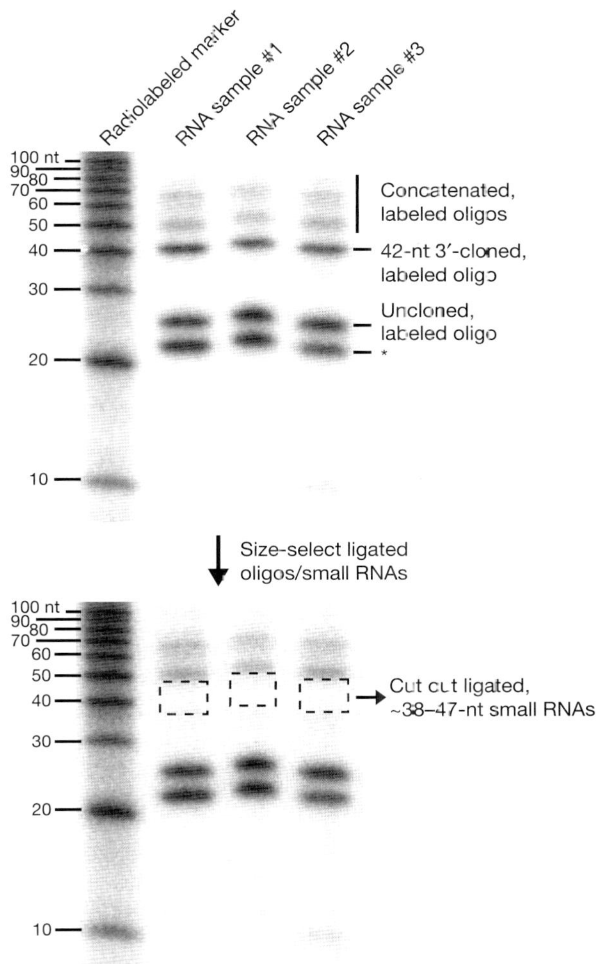

FIGURE 8-13. Size selection and gel extraction of 3′-ligated small RNAs. Radioactively labeled markers and ligated oligonucleotides were resolved on a 10% polyacrylamide gel. The 20–23-nucleotide small RNAs (including the 24-nucleotide radiolabeled oligonucleotide) now have an 18-nucleotide linker ligated to their 3′ termini, leading to a predicted size shift of 18 nucleotides. Here, ~38–47-nucleotide small RNAs (including the ligated, radiolabeled 24-nucleotide oligonucleotide, now running at 42 nucleotides) have been cut from the gel. The gel was then reexposed, showing the removal of radiolabeled oligonucleotides (and underlying small RNAs). Although we have not characterized additional products on the gel, based upon size, we have indicated the likely origin of labeled species. Asterisk indicates bands of unknown production.

39. Incubate overnight at room temperature with agitation (secure the sample on a vortex).
40. Centrifuge the gel slice homogenate through a micropore filter at full speed for 1 min at room temperature.
41. Transfer the filtrate to a microcentrifuge tube.
42. Add 20 µg of GlycoBlue, mix, and add 2.5 volumes of 100% ethanol.
43. Incubate the mixture for 3–6+ h at –20°C.
44. Centrifuge at full speed for 30 min at 4°C.
45. Wash the pellet with 70% ethanol and then remove *all* of the ethanol.
46. Air-dry the pellet for <5 min and then resuspend it in 13 µL of DEPC-Milli-Q H_2O. 3′-End cloned small RNA molecules have now been purified.

Section IV: 5′ Linker Ligation

47. Set up the following reaction and incubate for 1 h at 37°C.

Ligated RNA product (in H_2O)	13 µL
T4 RNA ligase buffer (10x)	2 µL
DMSO	2 µL
Solexa linker (50 µM)	1 µL
T4 RNA ligase	2 µL
Total	20 µL

48. Add 20 µL of Gel-loading buffer II and the heat-inactivate for 5 min at 95°C. Proceed to Section V.

Section V: Gel Purification of 5′- and 3′-ligated RNA Product

49. Heat the samples from Step 48 for 5 min at 95°C and then chill for 1 min.
50. Load 2.5 ng of ^{32}P-labeled Decade Marker as a size marker.
51. Prepare a 1.0-mm 15% vertical polyacrylamide/urea gel.
52. Load the sample(s) and run the gel at a constant 10W for 2+ h (until the first dye front passes the bottom of the gel).
53. Once the run has completed, pull one glass plate away from the gel and place it face up. Then, blot a tiny amount of radioactivity onto small cut-out paper triangles and place them at three points on the face of the gel. These will serve as references when lining up the gel on the radioactive image and align where the bands should be cut out.
54. Expose a phosphorimager screen over the gel for 1+ h (time will vary) (Fig. 8-14). Print the gel image at actual size so that it can be placed under the glass plate and gel to use as a reference to cut out small RNAs of desired size in the following step.
55. Excise the precise band corresponding to the desired size of small RNAs (including the ligated radiolabeled oligonucleotide[s]) into a microcentrifuge tube.

    Selection of 69–79-nucleotide-sized small RNAs is shown in Figure 8-14.
56. Centrifuge the gel slice briefly at maximum speed for 1 min at room temperature.
57. Carefully grind the gel slices using a pestle.
58. Centrifuge the microfuge tube for 1 min at room temperature to pellet the ground gel to the bottom of the tube.
59. Add 420 µL of 0.4 M NaCl to the sample.
60. Quickly freeze the samples in a bath of ethanol and dry ice for 1 min.
61. Incubate overnight at room temperature with agitation (secure the sample on a vortex).
62. Centrifuge the gel slice homogenate through a micropore filter at full speed for 1 min at room temperature.
63. Transfer the eluant filtrate to a microcentrifuge tube.
64. Add 20 µg of GlycoBlue, mix, and add 2.5 volumes of 100% ethanol.
65. Incubate the mixture for 3–6+ h at –20°C.
66. Centrifuge for 30 min at 4°C.
67. Wash the pellet with 70% ethanol and then remove *all* of the ethanol.
68. Air-dry for <5 min and resuspend the pellet in 6.3 µL of DEPC-Milli-Q H_2O. 3′- and 5′-End cloned small RNA molecules have now been purified.

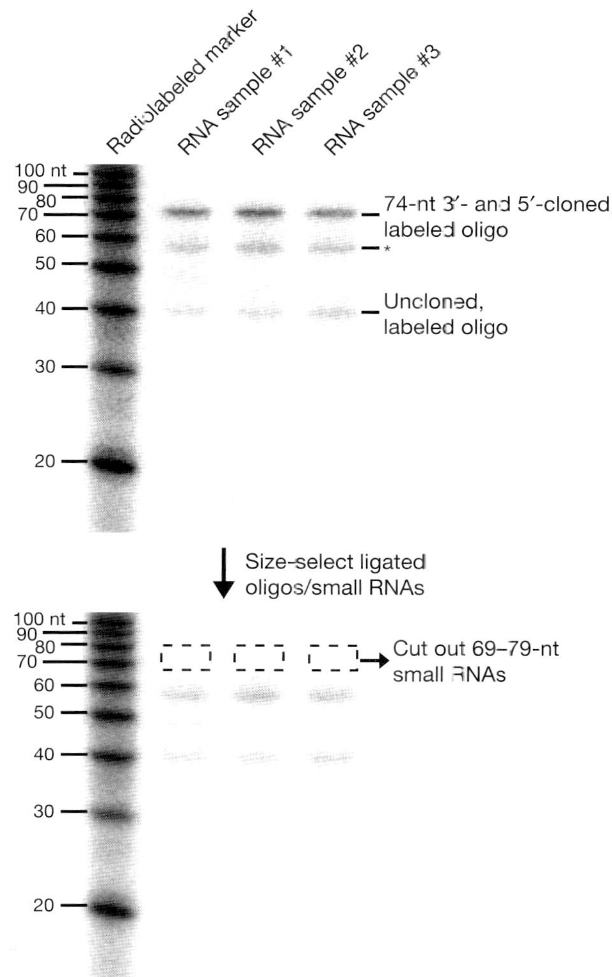

FIGURE 8-14. Size selection and gel extraction of 3′- and 5′-ligated small RNAs. Radioactively labeled markers and ligated oligonucleotides are resolved on a 10% polyacrylamide gel. The ~38–47-nucleotide small RNAs (including the 42-nucleotide radiolabeled oligonucleotide) now have a 32-nucleotide linker ligated to their 5′ termini, leading to a predicted size shift of 32 nucleotides. Here, ~69–79-nucleotide small RNAs (including the ligated, radiolabeled 24-nucleotide oligonucleotide, now running at 74 nucleotides) have been cut from the gel. The gel was then reexposed, showing the removal of radiolabeled oligonucleotides (and underlying small RNAs). Although we have not characterized additional products on the gel, based on size, we have indicated the likely origin of labeled species. Asterisk indicates bands of unknown production.

Section VI: Reverse Transcription

69. Set up the following reaction in a 150-µL PCR thermocycler tube and incubate for 2 min at 72°C.

Ligated RNA product (in H_2O)	6.3 µL
BanOne primer (5 µM)	4.2 µL

70. Centrifuge for 1 min at room temperature.
71. Cool for 2 min on ice.
72. Add 8.4 µL of Reverse transcriptase (RT) mix.
73. Split the mixture into two tubes (9 µL each).
74. Add either 1 µL SuperScript III RT (Invitrogen) (+RT) or 1 µL DEPC–Milli-Q H_2O (–RT control).

75. Incubate for 1 h at 50°C and then heat for 15 min to 70°C.
76. Store at –20°C until use or proceed directly to Section VII.

Section VII: PCR Amplification of cDNA

77. Set up two PCRs for each sample (+RT and –RT) in 150-μL PCR thermocycler tubes.

First-strand cDNA (+RT) or control (–RT) (from Step 76)	5 μL
PCR buffer + $MgCl_2$ (10x)	10 μL
dNTP mix (10 mM of each)	2 μL
Sol_5_SBS3 (100 μM)	1 μL
Sol_3_Modban (100 μM)	1 μL
Milli-Q H_2O	80 μL
Taq polymerase (5 units/μL)*	1 μL
Total	100 μL

*High-fidelity polymerase can be used.

78. Place the reaction in a thermal cycler and run the following PCR program:

    Repeat 5 times: 94°C for 2 min, 94°C for 15 sec, 54°C for 30 sec, 72°C for 30 sec

    Repeat between 15 and 25 times: *94°C for 15 sec, 60°C for 30 sec, 72°C for 30 sec

    72°C for 2 min, 4°C for ∞

    *This range depends on the amount of starting RNA, small RNA abundance, and ligation/elution efficiencies; we recommend as few repeats as possible.

79. Transfer the +RT samples to fresh microcentrifuge tubes and add 200 μL of 0.5 M NaCl.
80. Add 300 μL of PCA and vortex for 1 min.
81. Centrifuge at maximum speed for 5 min at room temperature.
82. Transfer the aqueous layer to 900 μL (or 2.5–3 volumes) of 100% ethanol and 20 μg of GlycoBlue (optional).
83. Incubate the mixture for 3–6+ h at –20°C.
84. Centrifuge at maximum speed for 30 min at 4°C.
85. Wash the pellet with 70% ethanol and then remove all of the ethanol.
86. Wash the pellet with 100% ethanol and then remove all of the ethanol.
87. Air-dry the pellet and then resuspend it in 23.7 μL of H_2O. Your cDNA-converted small RNA library is now amplified.

Section VIII: PmeI Digestion of Radiolabeled Oligonucleotides

88. Set up the following reaction and incubate for 2–4 h at 37°C:

PCR products (in H_2O) from Step 87	23.7 μL
NEB buffer 4 (10x)	3 μL
PmeI	3 μL
BSA (100x)	0.3 μL
Total	30 μL

89. Add 6 μL of fresh 6x Gel-loading dye. All RNA size markers containing a PmeI restriction enzyme site are now cleaved away.
90. Proceed to Section IX.

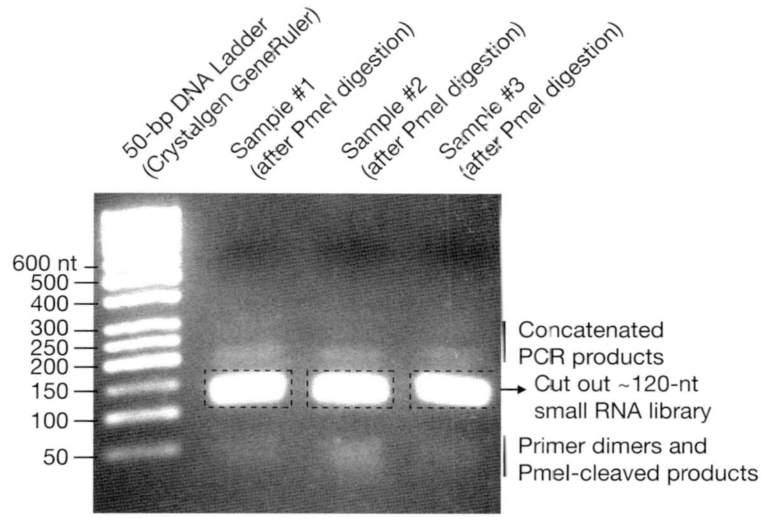

FIGURE 8-15. Size selection and gel extraction of small RNA libraries. Processed libraries are resolved by UV light on a 2% low-melting agarose gel stained with ethidium bromide. A 50-bp size marker and PmeI-digested PCR products of each library are shown. A primary band at ~120 nucleotides is indicated, which was carefully cut and extracted from the gel (see Section IX). Although we have not characterized all products on the gel, based on size, we have indicated the likely origin of additional species.

Section IX: Gel Purification of Amplified cDNA

91. Prepare a 2% low-melt agarose gel (using Seakem GTG agarose) in TAE, according to the manufacturer's instructions, including a minimal amount of ethidium bromide.
92. Load 8–12 µL of 50-bp DNA ladder, PCR-amplified library from cDNA (+RT), and control (–RT) (from Step X). Run the gel at a constant 80V until the bromophenol blue band is two-thirds through the gel.
93. View the PCR products with a long-wave UV source (Fig. 8-15).

    To maintain the integrity of the libraries, take all precautions to minimize the amount of time that the samples are exposed to UV light.
94. Excise the DNA band with a clean razor blade and transfer the gel slice into a preweighed microcentrifuge tube.
95. Centrifuge at full speed for 1 min at room temperature.
96. Add 0.5 M NaCl to a total volume 500 µL .
97. Melt the agarose in the solution for 10 min at 70°C, flicking tube the every couple of minutes.
98. Place microcentrifuge tubes with 500 µL of phenol (equilibrated, pH 8) at 70°C.
99. Add hot phenol to the tubes with melted agar and vortex immediately for 1 min.

    Perform under a chemical fume hood.
100. Centrifuge the mixture at maximum speed for 5 min at room temperature.
101. Transfer the aqueous layer to 1 volume of PCA, vortex, and centrifuge at maximum speed for 5 min at room temperature.
102. Transfer the aqueous layer to 1 volume of chloroform, vortex, and centrifuge at maximum speed for 5 min at room temperature.
103. Transfer the aqueous layer to 2.5–3 volumes of 100% ethanol.
104. Incubate for 2 hr to overnight at –20°C.

105. Centrifuge at maximum speed for 15 min at 4°C.
106. Wash the pellet with 70% ethanol and then remove all of the ethanol.
107. Wash pellet with 100% ethanol and then remove all of the ethanol.
108. Air-dry the pellet and then resuspend it in 20 μL of H_2O.
109. Determine sample concentration and dilute to 10 nM before sequencing.

TIPS AND TROUBLESHOOTING

For a higher yield of RNA from polyacrylamide gel slices, perform dialysis/electrophoresis extraction rather than grinding the slices and extracting by diffusion.

SOLUTION AND REAGENT RECIPES

CAUTION: See the Cautions Appendix for appropriate handling of material marked with <!>.

Reverse Transcriptase Mix Stock

Reagent	Quantity (for 60 μL)	Final concentration
First-strand buffer (5x)	12 μL	2.5 mM
dNTP (10 mM each)	15 μL	2.5 mM
DTT (0.1 M) <!>	3 μL	5 mM
Milli-Q H_2O	30 μL	

REFERENCE

Lau NC, Lim LP, Weinstein EG, Bartel DP. 2001. An abundant class of tiny RNAs with probable regulatory roles in *Caenorhabditis elegans*. *Science* **294:** 858–862.

Protocol 7

Large-scale Immunopurification of RNP Complexes from *Drosophila* Nucleoplasmic Extracts for Tiling Microarrays (RNA Immunoprecipitation Microarray; RIP-Chip)

It is of interest to be able to define sets of cellular RNAs associated with specific RNA-binding proteins. This "guilt by association" can lead to new insights into how RNA-binding proteins can modulate posttranscriptional gene expression of specific target RNAs. Immunopurification of mRNAs associated with specific RNA-binding proteins and their identification using microarrays was first described in 2000 and extended to nuclear ribonucleoprotein (RNP) fractions using spotted expressed sequence tag (EST) cDNA microarrays. Several drawbacks of this method are that the investigator typically must generate or obtain an antibody (or generate cells expressing an epitope-tagged version of the gene) to retrieve the protein-associated RNAs. The method gives an enrichment relative to the total RNA (the starting RNP fraction), but it can also result in false positives if the antibody cross-reacts with additional epitopes besides the protein of interest or if the RNA-binding protein of interest rebinds new RNAs after extract preparation. Some results indicate that this is not a significant problem (Keene et al. 2006), but one example shows that this can happen in at least one case for the human HuR protein (Mili and Steitz 2004). Some procedures use in vivo protein–RNA cross-linking—either formaldehyde or UV cross-linking (Ule et al. 2003; Licatalosi et al. 2008; Sanford et al. 2009; Xue et al. 2009; Yeo et al. 2009)—although other procedures have not used these cross-linking methods to stabilize RNA–protein complexes before cell lysis (Keene et al. 2006).

This method was developed to identify and localize binding regions on nuclear transcripts for the *Drosophila* heterogeneous nuclear RNP (hnRNP) proteins hrp48, hrp40, hrp38, and hrp36 (Olson et al. 2007; Blanchette et al. 2009). The method uses nucleoplasmic RNP-enriched fractions from a procedure originally described for mammalian cells. At least for several *Drosophila* RNA-binding proteins and splicing factors (Gerber et al. 2006; Olson et al. 2007; Blanchette et al. 2009; Sanford et al. 2009) and certain mammalian RNA-binding proteins (Keene et al. 2006), no cross-linking was necessary to retrieve discrete protein-bound mRNA and pre-mRNA fragments.

MATERIALS

The recipes for the items marked with <R> are at the end of the protocol or in the Common Recipes Appendix.

CAUTION: See the Cautions Appendix for appropriate handling of materials marked with <!>.

Reagents

Affinity-purified antibody to the protein of interest
Buffer RCT + β-mercaptoethanol <!> (from QIAGEN RNeasy Minikit)
Buffer RPE (comes with RNeasy kit)
Embryonic or cell culture nucleoplasmic RNP-enriched fraction (see Appendix 3, Protocol 7)
Ethanol (100%) <!>
IPB1 buffer <R>
IPB2 buffer <R>
Protein A–Sepharose CL-4B (Sigma-Aldrich or GE Healthcare 17-0780-01) or Protein A magnetic beads (Dynal, Invitrogen)

QIAGEN RNeasy Minikit
Rabbit IgG, nonimmune control (Sigma-Aldrich or Zymed/Invitrogen)
RiboGreen (Molecular Probes)

Equipment

Falcon tube (12.5 mL, no. 2059)
Microcentrifuge (Beckman GS-6R tabletop)
Millex-GV syringe filter (0.2 mm, 4 mm; Millipore SLGVL 040S)
NanoDrop spectrophotometer
Polyethylene tubing (0.062 inch/0.082 inch ID/OD; Clay Adams 427446, Fisher 14-170-12G)
Poly-Prep Econo-Columns (Bio-Rad 731-1550)
Polypropylene snap-cap tubes (75 x 100 mm; Falcon 2059)
RiboGreen
Rotator, motorized (Scientific Industries multipurpose rotator model 151)
Vortex machine

OVERVIEW

The antibody is added to the RNP extract and incubated to allow antigen–antibody complexes to form. The antibody–antigen complexes are then retrieved by binding of the antibody constant region to immobilized staphylococcal Protein A immobilized on Sepharose beads. The bead-immobilized complexes are then washed and RNA is prepared. The RNA is used to generate random-primed cDNA, cRNA, and biotinlyated cDNA probes for use on Affymetrix whole-genome *Drosophila* tiling arrays.

METHOD

Immunopurification

1. Centrifuge 1 mL of thawed embryonic or cell culture nucleoplasmic RNP fraction in a microcentrifuge at 14,000 rpm (16,000*g*) for 10 min at 4°C.
2. While the centrifuge is still running, mix the antibody and IPB1 buffer on ice in a 12.5-mL Falcon tube.
   i. Add antibody (affinity purified, preferably) to give sufficient depletion (typically 0.1–1 mg of affinity-purified antibody).

      The amount of antibody used must be determined empirically by titration for each antibody and target protein (see Tips and Troubleshooting).

   ii. Add cold IPB1 buffer to 9 mL.
   iii. Add 1 mL of RNP extract from Step 1 (see Appendix 3, Protocol 7).
   iv. Rotate the mixture on the rotator for 20 min at 4°C in the cold room.
3. While rotating the antibody–RNP extract in the cold room, place a fresh Falcon 2059 polypropylene tube on ice containing 0.25 mL of a 50% slurry Protein A–Sepharose CL-4B in IPB2 buffer without Triton X-100.

   This can be prepared in advance and stored for years with 0.2% azide.

   The Protein A–Sepharose CL-4B comes as a dry powder that must be hydrated overnight before use.

4. Filter the antibody–RNP extract with a 0.22-µm Millipore Millex GV 4-mm syringe filter attached to a 10-mL syringe directly on top of the Protein A–Sepharose resin in the Falcon 2059 tube.
5. Seal the tube and rotate on a rotator for 20 min at 4°C in the cold room.

6. Pour the bead slurry on top of a 10-mL Poly-Prep Econo-Column (Bio-Rad) connected to ~20 cm of 0.062-inch/0.082-inch ID/OD polyethylene tubing. Recover the flowthrough fraction in a Falcon 2059 polypropylene tube.
7. Wash the column five times with 10 mL of cold IPB2 buffer in the cold room. Do not save the washes.

RNA Isolation

8. Close the column outlet with the yellow cap that comes with the column.
   i. Add 350 µL of buffer RLT + β-mercaptoethanol (from a QIAGEN RNeasy Minikit) and vortex.
   ii. Add 250 µL of 100% ethanol and vortex.
9. Remove the yellow cap and rapidly transfer the column to a fresh Falcon 2059 17 x 100-mm polyproylene tube. Centrifuge at 3800 rpm (3000*g*) in a Beckman GS-6R tabletop centrifuge at 4°C.
10. Recover the RNA following the QIAGEN RNeasy protocol as follows:
    i. Apply the 700 µL of RNA mixture to the column.
    ii. Centrifuge at the maximum speed of 14,000 rpm (16,000*g*) for 15 sec.
    iii. Transfer the column to a new recovery tube.
    iv. Wash the column with 500 µL of buffer RPE with ethanol added to it.
    v. Centrifuge at the maximum speed of 14,000 rpm (16,000*g*) for 15 sec.
    vi. Wash the column again with 500 µL of buffer RPE with ethanol added to it.
    vii. Centrifuge at the maximum speed of 14,000 rpm (16,000*g*) for 2 min.
    viii. Transfer the column to the RNase-free Eppendorf microcentrifuge tube (comes with the QIAGEN kit).
    ix. Add 50 µL of RNase-free H_2O to elute the RNA from the column (comes with the QIAGEN kit).
    x. Centrifuge at the maximum speed of 14,000 rpm (16,000*g*) for 30 sec.
11. Dilute 0.5 µL of the eluted RNA in 200 µL of RiboGreen diluted 1:4000 to quantitate the RNA on a fluorometer *or*, alternatively, assay the concentration of the purified RNA directly using a NanoDrop 1000 spectrophotometer. Typically, 50–500 ng of RNA is retrieved in this procedure.

cDNA Synthesis and Affymetrix Tiling Array Hybridization

12. Amplify and label total RNA purified from a single large-scale immunopurification.

    Typically, the Affymetrix GeneChip WT Amplified Double-stranded cDNA Synthesis Kit is used. This kit takes the isolated RNA, converts it to cDNA using random hexamer priming, makes cRNA using T7 RNA polymerase, converts the cRNA to cDNA, and uses incorporation of dUTP into the final cDNA to allow fragmentation using uracil *N*-glycosidase (see Affymetrix website).

13. Hybridize the labeled cDNA on a single GeneChip *Drosophila* Tiling 1.0R Array whole-genome tiling array using the 2-µm scanner and hybridization station from Affymetrix.

    Typically, for Affymetrix tiling arrays where internal control probes are provided during the labeling protocol, individual biological duplicated or triplicates of immunopurified RNA samples are hybridized for each individual RNA-binding protein of interest and compared to 1 µg of total RNA purified directly from the starting RNP fraction and amplified, labeled, and hybridized using the same procedure described above.

14. Perform downstream data analysis using the TiMAT application suite (http://bdtnp.lbl.gov/TiMAT/). Visualize the data using the Affymetrix Integrated Genome Browser (http://www.affymetrix.com/support/developer/tools/download_igb.affx).

TIPS AND TROUBLESHOOTING

1. It may be more effective to preload the Protein A–Sepharose (or Protein A magnetic beads) with antibody before incubation with the RNP extract. For pilot immunoprecipitations, adding increasing amounts of affinity-purified antibody to a constant amount of extract may be easier, but once an amount of antibody is found that efficiently depletes the protein of interest, preloading the antibody onto the Protein A affinity matrix may give improved depletion of the target antigen from the extract.
2. It is critical to determine the amount of antibody necessary to deplete the protein of interest from the RNP fraction. This is most easily done using a fixed amount of extract with increasing amounts of antibody on small samples (50–100-mL extract) and 20 mL of Protein A–Sepharose. After mixing and removing the Protein A–Sepharose by centrifugation, the supernatant is analyzed by immunoblotting to detect depletion of the target protein. If the protein is completely removed by a certain concentration of antibody, it should be possible to scale up the reaction directly for a preparative purification. It is *essential* to know how much antibody to use for the success of this protocol.
3. Protein A magnetic beads (Dynal/Invitrogen) may give less background binding than Protein A–Sepharose CL4B. These have been used extensively for chromatin immunoprecipitations. However, the binding capacity of the magnetic beads is much lower than the Protein A–Sepharose.

SOLUTION AND REAGENT RECIPES

CAUTION: See the Cautions Appendix for appropriate handling of materials marked with <!>.

IPB1 Buffer

Reagent	Quantity (for 100 mL)	Final concentration
Tris-HCl (1 M, pH 7.5)	1 mL	10 mM
NaCl (5 M)	3 mL	150 mM
$MgCl_2$ (1 M) <!>	0.25 mL	2.5 mM
Triton X-100 (10%) <!>	0.5 mL	0.05%
PMSF (0.1 M) <!> (see Protocol 8)	0.1 mL	0.5 mM
Protease inhibitor cocktail (1000x) (see Protocol 8)	0.1 mL	1x
Cold ddH_2O	add to 100 mL	

Store at 4°C.

IPB2 Buffer

Reagent	Quantity (for 500 mL)	Final concentration
Tris-HCl (1 M, pH 7.5)	5 mL	10 mM
NaCl (5 M)	15 mL	150 mM
$MgCl_2$ (1 M) <!>	1.25 mL	2.5 mM
Triton X-100 (10%) <!>	12.5 mL	0.25%
Cold ddH_2O	add to 100 mL	

Store at 4°C.

REFERENCES

Blanchette M, Green RE, MacArthur S, Brooks AN, Brenner SE, Eisen MB, Rio DC. 2009. Genome-wide analysis of alternative pre-mRNA splicing and RNA binding specificities of the *Drosophila* hnRNP A/B family members. *Mol Cell* **33:** 438–449.

Gerber AP, Luschnig S, Krasnow MA, Brown PO, Herschlag D. 2006. Genome-wide identification of mRNAs associated with the translational regulator PUMILIO in *Drosophila melanogaster*. *Proc Natl Acad Sci* **103:** 4487–4492.

Keene JD, Komisarow JM, Friedersdorf MB. 2006. RIP-Chip: The isolation and identification of mRNAs, microRNAs and protein components of ribonucleoprotein complexes from cell extracts. *Nat Protoc* **1:** 302–307.

Licatalosi DD, Mele A, Fak JJ, Ule J, Kayikci M, Chi SW, Clark TA, Schweitzer AC, Blume JE, Wang X, et al. 2008. HITS-CLIP yields genome-wide insights into brain alternative RNA processing. *Nature* **456:** 464–469.

Mili S, Steitz JA. 2004. Evidence for reassociation of RNA-binding proteins after cell lysis: Implications for the interpretation of immunoprecipitation analyses. *RNA* **10:** 1692–1694.

Olson S, Blanchette M, Park J, Savva Y, Yeo GW, Yeakley JM, Rio DC, Graveley BR. 2007. A regulator of Dscam mutually exclusive splicing fidelity. *Nat Struct Mol Biol* **14:** 1134–1140.

Sanford JR, Wang X, Mort M, Vanduyn N, Cooper DN, Mooney SD, Edenberg HJ, Liu Y. 2009. Splicing factor SFRS1 recognizes a functionally diverse landscape of RNA transcripts. *Genome Res* **19:** 381–394.

Ule J, Jensen KB, Ruggiu M, Mele A, Ule A, Darnell RB. 2003. CLIP identifies Nova-regulated RNA networks in the brain. *Science* **302:** 1212–1215.

Xue Y, Zhou Y, Wu T, Zhu T, Ji X, Kwon YS, Zhang C, Yeo G, Black DL, Sun H, et al. 2009. Genome-wide analysis of PTB-RNA interactions reveals a strategy used by the general splicing repressor to modulate exon inclusion or skipping. *Mol Cell* **36:** 996–1006.

Yeo GW, Coufal NG, Liang TY, Peng GE, Fu XD, Gage FH. 2009. An RNA code for the FOX2 splicing regulator revealed by mapping RNA-protein interactions in stem cells. *Nat Struct Mol Biol* **16:** 130–137.

Protocol 8

CLIP (Cross-linking and Immunoprecipitation) Identification of RNAs Bound by a Specific Protein

Guilt by association remains a powerful argument for assigning function in biological processes: Proteins that bind a specific RNA are strongly implicated in modulating the fate and action of that RNA. Detecting such associations specifically is therefore a major goal in the understanding of how RNA function is regulated. The CLIP method allows assignments of RNA-binding proteins to their RNA targets in a broad sense by using UV cross-linking to covalently capture close (near-covalent bond distances) association of RNA with protein, followed by immunopurification of the protein partner with extraction and subsequent characterization of the RNA partner sequence(s). This allows complex pools of target sequences to be captured and their genomic origins studied so that regulatory and other functional relationships between the protein and its RNA targets can be inferred (Ule et al. 2005; Jensen and Darnell 2008; Sanford et al. 2009; Xue et al. 2009; Yeo et al. 2009). The advantage of this protocol appears to be that it only captures very intimately associated RNAs and proteins, and so is expected to be highly specific. Some investigators report capturing cross-links between rRNA and nonribosomal proteins in cases where ribosomes are not efficiently cleared from the lysate before immunopurification of the target protein, so the limits of specificity for the method are yet to be fully explored. Its main disadvantages are its many steps and that it may fail to work for all proteins or capture all legitimate targets of any given protein if the necessary chemical groups are not optimally arranged for photocross-linking in the complex. However, properly performed, this method should provide many useful insights into the nature of the RNA ligands for an RNA-binding protein of interest.

MATERIALS

CAUTION: See the Cautions Appendix for appropriate handling of materials marked with <!>.

Reagents

3′ RNA linker, gel purified, L3: 5′-P-GUGUCAGUCACUUCCAGCGG-3′-puromycin
5′ RNA linker, gel purified, L5: 5′-OH-AGGGAGGACGAUGCGG-3′-OH
Accuprime Pfx Supermix (Invitrogen 12344-040)
Acrylamide:bis-acrylamide (19:1) <!>
Antibody against the protein of interest
APS (10%)
ATP (1 mM, 10 mM)
Bovine serum albumin (BSA, 100 μg/mL) in RNase-free water
Calf intestinal alkaline phosphate (CIP)
Chloroform:isoamyl alcohol (49:1) <!>
Dephosphorylation buffer (10x; Roche 712023)
Dithiothreitol (DTT) (0.1 M) <!>

We thank Robert Darnell, The Rockefeller University, for supplying this protocol. Modifications and tips were provided by Jeremy Sanford (University of California, Santa Cruz) and Xiang-Dong Fu (University of California, San Diego) have also been incorporated throughout.

DNA P3-454 capture primer: 5′ GCC TTG CCA GCC CGC TCA GCC GCT GGA AGT GAC TGA CAC 3′ (100 μM)
DNA P3 primer: 5′-CCGCTGGAAGTGACTGACAC-3′ (30 pmol/μL)
DNA P5-454 capture primer: 5′ GCC TCC CTC GCG CCA TCA GAG GGA GGA CGA TGC GG 3′ (100 μM)
DNA P5 primer: 5′-AGGGAGGACGATGCGG-3′ (30 pmol/μL)
DNase I buffer
dNTP (3 mM)
EDTA <!>
Ethanol (75%) <!>
Ethanol:isopropanol (1:1) <!>
[γ-^{32}P-]ATP (10 μCi/μL) <!>
Glycogen (Ambion 9510)
Hanks' balanced salt solution (HBSS), Ca-Mg free (GIBCO 14186-012), 10 mM HEPES (pH 7.3)
Low-molecular-weight marker (e.g., Amplisize Molecular Ruler, Bio-Rad)
Micrococcal nuclease, nuclease S7 (Roche), dissolved in water at ~10 units/μL
Micrococcal nuclease reaction buffer (MNase reaction buffer) (50 mM Tris-Cl at pH 7.9, 5 mM $CaCl_2$)
NaCl (5 M)
NaOAc (3 M at pH 5.2; 0.15 M at pH 5.2)
Novex loading buffer (Invitrogen)
Novex NuPAGE precast 10% Bis-Tris gel
NuSieve agarose (2%)
Phenol:chloroform (1:1) <!>
Phosphate-buffered saline (1x PBS) (137 mM NaCl, 2.7 mM KCl, 10 mM Na_2HPO_4, 5 mM KH_2PO_4 at pH 7.4)
Polyacrylamide gel (10% denaturing) <!>
PK buffer (100 mM Tris-Cl at pH 7.5, 50 mM NaCl, 10 mM EDTA <!>)
PK buffer + 7 M urea (must be fresh)
PNK (polynucleotide kinase) buffer (50 mM Tris-Cl at pH 7.4, 10 mM $MgCl_2$, 0.5% NP-40, protease inhibitor cocktail)
PNK + EGTA buffer (50 mM Tris-Cl at pH 7.4, 20 mM EGTA, 0.5% NP-40, protease inhibitor cocktail)
Protein A Dynabeads (Dynal 100.02)
Proteinase K in PK buffer (4 mg/mL)
PXL buffer (1x PBS [tissue culture grade; no Mg^{++}, no Ca^{2+}, see recipe above], plus 0.1% SDS, 0.5% deoxycholate, 0.5% NP-40, protease inhibitor cocktail)

Some antibodies do poorly with deoxycholate; if so, omit it.

PXL100 buffer (1x PBS [tissue culture grade; no Mg^{++}, no Ca^{2+}], 100 mM additional NaCl over that present in PBS, 0.1% SDS <!>, 0.5% deoxycholate <!>, 0.5% NP-40, protease inhibitor cocktail)

Some antibodies do poorly with deoxycholate; if so, omit it.

PXL400 buffer (1x PBS [tissue culture grade; no Mg^{++}, no Ca^{2+}], 400 mM NaCl over that present in PBS, 0.1% SDS <!>, 0.5% deoxycholate <!>, 0.5% NP-40, protease inhibitor cocktail)

Some antibodies do poorly with deoxycholate; if so, omit it.

QIAEX II Gel Extraction Kit (QIAGEN)
Rainbow marker (Amersham RPN800)
RNase A (USB 70194Y) <!>
RNasin (Promega N2515) or RNaseOUT (Invitrogen)
RSB100 cell lysis buffer (10 mM Tris-Cl at pH 7.4, 100 mM NaCl, 2.5 mM $MgCl_2$, 0.5% NP-40, 0.5% Triton X-100 <!>)

RQ1 DNase (Promega M6101)
Sodium phosphate (0.1 M, pH 8.1)
Source material (Option 1, tissue; Option 2, cultured cells)
SuperScript III (Invitrogen 18080-044)
SuperScript RT buffer (5×)
SYBR Gold (Molecular Probes)
T4 PNK enzyme (New England Biolabs M0201L)
T4 RNA ligase (Fermentas)
T4 RNA ligase buffer (10×; Fermentas)
TBE (5×)
TEMED <!>
Urea (Fisher U15-3) <!>

Equipment

Autoradiography equipment and X-ray film
Beckman TLA 120.2 rotor with polycarbonate tubes
Beckman ultramicrocentrifuge
Freezer
Gel electrophoresis apparatus and power supply
Ice
Kimwipes
Luminescent stickers
Microcentrifuge tubes (1.5 mL)
Nitrocellulose (BA85, S&S)
Novex NuPage precast gels (10% Bis-Tris; Invitrogen)
Novex wet-transfer apparatus
Plastic pipette (5 or 10 mL)
Plastic wrap
Pipettor tip (1 mL, 200 μL)
Rotating wheel
Scalpel
Scintillation counter
Sonicator
SpeedVac
Thermal cycler
Thermomixer R (Eppendorf)
Tissue culture plates
UV box (e.g., Stratalinker 2400, Stratagene)

OVERVIEW

The protocol takes several days and comprises more than 100 steps. As written, this protocol incorporates two points at which options are available depending on the starting material (Starting materials Option 1: tissue, or Option 2: cultured cells) as well as the nuclease used for trimming the cross-linked RNA (Nuclease Option 1: mixture of RNase A and RNase T1 before immunoprecipitation, or Nuclease Option 2: micrococcal nuclease on beads after immunoprecipitation) (Fig. 8-16). In all cases, however, the starting material is first irradiated with UV to cross-link proteins to nucleic acids in vivo. Second, a lysate is prepared from the cross-linked material, and aliquots are treated with dilutions of nuclease to trim the cross-linked RNAs to a size of ~50–150 nucleotides. The lysates are cleared of ribosomes and then incubated with beads bearing antibody against the protein of interest. After binding of the protein of interest (plus its covalently cross-linked RNA target fragments) to the antibody-bead matrix,

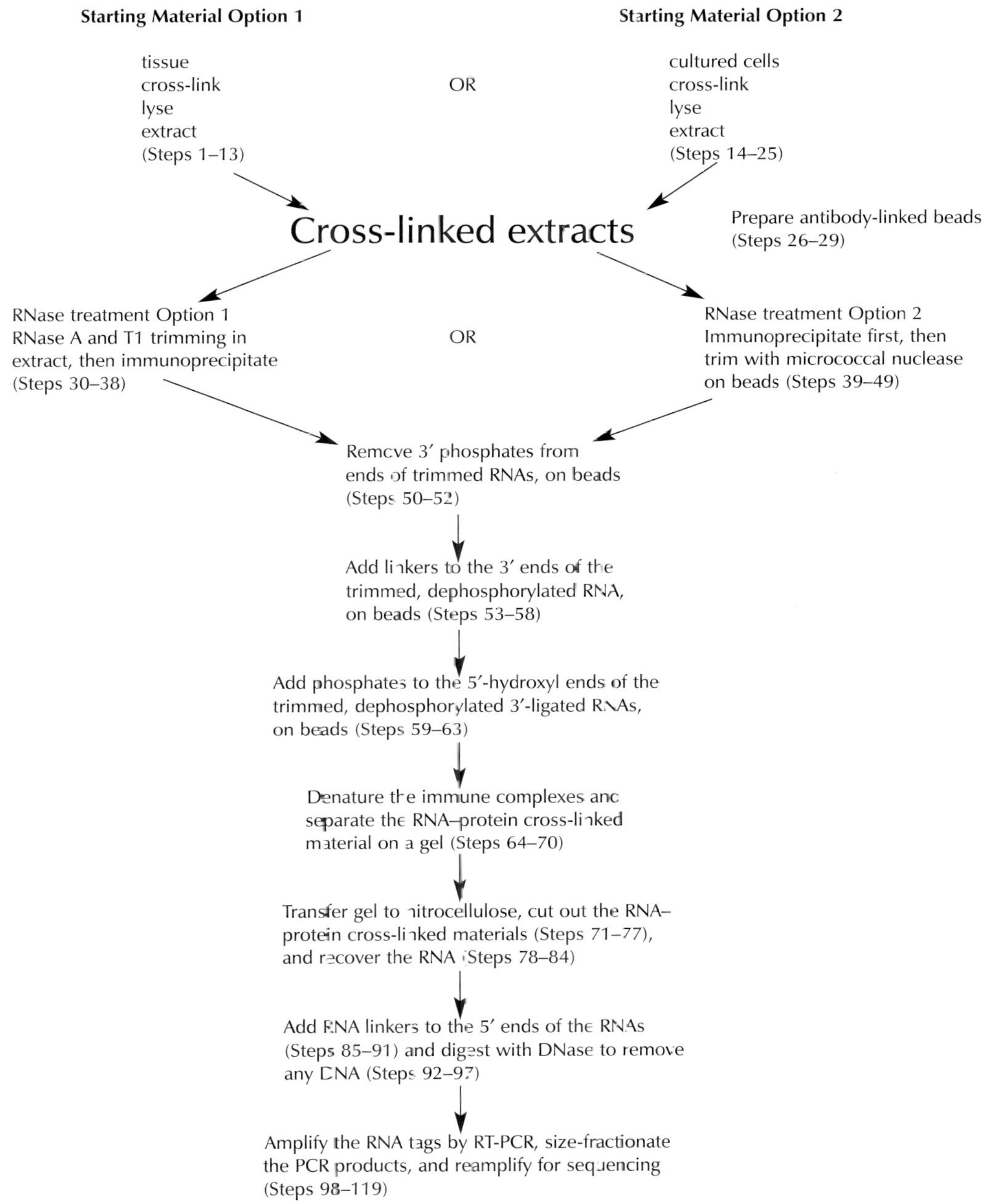

FIGURE 8-16. Outline of CLIP protocol.

unbound proteins are washed off and a series of reactions is performed on the cross-linked RNAs while they are still attached to the protein and bound to the beads. The cross-linked RNA is dephosphorylated (at both its 3′ and 5′ ends) with phosphatase, and an RNA linker (with a blocked 3′ end to prevent multiple ligation events) is attached to the 3′ end of the cross-linked fragments using RNA ligase. The 5′ end of the cross-linked RNA is then phosphorylated and labeled by polynucleotide kinase in the presence of [γ-^{32}P]ATP. The labeled RNA–protein complexes are resolved on a denaturing protein gel and radioactive material migrating ~20 kDa more slowly than the native uncross-linked proteins is isolated, ligated to a 5′ RNA linker, purified, and then RT-PCR–amplified using primers for the linker sequences. These products can then be cloned or subjected to sequencing by further steps. The derived sequences identify the genomic source(s) of the cross-linked RNA.

METHOD

Starting Material Option 1: UV Cross-linking of Tissues

1. Harvest tissue (e.g., material equivalent to one mouse brain [~500 mg] suffices for neuronal RNA-binding proteins). If necessary, allow tissue to sit in ice-cold PBS (alternatively, HBSS can be used instead for brain tissue) until harvest is complete.
2. Disrupt the tissue. Add 10 tissue sample volumes of PBS (or HBSS) and break up the tissue by passage into and out of a 5- or 10-mL plastic pipette. Using a 5-mL plastic pipette to which a 1-mL pipettor tip has been attached, disrupt further, and then finish by using a 200-μL pipettor tip attached to a 5-mL pipette. Only very small pieces of tissue should remain. Avoid bubbles.
3. Dispense the suspension into tissue culture plates (placing 10 mL per each 10-cm tissue culture plate).
4. Irradiate the dish (with the cover off!) three times with 100–400 mJ/cm^2 in a UV box. Mix the suspension between each irradiation.

   Because UV light can penetrate a few cell layers, stringent disruption of the tissue to the single-cell level is not necessary.

   The amount of cross-linking for each protein should be optimized separately, because each RNA-binding protein will cross-link with different efficiencies. For a preliminary experiment, try 100, 200, and 400 mJ/cm^2 and then use the shortest condition that gives >70% of the maximum signal.
5. Collect the suspension, pellet the cells at 2500 rpm for 3 min at 4°C, resuspend pellet in ~3x pellet volumes of PBS, and distribute 1 mL of suspension to microcentrifuge tubes.
6. Centrifuge quickly at 4°C, remove the supernatant, and freeze the cell pellets at –80°C until use. Each tube should have ~200 μL of cells.
7. When ready to continue, thaw each tube of cross-linked cells up to ~1 mL using PXL buffer. Allow the tube to sit on ice for 10 min to lyse cells and solubilize complexes.

   If tissue is resistant to lysis, as is, e.g., adult brain tissue, sonication may be necessary at this step.
8. Add 11 μL of 0.1 M DTT to each tube. Then add 30 μL of RNaseOUT or RNasin and 60 μL of RQ1 DNase to each tube. Incubate on the Thermomixer at 1200 rpm for 15 min at 37°C. Invert the tubes after 8 min.

   It is critical to add DTT to the solution because the RNase inhibitors require a minimum of 1 mM DTT to maintain activity.
9. Incubate for 5 min on ice.
10. Add EDTA to a final concentration of 20 mM.
11. Centrifuge the lysates in a prechilled ultramicrocentrifuge (polycarbonate tubes in TLA 120.2 rotor) at 49,000 rpm for 20 min at 4°C.

    If the protein of interest sticks to ribosomes under these conditions, or ribosomal complexes or other large RNPs are being studied, this step should be modified so that the target protein stays in the supernatant, or the step should be skipped. If so, beware that spurious cross-linking to rRNA may be observed.
12. Transfer 120–140-μL aliquots of the supernatants to new 1.5-mL microcentrifuge tubes.
13. Proceed to the Prepare the Dynabead–Antibody Complexes section (Step 26).

Starting Material Option 2: UV Cross-linking of Cultured Cells

If Starting Material Option 1 was used, proceed directly to Prepare the Dynabead–Antibody Complexes, Step 26.

14. Grow the cells in a 100- or 150-mm culture plates. For a typical experiment, set up three plates and irradiate two, reserving the other as a "no UV" control.

    Cell culture is a useful way to study proteins for which no suitable antibody is readily available. By expressing epitope-tagged proteins for which good antibodies specific for the tag are available, different proteins can be isolated in parallel experiments using the same antibody. The use of some tags (or their antibodies) has been observed to lead to copurification of nontarget RNA-binding proteins, so we recommend that control purifications on untagged extracts be performed as well.

15. Rinse cell layers once with PBS and place the plates in a Stratalinker (with the cover off!). Irradiate one time for 100–400 mJ/cm^2 in the Stratalinker.

    Because adherent cells grow in a single layer, a single UV exposure may be sufficient for uniform cross-linking.

    The amount of cross-linking for each protein should be optimized separately, because each RNA-binding protein will cross-link with different efficiencies. For a preliminary experiment, try 100, 200, and 400 mJ/cm^2 and then use the shortest condition that gives >70% of the maximum signal.

16. Wash the cells twice with 5 mL of cold PBS.

17. Add 400 µL of a cell lysis buffer (e.g., RSB100) per plate. Scrape the cells from the dish and collect into a microfuge tube.

    RSB100 contains NP-40 and Triton X-100 and is useful for HEK293 cells, but other detergents may work better for other cell lines and experimentation may be necessary.

18. Place the cells on ice for 5 min so that lysis becomes complete.

19. Centrifuge the lysate in a microcentrifuge at 10,000 rpm for 10 min at 4°C.

20. Transfer the supernatant to a new 1.5-mL microcentrifuge tube. Combine the +UV supernatants at this point.

21. Add 11 µL of 0.1 M DTT. Then add 30 µL of RNaseOUT or RNasin and 60 µL of RQ1 DNase to each tube. Incubate on the Thermomixer at 1200 rpm for 15 min at 37°C. Invert the tubes after 8 min.

    It is critical to add DTT to the solution because the RNase inhibitors require a minimum of 1 mM DTT to maintain activity.

22. Incubate for 10 min at 37°C.

23. Add EDTA to a final concentration of 20 mM.

24. Centrifuge the lysates in a prechilled ultramicrocentrifuge (polycarbonate tubes in TLA 120.2 rotor) at 49,000 rpm for 20 min at 4°C.

    If the protein of interest sticks to ribosomes under these salt conditions, or ribosomal complexes or other large RNPs are being studied, this step should be modified so that the target protein stays in the supernatant, or the step should be skipped. If so, be aware that spurious cross-linking to rRNA may be observed.

25. Transfer 120- to 140-µL aliquots of the supernatants to new 1.5-mL microcentrifuge tubes.

Prepare the Dynabead–Antibody Complexes

26. For each 120–140 µL of lysate, use 400 µL of Protein A–Dynabead suspension.

    Dynabeads are paramagnetic beads that can be collected against the side of the tube using a magnetic rack, or they can be centrifuged briefly in a microcentrifuge (2 min at 10,000 rpm). An advantage of the magnetic rack is that supernatant fluid can be removed while the magnetic force is in place, preventing loss of the beads as occurs during resuspension of pelleted beads.

27. Wash the beads three times with 0.1 M sodium phosphate (pH 8.1).

    The high pH buffer increases antibody capture by Protein A. However, any or all of the immunoprecipitation buffers may need to be optimized for each individual antibody used. For example, some antibodies do not work with the deoxycholate in the wash buffer. If the antibody to be used is mouse IgG_1, which interacts poorly with Protein A, the beads can be preincubated with rabbit antimouse IgG. The rabbit antimouse antibody will act as an adapter, binding to the Protein A by its constant region and then to the constant region of the mouse IgG_1.

28. Resuspend the beads in 320 µL of 0.1 M sodium phosphate (pH 8.1) and add 80 µL of an appropriate concentration of antibody.
29. Rotate the tubes for 30–45 min at room temperature. Wash three times with PXL buffer. If not yet ready to add cross-linked lysate, leave beads in the last wash until ready, but use them the same day as they are made.

RNase Treatment Option 1: Trim the RNA Tag with RNase A and T1 Mix

This option uses a mixture of two difficult-to-control nucleases in solution before the cross-linked material is immunoprecipitated. RNase treatment Option 2 uses micrococcal nuclease, applied to the immunoprecipitate after capture of the cross-linked material. This enzyme is easy to control with the use of EGTA because the enzyme is strictly calcium dependent. If using RNase treatment Option 2, skip ahead to Step 39.

30. Prepare the following RNase A/T1 solutions (Ambion, cocktail of RNase A and T1):

1x	neat
1:10	1 µL RNase:9 µL PXL
1:50	1 µL RNase:49 µL PXL
1:500	1 µL RNase:499 µL PXL

Optimal RNase concentrations must be determined empirically for each RNA-binding protein of interest. The conditions above represent a good starting point. Ideally a UV-dependent, RNase-sensitive complex will be observed following immunoprecipitation and autoradiography.

31. Bring each tube of cross-linked lysate from either Step 12 (Option 1: tissues) or Step 25 (Option 2: cell lines) up to ~1 mL using PXL buffer. Remove 10 µL for immunoblot analysis. Save another 20 µL for Total control (this will be used to clone or sequence the control RNA tags present in the total pool of RNA-binding proteins).
32. Set up the following six reactions by adding 1 µL of the appropriate RNase dilutions per 100 µL of the appropriate extract and incubate for 10 min at 37°C. Be sure to include the "no UV" and "no antibody" controls:

Standard Controls and RNase Dilutions for Typical CLIP

[RNase]	1:10	1:10	Undiluted	1:10	1:50	1:500
Antibody	+	–	+	+	+	+
Beads	+	+	+	+	+	+
UV	–	+	+	+	+	+
Tube	**1**	**2**	**3**	**4**	**5**	**6**

33. Add an equal volume of 2x PXL to each tube to denature the extract and inactivate RNases.
34. Add 450 µL of PXL and 200 µL of antibody–beads to each tube, as indicated in the table above.
35. Incubate the tubes for 1 hr at 4°C.
36. Recover the beads using the magnetic rack and remove the lysate. Save an aliquot for a western blot to determine the efficiency of immunoprecipitation.
37. Wash the beads with 1000 µL of the following ice-cold buffers:

    Twice with PXL

    Twice with PXL-400

    Twice with 1x PNK buffer
38. Proceed to the Remove Phosphates from the RNA Ends section below.

RNase Treatment Option 2: Trim the RNA Tag with Micrococcal Nuclease (on Bead)

This option trims the RNA in the cross-linked material after it is captured on the beads. It may be easier to control digestion with this enzyme as compared to RNases A and T1.

If RNase treatment Option 1 was used, skip this step and proceed to Remove Phosphates from the RNA Ends, Step 50.)

39. Adjust each tube of cross-linked lysate from Step 12 (Option 1: tissues) or Step 25 (Option 2: cultured cells) to ~1 mL using PXL buffer. Remove 10 µL for immunoblot analysis. Save another 20 µL for Total control (this will be used to clone or squence the control RNA tags present in the total pool of RNA-binding proteins).
40. Aliquot the antibody–beads (from Step 29) and remove the last wash from each tube. Then, add each aliquot of lysate to one tube of antibody-coated beads. Rotate beads–lysate mix for 1 h at 4°C.
41. Collect the beads in the magnetic rack. Remove the supernatant and save 10 µL for immunoblot analysis: This aliquot should be depleted of the target protein.
42. Wash beads twice with ice-cold PXL buffer.
43. Wash twice with ice-cold PXL100 buffer.
44. Wash twice with ice-cold PXL400 buffer.

    If this step decreases the efficiency of immunoprecipitation, it can be omitted.
45. Wash twice with ice-cold 1x PNK buffer.
46. Dilute micrococcal nuclease with 1x MNase reaction buffer. Optimize the reaction by making a series of dilutions of 15, 2, 0.2, and 0.02 units of MNase in 500-µL aliquots of MNase reaction buffer. The 0.2-unit dilution seems to work best for cultured cell extracts; however, this may vary with the protein being studied. The 15-unit dilution will be the high (complete digestion) MNase control.
47. Add 500 µL of diluted MNase to the washed beads containing the antibody cross-linked RNA–protein complexes from Step 45.
48. Incubate the mixture at 1200 rpm on a Thermomixer for 1 min at 37°C and then place the mixture on ice. Approximately every 3 min for the next 10 min, place the mixture on the Thermomixer for 15 sec, returning the tube to the ice between mixing.
49. Wash the beads with the following series of ice-cold buffers:

    Twice with 1x PNK + EGTA buffer (EGTA will chelate the Ca^{2+} and quench the MNase activity)

    Twice with PXL buffer

    Twice with 1x PNK buffer

Remove Phosphates from the RNA Ends (on Bead)

50. Set up the reaction with calf intestinal alkaline phosphatase (CIP) as follows:

Dephosphorylation buffer (10x)	8 µL
CIP	3 µL
H_2O	69 µL
Total	80 µL

51. Add the mixture to the beads and incubate in the Thermomixer for 10 min at 37°C. Set the mixer to shake at 1000 rpm for 15 sec every 3 min.
52. Wash the beads with the following series of ice-cold buffers:

Twice with PNK + EGTA buffer
Twice with PNK buffer
Twice with 0.1 mg/mL BSA in RNase-free water

BSA enhances T4 RNA ligase efficiency and reduces aggregation of the beads.

Add RNA Linkers to the 3′ End of the Fragments with RNA Ligase (on Bead)

53. Make up the linker mix by adding

3′ RNA linker at 20 pmol/μL	8 μL
H_2O	32 μL
Total	40 μL

54. Add 40 μL of the linker mix to each tube of beads.
55. Make up the RNA ligase mix by adding

T4 RNA ligase buffer (10×)	8 μL
BSA (0.2 μg/μL)	8 μL
ATP (10 mM)	8 μL
T4 RNA ligase	2 μL
H_2O	14 μL
Total	40 μL

56. Add 40 μL of ligase mix to each tube of beads (final linker concentration is 2 pmol/μL).
57. Incubate the mixture in the Thermomixer overnight at 16°C. Set the mixer to shake at 1000 rpm for 15 sec every 5 min.

    The linker is blocked at its 3′ end with a puromycin moiety and thus only a single linker can be added to a given RNA.

58. Wash the beads three times with PNK buffer.

Phosphorylate the 5′ Ends of the RNAs with PNK (on Bead)

59. Make the polynucleotide kinase (PNK) mix by adding

PNK buffer (10×)	8 μL
[γ-^{32}P]ATP (10 μCi/μL)	2 μL
T4 PNK enzyme	4 μL
H_2O	66 μL
Total	80 μL

60. Add 80 μL of PNK mix to each sample and incubate in the Thermomixer for 10 min total at 37°C. Set the mixer to shake at 1000 rpm for 15 sec every 4 min.
61. Add 10 μL of 1 mM ATP (nonradioactive) and allow the reaction to proceed for an additional 5 min.
62. Wash three times with PNK + EGTA.

    CAUTION: These washes are radioactive and should be disposed of properly!

63. Resuspend the beads in 30 μL of PNK + EGTA.

Resolve the RNA–Protein Cross-linked Complexes on a Gel

This step makes use of precast gels sold under the name Novex NuPAGE (Invitrogen). They are different from the commonly used "Laemmli" SDS polyacrylamide gels in several ways. The most important difference is that the buffer system does not produce a high pH wave through the sample during electrophoresis, thus protecting the RNA tags from alkaline hydrolysis.

Depending on the size of the protein of interest, the presence of antibody in the sample may interfere with resolution of the protein and its RNA tags. The antibody heavy chain (~50 kDa) and light chain (~25 kDa) can be separated and resolved from each other in the presence of a reducing agent such as β-mercaptoethanol or DTT. If the protein of interest is larger than and can be separated from the 50-kDa heavy chain, we recommend that a reducing agent be added. If the protein of interest is the same size or smaller than 50 kDa, we do not recommend that the sample be reduced; this will leave the disulfide bridges in the antibody protein intact and it will not migrate near the bands of interest in the gel.

64. Add to the beads 30 μL of Novex loading buffer (with or without reducing agent, depending on the size of the protein of interest).
65. Mix the 20 μL of Total control (supernatant removed from lysate before the immunoprecipitation at Steps 31 or 39, depending on which nuclease was used) and 10 μL of Novex loading buffer.
66. Incubate the mixture in the Thermomixer at 1000 rpm for 10 min at 70°C. Collect the beads against the side of the tube and remove the supernatant for loading.
67. Split and load the sample from each tube of beads into two wells of a 10-well precast Novex NuPAGE 10% Bis-Tris gel.
68. Load the prestained markers in a side well of the gel.

    Some prestained molecular-weight markers may run differently on Novex NuPAGE gels. We recommend using the Rainbow Marker (Amersham RPN800), which runs at the expected molecular weights on this gel system.

69. When the dye front has migrated to the end of the gel, disassemble the gel and transfer it to S&S BA-85 nitrocellulose using the Novex wet-transfer apparatus.

    BA-85 is pure nitrocellulose and a little fragile, but it works better for the extraction step later.

    Some of the radioactivity in the loaded sample will not be retained on the gel. Some will be in the lower buffer, which results from carryover of free ATP and small free RNAs that were not washed from the beads after the kinase step. Free RNA <100 nucleotides long will migrate faster than the 30-kDa marker.

70. After transfer, gently rinse the nitrocellulose filter in 1x PBS and carefully blot the side that was away from the gel on Kimwipes. Wrap the membrane in plastic wrap and expose to X-ray film for autoradiography (see Fig. 8-17).

    Free RNA will not bind to the membrane and will be lost in the transfer step. The film must be aligned to the blot so that the proper material can be cut out. Apply luminescent stickers to ensure that this alignment is as accurate as possible.

    Depending on the protein and amount of starting material, a strong signal may be seen on the autoradiogram after 30 min of exposure. If much less material is used, an overnight exposure will sometimes be necessary to see a decent signal.

Recovering the Cross-linked RNA

71. Go to the next step if the overdigested (highest nuclease concentration) sample lane has a band close to the expected molecular mass of the protein of interest (as in Fig. 8-17).
72. Calculate the distance between the protein of interest and the closest contaminating bands and estimate the relative signal strength of the protein of interest relative to other bands.

    It should be possible to purify RNAs specific for the protein of interest if it migrates >10 kDa away from any other strong protein band and its signal is as strong as the other bands on the gel.

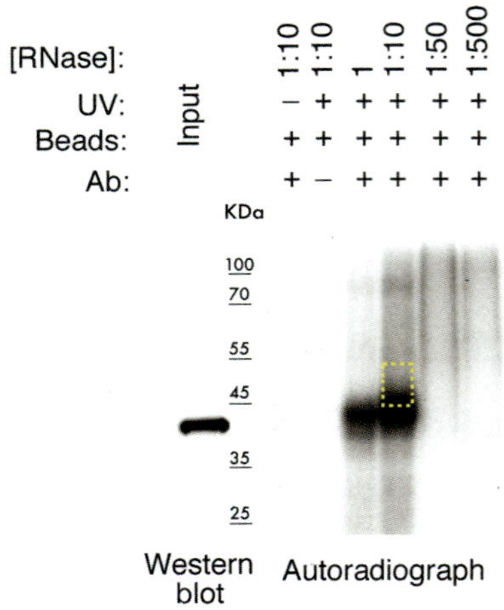

FIGURE 8-17. A sample CLIP experiment for hnRNP A1, a 36-kDa RNA-binding protein. The strong cross-linked bands at 42 kDa in lanes 3 and 4 are dependent on UV treatment and antibody in the immunoprecipitation as well as a proper level of ribonuclease activity. The region shown in the autoradiogram (fourth lane from the left, white dashed square) should be excised from the gel and treated to continue isolation of the RNA. (Courtesy of Jeremy Sanford, University of California, Santa Cruz.)

73. Estimate the position of migration of the protein of interest cross-linked to an original trimmed tag of 50 nucleotides, using the following considerations: The average molecular mass of 50 nucleotides of RNA is 16.2 kDa. Because each original tag now contains a 20-nucleotide-long linker (L3), the average size of RNA desired will thus be 70 nucleotides, which should cause the protein to migrate on average ~20 kDa more slowly than the molecular mass of the free protein. Because the nuclease trimming of the tags generates heterogeneous ends, the tag sizes will vary from ~10 to 150 nucleotides and thus the complex will appear as diffuse radioactivity.
74. Use a clean scalpel blade to cut a band from the nitrocellulose spanning ~10–20 kDa above the molecular mass of the free protein of interest.
75. In addition, cut two bands below and above this band, with ~15 kDa distance between the target band and these side fractions.

    With these samples, it may be possible to determine whether any other RNA-binding proteins are contaminating the sample.
76. From the Total control lane, cut out a very wide band from 30 kDa to almost the top of the gel.

    This control should contain RNA tags bound to all the cellular RNA-binding proteins with size from 10 to 200 kDa. It should not contain free RNA with a size <100 kDa, because such RNAs migrate lower than 30 kDa. This control tag can later be compared with the tags that bound to the protein of interest.
77. Cut each band into little pieces and place the pieces from each band into a microcentrifuge tube. Count radioactivity in a scintillation counter to determine the yield of each.

Isolate the RNA from the Nitrocellulose

78. Make a 4 mg/mL proteinase K stock solution in 1x PK buffer. Preincubate this stock for 20 min at 37°C to ensure digestion of any RNases.
79. Add 200 µL of proteinase K solution to each tube of isolated nitrocellulose pieces. Incubate on the Thermomixer at 1000 rpm for 20 min at 37°C .
80. Add 200 µL of fresh PK + 7 M urea solution. Incubate on the Thermomixer at 1000 rpm for another 20 min at 37°C.
81. Add 400 µL of phenol equilibrated with 0.15 M NaOAc (pH 5.2) and 130 µL of $CHCl_3$:isoamyl alcohol (49:1) to the suspension. Incubate on the Thermomixer at 1000 rpm for 20 min at 37°C.
82. Centrifuge the tubes in a microcentrifuge at 12,000 rpm for 5 min at room temperature. Remove the aqueous phase to a new tube.

83. Add 50 μL of 3 M NaOAc (pH 5.2), 0.5 μL of 5 mg/mL glycogen, and 1 mL of ethanol:isopropanol (1:1) to the aqueous phase. Mix well.

Glycogen helps to precipitate small quantities of RNA, but too much may inhibit RNA ligase.

84. Precipitate overnight at –20°C.

Ligate RNA Linkers to the 5′ Ends of the Tags

85. Centrifuge the RNA in a microcentrifuge at 12,000 rpm for 10 min at 4°C. To determine if the precipitation was successful, use a Geiger counter to detect evidence of radioactivity.

86. Rinse the pellet twice with 150 μL of ice-cold 75% ethanol, centrifuging the precipitate after each rinse at 12,000 rpm for 5 min at 4°C.

This should minimize loss of the precipitate and ensure removal of salt.

87. Dry *briefly* (<2 min) in the SpeedVac. Count the RNA in a scintillation counter.

88. Resuspend the pellet in 6 μL of H_2O.

89. Set up the RNA ligation premix:

T4 RNA ligase buffer (10×)	1 μL
BSA (0.2 μg/μL)	1 μL
ATP (10 mM)	1 μL
T4 RNA ligase (3 units)	0.1 μL
L5 RNA linker at 20 pmol/μL	1 μL
Total	4.1 μL

90. Add 5.9 μL of RNA resuspended in H_2O to the premix. The final volume is 10 μL total.

91. Incubate the mixture for >2 h at 16°C. Try to avoid overnight.

Ligation can only occur between the phosphorylated 5′ end of the RNA tag and the 3′ OH of the 5′ linker. The 5′ linker itself cannot dimerize or circularize (it has a 5′ OH that is not reactive), and the CLIP tag-3′ linker cannot dimerize or circularize (it is blocked by puromycin at its 3′ end). Extended incubation may lead to RNA fragmentation and undesired ligation events. Do not heat inactivate ligase, because RNA may be degraded with Mg^{++} present. Do not inactivate with EDTA, which will bind Mg^{++}, inhibiting the DNase treatment.

Remove Any Contaminating DNA from the RNA Tags

92. Add to the RNA ligation reaction:

H_2O	78 μL
DTT (0.1 M)	1 μL
DNase I buffer 1 (10×)	1 μL
RNasin	5 μL
RQ1 DNase	5 μL

93. Incubate the mixture for 20 min at 37°C.

This DNase step reduces the production of spurious PCR products in subsequent amplification rounds and ensures that amplicons are derived from RNA by RT-PCR.

94. Extract the RNA as follows: To the DNase reaction, add 300 μL of H_2O, 300 μL of phenol equilibrated with 0.15 M NaOAc (pH 5.2), and 100 μL of chloroform:isoamyl alcohol (49:1).

95. Vortex the mixture to emulsify and then centrifuge at 10,000 rpm for 5 min at room temperature to separate the phases. Remove the aqueous layer to a new tube.

96. Precipitate the RNA from the aqueous layer by adding 50 μL of 3 M NaOAc (pH 5.2), 1 μL of 5 mg/mL of glycogen, and 1 mL of ethanol:isopropanol (1:1).

97. Precipitate overnight (or <2 h) at –20°C.

Amplify the Linker-ligated Tags by RT-PCR

98. Centrifuge the RNA in a microcentrifuge at 12,000 rpm for 10 min at 4°C. To determine if the precipitation was successful, use a Geiger counter to detect evidence of radioactivity in the pellet.
99. Rinse the pellet twice with 150 μL of ice-cold 75% ethanol, centrifuging the precipitate after each rinse at 12,000 rpm for 5 min at 4°C.

    This should minimize loss of the precipitate and ensure removal of salt.
100. Dry the pellet *briefly* (<2 min) in the SpeedVac. Count the RNA in a scintillation counter.
101. Resuspend the pellet in 8 μL of H_2O. This is the ligated RNA.
102. Mix the 8 μL of ligated RNA with 2 μL of DNA primer P3 (5 pmol/μL).
103. Heat the mixture for 5 min at 65°C. Then, chill on ice and centrifuge quickly to collect any droplets that may have formed by condensation.
104. To the RNA + primer mix add

dNTP (3 mM each)	3 μL
DTT (0.1 M)	1 μL
SuperScript RT buffer (5x)	4 μL
RNasin	1 μL
SuperScript III (Invitrogen 18080-044)	1 μL
Total	20 μL

105. Incubate the mixture for 30 min at 50°C and then for 5 min at 90°C. Finally, adjust the temperature to 4°C.
106. Set up the PCR by mixing

Accuprime Pfx Supermix	27 μL
P5 primer, 30 pmol/μL	0.5 μL
P3 primer, 30 pmol/μL	0.5 μL
RT reaction from Step 106	2 μL
Total	30 μL

107. Cycle as follows in a thermal cycler:

    Step 1: 95°C for 5 min

    Step 2: Cycle 25–35 times (depending on the amount of RNA): 95°C for 20 sec, 61°C for 30 sec, and then 68°C for 20 sec

    Step 3: 68°C for 5 min

    Step 4: Hold at 4°C

Purify the PCR-amplified Tag Sequences

108. Pour a 10% denaturing polyacrylamide gel. Amounts may vary depending on size. For 20 mL of the gel mix, add

Urea (Fisher U15-3)	8.4 g
40% acrylamide:bis-acrylamide (19:1)	5 mL
TBE (5x)	4 mL
H_2O	to 20 mL total

109. Immediately before pouring the gel, add 200 μL of freshly made 10% APS and 7.5 μL of TEMED.
110. Pour the gel and insert the comb. Allow the gel to polymerize.
111. Run the entire PCR on one lane of the gel. Use low-molecular-weight markers (e.g., 3 μL per lane Amplisize Molecular Ruler [Bio-Rad]). Do not heat.

112. Visualize DNA by immersing the gel for 10 min in a 10,000-fold dilution of SYBR Gold in TBE.

   In the case of the nitrocellulose band representing material migrating ~20 kDa above the free molecular mass of the protein of interest, the DNA amplified from the RNA tags will appear as a diffuse band migrating at ~80–90 nucleotides (average RNA tag size of ~50 nucleotides + 36 nucleotides for the linkers). There may be a smear above this size as well, indicating the presence of longer tags. Sharp bands as well as any obvious primer dimers should be avoided.

113. Cut out the DNA of 80–90 nucleotides or greater if present from the gel, avoiding sharp bands. Go for the heterogeneous material. Extract the DNA with the QIAEX II Kit (follow protocol for polyacrylamide gel) and resuspend the extracted DNA in 20 µL of H_2O.

   This is the amplicon library that represents the pool of RNA sequences that cross-link to the protein of interest. At this point, the library can be cloned and sequenced or sequenced directly using high-throughput sequencing methods. These methods require the addition of specific capture sequences, such as that described below for 454 sequencing.

Add 454 Capture Linkers via Second PCR Step

114. Set up the PCR by mixing

Accuprime Supermix Pfx	95 µL
P5-454 (100 µM)	1 µL
P3-454 (100 µM)	1 µL
QIAEX-II-purified RT-PCR product from Step 113	3 µL
Total	100 µL

115. Cycle as follows in a PCR machine:

   Step 1: 95°C for 5 min

   Step 2: Cycle 15–20 times (depending on the amount of RNA): 95°C for 15 sec, 60°C for 30 sec, and then 68°C for 30 sec

   Step 3: 68°C for 5 min

   Step 4: Hold at 4°C

116. Add 200 µL of phenol:chloroform (Sigma-Aldrich P3803), vortex briefly, and then centrifuge at 12,000 rpm for 5 min at room temperature.

117. Collect the aqueous fraction and add 200 µL chloroform:isoamyl alcohol (49:1). Vortex briefly and then centrifuge at 12,000 rpm for 5 min at room temperature.

118. Collect the aqueous fraction. Add 14 µL of 5 M NaCl (0.1 M final) and 500 µL of ethanol (2.5 volumes). Precipitate the DNA for 2 h (or overnight) at –20°C.

119. Separate the amplicons on a 2% NuSieve agarose and then gel-purify them. PCR products should be ~125 bp in length. Extract the DNA with the QIAEX II Kit (follow protocol for agarose gel) and resuspend it in 20 µL of H_2O.

   Small amounts of amplicons are required for the sequencing (titration runs and full runs). Consult with the sequencing facility to find out how much is required for 454 sequencing, but emphasize that the capture sequences are already present in the library of amplicons if this step has been performed!

TIPS AND TROUBLESHOOTING

Because of the many steps in this protocol, we have inserted numerous tips and troubleshooting comments just below the appropriate steps throughout. In general, however, the overall success of this procedure is likely to depend on several main factors listed below.

1. The length of time of UV exposure and the yield of cross-links, which depends on the intrinsic geometries of the population of RNA–protein complexes of interest, may vary.
2. Calibrating the trimming of the RNA ends is key to obtaining useful tags for sequencing.
3. The antibody will have characteristic properties that must be considered, especially if comparing CLIP results using different antibodies. Use of tagged proteins of interest and the same antitag antibody may circumvent some of this variation.
4. Cutting the correct region of the nitrocellulose above the unreacted protein band will contribute to the size and complexity of the recovered RNA population.

REFERENCES

Jensen KB, Darnell RB. 2008. CLIP: Crosslinking and immunoprecipitation of in vivo RNA targets of RNA-binding proteins. *Methods Mol Biol* **488:** 85–98.

Sanford JR, Wang X, Mort M, Vanduyn N, Cooper DN, Mooney SD, Edenberg HJ, Liu Y. 2009. Splicing factor SFRS1 recognizes a functionally diverse landscape of RNA transcripts. *Genome Res* **19:** 381–394.

Ule J, Jensen K, Mele A, Darnell RB. 2005. CLIP: A method for identifying protein-RNA interaction sites in living cells. *Methods* **37:** 376–386.

Xue Y, Zhou Y, Wu T, Zhu T, Ji X, Kwon YS, Zhang C, Yeo G, Black DL, Sun H, et al. 2009. Genome-wide analysis of PTB-RNA interactions reveals a strategy used by the general splicing repressor to modulate exon inclusion or skipping. *Mol Cell* **36:** 996–1006.

Yeo GW, Coufal NG, Liang TY, Peng GE, Fu XD, Gage FH. 2009. An RNA code for the FOX2 splicing regulator revealed by mapping RNA-protein interactions in stem cells. *Nat Struct Mol Biol* **16:** 130–137.

Protocol 9

Preparation of mRNA-Seq Libraries from Poly(A)$^+$ mRNA for Illumina Transcriptome High-throughput Sequencing

Many investigators who do high-throughput sequencing of mRNA (mRNA-Seq) use kits for library preparation purchased from Illumina. Recognizing that these kits are continually being updated and improved, we provide here some background information and a list of Tips and Troubleshooting for the currently available kits. For information on exact protocols for library preparation and sequencing, see http://www.illumina.com/.

Preparation of cDNA libraries from poly(A)$^+$ mRNA for high-throughput sequencing (HTS) or mRNA-Seq is becoming a routine method for gene expression profiling (Mortazavi et al. 2008; Wang et al. 2009; www.illumina.com) and studies of alternative splicing patterns (Pan et al. 2008; Sultan et al. 2008; Wang et al. 2008). There are several different approaches to generating cDNA libraries compatible with short-read Illumina HTS (Mortazavi et al. 2008; www.illumina.com). One of the best approaches, in terms of more complete representation of all regions of a given transcript, is to fragment the mRNA using metal-ion–mediated cleavage before random-primed cDNA synthesis (Mortazavi et al. 2008; Wilhelm and Landry 2009; Nagalakshmi et al. 2010; Wilhelm et al. 2010; www.illumina.com). The procedure discussed here is adapted from the Illumina mRNA-Seq protocol (Mortazavi et al. 2008; www.illumina.com) to generate short cDNA libraries from fragmented poly(A)$^+$ mRNA suitable for HTS on an Illumina Genome Analyzer. Bioinformatic analysis of the HTS data is a critical aspect of this procedure once the cDNA libraries generated by this protocol are carried out (Pepke et al. 2009). (See also Protocols 10 and 11.)

OVERVIEW

This procedure is based on the RNA shatter method (Mortazavi et al. 2008; Wilhelm and Landry 2009; www.illumina.com) using metal-ion–induced mRNA fragmentation before cDNA synthesis. The method starts with ~10 µg of total cellular RNA and uses two cycles of oligo(dT) magnetic bead purification of poly(A)$^+$ mRNA. The purified mRNA is then fragmented to ~200 nucleotides, and random-primed cDNA fragments are generated. These cDNA pools are made double-stranded by use of RNase H and *Escherichia coli* DNA polymerase. The cDNA ends are then repaired with T4 DNA polymerase, Klenow fragment DNA polymerase, and T4 polynucleotide kinase. Nontemplated A residues are added to the 3′ ends of the double-stranded cDNA and then special Illumina adapters are ligated onto the cDNA pools, PCR-amplified, and submitted to Illumina sequencing on the Illumina Genome Analyzer.

Note that the size of the fragments generated can be controlled by the length of the incubation. For instance, a 2-min incubation yields fragments of up to 700 nucleotides, a 3-min incubation gives fragments of ~100–500 nucleotides, and a 5-min incubation gives fragments of ~200 nucleotides. The fragmentation size should match the purpose of the experiment.

Notes and Considerations before Beginning

For RNA-Seq, one must consider the purpose of the experiment when deciding how to make the libraries. For example, single reads and short reads (~36 nucleotides) are typically sufficient for performing gene expression analysis. In contrast, deeper coverage (in terms of total reads) and longer,

Tips and troubleshooting notes were supplied by Brent Graveley, University of Connecticut.

paired-end reads significantly enhance the ability to detect alternative splicing events. Moreover, greater depth is necessary for discovery of rare RNA forms than for quantitation of well-expressed genes. It is also important to keep in mind that although paired-end reads definitely enhance the analysis of alternative splicing events, it complicates the downstream bioinformatics analysis because the mate pairs need to be tracked. In addition, the accuracy of using the insert size to infer splicing events between the two mate pairs depends entirely on how tightly selected the insert size. There is also a balance between the cost of generating longer or paired-end sequences and whether there is likelihood that the read can be mapped. For example, with 50-nucleotide reads, fewer will map uniquely to the human genome than to the *Drosophila* genome due to the size of the fly genome.

As a guideline, when using a 76-nucleotide read length, between 10% and 13% of the reads span a splice junction. Furthermore, ~20 million uniquely mapped reads provide sufficiently robust coverage of splice junctions to obtain statistically relevant results for well-expressed genes when using cell lines. Obviously, for complex samples such as tissues or whole *Drosophila* or *Caenorhabditis elegans* animals, more reads are necessary and it is difficult to determine exactly the depth needed.

It is important to note that libraries prepared using paired-end linkers and polymerase chain reaction (PCR) primers can be sequenced using either single or paired-end reads. As a result, paired-end libraries yield the versatility of performing either type of sequencing.

It is also advisable to use a set of spike-in RNAs (Mortazavi et al. 2008) that cover a broad range of concentrations. This can ensure the quality of the library and, in samples such as cell lines, the possibility of quantitating the number of molecules of each mRNA in a cell. These should be added to constitute between 0.1% and 1% of the library and can be added to the total RNA if they are polyadenylated or to the mRNA after the poly(A)$^+$ selection.

Finally, it is advisable to generate independent libraries from biological replicate samples. After sequencing and determining the correlation between the two samples at the gene level, it is common practice to pool the reads from the replicates, provided that they have an $r^2 > 0.9$.

One can routinely begin with only 1 µg of total RNA and get great libraries. Starting with 10 µg is ideal if one has the luxury of generating copious quantites of RNA, but it is certainly possible to start with less.

TIPS AND TROUBLESHOOTING

General

1. Optimization of each step in this procedure is critical for full transcriptome representation. This has been done for genomic DNA sequencing at the Sanger Centre, Cambridge, UK and the reader is referred to this protocol for additional reagent troubleshooting of individual steps in library preparation (Quail et al. 2008, 2009).

Adapter Ligation

2. Residual ethanol in the samples is a problem at this step. If there is residual ethanol, the samples will float out of the gel and cause the loss of all of the DNA. To solve this problem, open the caps of the column tubes before the elution step and let them air-dry for ~5 min. Then, elute with 10 µL of QIAGEN Elution Buffer.

First Gel Purification

3. Add a large amount of loading dye to each sample to prevent them from floating away. Generally, 5 µL of 6x loading dye works well. Dilute the 6x loading dye with clear loading dye to decrease the amount of dye in the gel.

4. When loading the sample, leave a space, load the ladder, leave a space, etc., so that there is a ladder on each side of the sample.
5. For both gel purification steps, it is important to use as long a gel as possible and run the gel until the 100-bp band is at the bottom. The longer the gel, the tighter the insert size region, which is particularly important for paired-end applications. A tight insert size region will provide greater accuracy in inferring splicing of exons located between the mate pair reads, whereas a broad insert size will make this difficult.
6. Be sure to add isopropanol in equal volume to the gel slice to the QIAGEN QG solution (comes with QIAGEN columns) after the gel slice is dissolved (see QIAGEN protocol for small [<500 bp] fragments.

PCR Enrichment

7. Again, be careful with the residual ethanol (see above), which will cause the sample to float out of the well when the gel is loaded below.
8. It is common to skip the column purification step here and instead load the 50 µL of PCRs directly onto agarose gels. It seems redundant to remove the polymerase and dNTPs before gel-purifying.

Second Gel Purification

9. The size of the PCR fragment is ~100–125 bp larger than the insert size of the library (depending on exactly which linkers and PCR primers are used). Thus, be sure to include the size of the linkers when determining the size fragment to cut out of the gel. It may be better to use MinElute columns and elute the library in 10 µL of QIAGEN Elution buffer to give a higher library cDNA concentration, which generally gives better NanoDrop readings.
10. It is also possible to also use the QIAGEN MinElute gel extraction kit. Regardless of using Qiaquick or MinElute columns, it is useful to elute with only 10 µL of QIAGEN Elution Buffer. Again, add isopropanol to the QIAGEN QG solution for a small fragment size. The smaller the eluted volume, the more concentrated the sample, and it will be easier and accurate to quantify using the NanoDrop spectrophotometer.
11. Typically, we subject the eluted library DNA to NanoDrop spectrophotometry. NanoDrop readings below ~10–20 ng/µL are very unreliable, and thus, libraries may need to be sequenced more than once.

REFERENCES

Mortazavi A, Williams BA, McCue K, Schaeffer L, Wold B. 2008. Mapping and quantifying mammalian transcriptomes by RNA-Seq. *Nat Methods* **5:** 621–628.

Nagalakshmi U, Waern K, Snyder M. 2010. RNA-Seq: A method for comprehensive transcriptome analysis. *Curr Protoc Mol Biol* Chapter 4: Unit 4.11.1–13.

Pan Q, Shai O, Lee LJ, Frey BJ, Blencowe BJ. 2008. Deep surveying of alternative splicing complexity in the human transcriptome by high-throughput sequencing. *Nat Genet* **40:** 1413–1415.

Pepke S, Wold B, Mortazavi A. 2009. Computation for ChIP-seq and RNA-seq studies. *Nat Methods* **6:** S22–S32.

Quail MA, Kozarewa I, Smith F, Scally A, Stephens PJ, Durbin R, Swerdlow H, Turner DJ. 2008. A large genome center's improvements to the Illumina sequencing system. *Nat Methods* **5:** 1005–1010.

Quail MA, Swerdlow H, Turner DJ. 2009. Improved protocols for the Illumina Genome Analyzer sequencing system. *Curr Protoc Hum Genet* Chapter 18: Unit 18.2.

Sultan M, Schulz MH, Richard H, Magen A, Klingenhoff A, Scherf M, Seifert M, Borodina T, Soldatov A, Parkhomchuk D, et al. 2008. A global view of gene activity and alternative splicing by deep sequencing of the human transcriptome. *Science* **321:** 956–960.

Wang Z, Gerstein M, Snyder M. 2009. RNA-Seq: A revolutionary tool for transcriptomics. *Nat Rev Genet* **10:** 57–63.
Wang ET, Sandberg R, Luo S, Khrebtukova I, Zhang L, Mayr C, Kingsmore SF, Schroth GP, Burge CB. 2008. Alternative isoform regulation in human tissue transcriptomes. *Nature* **456:** 470–476.
Wilhelm BT, Landry JR. 2009. RNA-Seq-quantitative measurement of expression through massively parallel RNA-sequencing. *Methods* **48:** 249–257.
Wilhelm BT, Marguerat S, Goodhead I, Bähler J. 2010. Defining transcribed regions using RNA-seq. *Nat Protoc* **5:** 255–266.

WEB RESOURCE

http://www.illumina.com/ Illumina, Inc. home page

BIOINFORMATIC ANALYSIS

With the development of whole-genome-level methods for analyzing gene expression, biologists have recently enjoyed coming to grips with spectacular amounts of data. Data, no matter the quantity, can be of good quality or bad quality, and determining the difference will always be key to making sound conclusions. In the next two sections, we discuss analysis of microarray data (Protocol 11) and high-throughput sequencing data (Protocol 12). An in-depth discussion of the strengths and weaknesses of the many methods available is beyond the scope of this book. Thus, the protocols we provide here are really higher-level "metaprotocols"—sets of general practices and pointers to available programs that facilitate analysis.

Biologists without experience in collaborating with statisticians, computer scientists, and other bioinformaticians may find making the transition to using genome-level methods a daunting prospect. Many of us have forgotten what little training in statistics we might have had because of the "either you see a band or you don't" nature of molecular biology data during the last 30 years. Moving past one-gene-at-a-time thinking requires statistics, necessitating both mid-career retraining for faculty and modification of current training practices for students. The goal of these protocols is to give biologists a familiarity with the language and strategies of bioinformatics. Thus armed, they can better help their statistical collaborators to use trustworthy bioinformatics rather than "bioinfauxmatics," or worse, "biodisinformatics." It is worth remembering that the results from computational analyses can trigger expensive, labor-intensive wet lab validation and hypothesis testing. Spending extra attention and time getting the analysis right can save time and other resources during the next cycle of biological experimentation.

It is not always practical or necessary for the biologist to understand every statistical method in fundamental theoretical mathematical detail. Some of these can be quite advanced and complex, whereas others are straightforward. But it can be important to understand what is being filtered out in the process. Usually, a statistical test at a filtering or other decision point is set to exclude the hypothesis that the observed variation in the data is due to chance sampling variation (the null hypothesis). Excluding the null hypothesis does not explain which of the non-null hypotheses might be correct. Even so, preprocessing and filtering the data, as well as the initial experimental setup itself, can contribute acquisition bias that dooms the chances of properly disproving a correct null hypothesis. Once this happens, confirmation bias can take over, leading to acceptance of an illusory hypothesis and a trip down the garden path. It is up to the biologist to consider the experimental biases going in and the interpretational biases coming out, and to help bioinformatics collaborators apply the statistical tests in a principled fashion in between.

Finally, it must be emphasized that statistical significance does not translate directly to biological significance. Scientists are attracted by large changes in response to an independent variable. Although these easily pass statistical tests, many are peripheral to biological function. Some very small-fold changes (e.g., a 1.1-fold increase) may pass statistical tests with ease, but their effect on the biology may be miniscule. On the other hand, changes of twofold can have powerful biological consequences (e.g., as in dosage compensation), but many of these may not be labeled significant due to sampling error or biological noise that prevents a robust formal rejection of the null hypothesis. Computational collaborators may miss such subtleties, especially if their training predisposes them to think of biological systems as "engineered." But biological systems are the products of evolution, not design. They are incompletely "optimized" through a process of mutation, reproduction, and selection, with a nearly frozen set of historical constraints and a long experience with wobbly selection pressures and waves of mutation, as well as plenty of stochasticity and luck thrown into the mix. A better understanding of biological systems will require both an appreciation for their organic and emergent properties and an adherence to the principles of reducing bias and understanding chance variation in the data. Hopefully, such an understanding can arise naturally from the friendly mixing of the disciplines needed for the application of these analysis methods.

Protocol 10

Methods for Processing Microarray Data

As described in the introduction of this chapter, there are several platforms for acquiring microarray data. Different methods are needed depending on whether the single-label (intensity based) or double-label (ratio based) strategy is used. There are several steps in the analysis of microarray experiments. Because the utility of the results depends directly on experimental design, the first critical step is to design the experiment so that it can be properly analyzed and interpreted. What is the biological question? What is the best way to do the experiment? How many replicates will be required to obtain the desired statistical resolution? Next, the samples must be prepared, pass quality controls for integrity and representation, and be hybridized and scanned (Protocols 3 and 4). Another quality-control step is necessary to begin the analysis: Slides with defects, missing data, high background, or weak signal must be rejected. Data from individual slides must be normalized and combined so that the data are as free of systematic bias as possible. The third phase is to apply statistical filters and tests to the data to determine genes (1) expressed above background, (2) whose expression level changes in different samples, and (3) whose RNA-processing patterns or protein associations change. Next, a subset of the data should be validated by an alternative method, such as reverse transcription–polymerase chain reaction (RT-PCR). Provided that this endorses the general conclusions of the array analysis, gene sets whose expression, splicing, polyadenylation, protein binding, and the like change in different samples can be classified with respect to function, sequence motif properties, as well as other categories to extract hypotheses for their biological roles and regulatory logic.

MATERIALS

Equipment

Account permissions to install open-source software (e.g., the R statistics package)
Desktop computer
Electricity
Good chair
Internet service
Pencils (with erasers)
Scanned array files in appropriate file format(s)
Scratch paper
Software for web browsing
Unix-based server

METHOD

Experimental Design

Before beginning an array analysis, it is advisable to determine whether the experiment in mind will be adequate to produce the desired result. The following questions represent many, but not necessarily all, that need to be answered. If the data are to be published or archived, it is important to comply with the "Minimum Information About a Microarrray Experiment" (MIAME) conventions (Brazma et al. 2001; Rayner et al. 2006). There are also open-source resources for array analysis that are well documented and widely used, in particular the TM4 suite (Saeed et al. 2006) and Bioconductor (Gentleman et al. 2004; Reimers and Carey 2006).

1. How much biological starting material will be needed? Depending on method, this will vary from a few tens of nanograms to tens of micrograms per array hybridization. If amplification is to be used to increase the amount of material to be hybridized, all samples to be compared must be handled the same way. For example, if an RNA immunoprecipitation is performed, there will be far less RNA in the immunoprecipitation sample than in the input sample. To label the two samples for comparison, it is better to start with equivalent amounts of RNA from each and amplify them the same way than to amplify the immunoprecipitated RNA and compare it to unamplified input RNA. Although many amplification protocols claim to provide uniform amplification of every sequence in the mixture, this claim cannot be taken seriously, because it is usually based on very sparse testing and a great deal of bravado.
2. What are the proper controls for statistical comparison? Binary comparisons between two sets of samples that differ by one variable are the easiest for most investigators to design, think about, and analyze, but there are many other considerations (Churchill 2002). In a simple example where a treatment is compared to an untreated control, or a mutant is compared to wild type, the comparison is obvious. In the case where multiple treatments or mutants will be tested, they should all be compared to a common control. For example, treatment of a cell culture with two different drugs each dissolved in dimethylsulfoxide (DMSO) could be compared to cultures treated with DMSO alone for the same amount of time. Choice of an appropriate reference sample, especially with respect to time series experiments when expression of many genes is changing dramatically, can influence the quality of the data (Quackenbush 2002). Once the experimental and control samples are identified, they can be compared by either single-label arrays using statistical comparisons or directly by the double-label (ratio based) approach.
3. How will arrays be normalized so that they can be compared to one another? It is commonly the case that arrays vary in their overall intensities after hybridization for a variety of difficult to control reasons. Normalization allows simple intensities and intensity ratios (in the case of double-label experiments) obtained on different arrays to be compared (Quackenbush 2002). If the number of genes or events to be tested is large (e.g., all genes in the genome) and only a small fraction of all genes is expected to change between the samples and controls, then it is possible to normalize over the entire array. If there are few genes represented on the array or a large fraction of them are expected to change with the treatment, other methods will be needed. One possibility is to use a set of genes whose expression is known not to change in the experiment, and use those genes for normalization. Another possibility is to use "spike-in" control RNAs that are added to the samples in the same known concentrations to normalize. This latter approach requires that the arrays be manufactured with probes for the spike-in control sequences.
4. How many replicates will be needed to identify differences in gene expression or RNA processing with statistical confidence? Answering this question requires an understanding of the variation intrinsic to the biological system and the methods used (Churchill 2002). The ability to detect small statistical differences between two broad distributions requires more samples. Thus, the more intrinsic the variation in the system, the more replicate experiments needed to identify accurately experimentally induced changes. A power analysis method for microarrays has been developed and can be used to analyze existing data sets or to upload and analyze unpublished data (Page et al. 2006).

   It is important to understand the sources of variation in the experiment and to address them in one of two ways: Control variation that can be controlled and randomize variation that cannot be controlled. To do this, it is useful to think about different kinds of "replicate" experiments, some of which capture only part of the variation in an experimental protocol. For example, a "technical replicate" would involve taking the very same RNA sample, splitting it into three aliquots, labeling each for array analysis, and applying each to one of three separate arrays. Variation in such an experiment should arise from the labeling, hybridization, and array manufacture and handling properties of the process. Variation in this part of the experiment should be very low in any case, indicating that it is well controlled (hopefully).

Another set of replicates might involve splitting a cell culture into three aliquots, extracting RNA separately, and then analyzing each RNA sample. These replicates will capture variability in the RNA extraction steps, plus that captured in the previous example. Because RNA extraction should be well controlled, these kinds of replicates are not as useful either. If, on the other hand, separate cell cultures were grown and then each separately extracted and tested, the experiment would detect variability in cell growth, which might be harder to control. Although it is tempting to avoid variability that cannot be controlled in experimental design to reduce noise or because it requires more effort, the goal of the statistical analysis should be to reveal changes that occur in response to the experimental manipulation, against the background of uncontrolled variation in cell culture conditions and RNA preparation, labeling, and hybridization. Some factors are easier to control than others (e.g., genetic background, growth medium, cell handling, sample processing), and still many others are unknown and can only be captured by replication.

Array Quality Control, Data Extraction, and Normalization

5. Because arrays are subject to manufacturing and handling defects that may only become apparent after hybrization, it is important to visually inspect each scan for defects, including scratches, blurred-together spots, and other defects. Some hybridizations fail as well, giving high background over some or all of the array, blank spots (usually due to bubbles), or generally low signal. Such arrays should be eliminated from analysis. Because this happens at a reasonable rate, it is useful to plan to be ready to do 12–13 arrays for every 10 needed, so that failed experiments can be repeated.
6. Data extraction from arrays that pass this simple check is highly automated in the case of Affymetrix arrays (see Protocol 4) and is described for spotted arrays in Protocol 3. In these cases, in general, the intensity values of each pixel in the array feature corresponding to a spot or cell are identified in the image file (e.g., a ".tiff" or ".tif" image file, or Affymetrix ".DAT" file), as are background and control feature intensities, and then collected to produce a table of data (e.g., GenePix ".gpr" file or Affymetrix ".CEL" file).

Normalization and Scaling

7. The rationale and methods used to perform the key normalization steps in typical experiments have been reviewed previously (Quackenbush 2002). A main concern in comparing different arrays to one another is that differences in overall intensities can arise from variation in sample preparation, labeling, hybridization volume, temperature fluctuations, and so forth. Normalization and scaling processes bring a group of arrays into alignment with one another with respect to the magnitude of intensities and their distributions (Quackenbush 2002). In the case of double-labeled experiments, the ratios of the two intensity channels are also normalized by location across the slide, usually using the print tip origin group to approximate location. This removes common artifacts of hybridization in which the background red/green ratio changes across the slide. Second, the intensity of the signals appears to have an influence on the ratio, necessitating Lowess normalization in which intensity-based adjustments in the normalization factor are made (Quackenbush 2002). The Bioconductor project (http://bioconductor. org/) is the best place to learn about and download software for statistical analysis of data from microarrays (Gentleman et al. 2004; Reimers and Carey 2006).

   Affymetrix arrays and other single-label array methods do not have direct ratio data requiring normalization of intensity ratios, but scaling of the data and reshaping the distribution of intensities (also called normalization) are necessary to allow statistical comparison of the arrays. There are many ways to do this, including the commonly used quantile normalization, which adjusts the intensity distributions of two or more arrays so that they appear to be identical. Intensities tend to be higher for GC-rich oligonucleotide probe sequences, and this effect can be reduced by correction factors applied over probes with respect to their GC contents. Again, numerous approaches and software packages are available at the Bioconductor web site.

Deciding Whether a Transcript Is Detected and Identifying Changes in Transcript Levels

8. Following normalization and scaling, a set of matched similarly treated array data files are ready for analysis. Here, there may be some need to consider the relationship of the probes on the array to the genes in the genome. For many commercial arrays or general-use arrays for gene expression, this means simply assigning each array probe to a gene. For tiling arrays, probes are defined by their genome coordinates, and some additional layer of annotation is necessary to know to which gene (if any) a probe belongs. If a gene expression level is desired from a tiling array, the method must include a way of pooling or summarizing the signals of those probes into an expression signal. Finally, for more sophisticated analysis such as is necessary for splicing or polyadenlation changes, the probes might need to be annotated with respect to alternative mRNA forms in addition to belonging to a gene. Alternatively, probes that represent short regions over which mutually exclusive splicing choices differ should be related to one another so that changes can be detected. This annotation-level problem is addressed by summarizing probe signals together as genes, alternative mRNA forms, or alternative splice sites and then creating derived files of the summarized data (if needed) for statistical comparison.
9. The next step is to decide whether an RNA transcript (or a gene) is being detected, i.e., is the signal associated with this gene above background with the likelihood that the signal detected represents expression? There are numerous methods for setting background levels. A common approach is to call any signal that is more than two standard deviations above mean background "expressed." For spotted arrays, background can be estimated using intensity of pixels near but not within spots. Alternatively, the intensity of pixels in spots designed not to hybridize any sequence found in the sample can be taken as background. For high-density arrays that may possess many spots that do not hybridize anything in the sample, the bottom 1% of the distribution of all probe signals can be taken as the intensity level below which the signal is unlikely to represent true detection of expression. Most of the array preprocessing packages have methods to determine expression (Gentleman et al. 2004).
10. Once values significantly above background have been obtained for each array, and the arrays have been grouped for comparison (e.g., treatment vs. control), statistical tests can be applied to determine whether the values in one group are likely to come from a different distribution than the corresponding values in the comparison group. Here, the impact of variability in the samples and the shape of the distributions as represented by the replicate samples come into play. There are numerous approaches to this question of determining whether a change has occurred. One commonly used method is SAM (significance analysis of microarrays), which identifies where sampled values differ significantly and estimates false discovery rates (Tusher et al. 2001; Zhang 2007). Many other statistical methods have also been applied. Search the Bioconductor site for tested methods and packages for running these analyses.

    After this step is performed, a list of gene expression or RNA-processing changes by gene or genomic location, along with the magnitude of each change and some sort of measure of the likelihood that the change is not due to chance, are obtained. These represent the statistically valid set of changes that might be due to experimental manipulation or other feature that distinguishes the groups biologically. It now becomes important to determine how these data stand up to an orthogonal method of measurement.

Validation

11. Microarray data are useful for general characterization of larger groups of genes, but the data for any individual gene picked at random are less reliable. Thus, there are two philosophical reasons for performing validation experiments. Validation is useful to biologists to find out if a favorite gene that came up in the list is truly changed in the experiment, because they are searching for specific biological insights rather than higher-level regulatory logic or general regulatory trends. On the other hand, bioinformaticists may need to know how many of the results are true and to understand where in the ranked list of best-scoring genes or events the predictions become incorrect to identify the scores that provide the best receiver operator characteristics (ROC) curve. This will be important because bioinformaticists require a large high-confidence set of genes for further analysis. So a compromise validation list must be developed for maximum progress and happiness of both collaborators. Once settled, quantitative RT-PCR or semi-quantitative RT-PCR or other nonmicroarray detection and measurement methods should be used to determine how well the array measurements predict the true changes as measured by the more labor-intensive approach.

Classification of Changes and Their Associated Biology

12. After validation, a high-confidence list of gene expression changes or RNA-processing changes is now in hand. It is time to search for biological function and genetic sequence associations (to name two common examples) that appear enriched in the high-confidence group relative to a randomly selected comparison groups ("background sets"), either the genome in its entirety or the genes expressed in the cell type under study, or some other group of genes selected under a different experiment. In some cases, the list of genes may provide obvious clues to the biology, which will depend on the mind-set and the experience of the investigator who is looking at the list. It is advisable to ask the question in a more open way using Gene Ontology (GO) tools (Gene Ontology Consortium 2006) available at Bioconductor. These tools have data that link gene names to terms for cellular components, biological processes, and molecular functions that have been assigned to gene products by various methods. GO has a structured language and seeks to formalize the meaning of terms and connections. In a statistical fashion, the list of genes obtained in the experiment is used as the input, and enrichment of GO terms associated with that group of genes is evaluated compared to the distribution of GO terms among the genes in the background set. If the biological treatment in the experiment impacts expression of a functionally distinct group of genes in the cell, then terms associated with this function should appear to be enriched.

    Sequences in promoters, near splice sites, or within untranslated mRNA regions may also be enriched in the selected gene set. Finding sequence motifs is more complex, and there are numerous approaches to this question. A good starting point is the "GREAT" program (McLean et al. 2010), which in addition to identifying GO term enrichment, allows identification of conserved promoter elements associated with an input gene list. These and other downstream analysis methods aid in hypothesis development for further experimentation. In the end, the arrays are just the beginning, and it is likely that many follow-up experiments will be necessary to close the loop and obtain a deeper understanding of the biological impact of the original experiment.

Visualization

13. Data that arise from genomic-level experiments usually exist in large tables that are not easy for humans to grasp and digest. A large variety of visualization tools are available at Bioconductor (http://bioconductor.org/). Also useful is the UCSC Genome Browser (http://genome.ucsc.edu/);

see also Mangan et al. (2008) and Zweig et al. (2008). The Galaxy website (http://galaxyproject.org/) is very useful to biologists who may have limited experience writing scripts or ensuring file format integrity (Blankenberg et al. 2010). This site is especially useful for creating track files for the UCSC Genome Browser. Finally, the display of results in color figures can create concern, especially considering the frequency of red–green color blindness in the human population. An excellent discussion of this point may be found at http://jfly.iam.u-tokyo.ac.jp/color/.

TIPS AND TROUBLESHOOTING

1. Problems may arise as a consequence of "interdisciplinary strain," in which collaborators struggle to communicate their needs and perspectives on the work across the gulf that separates them. In such cases, talking is best, with the recognition that goals, biases about what is important, and reward systems differ in different disciplines and that it is best if all of these can be recognized and respected as valid concerns.
2. File formats may become corrupt or incompatible. Check Galaxy for file conversion and quality-checking methods.
3. Dedicated computer time and space may become a problem. It is important for the project to have adequate resources for the computational parts.
4. Quality control must be maintained at every step from RNA isolation through statistical evaluation, preferably by a small number of investigators who apply consistent standards. If RNA quality is variable or poor, little in the subsequent analysis will be able to improve things. Likewise, errors in analysis can crop up, so spot validation and visual inspection of results or other more elaborate "sanity checks" may be needed to confirm that the analysis is proceeding well.

REFERENCES

Blankenberg D, Von Kuster G, Coraor N, Ananda G, Lazarus R, Mangan M, Nekrutenko A, Taylor J. 2010. Galaxy: A web-based genome analysis tool for experimentalists. *Curr Protoc Mol Biol* **2010:** Unit 19.10.1–21.

Brazma A, Hingamp P, Quackenbush J, Sherlock G, Spellman P, Stoeckert C, Aach J, Ansorge W, Ball CA, Causton HC, et al. 2001. Minimum information about a microarray experiment (MIAME) toward standards for microarray data. *Nat Genet* **29:** 365–371.

Churchill GA. 2002. Fundamentals of experimental design for cDNA microarrays. *Nat Genet* (suppl.) **32:** 490–495.

Gene Ontology Consortium. 2006. The Gene Ontology (GO) project in 2006. *Nucleic Acids Res* **34** (Database issue): D322–D326.

Gentleman RC, Carey VJ, Bates DM, Bolstad B, Dettling M, Dudoit S, Ellis B, Gautier L, Ge Y, Gentry J, et al. 2004. Bioconductor: Open software development for computational biology and bioinformatics. *Genome Biol* **5:** R80.

Mangan ME, Williams JM, Lathe SM, Karolchik D, Lathe WC III. 2008. UCSC genome browser: Deep support for molecular biomedical research. *Biotechnol Annu Rev* **14:** 63–108.

McLean CY, Bristor D, Hiller M, Clarke SL, Schaar BT, Lowe CB, Wenger AM, Bejerano G. 2010. GREAT improves functional interpretation of *cis*-regulatory regions. *Nat Biotechnol* **28:** 495–501.

Page GP, Edwards JW, Gadbury GL, Yelisetti P, Wang J, Trivedi P, Allison DB. 2006. The PowerAtlas: A power and sample size atlas for microarray experimental design and research. *BMC Bioinformatics* **7:** 84.

Quackenbush J. 2002. Microarray data normalization and transformation. *Nat Genet* (suppl.) **32:** 496–501.

Rayner TF, Rocca-Serra P, Spellman PT, Causton HC, Farne A, Holloway E, Irizarry RA, Liu J, Maier DS, Miller M, et al. 2006. A simple spreadsheet-based, MIAME-supportive format for microarray data: MAGE-TAB. *BMC Bioinformatics* **7:** 489.

Reimers M, Carey VJ. 2006. Bioconductor: An open source framework for bioinformatics and computational biology. *Methods Enzymol* **411:** 119–134.

Saeed AI, Bhagabati NK, Braisted JC, Liang W, Sharov V, Howe EA, Li J, Thiagarajan M, White JA, Quackenbush J. 2006. TM4 microarray software suite. *Methods Enzymol* **411:** 134–193.

Tusher VG, Tibshirani R, Chu G. 2001. Significance analysis of microarrays applied to the ionizing radiation response. *Proc Natl Acad Sci* **98:** 5116–5121.

Zhang S. 2007. A comprehensive evaluation of SAM, the SAM R-package and a simple modification to improve its performance. *BMC Bioinformatics* **8:** 230.

Zweig AS, Karolchik D, Kuhn RM, Haussler D, Kent WJ. 2008. UCSC genome browser tutorial. *Genomics* **92:** 75–84.

WEB RESOURCES

http://bioconductor.org/ Bioconductor, an open-source software for bioinformatics.
http://galaxyproject.org/ Galaxy, an easy-to-use, open-source, scalable framework for tool and data integration
http://genome.ucsc.edu/ UCSC Genome Browser
http://jfly.iam.u-tokyo.ac.jp/color/ Color Universal Design (CUD)

Protocol 11

Methods for Processing High-throughput Sequencing Data

High-throughput RNA sequencing ("RNA-Seq") methods are gaining rapid application to many experimental situations in which an RNA population must be characterized. Whether the goal is capturing transcriptome complexity from a tissue or identifying small fragments of RNA cross-linked to a protein of interest, conversion of the RNA to cDNA followed by direct sequencing using the latest methods is a developing practice, with new technical modifications and applications appearing every day. Even more rapid is the development and improvement of methods for analysis of the very large amounts of data that arrive at the end of an RNA-Seq experiment. This metaprotocol is designed to review and emphasize a pathway of analysis that is likely to be successful, with the recognition that better methods are right around the corner.

As with the analysis of microarray experiments, the steps in preparing for and thinking about an RNA-Seq experiment require thought and planning, especially because the expense in time and materials is currently higher and the protocols are far less routine than those used for microarrays. As always, good experimental design will make analysis and interpretation easier. Having a clear biological question, an idea about the best way to do the experiment, and an understanding of the number of replicates needed will make the entire process more satisfying.

Once an appropriate number of high-quality samples are prepared, libraries are prepared by a variety of methods (e.g., Protocols 5, 6 and 10, as well as others [Nagalakshmi et al. 2010], including those provided by commercial kit suppliers). Most researchers have sequencing done at a core facility where they send RNA samples for library construction, quality control, and sequencing and receive in return a large amount of data (on the order of several gigabytes delivered either by file transfer or on a portable disk drive), per experiment. Some core facilities will also perform bioinformatics analysis; however, at some point, the biologist will be handed a large amount of data that must be stored, because most core facilities lack the disk space to store data that they produce for very long. At that time, it will be important for the biologist to understand what exactly has been delivered (raw reads? mapped reads? filtered reads?) and how they have been processed and to determine what sort of additional manipulations will be needed to extract the information necessary to draw conclusions from the experiment.

Once reads are mapped to the genome, depending on the research goals, transcript (isoform) assembly can lead to discovery of new genes or discovery of new RNA isoforms of known genes. Reads can be assigned to genes (or RNA isoforms) and counted to measure expression levels in each sample and, more importantly, changes in expression level, promoter usage, splicing, and polyadenylation between samples. Unlike analog microarray signals, sequence read data are digital and subject to analysis using event-based statistical methods. As for any highly parallel analysis, a subset of the data including key findings should be validated by alternative methods. Once the analysis is validated, gene sets whose expression, splicing, polyadenylation, protein binding, or other properties of interest can be tested for association with genomic data and classified with respect to function, sequence motif properties, and other features to extract conclusions and develop refined hypotheses.

We note that both array-based and HTS analyses generate incredibly large data sets in that identification of considerations regarding reproducibility, validation, visualization, and interpretation arise from each approach. Accordingly, treatment of these issues is reproduced verbatim in Protocols 10 and 11.

MATERIALS

Equipment

Account permissions to install open source software
Desktop computer
Electricity
Good chair
Internet service
Pencils (with erasers)
Scratch paper
Sequence reads in appropriate file format(s)
Several Terabytes of disk storage space (or more)
Software for web browsing
Unix-based server

METHOD

Experimental Design

Before applying high-throughput RNA sequencing methods, it is important to understand whether the method is appropriate to the cost and scale of the experiment. The outline below may serve as a guide to these considerations. If the data are to be published or archived, it is important to review the requirements for submission to the Gene Expression Omnibus (Barrett et al. 2009). There is a rapidly growing collection of open-source resources for analyzing high-throughput sequencing data (Bateman and Quackenbush 2009). As for other bioinformatics methods, many of these are downloadable from the web resources of Bioconductor (Gentleman et al. 2004; Reimers and Carey 2006).

1. How much starting material will be needed to obtain sufficient reads? RNA sequencing methods can be used to detect RNAs in very small highly purified pools of RNA, such as those cross-linked to a protein of interest in a "CLIP-Seq" experiment (Protocol 10). For transcriptome studies, a few hundred nanograms of oligo(dT)-selected RNA (Chapter 2, Protocol 10) can serve as the starting material. Shotgun approaches can proceed with as little as 10 ng, but beware of contaminants such as mycoplasma. Thus, only small amounts of input RNA are required to generate millions of reads. Unlike microarrays, which have specific probes that capture a target sequence with high specificity in the presence of a large excess of other sequences in the sample, RNA sequencing captures sequences mostly with respect to their presence in the sample. A total RNA sample will provide rRNA reads at higher than a 90% rate, leaving some 10% of the reads to be split among the mRNAs and other RNAs in the sample, thus reducing the sensitivity of detection and the accuracy of quantification in particular for lower-abundance transcripts. To improve collection of reads for low-abundance mRNAs, it is advisable to reduce the amount of rRNA in the sample by using oligo(dT) selection (Chapter 2, Protocol 10) or by oligonucleotide selection methods (Chapter 2, Protocol 9). Despite these and more extreme efforts, including sequential oligo(dT) selections or rRNA depletions, large amounts of rRNA will remain in the starting RNA and reduce the numbers of useful reads.
2. How will the library preparation method influence the data?
   i. ***Priming or Fragmentation?*** A question that influences the outcome has to do with the method of library preparation and the different biases that this choice can create. Many transcriptome libraries are prepared from cDNA that is primed with random hexamers. This can create two kinds of biases in the read distribution. One is a consequence of the

fact that cDNA is synthesized in the 5′ to 3′ direction, leading to higher representation of reads at the 3′ end of the transcript than at the 5′ end (Wilhelm and Landry 2009; Wilhelm et al. 2010). A second bias that has been noted appears to be caused by depletion of some primers in the hexamer mix probably due to noncorrespondence between the representation of hexamers in the synthetic primer mix as compared to the representation of six nucleotide primer-binding sites in natural RNA. A method to address this bias has been developed (Armour et al. 2009).

Rather than priming RNA directly to make cDNA, some library preparation protocols fragment the RNA (by a variety of enzymatic or chemical methods) and then add RNA linkers directly to the ends of the RNA fragments using RNA ligase. Depending on the method used, this may require removal of phosphates from the 3′ end and the addition of phosphate to the 5′ end to achieve ligation after fragmentation. These manipulations can create biases in the library due to variation in efficiency of RNA strand cleavage at different sites and efficiencies of phosphatase, kinase, and RNA ligase for different RNA ends. These biases have not been studied systematically and are likely reduced (except for fragmentation bias) by ensuring that the end-treatment reactions go to completion. Library preparation can be modified to capture transcript ends selectively, in order to capture reads that represent polyadenylated 3′ ends (Yoon and Brem 2010) or capped 5′ ends (Fejes-Toth et al. 2009). By using structure-sensitive nucleases for fragmentation, it may be possible to map RNA structure in a high-throughput fashion using read ends to identify sites of nuclease sensitivity (J Underwood, pers. comm.).

ii. ***Amplification.*** Library preparation requires amplification of an originating set of RNA-derived cDNA fragments by PCR. This can create bias in the library if some fragments amplify more efficiently than others. This problem may be more severe for longer-fragment libraries. Size selection before amplification and a check that the amplified product remains distributed around the original selected size help to ensure that the population of fragments is efficiently replicated; however, it does not prevent loss or overamplification of subsets of product. Overamplified products can be identified at the mapping stage by the overrepresentation of fragments with identical ends, and these can be filtered out (Pepke et al. 2009). Unfortunately, lost fragments will not appear in the data, but their existence should remain somewhere in the back of the investigator's mind.

iii. ***Directionality (Strand Selection).*** Because RNA transcription is directional, it is important to know the genomic strand of origin of the RNA sequence that has been captured. A number of strand selection strategies are available. These take advantage either of the specific ligation of 3′ blocked linkers to the 3′ ends of RNA fragments or of the 5′ to 3′ directionality of reverse transcription to preserve information on the polarity of the original RNA fragment (e.g., see Ingolia et al. 2009; Yoon and Brem 2010).

iv. ***Paired Ends.*** One option available under most sequencing protocols is to obtain "paired-end" reads. This requires that the library be constructed with the appropriate linkers and involves capturing read data from individual beads or amplification centers by first using a sequencing primer from one end of each fragment and then clearing and regenerating single-stranded DNA and sequencing a second time with a primer from the other end of the fragment. Thus, two sequence reads, one on either strand, can be obtained from an individual seed molecule in the library. Coupled with an idea of how long the average fragment is in the library (obtained during amplification while preparing the library), paired-end data can greatly improve mapping, especially if one end of the fragment lies in a repeated region or the ends of the fragments lie in two different exons of a spliced RNA. Mapping methods are available for capturing splicing in paired-end RNA sequencing data (Au et al. 2010; Hu et al. 2010; Trapnell et al. 2010). Core facilities will usually offer paired-end sequencing; however, due to the extra reagents needed for the second round of sequencing, this option will be more expensive (although it should not be twice as expensive).

v. ***Read Length.*** In the early days of high-throughput sequencing, read lengths were short, offering perhaps 25–32 high-confidence base calls with which to determine the genomic origin of the sequence. In even moderately repeated regions of the genome, including gene families and dispersed repeats, the short length of reads meant that many could not be mapped to a single region. This is even more complicated than it might first seem, because there is some error rate in the reads that confounds mapping. In addition, there are differences in the "mappability" of genomic sites due to their proximity in sequence space to other genomic sequences, as well as the existence of single-nucleotide polymorphisms in populations relative to the reference genome used for mapping. The bottom line is that all of these problems fall away rapidly as read length increases; in particular, the identification of splice junction reads improves greatly even in the absence of paired-end data (Au et al. 2010; Hu et al. 2010; Trapnell et al. 2010). Most core facilities will offer a variety of read-length options under different platforms, and these have corresponding cost increases due to the reagents needed for additional sequencing cycles. Thus, a decision with respect to the need for long reads and the cost must be made.

vi. ***Bar Coding.*** The bar-coding option allows multiple libraries to be sequenced in the same sequencing lane or run as a mixture. During individual library construction, a distinct linker sequence is attached to each library. These "bar codes" constitute a very few nucleotides that can be read with high confidence early in the read such that the library origin of any individual read is immediately identifiable. This allows cost-effective sequencing of multiple libraries at once while reducing the depth at which any of the libraries can be read (see below). When multiple samples do not need to be read at great depth, bar coding is an extremely useful method. Most cores offer this service and most of the commercial kits have bar-coded linker sets that can be used in their protocols.

3. How many reads are needed to determine transcript structure with confidence? Like genomic tiling arrays, an RNA-Seq experiment can capture evidence for previously unannotated genes and RNA isoforms. If this is a goal, a certain number of reads, but not many sample replicates, will be needed to convince, because a clear qualitative comparison of transcript sequence content to the genome is required rather than quantitative comparison between groups of samples. The utility of RNA sequencing to uncover new transcripts is well documented (Mortazavi et al. 2008; Pepke et al. 2009; Ramskold et al. 2009; Wilhelm and Landry 2009; Hu et al. 2010; Wilhelm et al. 2010). The relationship between read density and transcript discovery has been studied, in particular with respect to alternatively spliced isoforms. Clearly, more data are always better, but there is a relationship between transcript abundance and read density that allows determination of the likelihood that a transcript of a particular abundance will be detected (Li et al. 2008; Sultan et al. 2008; Wang et al. 2008; Yeo et al. 2009; Trapnell et al. 2010). This problem is particularly acute for alternative isoforms of low-abundance mRNAs because, depending on the read length, reads that cross splice junctions are rare.

4. How many replicate libraries will be needed to identify differences in gene expression or RNA processing with statistical confidence? If the goal is to establish whether a change in mRNA levels occurs between biological states (growth conditions, drug treatments, tissues, knock down, time points), biological sampling will still be necessary. As with any method, the resolution of the experiment to see a change will depend on the sampling error or noise of both the combined technical steps and the underlying biology; with sufficient replicates, the biological variation can be accessed (Pickrell et al. 2010). Simple experiments will involve comparisons between two sets of samples that differ by one variable. In the case of multiple different treatments that must be compared with one another, a common set of control replicates makes the best reference. In either case, as with microarrays, it is important to keep technical variation to a minimum to obtain strong statistical

support for biological variation in the system under study. Several laboratories have provided evidence that library preparation and sequencing sets are technically very reproducible (e.g., see Marioni et al. 2008); however, this may depend on the details of the exact method used given the many options for library preparation (see above). Thus, adequate replication of experiment and controls will be needed to show that a result is trustworthy, at a given desired resolution.

It is key to consider and address the sources of variation in the experiment by controlling the variation that can be controlled (technically) and randomizing the variation that cannot be controlled (usually this will be intrinsic to the biology). There are different kinds of "replicate" experiments, some of which capture only part of the variation in an experimental protocol. A so-called "technical replicate" involves taking the very same RNA sample, splitting it into three aliquots, and making three separate fragment libraries. Variation in such an experiment should arise from the fragmentation or priming method, ligation efficiencies, amplification, and other variables in the technical steps of library construction and amplification. Variation in these steps should be held very low if technical replicates are expected to reproducibly if not exactly represent the original samples.

Another set of replicates might involve splitting a cell culture into three aliquots, extracting RNA separately, and then analyzing each RNA sample. These replicates will capture variability in the RNA extraction steps, plus that captured in the previous example. Because RNA extraction should be well controlled, these kinds of replicates are not as useful either. If, on the other hand, separate cell cultures were grown and then each separately extracted and tested, the experiment would detect variability in cell growth, which might be harder to control. Although it is tempting to avoid the variability that cannot be controlled in experimental design to reduce noise or because it requires more effort, the goal of the statistical analysis should be to reveal changes that occur in response to the experimental manipulation, against the background of uncontrolled variation in cell culture conditions, RNA preparation, labeling, and hybridization. Some factors are easier to control than others (e.g., genetic background, growth medium, cell handling, and sample processing) and still many others are unknown and can only be captured by replication.

5. How will sequencing experiments be normalized so that they can be compared to one another? The admonition to perform replicate library preparation on different biological replicates to obtain reliable measurements of the change in gene expression may require normalization or scaling of libraries with one another. Some libraries are deeper than others. Stochastic effects, especially on rare transcripts, may lead to poor library overlap with respect to recovery of rare sequences. Methods are being evaluated for replicate library comparison and normalization (Robinson and Oshlack 2010). In addition, internal controls can be created such as "spike-in" control RNAs added to the samples in a series of known concentrations for use in evaluating library scale and the efficiency of recovery of different abundance groups. Correction for library depth may also be applied at a postmapping step to create a comparable derived metric such as "RPKM" or "reads per kilobase per million" mapped reads (Mortazavi et al. 2008; Pepke et al. 2009). A sequence derived from a pair of matched-end reads (two paired-end reads) is called a "fragment" and is counted as FPKM or fragments per kilobase per million (Trapnell et al. 2010). The measures (see below) normalize for both gene size (more reads or fragments can be mapped to larger genes) and the total number of reads or fragments (per million mapped). However, (obviously) the RPKM value for a gene from a deep library may have more statistical meaning than an equivalent value from a more shallow library (Bullard et al. 2010).

   For this and other aspects of this rapidly evolving area of research, it is advisable to search the Bioconductor site at http://bioconductor.org/ (Gentleman et al. 2004; Reimers and Carey 2006) and to check a continuingly updated site associated with recent publications of interest in this area (Bateman and Quackenbush 2009). Once these questions are answered, RNA can be sent to the core facility with detailed and specific instructions. Alternatively, library preparation can proceed, and the fragment libraries can be sequenced. In either case, the next phase is data analysis.

Processing and Mapping Sequencing Reads

6. ***Processing Raw Reads.*** High-throughput sequencing machines have internal computational methods for determining base identity. These depend in some cases on the method (e.g., SOLiD sequencing produces a read in "color space" due to the fact that each ligation step reads a dinucleotide and overlapping dinucleotide reads are used to identify the base at each position. The color space data must be converted to a sequence, a process thought to add fidelity to the sequence determination but that is not needed for the other platforms (see introduction to this chapter). In the end, each platform produces a file that includes a base call at each read position plus a quality score indicating the confidence that the base call at that position is correct (a so-called "raw read"). Depending on the platform and the sequence, raw reads have distinct kinds of error profiles that are characteristic of the sequencing method used to obtain them. Raw reads are trimmed so that low-confidence bases, usually at the end of the read, are removed. If the sequencing run has been derived from a mixture of bar-coded libraries, the next step is to sort the reads by their bar codes into separate files that contain only reads from a given input library. Once this is accomplished, the bar codes and other recognizable linker sequences are trimmed to produce the processed reads. These are labeled by library origin from the bar code, are free of linker sequence, and retain their quality scores.
7. ***Mapping.*** Once the processed reads are obtained, they must be mapped to a reference genome or, if no genome is available, assembled de novo. A large number of methods and approaches exist for this step, and many recent reviews have appeared (Mortazavi et al. 2008; Bateman and Quackenbush 2009; Langmead et al. 2009; Pepke et al. 2009; Trapnell et al. 2009; Wilhelm and Landry 2009; Trapnell et al. 2010; Wilhelm et al. 2010). In addition, and because the developments in this field are rapidly being converted to functional and available software packages, the Bioconductor website should be searched for the latest developments (Gentleman et al. 2004; Reimers and Carey 2006). Extra efforts may be necessary to capture and map splice junction reads (Trapnell et al. 2009; Ameur et al. 2010; Au et al. 2010; Trapnell et al. 2010).
8. ***Transcript (Isoform) Assembly.*** For discovery of novel transcripts, mapped reads must be assembled into transcripts. This problem has also been widely addressed in recent reviews (Bateman and Quackenbush 2009; Pepke et al. 2009). Multiple packages exist for carrying out this process. A commonly used approach uses three programs in sequence. The first is Bowtie (Langmead et al. 2009), which maps reads that do not span junctions and assembles them into "transfrags" that often represent individual exons. Reads that do not map in this initial step are retained and fed to Tophat (Trapnell et al. 2009), which finds and evaluates reads that connect the transfrags produced by Bowtie. Finally, Cufflinks (Trapnell et al. 2010) is used to evaluate the joining solutions created by Tophat to create transcript models that incorporate alternative splicing patterns observed. Other packages can perform similar processing (Reimers and Carey 2006; Bateman and Quackenbush 2009; Pepke et al. 2009).

Measuring Expression Level and Changes in Expression

9. ***Expression Level from Reads, Changes in Expression.*** As for the mapping problem, methods for determining expression levels are also evolving (Reimers and Carey 2006; Bateman and Quackenbush 2009; Pepke et al. 2009). A popular metric is RPKM, which allows a provisionally comparable number to be extracted from different libraries in which the mapping depth varies, adjusted for genes of different sizes (Mortazavi et al. 2008; Pepke et al. 2009; Trapnell et al. 2010). Although this measure is convenient, it has not yet received exhaustive scrutiny, and there is a suggestion that it could be attached to statistical quality metrics in order to become more reliable (Bullard et al. 2010; Robinson and Oshlack 2010; Young et al. 2010), especially if it is to be used to detect and measure changes in gene expression. After expression is detected

and compared, a list of gene expression or RNA-processing changes by gene or genomic location, along with the magnitude of each change and some sort of measure of the likelihood that the change is not due to chance, is obtained. These represent the statistically valid set of changes that might be due to the experimental manipulation or other feature that distinguishes the groups biologically. It now becomes important to determine how these data stand up to an orthogonal method of measurement.

Validation

10. ***Validation.*** For any high-throughput data acquisition method, a result for any individual gene of special interest should be validated by a direct experiment. There are two philosophical reasons for performing validation experiments. Validation is useful to biologists to find out if a favorite gene that came up in the list is truly changed in the experiment, because their goal is more likely to be biological insights in a specific process of interest. On the other hand, bioinformaticists may need to know how many of the results are true and to understand where in the ranked list of best-scoring genes or events the predictions become incorrect to identify the scores that provide the best ROC curve. This will be important because bioinformaticists will require a large high-confidence set of genes for further analysis. Thus, a compromise validation list must be developed for maximum progress and happiness of both collaborators. Once settled, quantitative RT-PCR, semiquantitative RT-PCR, or other independent measurement methods should be used to determine how well the sequencing data predict the true changes as measured by the more labor-intensive approach. A number of studies have been done that indicate good agreement between RNA sequencing and microarrays (e.g., see Marioni et al. 2008), suggesting they can be used to cross-validate each other.

Classification of Changes and Their Associated Biology

11. ***Classification and Association.*** After validation, a high-confidence list of gene expression changes or RNA-processing changes is now in hand. It is time to search for biological function and genetic sequence associations (to name two common examples) that appear to be enriched in the high-confidence group relative to a randomly selected comparison groups ("background sets"), either the genome in its entirety or the genes expressed in the cell type under study, or some other group of genes selected under a different experiment. In some cases, the list of genes may provide obvious clues to the biology. This will depend on the mind-set and the experience of the investigator looking at the list. It is advisable to ask the question in a more open way using GO tools (Gene Ontology Consortium 2006) available at Bioconductor. These tools have data that link gene names to terms for cellular components, biological processes, and molecular functions that have been assigned to gene products by various methods. GO has a structured language and seeks to formalize the meaning of terms and connections. In a statistical fashion, the list of genes obtained in the experiment is used as the input, and enrichment of GO terms associated with that group of genes is evaluated compared to the distribution of GO terms among the genes in the background set. If the biological treatment in the experiment impacts expression of a functionally distinct group of genes in the cell, terms associated with this function should appear to be enriched. Recently, it has been discovered that gene lists from RNA sequencing data sets may suffer from a selection bias that overrepresents long and highly expressed genes (Young et al. 2010). It may be important to take this into account.

    Sequences in promoters, near splice sites, or within untranslated mRNA regions may also be enriched in the selected gene set. Finding sequence motifs is more complex, and there are numerous approaches to this question. A good starting point is the "GREAT" program (McLean et al. 2010), which, in addition to identifying GO term enrichment, allows identification of conserved promoter elements associated with an input gene list. These and other downstream analysis methods aid in hypothesis development for further experimentation. In the end, the sequencing data are just the beginning, and it is likely that many follow-up experiments will be necessary to close the loop and obtain a deeper understanding of the biological impact of the original experiment.

Visualization

12. ***Presenting the Data.*** Data that arise from genomic-level experiments usually exist in large tables that are not easy for humans to grasp and digest. A large variety of visualization tools are available at Bioconductor (http://bioconductor.org/). Also useful is the UCSC Genome Browser (http://genome.ucsc.edu/); see also Mangan et al. (2008) and Zweig et al. (2008). The Galaxy website (http://galaxyproject.org/) is very useful to the biologist who may have limited experience writing scripts or ensuring file format integrity (Blankenberg et al. 2010). This site is especially useful for creating track files for the UCSC Genome Browser. Finally, the display of results in color figures can create concern, especially considering the frequency of red–green color blindness in the human population. An excellent discussion of this point may be found at http://jfly.iam.u-tokyo.ac.jp/color/.

TIPS AND TROUBLESHOOTING

1. Problems may arise as a consequence of "interdisciplinary strain," in which collaborators struggle to communicate their needs and perspectives on the work across the gulf that separates them. In such cases, talking is best, with the recognition that goals, biases about what is important, and reward systems differ in different disciplines and that it is best if all of these can be recognized and respected as valid concerns.
2. File formats may become corrupt or incompatible. Check Galaxy for file conversion and quality-checking methods.
3. Dedicated computer time and space may become a problem. It is important for the project to have adequate resources for the computational parts.
4. Quality control must be maintained at every step from RNA isolation through statistical evaluation, preferably by a small number of investigators who apply consistent standards. If RNA quality is variable or poor, little in the subsequent analysis will be able to improve things. Likewise, errors in analysis can crop up, so spot validation and visual inspection of results or other more elaborate "sanity checks" may be needed to confirm that the analysis is proceeding well.

REFERENCES

Ameur A, Wetterbom A, Feuk L, Gyllensten U. 2010. Global and unbiased detection of splice junctions from RNA-seq data. *Genome Biol* **11:** R34.

Armour CD, Castle JC, Chen R, Babak T, Loerch P, Jackson S, Shah JK, Dey J, Rohl CA, Johnson JM, et al. 2009. Digital transcriptome profiling using selective hexamer priming for cDNA synthesis. *Nat Methods* **6:** 647–649.

Au KF, Jiang H, Lin L, Xing Y, Wong WH. 2010. Detection of splice junctions from paired-end RNA-seq data by SpliceMap. *Nucleic Acids Res* **38:** 4570–4578.

Barrett T, Troup DB, Wilhite SE, Ledoux P, Rudnev D, Evangelista C, Kim IF, Soboleva A, Tomashevsky M, Marshall KA, et al. 2009. NCBI GEO: Archive for high-throughput functional genomic data. *Nucleic Acids Res* (database issue) **37:** D885–D890.

Bateman A, Quackenbush J. 2009. Bioinformatics for next generation sequencing. *Bioinformatics* **25:** 429.

Blankenberg D, Von Kuster G, Coraor N, Ananda G, Lazarus R, Mangan M, Nekrutenko A, Taylor J. 2010. Galaxy: A web-based genome analysis tool for experimentalists. *Curr Protoc Mol Biol* **2010:** Unit 19.10.1–21.

Bullard JH, Purdom E, Hansen KD, Dudoit S. 2010. Evaluation of statistical methods for normalization and differential expression in mRNA-Seq experiments. *BMC Bioinformatics* **11:** 94.

Fejes-Toth K, Sotirova V, Sachidanandam R, Assaf G, Hannon G, Kapranov P, Foissac S, Willingham A, Duttagupta R, Dumais E, et al. 2009. Post-transcriptional processing generates a diversity of 5′-modified long and short RNAs. *Nature* **457:** 1028–1032.

Gene Ontology Consortium. 2006. The Gene Ontology (GO) project in 2006. *Nucleic Acids Res* (database issue) **34:** D322–D326.

Gentleman RC, Carey VJ, Bates DM, Bolstad B, Dettling M, Dudoit S, Ellis B, Gautier L, Ge Y, Gentry J, et al. 2004. Bioconductor: Open software development for computational biology and bioinformatics. *Genome Biol* **5:** R80.

Hu Y, Wang K, He X, Chiang DY, Prins JF Liu J. 2010. A probabilistic framework for aligning paired-end RNA-seq data. *Bioinformatics* **26:** 1950–1957.

Ingolia NT, Ghaemmaghami S, Newman JR, Weissman JS. 2009. Genome-wide analysis in vivo of translation with nucleotide resolution using ribosome profiling. *Science* **324:** 218–223.

Langmead B, Trapnell C, Pop M, Salzberg SL. 2009. Ultrafast and memory-efficient alignment of short DNA sequences to the human genome. *Genome Biol* **10:** R25.

Li H, Lovci MT, Kwon YS, Rosenfeld MG, Fu XD, Yeo GW. 2008. Determination of tag density required for digital transcriptome analysis: Application to an androgen-sensitive prostate cancer model. *Proc Natl Acad Sci* **105:** 20179–20184.

Mangan ME, Williams JM, Lathe SM, Karolchik D, Lathe WC III. 2008. UCSC genome browser: Deep support for molecular biomedical research. *Biotechnol Annu Rev* **14:** 63–108.

Marioni JC, Mason CE, Mane SM, Stephens M, Gilad Y. 2008. RNA-seq: An assessment of technical reproducibility and comparison with gene expression arrays. *Genome Res* **18:** 1509–1517.

McLean CY, Bristor D, Hiller M, Clarke SL, Schaar BT, Lowe CB, Wenger AM, Bejerano G. 2010. GREAT improves functional interpretation of *cis*-regulatory regions. *Nat Biotechnol* **28:** 495–501.

Mortazavi A, Williams BA, McCue K, Schaeffer L, Wold B. 2008. Mapping and quantifying mammalian transcriptomes by RNA-Seq. *Nat Methods* **5:** 621–628.

Nagalakshmi U, Waern K, Snyder M. 2010. RNA-Seq: A method for comprehensive transcriptome analysis. *Curr Protoc Mol Biol* **2010:** Unit 4.11.1–13.

Pepke S, Wold B, Mortazavi A. 2009. Computation for ChIP-seq and RNA-seq studies. *Nat Methods* **6** (suppl 11)**:** S22–S32.

Pickrell JK, Marioni JC, Pai AA, Degner JF, Engelhardt BE, Nkadori E, Veyrieras JB, Stephens M, Gilad Y, Pritchard JK. 2010. Understanding mechanisms underlying human gene expression variation with RNA sequencing. *Nature* **464:** 768–772.

Ramskold D, Wang ET, Burge CB, Sandberg R. 2009. An abundance of ubiquitously expressed genes revealed by tissue transcriptome sequence data. *PLoS Comput Biol* **5:** e1000598.

Reimers M, Carey VJ. 2006. Bioconductor: An open source framework for bioinformatics and computational biology. *Methods Enzymol* **411:** 119–134.

Robinson MD, Oshlack A. 2010. A scaling normalization method for differential expression analysis of RNA-seq data. *Genome Biol* **11:** R25.

Sultan M, Schulz MH, Richard H, Magen A, Klingenhoff A, Scherf M, Seifert M, Borodina T, Soldatov A, Parkhomchuk D, et al. 2008. A global view of gene activity and alternative splicing by deep sequencing of the human transcriptome. *Science* **321:** 956–960.

Trapnell C, Pachter L, Salzberg SL. 2009. TopHat: Discovering splice junctions with RNA-Seq. *Bioinformatics* **25:** 1105–1111.

Trapnell C, Williams BA, Pertea G, Mortazavi A, Kwan G, van Baren MJ, Salzberg SL, Wold BJ, Pachter L. 2010. Transcript assembly and quantification by RNA-Seq reveals unannotated transcripts and isoform switching during cell differentiation. *Nat Biotechnol* **28:** 511–515.

Wang ET, Sandberg R, Luo S, Khrebtukova I, Zhang L, Mayr C, Kingsmore SF, Schroth GP, Burge CB. 2008. Alternative isoform regulation in human tissue transcriptomes. *Nature* **456:** 470–476.

Wilhelm BT, Landry JR. 2009. RNA-Seq-quantitative measurement of expression through massively parallel RNA-sequencing. *Methods* **48:** 249–257.

Wilhelm BT, Marguerat S, Goodhead I, Bahler J. 2010. Defining transcribed regions using RNA-seq. *Nat Protoc* **5:** 255–266.

Yeo GW, Coufal NG, Liang TY, Peng GE, Fu XD, Gage FH. 2009. An RNA code for the FOX2 splicing regulator revealed by mapping RNA-protein interactions in stem cells. *Nat Struct Mol Biol* **16:** 130–137.

Yoon OK, Brem RB. 2010. Noncanonical transcript forms in yeast and their regulation during environmental stress. *RNA* **16:** 1256–1267.

Young MD, Wakefield MJ, Smyth GK, Oshlack A. 2010. Gene ontology analysis for RNA-seq: Accounting for selection bias. *Genome Biol* **11:** R14.

Zweig AS, Karolchik D, Kuhn RM, Haussler D, Kent WJ. 2008. UCSC genome browser tutorial. *Genomics* **92:** 75–84.

WEB RESOURCES

http://bioconductor.org/ Bioconductor, an open-source software for bioinformatics
http://galaxyproject.org/ Galaxy, an easy-to-use, open-source, scalable framework for tool and data integration
http://genome.ucsc.edu/ UCSC Genome Browser
http://jfly.iam.u-tokyo.ac.jp/color/ Color Universal Design (CUD)

APPENDIX

1 Common Recipes

Annealing Buffer (5x)

Potassium acetate (500 mM)
HEPES-KOH (150 mM, pH 7.4) <!>
Magnesium acetate (10 mM)

Church and Gilbert Hybridization Buffer (for 100 mL)

	High	Moderate	Low
BSA (10 mg/mL final)	1 g	1 g	1 g
H_2O	30 mL	—	15 mL
NaP (1 M) <!>	20 mL	50 mL	50 mL
Formamide <!>	15 mL (15% final)	15 mL (15% final)	—
0.5 M EDTA (1 mM) <!>	0.2 mL	0.2 mL	0.2 mL
20% SDS (7% final) <!>	35 mL	35 mL	35 mL

The preparation of this buffer for Church and Gilbert hybridization requires that dibasic sodium phosphate be pHed with phosphoric acid.

Formamide/Dye Loading Buffer

Reagent	Quantity (for 10 mL)	Final concentration
Deionized formamide <!>	9.5 mL	95%
Bromophenol blue <!>	2.5 mg	0.025% (w/v)
Xylene cyanol FF <!>	2.5 mg	0.025% (w/v)
EDTA (0.5 m, pH 8.0) <!>	100 mL	5 mm
H_2O	400 mL	

Store in aliquots at –20°C.

Gel Elution Buffer

Reagent	Quantity (for 500 mL)	Final concentration
Tris-HCl (1 M, pH 7.5)	10 mL	20 mM
Sodium acetate (2.5 M, prepared without pH adjustment)	50 mL	0.25 M
EDTA (0.5 M, pH 8.0) <!>	1 mL	1 mM
SDS (10%, w/v) <!>	12.5 mL	0.25%
H_2O	426.5 mL	

If the SDS precipitates, warm to 37°C until the precipitate disappears. Store indefinitely at room temperature.

Gel Elution Buffer (no SDS)

Reagent	Quantity (for 500 mL)	Final concentration
Tris-HCl (1 M, pH 7.5)	10 mL	20 mM
Sodium acetate (2.5 M, unpH'ed)	50 mL	0.25 M
EDTA (0.5 M, pH 8.0) <!>	1 mL	1 mM
H_2O	to 500 mL	

Store indefinitely at 4°C.

Gel Elution Buffer (with SDS)

Reagent	Quantity (for 500 mL)	Final concentration
Sodium acetate (2.5 M, unpH'd)	50 mL	250 mM
Tris-HCl (1 M, pH 7.5)	10 mL	20 mM
EDTA (0.5 M, pH 8.0) <!>	1 mL	1 mM
SDS (10%, w/v) <!>	12.5 mL	0.25% (w/v)
H_2O	to 500 mL	

Store indefinitely at room temperature.

Klenow Buffer (5x)

Reagent	Quantity (for 1 mL)	Final concentration
NaCl (5 M)	50 µL	250 mM
Tris-HCl (1 M, pH 7.9)	50 µL	50 mM
DTT (1 M) <!>	5 µL	5 mM
H_2O	895 µL	

Store for 6 mo at –20°C.

Ligation Buffer (10x)

Reagent	Quantity (for 1 mL)	Final concentration
Tris-HCl (1 M, pH 7.5)	0.5 mL	500 mM
$MgCl_2$ (1 M) <!>	0.1 mL	100 mM
DTT (1 M) <!>	0.1 mL	100 mM
ATP (100 mM)	0.1 mL	10 mM
H_2O	0.2 mL	

Store for 6 mo at –20°C.

Luria Broth (LB)

Add the following to 800 mL of H_2O:

Bacto-tryptone	10 g
Yeast extract	5 g
NaCl	10 g

Adjust the pH to 7.5 with NaOH and then adjust the volume to 1 L with dH_2O. Sterilize by autoclaving.

M9+ Medium

1x M9 stock plus:

Cholesterol	10 µg/mL
Ampicillin <!>	50 µg/mL
Tetracycline <!>	12 µg/mL
IPTG <!>	200 µg/mL
Fungizone	0.1 µg/mL

M9 Stock (1x)

KH_2PO_4	3.0 g
Na_2HPO_4	6.0 g
NaCl	5.0 g
$MgSO_4$ (1 M) <!>	1.0 mL
dH_2O	to 1 L

Native Gel-loading Buffer

Reagent	Quantity (for 10 mL)	Final concentration
TBE (10x)	2.5 mL	2.5x
Glycerol (100%)	5 mL	50%
Bromophenol blue (1%) <!>	1 mL	0.1%
H_2O	1.5 mL	

Store at room temperature.

NGM Plates

Bacto-peptone	2.5 g
NaCl	3 g
Bacto-agar	18 g
dH_2O	to 975 mL

Autoclave. After cooling, add

$CaCl_2$ (1 M) <!>	1 mL
$MgSO_4$ (1 M) <!>	1 mL
Cholesterol (5 mg/mL) (dissolved in 100% ethanol <!>)	1 mL
$NaPO_4$ (1 M, pH 6) <!>	25 mL

OptiKinase Buffer (10x)

Reagent	Quantity (for 1 mL)	Final concentration (10x)
Tris-HCl (1 M, pH 7.5)	500 µL	0.5 M
$MgCl_2$ (1 M) <!>	100 µL	100 mM
DTT (1 M) <!>	50 µL	50 mM
H_2O	350 µL	

Store for up to 6 mo at –20°C.

PBS/EDTA/EGTA

Reagent	Quantity (for 500 mL)	Final concentration
PBS (10x)	50 mL	1x
EDTA (0.5 M, pH 8.0) <!>	1 mL	1 mM
EGTA (0.1 M, pH 8.0)	5 mL	1 mM
H_2O	to 500 mL	

PCR Buffer (10x)

Reagent	Quantity (for 1 mL)	Final concentration (10x)
Tris-HCl (1 M, pH 8.3)	100 µL	100 mM
KCl (2 M) <!>	250 µL	500 mM
$MgCl_2$ (1 M) <!>	15 µL	15 mM
H_2O	635 µL	

Store for up to 1 yr at –20°C.

PCR Buffer (10x)

Reagent	Quantity (for 1 mL)	Final concentration
Tris-HCl (1 M, pH 8.6)	100 µL	100 mM
KCl (2 M) <!>	250 µL	500 mM
$MgCl_2$ (1 M) <!>	15 µL	15 mM
H_2O	635 µL	

Store for 6 mo at –20°C.

Phosphatase Buffer (10x)

Reagent	Quantity (for 1 mL)	Final concentration (10x)
Tris-HCl (1 M, pH 7.9)	0.5 mL	0.5 M
$MgCl_2$ (1 M) <!>	0.1 mL	100 mM
NaCl (5 M)	0.2 mL	1 M
DTT (1 M) <!>	10 µL	10 mM
H_2O	0.19 mL	

Store for up to 6 mo at –20°C.

Phosphate-buffered Saline (PBS)

Reagent	Quantity (for 1 L)	Final concentration
NaCl	8 g	137 mM
KCl <!>	0.2 g	2.7 mM
Na_2HPO_4	1.44 g	10 mM
KH_2PO_4	0.24 g	2 mM

Dissolve in 800 mL of Milli-Q H_2O. Adjust the pH to 7.4 and then add Milli-Q H_2O to 1 L. Autoclave or filter to sterilize. Store at room temperature. Alternatively, buy or make 10x PBS.

PK Buffer (2x)

Reagent	Quantity (for 100 mL)	Final concentration
Tris-HCl (1 M, pH 7.5)	20 mL	0.2 M
EDTA (0.5 M, pH 8.0) <!>	10 mL	25 mM
NaCl (5 M)	6 mL	0.3 M
SDS (10%, w/v) <!>	20 mL	2%
H_2O	to 100 mL	

Store indefinitely at room temperature.

Poly(A) Polymerase Buffer (5x)

Reagent	Quantity (for 1.0 mL)	Final concentration (5x)
Tris-HCl (1 M, pH 8.0)	100 µL	0.1 M
$MnCl_2$ (0.3 M) <!>	10 µL	3 mM
EDTA (0.1 M) <!>	1 µL	0.1 mM
DTT (0.5 M) <!>	2 µL	1 mM
Bovine serum albumin (BSA), acetylated (50 mg/mL)	10 µL	500 µg/mL
H_2O	877 µL	

Store indefinitely at –20°C.

Proteinase K Buffer (2×)

Reagent	Quantity (for 50 mL)	Final concentration
SDS (10%) <!>	2.5 mL	0.5%
EDTA (0.5 M, pH 8.0) <!>	0.4 mL	4 mM
H_2O	47.1 mL	

Store indefinitely at room temperature.

Rapid Ligation Mix (2×)

Reagent	Quantity (for 1 mL)	Final concentration (2×)
Tris-HCl (1 M, pH 7.5)	132 µL	132 mM
$MgCl_2$ (1 M) <!>	20 µL	20 mM
DTT (1 M) <!>	2 µL	2 mM
ATP (100 mM)	20 µL	2 mM
Polyethylene glycol (PEG) (50%, w/v)	300 µL	15% (w/v)
H_2O	526 µL	

Store for up to 6 mo at –20°C. Note that rapid ligation buffers are available from many commercial suppliers.

Reverse Transcriptase Buffer (2.5×)

Reagent	Quantity (for 1 mL)	Final concentration
Tris (1 M, pH 8.3) <!>	50 µL	50 mM
DTT (1 M) <!>	50 µL	50 mM
$MgCl_2$ (1 M) <!>	20 µL	20 mM
dATP (10 mM)	40 µL	0.4 mM
dCTP (10 mM)	40 µL	0.4 mM
dGTP (10 mM)	40 µL	0.4 mM
dTTP (10 mM)	40 µL	0.4 mM
H_2O	720 µL	

Store indefinitely at –20°C.

Reverse Transcription Buffer (10×)

Reagent	Quantity (for 1 mL)	Final concentration
Tris-HCl (1 M, pH 8.3)	500 µL	500 mM
KCl (2 M) <!>	200 µL	400 mM
$MgCl_2$ (1 M) <!>	60 µL	60 mM
DTT (1 M) <!>	10 µL	10 mM
H_2O	230 µL	

Store for 6 mo at –20°C.

RNA Gel-loading Buffer (1.5x)

Formamide + dyes

Reagent	Quantity (for 10 mL of 1.5x)	Final concentration
Ultrapure formamide <!>	9.5 mL	95%
Bromophenol blue (2.5%, w/v) <!>	100 μL	0.025%
Xylene cyanol FF (2.5%, w/v) <!>	100 μL	0.025%
EDTA (0.25 M, pH 8.0) <!>	200 μL	5 mM EDTA

Use 5 μL for a 2.5-μL sample. Purchase a distilled, deionized preparation of formamide and the above loading dyes. Store in small aliquots at –20°C. This solution is available commercially (Ambion) and is recommended over homemade. Make 1-mL aliquots. Store for up to 1 yr at –20°C.

RNase H Buffer (5x)

Reagent	Quantity (for 10 mL)	Final concentration (5x)
Tris-HCl (1 M, pH 7.5)	1 mL	100 mM
KCl (2 M) <!>	2.5 mL	500 mM
$MgCl_2$ (1 M) <!>	0.5 mL	50 mM
DTT (1 M) <!>	0.5 mL	50 mM
Sucrose	2.5 g	25 (w/v)
H_2O	to 10 mL	

If this buffer is used infrequently, prepare a stock without DTT, aliquot into 1-mL aliquots, and add 50 μL of 1 M DTT to each aliquot before use. Store working aliquot with DTT for up to 1 mo at –20°C.

S1 Digestion Buffer

Reagent	Quantity (for 100 mL)	Final concentration
NaCl (5 M)	5.6 mL	0.28 M
Sodium acetate (0.5 M, pH 4.6)	10 mL	0.05 M
Zinc acetate (450 mM)	1 mL	4.5 mM

Filter with a 0.2-μm filter unit. Store for up to 1 yr at 4°C.

S1 Hybridization Buffer (5x)

Reagent	Quantity (for 100 mL)	Final concentration (5x)
PIPES, free acid	6.04 g	0.2 M
NaCl (5 M)	40 mL	2 M
EDTA (250 mM, pH 8.0) <!>	2 mL	5 mM
H_2O	to 100 mL	

Before bringing the final volume to 100 mL with H_2O, adjust the pH to 6.4 with NaOH <!>. Filter with a 0.2-μm filter unit. Store for up to 1–2 yr at room temperature.

S1 Termination Mix

Reagent	Quantity (for 100 mL)	Final concentration
Ammonium acetate	30.83 g	4 M
Na_2EDTA <!>	3.72 g	0.1 M

Filter with a 0.2-µm filter unit. Store for up to 1–2 yr at room temperature.

SDS Extraction Buffer

Reagent	Quantity (for 100 mL)	Final concentration
Tris (1 M, pH 7.5) <!>	2 mL	20 mM
EDTA (0.5 M, pH 8.0) <!>	0.4 mL	1 mM
SDS (10%, w/v) <!>	5 mL	0.5% (w/v)
H_2O	92.4 mL	

Store indefinitely at room temperature.

Sodium Phosphate Buffer (1 M, pH 7.2)

Reagent	Quantity (for 1 L)	Final concentration (10x)
$Na_2HPO_4 \cdot 7H_2O$	134 g	500 mM
H_3PO_4 (85%) <!>	4 mL	0.34%
H_2O	to 1 L	

The pH should be at 7.2. Store indefinitely at room temperature.

Splicing Master Mix (4x)

Reagent	Quantity (for 62.2 µL)	Final concentration (1x splicing reaction)
Creatine phosphate (0.5 M)	10 µL	20 mM
ATP (100 mM)	5 µL	2 mM
KCl (3 M) <!>	2.5 µL	30 mM*
$MgCl_2$ (0.5 M) <!>	2.1 µL	4.2 mM
DTT (1 M) <!>	1 µL	4 mM
Tris (1 M, pH 8.0) <!>	1 µL	16 mM
Aprotinin and leupeptin <!>	1 µL	
Polyethyleneglycol (PEG 8000; 20%, w/v)	37.5 µL	3%
H_2O	2.1 µL	

Prepare fresh and keep on ice.

*The final KCl concentration is usually between 50 and 80 mM. The extract also contains KCl, which contributes to the final concentration. Adjust according to the concentration in dialysis buffer and the amount of extract used in the reaction.

T1 Sequencing Buffer

Reagent	Quantity (for 1 mL)	Final concentration
Sodium acetate (3 M, pH 5.3)	6.6 µL	20 mM
Urea (Molecular Biology grade) <!>	0.42 g	7 M
EDTA (0.5 M, pH 8.0) <!>	2 µL	1 mM
H_2O	adjust to 1 mL	

Store for 6 mo at room temperature.

T7 Transcription Buffer (10x)

Reagent	Quantity (for 1 mL)	Final concentration
Tris-HCl (1 M, pH 8.0)	400 µL	400 mM
$MgCl_2$ (1 M) <!>	200 µL	200 mM
DTT (1 M) <!>	50 µL	50 mM
H_2O	350 µL	

Store for 6 mo at –20°C.

TBE Electrophoresis Buffer (10x)

Reagent	Quantity (for 1 L)	Final concentration
Tris base	121.1 g	1 M
Boric acid <!>	61.8 g	1 M
EDTA (disodium salt) <!>	7.4 g	0.02 M

Dilute 50 mL to 1 L to make gel running buffer. Store for up to 6 mo at room temperature.

TE Buffer

Reagent	Quantity (for 500 mL)	Final concentration
Tris (1 M, pH 7.5) <!>	5 mL	10 mM
EDTA (0.5 M, pH 8.0) <!>	1 mL	1 mM
H_2O	to 500 mL	

Use Tris at pH 8.0 if desired. Store at 4°C.

Urea/Dye Loading Buffer

Reagent	Quantity (for 10 mL)	Final concentration
Urea <!>	4.8 g	8 M
Tris-HCl (1 M, pH 8.0)	200 µL	20 mM
EDTA (0.25 M, pH 8.0) <!>	40 µL	1 mM
Bromophenol blue (0.5%) <!>	1 mL	0.05%
Xylene cyanol (0.5%) <!>	1 mL	0.05%
H_2O	to 10 mL	

Urea Loading Dye

Reagent	Quantity (for 10 μL)	Final concentration
Urea (ultrapure) <!>	5.4 g	9 M
TBE*		
Xylene cyanol <!>/ bromophenol blue <!>	0.05% w/v	
(1%, w/v) (optional)	500 μL	0.5%

*Add enough to reach the 10-mL mark in a 15-mL conical tube containing urea. Heat at 50°C until dissolved. Check before using to make sure urea is still in solution.

Store at room temperature and make sure urea is in solution before using.

APPENDIX 2

Quick Reference for Enzymes Commonly Used in RNA Research

Ribonucleases

Enzyme	Endo or exo	Specificity	Termini	Special
RNase T1	endo	3′ of G single strand	3′P 5′OH	very specific
RNase A	endo	3′ of pyrimidiness single strand	3′P 5′OH	hard to inactivate
RNase 1	endo	single strand no base specificity	3′P 5′OH	
RNase T_2	endo	single strand no base specificity	3′P 5′OH	
RNase U2	endo	3′ of A single strand	3′P 5′OH	moderate specificity A>G>>U,C
RNase VI	endo	double strand or structure specific no base preference	3′OH 5′P	
RNase P1	endo	single strand no base preference	3′OH 5′P	
RNase H	endo	digests RNA in RNA/DNA hybrids	3′OH 5′P	very specific
Micrococcal nuclease	endo	digests RNA, DNA (both double stranded and single stranded)	3′P 5′OH	requires calcium inactivated by removal of Ca^{2+} by EGTA
S1 nuclease	endo	digests single-stranded RNA or DNA	3′OH 5′P	requires Zn^{++} ions for activity
RNase III	endo	extended double-stranded regions	3′OH 5′P	can detect long double-stranded regions or fragment single-stranded RNA under released conditions
Terminator RNase	exo	digests single-stranded RNA with 5′P terminus	3′OH 5′P	remove rRNA
Tobacco acid phosphatase (TAPase)	endo	digests pyrophosphate in linkages	5′P	remove caps

DNases

Enzyme	Endo or exo	Specificity	Termini	Special
DNase 1	endo	double-stranded or single-stranded DNA no base specificity	3′OH 5′P	easily denatured
ExoVIII	exo	single-stranded DNA either 5′ →3′ or 3′ → 5′	3′OH 5′P	effective at removing PCR primers
Micrococcal nuclease	endo	double-stranded or single-stranded DNA no base specificity	3′P 5′OH	requires Ca^{2+}
S1 nuclease	endo	single-stranded DNA no base specificity	3′OH 5′P	requires Zn^{++}

For detailed descriptions of uses and conditions, see individual protocols. Also see Enzymes used in Appendix 4 of Sambrook and Russell (2001) and supplier's instructions.

REFERENCE

Sambrook JS, Russell DW. 2001. *Molecular cloning: A laboratory manual*, 3rd edition. Cold Spring Harbor Laboratory Press, Cold Spring Harbor, N.Y.

APPENDIX

3 Protocols

1 RNA-friendly Plasmid Preparation, 526

2 Preparing Cellular DNA from Nuclei or Whole Cells, 532

3 Expression and Purification of Active Recombinant T7 RNA Polymerase from *E. coli*, 535

4 Preparing Size Markers for Gel Electrophoresis, 541

5 Toeprinting, 544

6 Cell Fractionation, 547

7 hnRNP-Enriched Nucleoplasmic Extracts from *Drosophila* or Mammalial Tissue Culture Cells, 548

8 Measuring the Length of Poly(A) Tails, 554

Method 1: Low-resolution Measurement of Poly(A) Tails Using Northern Blotting, 555

Method 2: High-resolution Measurement of Poly(A) Tails Using Northern Blotting, 555

Method 3: PCR-based and Ligation-mediated Approaches for Measuring Poly(A) Tail Length, 557

Protocol 1

RNA-friendly Plasmid Preparation

With the advent of miniprep kits for the preparation of plasmids, we have noticed that many investigators use plasmids prepared by these methods as transcription templates or for cell transfection experiments. This can be quite problematic because miniprep (or maxiprep) DNA can often be contaminated with RNases. Indeed, most DNA preparation kits include pancreatic RNase. This RNase is extremely difficult to inactivate and can contaminate benchtops and pipetman. We highly recommend that plasmid DNAs, especially those that will be used repeatedly, be purified using traditional CsCl density gradient methods that do not involve the use of RNase. We have found that the protocol below yields highly pure, noncontaminated plasmid DNA suitable for in vitro transcription and/or transfection analyses. An added benefit to this procedure is that it will yield enough plasmid DNA for many, many experiments.

MATERIALS

The recipes for the items marked with <R> are at the end of the protocol or in the Common Recipes Appendix.

CAUTION: See the Cautions Appendix for appropriate handling of materials marked with <!>.

Reagents

Acetic acid (glacial) <!>
Alkaline SDS solution <R>
Ammonium acetate (7.5 M)
Antibiotic
Cell culture
Cesium chloride balance solution <R> <!>
Culture medium (LB or BHI) <R>
EDTA (0.5 M, pH 8.0) <!>
Ethanol (100%) <!>
Ethidium bromide (10 mg/mL) <!>
Glucose
Isoamyl alcohol (3-methyl 1-butanol) <!>
Lysis buffer <R>
Lysozyme (40 mg/mL) <!>
NaOH (2 N), newly prepared <!>
Potassium acetate solution <R>
Sarkosyl (10%, w/v) <!>
SDS (10%, w/v) <R> <!>
Sodium acetate (3 M, pH 5.2)
TE buffer <R>
Tris (1 M, pH 8.0) <!>

Equipment

Centrifuge and rotors for 200-mL bottles and/or 40-mL Oak Ridge tubes
Centrifuge (low speed; Beckmann model J-6B)

Centrifuge tubes (polypropylene, 12 mL)
Conical tubes (15 mL, 50 mL)
Dry ice <!>
Flasks (500 mL, sterile for bacterial cultures)
Ice
Incubator/shaker (37°C)
Kimwipes
Pasteur pipettes
Rotors (Sorvall RC-5B GSA and RC-5B SS-34, and VTi 90)
Scissors or razor blade
Syringes (3 mL, 10 mL) and 18-gauge needles
Ultracentrifuge
Vertical rotor, tubes (sealable), and rack
Vortex mixer

METHOD

Culture

1. Inoculate 5 mL of culture medium + antibiotic and incubate overnight in a 37°C shaker.
2. Prepare 100 mL of culture medium + antibiotic for each plasmid.
3. Inoculate 100 mL of culture medium with 0.5 mL of overnight culture and incubate overnight in a 37°C shaker.

Cell Lysis

4. Pour the culture into a 200-mL centrifuge bottle and pellet the cells in a Sorvall model RC-5B, GSA rotor at 5000 rpm for 10 min at 4°C.
5. Drain the pellet by inverting the bottle and wipe excess liquid off with a Kimwipe.
6. Resuspend the cell pellet in 30 mL of TE buffer and then transfer to an Oak Ridge tube.
7. Centrifuge the pellet in a Sorvall model RC-5B, SS-34 rotor at 5000 rpm for 10 min at 4°C.
8. While centrifuging, prepare 40 mg/mL of lysozyme in lysis buffer.
9. Pour off the supernatant from Step 7 and place the cell pellet on ice.
10. Resuspend the pellet in 4 mL of cold lysis buffer by pipetting up and down and vortexing.
11. Add 400 μL of lysozyme (40 mg/mL) to the resuspended pellet. Mix and incubate for 30 min on ice. The solution should become fairly viscous.
12. While the pellet is incubating, make the alkaline SDS solution.
13. Add 8 mL of alkaline SDS solution to each tube and sharply invert the tube 20 times to mix. Then, incubate the pellet for 5 min on ice. The solution should become fairly clear.
14. Add 6 mL of cold potassium acetate solution and gently invert the tube 20 times to mix. Then, incubate for 10 min on ice. The chromosomal DNA should precipitate.
15. Centrifuge in an SS-34 rotor at 15,000 rpm for 20 min at 4°C. Save the supernatant in a 50-mL conical tube.

Ethanol/Isopropanol Precipitation

16. Add 2 volumes of 100% ethanol to the supernatant from Step 15 and then invert to mix.

17. Incubate the tube on dry ice for more than 30 min or overnight at –20°C.
18. Transfer the contents to a 40-mL Oak Ridge tube and centrifuge in a Sorvall model RC-5B, SS-34 rotor at 11,500 rpm for 10 min at 4°C.

    Two centrifugations are necessary because the tubes cannot hold the entire volume.
19. Pour off the supernatant and discard.
20. Resuspend pellet in 5 mL of TE buffer.
21. Add 2.67 mL of 7.5 M ammonium acetate (NH_4OAc) to the pellet and then add 20 mL of isopropanol (2-propanol).
22. Incubate for 10 min at room temperature.
23. Centrifuge in a Sorvall model RC-5B, SS-34 rotor at 11,500 rpm for 10 min at 4°C.
24. Pour off the supernatant.
25. Centrifuge again in an SS-34 rotor at 11,500 rpm for 5 min at 4°C to collect the residual supernatant into the bottom of the tube.
26. Pipette off the remaining supernatant and resuspend the pellet in 4 mL of TE buffer.

Cesium Chloride Gradient

27. Transfer the pellet to a 15-mL conical tube.
28. Add 4.4 g of cesium chloride (1.1 g of CsCl/mL TE solution), 200 µL of 10% Sarkosyl, and 240 µL 10 mg/mL of ethidium bromide (0.6 mL/mL TE buffer) to the pellet.
29. Transfer the samples to sealable, vertical rotor centrifuge tubes using a 10-mL syringe with an 18-gauge needle to funnel liquid through the neck of the tube. If tubes are not full or if a "balance tube" is needed, fill tubes up to the neck with CsCl balance solution.
30. Remove liquid from the neck of tube using a Kimwipe. Seal the tubes as recommended by the manufacturer.

    The tubes must be balanced to within 0.5 g of each other.
31. Place tubes in a VTi 90 rotor. Be sure to take note of the loading order. Centrifuge in the ultracentrifuge, at 23°C at 60,000 rpm, decel 9, and hold for 16 h.
32. Remove the tubes to a stable rack that provides access to the side of the tube. The tube should have an area of very dark red (protein) at the top along with a dark red ribbon (RNA) running down the side. There should be an orange band (the DNA sample) near the center of the tube. A smaller band above is residual chromosomal DNA (Fig. A-1).
33. Cut off the top of the tube with scissors or a razor blade.
34. Extract the orange plasmid band using a 3-mL syringe with an 18-gauge needle. Insert the needle bevel-side up at the bottom of the band to remove the band from below. Transfer the liquid from the syringe to a 15-mL conical tube.

    If the application for the plasmid requires extreme purity, another cesium chloride spin may be required. In this case, once the band has been removed, place it into a new 5-mL ultracentrifuge tube topped off with CsCl balance solution. Repeat the centrifugation as in Step 31 above and extract the band in the same manner.

Ethidium Bromide Extraction

35. Adjust the volume in the conical tube to 4 mL with TE buffer.
36. Remove the ethidium bromide by extraction with 1 volume (4 mL) of isoamyl alcohol (3-methyl 1-butanol) in a chemical fume hood. Vortex well to mix.

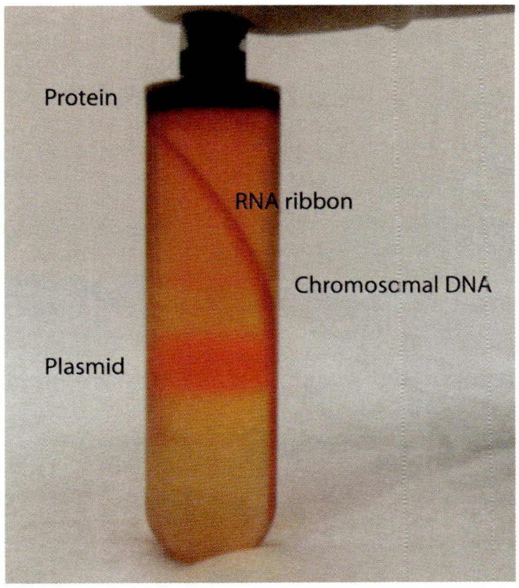

FIGURE A-1. Photograph of a plasmid preparation after CsCl centrifugation. The bright orange band in the middle of the gradient is the plasmid. The light orange band above the plasmid is chromosomal DNA and the dark material (called RNA ribbon in the photograph) is the RNA that pellets during centrifugation. Protein floats on these gradients and is indicated in the photograph. The plasmid is harvested by puncturing the tube (after the top is snipped off) with a syringe. (Nilsen lab, unpubl.)

37. Let the tubes stand in a rack in the chemical fume hood to allow the phases to separate or centrifuge briefly in a low-speed centrifuge at 2000 rpm.
38. Remove and discard top (pink) phase using a pasteur pipette. DNA is in the heavier bottom phase containing buffer and CsCl.
39. Repeat the extractions until the color is gone from both phases and then perform one extra extraction. On the last extraction, remove as much top phase as possible.
40. Use a fresh pipette to remove the bottom phase and transfer it to a new 12-mL polypropylene centrifuge tube. Take care to avoid the alcohol phase.

Ethanol Precipitation

41. Add 0.3 M NaOAc (pH 5.2) and 8 mL of 100% ethanol and vortex or invert to mix. Make sure to mix completely.
42. Precipitate for ~45 min on dry ice or overnight at –20°C.
43. Centrifuge in a Sorvall model RC-5B, SS-34 for at 10,000 rpm 20 min at 4°C. Remove the ethanol.
44. Resuspend the pellet in 2 mL of DNase-free, RNase-free H_2O. Add 200 µL of 3 M NaOAc (pH 5.2) and 5 mL of 100% ethanol to reprecipitate.
45. Place on dry ice for ~45 min.
46. Thaw and centrifuge in a SS-34 rotor at 10,000 rpm for 20 min at 4°C.
47. Remove the ethanol.

    If the pellet still looks white, additional precipitations may be required to remove residual cesium chloride. Repeat as needed. The pellet should have a slightly blue tint to it.
48. Resuspend the pellet after the final precipitation with 150 µL of H_2O and quantitate the DNA.

SOLUTION AND REAGENT RECIPES

CAUTION: See the Cautions Appendix for appropriate handling of materials marked with <!>.

Cesium Chloride Balance Solution

Mix together 4 mL of TE buffer, 4.4 g of CsCl <!>, 200 µL of 10% Sarkosyl <!>, and 240 µL of ethidium bromide <!> (10 mg/mL).

BHI Medium

Add 3.7 g of BHI to 100 mL of H_2O. Autoclave. Add antibiotic after cooling.

LB Medium

Reagent	Quantity (for 100 mL)
Tryptone	1 g
Yeast extract	0.5 g
NaCl	1 g
ddH_2O	to 100 mL

Autoclave, cool, and add antibiotic at the time of use.

TE Buffer

Reagent	Quantity (for 1 L)	Final concentration
Tris (1 M, pH 7.5) <!>	10 mL	10 mM
EDTA (0.5 M) <!>	2 mL	1 mM
ddH_2O	to 1 L	

Store at 4°C.

Lysis Buffer

Reagent	Quantity (for 500 mL)	Final concentration
Tris (1 M, pH 8.0) <!>	12.5 mL	25 mM
EDTA (0.5 M) <!>	10 mL	5 mM
Glucose	4.5 g	
H_2O	to 500 mL	

Store at 4°C.

Alkaline SDS Solution

Reagent	Quantity (for 10 mL)	Final concentration
NaOH (2 N) <!>	0.5 mL	0.1 N
SDS (10%, w/v) <!>	1 mL	1% (w/v)
ddH_2O	8.5 mL	

Prepare fresh for each use.

Potassium Acetate Solution (5 M)

Reagent	Quantity (for 100 mL)	Final concentration
Potassium acetate (5 M)	60 mL	3 M
Acetic acid (glacial) (17 M) <!>	11.5 mL	~2 M
ddH_2O	28.5 mL	

Store at room temperature.

Protocol 2

Preparing Cellular DNA from Nuclei or Whole Cells

It is often desirable to have cellular DNA on hand. This DNA can be used for a variety of purposes, including as a template for polymerase chain reaction (PCR) (to isolate specific fragments or to prepare probes or transcription templates). DNA is stable and can be maintained intact for many years. It is clearly the preferred template choice for PCR or PCR mutagenesis because reverse transcription is not a required step. Moreover, cellular DNA can be used for Southern blots and to determine sites of integration of DNA fragments (e.g., in stable cell lines). This method is for preparation of DNA suitable for PCRs and Southern blots of genomic DNA to determine copy number and sites of integration of plasmids in stable cell lines. Quantitation of DNA may not be exact because RNA is not completely removed. This method can also be used on whole cells, but there will be more RNA contamination.

MATERIALS

The recipes for the items marked with <R> are at the end of the protocol or in the Common Recipes Appendix.

CAUTION: See the Cautions Appendix for appropriate handling of materials marked with <!>.

Reagents

Chloroform <!>
Ethanol (70% and 100%), ice cold <!>
Nuclei (separated from cytoplasmic extract) (~0.2 mL)
Phenol <!>
Proteinase K (10 mg/mL) <!>
Resuspension buffer <R>
SDS extraction buffer <R>
Sodium acetate (3 M, pH 5.2)
T10E0.1 buffer <R>
Triton X-100 (10%, w/v) <!>

Equipment

Bunsen burner
Centrifuge
Centrifuge tubes (15 and 50 mL, conical)
Pasteur pipettes
Pipettes (10 mL)
Vortex mixer
Water bath (42°C)

METHOD

Preparation of DNA from Nuclei after Cytoplasmic Extract Has Been Removed

1. Start with 0.2 mL of nuclei in a 15-mL conical tube (after removing cytoplasmic extract) (see Chapter 2). Resuspend in 3 mL of resuspension buffer.
2. Centrifuge the nuclei at 1500 rpm for 5 min at 4°C and discard the supernatant.
3. Lyse the nuclei with 3 mL of SDS extraction buffer.
4. Mix and resuspend the lysed nuclei with a pasteur pipette or 10-mL pipette.
5. Add 100 µg/mL of proteinase K and digest for 20 min at 42°C.
6. Add NaOAc (pH5.2) to 0.3 M and extract with 2 volumes of phenol:chloroform (1:1). Mix well, but not vigorously, for 5 min.
7. Centrifuge at 2000 rpm for 5 min at room temperature to separate the phases.
8. Remove the aqueous phase with a pipette with a large opening and transfer this phase to a fresh tube.
9. Repeat extraction of the aqueous phase with phenol:chloroform, centrifuge at 2000 rpm for 5 min at room temperature, and transfer the aqueous phase to a fresh tube.
10. Extract with 2 volumes of chloroform:isoamyl alcohol (24:1) (mixing for 5 min) and centrifuge at 2000 rpm for 5 min at room temperature.
11. Transfer the aqueous phase to a 50-mL conical tube and slowly add 2.5 volumes of ice-cold ethanol on top of the liquid.
12. Gently swirl until the ethanol and aqueous phase mix and DNA strands become visible.
13. Continue to swirl until the phases are completely mixed and DNA is visible as a viscous ball of threads.
14. With a Bunsen burner, carefully flame closed the end of a pasteur pipette and then melt and bend the end to make a hook. Use the hook to catch and remove the DNA. Allow the excess ethanol containing RNA to drip off.
15. Rinse the DNA, while it is still attached to the glass hook, in a tube containing cold 70% ethanol. Then, transfer the DNA to a tube containing TE buffer (~400 µL). Allow DNA to dissolve overnight at 4°C and quantitate by A_{260}.

SOLUTION AND REAGENT RECIPES

CAUTION: See the Cautions Appendix for appropriate handling of materials marked with <!>.

Resuspension Buffer

Reagent	Quantity (for 100 mL)	Final concentration
Tris (1 M, pH 8.0) <!>	2 mL	20 mM
EDTA (0.5 M, pH 8.0) <!>	0.2 mL	1 mM
NaCl (5 M)	0.2 mL	10 mM
Triton X-100 (10%, w/v) <!>	1 mL	0.1% (w/v)
H_2O	96.6 mL	

Store indefinitely at 4°C.

SDS Extraction Buffer

Reagent	Quantity (for 500 mL)	Final concentration
Tris (1 M, pH 7.5) <!>	10 mL	20 mM
EDTA (0.5 M, pH 8.0) <!>	2 mL	21 mM
SDS (10%, w/v) <!>	25 mL	0.5% (w/v)
H_2O	463 mL	

Store indefinitely at room temperature.

T10E0.1 Buffer

Reagent	Quantity (for 100 mL)	Final concentration
Tris (1 M, pH 8.0) <!>	1 mL	10 mM
EDTA (0.1 M, pH 8.0) <!>	0.1 mL	0.1 mM
H_2O	98.9 mL	

Store indefinitely at 4°C.

Protocol 3

Expression and Purification of Active Recombinant T7 RNA Polymerase from *E. coli*

Often for large-scale transcription reactions or for cost savings, a lab may want to prepare its own recombinant T7, SP6, or T3 phage RNA polymerases. It is convenient to perform this prep every 2–3 years and have a consistent and reliable source of phage RNA polymerase for many in vitro transcription reactions (Studier et al. 1990; Ichetovkin et al. 1997). The purified enzyme is stored for months or years at –20°C.

MATERIALS

The recipes for the items marked with <R> are at the end of the protocol or in the Common Recipes Appendix.

CAUTION: See the Cautions Appendix for appropriate handling of materials marked with <!>.

Reagents

Bacterial expression strain: BL21 (DE3; Novagen)
BL21(DE3)-competent cells
Elution buffer <R>
Imidazole (0.5 M, pH 8.0) <!>
Isopropylthiogalactoside (IPTG, 1.0 M) <!>
L broth + ampicillin plates
Leupeptin (5 mg/mL) <!>
Lysis buffer (1x, 5x) <R>
Lysozyme (10 mg/mL) <!>
M9TB expression medium <R>
Ni-NTA agarose (QIAGEN 1000630) <!>
Phenylmethysulfonyl fluoride (PMSF, 20 mg/mL) <!>
Potassium phosphate (0.2 M, pH 7.5)
SDS polyacrylamide gel
Sodium deoxycholate (8%) <!>
Storage buffer
T7 RNA polymerase expression plasmid

- pT7-911Q obtained from Thomas E. Shrader, Albert Einstein College of Medicine
- pT7-911Q = pQE8 (QIAGEN) with lacIQ in XbaI site and His_6T7-RNAP in BamHI site
- PLac promoter/operator; Amp^R

Wash buffer <R>

Equipment

Bio-Rad Econo-Column (737-4251; 2.5-cm diameter x 10-cm length)
Bottles (250 mL, 1 L)
Centrifuge (e.g., Sorvall RC-3B or equivalent with SS-34 rotor or Sorvall RC-5B with SLA-1500 rotor)
Centrifuge tubes (50 mL; Nalgene)
Cold room

Dialysis equipment
Falcon tubes (50 mL)
Fernbach flasks for cell culture
Ice
Incubator
Oak Ridge tubes (50 mL)
Rocker
Screw-cap tubes (1.5 mL)
Shaker (37°C)
Sonicator

OVERVIEW

The recombinant plasmid expressing T7 RNA polymerase as a his6-tagged molecule is under an IPTG-inducible promoter. The bacteria are lysed by sonication, the his_6-tagged protein in the bacterial lysate is purified by binding to Ni-NTA agarose, and the resin is then extensively washed and eluted with imidazole. The protein is dialyzed against a glycerol-containing storage buffer and then stored at –20°C.

METHOD

Bacterial Protein Expression

Day 1

1. Transform pT7-911Q into BL21-(DE3) competent cells. Plate on L broth + ampicillin plates.

Day 2

2. Prepare M9TB expression medium.
3. Prepare the inoculum by resuspending two to three fresh bacterial colonies (~1 mm in diameter) in 1.0 mL of culture medium.
4. Inoculate the Fernbach flasks at the end of the day, using 100 µL of inoculum per liter culture. Incubate the flasks overnight in a 37°C shaker.

Day 3

5. Check the OD_{600} of the cells in the flasks: The OD_{600} should be 0.5–0.6.
6. Induce cells in the flasks at OD_{600} = 0.5–0.6 with 0.5 mL of 1 M IPTG per flask.
7. Return the flasks to a 37°C shaker for 3 h.
8. Harvest 1 L of cell culture by centrifugation (10 min in a G53 rotor, in 1-L bottles, at 4000 rpm, or at 7000 rpm in an SLA-1500 rotor, in 250-mL bottles, in a Sorvall RC-5B centrifuge).

Bacterial Lysis and Nickel-Agarose Affinity Purification

Day 3

9. Resuspend the cells in 25 mL of lysis buffer per pellet. Transfer the cells to 50-mL Falcon tubes.

10. Add 1.0 mL of 10 mg/mL lysozyme, 50 μL of 20 mg/mL PMSF, and 20 μL of 5 mg/mL leupeptin.
11. Incubate the cells with rocking for 20 min at 4°C.
12. Add 0.25 mL of 8% sodium deoxycholate and incubate with rocking for 20 min at 4°C.
13. Sonicate each pellet four times for 15 sec with a sonicator until broken. Keep the samples in ice during sonication.
14. Transfer samples to 50-mL Nalgene centrifuge tubes. Centrifuge in a Sovall SS-34 rotor at 15,500 rpm for 30 min at 4°C.
15. Decant the supernatant into a clean bottle or flask. Keep it on ice until ready for the next step.
16. Make a column of Ni-NTA agarose (10 mL, 50% slurry) for each liter of starting cell culture. If preparing 2 L, then use 20 mL, 50% slurry.
17. Equilibrate the resin with at least 10 column volumes of lysis buffer.
18. Load the protein sample by gravity flow in a Bio-Rad Econo-Column and collect the flowthrough.
19. Wash the resin with 3 column volumes of lysis buffer followed by 10 column volumes of wash buffer. Collect the flowthrough.
20. Elute T7-RNAP with 10 column volumes of elution buffer. Collect in 1 column volume aliquots.
21. Check all of the fractions from Step 20 on an SDS-polyacrylamide gel (use 5 μL of each fraction to test on a 7.5% gel).
22. Pool the peak fractions containing T7-RNAP (the protein runs at ~100 kD).
23. Dialyze into storage buffer for 24 h, three changes of 100 volumes each.

Day 4

24. Quantitate the purified T7-RNAP by A_{280}. The estimated mg/mL = A_{280} × (dilution factor/1.3). Then, aliquot the enzyme in ~100–200-μL aliquots into 1.5-mL screw-cap tubes on ice in the cold room.
25. Store the purified T7-RNAP at –20°C.
26. A practical way to determine the activity of the purified T7-RNAP is to perform in vitro transcription reactions with serially diluted enzyme from your preparation versus several amounts of commercially available enzyme and assay the incorporation of [α-^{32}P]UTP by TCA precipitation or DEAE filter retention.

TIPS AND TROUBLESHOOTING

1. It is important to use freshly transformed bacterial colonies obtained right after a transformation into BL21(DE3)-competent cells. Do not store the plates at 4°C.
2. To remove all of the bacterial nuclease contamination for your purified protein, it critical that you wash the column extensively with at least 10 column volumes of wash buffer, before the imidazole elution.

SOLUTION AND REAGENT RECIPES

CAUTION: See the Cautions Appendix for appropriate handling of materials marked with <!>.

Bacterial Media (M9 Tryptone Broth + Supplements)

M9 Salts (10x)

For 900 mL: Mix 68 g of Na_2HPO_4, 30 g of KH_2PO_4, and 10 g of NH_4Cl <!> and adjust to 900 mL with ddH_2O. Filter through a 0.2-µm filter; aliquot in sterile bottles.

Tryptone Broth

For 900 mL: Mix 10 g of tryptone, 5.5 g of NaCl, and 900 mL of ddH_2O in a 2-L Fernbach flask. Autoclave for 20 min on a slow exhaust, liquid cycle.

Other Components for Medium (Prepare by Filtration through a 0.2-µm Filter)

Ampicillin (Sodium Salt) (50 mg/mL)

For 10 mL: To 0.5 g ampicillin sodium salt <!> (Sigma-Aldrich A-9518-100G), add ddH_2O to 100 mL. Filter through a 0.2-µm filter. Store at –20°C.

$CaCl_2$ (0.1 M)

For 100 mL: To 1.47 g of ($CaCl_2$•$2H_2O$), add ddH_2O to 100 mL. Filter through a 0.2-µm filter.

Glycerol (25%)

For 100 mL: To 25 mL of glycerol, add ddH_2O to 100 mL. Autoclave.

M9TB Expression Medium

For 1 L: Mix 900 mL of tryptone broth, 100 mL of 10x M9 salts, 1.0 mL of 0.1 M $CaCl_2$, 1 mL of 1 M $MgSO_4$, 8 mL of 25% glycerol, and 0.8 mL of 50 mg/mL ampicillin <!>.

$MgSO_4$ (1 M)

For 100 mL: To 20.33 g of ($MgCl_2$•$6H_2O$), add ddH_2O to 100 mL. Autoclave.

Other Buffers and Reagents

Elution Buffer

Reagent	Quantity (for 200 mL)	Final concentration
Lysis buffer (5x)	40 mL	1x
β-Mercaptoethanol (14.1 M) <!>	71 µL, add fresh	
Imidazole (1 M, pH 8.0) <!>	100 mL	500 mM
H_2O	to 200 mL	

IPTG (1.0 M)

For 5 mL: Dissolve 1.69 g of isopropyl-β-D-thiogalactopyranoside <!> (FW 238.29; Gold BioTechnology I2481C50) in ddH_2O.

Leupeptin (5 mg/mL)

For 10 mL: Dissolve 50 mg of leupeptin <!> (Chemicon E 18) in 100% DMSO <!>. Store in a polypropylene tube at –20°C.

Lysis Buffer (1X)

Reagent	Quantity (for 200 mL)	Final concentration
β-Mercaptoethanol (14.1 M) <!>	4 mL	1X
Imidazole (1 M, pH 8.0) <!>	71 μL, add fresh	1 mM (pH 8.0)
ddH_2O	to 200 mL	

Add β-mercaptoethanol before use or add to diluted buffers.

Lysis Buffer (5X)

Reagent	Quantity (for 200 mL)	Final concentration
Tris-HCl (1 M, pH 8.0)	50 mL	50 mM
NaCl (5 M)	20 mL	500 mM
Glycerol (100%)	50 mL	25% (v/v)

Autoclave. Store for 6 mo at 4°C.

Lysozyme (10 mg/mL)

For 10 mL: Mix 100 mg of chicken egg white (Sigma-Aldrich L-6876) with ddH_2O to 10 mL. Prepare fresh.

PMSF (20 mg/mL)

For 100 mL: Dissolve 2 g of PMSF <!> (Sigma-Aldrich P-7626) in isopropanol <!>. Store in a dark glass bottle at –20°C.

Potassium Phosphate (0.2 M, pH 7.5)

For 1 L: To 27.2 g K_2HPO_4 (monobasic; FW 136.01), add ddH_2O to 950 mL. Adjust pH to 7.5 with 10 M KOH <!>. Autoclave.

Sodium Deoxycholate (8%)

For 10 mL: Dissolve 0.8 g of deoxycholic acid <!> and sodium salt (Sigma-Aldrich D-6750) in ddH_2O. Prepare fresh.

Storage Buffer

Reagent	Quantity (for 1 mL)	Final concentration
Potassium phosphate (0.2 M, pH 7.5)	100 mL	20 mM
NaCl (5 M)	20 mL	100 mM
Glycerol	500 mL	50%
Dithiothreitol (DTT, 0.5 M) <!>	20 mL	10 mM
EDTA (0.25 M, pH 8.0) <!>	0.4 mL	0.1 mM
NaN_3 powder (Sigma-Aldrich) <!>	2 g	0.2%

Wash Buffer

Reagent	Quantity (for 100 mL)	Final concentration
Lysis buffer (5x)	100 mL	1x
β-Mercaptoethanol (14.1 M) <!>	177.5 μL, add fresh	5 mM
Imidazole (1 M, pH 8.0) <!>	5 mL	10 mM
H_2O	to 500 mL	

REFERENCES

Ichetovkin IE, Abramochkin G, Shrader TE. 1997. Substrate recognition by the leucyl/phenylalanyl-tRNA-protein transferase. Conservation within the enzyme family and localization to the trypsin-resistant domain. *J Biol Chem* **272:** 33009–33014.

Studier FW, Rosenberg AH, Dunn JJ, Dubendorff JW. 1990. Use of T7 RNA polymerase to direct expression of cloned genes. *Methods Enzymol* **185:** 60–89.

Protocol 4

Preparing Size Markers for Gel Electrophoresis

In any procedure involving gel electrophoreses, it is necessary to run appropriate size markers in parallel. The migration of these markers tells one if the gel has run correctly, i.e., achieved the desired separation. Selection of size markers depends on the size of the expected (experimental) material. If a 20-mer is expected, then size markers with a range of from 100 kb to 1 kb is clearly inappropriate. Similarly, if a band of 1 kb is expected, size markers from 10 nucleotides to 100 nucleotides are also inappropriate. A good rule of thumb is to choose markers that cover the expected size; ideally, one marker should be identical or nearly identical in migration to the expected band. Happily, there is now a huge range of size markers available from commercial suppliers. These can be single-stranded or double-stranded DNAs and RNAs. Labeling of these markers is easily achievable in most cases by kinasing (see Protocol 1, Chapter 3). When using labeled markers, it is important to load the appropriate amount of markers, i.e., such that the appearance of marker bands is approximately the same as that anticipated for the experimental loading. Overloading or underloading can be problematic when attempting to achieve proper exposures.

Here, we present two simple methods for preparing markers that we have found to be useful for a number of applications. The first is alkaline ladders of end-labeled RNA. We note that there are many ways to produce good alkaline ladders. In addition to the one presented here, other methods are presented in Chapter 5, Protocols 6 and 8. These ladders are useful in the size range of 1 to 100 nucleotides and provide single-base resolution. The second method is the labeling of DNA fragments produced by digestion of pBR322 by the restriction enzyme MspI. These markers are useful across a wide size range of from 15 nucleotides to 622 nucleotides.

MATERIALS

The recipes for the items marked with <R> are at the end of the protocol or in the Common Recipes Appendix.

CAUTION: See the Cautions Appendix for appropriate handling of materials marked with <!>.

Reagents

[α-^{32}P]dCTP (3000 Ci/mmol) <!>
dATP, dGTP, dTTP (2 mM)
ddH_2O
DNA (MspI digested)
EDTA (1 mM) <!>
Ethanol (100%) <!>
Labeling buffer (10x) <R>
NaOAc
Phenol:chloroform:isoamyl alcohol (25:24:1) <!>
RNA (^{32}P-end labeled) <!>
RNA loading dye <!>
SDS extraction buffer <R>
Sodium bicarbonate (50 mM, pH 9.0)
Tris (10 mM) <!>
tRNA (10 μg)

Equipment

Dry ice <!>
Incubator
Microcentrifuge
SpeedVac

METHODS

RNA Ladder by Alkaline Hydrolysis

This procedure is useful for generating an RNA marker ladder with ^{32}P 5′- or 3′-end-labeled RNA.

1. Ethanol-precipitate the end-labeled RNA with 10 μg tRNA (see Chapter 2, Protocol 3).
2. Resuspend the RNA pellet in 10 μL of 50 mM sodium bicarbonate (pH 9.0, adjust pH with NaOH).
3. Incubate the pellet for 5–15 min at 90°C or for 20 min at 65°C. Check small aliquots to find the best conditions for the desired size range of the fragments.
4. Terminate the hydrolysis reaction by the addition of 10 μL of 50 mM Tris, pH 7.0, and 20 μL of RNA loading dye.
5. Store the markers at –20°C.
6. Add 10 μg of tRNA to the tube to prevent sticking.

Preparation of Labeled MspI-digested pBR322 Fragments

1. Set up the following on ice:

MspI-digested DNA (1 μg/μL)	2 μL
Labeling buffer (10x)	2 μL
H_2O	8 μL
[α-^{32}P]dCTP (3000 Ci/mmol) followed with 10 μCi/μL	5 μL
dATP, dGTP, dTTP (2 mM each)	2 μL
Klenow (1 unit/uL)	1 uL

Sizes in Base Pairs of MspI-digested Fragments

622	238	160	76
527	217	147	67
404	201	123	34
307	190	110	26
242	180	90	15

2. Incubate the mixure for 30 min at room temperature.
3. Add 2 μL of 2 mM dCTP.
4. Incubate the mixture for 30 min at room temperature.
5. Dilute with 250 μL of SDS extraction buffer and then add NaOAc to 0.3 M.
6. Extract with phenol:chloroform:isoamyl alcohol and precipitate with 10 μg of tRNA and 600 μL of ethanol (see Chapter 2, Protocol 1).
7. Freeze on dry ice, thaw, and centrifuge at maximum speed in a microcentrifuge for 10 min.
8. Carefully remove and dispose of the ethanol in radioactive waste.
9. Resuspend the pellet in 30 μL of TE buffer. Then, dilute 1 μL of the 30 μL to ~200 μL in TE buffer with 20 μg of tRNA. Store at 4°C.

   tRNA helps prevent the small amounts of DNA in the diluted markers from sticking to the sides of the tube. These markers will last for many weeks. Adjust the amount of dilution according to the age of the [α-^{32}P]dCTP.

SOLUTION AND REAGENT RECIPES

CAUTION: See the Cautions Appendix for appropriate handling of materials marked with <!>.

Labeling Buffer (10x)

Reagent	Quantity (for 1 mL)	Final concentration
NaCl (5 M)	100 µL	500 mM
$MgCl_2$ (1 M) <!>	100 µL	100 mM
Tris (1 M, pH 7.5) <!>	100 µL	100 mM
DTT (1 M) <!>	10 µL	10 mM
H_2O	690 µL	

Store for 1 yr at –20°.

SDS Extraction Buffer

Reagent	Quantity (for 100 mL)	Final concentration
Tris (1 M, pH 7.5) <!>	1 mL	10 mM
EDTA (0.5 M, pH 8.0) <!>	0.2 mL	1 mM
SDS (10%, w/v) <!>	5 mL	0.5%
H_2O	93.8 mL	

Protocol 5

Toeprinting

Toeprinting is a technique that was developed in the translation field to measure precisely the position of ribosomes (or ribosomal subunits) on mRNAs. It relies on the fact that reverse transcriptase is "stopped" when it encounters a blocking complex (RNP or protein) on the RNA. The length of the primer extension product produced when the RNA is occupied by such a complex compared to the full-length primer extension product indicates the 3′ position of the complex of interest. Although primarily used in measuring ribosome occupancy and position, toeprinting can be used in a variety of other contexts. The procedure described below was developed to measure the position (and presence) of the exon junction complex (EJC) deposited on mRNAs as a consequence of splicing. Toeprinting can, in principle, be used whenever the complex of interest is bound stably enough to withstand displacement by transcribing reverse transcriptase.

MATERIALS

The recipe for the item marked with <R> is at the end of the protocol or in the Common Recipes Appendix.

CAUTION: See the Cautions Appendix for appropriate handling of materials marked with <!>.

Reagents

Chloroform:isoamyl alcohol <!> (24:1)
Denaturing loading dyes <!>
Denaturing polyacrylamide gel reagents
DNA marker (^{32}P labeled)
dNTPs (10 mM)
Dithiothreitol (DTT, 0.1 M) <!>
Ethanol (100%) <!>
GlycoBlue (10 μg/mL)
HCl (optional; see Step 3.iv) <!>
$MgCl_2$ (100 mM) <!>
NaOAc (optional; see Step 3.iv)
Phenol <!>
Primer (5′-end labeled for primer extension)
Reverse transcriptase (SuperScript II) 200 units/μL
RNase inhibitor (10 units/μL)
RNA substrate (trace labeled, low specific activity, see Chapter 4, Protocol 1)
SDS extraction buffer <R>
TBE buffer

Equipment

Gel dryer
Gel electrophoresis equipment and power supply to run a 6%–8% denaturing polyacrylamide gel
Microfuge tubes (1.5 mL)
Phosphorimaging or autoradiography equipment
Water baths (37°C and 65°C) (optional)

METHOD

For best results, complex formation must have greater than 50% efficiency. If this is not the case, the complex must be enriched, e.g., by gradient fractionation.

1. Split the complexed RNA (or gradient fraction containing enriched RNA complex) into two parts. Extract one part with phenol:chloroform:isoamyl alcohol (25:24:1), add 20 µg GlycoBlue, and precipitate with ethanol (see Chapter 2).

   This RNA (designated control) will be used as a marker for uncomplexed substrate or product and will also show any nicks or primer extension stops that are not the result of protein binding.

2. Resuspend the control RNA in its original volume of reaction buffer or gradient buffer so that it is equivalent to the nonextracted portion.
3. Set up primer extension reactions on control RNA and complexes.
   i. Mix together the following:

Reagent	Quantity	Final concentration
Control or complex in reaction or gradient buffer	35 µL	50–100 mM NaCl
RNase inhibitor	1 µL	0.8 units/µL
dNTPs (10 mM)	2.5 µL	0.5 mM each dCTP, dATP, dGTP, dTTP
$MgCl_2$ (100 mM)	*µL	5 mM total
DTT (100 mM)	5 µL	10 mM
5′-End-labeled primer	2.5 µL	~2 ng or 100,000 cpm

*Samples in gradient buffer or reaction buffer will contain NaCl or KCl as well as Mg and buffer. Include these contributions when calculating final concentrations in the reverse-transcriptase reactions. Make sure that the $MgCl_2$ concentration is 3 mM greater than the dNTP concentration. Total dNTP concentration is 2 mM, so $MgCl_2$ should be 5 mM.

   ii. Incubate the reaction for 10 min at room temperature.

   iii. Add 1 µL of reverse transcriptase (SuperScript II) and incubate for 45 min at 37°C.

   If complexes are not stable, try lower temperatures starting with 30°C.

   iv. If the labeled RNA in the complexes interferes with potential primer extension products, add a base hydrolysis step before extraction as follows:

   a. Add 1 µL of 2 N NaOH and incubate for 10 min at 65°C.

   b. Neutralize with HCl. Continue with Step v.

   v. Dilute the reaction with 0.2 mL of SDS extraction buffer and add NaOAc to 0.3 M. Extract with phenol:chloroform:isoamyl alcohol, add 20 µg of GlycoBlue, and precipitate with ethanol.

   vi. Resuspend the primer extensions in 10 µL of H_2O and analyze 2.5–5 µL on a 6%–8% denaturing polyacrylamide gel.

   Sequencing ladders made using the same primer and substrate are the best markers, but if this is not possible, an end-labeled 10-nucleotide ladder is acceptable.

   vii. Dry gel and visualize bands by phosphorimagery or autoradiography (Fig. A-2).

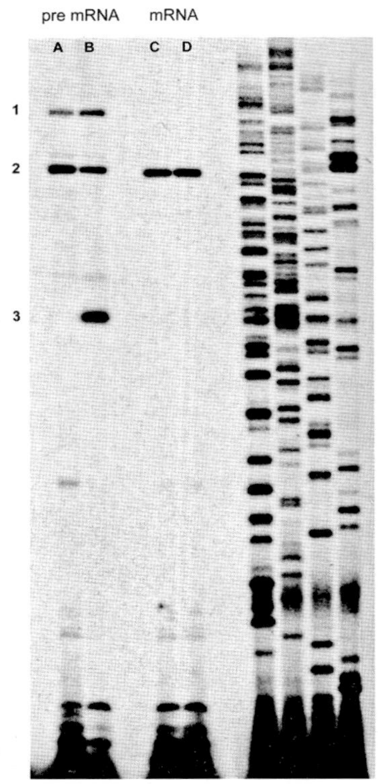

FIGURE A-2. Toeprint of EJC on spliced mRNA substrate. (Lanes *A,C*) Control samples that were deproteinized before primer extension; (Lanes *B,D*) complexes purified on sucrose gradients. Bands caused by blocks to primer extension are from (*1*) inspliced substrate, (*2*) spliced mRNA, and (*3*), the 3′ boundary of the "toeprint" of EJC on the spliced mRNA substrate. The markers are a dideoxy sequencing ladder by the same primer and substrate. (Nilsen lab, unpubl.)

SOLUTION AND REAGENT RECIPES

CAUTION: See the Cautions Appendix for appropriate handling of materials marked with <!>.

SDS Extraction Buffer

Reagent	Quantity (for 100 mL)	Final concentration
Tris (1 M, pH 7.5) <!>	2 mL	20 mM
EDTA (0.5 M, pH 8.0) <!>	0.2 mL	1 mM
SDS (10%, w/v) <!>	5 mL	0.5% (w/v)
H_2O	92.8 mL	

Store indefinitely at room temperature.

READING LIST

Hartz D, McPheeters DS, Traut R, Gold L. 1988. Extension inhibition analysis of translation initiation complexes. *Methods Enzymol* **164:** 419–425.

Sachs MS, Wang Z, Gaba A, Fang P, Belk J, Ganesan R, Amrani N, Jacobson A. 2002. Toeprint analysis of the positioning of translation apparatus components at initiation and termination codons of fungal mRNAs. *Methods* **26:** 105–114.

Protocol 6

Cell Fractionation

It is often desirable to determine the subcellular localization (nuclear or cytoplasmic) of particular RNAs (or proteins). This is often done by immunological-based (e.g., immunofluorescence) or hybridization-based (e.g., in situ hybridization) methods. However, these methods require sophisticated equipment and are often not quantitative. Biochemical fractionation offers a simple, reliable, and, in most cases, quantitative alternative to these methods. It is important to realize that preparation of "pure" nuclear fractions is usually incompatible with the simultaneous preparation of "pure" cytoplasmic fractions. Accordingly, the two fractions should be prepared in parallel. For preparation of clean nuclei, we recommend Dounce homogenization, followed by treatment with nonionic detergent (e.g., NP-40 or Triton X-100), and centrifugation through a 30% sucrose cushion. When using this procedure, it is important that all intact cells be broken (some nuclei will inevitably be damaged). The remaining nuclei will be stripped of any residual cytoplasmic material by the detergent, and the sucrose cushion effectively separates intact nuclei from cytoplasmic debris.

The opposite approach is taken to prepare clean cytoplasmic material. Here, one wants to only mildly break the cells by Dounce homogenization (≥50%–60%), under conditions where nuclear breakage is minimized. Again, the extract is treated with nonionic detergent (to recover material that may be associated with the nuclear membrane) and the remaining intact cells and nuclei are removed by centrifugation through a sucrose cushion. The material above the cushion is clean cytoplasm. These methods are quite reliable, except in cases where nuclear material "leaks" into the cytoplasmic fraction. Such cases are impossible to overcome by biochemical techniques unless anhydrous methods are used, the description of which is beyond the scope of this manual.

When suitable fractions have been prepared, it is necessary to assess their "purity." This can be done either by western blotting (for proteins) by or northern blotting for RNA. A good protein marker for the cytoplasm is tubulin—it should be abundant in the cytoplasm and absent from the nucleus. Proliferating cell nuclear antigen (PCNA) is a good protein marker for the nucleus—it should be present in nuclear material and absent from the cytoplasm. Alternatively, depending on the cell type, transcription factors can also be good nuclear markers.

For RNAs, U6 snRNA is found almost exclusively in the nucleus and small cytoplasmic RNAs (e.g., rho) are present exclusively in the cytoplasm. These simple controls will show whether the fractionation was successful.

Protocol 7

hnRNP-Enriched Nucleoplasmic Extracts from *Drosophila* or Mammalian Tissue Culture Cells

This procedure is based on the preparation of nucleoplasmic extracts from mammalian cells (Pinol-Roma et al. 1990a,b; Swanson and Dreyfuss 1999). In these fractions, heterogeneous nuclear ribonucleoprotein (hnRNP) complexes are enriched and can be isolated using antibodies to specific RNA-binding proteins, associated RNA isolated and used for microarray analysis (also known as RIP-Chip) (Labourier et al. 2002; Blanchette et al. 2004, 2009; Olson et al. 2007) or HTS (high-throughput RNA sequencing) (RIP-Seq).

MATERIALS

The recipes for the items marked with <R> are at the end of the protocol or in the Common Recipes Appendix.

CAUTION: See the Cautions Appendix for appropriate handling of material marked with <!>.

Reagents

Buffer A <R>
Drosophila S2 (L2, Schneider Line 2) cells (4–16 L)
HNEB2 buffer <R>
HNEB2–30% sucrose <R>
Liquid nitrogen <!>
Phosphate-buffered saline (1×) + $MgCl_2$ <!> <R>
Phosphate-buffered saline (PBS; 10×) <R>

Equipment

Beckman Accuspin GS-6R tabletop refrigerated centrifuge (for 50-mL Corning tubes)
Centrifuge tubes (3 × 1-1/2 inch ultraclear; Beckman)
Centrifuge tubes (round-bottomed polycarbonate Oak Ridge screw cap)
Cold room
Conical tubes (50 mL, Corning orange cap)
Dounce homogenizer (40 mL; Bellco 1984-10040) with A (loose) pestle
Eppendorf tubes (1.5 mL for 1-mL aliquots)
Forceps or hemostat
Ice
Microscope (phase contrast)
Pipettes (two, wide-bore, 10 mL; Bellco Glass)
Plastic pipettes (10 mL, 5 mL)
Sonicator (e.g., Branson Sonifier 450)
Sorvall RC3 low-speed centrifuge (for 1-L bottles)
Sorvall RC5C superspeed centrifuge (for 50-mL polycarbonate tubes)
Sorvall SS-34 fixed-angle rotor
Spatula (curved, metal)
SW28 rotor and buckets (Beckman)
Tube clamp

OVERVIEW

The procedure starts by harvesting suspension cultures of *Drosophila* L2 (S2; Schneider Line 2) cells and washing them in PBS + $MgCl_2$. The cells are resuspended in hypotonic buffer (Buffer A) and then homogenized. Nuclei are isolated by centrifugation and then sonicated. The nuclear sonicate is then placed on a 30% sucrose cushion and sedimented. The soluble nuclear ribonucleoprotein (RNP) complexes remain in the supernatant and the nuclear membrane fragments and chromatin pellet through the sucrose cushion.

METHOD

1. Harvest 4–16 L of suspension *Drosophila* L2 cells by centrifugation in a Sorvall RC3 centrifuge in 1-L bottles at 4000 rpm (~3000*g*) for 10 min at 4°C.
2. Use a 10-mL wide-bore pipette to resuspend the cells in 1x PBS + $MgCl_2$. Transfer the mixture to a Corning orange cap conical tube.
3. Centrifuge the cells in a Beckman GS-6R tabletop centrifuge at 2800 rpm (1800*g*) for 10 min at 4°C.
4. Record the packed cell volume (PCV). The PCV is the volume of the cells, once washed with PBS and centrifuged. Use the graduations on the 50-mL Corning tubes to determine and record the PCV. Remove the supernatant by decanting or use a pipette.

   The supernatant may be a little cloudy.
5. Resuspend the cell pellet in 5x PCV of hypotonic Buffer A. Resuspend gently with a 10-mL wide-bore pipette. Make sure to resuspend the cell pellet completely, but do it gently, because the cells can lyse from too vigorous mixing.
6. Centrifuge the cells in a Beckman GS-6R tabletop centrifuge at 2800 rpm (1800*g*) for 10 min at 4°C.
7. Resuspend the cell pellet in 2x PCV of hypotonic Buffer A. Resuspend the pellet gently with a 10-mL wide-bore pipette. Make sure to resuspend the cell pellet completely, but do it gently, because the cells can lyse from too vigorous mixing.
8. Allow the cells to swell for 10 min on ice in the cold room. Pour the cells from the 50-mL Corning tube into a 40-mL Dounce homogenizer on ice.
9. Use 15 firm strokes to homogenize the cell suspension in a Dounce homogenizer with the loose A pestle. Use a strong downstroke and a slow upstroke, so as not to implode the bottom of the homogenizer.
10. Check the fraction of lysed cells on a phase-contrast microscope. If the cells are not >90% lysed, use five more strokes.
11. Transfer the cell lysate to 50-mL clear polycarbonate Oak Ridge screw-cap centrifuge tubes and centrifuge the nuclei in a Beckman GS-6R tabletop centrifuge at 3000 rpm (2000*g*) for 10 min at 4°C.
12. Remove the cloudy supernatant with a 10-mL plastic pipette. The nuclear pellet will be loose but remove as much of the cloudy supernatant as possible.
13. Transfer the tubes to the Sorvall SS-34 rotor and centrifuge at 15,000 rpm (27,000*g*) in the Sorvall 5B centrifuge for 10 min at 4°C. Remove the supernatant with a 5-mL plastic pipette.
14. Add 2 PCV of HNEB2 buffer to the tubes. Dislodge the nuclear pellet with a curved metal spatula and use a 10-mL wide-bore pipette to transfer the nuclei (in chunks) to a 40-mL Dounce homogenizer on ice. Resuspend the nuclear pellet with a few strokes of the tight B pestle. Transfer the nuclear suspension to a 50-mL Corning orange cap tube.

15. Sonicate the nuclear suspension for 3 x 30 sec on the high setting and allow 30 sec on ice between bursts to cool.
16. Gently load the nuclear lysate onto an 18-mL 30% sucrose cushion (in HNEB2 buffer) in a 3 x 1-1/2 inch Beckman ultraclear centrifuge tube. Use forceps or a hemostat to transfer the tubes to precooled Beckman SW28 rotor buckets and close the caps.
17. Centrifuge the tubes at 4800 rpm (~4000*g*) in the SW28 rotor for 15 min at 4°C. Set the centrifuge controls to slow acceleration and brake off.
18. Use forceps or a hemostat to remove the tubes and place into a tube clamp in the cold room. Remove the top half of the liquid (i.e., the light phase) and use a 10-mL pipette to aliquot into 1-mL fractions in cold 1.5-mL microcentrifuge tubes.
19. Flash-freeze the tubes in liquid nitrogen. Store at –80°C.

SOLUTION AND REAGENT RECIPES

CAUTION: See the Cautions Appendix for appropriate handling of materials marked with <!>.

PBS (1x) + $MgCl_2$

Reagent	Quantity (for 1 L)	Final concentration
PBS (10x)	100 mL	1x
ddH_2O	900 mL	
$MgCl_2$ <!>•$7H_2O$	1 g	

Store for 6 mo to 1 yr at 4°C.

Buffer A

Reagent	Quantity (for 250 mL)	Final concentration
HEPES-KOH <!> (1 M, pH 7.6)	0.5 mL	10 mM
$MgCl_2$ (1 M) <!>	375 µL	1.5 mM
KCl (1 M) <!>	2.5 mL	10 mM
DTT (0.5 M) <!>	250 µL (added fresh)	0.5 mM
ddH_2O	to 250 mL	

Store for 6 mo to 1 yr at 4°C.

HNEB1 Buffer

Reagent	Quantity (for 1 L)	Final concentration
HEPES-KOH <!> (1 M, pH 7.6)	15 mL	15 mM
KCl (3 M) <!>	3.3 mL	10 mM
$MgCl_2$ (1 M) <!>	2.5 mL	2.5 mM
Sucrose	119.8 g	350 mM
EDTA (0.5 M) <!>	0.2 mL	0.1 mM

(*continued*)

HNEB1 Buffer (continued)

Reagent	Quantity (for 1 L)	Final concentration
DTT (1 M) <!>	add 1 mL just before use	1 mM
PMSF (0.2 M) <!> (see recipe below)	add 1 mL just before use	0.2 mM
ddH_2O	adjust to 1 L and filter with 0.2-mm filter	

Store at 4°C.

HNEB2 Buffer

Reagent	Quantity (for 250 mL)	Final concentration
Tris-HCl (1 M, pH 7.5)	2.5 mL	10 mM
NaCl (5 M)	5 mL	100 mM
$MgCl_2$ (1 M) <!>	625 µL	2.5 mM
Triton X-100 (20%) <!>	6.25 mL	0.5%
PMSF (0.2 M) <!> (see recipe below)	add 250 µL just before use	0.2 mM
Protease inhibitor cocktail (1000x) (see recipe below)	add 250 µL just before use	1x
ddH_2O	adjust to 250 mL and filter with 0.2-mm filter	

Store at 4°C.

HNEB2–30% Sucrose

Reagent	Quantity (for 250 mL)	Final concentration
Tris-HCl (1 M, pH 7.5)	2.5 mL	10 mM
NaCl (5 M)	5 mL	100 mM
$MgCl_2$ (1 M) <!>	625 µL	2.5 mM
Triton X-100 (20%) <!>	6.25 mL	0.5%
Ultrapure sucrose (MP Biomedicals)	75 g	30% sucrose (w/v)
PMSF (0.2 M) <!> (see recipe below)	add 250 µL just before use	0.2 mM
Protease inhibitor cocktail (1000x) (see recipe below)	add 250 µL just before use	1x
ddH_2O	adjust to 250 mL and filter with 0.2-mm filter	

Store for 6 mo to 1 yr at 4°C.

PBS (10x) (Tissue Culture)

Reagent	Quantity (for 1 L)
NaCl	80 g
KCl <!>	2 g
Na_2HPO_4 (anhydrous dibasic) *or* $Na_2HPO_4 \cdot 7H_2O$	9.2 g 17.38 g
KH_2PO_4 (monobasic)	2 g
ddH_2O	to 1 L

Autoclave and then store for 6 mo to 1 yr at room temperature.

Protease Inhibitor Cocktail (1000x) (1x = 0.5 µg/mL)

Reagent	Quantity (for 10 mL)
Aprotinin <!>	5 mg
Leupeptin <!>	5 mg
Pepstatin	5 mg
Chymotrypsin	5 mg
Antipain <!>	5 mg
DMSO <!>	7 mL
ddH_2O	3 mL

Aprotinin, leupeptin, pepstatin, chymotrypsin, and antipain in 70% DMSO, 0.5 mg/mL stock = 1000x. Combine in a 15-mL screw-cap polypropylene tube and mix overnight at room temperature. Make 100-µL aliquots in 0.5-mL Eppendorf tubes and store at –20°C.

PMSF (Phenylmethylsulfonylfluoride; 100 mM, MW 174.2)

Reagent	Quantity (for 10 mL)
PMSF (Sigma-Aldrich P-7626) <!>	1.74 g
Isopropanol <!>	to 100 mL

Use gloves when preparing. Store at –20°C in a dark glass bottle. This compound is unstable in H_2O.

READING LIST

Blanchette M, Labourier E, Green RE, Brenner SE, Rio DC. 2004. Genome-wide analysis reveals an unexpected function for the *Drosophila* splicing factor U2AF50 in the nuclear export of intronless mRNAs. *Mol Cell* **14:** 775–786.

Blanchette M, Green RE, MacArthur S, Brooks AN, Brenner SE, Eisen MB, Rio DC. 2009. Genome-wide analysis of alternative pre-mRNA splicing and RNA binding specificities of the *Drosophila* hnRNP A/B family members. *Mol Cell* **33:** 438–449.

Kaneko S, Manley JL. 2005. The mammalian RNA polymerase II C-terminal domain interacts with RNA to suppress transcription-coupled 3′ end formation. *Mol Cell* **20:** 91–103.

Labourier E, Blanchette M, Feiger JW, Adams MD, Rio DC. 2002. The KH-type RNA-binding protein PSI is required for *Drosophila* viability, male fertility, and cellular mRNA processing. *Genes Dev* **16:** 72–84.

Niranjanakumari S, Lasda E, Brazas R, Garcia-Blanco MA. 2002. Reversible cross-linking combined with immunoprecipitation to study RNA-protein interactions in vivo. *Methods* **26:** 182–190.
Olson S, Blanchette M, Park J, Savva Y, Yeo GW, Yeakley JM, Rio DC, Graveley BR. 2007. A regulator of Dscam mutually exclusive splicing fidelity. *Nat Struct Mol Biol* **14:** 1134–1140.
Pinol-Roma S, Choi YD, Dreyfuss G. 1990a. Immunological methods for purification and characterization of heterogeneous nuclear ribonucleoprotein particles. *Methods Enzymol* **181:** 317–325.
Pinol-Roma S, Swanson MS, Matunis MJ, Dreyfuss G. 1990b. Purification and characterization of proteins of heterogeneous nuclear ribonucleoprotein complexes by affinity chromatography. *Methods Enzymol* **181:** 326–331.
Swanson MS, Dreyfuss G. 1999. Preparation of heterogeneous nuclear ribonucleoprotein complexes. *Methods Mol Biol* **118:** 299–308.
Vasudevan S, Steitz JA. 2007. AU-rich-element-mediated upregulation of translation by FXR1 and Argonaute 2. *Cell* **128:** 1105–1118.

Protocol 8

Measuring the Length of Poly(A) Tails

Adenylation status has an important role in the regulation of mRNA metabolism: mRNAs are deadenylated before degradation, microRNAs (miRNAs) can cause deadenylation, and the poly(A) length of certain mRNAs is regulated during development. The three methods described here can be used to measure the poly(A) tail length of specific mRNAs. These methods include, in order of increasing sensitivity, (1) northern blotting of intact and experimentally deadenylated mRNAs, (2) northern blotting of intact and experimentally deadenylated mRNA fragments that have been cleaved near the 3′ end with RNase H, and (3) PCR-based poly(A) length measurement.

MATERIALS

The recipes for the items marked with <R> are at the end of the protocol or in the Common Recipes Appendix.

CAUTION: See the Cautions Appendix for appropriate handling of material marked with <!>.

Reagents

Acrylamide (6%), 7 M urea gel
Agarose gel
Bromophenol blue <!>
For the northern-blotting methods only (see Chapter 3, Protocols 7–9):
  Gene-specific oligonucleotide (for high-resolution method; see Steps 4–6)
  Oligo(dT)$_{16-18}$
  Probe for northern blot
    For Method 1, use asymmetric PCR to synthesize a probe that hybridizes to the coding region. For Method 2, make sure to design the probe to the region of the mRNA 3′ to the site of RNase H cleavage in the untranslated region (UTR); coding-region probes for standard northern blots will not recognize the 3′UTR fragments produced during this procedure in Method 2.
  RNase H (1 unit/μL)
  RNase H buffer (5×) <R>
  RNasin
Reagents for phenol:chloroform extraction and ethanol precipitation <!>
RNA of interest
RNase H buffer <R>
Transfer buffer <R>

Equipment

Blot block paper
Cafeteria tray (large)
Gel boxes (agarose or acrylamide) and appropriate power supplies
GeneScreen Plus
Ice
Incubators, temperature blocks, or water baths (37°C and 95°C)
Materials for northern hybridization
Microcentrifuge
Whatman 3MM paper

OVERVIEW

In the first method, the sizes of intact mRNA and experimentally deadenylated mRNA are compared by northern hybridization; this approach can be used to estimate the size of a poly(A) tail on small mRNAs up to 1–2 kb. The second method involves the use of a gene-specific oligonucleotide that directs RNase H cleavage 100–300 nucleotides upstream of the 3′ end (the polyadenylation site) of the mRNA of interest. Cleavage by RNase H releases a short fragment of mRNA that can be visualized on a high-resolution acrylamide northern blot. The third method relies on the ability of yeast poly(A) polymerase to add a limited stretch of G or I residues to the 3′ end of an RNA. One primer is designed to anneal to these residues (and part of the adjoining poly[A] tail), and another primer is designed 150–200 bases upstream of the polyadenylation site. The method can be used to measure the poly(A) tail length of any mRNA.

Before beginning any of these procedures, decide which method is most appropriate. It is unlikely that the first procedure will provide sufficient resolution if the mRNA of interest is >1 kb. In contrast, the second northern blotting procedure is valuable for longer RNAs. Both northern blotting procedures are only practical for mRNAs of moderate to high abundance. The PCR method can be applied to all mRNAs, regardless of size and abundance.

METHODS

Method 1: Low-resolution Measurement of Poly(A) Tails Using Northern Blotting

1. Divide the purified RNA sample into two aliquots for subsequent analysis. Leave one aliquot as an untreated control.
2. Deadenylate the RNA of interest.
   i. For a 50-µL final reaction volume, combine the following:

RNA (purified)	5–20 µg
Oligo(dT)$_{16–18}$	100 ng
H_2O	to 39 µL

   ii. Heat the mixture for 1 min at 95°C and then transfer to ice.
   iii. Add 10 µL of 5x RNase H buffer and 1 µL of 1 unit/µL RNase H.
   iv. Incubate for 45 min at 37°C.
   v. Dilute reaction to 200 µL and then extract the RNA with phenol:chloroform. Then, precipitate the RNA with ethanol and 20 µg of GlycoBlue as described in Chapter 2.
3. Resuspend the pellet in 10 µL of H_2O. Run equal aliquots (5–10 µL) of untreated and deadenylated mRNAs on an agarose gel and detect by northern blotting as described in Chapter 3, Protocols 9 and 10. Compare the sizes of the fragments (Fig. A-3).

   The difference in size gives an approximation of the size of the poly(A) tail. If the mRNA of interest is deadenylated or not adenylated, there will be no difference between the size of the intact RNA and the RNA that has been experimentally deadenylated. See Tips and Troubleshooting.

Method 2: High-resolution Measurement of Poly(A) Tails Using Northern Blotting

1. Divide the purified RNA sample into four aliquots for subsequent analysis. Leave one aliquot (the fourth aliquot) as an untreated control.
2. Add 50 ng of the gene-specific oligonucleotide (designed to be complementary to the mRNA ~200 nucleotides upstream of the 3′ end) to each of the first two aliquots of RNA. In addition,

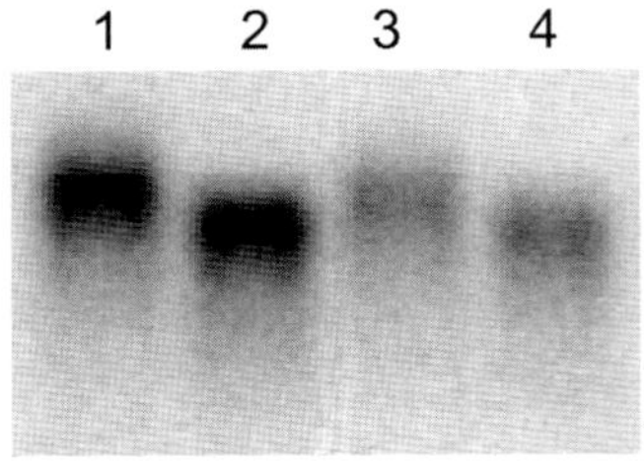

FIGURE A-3. Low-resolution northern blot. Low-resolution measurement of a deadenylation status. (Lanes *1,3*) Intact mRNA was detected by northern blotting, (lanes *2,4*) mRNA was deadenylated via the use of oligo(dT) and RNase H. The reduction in size (compare lanes *1* and *2* and *3* and *4* after deadenylation) shows that the intact mRNA was polyadenylated. (Adapted from Behm-Ansmant et al. 2006.)

add 100 ng of oligo$(dT)_{16-18}$ to the second aliquot. Digest both samples with RNase H exactly as described in Method 1, Step 2.

3. Fractionate the samples on an acrylamide gel and detect by northern blotting as described in Chapter 3, Protocol 11 or by continuing with the northern blot protocol below. Compare the sizes of the fragments.

Alternative Protocol for Polyacrylamide Northern Blots

This method works well for RNAs <600 nucleotides in length.

1. Denature and run the samples on a 0.4-mm-thick 6% acrylamide, 7 M urea gel.
2. Run until the Bromophenol blue is 22 cm from the wells.
3. Disassemble the gel plates and leave the gel on the large back plate.
4. Trim the gel. Only the part containing samples will be transferred.
5. Transfer of the gel will be from the bottom up. Remember to mark the membrane for orientation.
   - i. Cut to the size of the gel six pieces of Whatman 3MM filter paper, four pieces of blot block paper (or eight pieces of Whatman 3MM paper), and one piece of GeneScreen Plus for each gel to be transferred. Presoak the GeneScreen Plus membrane and a four-piece stack of blot block paper in transfer buffer.
   - ii. Set up transfer in a large cafeteria tray. Place four pieces of blot block as the bottom layer. Wet with enough transfer buffer to saturate the paper plus have extra buffer in the tray.
   - iii. Pick up the gel with a dry piece of Whatman 3MM paper and place it gel-side up on the wet filter paper.
   - iv. Lay the prewet GeneScreen Plus membrane on top of the gel. Be careful to line it up correctly before setting it down. Do not pick it up once it is put down.
   - v. Put five dry pieces of Whatman 3MM paper on top of the GeneScreen Plus membrane. Then place 1/2 inch of paper towels on top of the Whatman 3MM paper. Do not let the dry filter paper or towels get wet. Use plastic wrap to keep them out of the buffer tray.
   - vi. Place a glass plate on top of the stack and then a weight (such as a 500-mL bottle filled with H_2O). Transfer overnight.
   - vii. After transfer, disassemble and process the blot as described in Chapter 3, Protocol 8 (Fig. A-4A,B).

TIPS AND TROUBLESHOOTING

1. The untreated control should be blank in the region of the RNA fragment because the intact mRNA usually does not reach that part of the gel.

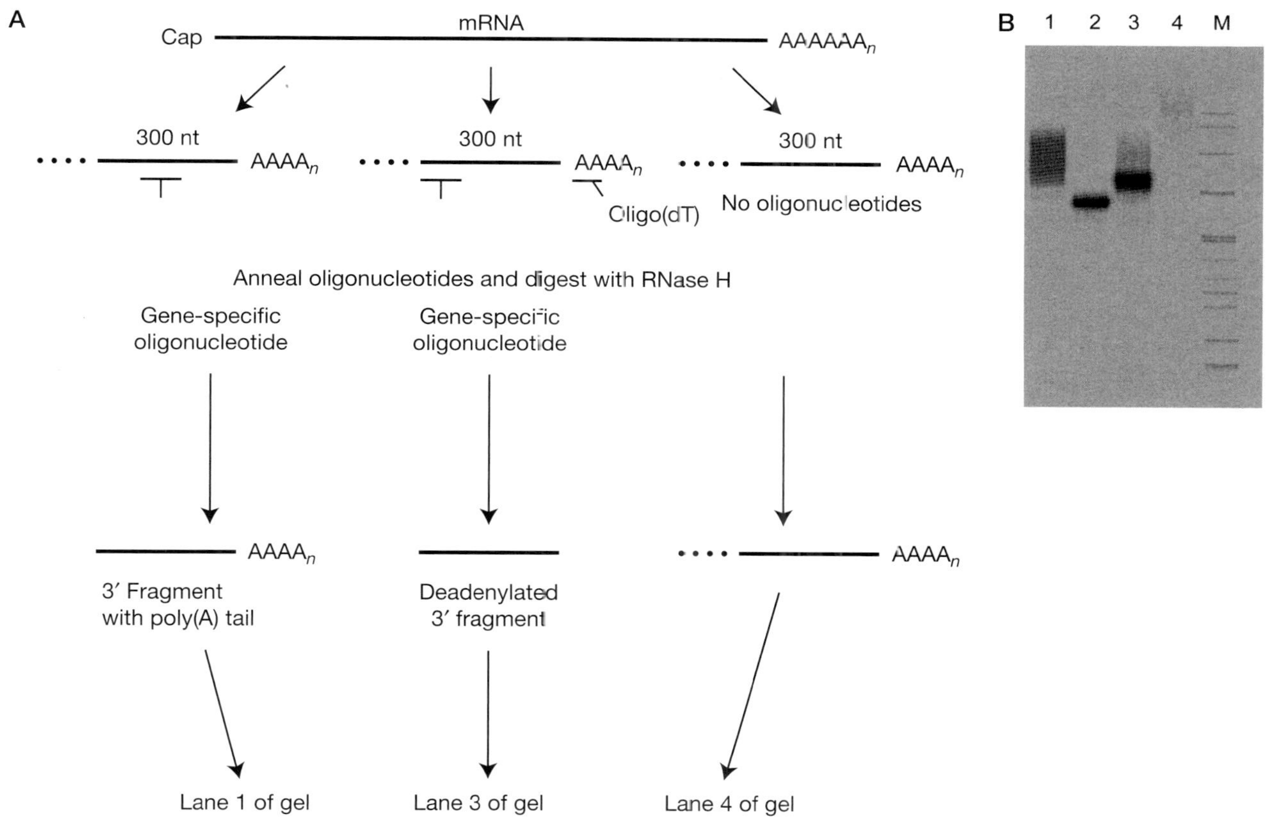

FIGURE A-4. (*A*) Schematic illustration of high-resolution northern blot technique to measure poly(A) tail length. (*B*) High-resolution northern blot. High-resolution measurement of poly(A) tail length by northern blotting. (Lane *1*) Signal obtained when the mRNA was cleaved with RNase H ~300 nucleotides upstream of the 3′ end, (lane *3*) same mRNA after deadenylation, with olido(dT) and RNase H, (lane *2*) same mRNA cleaved by two oligonucleotides (300 nucleotides upstream of the 3′ end (see note 2, below) and 20 nucleotides upstream of the 3′ end). The heterogeneous distribution of bands in lane 1 is expected because the poly(A) tail undergoes progressive deadenylation in the cell. (Lane *4*) Uncleaved mRNA, (lane *M*) labeled markers. (See Protocol 4 in Protocols Appendix.) (Nilsen lab, unpubl.)

2. In some cases, it is desirable to use a second oligonucleotide that is complementary to the last 20 bases of the 3′UTR in Method 2, Step 2. The use of both oligonucleotides with RNase H will produce a fragment of a discrete size and ensures that the experiment was carried out as planned. It will serve as a marker for calculating the poly(A) tail length.

Method 3: PCR-based and Ligation-mediated Approaches for Measuring Poly(A) Tail Length

For low-abundance mRNAs, it is very difficult (if not impossible) to measure poly(A) tail length by northern blot. Two alternative approaches are available for such cases. The first is a PCR-based approach outlined schematically in Figure A-5. Although this approach can be performed with commercially available reagents, we recommend starting with a kit such as the one described in Figure A-5 that contains all of the required components. We have used this kit and found that it performs well.

An alternative to the Affymetrix/USB kit is a kit from BioScientific (All-Tail Kit). Instead of extension with G/I, this kit uses T4 RNA ligase 2 to attach a linker to the 3′ end. Following reverse transcription, a primer complementary to the linker and a gene-specific primer are used to measure poly(A) tail length. We have not tested this particular kit and therefore cannot comment on its performance.

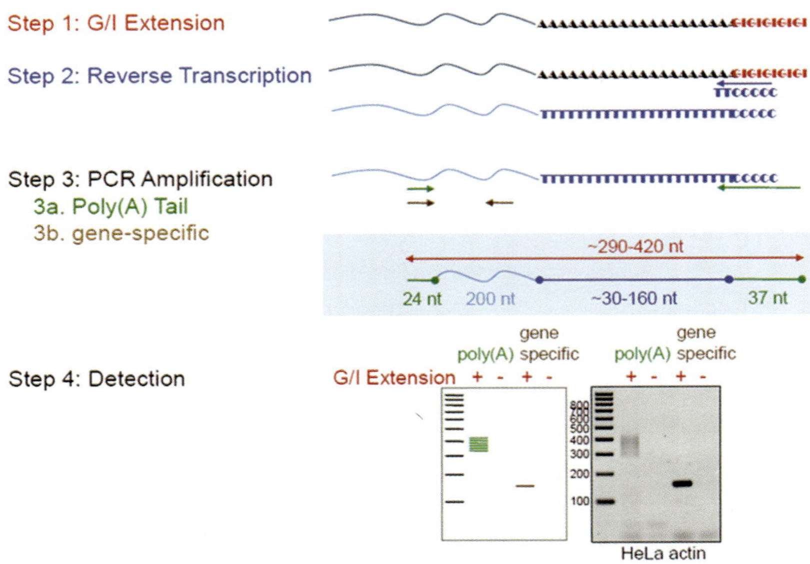

FIGURE A-5. mRNA poly(A) tail length measurement by PCR. (Courtesy of Sangpen Chamnongpol, Affymetrix/USB. Redrawn, with permission of USB Products from Affymetrix, Inc.; http://www.usbweb.com/proto/ 76450a.pdf.)

SOLUTION AND REAGENT RECIPES

CAUTION: See the Cautions Appendix for appropriate handling of material marked with <!>.

RNase H Buffer (5x)

Reagent	Quantity (for 10 mL)	Final concentration (5x)
Tris-HCl (1 M, pH 7.5)	1 mL	100 mM
KCl (2 M) <!>	2.5 mL	0.5 M
$MgCl_2$ (1 M) <!>	0.5 mL	50 mM
Dithiothreitol (DTT, 1 M) <!>	0.5 mL	50 mM
Sucrose	2.5 g	25% (w/w)
H_2O	to 10 mL	

Prepare a stock without DTT for long-term storage. Add DTT to aliquots as needed.

Transfer Buffer (10x SSC)

Reagent	Quantity (for 1 L)	Final concentration
NaCl	87.6 g	1.5 M
Sodium citrate <!>	44.1g	0.15 M
H_2O	to 1 L	

REFERENCE

Behm-Ansmant I, Rehwinkel J, Doerks T, Stark A, Bork P, Izaurralde E. 2006. mRNA degradation by miRNAs and GW182 requires both CCR4:NOT deadenylase and DCP1:DCP2 decapping complexes. *Genes Dev* **20:** 1885–1898.

APPENDIX

4 Cautions

GENERAL CAUTIONS

Please note that the Cautions Appendix in this manual is not exhaustive. Readers should always consult individual manufacturers and other resources for current and specific product information. Chemicals and other materials discussed in text sections are not identified by the icon <!> used to indicate hazardous materials in the protocols. However, without special handling, they may be hazardous to the user. Please consult your local safety office or the manufacturer's safety guidelines for further information.

The following general cautions should always be observed.

- **Before beginning the procedure,** become completely familiar with the properties of substances to be used.
- **The absence of a warning** does not necessarily mean that the material is safe, because information may not always be complete or available.
- **If exposed to toxic substances,** contact your local safety office immediately for instructions.
- **Use proper disposal procedures** for all chemical, biological, and radioactive waste.
- **For specific guidelines on appropriate gloves to use**, consult your local safety office.
- **Handle concentrated acids and bases** with great care. Wear goggles and appropriate gloves. A face shield should be worn when handling large quantities.

  Do not mix strong acids with organic solvents because they may react. Sulfuric acid and nitric acid especially may react highly exothermically and cause fires and explosions.

  Do not mix strong bases with halogenated solvent because they may form reactive carbenes which can lead to explosions.
- **Handle and store pressurized gas containers** with caution because they may contain flammable, toxic, or corrosive gases; asphyxiants; or oxidizers. For proper procedures, consult the Material Safety Data Sheet that must be provided by your vendor.
- **Never pipette** solutions using mouth suction. This method is not sterile and can be dangerous. Always use a pipette aid or bulb
- **Keep halogenated and nonhalogenated** solvents separately (e.g., mixing chloroform and acetone can cause unexpected reactions in the presence of bases). Halogenated solvents are organic solvents such as chloroform, dichloromethane, trichlorotrifluoroethane, and dichloroethane. Nonhalogenated solvents include pentane, heptane, ethanol, methanol, benzene, toluene, *N,N*-dimethylformamide (DMF), dimethyl sulfoxide (DMSO), and acetonitrile.
- **Laser radiation**, visible or invisible, can cause severe damage to the eyes and skin. Take proper precautions to prevent exposure to direct and reflected beams. Always follow the manufactur-

er's safety guidelines and consult your local safety office. See caution below for more detailed information.

- **Flash lamps**, due to their light intensity, can be harmful to the eyes. They also may explode on occasion. Wear appropriate eye protection and follow the manufacturer's guidelines.
- **Photographic fixatives, developers, and photoresists** also contain chemicals that can be harmful. Handle them with care and follow the manufacturer's directions.
- **Power supplies and electrophoresis equipment** pose serious fire hazard and electrical shock hazards if not used properly.
- **Microwave ovens and autoclaves** in the lab require certain precautions. Accidents have occurred involving their use (e.g., when melting agar or bacto-agar stored in bottles or when sterilizing). If the screw top is not completely removed and there is inadequate space for the steam to vent, the bottles can explode and cause severe injury when the containers are removed from the microwave or autoclave. Always completely remove bottle caps before microwaving or autoclaving. An alternative method for routine agarose gels that do not require sterile agar is to weigh out the agar and place the solution in a flask.
- **Ultrasonicators** use high-frequency sound waves (16–100 kHz) for cell disruption and other purposes. This "ultrasound," conducted through air, does not pose a direct hazard to humans, but the associated high volumes of audible sound can cause a variety of effects, including headache, nausea, and tinnitus. Direct contact of the body with high-intensity ultrasound (not medical imaging equipment) should be avoided. Use appropriate ear protection and display signs on the door(s) of laboratories where the units are used.
- **Use extreme caution when handling cutting devices,** such as microtome blades, scalpels, razor blades, or needles. Microtome blades are extremely sharp! Use care when sectioning. If unfamiliar with their use, have an experienced user demonstrate proper procedures. For proper disposal, use the "sharps" disposal container in the lab. Discard used needles *unshielded*, with the syringe still attached. This prevents injuries and possible infections when manipulating used needles because many accidents occur while trying to replace the needle shield. Injuries may also be caused by broken Pasteur pipettes, coverslips, or slides.
- **Procedures for the humane treatment of animals** must be observed at all times. Consult your local animal facility for guidelines. Animals, such as rats, are known to induce allergies that can increase in intensity with repeated exposure. Always wear a lab coat and gloves when handling these animals. If allergies to dander or saliva are known, wear a mask.

GENERAL PROPERTIES OF COMMON CHEMICALS

The hazardous materials list can be summarized in the following categories.

- Inorganic acids, such as hydrochloric, sulfuric, nitric, or phosphoric, are colorless liquids with stinging vapors. Avoid spills on skin or clothing. Spills should be diluted with large amounts of water. The concentrated forms of these acids can destroy paper, textiles, and skin and cause serious injury to the eyes.
- Inorganic bases, such as sodium hydroxide, are white solids that dissolve in water and under heat development. Concentrated solutions will slowly dissolve skin and even fingernails.
- Salts of heavy metals are usually colored, powdered solids that dissolve in water. Many of them are potent enzyme inhibitors and therefore toxic to humans and the environment (e.g., fish and algae).
- Most organic solvents are flammable volatile liquids. Avoid breathing the vapors, which can cause nausea or dizziness. Also avoid skin contact.

- Other organic compounds including organosulfur compounds, such as mercaptoethanol or organic amines, can have very unpleasant odors. Others are highly reactive and should be handled with appropriate care.
- If improperly handled, dyes and their solutions can stain not only your sample, but also your skin and clothing. Some are also mutagenic (e.g., ethidium bromide), carcinogenic, and toxic.
- Nearly all names ending with "ase" (e.g., catalase, β-glucuronidase, or zymolyase) refer to enzymes. There are also other enzymes with nonsystematic names such as pepsin. Many of them are provided by manufacturers in preparations containing buffering substances, etc. Be aware of the individual properties of materials contained in these substances.
- Toxic compounds are often used to manipulate cells. They can be dangerous and should be handled appropriately.
- Be aware that several of the compounds listed have not been thoroughly studied with respect to their toxicological properties. Handle each chemical with appropriate respect. Although the toxic effects of a compound can be quantified (e.g., LD_{50} values), this is not possible for carcinogens or mutagens where one single exposure can have an effect. Also realize that dangers related to a given compound may also depend on its physical state (fine powder vs. large crystals/diethylether vs. glycerol/dry ice vs. carbon dioxide under pressure in a gas bomb). Anticipate under which circumstances during an experiment exposure is most likely to occur and how best to protect yourself and your environment.

HAZARDOUS MATERIALS

Acetic acid (concentrated) must be handled with great care. It may be harmful by inhalation, ingestion, or skin absorption. Wear appropriate gloves and goggles and use in a chemical fume hood.

Acetic acid (glacial) is highly corrosive and must be handled with great care. It may be a carcinogen. Liquid and mist cause severe burns to all body tissues. It may be harmful by inhalation, ingestion, or skin absorption. Wear appropriate gloves and goggles and use in a chemical fume hood. Keep away from heat, sparks, and open flame.

Acrylamide (unpolymerized) is a potent neurotoxin and is absorbed through the skin (the effects are cumulative). Avoid breathing the dust. Wear appropriate gloves and a face mask when weighing powdered acrylamide and methylene-bisacrylamide. Use in a chemical fume hood. Polyacrylamide is considered to be nontoxic, but it should be handled with care because it might contain small quantities of unpolymerized acrylamide.

Actinomycin D is a teratogen and a carcinogen. It is highly toxic and may be fatal if inhaled, ingested, or absorbed through the skin. It may also cause irritation. Avoid breathing the dust. Wear appropriate gloves and safety glasses and always use in a chemical fume hood. Solutions of actinomycin D are light-sensitive.

Aminomethyltrioxalen causes severe skin burns and eye damage. It is very toxic and harmful by inhalation, ingestion, and skin absorption. Wear appropriate gloves and safety goggles. Do not breathe the dust.

Ammonium chloride, NH_4Cl, may be harmful by inhalation, ingestion, or skin absorption. Wear appropriate gloves and safety glasses and use in a chemical fume hood.

Ammonium persulfate, $(NH_4)_2S_2O_8$, is extremely destructive to tissue of the mucous membranes and upper respiratory tract, eyes, and skin. Inhalation may be fatal. Wear appropriate gloves, safety glasses, and protective clothing and use only in a chemical fume hood. Wash thoroughly after handling.

Ampicillin may be harmful by inhalation, ingestion, or skin absorption. Wear appropriate gloves and safety glasses and use in a chemical fume hood.

Aniline is a carcinogen, mutagen, highly toxic, and combustible. It is harmful by inhalation, ingestion, or skin absorption. Wear appropriate gloves and safety goggles and use in a chemical fume hood. Do not inhale the vapor or mist. Keep away from heat, sparks, and open flame.

Antipain may be harmful by inhalation, ingestion, or skin absorption. Wear appropriate gloves and safety glasses and use in a chemical fume hood.

Aprotinin may be harmful by inhalation, ingestion, or skin absorption. It may also cause allergic reactions. Exposure may cause gastrointestinal effects, muscle pain, blood pressure changes, or bronchospasm. Wear appropriate gloves and safety glasses and use only in a chemical fume hood. Do not breathe the dust.

Note: In general, proprietary materials are not listed here. Kits and other commercial items as well as most anesthetics, sedatives, dyes, fixatives, embedding media, stains, herbicides, and fungicides are also not included. Anesthetics and antibiotics also require special care. Follow the manufacturer's safety guidelines that accompany these products.

Aurintricarboxylic acid is an irritant and is harmful by inhalation, ingestion, or skin absorption. Wear appropriate gloves and safety glasses. Do not breathe the dust.

Bisacrylamide is a potent neurotoxin and is absorbed through the skin (the effects are cumulative). Avoid breathing the dust. Wear appropriate gloves and a face mask when weighing powdered acrylamide and methylene-bisacrylamide.

Bis-Tris is an irritant and may be harmful by inhalation, ingestion, or skin absorption. Wear appropriate gloves and safety glasses. Do not breathe the dust

$CaCl_2$, *see* **Calcium chloride**

Cacodylate contains arsenic, is highly toxic, and may be fatal if inhaled, ingested, or absorbed through the skin. Wear appropriate gloves and safety glasses and use in a chemical fume hood. *See also* **Potassium cacodylate; Sodium cacodylate.**

Calcium chloride, $CaCl_2$, is hygroscopic and may cause cardiac disturbances. It may be harmful by inhalation, ingestion, or skin absorption. Do not breathe the dust. Wear appropriate gloves and safety goggles.

Carbon dioxide, CO_2, in all forms may be fatal by inhalation, ingestion, or skin absorption. In high concentrations, it can paralyze the respiratory center and cause suffocation. Use only in well-ventilated areas. In the form of dry ice, contact with carbon dioxide can also cause frostbite. Do not place large quantities of dry ice in enclosed areas such as cold rooms. Wear appropriate gloves and safety goggles.

Chloroform, $CHCl_3$, is irritating to the skin, eyes, mucous membranes, and respiratory tract. It is a carcinogen and may damage the liver and kidneys. It is also volatile. Avoid breathing the vapors. Wear appropriate gloves and safety glasses and always use in a chemical fume hood.

Chloroquine may be harmful by inhalation, ingestion, or skin absorption. Prolonged exposure can lead to permanent eye damage. Wear appropriate gloves and safety goggles.

Citric acid is an irritant and may be harmful by inhalation, ingestion, or skin absorption. It poses a risk of serious damage to the eyes. Wear appropriate gloves and safety goggles. Do not breathe the dust.

CMC/CMCT, *see* **1-Cyclohexyl-3-(2-morpholino-ethyl)carbodiimide metho-*p*-toluene sulfonate)**

Cobalt chloride, $CoCl_2$, is toxic, a possible carcinogen, and is dangerous to the environment. It can cause burns and may be harmful by inhalation, ingestion, or skin absorption. Wear appropriate gloves and safety glasses.

$CoCl_2$, *see* **Cobalt chloride**

Coomassie brilliant blue (CBR) may be harmful by inhalation, ingestion, or skin absorption. Wear appropriate gloves and safety glasses.

1-Cyclohexyl-3-(2-morpholino-ethyl)carbodiimide metho-*p*-toluene sulfonate (CMC/CMCT) may be harmful by inhalation, ingestion, or skin absorption. Wear appropriate gloves and safety glasses. Do not breathe the dust.

DEPC, *see* **Diethyl pyrocarbonate**

Diethyl pyrocarbonate (DEPC) is a potent protein denaturant and is a suspected carcinogen. Aim bottle away from you when opening it; internal pressure can lead to splattering. Wear appropriate gloves, safety goggles, and lab coat and use in a chemical fume hood.

Dimethyl sulfate (DMS), $(CH_3)_2SO_4$, is extremely toxic and is a carcinogen. Avoid breathing the vapors. Wear appropriate gloves and safety glasses and use only in a chemical fume hood. Dispose of solutions containing DMS by pouring them slowly into a solution of sodium hydroxide or ammonium hydroxide and allowing them to sit overnight in the chemical fume hood. Contact your local safety office before reentering the lab to clean up a spill.

Dimethyl sulfoxide (DMSO) may be harmful by inhalation or skin absorption. It easily penetrates the skin and anything dissolved or mixed with it will be absorbed. Wear appropriate gloves and safety glasses and use in a chemical fume hood. DMSO is also combustible. Store in a tightly closed container. Keep away from heat, sparks, and open flame.

Dithiothreitol (DTT) is a strong reducing agent that emits a foul odor. It may be harmful by inhalation, ingestion, or skin absorption. When working with the solid form or highly concentrated stocks, wear appropriate gloves and safety glasses and use in a chemical fume hood.

DMSO, *see* **Dimethyl sulfoxide**

Dry ice, *see* **Carbon dioxide**

DTT, *see* **Dithiothreitol**

Dyes, *follow* **manufacturer's safety guidelines**

EDTA, *see* **Ethylenediamenetetraacetic acid**

Empigen detergent may be harmful by inhalation, ingestion, or skin absorption. Wear appropriate gloves and safety glasses. Do not breathe the vapor, mist, or gas.

Ethanol (EtOH), CH_3CH_2OH, is highly flammable and may be harmful by inhalation, ingestion, or skin absorption. Wear appropriate gloves and safety glasses. Keep away from heat, sparks, and open flame.

Ethanolamine, $HOCH_2CH_2NH_2$, is toxic and harmful by inhalation, ingestion, or skin absorption. Handle with care and avoid any contact with the skin. Wear appropriate gloves and goggles and use in a chemical fume hood. Ethanolamine is highly corrosive and reacts violently with acids.

Ethidium bromide is a powerful mutagen, is toxic and may cause irreversible effects. Consult the local institutional safety officer for specific handling and disposal procedures. Avoid breathing the dust. Wear appropriate gloves when working with solutions that contain this dye.

Ethylenediamenetetraacetic acid (EDTA) may be harmful by inhalation, ingestion, or skin absorption. Wear appropriate gloves and safety glasses. Severe overexposure can result in death.

Formaldehyde, HCHO, is highly toxic and volatile. It is also a possible carcinogen. It is readily absorbed through the skin and is irritating or destructive to the skin, eyes, mucous membranes, and upper respiratory tract. Avoid breathing the vapors. Wear appropriate gloves and safety glasses and always use in a chemical fume hood. Keep away from heat, sparks, and open flame.

Formamide is teratogenic. The vapor is irritating to the eyes, skin, mucous membranes, and upper respiratory tract. It may be harmful by inhalation, ingestion, or skin absorption. Wear appropriate gloves and safety glasses and always use in a chemical fume hood when working with concentrated solutions of formamide. Keep working solutions covered as much as possible.

Gamma rays, *see* **Radioactive substances**

Glyoxal is highly toxic and may cause irreversible effects. It may be harmful by inhalation, ingestion, or skin absorption. Wear gloves and safety glasses. Do not breathe the vapor.

HCl, *see* **Hydrochloric acid**

Heparin is an irritant and may act as anticoagulant subcutaneously or intravenously. It may be harmful by inhalation, ingestion, or skin absorption. Wear appropriate gloves and safety glasses.

H_2O_2, *see* **Hydrogen peroxide**

Hydrazine, N_2H_4, is highly toxic and explosive in the anhydrous state. It may be harmful by inhalation, ingestion, or skin absorption. Avoid breathing the vapors. Wear appropriate gloves, goggles, and protective clothing and use only in a chemical fume hood. Dispose of solutions containing hydrazine in accordance with MSDS recommendations. Keep away from heat, sparks, and open flame.

Hydrochloric acid, HCl, is volatile and may be fatal if inhaled, ingested, or absorbed through the skin. It is extremely destructive to mucous membranes, upper respiratory tract, eyes, and skin. Wear appropriate gloves and safety glasses and use with great care in a chemical fume hood. Wear goggles when handling large quantities.

Hydrogen peroxide, H_2O_2, is corrosive, toxic, and extremely damaging to the skin. It may be harmful by inhalation, ingestion, and skin absorption. Wear appropriate gloves and safety glasses and use only in a chemical fume hood.

Hygromycin B is highly toxic and may be fatal if inhaled, ingested, or absorbed through the skin. Wear appropriate gloves and safety goggles and use only in a chemical fume hood. Do not breathe the dust.

Imidazole is corrosive and may be harmful by inhalation, ingestion, or skin absorption. Wear appropriate gloves and safety glasses and use in a chemical fume hood.

Iodine crystals are volatile at room temperature and form a purple vapor that is intensely irritating to the eyes and mucous membranes. Always wear appropriate gloves and safety glasses when handling iodine and use in a chemical fume hood.

2-Iodoethanol is fatal if swallowed and is combustible. It is harmful by inhalation, ingestion, or skin absorption and is flammable. Wear appropriate gloves and safety goggles and use only in a chemical fume hood. Do not breathe the vapors, mist, or gas.

IPTG, *see* **Isopropyl-β-D-thiogalactopyranoside**

Isoamyl alcohol (IAA) may be harmful by inhalation, ingestion, or skin absorption, and presents a risk of serious damage to the eyes. Wear appropriate gloves and safety goggles. Keep away from heat, sparks, and open flame.

Isopropanol is flammable and irritating. It may be harmful by inhalation, ingestion, or skin absorption. Wear appropriate gloves and safety glasses. Do not breathe the vapor. Keep away from heat, sparks, and open flame.

Isopropyl-β-D-thiogalactopyranoside (IPTG) may be harmful by inhalation, ingestion, or skin absorption. Wear appropriate gloves and safety glasses.

KCl, *see* **Potassium chloride**

KOH, *see* **Potassium hydroxide**

Leupeptin (or its **hemisulfate**) may be harmful by inhalation, ingestion, or skin absorption. Wear appropriate gloves and safety glasses and use in a chemical fume hood.

LiCl, *see* **Lithium chloride**

Liquid nitrogen (LN_2) can cause severe damage due to extreme temperature. Handle frozen samples with extreme caution. Do not breathe the vapors. Seepage of liquid nitrogen into frozen vials can result in an exploding tube upon removal from liquid nitrogen. Use vials with O-rings when possible. Wear cryo-mitts and a face mask. Do not allow the liquid nitrogen to spill onto your clothes. Do not breathe the vapors.

Lithium chloride, LiCl, is an irritant to the eyes, skin, mucous membranes, and upper respiratory tract. It may be harmful by inhalation, ingestion, or skin absorption. Wear appropriate gloves and safety goggles and use in a chemical fume hood. Do not breathe the dust.

Lysozyme is caustic to mucous membranes. Wear appropriate gloves and safety glasses.

Magnesium chloride, $MgCl_2$, may be harmful by inhalation, ingestion, or skin absorption. Wear appropriate gloves and safety glasses and use in a chemical fume hood.

Magnesium sulfate, $MgSO_4$, presents chronic health hazards and affects the central nervous system and the gastrointestinal tract. It may be harmful by inhalation, ingestion, or skin absorption. Wear appropriate gloves and safety glasses and use in a chemical fume hood.

β-Mercaptoethanol (2-Mercaptoethanol), $HOCH_2CH_2SH$, may be fatal if inhaled or absorbed through the skin and is harmful if ingested. High concentrations are extremely destructive to the mucous membranes, upper respiratory tract, skin, and eyes. β-Mercaptoethanol has a very foul odor. Wear appropriate gloves and safety glasses and always use in a chemical fume hood.

Methanol, MeOH or H_3COH, is toxic and can cause blindness, and is highly flammable. It may be harmful by inhalation, ingestion, or skin absorption. Adequate ventilation is necessary to limit exposure to vapors. Avoid inhaling these vapors. Wear appropriate gloves and safety goggles and use only in a chemical fume hood.

Methylene blue is irritating to the eyes and skin. It may be harmful by inhalation, ingestion, or skin absorption. Wear appropriate gloves and safety glasses.

$MgCl_2$, *see* **Magnesium chloride**

$MgSO_4$, *see* **Magnesium sulfate**

3-(*N*-Morpholino)propanesulfonic acid (MOPS) may be harmful by inhalation, ingestion, or skin absorption. It is irritating to mucous membranes and upper respiratory tract. Wear appropriate gloves and safety glasses and use in a chemical fume hood.

MOPS, *see* **3-(*N*-Morpholino)-propanesulfonic acid**

NaOH, *see* **Sodium hydroxide**

$NaPO_4$, *see* **Sodium phosphate**

Neomycin may be harmful by inhalation, ingestion, or skin absorption. Wear appropriate gloves and safety glasses.

Nitrogen (gaseous or liquid) may be harmful by inhalation, ingestion, or skin absorption. Wear appropriate gloves and safety glasses. Consult your local safety office for proper precautions.

Nonidet P40 (NP-40) causes severe eye irritation and burns. It may be harmful by inhalation, ingestion, or skin absorption. Wear appropriate gloves and safety goggles.

NP-40, *see* **Nonidet P40**

Nuclease S1 is an irritant and may be harmful by inhalation, ingestion, or skin absorption. Wear appropriate gloves and safety glasses. Do not breathe the vapor.

Phenol is extremely toxic, highly corrosive, and can cause severe burns. It may be harmful by inhalation, ingestion, or skin absorption. Wear appropriate gloves, goggles, and protective clothing and always use in a chemical fume hood. Rinse any areas of skin that come in contact with phenol with a large volume of water and wash with soap and water; do not use ethanol!

Phenylmethylsulfonyl fluoride (PMSF), $C_7H_7FO_2S$, is a highly toxic cholinesterase inhibitor. It is extremely destructive to the mucous membranes of the respiratory tract, eyes, and skin. It may be fatal by inhalation, ingestion, or skin absorption. Wear appropriate gloves and safety glasses and always use in a chemical fume hood. In case of contact, immediately flush eyes or skin with copious amounts of water and discard contaminated clothing.

PMSF, *see* **Phenylmethylsulfonyl fluoride**

Polyacrylamide is considered to be nontoxic, but it should be treated with care because it may contain small quantities of unpolymerized material (*see* **Acrylamide**).

Potassium cacodylate, *see* **Cacodylate**

Potassium chloride, KCl, may be harmful by inhalation, ingestion, or skin absorption. Wear appropriate gloves and safety glasses.

Potassium hydroxide, KOH, and **KOH/methanol** are highly toxic and may be fatal if swallowed. It may be harmful by inhalation, ingestion, or skin absorption. Solutions are corrosive and can cause severe burns. It should be handled with great care. Wear appropriate gloves and safety goggles.

Proteinase K is an irritant and may be harmful by inhalation, ingestion, or skin absorption. Wear appropriate gloves and safety glasses.

Radioactive substances: When planning an experiment that involves the use of radioactivity, include the physicochemical properties of the isotope (half-life, emission type, and energy), the chemical form of the radioactivity, its radioactive concentration (specific activity), total amount, and its chemical concentration. Order and use only as much as really needed. Always wear appropriate gloves, lab coat, and safety goggles when handling radioactive material. **X-rays** and **gamma rays** are electromagnetic waves of very short wavelengths either generated by technical devices or emitted by radioactive materials. They may be emitted isotropically from the source or may be focused into a beam. Their potential dangers depend on the time period of exposure, the intensity experienced, and the wavelengths used. Be aware that appropriate shielding is usually of lead or other similar material. The thickness of the shielding is determined by the energy(s) of the X-rays or gamma rays. Consult the local safety office for further guidance in the appropriate use and disposal of radioactive materials. Always monitor thoroughly after using radioisotopes. A convenient calculator to perform routine radioactivity calculations can be found at http://graphpad.com/quickcalcs.

RNase A is an irritant and may be harmful by inhalation, ingestion, or skin absorption. Wear appropriate gloves and safety glasses. Do not breathe the dust.

RNase T1 affects the kidneys. It may be harmful by inhalation, ingestion, or skin absorption. Wear appropriate gloves and safety glasses.

Sarkosyl, *see* **Sodium *N*-lauroylsarcosinate**

Sodium azide, NaN_3, is highly poisonous. It blocks the cytochrome electron transport system. Solutions containing sodium azide should be clearly marked. It may be harmful by inhalation, ingestion, or skin absorption. Wear appropriate gloves and safety goggles and handle it with great care. Sodium azide is an oxidizing agent and should not be stored near flammable chemicals.

Sodium borohydride, $NaBH_4$, is corrosive and causes burns. It may be harmful by inhalation, ingestion, or skin absorption. Wear appropriate gloves and safety goggles and use in a chemical fume hood.

Sodium cacodylate may be carcinogenic and contains arsenic. It is highly toxic and may be fatal by inhalation, ingestion, or skin absorption. It also may cause harm to the unborn child. Effects of contact or inhalation may be delayed. Do not breathe the dust. Wear appropriate gloves and safety goggles and use only in a chemical fume hood. *See also* **Cacodylate.**

Sodium citrate, *see* **Citric acid**

Sodium deoxycholate is irritating to mucous membranes and the respiratory tract and may be harmful by inhalation, ingestion, or skin absorption. Wear appropriate gloves and safety glasses when handling the powder. Do not breathe the dust.

Sodium deoxycholic acid, *see* **Sodium deoxycholate**

Sodium dodecyl sulfate (SDS) is toxic, an irritant, and poses a risk of severe damage to the eyes. It may be harmful by inhalation, ingestion, or skin absorption. Wear appropriate gloves and safety goggles. Do not breathe the dust.

Sodium hydroxide, NaOH, and **solutions containing NaOH** are highly toxic and caustic and should be handled with great care. Wear appropriate gloves and a face mask. All other concentrated bases should be handled in a similar manner.

Sodium *N*-lauroylsarcosinate may be harmful by inhalation, ingestion, or skin absorption. Wear appropriate gloves and safety glasses. Do not breathe the dust.

Sodium phosphate, $NaPO_4$, is corrosive and causes burns. It may be harmful by inhalation, ingestion, or skin absorption. Material is irritating to mucous membranes, upper respiratory tract, eyes, and skin. Wear appropriate gloves and safety goggles.

Sodium pyrophosphate is an irritant and may be harmful by inhalation, ingestion, or skin absorption. Wear appropriate gloves and safety glasses. Do not breathe the dust.

Spermidine may be corrosive and cause severe eye and skin burns. It may be harmful by inhalation, ingestion, or skin absorption. Effects may be delayed. Wear appropriate gloves and safety goggles and use in a chemical fume hood.

Streptomycin is toxic and a suspected carcinogen and mutagen. It may cause allergic reactions. It may be harmful by inhalation, ingestion, or skin absorption. Wear appropriate gloves and safety glasses.

SYBR GREEN/GOLD is supplied by the manufacturer as a 10,000-fold concentrate in DMSO which transports chemicals across the skin and other tissues. Wear appropriate gloves and safety glasses and decontaminate according to your safety office guidelines. *See* **DMSO**.

TE buffer contains **Tris-acetate.**

TEMED, *see* ***N,N,N′,N′*-Tetramethylethylenediamine**

Tetracycline may be harmful by inhalation, ingestion, or skin absorption. Wear appropriate gloves and safety glasses and use in a chemical fume hood.

***N,N,N′,N′*-Tetramethylethylenediamine (TEMED)** is highly caustic to the eyes and mucous membranes and may be harmful by inhalation, ingestion, or skin absorption. Wear appropriate gloves and tightly sealed safety goggles.

Thiourea may be carcinogenic and may be harmful by inhalation, ingestion, or skin absorption. Wear appropriate gloves and safety glasses and use in a chemical fume hood.

Trichloroacetic acid (TCA) is highly caustic. Wear appropriate gloves and safety goggles.

Tris is an irritant and may be harmful by inhalation, ingestion, or skin absorption. Wear appropriate gloves and safety glasses.

Trisodium citrate, *see* **Citric acid**

Triton X-100 causes severe eye irritation and burns. It may be harmful by inhalation, ingestion, or skin absorption. Wear appropriate gloves and safety goggles. Do not breathe the vapor.

TRIzol may be fatal if absorbed through the skin, inhaled, or swallowed. It can also cause severe burns. Wear appropriate gloves, safety goggles, and protective clothing and always use in a chemical fume hood. Rinse any areas of skin that come in contact with TRIzol with a large volume of water and wash with soap and water; do not use ethanol!

Trypsin may cause an allergic respiratory reaction. It may be harmful by inhalation, ingestion, or skin absorption. Do not breathe the dust. Wear appropriate gloves and safety goggles. Use with adequate ventilation.

Urea may be harmful by inhalation, ingestion, or skin absorption. Wear appropriate gloves and safety glasses.

UV light and/or **UV radiation** is dangerous and can damage the retina. Never look at an unshielded UV light source with naked eyes. Examples of UV light sources that are common in the laboratory include hand-held lamps and transilluminators. View only through a filter or safety glasses that absorb harmful wavelengths. UV radiation is also mutagenic and carcinogenic. To minimize exposure, make sure that the UV light source is adequately shielded. Wear protective appropriate gloves when holding materials under the UV light source.

Viruses may pose special hazards, and it is recommended that all researchers working with them consult their local safety office and the handbook edited by J.Y. Richmond and R.W. McKinney. 1999. *Biosafety in microbiological and biomedical laboratories (BMBL)*, 4th edition. U.S. Department of Health and Human Services, Centers for Disease Control at http://www.cdc.gov/od/ohs/biosfty/biosfty.htm.

Xylene cyanol is an irritant and may be harmful by inhalation, ingestion, or skin absorption. Wear appropriate gloves and safety glasses. Do not breathe the dust.

Zinc chloride, $ZnCl_2$, is corrosive and poses possible risk to the unborn child. It may be harmful by inhalation, ingestion, or skin absorption. Wear appropriate gloves and safety glasses. Do not breathe the dust.

$ZnCl_2$, *see* **Zinc chloride**

Zymolyase may be harmful by inhalation, ingestion, or skin absorption. Wear appropriate gloves and safety glasses.

Index

A

- A > G Reaction buffer (recipe), 197
- A + G Reaction buffer (recipe), 197, 297
- Absorbance at 260 nm, 8, 67–68
- Acrylamide, 60
- Acrylamide stock (40%; 19:1 acrylamide:bis-acrylamide), 65
- Actinomycin D, 144
- Adenosine triphosphate (ATP)
  - DNA oligonucleotide radiolabeling by terminal deoxynucleotidyl transferase (TdT) (protocol), 92–94
    - materials, 92
    - method, 92–93
    - overview, 92
    - recipes, 93
    - tips and troubleshooting, 93
  - 5′-end labeling of RNA with $[\gamma\text{-}^{32}P]$ATP and T4 polynucleotide kinase (protocol), 231–233
    - materials, 231
    - method, 232
    - overview, 231
    - recipes, 232
    - tips and troubleshooting, 232
  - labeling of oligonucleotide probes (DNA, LNA, RNA) by polynucleotide kinase and $[\gamma\text{-}^{32}P]$ATP (protocol), 89–91, 89f
    - materials, 89–90
    - method, 90
    - overview, 90
    - recipes, 91
    - tips and troubleshooting, 90
- AE buffer (recipe), 42
- Affinity-based techniques
  - basic affinity selection methods, 314–320
    - recipes, 319–320
    - selection of ribonucleoprotein using complementary biotinylated oligonucleotide, 317–319
    - selection with beads, 314–316
    - selection with biotinylated oligonucleotide, 316–317
  - description, 249–252
  - nickel-agarose affinity purification, 536–537
  - problems with, 251–252
  - tagged proteins, 251
    - epitope tags, 251
    - functional tags, 251
  - tagged RNAs, 250–251
- Affymetrix, Inc. microarrays, 422–423, 451–456, 473, 498
- Agarose beads, affinity selection using, 314–316
- Agarose gel electrophoresis
  - formaldehyde-agarose gel, preparation of, 107
  - gel purification of amplified cDNA, 469–470, 469f
  - normalizing different RNA preparations with, 9
  - protocols
    - general protocol, 63–66
      - materials, 64
      - method, 64–65
      - overview, 64
      - recipes, 65–66
      - tips and troubleshooting, 65
    - northern blots: denaturation and electrophoresis of RNA with formaldehyde, 106–109
      - materials, 106–107
      - method, 107–108
      - overview, 107
      - recipes, 108–109
      - tips and troubleshooting, 108
    - northern blots: denaturation and electrophoresis of RNA with glyoxal, 110–113
      - materials, 110–111
      - method, 111–112
      - overview, 111
      - recipes, 112–113
  - quality of RNA purification assessed by, 9
  - "ugly gels," avoiding, 9
- Agarose–acrylamide composite gels, analysis of splicing complexes on, 352–354
- Agilent, Inc. microarrays, 422
- Alkaline hydrolysis, of RNA, 2, 3f
- Alkaline phosphatase
  - in 5′ rapid amplification of cDNA ends (RACE) protocol, 174, 187
  - in CLIP protocol, 483
  - dephosphorylation of RNA using calf intestine alkaline phosphatase or shrimp alkaline phosphatase (protocol), 103–105
- Alkaline phosphatase buffer (10×) (recipe), 177, 188
- Alkaline SDS solution (recipe), 530
- Ambion MEGAscript T7 Kit, 360
- 4′-Aminomethyl trioxsalen (AMT), 249
- Ammonium persulfate, 61
- Amphotropic virus, 399
- Ampicillin (recipe), 538
- AmpliTaq DNA polymerase (Applied Biosystems), 155, 160–161
- AMT (4′-aminomethyl trioxsalen), 249
- Aniline (1 M) (recipe), 274
- Aniline cleavage, 272, 293
- Aniline/acetate buffer (recipe), 197
- Annealing buffer (2×) (recipe), 147, 191, 274
- Annealing buffer (5×) (recipe), 143, 151, 397, 513
- Annealing buffer (6×) (recipe), 237
- Annealing buffer (10×) (recipe), 202
- Annealing buffer (recipe), 391
- Antireverse cap analog (ARCA), 214
- Anti-2,2,7-trimethylguanosine (anti-TMG) antibody, 10
- Aprotinin and leupeptin (recipe), 333
- Argonaute proteins, 356–357, 430
- ArrayScript (Ambion), 156
- Asymmetric polymerase chain reaction to generate single-stranded probes (protocol), 95–96
  - materials, 95
  - method, 96
  - overview, 95
  - recipes, 96
  - tips and troubleshooting, 96
- ATP. *See* Adenosine triphosphate
- Autoclaving, resistance of ribonucleases (RNases) to, 3

B

- Bacterial RNA isolation from *E. coli* (protocol), 44–47
  - discussion, 46–47
  - materials, 44–45
  - method, 45–46
  - overview, 45
  - tips and troubleshooting, 46

Page references followed by f denote figures; those followed by t denote tables.

Bar coding, 506
β-Mercaptoethanol (recipe), 414
BHI medium (recipe), 530
Binding buffer (recipes), 55, 260, 320
Binding master mix (recipe), 256
Bind/wash buffer (recipe), 320
Bioanalyzer (Aligent), RNA quality assessment using, 41–42, 42f, 70–72, 71f
Bioconductor project, 496, 498–500, 504, 507, 509
Bioinformatic analysis
    methods for processing high-throughput sequencing data, 503–510
        classification of changes and associated biology, 509
        experimental design, 504–507
        materials, 503–504
        measuring expression level and changes in expression, 508
        processing and mapping sequencing reads, 507–508
        tips and troubleshooting, 510
        validation, 509
        visualization, 509–510
    methods for processing microarray data, 496–502
        array quality control, data extraction, and normalization, 498
        classification of changes and associated biology, 500
        experimental design, 496–498
        materials, 496
        normalization and scaling, 498–499
        tips and troubleshooting, 501
        transcript detection and change identification in transcript levels, 499
        validation, 500
        visualization, 500–501
    overview, 495
Biotinylated nucleotide incorporation, 213–214
Biotinylated oligonucleotides
    affinity selection with, 316–319
    for rRNA removal by hybrid selection, 51–52
Blocking buffer (recipe), 445
Blocking slides, 441, 442, 442f
Body labeling, 207
Borohydride stop mix (recipe), 198
Boundary analysis
    description, 246, 246f
    example, 300f
    protocol, 298–300
        materials, 298–299
        method, 299–300, 300f
        overview, 299
BPTE buffer (10×) (recipe), 112
Bromophenol blue, 62
Bubbles, micropipettor use and, 13
Buffer A (recipes)
    formaldehyde cross-linking protocol, 313
    hnRNP-enriched nucleoplasmic extract preparation, 550
    hypotonic for nuclear extract preparation, 328
    nuclear extract preparation, 337, 346
    RNA purification protocols, 35, 37
Buffer B (recipes)
    low salt for nuclear extract preparation, 328
    nuclear extract preparation, 337
Buffer C (recipes)
    high salt for nuclear extract preparation, 328
    nuclear extract preparation, 337
Buffer D (recipes)
    dialysis for nuclear extract preparation, 329
    for nuclear extract preparation, 337, 346

C

C > U Reaction buffer (recipe), 199
Ca^{2+}, chelation with EGTA, 3
$CaCl_2$ (recipe), 538
Caenorhabditis elegans, RNA interference in
    discovery of RNAi, 355
    experimental design basics, 358
    overview, 369
    protocols
        dsRNA-induced RNAi in *C. elegans* by feeding with dsRNA-expressing *E. coli,* 370–373
            materials, 370–371
            methods, 371–372
            overview, 371
            pL4440 expression vector, 370f
            recipes, 372–373
            RNAi of L1 larvae in 96-well liquid cultures, 371
            RNAi on plates, 372
        RNAi in *C. elegans* by injection of dsRNA, 374–377
            materials, 374
            method, 374–375
            overview, 374
            recipes, 376–377
            tips and troubleshooting, 375–376
    screens, 416
    targeting multiple mRNA, 375–376
Caenorhabditis elegans Genetics Center, 375
Calcium phosphate transfection, 399–401
Calf intestine alkaline phosphatase (CIP)
    in CLIP protocol, 483
    dephosphorylation of RNA using (protocol), 103–105
        materials, 103
        method, 104
        overview, 103
        recipes, 105
        tips and troubleshooting, 104
Capillary transfer, in northern blot protocols
    alternative method for processing northern blots after capillary transfer, 121–125
        materials, 121
        method, 122–124
        overview, 122
        recipes, 124
    capillary transfer of RNA from agarose gels and filter hybridization using standard stringency hybridization conditions, 114–120
        materials, 114–115
        method, 115–117, 116f
        overview, 115
        recipes, 118–120
        tips and troubleshooting, 118
Capping of in-vitro-synthesized transcripts, 213
Carriers
    colored, 7
    for ethanol precipitation, 7, 27–29
    GlycoBlue (Ambion), 7, 25, 29
    glycogen, RNase-free, 29
    transfer RNA (tRNA), 7, 29
Cautions, 559–565
cDNA
    PCR amplification of, 468
    preparation of fluorescent-dye-labeled cDNA from RNA for microarray hybridization (protocol), 435–439
        discussion, 438–439
        materials, 435–436
        recipes, 439
        tips and troubleshooting, 438
    preparation of small RNA libraries for high-throughput sequencing (protocol)
        gel purification of amplified cDNA, 469–470, 469f
        PCR amplification of cDNA, 468
    purification, 437
    rapid amplification of cDNA ends (RACE)
        3′ rapid amplification of cDNA ends (RACE)
            description, 83, 84f, 85
            protocol, 180–185, 180f, 181f
        5′ rapid amplification of cDNA ends (RACE)
            description, 83, 84f, 85
            protocol, 171–179, 172f
Cell fractionation (protocol), 547
Cell-free systems, 321–354
    developing from poorly studied organisms, 323
    guidelines for testing and optimizing extracts, 322
    guidelines for the preparation of active, 321–322
    protocols, 325–354
        analysis of pre-mRNA splicing using HeLa cell nuclear extracts, 330–333
            materials, 330–331
            method, 331–332, 332f
            overview, 331
            recipes, 332–333
            tips and troubleshooting, 332
        analysis of splicing complexes on native gels, 352–354
            discussion, 354
            materials, 352
            method, 353
            recipes, 354
            tips and troubleshooting, 353–354
        analysis of splicing in vitro using extracts of *Saccharomyces cerevisiae,* 348–351
            ATP depletion, 350
            discussion, 350
            materials, 348–349
            method, 349–350
            overview, 349

recipes, 351
tips and troubleshooting, 350
preparation and analysis of cell-free splicing extracts from *Saccharomyces cerevisiae,* 343–347
discussion, 345
materials, 343–344
method, 344–345
overview, 344
recipes, 346–347
tips and troubleshooting, 345
preparation of *Drosophila* Kc cell nuclear extracts for in vitro splicing, 334–338
materials, 334
method, 335–336
overview, 335
recipes, 337–338
tips and troubleshooting, 336
preparation of nuclear extracts from HeLa cells, 325–329
materials, 325–326
method, 326–327
overview, 326
recipes, 328–329
tips and troubleshooting, 328
in vitro splicing reactions in *Drosophila* Kc cell nuclear extracts, 339–342
materials, 339
method, 340
overview, 340
recipes, 341–342
tips and troubleshooting, 341
Cellular DNA, preparing from nuclei and whole cells (protocol), 532–534
Cesium chloride balance solution (recipe), 530
Cesium chloride (CsCl) centrifugation, after guanidine isothiocyanate solubilization, 5
Cesium chloride gradient, 528, 529f
ChargeSwitch (Invitrogen), 31
Chemical sequencing of end-labeled RNA (protocol), 192–199, 197f
materials, 192–193
method, 193–196
overview, 193
recipes, 197–199
tips and troubleshooting, 196
Chemical structure, of RNA, 1–2, 2f
Chloroform. *See also* PCA (phenol:chloroform: isoamyl alcohol) extraction
in TRIzol extraction procedure, 24
Chloroform:isoamyl alcohol, 19, 21
Church and Gilbert hybridization buffer (recipe), 124, 513
CIP. *See* Calf intestine alkaline phosphatase
CLIP (cross-linking and immunoprecipitation)
identification of RNAs bound by specific protein, 476–490
materials, 476–478
method, 480–489
addition of 454 capture linkers via second PCR step, 489
Dynabead–antibody complex preparation, 481–482
isolation of RNA from nitrocellulose, 486–487
ligation of linkers to 5′ ends of tags, 487
phosphate removal, 483–484
phosphorylation of 5′ ends, 484
purification of PCR-amplified tag sequences, 488–489
recovery of cross-linked RNA from gel, 485–486
removal of contaminating DNA from RNA tags, 487
resolution of RNA–protein cross-linked complexes on a gel, 484–485
RNA linker addition to 3′ ends, 484
RNase treatment (micrococcal nuclease), 483
RNase treatment (RNase A and T1 mix), 482
RT-PCR amplification of linker-ligated tags, 488
UV cross-linking of cultured cells, 480–481
UV cross-linking of tissues, 480
overview, 478–479, 479f
tips and troubleshooting, 489–490
overview, 429, 429f
CMCT, 244, 271, 279
CodeLink slides, 440–445
Colored carrier, 7
Column chromatography, 6
Concentration, calculation of RNA, 361
Coomassie blue stain (recipe), 303, 306
Cordycepin 5′-[α-^{32}P]triphosphate, 228–230
Cross-linking
advantages and disadvantages of methodologies, 250t
CLIP (cross-linking and immunoprecipitation)
identification of RNAs bound by specific protein, 476–490
materials, 476–478
method, 480–489
overview, 478–479, 479f
tips and troubleshooting, 489–490
overview, 429, 429f
formaldehyde, 249, 311–313
materials, 311
method, 312
overview, 311
recipes, 313
tips and troubleshooting, 312–313
RNA–protein interaction detection
description, 248–249
photocross-linking using uridine analogs (protocol), 304–306
materials, 304
method, 305
overview, 305
recipes, 306
tips and troubleshooting, 305–306
ultraviolet photochemical cross-linking to detect RNA-binding proteins (protocol), 301–303
materials, 301
method, 302–303
overview, 302
recipes, 303
tips and troubleshooting, 303
UV cross-linking using RNAs substituted with photoactivatable nucleosides, 249
RNA–RNA interaction detection using psoralen derivatives, 307–310
materials, 307
method, 308–309, 309f
overview, 308
recipes, 310
tips and troubleshooting, 309
Cy3 and Cy5 dyes, 438
Cycle threshold, 168–169, 168f
Cytomegalovirus (CMV) promoter, 393, 401
Cytoplasmic RNA, preparation from tissue culture cells (protocol), 33–35

D

Degradation of RNA
assessment
Bioanalyzer trace, 41–42, 42f
molar ratio of 28S–18S rRNA, 41
causes
metal ions, 2–3
pH, 2, 3f
ribonucleases, 3–4
Deionized glyoxal (recipe), 113
Denaturing stop buffer (recipe), 351
Denhardt's solution (100×) (recipe), 118, 129
DEPC, 3, 247–248, 271
Dephosphorylation
in 5′ rapid amplification of cDNA ends (RACE) protocol, 174, 187
of cleaved RNA in site-specific labeling protocol, 236
enzymatic of RNA using calf intestine alkaline phosphatase or shrimp alkaline phosphatase (protocol), 103–105
materials, 103
method, 104
overview, 103
recipes, 105
tips and troubleshooting, 104
Detection and characterization of RNAs, 75–203
overview of techniques, 75–88
northern blotting, 77
nuclease protection, 77f, 78, 78f
PCR-based approaches, 80–86
end-point PCR, 81
quantitative PCR, 82–83, 82f, 83f
RACE, 83, 84f, 85
semiquantitative PCR, 81–82, 81f
sequencing, 85
primer extension, 78–79, 79f
probe preparation, 75–77, 76t
RNase H digestion, 80, 80f
for small RNAs
gel-shift assay, 87, 87f
native gel electrophoresis, 87, 87f

Detection and characterization of RNAs (*continued*)
splinted ligations, 86, 86f
stem-loop PCR, 87, 88f
protocols, 89–203
3′ rapid amplification of cDNA ends (RACE), 180–185, 180f, 181f
materials, 180–182
method, 182–184
overview, 182
recipes, 184–185
tips and troubleshooting, 184
5′ rapid amplification of cDNA ends (RACE), 171–179, 172f
materials, 171–173
method, 174–177
overview, 173–174
recipes, 177–178
tips and troubleshooting, 177
asymmetric polymerase chain reaction to generate single-stranded probes, 95–96
materials, 95
method, 96
overview, 95
recipes, 96
tips and troubleshooting, 96
detection of RNAs by 3′-end labeling and RNase H digestion, 152–154
materials, 152–153
method, 153
overview, 153
recipes, 154
tips and troubleshooting, 154
direct chemical sequencing of end-labeled RNA, 192–199, 197f
materials, 192–193
method, 193–196
overview, 193
recipes, 197–199
tips and troubleshooting, 196
DNA oligonucleotide radiolabeling by terminal deoxynucleotidyl transferase (TdT), 92–94
materials, 92
method, 92–93
overview, 92
recipes, 93
tips and troubleshooting, 93
enzymatic dephosphorylation of RNA using calf intestine alkaline phosphatase or shrimp alkaline phosphatase (SAP), 103–105
materials, 103
method, 104
overview, 103
recipes, 105
tips and troubleshooting, 104
labeling of oligonucleotide probes (DNA, LNA, RNA) by polynucleotide kinase and [γ-^{32}P]ATP, 89–91, 89f
materials, 89–90
method, 90
overview, 90
recipes, 91
tips and troubleshooting, 90
nick translation of double-stranded DNA for the preparation of ^{32}P-labeled hybridization probes, 100–102
materials, 100
method, 101
overview, 100
recipes, 101–102
tips and troubleshooting, 101
northern blots: alternative method for processing northern blots after capillary transfer, 121–125
materials, 121
method, 122–124
overview, 122
recipes, 124
northern blots: capillary transfer of RNA from agarose gels and filter hybridization using standard stringency hybridization conditions, 114–120
materials, 114–115
method, 115–117, 116f
overview, 115
recipes, 118–120
tips and troubleshooting, 118
northern blots: denaturation and electrophoresis of RNA with formaldehyde, 106–109
materials, 106–107
method, 107–108
overview, 107
recipes, 108–109
tips and troubleshooting, 108
northern blots: denaturation and electrophoresis of RNA with glyoxal, 110–113
materials, 110–111
method, 111–112
overview, 111
recipes, 112–113
northern blots for small RNAs and microRNAs, 126–130
materials, 126–127
method, 127–128
overview, 127
recipes, 129–130
nuclease S1 protection mapping, 136–140, 136f
materials, 137
method, 138
overview, 137–138
recipes, 139–140
tips and troubleshooting, 139
poisoned primer extension, 149–151
materials, 149
method, 150
overview, 149–150
recipes, 151
tips and troubleshooting, 150
primer extension analysis of RNA, 141–144
materials, 141–142
method, 142–143
overview, 142
recipes, 143–144
tips and troubleshooting, 143
quantitative PCR using real-time fluorescence measurements, 165–170
discussion, 168–169, 168f, 169t, 170f
materials, 165–166
melting cycle, 169, 170f
method, 166–167
overview, 166
threshold cycle, 168–169, 168f, 169t
tips and troubleshooting, 168
RACE on uncapped or nonpolyadenylated RNAs, 186–188
materials, 186
method, 187
overview, 186
recipes, 188
tips and troubleshooting, 188
random hexamer ^{32}P radiolabeling of DNA fragments as hybridization probes, 97–99
materials, 97
method, 98
overview, 97
recipes, 99
tips and troubleshooting, 98
reverse primer extension, 145–148, 145f
materials, 145–146
method, 146–147
overview, 146
recipes, 147–148
tips and troubleshooting, 147
reverse transcription–polymerase chain reaction, 155–164, 157f, 163f
discussion, 161–162, 163f
exponential range, determining, 162, 163f
hot-start PCR, 162
materials, 155–156
method, 158–161
overview, 156, 157f
primer design, 161–162
recipes, 164
tips and troubleshooting, 161
RNA sequencing by primer extension, 189–191
materials, 189
method, 190
overview, 189
recipes, 191
tips and troubleshooting, 190
RNase protection assay, 131–135
materials, 131–132
method, 132–133
overview, 132
recipes, 134
tips and troubleshooting, 133–134
splinted ligation method to detect small RNAs, 200–203, 200f
materials, 200–201
method, 201–202
overview, 201
recipes, 202–203
tips and troubleshooting, 202
DGCR-8/Pasha, 399

Dicer, 356–358, 356f, 384
Diethyl pyrocarbonate (DEPC), 3, 247–248, 271
Direct chemical sequencing of end-labeled RNA (protocol), 192–199, 197f
materials, 192–193
method, 193–196
overview, 193
recipes, 197–199
tips and troubleshooting, 196
Divalent metal ions
chelating with EDTA, 2–3
as contaminants of buffer components, 2
RNA degradation and, 2–3
DMS, 247, 270–271, 279
DNA, preparing cellular from nuclei and whole cells (protocol), 532–534
DNA labeling
oligonucleotide probes by polynucleotide kinase and [γ-^{32}P]ATP (protocol), 89–91, 89f
materials, 89–90
method, 90
overview, 90
recipes, 91
tips and troubleshooting, 90
oligonucleotide radiolabeling by terminal deoxynucleotidyl transferase (TdT) (protocol), 92–94
materials, 92
method, 92–93
overview, 92
recipes, 93
tips and troubleshooting, 93
DNA polymerase buffer (5×) (recipe), 147
DNA removal from RNA (protocol), 56–57
materials, 56
method, 56–57
overview, 56
recipes, 57
tips and troubleshooting, 57
DNase
nick translation of double-stranded DNA for the preparation of ^{32}P-labeled hybridization probes (protocol), 100–102
on-column treatment, 31
quick reference guide to DNases, 523
removal of contaminating DNA from RNA tags, 487
treatment of purified RNA sample with, 8, 56–57, 67
DNase buffer (10×), 56
DNase I buffer (recipe), 320
DNase I storage buffer (1×) (recipe), 101
Dot blot, 420
Double-stranded RNA (dsRNA)
Dicer processing of, 356, 358
protein kinase R response to, 358
purification of, 360–361
RNA interference, 359–361
dsRNA synthesis, 360–361
protocols
dsRNA-induced RNAi in *C. elegans* by feeding with dsRNA-expressing *E. coli*, 370–373
dsRNA-induced RNAi in *Drosophila* cells by soaking, 363–365
dsRNA-induced RNAi in *Drosophila* cells by transfection, 366–369
RNAi in *C. elegans* by injection of dsRNA, 374–377
transcription template generation, 359–360
trigger length, 360
trigger selection, 359–360
storage of, 361
synthesis of, 360–361
Dounce homogenization, 322
Drosha, 356f, 357, 384, 399
Drosophila
hnRNP-enriched nucleoplasmic extracts, 548–553
large-scale immunoprecipitation of RNP complexes from *Drosophila* nucleoplasmic extracts for tiling microarrays (RNA immunoprecipitation microarray; RIP-Chip), 471–475
materials, 471–472
method, 472–473
overview, 472
recipes, 474
tips and troubleshooting, 474
preparation of *Drosophila* Kc cell nuclear extracts for in vitro splicing, 334–338
materials, 334
method, 335–336
overview, 335
recipes, 337–338
tips and troubleshooting, 336
RNA interference (RNAi)
encoded triggers, 383–385, 384f, 385f
experimental design basics, 358–359
protocols
construction of transgenic *Drosophila* expressing shRNAs in the miR-1 backbone, 387–392
dsRNA-induced RNAi in *Drosophila* cells by soaking, 363–365
dsRNA-induced RNAi in *Drosophila* cells by transfection, 366–369
RNAi screens, 416
in vitro splicing reactions in *Drosophila* Kc cell nuclear extracts, 339–342
materials, 339
method, 340
overview, 340
recipes, 341–342
tips and troubleshooting, 341
Drosophila RNAi Screening Center (DRSC), 416
DRSC (*Drosophila* RNAi Screening Center), 416
dsRNA. *See* Double-stranded RNA

E

Ecotropic virus, 399, 401
EDTA
chelating metal ions with, 2–3
NaOH (0.1 M)/EDTA (1 mM) (recipe), 140
sodium acetate (0.3 M)/EDTA solution (0.1 mM) (recipe), 199
Tris borate, EDTA (TBE) electrophoresis buffer, 63–64, 66
EDTA/glycogen mixture (recipe), 93
Effectene, 366, 367
EGTA, for Ca^{2+} chelation, 3
Electrophoresis. *See* Gel electrophoresis
Electrophoretic mobility shift assay (EMSA), 257–261
materials, 257–258
overview, 258–259
recipes, 260–261
tips and troubleshooting, 260
Electrophoretic transfer buffer (20×) (recipe), 129
Electroporation, of transgenes into KH2 ES cells, 410–411
Elution buffer (recipes), 55, 320, 538
Emetine, 35
Empigen (recipe), 313
EMSA. *See* Electrophoretic mobility shift assay (EMSA)
3′-End-labeling protocols
detection of RNAs by labeling and RNase H digestion, 152–154
materials, 152–153
method, 153
overview, 153
recipes, 154
tips and troubleshooting, 154
in vitro transcription
labeling of RNA with [^{32}P]pCp and T4 RNA ligase 1, 225–227
materials, 225
method, 226
overview, 226
recipes, 227
tips and troubleshooting, 226
labeling of RNA with yeast poly(A) polymerase and 3′-deoxyadenosine 5′-[α-^{32}P]triphosphate (cordycepin 5′-[α-^{32}P]triphosphate), 228–230
materials, 228
method, 229
overview, 228–229
recipes, 229–230
overview, 208
5′-End-labeling protocols
labeling of RNA with [γ-^{32}P]ATP and T4 polynucleotide kinase, 231–233
materials, 231
method, 232
overview, 231
recipes, 232
tips and troubleshooting, 232
overview, 208
End-labeled RNA
direct chemical sequencing of (protocol), 192–199, 197f
materials, 192–193
method, 193–196
overview, 193
recipes, 197–199
tips and troubleshooting, 196
in vitro–synthesized RNAs, 208

End-point PCR, for detection and characterization of RNAs, 81
Enzymatic dephosphorylation of RNA using calf intestine alkaline phosphatase or shrimp alkaline phosphatase (protocol), 103–105
  materials, 103
  method, 104
  overview, 103
  recipes, 105
  tips and troubleshooting, 104
Enzymes. *See also specific enzymes*
  DNases, 523
  exonucleases for rRNA removal from RNA preparation, 10, 48–49
  importance of K_m, 13
  ribonucleases, 523
Epitope tags, 251
Eri-6/7 locus, 376
Escherichia coli, RNA isolation from (protocol), 44–47
  discussion, 46–47
  materials, 44–45
  method, 45–46
  overview, 45
  tips and troubleshooting, 46
Escherichia coli RNase III, fragmentation of whole-transcriptome RNA using, 457–459
"Ethanol grabbers," 8
Ethanol precipitation
  guidelines, 6–7
    colored carrier use, 7
    procedure, 7
    recovery by centrifugation, 7, 7f
    removal of excess ethanol, 7–8
    RNA concentration required for, 7
    salt concentration, 6
  protocol, 27–29
    expected yield from various tissues and cells, 28t
    materials, 27
    method, 28–29
    overview, 27
    tips and troubleshooting, 29
Ethanolamine, 441, 445
Ethidium bromide, 70
Exonuclease
  for rRNA removal from RNA preparation, 10, 48–49
  Terminator (Epicentre), 48–49

F

Feeder medium (recipe), 415
Fe(II)-EDTA, 276, 278
Filter-binding assay
  description, 241–242
  protocol, 254–256
    materials, 254
    method, 255
    overview, 254–255
    recipes, 256
    tips and troubleshooting, 256
First-strand buffer (5×) (recipe), 164
Flock house virus, 367
Flp-In vector, 407, 407f
Fluorescent dye binding, for RNA and DNA quantification, 68–70
Fluorescent-dye-labeled cDNA, preparation of, 435–439
Fluorometry, 68–70
Footprinting
  description, 244–245, 245f
  mapping sites using chemical methods, 276–283
    CMCT, 279
    DMS, 279
    Fe(II)-EDTA, 276, 278
    kethoxal, 279
    phosphorothioate-substituted RNA, 280–283
  reverse, 245–246, 245f, 288–289, 289f
  RNase, 284–287, 286f
Formaldehyde
  cross-linking, 249, 311–313, 430
    advantages and disadvantages, 250t
    materials, 311
    method, 312
    overview, 311
    recipes, 313
    tips and troubleshooting, 312–313
  northern blots: denaturation and electrophoresis of RNA with, 106–109
    materials, 106–107
    method, 107–108
    overview, 107
    recipes, 108–109
    tips and troubleshooting, 108
Formaldehyde (10%) (recipe), 313
Formamide hybridization buffer (recipe), 134
Formamide/dye loading buffer (recipe), 129, 144, 191, 224, 513
Fragmentation of whole-transcriptome RNA using *E. coli* RNase III, 457–459
Freezing medium (recipe), 415

G

G reaction buffer (recipe), 198
G reaction stop buffer (recipe), 198
Gal4 drivers, 359, 387, 416
Galaxy website, 501, 510
Gel electrophoresis
  detection and characterization of RNAs, 87, 87f
  gel purification of amplified cDNA, 469–470, 469f
  native polyacrylamide-gel-electrophoresis-binding assay for RNA–protein complexes, 257–261
  northern blots for small RNAs and microRNAs (protocol), 126–130
  preparing size markers for, 541–543
  protocols, 58–66
    agarose (nondenaturing), 63–66
    northern blots: denaturation and electro phoresis of RNA with formaldehyde, 106–109
      materials, 106–107
      method, 107–108
      overview, 107
      recipes, 108–109
      tips and troubleshooting, 108
    northern blots: denaturation and electrophoresis of RNA with glyoxal, 110–113
      materials, 110–111
      method, 111–112
      overview, 111
      recipes, 112–113
    polyacrylamide, 58–63, 60f
    purification of in vitro transcripts, 222–224
      materials, 222
      method, 223–224
      overview, 223
      recipes, 224
  recovery of cross-linked RNA from gel, 485–486
  resolution of RNA–protein cross-linked complexes on a gel, 484–485
  yield determination by, 70–72
Gel elution buffer (recipe), 224, 514
  no SDS, 238, 514
  with SDS, 514
Gel mobility shift assay
  description, 242–243
  protocol, 257–261
    materials, 257–258
    overview, 258–259
    recipes, 260–261
    tips and troubleshooting, 260
Gel-running buffer (recipe), 112
Gel-shift assay, 87, 87f
Gene Expression Omnibus, 504
Gene Ontology, 500, 509
GenePix, 446–449
Genomics
  bioinformatic analysis, 495
  data analysis overview, 431–433, 432f
  high-throughput RNA sequencing (HTS)
    applications, 427–428
    overview, 424–428
    seed molecule library creation, 427
    systems
      Illumina, 425, 425f
      pyrosequencing ("454"), 425–426, 426f
      SOLiD, 426–427, 427f
  identifying RNAs associated with specific proteins, 428–431
    antibodies, 430–431
    cross-linking, 428–430
      CLIP, 429, 429f
      formaldehyde, 430
      UV, 428–430
    extract preparation, 431
  microarrays
    array platforms
      Affymetrix, Inc., 422–423
      Agilent, Inc., 422
      "ink-jet" printed arrays, 422
      Nimblegen, Inc., 423, 423f
      pin-printed arrays, 421–422, 422f
    commercial providers of, 420–421
    overview, 419, 420–424

double-label cohybridization of two samples, 421
limitations of arrays, 423–424
single-label one-sample-per-array strategy, 421
tiling arrays, 420
protocols, 435–510
CLIP (cross-linking and immunoprecipitation) identification of RNAs bound, 476–490
materials, 476–478
method, 480–489
overview, 478–479, 479f
tips and troubleshooting, 489–490
fragmentation of whole-transcriptome RNA using *E. coli* RNase III, 457–459
materials, 457
method, 458
overview, 458
recipe, 459
tips and troubleshooting, 458
large-scale immunoprecipitation of RNP complexes from *Drosophila* nucleoplasmic extracts for tiling microarrays (RNA immunoprecipitation microarray; RIP-Chip), 471–475
materials, 471–472
method, 472–473
overview, 472
recipes, 474
tips and troubleshooting, 474
methods for processing high-throughput sequencing data, 503–510
classification of changes and associated biology, 509
experimental design, 504–507
materials, 503–504
measuring expression level and changes in expression, 508
processing and mapping sequencing reads, 507–508
tips and troubleshooting, 510
validation, 509
visualization, 509–510
methods for processing microarray data, 496–502
array quality control, data extraction, and normalization, 498
classification of changes and associated biology, 500
experimental design, 496–498
materials, 496
normalization and scaling, 498–499
tips and troubleshooting, 501
transcript detection and change identification in transcript levels, 499
validation, 500
visualization, 500–501
microarray slide hybridization, 440–445
materials, 440–441
method, 442–445
overview, 441
recipes, 445
tips and troubleshooting, 445
preparation of fluorescent-dye-labeled cDNA from RNA for microarray hybridization, 435–439
discussion, 438–439
materials, 435–436
recipes, 439
tips and troubleshooting, 438
preparation of mRNA-Seq libraries from poly(A)$^+$ mRNA for Illumina transcriptome high-throughput sequencing, 491–494
overview, 491–492
tips and troubleshooting, 492–493
preparation of small RNA libraries for high-throughput sequencing, 460–470
gel purification of 5′- and 3′-ligated RNA product, 466, 467f
gel purification of amplified cDNA, 469–470, 469f
gel purification of 3′-ligated RNA product, 464–465, 465f
linker ligation, 3′, 464
linker ligation, 5′, 466
materials, 460–461
method, 461–470
PCR amplification of cDNA, 468
PmeI digestion of radiolabeled oligonucleotides, 463
recipes, 470
reverse transcription, 467–468
size selection and gel purification of RNA sample, 461–464, 463f
tips and troubleshooting, 470
scanning microarray slides, 446–450
discussion, 449
materials, 446
method, 447–448
overview, 447
tips and troubleshooting, 449–450, 450f
tips on hybridizing, washing, and scanning Affymetrix microarrays, 451–456
materials, 451
method, 452–456
overview, 451–452
tips and troubleshooting, 456
Glass beads, vortexing with, 40–41
Glycerol, 12
Glycerol (25%) (recipe), 538
GlycoBlue (Ambion), 7, 25, 29
Glycogen
EDTA/glycogen mixture (recipe), 93
RNase-free, 29
Glyoxal, denaturation and electrophoresis of RNA with, 110–113
materials, 110–111
method, 111–112
overview, 111
recipes, 112–113
Glyoxal denaturation buffer (recipe), 113
GPS (glutamine/penicillin/streptomycin) solution (recipe), 414
GR1379 strain, 375
"GREAT" program, 500, 509
Guanidine isothiocyanate, solubilization with, 5
Guide strand, siRNA, 357, 378

H

Hairpin, 383, 384f. *See also* shRNA
Hazardous materials
general properties, 560–561
list of materials, 561–565
HBS solution (2x) (recipe), 402
HeLa cells
analysis of pre-mRNA splicing using HeLa cell nuclear extracts, 330–333
materials, 330–331
method, 331–332, 332f
overview, 331
recipes, 332–333
tips and troubleshooting, 332
history of, 325
preparation of nuclear extracts from HeLa cells, 325–329
materials, 325–326
method, 326–327
overview, 326
recipes, 328–329
tips and troubleshooting, 328
Herring testis carrier DNA (recipe), 119, 129
Hexamer mix (recipe), 99
High-throughput sequencing (HTS)
applications, 427–428
bar coding, 506
data analysis, 431–433, 432f
fragmentation of whole-transcriptome RNA using *E. coli* RNase III, 457–459
materials, 457
method, 458
overview, 458
recipe, 459
tips and troubleshooting, 458
methods for processing high-throughput sequencing data, 503–510
classification of changes and associated biology, 509
experimental design, 504–507
materials, 503–504
measuring expression level and changes in expression, 508
processing and mapping sequencing reads, 507–508
tips and troubleshooting, 510
validation, 509
visualization, 509–510
overview, 424–428
preparation of mRNA-Seq libraries from poly(A)$^+$ mRNA for Illumina transcriptome high-throughput sequencing, 491–494
overview, 491–492
tips and troubleshooting, 492–493
preparation of small RNA libraries for high-throughput sequencing, 460–470
gel purification of 5′- and 3′-ligated RNA product, 466, 467f

High-throughput sequencing (HTS) (*continued*)
gel purification of amplified cDNA, 469–470, 469f
gel purification of 3′-ligated RNA product, 464–465, 465f
linker ligation, 3′, 464
linker ligation, 5′, 466
materials, 460–461
method, 461–470
PCR amplification of cDNA, 468
PmeI digestion of radiolabeled oligonucleotides, 468
recipes, 470
reverse transcription, 467–468
size selection and gel purification of RNA sample, 461–464, 463f
tips and troubleshooting, 470
read length, 505–506
seed molecule library creation, 427
systems
Illumina, 425, 425f
pyrosequencing ("454"), 425–426, 426f
SOLiD, 426–427, 427f
High-yield transcription buffer (10×) (recipe), 218
HNEB1 buffer (recipe), 550–551
HNEB2 buffer (recipe), 551
HNEB2–30% sucrose (recipe), 551
hnRNP-enriched nucleoplasmic extracts from *Drosophila* or mammalian tissue culture cells, 548–553
HotStarTaq DNA polymerase, 162
Housekeeping messenger RNA (mRNA), 72
HTS. *See* High-throughput sequencing (HTS)
Hybrid selection, for rRNA removal from RNA preparation, 10, 51–52
Hybridization. *See* Microarrays
Hybridization buffer (5×) (recipe), 134
Hybridization probes
nick translation of double-stranded DNA for the preparation of ^{32}P-labeled hybridization probes (protocol), 100–102
materials, 100
method, 101
overview, 100
recipes, 101–102
tips and troubleshooting, 101
random hexamer ^{32}P radiolabeling of DNA fragments as (protocol), 97–99
materials, 97
method, 98
overview, 97
recipes, 99
tips and troubleshooting, 98
Hydrazine 50% (U reaction buffer) (recipe), 297
Hydrogen peroxide (2.5%, w/v) (recipe), 289
Hydrophilicity of RNA, 27

I

Illumina sequencing process, 425, 425f, 491
Immunoprecipitation
CLIP (cross-linking and immunoprecipitation) identification of RNAs bound by specific protein, 476–490
materials, 476–478
method, 480–489
overview, 478–479, 479f
tips and troubleshooting, 489–490
overview, 429, 429f
for enrichment of RNA species, 10
large-scale immunoprecipitation of RNP complexes from *Drosophila* nucleoplasmic extracts for tiling microarrays (RNA immunoprecipitation microarray; RIP-Chip), 471–475
materials, 471–472
method, 472–473
overview, 472
recipes, 474
tips and troubleshooting, 474
In vitro transcription, 205–239
basics of, 205–206
failure, reasons for, 205–206
labeling of RNAs, 207–209
overview
5′ and 3′ labeling, 208
body labeling, 207
site-specific labeling, 208–209, 209f
protocols
5′-end labeling of RNA with [γ-^{32}P]ATP and T4 polynucleotide kinase, 231–233
3′-end labeling of RNA with [^{32}P]pCp and T4 RNA ligase 1, 225–227
3′-end labeling of RNA with yeast poly(A) polymerase and 3′-deoxyadenosine 5′-[α-^{32}P]triphosphate (cordycepin 5′-[α-^{32}P]triphosphate), 228–230
site-specific labeling and substitution of RNA, 234–239
synthesis, labeling, and substitution, 210–215
protocols, 210–239
determining yield of RNA synthesized in vitro, 220–221
materials, 220
method, 220–221
overview, 220
tips and troubleshooting, 221
5′-end labeling of RNA with [γ-^{32}P]ATP and T4 polynucleotide kinase, 231–233
materials, 231
method, 232
overview, 231
recipes, 232
tips and troubleshooting, 232
3′-end labeling of RNA with [^{32}P]pCp and T4 RNA ligase 1, 225–227
materials, 225
method, 226
overview, 226
recipes, 227
tips and troubleshooting, 226
3′-end labeling of RNA with yeast poly(A) polymerase and 3′-deoxyadenosine 5′-[α-^{32}P]triphosphate (cordycepin 5′-[α-^{32}P]triphosphate), 228–230
materials, 228
method, 229
overview, 228–229
recipes, 229–230
gel purification of RNA, 222–224
materials, 222
method, 223–224
overview, 223
recipes, 224
high-yield synthesis of RNA using T7 RNA polymerase and plasmid DNA or oligonucleotide templates, 216–219
materials, 216
method, 217
overview, 216–217
recipes, 218
tips and troubleshooting, 217–218
site-specific labeling and substitution of RNA, 234–239
cleavage of RNA with chimeric oligonucleotide, 235–236
dephosphorylation of cleaved RNA, 236
ligation of RNA using a bridge oligonucleotide, 236–237
materials, 234–235
method, 235–237
overview, 235
phosphorylation of RNA with [γ-^{32}P]ATP, 236
recipes, 237–239
tips and troubleshooting, 237
synthesis, labeling, and substitution, 210–215
capping and substitution of transcripts, 213
materials, 210
method, 211–214
overview, 211
random incorporation of modified nucleotides, 213–214
recipes, 215
tips and troubleshooting, 214–215
synthesis with modified nucleotides, 207
yield of RNA transcripts, 206
Injection buffer (1×) (recipe), 391
Inorganic pyrophosphatase (IPP) stock and storage buffer (recipe), 218
Integrity, assessing RNA preparations, 8–9
Inverted repeats, as silencing triggers for RNAi in *Drosophila,* 383–385, 384f, 385f
Iodoethanol, 280–283
Iodoethanol (recipe), 289, 297
IPB1 buffer (recipe), 474
IPB2 buffer (recipe), 474
IPTG (recipe), 373, 539
Isoamyl alcohol. *See* PCA (phenol:chloroform: isoamyl alcohol) extraction
Isoforms, RNA, 503, 508
Isopropanol precipitation, 24–25

K

Kc splicing mix (4×) (recipe), 341
Kethoxal, 244, 279
Kinase, 5′-end labeling of RNA with, 208, 231–233
Kinase buffer (10×) (recipe), 143, 203
Kits. *See* Purification kits
Klenow buffer (5×) (recipe), 267, 514
Klenow DNA polymerase, 97–98
K_m, importance of, 13

L

Labeling. *See also* Radiolabeling
  3′-end labeling (protocol), 152–154
    materials, 152–153
    method, 153
    overview, 153
    recipes, 154
    tips and troubleshooting, 154
  of oligonucleotide probes (DNA, LNA, RNA) by polynucleotide kinase and [γ-^{32}P]ATP (protocol), 89–91, 89f
    materials, 89–90
    method, 90
    overview, 90
    recipes, 91
    tips and troubleshooting, 90
  in vitro transcription, 206–209
    5′ and 3′ labeling, 208
    body labeling, 207
    site-specific labeling, 208–209
Labeling buffer (10×) (recipe), 543
Large RNAs, selective precipitation with LiCl or PEG/NaCl, 49–50
LB medium (recipe), 530
Lentivirus, 398, 401
Leupeptin (recipe), 539
Library, for high-throughput sequencing
  protocols
    preparation of mRNA-Seq libraries from poly(A)$^+$ mRNA for Illumina transcriptome high-throughput sequencing, 491–494
      overview, 491–492
      tips and troubleshooting, 492–493
    preparation of small RNA libraries for high-throughput sequencing, 460–470
      gel purification of 5′- and 3′-ligated RNA product, 466, 467f
      gel purification of amplified cDNA, 469–470, 469f
      gel purification of 3′-ligated RNA product, 464–465, 465f
      linker ligation, 3′, 464
      linker ligation, 5′, 466
      materials, 460–461
      method, 461–470
      PCR amplification of cDNA, 468
      PmeI digestion of radiolabeled oligonucleotides, 468
      recipes, 470
      reverse transcription, 467–468
      size selection and gel purification of RNA sample, 461–464, 463f
      tips and troubleshooting, 470
  seed molecule library creation, 427
Ligase, T4 RNA, 152–153
Ligation buffer (2×) (recipe), 203
Ligation buffer (10×) (recipe), 391, 397, 515
Lin-29, 376
Lipofectamine 2000, 380–381
Liquid nitrogen, snap-freezing tissue for RNA extraction, 6
Lithium chloride (LiCl), selective precipitation of large RNAs with, 49–50
LNA labeling of probes by polynucleotide kinase and [γ-^{32}P]ATP (protocol), 89–91, 89f
  materials, 89–90
  method, 90
  overview, 90
  recipes, 91
  tips and troubleshooting, 90
Loading buffer (recipe), 260
Loading dye (recipe), 198, 274
Luria broth (LB) (recipe), 373, 515
Lysis buffer (recipes), 43, 530, 539
Lysozyme
  recipe, 539
  use in RNA isolation from bacteria, 45

M

M9+ medium (recipe), 372, 515
M15 medium (recipe), 415
M15 medium + hygromycin (recipe), 415
M9 salts (recipe), 538
M9 stock (1×) (recipe), 372, 376, 515
Magnetic beads
  affinity selection using, 314–316
  protein A, 474
Mammalian cells, RNA interference in
  encoded triggers, 393–394
  experimental design basics, 359
  protocols
    creating an miR30-based shRNA vector, 395–397
      materials, 395–396
      method, 396–397
      recipes, 397
    creating transgenic shRNA mice by recombinase-mediated cassette exchange, 407–417
      discussion, 415–417
      materials, 407–409
      method, 409–414
      overview, 409
      recipes, 414–415
      tips and troubleshooting, 414
    infection of mammalian cells with retroviral shRNAs, 403–405
      materials, 403
      method, 403–404
      overview, 403
      tips and troubleshooting, 404
    packaging shRNA retroviruses and host cell infection, 399–402
      materials, 399–400
      method, 400
      overview, 400
      recipes, 402
      tips and troubleshooting, 401
    RNAi in cultured mammalian cells using synthetic siRNAs, 380–382
      materials, 380
      method, 380–381
      overview, 380
      tips and troubleshooting, 381
  screens, 416
  synthetic siRNAs, 378–379
Master mixes, importance of, 11
Material safety data sheets (MSDSs), 30
Meiotic cells, isolation of total RNA from yeast, 40–41, 42f
Melting curve, 169–170
Messenger RNA (mRNA)
  enrichment of poly(A)$^+$ mRNA using immobilized oligo(dT) (protocol), 53–55
    materials, 53
    method, 54
    overview, 53
    recipes, 55
    tips and troubleshooting, 54
  enrichment using immobilized oligo(dT) cellulose, 10
  housekeeping, 72
  mRNA-Seq libraries, preparation of, 491–494
  percentage of total cellular RNA, 1
Metal ions
  chelating, 2–3
  RNA degradation and, 2–3
Mg^{++}
  chelation with EDTA, 3
  ribonucleoside triphosphates (rNTPs) requirement for, 206
$MgSO_4$ (recipe), 538
Microarrays
  blocking slides, 441, 442, 442f
  CodeLink slides, 440–445
  commercial providers of, 420–421
  data analysis, 431–433
  overview, 419, 420–424
    array platforms
      Affymetrix, Inc., 422–423
      Agilent, Inc., 422
      "ink-jet" printed arrays, 422
      Nimblegen, Inc., 423, 423f
      pin-printed arrays, 421–422, 422f
    double-label cohybridization of two samples, 421
    limitations of arrays, 423–424
    single-label one-sample-per-array strategy, 421
    tiling arrays, 420
  protocols
    large-scale immunoprecipitation of RNP complexes from *Drosophila* nucleoplasmic extracts for tiling microarrays (RNA immunoprecipitation microarray; RIP-Chip), 471–475

Microarrays (*continued*)
materials, 471–472
method, 472–473
overview, 472
recipes, 474
tips and troubleshooting, 474
methods for processing microarray data, 496–502
array quality control, data extraction, and normalization, 498
classification of changes and associated biology, 500
experimental design, 496–498
materials, 496
normalization and scaling, 498–499
tips and troubleshooting, 501
transcript detection and change identification in transcript levels, 499
validation, 500
visualization, 500–501
microarray slide hybridization, 440–445
materials, 440–441
method, 442–445
overview, 441
recipes, 445
tips and troubleshooting, 445
preparation of fluorescent-dye-labeled cDNA from RNA for microarray hybridization, 435–439
discussion, 438–439
materials, 435–436
recipes, 439
tips and troubleshooting, 438
scanning microarray slides, 446–450
discussion, 449
materials, 446
method, 447–448
overview, 447
tips and troubleshooting, 449–450, 450f
tips on hybridizing, washing, and scanning Affymetrix microarrays, 451–456
materials, 451
method, 452–456
overview, 451–452
tips and troubleshooting, 456
significance analysis of, 499
tiling, 420, 471–475
Micrococcal nuclease, 483
Micrococcal nuclease (recipe), 289
Microinjection, of dsRNA in *C. elegans,* 374–377
Micropipettor
accuracy, checking, 13
problematic solution, 12–13
proper use of, 11–13
RNase contamination of, 4
microRNA (miRNA)
isolation kits for, 31
miR-1, 384, 385, 387–392
miR-30, 384, 393, 395–397
northern blots for (protocol), 126–130
materials, 126–127
method, 127–128
overview, 127
recipes, 129–130
production from primary (pri-miRNAs), 357
regulation of natural, 384
stem-loop PCR, 87, 88f
Minimum Information About a Microarray Experiment (MIAME) conventions, 496
miR-1, 384, 385, 387–392
miR-30, 384, 393, 395–398
miRBase, 87
Mobility shift assay
description, 242–243
protocol, 257–261
materials, 257–258
overview, 258–259
recipes, 260–261
tips and troubleshooting, 260
Modification interference
description, 246–248, 247f
protocol, 291–294
materials, 291–292
method, 292–294
overview, 292
tips and troubleshooting, 294
Modification reaction buffer A (denaturing) (recipe), 275
Modification reaction buffer A (native) (recipe), 274
Modification reaction buffer B (denaturing) (recipe), 275
Modification reaction buffer B (native) (recipe), 275
Modified nucleotides, synthesis of RNA containing, 207, 213–214
Moloney murine leukemia virus (M-MuLV), 393, 398
MOPS buffer (10×) (recipe), 108
Mouse
creating transgenic shRNA mice by recombinase-mediated cassette exchange (protocol), 407–415
encoded RNAi in, 406
mRNA. *See* Messenger RNA (mRNA)
mRNA-Seq libraries, preparation of, 491–494
MSDSs (material safety data sheets), 30
MspI-digested pBR322 fragments, preparation of labeled, 542
M9TB expression medium (recipe), 538
MTT reduction assay, 416
Mut-16 gene, 376
Mycoplasma, infection of cells by, 326

N

NAIM. *See* Nucleotide analog interference mapping (NAIM)
NanoDrop (Thermo-Fisher), 67, 68, 438
Nanospectrophotometer, 40, 67, 68, 69f
NanoVue (GE Healthcare), 67, 68
NaOH (0.1 M)/EDTA (1 mM) (recipe), 140
Native gel-loading buffer (recipe), 354, 515
Native gels, analysis of splicing complexes on, 352–354
Native polyacrylamide-gel-electrophoresis-binding assay for RNA–protein complexes, 257–261
materials, 257–258
overview, 258–259
recipes, 260–261
tips and troubleshooting, 260
Native quench buffer (recipe), 351, 354
Nested PCR, 176–177, 184
NGM plates (recipe), 372–373, 376–377, 516
NGM RNAi plates (recipe), 373
Nhr-23, 376
N-hydroxysuccinimide (NHS) ester dyes, 435, 437, 438
Nick translation buffer (10×) (recipe), 102
Nick translation of double-stranded DNA for the preparation of ^{32}P-labeled hybridization probes (protocol), 100–102
materials, 100
method, 101
overview, 100
recipes, 101–102
tips and troubleshooting, 101
Nickel-agarose affinity purification, 536–537
Nickel-chelate chromatography, 431
Nimblegen, Inc. microarrays, 423, 423f
Nitrocellulose, isolation of RNA from, 486–487
NL2099 strain, 375
Nonpolyadenylated RNAs, RACE of (protocol), 186–188
materials, 186
method, 187
overview, 186
recipes, 188
tips and troubleshooting, 188
Northern blots, 420
advantages of, 77
description, 77
disadvantages of, 77
measuring length of poly(A) tails, 554–556, 556f, 557f
probe preparation, 76t
protocols
alternative method for processing northern blots after capillary transfer, 121–125
materials, 121
method, 122–124
overview, 122
recipes, 124
capillary transfer of RNA from agarose gels and filter hybridization using standard stringency hybridization conditions, 114–120
materials, 114–115
method, 115–117, 116f
overview, 115
recipes, 118–120
tips and troubleshooting, 118
denaturation and electrophoresis of RNA with formaldehyde, 106–109
materials, 106–107
method, 107–108
overview, 107
recipes, 108–109

tips and troubleshooting, 108
denaturation and electrophoresis of RNA with glyoxal, 110–113
materials, 110–111
method, 111–112
overview, 111
recipes, 112–113
for small RNAs and microRNAs, 126–130
materials, 126–127
method, 127–128
overview, 127
recipes, 129–130
for RNA integrity, 9, 73
NT2 buffer (recipe), 267
Nuclear extract preparation protocols. *See also* Cell-free systems
preparation and analysis of cell-free splicing extracts from *Saccharomyces cerevisiae,* 343–347
discussion, 345
materials, 343–344
method, 344–345
overview, 344
recipes, 346–347
tips and troubleshooting, 345
preparation of *Drosophila* Kc cell nuclear extracts for in vitro splicing, 334–338
materials, 334
method, 335–336
overview, 335
recipes, 337–338
tips and troubleshooting, 336
preparation of nuclear extracts from HeLa cells, 325–329
materials, 325–326
method, 326–327
overview, 326
recipes, 328–329
tips and troubleshooting, 328
Nuclear RNA, preparation from tissue culture cells (protocol), 33–35
Nuclease protection
breathing and end fraying, 78
description, 77f, 78, 78f
limitations of, 78
probe preparation, 76t
S1 protection mapping (protocol), 136–140, 136f
materials, 137
method, 138
overview, 137–138
recipes, 139–140
tips and troubleshooting, 139
Nucleotide analog interference mapping (NAIM)
description, 248
protocol, 294–297
materials, 291–292
method, 294–296, 295f, 296f
recipes, 297
tips and troubleshooting, 296
Nucleotide mix (50×) (recipe), 439
Nucleotides, synthesis of RNA containing modified, 207, 213–214

O

Oligo(dT) cellulose, for enrichment of mRNA, 10, 53–54
Oligohistidine, 251
On-Target (Dharmacon), 378
OptiKinase, 90, 187, 201, 236
OptiKinase buffer (10×) (recipe), 91, 188, 238, 516

P

P1 nuclease, 457
^{32}P radiolabeling
of DNA fragments as hybridization probes (protocol), 97–99
materials, 97
method, 98
overview, 97
recipes, 99
tips and troubleshooting, 98
5′-end labeling of RNA with [γ-^{32}P]ATP and T4 polynucleotide kinase (protocol), 231–233
materials, 231
method, 232
overview, 231
recipes, 232
tips and troubleshooting, 232
3′-end labeling of RNA with [^{32}P]pCp and T4 RNA ligase 1 (protocol), 225–227
materials, 225
method, 226
overview, 226
recipes, 227
tips and troubleshooting, 226
3′-end labeling of RNA with yeast poly(A) polymerase and 3′-deoxyadenosine 5′-[α-^{32}P]triphosphate (cordycepin 5′-[α-^{32}P]triphosphate) (protocol), 228–230
materials, 228
method, 229
overview, 228–229
recipes, 229–230
labeling of oligonucleotide probes (DNA, LNA, RNA) by polynucleotide kinase and [γ-^{32}P]ATP (protocol), 89–91, 89f
materials, 89–90
method, 90
overview, 90
recipes, 91
tips and troubleshooting, 90
nick translation of double-stranded DNA for the preparation of ^{32}P-labeled hybridization probes (protocol), 100–102
materials, 100
method, 101
overview, 100
recipes, 101–102
tips and troubleshooting, 101
random hexamer ^{32}P radiolabeling of DNA fragments as hybridization probes (protocol), 97–99
materials, 97
method, 98
overview, 97
recipes, 99
tips and troubleshooting, 98
PAGE. *See* Polyacrylamide gel electrophoresis
Passenger strand, siRNA, 357, 378
PAZ domain, 357
PB buffer (20×) (recipe), 112, 119
PBS (10×) (recipe), 552
PBS (1×) + $MgCl_2$ (recipe), 550
PBS/EDTA/EGTA (recipe), 365, 516
PCA (phenol:chloroform:isoamyl alcohol) extraction, 19–22, 38–41
PCR. *See* Polymerase chain reaction (PCR)
PCR buffer (10×) (recipe), 96, 178, 185, 268, 516
PEG/NaCl, selective precipitation of large RNAs with, 49–50
P element, 388
pH, RNA degradation and, 2, 3f
Phase Lock Gel, 38–39, 44–46
Phenol
contamination of RNA, 8
extraction, 19–22, 20f, 21f
Phosphatase buffer (10×) (recipe), 238, 517
Phosphate-buffered saline (PBS) (recipe), 364, 517
Phosphoramidite chemistry, 422
Phosphorothioate interference analysis, 294–295, 295f
Phosphorothioate-substituted RNA, footprinting with, 280–283
Phosphorylation of RNA with [γ-^{32}P]ATP, 236
PicoGreen, 69
Piperazine-*N,N*′-bis(2-ethanesulfonic acid) (PIPES), 31
Pipetman. *See* Micropipettor
PK buffer (2×) (recipe), 341, 517
PL4440 expression vector, 370f
Plasmid preparation, RNA friendly (protocol), 526–531
materials, 526–527
method, 526–529
cell lysis, 527
cesium chloride gradient, 528, 529f
culture, 527
ethanol precipitation, 529
ethanol/isopropanol precipitation, 527–528
ethidium bromide extraction, 528–529
recipes, 530–531
PmeI digestion of radiolabeled oligonucleotides, 468
PMSF (recipe), 338, 539, 552
Poisoned primer extension
description, 79, 79f
protocol, 149–151
materials, 149
method, 150
overview, 149–150
recipes, 151
tips and troubleshooting, 150

Poly(A)$^{+}$ mRNA enrichment using immobilized oligo(dT) (protocol), 53–55
materials, 53
method, 54
overview, 53
recipes, 55
tips and troubleshooting, 54
Polyacrylamide gel electrophoresis
gel purification of RNA (protocol), 222–224
materials, 222
method, 223–224
overview, 223
recipes, 224
protocol, 58–63, 60f
materials, 58–59
method, 59–62, 60f
cleaning gel plates, 59
disassembling the gel, 62
gel solution preparation, 60
loading the gel, 61–62
plate assembly, 60–61, 60f
pouring the gel, 61
running the gel, 62
setting up the gel, 61
northern blots for small RNAs and microRNAs, 126–130
overview, 59
tips and troubleshooting, 63
Polybrene solution (recipe), 404
Polymerase chain reaction (PCR). *See also* Reverse transcription–polymerase chain reaction
accumulation curves, 81, 81f
asymmetric polymerase chain reaction to generate single-stranded probes (protocol), 95–96
materials, 95
method, 96
overview, 95
recipes, 96
tips and troubleshooting, 96
in CLIP protocol, 488, 489
for detection and characterization of RNAs, 80–86
end-point PCR, 81
quantitative PCR, 82–83, 82f, 83f
RACE, 83, 84f, 85
semiquantitative PCR, 81–82, 81f
sequencing, 85
end point, 81
nested, 176–177, 184
poly(A) tail length measurement, 557, 558f
quantitative, 82–83, 82f, 83f, 165–170
RACE, 83, 84f, 85
RNA yield determination by quantitative or semiquantitative, 72
semiquantitative, 81–82, 81f
stem loop, 87, 88f
Polynucleotide kinase
in CLIP protocol, 484
labeling of oligonucleotide probes (DNA, LNA, RNA) by polynucleotide kinase and [γ-^{32}P]ATP (protocol), 89–91, 89f
Poly(A) polymerase, 187
3′-end labeling of RNA with yeast poly(A) polymerase and 3′-deoxyadenosine 5′-[α-^{32}P]triphosphate (cordycepin 5′-[α-^{32}P]triphosphate) (protocol), 228–230
Poly(A) polymerase buffer (5x) (recipe), 188, 229, 517
Poly(A) tails, measuring length of, 554–559
high-resolution measurement using northern blotting, 555–556, 557f
low-resolution measurement using northern blotting, 555, 556f
materials, 554
PCR based and ligation mediated, 557, 558
recipes, 558
Potassium acetate solution (recipe), 531
Potassium phosphate (recipe), 539
Precipitation. *See also* Ethanol precipitation
isopropanol, 24–25
selective of large RNAs with LiCl or PEG/NaCl, 49–50
Prehybridization/hybridization mix (recipe), 119, 130
Pre-miRNAs, 357
Pre-mRNA splicing. *See* Splicing
Primary microRNAs (pri-miRNAs), 357, 384, 385f
Primer3 (design tool), 360
Primer dimers, 162
Primer extension
description, 78–79, 79f
poisoned
description, 79, 79f
protocol, 149–151
materials, 149
method, 150
overview, 149–150
recipes, 151
tips and troubleshooting, 150
protocol, 141–144
materials, 141–142
method, 142–143
overview, 142
recipes, 143–144
tips and troubleshooting, 143
reverse
protocol, 145–148, 145f
materials, 145–146
method, 146–147
overview, 146
recipes, 147–148
tips and troubleshooting, 147
reverse primer extension, 79, 79f
RNA sequencing by (protocol), 189–191
materials, 189
method, 190
overview, 189
recipes, 191
tips and troubleshooting, 190
Primer extension (PE) mix (5x) (recipe), 144
Probes
asymmetric polymerase chain reaction to generate single-stranded probes (protocol), 95–96
materials, 95
method, 96
overview, 95
recipes, 96
tips and troubleshooting, 96
nick translation of double-stranded DNA for the preparation of ^{32}P-labeled hybridization probes (protocol), 100–102
materials, 100
method, 101
overview, 100
recipes, 101–102
tips and troubleshooting, 101
preparation, overview of, 75–77, 76t
random hexamer ^{32}P radiolabeling of DNA fragments as hybridization probes (protocol), 97–99
materials, 97
method, 98
overview, 97
recipes, 99
tips and troubleshooting, 98
Processive exonuclease, for rRNA removal from RNA preparation, 10
Promoter
cytomegalovirus, 393, 401
Gal4 responsive, 359
phage polymerase, 359–360, 359f
RNA polymerase II, 393
RNA polymerase III, 383
tissue specific, 393
Protease inhibitor cocktail (1000x) (recipe), 552
Protein A magnetic beads, 474
Protein kinase R (PKR), 357–358, 378
Proteinase K, 6, 20
Proteinase K buffer (2x) (recipe), 268, 518
Protein–RNA interactions. *See* RNA–protein interactions
Psoralens
advantages and disadvantages, 250t
detecting RNA–RNA interactions, 249, 307–310
materials, 307
method, 308–309, 309f
overview, 308
recipes, 310
tips and troubleshooting, 309
Purification
description of techniques, 5–6
column chromatography, 6
guanidine cesium chloride, 5
RNA extraction from tissues, 6
SDS/proteinase K/phenol:chloroform, 6
solubilizing and deproteinizing, 5
TRIzol, 5
overview
assessment of quantity and quality, 8–9
enrichment techniques, 10
mRNA enrichment using immobilized oligo(dT) cellulose, 10
precipitation guidelines, 6–8, 7f
resuspension of pelleted RNA, 8
ribosomal RNA (rRNA) removal, 9–10
technique descriptions, 5–6
protocols
bacterial RNA isolation *(E. coli)*, 44–47

discussion, 46–47
materials, 44–45
method, 45–46
overview, 45
tips and troubleshooting, 46
enrichment of poly(A)+ mRNA using immobilized oligo(dT), 53–55
materials, 53
method, 54
overview, 53
recipes, 55
tips and troubleshooting, 54
ethanol precipitation of RNA and use of carriers, 27–29
expected yield from various tissues and cells, 28t
materials, 27
method, 28–29
overview, 27
tips and troubleshooting, 29
gel electrophoresis, 58–66
agarose (nondenaturing), 63–66
polyacrylamide, 58–63, 60f
isolation of total RNA from yeast cell cultures, 38–43
method for meiotic cells, 40–41, 42f
method for vegetative cells, 38–40
quality control, 41–42, 42f
recipes, 42–43
overview, 17
preparation of cytoplasmic and nuclear RNA from tissue culture cells, 33–35
materials, 33
method, 34–35
overview, 33–34
recipes, 34
tips and troubleshooting, 35
purification kit use, 30–32
how kits work, 30–32
problems, 32
table of available kits, 31t
removal of DNA from RNA, 56–57
materials, 56
method, 56–57
overview, 56
recipes, 57
tips and troubleshooting, 57
removal of ribosomal subunits from cytoplasmic extracts before solubilization with SDS and deproteinization, 36–37
materials, 36
method, 37
overview, 36
recipes, 37
tips and troubleshooting, 37
removing rRNA from deproteinized, phenol-extracted total RNA, 48–52
enzymatic digestion of rRNA, 48–49
removal of rRNA by hybrid selection, 51–52
selective precipitation of large RNAs with LiCl or PEG/NaCl, 49–50
SDS solubilization and phenol extraction, 19–22
materials, 19
method, 20–21, 21f
overview, 19–20, 20f
recipes, 22
tips and troubleshooting, 22
strategy outline, 18f
using TRIzol (TRI reagent), 23–26
materials, 23
method, 24–25
overview, 24
recipes, 25
tips and troubleshooting, 25
yield and quality of purified RNA, determining, 67–73
by gel electrophoresis, 70–72
by northern blotting, 73
by PCR, quantitative or semiquantitative, 72
by spectrophotometry or fluorometry, 67–70, 69f
Purification kits, 30–32
how kits work, 30–32
problems with, 32
contamination with DNA or ethanol, 32
inadequate solubilization of starting material, 32
low or no yield, 32
table of available kits, 31t
Pyrosequencing ("454") method, 425–426, 426f

Q

QIAGEN RNeasy kit, 360
qPCR, RNA yield determination by, 72
Quality control
Bioanalyzer trace, 41–42, 42f, 70–72, 71f
molar ratio of 28S–18S rRNA, 41
spectrophotometry readings, 41, 67
Quality of purified RNA, determining, 8–9, 67–73
by gel electrophoresis, 70–72, 71f
by northern blotting, 73
by PCR, quantitative or semicuantitative, 72
by spectrophotometry or fluorometry, 67–70, 69f
Quantitative PCR
for detection and characterization of RNAs, 82–83, 82f, 83f
using real-time fluorescence measurements (protocol), 165–170
discussion, 168–169, 168f, 169t, 170f
materials, 165–166
melting cycle, 169, 170f
method, 166–167
overview, 166
threshold cycle, 168–169, 168f, 169t
tips and troubleshooting, 168
Quantity of purified RNA, determining, 8, 67–72
by gel electrophoresis, 70–72
by PCR, quantitative or semiquantitative, 72
by spectrophotometry or fluorometry, 67–70, 69f

R

RACE. *See* Rapid amplification of cDNA ends (RACE)
Radiolabeling
DNA oligonucleotide radiolabeling by terminal deoxynucleotidyl transferase (TdT) (protocol), 92–94
materials, 92
method, 92–93
overview, 92
recipes, 93
tips and troubleshooting, 93
labeling of oligonucleotide probes (DNA, LNA, RNA) by polynucleotide kinase and [γ-^{32}P]ATP (protocol), 89–91, 89f
materials, 89–90
method, 90
overview, 90
recipes, 91
tips and troubleshooting, 90
nick translation of double-stranded DNA for the preparation of ^{32}P-labeled hybridization probes (protocol), 100–102
materials, 100
method, 101
overview, 100
recipes, 101–102
tips and troubleshooting, 101
PmeI digestion of radiolabeled oligonucleotides, 468
probe preparation, 75–77, 76t
random hexamer ^{32}P radiolabeling of DNA fragments as hybridization probes (protocol), 97–99
materials, 97
method, 98
overview, 97
recipes, 99
tips and troubleshooting, 98
specific activity, 13–14
Rainex, 59
Random hexamer ^{32}P radiolabeling of DNA fragments as hybridization probes (protocol), 97–99
materials, 97
method, 98
overview, 97
recipes, 99
tips and troubleshooting, 98
Rapid amplification of cDNA ends (RACE)
3′ rapid amplification of cDNA ends (RACE)
description, 83, 84f, 85
protocol, 180–185, 180f, 181f
materials, 180–182
method, 182–184
overview, 182
recipes, 184–185
tips and troubleshooting, 184
5′ rapid amplification of cDNA ends (RACE)
description, 83, 84f, 85
protocol, 171–179, 172f
materials, 171–173

Rapid amplification of cDNA ends (RACE) (*continued*)
method, 174–177
overview, 173–174
recipes, 177–178
tips and troubleshooting, 177
description, 83, 84f, 85
on uncapped or nonpolyadenylated RNAs (protocol), 186–188
materials, 186
method, 187
overview, 186
recipes, 188
tips and troubleshooting, 188
Rapid ligation mix (2×) (recipe), 239, 518
Reaction buffer (10×) (recipe), 105
Reads per kilobase per million, 507
Real-time fluorescence measurements, quantitative PCR using (protocol), 165–170
discussion, 168–169, 168f, 169t, 170f
materials, 165–166
melting cycle, 169, 170f
method, 166–167
overview, 166
threshold cycle, 168–169, 168f, 169t
tips and troubleshooting, 168
Real-time PCR, 82
Recipes
A > G reaction buffer, 197
A + G reaction buffer, 297
acrylamide stock (40%; 19:1 acrylamide:bis-acrylamide), 65
AE buffer, 42
alkaline phosphatase buffer (10×), 177, 188
alkaline SDS solution, 530
ampicillin, 538
aniline (1 M), 274
aniline/acetate buffer (recipe), 197
annealing buffer, 391
annealing buffer (2×), 147, 191, 274
annealing buffer (5×), 143, 151, 397, 513
annealing buffer (6×), 237
annealing buffer (10×), 202
aprotinin and leupeptin, 333
β-mercaptoethanol, 414
BHI medium, 530
binding buffer, 55, 260, 320
binding master mix, 256
bind/wash buffer, 320
blocking buffer, 445
borohydride stop mix, 198
BPTE buffer (10×), 112
buffer A (formaldehyde cross-linking protocol), 313
buffer A (hnRNP-enriched nucleoplasmic extraction), 550
buffer A (hypotonic for nuclear extract preparation), 328
buffer A (nuclear extract preparation), 337, 346
buffer A (RNA purification protocols), 35, 37
buffer B (low salt for nuclear extract preparation), 328
buffer B (nuclear extract preparation), 337
buffer C (high salt for nuclear extract preparation), 328
buffer C (nuclear extract preparation), 337
buffer D (dialysis for nuclear extract preparation), 329
buffer D (nuclear extract preparation), 337, 346
C > U reaction buffer, 199
$CaCl_2$, 538
cesium chloride balance solution, 530
Church and Gilbert hybridization buffer, 124, 513
common, 513–522
Coomassie blue stain, 303, 306
deionized glyoxal, 113
denaturing stop buffer, 351
Denhardt's solution (100×), 118, 129
DNA polymerase buffer (5×), 147
DNase I buffer, 320
DNase I storage buffer (1×), 101
EDTA/glycogen mixture, 93
electrophoretic transfer buffer (20×), 129
elution buffer, 55, 320, 538
Empigen, 313
feeder medium, 415
first-strand buffer (5×), 164
formaldehyde (10%), 313
formamide hybridization buffer, 134
formamide/dye loading buffer, 129, 144, 191, 224, 513
freezing medium, 415
G reaction buffer, 198
G reaction stop buffer, 198
gel elution buffer, 224, 514
gel elution buffer (no SDS), 238, 514
gel elution buffer (with SDS), 238, 514
gel-running buffer, 112
glycerol (25%), 538
glyoxal denaturation buffer, 113
GPS (glutamine/penicillin/streptomycin) solution, 414
HBS solution (2×), 402
herring testis carrier DNA, 119, 129
hexamer mix, 99
high-yield transcription buffer (10×), 218
HNEB1 buffer, 550–551
HNEB2 buffer, 551
HNEB2–30% sucrose, 551
hybridization buffer (5×), 134
hydrazine 50% (U reaction buffer), 297
hydrogen peroxide (2.5%, w/v), 289
injection buffer (1×), 391
inorganic pyrophosphatase (IPP) stock and storage buffer, 218
iodoethanol, 289, 297
IPB1 buffer, 474
IPB2 buffer, 474
IPTG, 373, 539
Kc splicing mix (4×), 341
kinase buffer, 143, 203
Klenow buffer (5×), 267, 514
labeling buffer (10×), 543
LB medium, 530
leupeptin, 539
ligation buffer (2×), 203
ligation buffer (10×), 391, 397, 515
loading buffer, 260
loading dye, 198, 274
Luria broth (LB), 373, 515
lysis buffer, 43, 530
lysis buffer (1×), 539
lysis buffer (5×), 539
lysozyme, 539
M9+ medium, 372, 515
M15 medium, 415
M15 medium + hygromycin, 415
M9 salts, 538
M9 stock (1×), 372, 376, 515
M9TB expression medium, 538
$MgSO_4$, 538
micrococcal nuclease, 289
modification reaction buffer A (denaturing), 275
modification reaction buffer A (native), 274
modification reaction buffer B (denaturing), 275
modification reaction buffer B (native), 275
MOPS buffer (10×), 108
NaOH (0.1 M)/EDTA (1 mM), 140
NaOH mix, 439
native gel-loading buffer, 354, 515
native quench buffer, 351, 354
NGM plates, 372–373, 376–377, 516
NGM RNAi plates, 373
nick translation buffer (10×), 102
NT2 buffer, 267
nucleotide mix (50×), 439
Optikinase buffer (10×), 91, 188, 238, 516
PB buffer (20×), 112, 119
PBS (10×), 552
PBS (1×) + $MgCl_2$, 550
PBS/EDTA/EGTA, 365, 516
PCR buffer (10×), 96, 178, 185, 268, 516
phosphatase buffer (10×), 238, 517
phosphate-buffered saline (PBS), 364, 517
PK buffer (2×), 341, 517
PMSF, 338, 539, 552
Polybrene solution, 404
poly(A) polymerase buffer (5×), 188, 229, 517
potassium acetate solution, 531
potassium phosphate, 539
prehybridization/hybridization mix, 119, 130
primer extension (PE) mix (5×), 144
protease inhibitor cocktail (1000×), 552
proteinase K buffer (2×), 268, 518
rapid ligation mix (2×), 239, 518
reaction buffer (10×), 105
resuspension buffer, 533
reverse transcriptase buffer (2.5×), 518
reverse transcriptase buffer (5×), 148, 178, 184
reverse transcriptase mix (10×), 151
reverse transcription buffer (5×), 275
reverse transcription buffer (10×), 268, 518
RNA gel-loading buffer (1.5×), 65, 519
RNA ligase buffer (10×), 227
RNase A/T1 solution, 134
RNase H buffer (5×), 154, 239, 290, 519, 558
RNase I dilution buffer, 290
RNase III buffer (10×), 459
RT buffer (2.5×), 191
RT mix stock, 470

S100 buffer (10×), 329
S1 digestion buffer, 139, 519
S1 hybridization buffer (5×), 139, 519
S1 termination mix, 139, 520
SDS buffer, 25, 148
SDS buffer (2×), 57
SDS buffer (10×), 22
SDS extraction buffer, 215, 230, 290, 297, 310, 333, 520, 534, 543, 546
SDS–PAGE protein sample buffer (2×), 303, 306
SET (20×), 119
sodium acetate (0.3 M, pH 5.2), 218
sodium acetate (0.3 M)/EDTA solution (0.1 mM), 199
sodium bicarbonate (1 M, pH 9.0), 439
sodium butyrate (1000×), 402
sodium deoxycholate, 539
sodium phosphate buffer, 124
sodium phosphate buffer (1 M, pH 7.2), 520
sorbitol buffer, 346
splicing buffer (5×), 351
splicing master mix (4×), 332–333, 520
splicing mix (4×), 261
SSC (20×), 118, 129, 319
stop buffer, 341
stop mix, 99
storage buffer, 540
T4 kinase buffer (10×), 232
T4 RNA ligase buffer (2×), 154
T4 RNA ligase buffer (10×), 178
T1 sequencing buffer, 275, 521
T7 transcription buffer (10×), 268, 521
tailing buffer, 93
TBE electrophoresis buffer (10×), 66, 521
TE buffer, 521, 530
T10E0.1 buffer, 534
TE buffer (10×), 105
tobacco acid pyrophosphatase (TAP) buffer (10×), 178
tracking dye, 109
transcription buffer (5×), 215
transfer buffer (10× SSC), 558
Tris glycine buffer (10×), 261
tryptone broth, 538
U reaction stop buffer, 198
urea loading dye, 290, 522
urea/dye loading buffer, 134, 521
urea/dyes, 342
wash buffer, 256, 320, 540
wash buffer A, 119
wash buffer B, 120
wash buffer C, 120
wash buffer D, 120
yeast medium, 346
zymolase buffer, 347
Recombinase-mediated cassette exchange (RMCE)
creating transgenic shRNA mice by (protocol), 407–415, 407f
materials, 407–409
method, 409–414
creation of transgenic mice, 413–414
culture of ES cells before electroporation, 410
introduction of transgenes into KH2 ES cells, 410–411
plating of irradiated feeder cells, 409–410
preparation of gelatin-coated ES culture plates, 409
selection of ES cell clones containing integrated shRNA cassettes, 411–412
testing for shRNA expression, 412
testing neomycin sensitivity, 412
validation of clones and preparation of frozen stocks, 412–413
validation of clones by Southern blotting, 413
recipes, 414–415
tips and troubleshooting, 414
description, 406, 407
Replicate, technical, 497, 507
Resuspending pellet RNA, 8
Resuspension buffer (recipe), 533
Retrovirus
infection of mammalian cells with retroviral shRNAs, 403–405
materials, 403
method, 403–404
overview, 403
tips and troubleshooting, 404
packaging shRNA retroviruses and host cell infection, 399–402
materials, 399–400
method, 400
overview, 400
recipes, 402
tips and troubleshooting, 401
Reverse footprinting
description, 245–246, 245f
example, 289f
protocol, 288–289
Reverse northern blot, 420
Reverse primer extension
description, 79, 79f
protocol, 145–148, 145f
materials, 145–146
method, 146–147
overview, 146
recipes, 147–148
tips and troubleshooting, 147
Reverse transcriptase buffer (2.5×) (recipe), 518
Reverse transcriptase buffer (5×) (recipe), 148, 178, 184
Reverse transcriptase mix (10×) (recipe), 151
Reverse transcription buffer (5×) (recipe), 275
Reverse transcription buffer (10×) (recipe), 268, 518
Reverse transcription–polymerase chain reaction (RT-PCR)
in CLIP protocol, 488
protocol, 155–164, 157f, 163f
discussion, 161–162, 163f
exponential range, determining, 162, 163f
hot-start PCR, 162
materials, 155–156
method, 158–161
PCR and detection of products, 160–161
reverse transcription (RT): cDNA priming and synthesis, 158–160
overview, 156, 157f
primer design, 161–162
recipes, 164
tips and troubleshooting, 161
in SELEX, 263, 266
RiboGreen, 68–69
Ribonucleases (RNases). *See also specific RNases*
quick reference guide to, 523
resistance to autoclaving, 3
RNA degradation from, 3–4
sources of, 3, 4
ubiquity of, 3
in vitro transcription failures and, 205–206
Ribonucleoside triphosphates (rNTPs)
label incorporation, determining transcript yield by, 206
Mg^{++} requirements of, 206
sources for in vitro transcription, 206
Ribosomal RNA (rRNA)
percentage of total cellular RNA, 1, 9
removing
from deproteinized, phenol-extracted total RNA (protocol), 48–52
enzymatic digestion of rRNA, 48–49
removal of rRNA by hybrid selection, 51–52
selective precipitation of large RNAs with LiCl or PEG/NaCl, 49–50
overview of procedures, 9–10
ribosomal subunits from cytoplasmic extracts before solubilization with SDS and deproteinization (protocol), 36–37
materials, 36
method, 37
overview, 36
recipes, 37
tips and troubleshooting, 37
Ribosomal subunits, removal from cytoplasmic extracts before solubilization with SDS and deproteinization (protocol), 36–37
Ribosome runoff, prevention of, 35
RIP-Chip. *See* RNA immunoprecipitation microarray
RISC (RNA-induced silencing complex)
miRNAs and, 357
shRNAs and, 383–384
siRNAs and, 356–357
RMCE. *See* Recombinase-mediated cassette exchange
RNA concentration, calculation of, 361
RNA gel-loading buffer (1.5×) (recipe), 65, 519
RNA immunoprecipitation microarray (RIP-Chip)
large-scale immunoprecipitation of RNP complexes from *Drosophila* nucleoplasmic extracts for tiling microarrays, 471–475
materials, 471–472
method, 472–473
overview, 472
recipes, 474
tips and troubleshooting, 474

RNA interference (RNAi), 355–418
  discovery of, 355
  experimental design basics
    in *C. elegans,* 358
    in *Drosophila,* 358–359
    encoded triggers, 358
    in mammalian cells, 359
    synthetic triggers, 357–358
  with long dsRNAs, 359–361
    dsRNA synthesis, 360–361
    transcription template generation, 359–360
    trigger length, 360
    trigger selection, 359–360
  off-target effects, 382
  overview of, 355–357
  pathway summary, 356f
  protocols, 363–417
    construction of transgenic *Drosophila* expressing shRNAs in the miR-1 backbone, 387–392
      materials, 388–389
      method, 389–390
      recipes, 391
      tips and troubleshooting, 391
    creating an miR30-based shRNA vector, 395–397
      materials, 395–396
      method, 396–397
      recipes, 397
    creating transgenic shRNA mice by recombinase-mediated cassette exchange, 407–417
      discussion, 415–417
      materials, 407–409
      method, 409–414
      overview, 409
      recipes, 414–415
      tips and troubleshooting, 414
    dsRNA-induced RNAi in *C. elegans* by feeding with dsRNA-expressing *E. coli,* 370–373
      materials, 370–371
      methods, 371–372
      overview, 371
      recipes, 372–373
      RNAi of L1 larvae in 96-well liquid cultures, 371
      RNAi on plates, 372
    dsRNA-induced RNAi in *Drosophila* cells by soaking, 363–365
      materials, 363
      method, 364
      overview, 363–364
      recipes, 364–365
    dsRNA-induced RNAi in *Drosophila* cells by transfection, 366–369
      materials, 366
      method, 367
      overview, 366–367
      tips and troubleshooting, 367–368
    infection of mammalian cells with retroviral shRNAs, 403–405
      materials, 403
      method, 403–404
      overview, 403
      tips and troubleshooting, 404
    packaging shRNA retroviruses and host cell infection, 399–402
      materials, 399–400
      method, 400
      overview, 400
      recipes, 402
      tips and troubleshooting, 401
    RNAi in *C. elegans* by injection of dsRNA
      materials, 374
      method, 374–375
      overview, 374
      recipes, 376–377
      tips and troubleshooting, 375–376
    RNAi in cultured mammalian cells using synthetic siRNAs, 380–382
      materials, 380
      method, 380–381
      overview, 380
      tips and troubleshooting, 381
  screens, 415–417
    in *C. elegans,* 416
    in *Drosophila,* 416
    in mammalian cells, 416
  silencing triggers
    encoded, 358, 383–385
      in *Drosophila,* 383–385
      in mammalian cell lines, 393–394
      in mice, 406
    synthetic, 357–358, 378–379, 382
      RNAi in cultured mammalian cells, 380–382
RNA ligase, 484, 487
RNA ligase buffer (10×) (recipe), 227
RNA Nano Chip, 71–72
RNA Pico Chip, 71
RNA polymerase
  expression and purification of active recombinant T7 RNA polymerase from *E. coli,* 535–540
  high-yield synthesis of RNA using T7 RNA polymerase and plasmid DNA or oligonucleotide templates (protocol), 216–219
    materials, 216
    method, 217
    overview, 216–217
    recipes, 218
    tips and troubleshooting, 217–218
  for in vitro transcription, 205–206
RNA processing
  analysis using cell-free systems, 321–354
    guidelines for testing and optimizing extracts, 322
    guidelines for the preparation of active, 321–322
    protocols, 325–354
      analysis of pre-mRNA splicing using HeLa cell nuclear extracts, 330–333
      analysis of splicing complexes on native gels, 352–354
      analysis of splicing in vitro using extracts of *Saccharomyces cerevisiae,* 348–351
      preparation and analysis of cell-free splicing extracts from *Saccharomyces cerevisiae,* 343–347
      preparation of *Drosophila* Kc cell nuclear extracts for in vitro splicing, 334–338
      preparation of nuclear extracts from HeLa cells, 325–329
      in vitro splicing reactions in *Drosophila* Kc cell nuclear extracts, 339–342
RNA sequencing by primer extension (protocol), 189–191
  materials, 189
  method, 190
  overview, 189
  recipes, 191
  tips and troubleshooting, 190
RNA shatter method, 491
RNA splicing. *See* Splicing
RNA structure probing
  description, 243–244, 244f
  protocol, 269–275
    recipes, 274–275
    using chemical methods, 269–272
      materials, 269–270
      method, 270–272
      overview, 270
      tips and troubleshooting, 272
    using nuclease digestion methods, 273
RNAi. *See* RNA interference
RNA-induced silencing complex. *See* RISC
RNALater (Ambion), 6, 44
RNAProtect (QIAGEN), 44–45
RNA-protective solutions, 44–45
RNA–protein interactions, 241–320
  detection methods
    filter-binding assays, 241–242, 254–256
    gel mobility shift assays, 242–243
    general overview, 241–243
    SELEX (systematic evolution of ligands by exponential enrichment), 243, 262–268
  mapping sites, 244–252
    affinity approaches, 249–252
      problems with, 251–252
      tagged proteins, 251
      tagged RNAs, 250–251
    boundary analysis, 246, 246f, 298–300
    chemical methods, 276–283
    cross-linking approaches, 248–249, 301–306, 311–313
      advantages and disadvantages, 250t
      UV cross-linking using RNAs substituted with photoactivatable nucleosides, 249
    footprinting, 244–245, 245f
      reverse, 245–246, 245f, 288–289, 289f
      RNase, 284–287, 286f
    modification interference, 246–248, 247f, 291–294
    nucleotide analog interference mapping, 248, 294–297
  protocols, 254–320

basic affinity selection methods, 314–320
with beads, 314–316
with biotinylated oligonucleotide, 316–317
recipes, 319–320
of ribonucleoprotein using complementary biotinylated oligonucleotide, 317–319
boundary analysis, 298–300
materials, 298–299
method, 299–300, 300f
overview, 299
cross-linking formaldehyde, 311–313
detecting RNA–RNA interactions using psoralen derivatives, 307–310
materials, 307
method, 308–309, 309f
overview, 308
recipes, 310
tips and troubleshooting, 309
filter-binding assay, 254–256
materials, 254
method, 255
overview, 254–255
recipes, 256
tips and troubleshooting, 256
mapping sites using chemical methods, 276–283
CMCT, 279
DMS, 279
Fe(II)-EDTA, 276, 278
kethoxal, 279
phosphorothioate-substituted RNA, 280–283
mapping sites using nuclease-based methods, 284–290
recipes, 289–290
reverse footprinting, 288–289, 289f
RNase footprinting, 284–287, 286f
modification interference, 291–294
materials, 291–292
method, 292–294
overview, 292
tips and troubleshooting, 294
native polyacrylamide-gel-electrophoresis-binding assay for RNA–protein complexes, 257–261
materials, 257–258
overview, 258–259
recipes, 260–261
tips and troubleshooting, 260
nucleotide analog interference mapping, 294–297
materials, 291–292
method, 294–296, 295f, 296f
recipes, 297
tips and troubleshooting, 296
photocross-linking using uridine analogs, 304–306
materials, 304
method, 305
overview, 305
recipes, 306
tips and troubleshooting, 305–306
RNA structure determination using chemical and nuclease digestion methods, 269–275
recipes, 274–275
using chemical methods, 269–272
using nuclease digestion methods, 273
SELEX, 262–268
materials, 262–263
method, 263–267, 264f
overview, 263
recipes, 267–268
ultraviolet photochemical cross-linking to detect RNA-binding proteins, 301–303
materials, 301
method, 302–303
overview, 302
recipes, 303
tips and troubleshooting, 303
RNA structure probing, 243–244, 244f
RNA–RNA interactions detecting using psoralen derivatives, 307–310
materials, 307
method, 308–309, 309f
overview, 308
recipes, 310
tips and troubleshooting, 309
RNase A, 287, 482
RNase assays, 4
RNase A/T1 solution (recipe), 134
RNase footprinting (protocol), 284–287, 286f
example, 286f
markers, 287
materials, 284–285
method, 285–287
overview, 285
tips for troubleshooting, 287
RNase VI, 273, 286–287
RNase H buffer (5x) (recipe), 154, 239, 290, 519, 558
RNase H digestion
description, 80, 80f
protocol, 152–154
materials, 152–153
method, 153
overview, 153
recipes, 154
tips and troubleshooting, 154
in reverse footprinting, 288, 289f
in site-specific labeling, 209, 209f, 235
RNase inhibitors, limitations of, 3
RNase I, 273, 285, 286f
RNase I dilution buffer (recipe), 290
RNase protection assay (protocol), 131–135
materials, 131–132
method, 132–133
overview, 132
recipes, 134
tips and troubleshooting, 133–134
RNase T1, 273, 287, 482
RNase test, 4
RNase III buffer (10x) (recipe), 459
RNase III, fragmentation of whole-transcriptome RNA using, 457–459
RNase-free water
preparation of, 3
storage of, 3
use in making RNase-free solutions, 3–4
RNasin (Promega), limitations of, 3
RNPs. *See* RNA–protein interactions
rNTFs. *See* Ribonucleoside triphosphates
ROSA-26 locus, 412
RQ1 DNase, 487
Rrf-1, 375
rRNA. *See* Ribosomal RNA
RT buffer (2.5x) (recipe), 191
RT mix stock (recipe), 470
RT-PCR. *See* Reverse transcription–polymerase chain reaction (RT-PCR)

S

S1 digestion buffer (recipe), 139, 519
S1 hybridization buffer (5x) (recipe), 139, 519
S1 nuclease protection mapping (protocol), 136–140, 136f
materials, 137
method, 138
overview, 137–138
recipes, 139–140
tips and troubleshooting, 139
S1 termination mix (recipe), 139, 520
S100 buffer (10x) (recipe), 329
Saccharomyces cerevisiae
analysis of splicing in vitro using extracts of *Saccharomyces cerevisiae*, 348–351
ATP depletion, 350
discussion, 350
materials, 348–349
method, 349–350
overview, 349
recipes, 351
tips and troubleshooting, 350
preparation and analysis of cell-free splicing extracts from *Saccharomyces cerevisiae*, 343–347
discussion, 345
materials, 343–344
method, 344–345
overview, 344
recipes, 346–347
tips and troubleshooting, 345
SCAN-ALIZE, 446
Schiff's base, 106
Schlieren line, 61
Schneider S2 cells, 363, 366
SDS, RNA storage in 0.5%, 8
SDS buffer recipes, 25, 148
2x buffer, 57
10x buffer, 22
SDS extraction buffer (recipe), 215, 230, 290, 297, 310, 333, 520, 534, 543, 546
SDS solubilization and phenol extraction (protocol), 19–22
materials, 19
method, 20–21, 21f
overview, 19–20, 20f
recipes, 22
tips and troubleshooting, 22

SDS–PAGE protein sample buffer (2X) (recipe), 303, 306
SDS/proteinase K/phenol:chloroform, 6
Selective precipitation of large RNAs with LiCl or PEG/NaCl, 49–50
SELEX (systematic evolution of ligands by exponential enrichment), 262–268
  description, 243
  protocol
    materials, 262–263
    method, 263–267, 264f
    overview, 263
    recipes, 267–268
Semiquantitative PCR
  for detection and characterization of RNAs, 81–82, 81f
  RNA yield determination by, 72
Sequencing. *See also* High-throughput sequencing (HTS)
  direct chemical sequencing of end-labeled RNA (protocol), 192–199, 197f
    materials, 192–193
    method, 193–196
    overview, 193
    recipes, 197–199
    tips and troubleshooting, 196
  RNA sequencing by primer extension (protocol), 189–191
    materials, 189
    method, 190
    overview, 189
    recipes, 191
    tips and troubleshooting, 190
SET (20X) (recipe), 119
SEVAG, 19
Shrimp alkaline phosphatase, dephosphorylation of RNA using (protocol), 103–105
  materials, 103
  method, 104
  overview, 103
  recipes, 105
  tips and troubleshooting, 104
shRNA
  design of expression vectors, 393–394
  RNA interference protocols
    construction of transgenic *Drosophila* expressing shRNAs in the miR-1 backbone, 387–392
      materials, 388–389
      method, 389–390
      recipes, 391
      tips and troubleshooting, 391
    creating an miR30-based shRNA vector, 395–397
      materials, 395–396
      method, 396–397
      recipes, 397
    creating transgenic shRNA mice by recombinase-mediated cassette exchange, 407–417
      discussion, 415–417
      materials, 407–409
      method, 409–414
      overview, 409
      recipes, 414–415
      tips and troubleshooting, 414
    infection of mammalian cells with retroviral shRNAs, 403–405
      materials, 403
      method, 403–404
      overview, 403
      tips and troubleshooting, 404
    packaging shRNA retroviruses and host cell infection, 399–402
      materials, 399–400
      method, 400
      overview, 400
      recipes, 402
      tips and troubleshooting, 401
  as silencing triggers for RNAi
    in *Drosophila,* 383–385, 384f, 385f
    in mammalian cells, 393–394
  strategies for integration of shRNA expression cassettes into genome, 398
  vector examples, 395f
Significance analysis of microarrays, 499
Silencing. *See* RNA interference
siRNAs. *See* Small interfering RNAs
Site-specific labeling, for in vitro transcription
  overview, 208–209
  protocol, 234–239
    cleavage of RNA with chimeric oligonucleotide, 235–236
    dephosphorylation of cleaved RNA, 236
    ligation of RNA using a bridge oligonucleotide, 236–237
    materials, 234–235
    method, 235–237
    overview, 235
    phosphorylation of RNA with [γ-^{32}P]ATP, 236
    recipes, 237–239
    tips and troubleshooting, 237
Site-specific substitution, 207, 234–239
Size markers for gel electrophoresis, 541–543
Small interfering RNAs (siRNAs)
  production by Dicer, 356, 358
  RISC and, 356–357
  RNAi in cultured mammalian cells using synthetic siRNAs (protocol), 380–382
    materials, 380
    method, 380–381
    overview, 380
    tips and troubleshooting, 381
  synthetic as silencing triggers, 358, 378–379
    custom, 378
    vendor designed, 378
  thermodynamic asymmetry, 378
Small nucleolar RNAs (snoRNAs), 10
Small RNAs
  detection and characterization of
    overview of techniques
      gel-shift assay, 87, 87f
      native gel electrophoresis, 87, 87f
      splinted ligations, 86, 86f
      stem-loop PCR, 87, 88f
  northern blots for (protocol), 126–130
    materials, 126–127
    method, 127–128
    overview, 127
    recipes, 129–130
SnapDragon (design tool), 360
snoRNAs (small nucleolar RNAs), 10
snRNAs, 10
Sodium acetate (0.3 M, pH 5.2) (recipe), 218
Sodium acetate (0.3 M)/EDTA solution (0.1 mM) (recipe), 199
Sodium bicarbonate (1 M, pH 9.0) (recipe), 439
Sodium butyrate (1000X) (recipe), 402
Sodium deoxycholate (recipe), 539
Sodium hydroxide mix (recipe), 439
Sodium phosphate buffer (recipe), 124, 520
SOLiD sequencing method, 426–427, 427f
Solubilization
  description of techniques, 5
  SDS solubilization and phenol extraction (protocol), 19–22
Sorbitol buffer (recipe), 346
Southern blotting, for validation clones, 413
SP6 RNA polymerase
  promoter, 360
  for in vitro transcription, 205
Specific activity
  defined, 13
  important considerations, 13–14
Spectrophotometry
  for assessing quantity of RNA, 8, 67–68
  contamination of RNA revealed by, 8, 67
SpeedVac Concentrator, 8
Spliceosome, 243, 259
Splicing, analysis using cell-free systems, 321–354
  guidelines for testing and optimizing extracts, 322
  guidelines for the preparation of active, 321–322
  protocols, 325–354
    analysis of pre-mRNA splicing using HeLa cell nuclear extracts, 330–333
    analysis of splicing complexes on native gels, 352–354
    analysis of splicing in vitro using extracts of *Saccharomyces cerevisiae,* 348–351
    preparation and analysis of cell-free splicing extracts from *Saccharomyces cerevisiae,* 343–347
    preparation of *Drosophila* Kc cell nuclear extracts for in vitro splicing, 334–338
    preparation of nuclear extracts from HeLa cells, 325–329
    in vitro splicing reactions in *Drosophila* Kc cell nuclear extracts, 339–342
Splicing buffer (5X) (recipe), 351
Splicing master mix (4X) (recipe), 332–333, 520
Splicing mix (4X) (recipe), 261
Splinted ligations, 200–203, 200f
  description, 86, 86f
  materials, 200–201
  method, 201–202
  overview, 201
  recipes, 202–203

tips and troubleshooting, 202
SSC (20×) (recipe), 118, 129, 319
Stem-loop PCR, 87, 88f
Stir bars, RNase contamination of, 4
Stock solutions, preparation of, 3–4
Stop buffer (recipe), 341
Stop mix (recipe), 99
Storage buffer (recipe), 540
Streptavidin beads, for rRNA removal by hybrid selection, 51–52
Streptavidin gel-shift assay, 87, 87f
Structure, of RNA, 1–2, 2f
Substitution of modified nucleotides, 213–214
SuperScript II reverse transcriptase (Invitrogen), 155–156, 158–160
SuperScript III, 173, 175, 181, 182–183
SYBR Gold, 70
SYBR Green, 82, 83f

T

T4 DNA ligase, in splinted ligation method, 200–202
T4 kinase buffer (10×) (recipe), 232
T4 polynucleotide kinase, 5′-end labeling of RNA with, 208, 231–233
T4 RNA ligase, 152–153, 175
3′-end labeling of RNA with [^{32}P]pCp and T4 RNA ligase 1 (protocol), 225–227
materials, 225
method, 226
overview, 226
recipes, 227
tips and troubleshooting, 226
T4 RNA ligase buffer (2×) (recipe), 154
T4 RNA ligase buffer (10×) (recipe), 178
T3 RNA polymerase
promoter, 360
for in vitro transcription, 205–206
T7 RNA polymerase
expression and purification of active recombinant from *E. coli,* 535–540
high-yield synthesis of RNA using T7 RNA polymerase and plasmid DNA or oligonucleotide templates (protocol), 216–219
materials, 216
method, 217
overview, 216–217
recipes, 218
tips and troubleshooting, 217–218
K_m of, 13
promoter, 359f, 360
for in vitro transcription, 205–206
T1 sequencing buffer (recipe), 275, 521
T7 transcription buffer (10×) (recipe), 268, 521
Tailing buffer (recipe), 93
Tandem affinity purification (TAP) tags, 251
Taq polymerase
in 3′ rapid amplification of cDNA ends (RACE), 181, 183–184
in 5′rapid amplification of cDNA ends (RACE), 176–177
hot-start PCR, 162
quantitative PCR, 82
RT-PCR, 155, 156, 160–161
TaqMan
quantitative PCR, 82, 82f, 83f
stem-loop PCR, 87, 88f
TBE electrophoresis buffer (10×) (recipe), 66, 521
T10E0.1 buffer (recipe), 534
TE buffer (recipe), 105, 521, 530
Technical replicate, 497, 507
TEMED, 61
Terminal deoxynucleotidyl transferase (TdT), DNA oligonucleotide radiolabeling by (protocol), 92–94
materials, 92
method, 92–93
overview, 92
recipes, 93
tips and troubleshooting, 93
Terminator exonuclease (Epicentre), 48–49
Tet-CMV promoter, 393
thread, 368
Tiling microarrays, 420, 471–475
Tissue culture
hnRNP-enriched nucleoplasmic extracts (protocol)
materials, 548
method, 549–550
recipes, 550–552
preparation of cytoplasmic and nuclear RNA from (protocol), 33–35
materials, 33
method, 34–35
breaking cells, 34–35
harvesting cells, 34
overview, 33–34
recipes, 34
tips and troubleshooting, 35
Tissues, RNA extraction from, 6
TM4 suite, 496
TNA (trinitriloacetate), Zn^{++}, chelation with, 3
Tobacco acid pyrophosphatase (TAP), 173, 174–175
Tobacco acid pyrophosphatase (TAP) buffer (10×) (recipe), 178
Toeprinting, 544–546, 546f
Total RNA isolation from yeast cell cultures (protocol), 38–43
Tracking dye, for polyacrylamide gel electrophoresis, 62
Tracking dye (recipe), 109
Transcription buffer (5×) (recipe), 215
Transfection, of dsRNA-induced RNAi in *Drosophila* cells (protocol), 366–368
Transfer buffer (10× SSC) (recipe), 558
Transfer RNA (tRNA), as carrier in ethanol precipitation, 7, 29
Trinitriloacetate (TNA), Zn^{++}, chelation with, 3
Tris borate, EDTA (TBE) electrophoresis buffer, 63–64, 66
Tris glycine buffer (10×) (recipe), 261
TRIzol
advantages and disadvantages of, 5
purification of RNA using (protocol), 23–26
materials, 23
method, 24–25
overview, 24
recipes, 25
tips and troubleshooting, 25
solubilization with, 5
Tryptone broth (recipe), 538

U

U reaction stop buffer (recipe), 198
U1 RNA, 258
UCSC Genome Browser, 500–501, 510
"Ugly gels," avoiding, 9
Uncapped or nonpolyadenylated RNAs, RACE of (protocol), 186–188
materials, 186
method, 187
overview, 186
recipes, 188
tips and troubleshooting, 188
Upstream activator sequences (UASs), 387–388, 388f
Urea loading dye (recipe), 290, 522
Urea/dye loading buffer (recipe), 134, 521
Urea/dyes (recipe), 342
Uridine analogs, photocross-linking using, 304–306
UV cross-linking, 249, 301–303, 428–430, 480–481

V

VALIUM vectors, 387–388, 388f
Vesicular stomatitis virus (VSV) G protein, 399, 401, 403
Volatile materials, as challenge for micropipettor use, 12–13

W

Wash buffer (recipe), 256, 320, 540
Wash buffer A (recipe), 119
Wash buffer B (recipe), 120
Wash buffer C (recipe), 120
Wash buffer D (recipe), 120
Water
for checking pipettor accuracy, 13
RNase-free, 3–4

X

Xylene cyanol, 62

Y

Yeast cell cultures, isolation of total RNA from (protocol), 38–43
method for meiotic cells, 40–41
method for vegetative cells, 38–40
recipes, 42–43
Yeast medium (recipe), 346

Yeast poly(A) polymerase, 228–230
Yield
determining of purified RNA (protocol)
by gel electrophoresis, 70–72
by northern blotting, 73
by PCR, quantitative or semiquantitative, 72
by spectrophotometry or fluorometry, 67–70, 69f
determining yield of RNA synthesized in vitro (protocol), 220–221
materials, 220
method, 220–221
overview, 220
tips and troubleshooting, 221

Z

Zn^{++}, chelation with TNA, 3
Zymolase buffer (recipe), 347